ADVANCES IN CHEMICAL PHYSICS

VOLUME L

ADVANCES IN CHEMICAL PHYSICS—VOLUME L

I. Prigogine and Stuart A. Rice—Editors

DYNAMICS OF THE EXCITED STATE

Edited by

K. P. LAWLEY

Department of Chemistry
University of Edinburgh

AN INTERSCIENCE® PUBLICATION

1807 1982

JOHN WILEY & SONS

CHICHESTER · NEW YORK · BRISBANE · TORONTO · SINGAPORE

British Library Cataloguing in Publication Data:

Dynamics of the excited state,
— (Advances in chemical physics; v. 50)
1. Molecules
2. Excited state chemistry
I. Lawley, K. P. II. Series
541.2′8   QD461.5

ISBN 0 471 10059 5

Photosetting by Thomson Press (India) Limited, New Delhi
Printed by The Pitman Press, Bath.

INTRODUCTION

Few of us can any longer keep up with the flood of scientific literature, even in specialized subfields. Any attempt to do more, and be broadly educated with respect to a large domain of science, has the appearance of tilting at windmills. Yet the synthesis of ideas drawn from different subjects into new, powerful, general concepts is as valuable as ever, and the desire to remain educated persists in all scientists. This series, *Advances in Chemical Physics*, is devoted to helping the reader obtain general information about a wide variety of topics in chemical physics, which field we interpret very broadly. Our intent is to have experts present comprehensive analyses of subjects of interest and to encourage the expression of individual points of view. We hope that this approach to the presentation of an overview of a subject will both stimulate new research and serve as a personalized learning text for beginners in a field.

ILYA PRIGOGINE

STUART A. RICE

CONTRIBUTORS TO VOLUME L

W. H. BRECKENRIDGE, Department of Chemistry, University of Utah, Salt Lake City, UT 84112, USA

T. A. BRUNNER, Research Laboratory of Electronics and Department of Physics, Massachusetts Institute of Technology, Cambridge, MA 02139, USA

M. A. A. CLYNE deceased, formerly Department of Chemistry, Queen Mary College, Mile End Road, London E1 4NS, UK

I. V. HERTEL, Institut für Molekülphysik, Fachbereich Physik der Freien Universität Berlin, Boltzmannstrasse 20, 1000 Berlin 33, West Germany

D. M. HIRST, Department of Chemistry and Molecular Sciences, University of Warwick, Coventry CV4 7AL, UK

D. S. KING, National Bureau of Standards, Division of Molecular Spectroscopy, Washington, DC 20234, USA

A. M. F. LAU, Corporate Research Science Laboratories, Exxon Research and Engineering Company, PO Box 45, Linden, NJ 07036, USA

S. R. LEONE, Joint Institute for Laboratory Astrophysics, National Bureau of Standards and University of Colorado; and Department of Chemistry, University of Colorado, Boulder, CO 80309, USA

I. S. MCDERMID, Molecular Physics and Chemistry Section, Jet Propulsion Laboratory, California Institute of Technology, 4800 Oak Grove Drive, Pasadena, CA 91109, USA

D. PRITCHARD, Research Laboratory of Electronics and Department of Physics, Massachusetts Institute of Technology, Cambridge, MA 02139, USA

M. QUACK, Institut für Physikalische Chemie der Universität, Tammannstrasse 6, D-3400 Göttingen, West Germany

H. UMEMOTO, Department of Chemistry, University of Utah, Salt Lake City, UT 84112, USA

CONTENTS

Dynamics of the Excited State
Edited by K. P. Lawley
© 1982 John Wiley & Sons Ltd.

LASER-INDUCED FLUORESCENCE: ELECTRONICALLY EXCITED STATES OF SMALL MOLECULES

THE LATE MICHAEL A. A. CLYNE

*Department of Chemistry, Queen Mary College,
Mile End Road, London E1 4NS, UK*

AND

I. STUART McDERMID

*Molecular Physics and Chemistry Section, Jet Propulsion
Loboratory, California Institute of Technology,
4800 Oak Grove Drive, Pasadena, CA 91109, USA*

CONTENTS

1

I. INTRODUCTION

The frequency of laser radiation may be tuned into coincidence with an absorption line of an atom or molecule. The resulting emission of radiation from the upper state of the transition is termed laser-induced fluorescence (LIF), and it provides a highly sensitive and specific means of detecting species in their ground states. Thus, LIF is very useful, for example, in the kinetic studies by Baronavski *et al.* (1978) of metastable-excited ground-state species, and as an analytical method of measuring small concentrations of certain trace species such as HCHO (Becker *et al.*, 1975) and OH (Wang and Davis, 1974; Davis *et al.*, 1979) radicals in the atmosphere. Suitable species for LIF studies are those whose excited electronic states are stable, i.e. not predissociated, and which possess intense transitions in an appropriate energy region.

When reasonably narrow-band excitation ($\leqslant 10\,\text{GHz}$) is used, LIF with dye lasers provides a means of determining vibrational and rotational populations in molecules. Following the pioneering work of Zare and his colleagues (Cruse *et al.*, 1973; Zare and Dagdigian, 1974; Zare, 1979), highly detailed and elegant studies are being carried out of initial energy distributions in molecules formed in elementary reactions. Many of these investigations are being undertaken under molecular beam conditions, in order to provide well determined collisions.

Application of LIF specifically to the identification of photofragments in beam experiments is another extremely promising application, which is discussed in the article by Leone in this volume. LIF provides a direct method of studying the dynamics of excited states. In these studies, a short (nanosecond) pulse of laser radiation is used to form the excited electronic state. The decay of fluorescence from the excited state is then measured in real time, giving kinetic information about collision-free or collisional elementary processes. Modern techniques for the direct study of fluorescence decay and the measurement of lifetimes and relaxation rate constants have recently been reviewed by McDermid (1981). Radiative and predissociative lifetimes may be determined, whilst collisional data of significance include the rate constants for electronic quenching, ro-vibrational energy transfer and collisional predissociation. Examples of species that have been studied in this way include the $B^3\Pi(0^+)$ states of halogens and interhalogens (Clyne and McDermid, 1978a), the OH radical (German, 1975a, b, 1976; McDermid and Laudenslager, 1980) and SO (Clyne and McDermid, 1979e) free radicals. It is a great simplification that initial excitation can be confined to a single ro-vibrational state (v', J') of a diatomic molecule, when lasers with relatively narrow bandwidths ($\leq 3\,GHz$) are used. Thus it is possible to determine state-to-state rate constants in work of this type.

Before considering laser-induced fluorescence for molecules in more detail, it is useful first to review briefly the conceptually simpler technique of atomic resonance fluorescence.

A. Atomic Resonance Fluorescence

The use of resonance fluorescence as a method of detecting atoms in defined quantum states was established many years ago, and was reviewed by Mitchell and Zemansky (1934) in their wellknown monograph *Resonance Radiation and Excited Atoms*. An atomic resonance line from a suitable lamp (e.g. Na) is absorbed by ground-state atoms, the subsequent re-emission of radiation being an extremely sensitive and specific indicator of the atoms in question. Quantitative calibration of the fluorescence intensity in terms of atom concentration was seldom achieved in earlier work, because the intensities of absorption and fluorescence are dependent on the lineshape of the emitter in the lamp, and on that of the absorbing atoms. The absorber normally has a well defined lineshape, which is typically a Doppler, Gaussian profile at low pressures. However, the lamp emitter normally is strongly self-reversed and has a translational temperature that is not well defined but is above 300 K; values of 500–1000 K are typical for a low-power (50 W) microwave discharge operating at 2·45 GHz.

Recently, atomic resonance fluorescence has been developed as a quantitative method of measuring atom concentrations (Braun and Carrington, 1969; Anderson and Kaufman, 1972; Clyne and Nip, 1979). The use of a tunable dye laser exhibiting narrow bandwidth has obvious advantages over the earlier resonance lamps. When a continuous-wave (CW) dye laser is set up with several intracavity etalons, the laser oscillates on a single longitudinal mode whose bandwidth is much less than the Doppler linewidth ($- 3\,GHz$) of a typical atom or molecule at 300 K (Schafer, 1973). With stabilization of laser frequency and intensity, extremely low sodium atom concentrations (< 1 atom cm^{-3}) can be detected at 570 nm (Kuhl and Marowsky, 1971). In addition, the narrow bandwidth of CW dye lasers has opened up a vast new field of sub-Doppler atomic spectroscopy that encompasses the various types of hyperfine splitting of energy levels. This field has been reviewed in books edited by Walther (1976) and by Shimoda (1976) and will not be covered in the present work.

Unfortunately, the scope for detecting a variety of atoms using laser resonance fluorescence is rather limited. This limitation arises from the restricted range of wavelengths available from dye lasers, particularly from CW dye lasers which operate most easily at wavelengths above 550 nm. Pulsed dye lasers are more useful, since their fundamental wavelength ranges extend routinely below 400 nm, and frequency doubling or frequency mixing leads to usable ultraviolet (UV) energies below 260 nm; see, for example, Byer and Herbst (1978) and Byer (1980).

For the detection of most non-metallic atoms, resonant radiation in the vacuum ultraviolet spectrum ($\lambda < 200$ nm) is required. Lasers capable of operating in this wavelength region are beginning to become available, and two-photon excitation with conventional UV photons has been demonstrated by Hansch et al. (1975) for the $1s$–$2s$ transition of H near 121·6 nm. However, at the present time, conventional microwave-excited resonance lamps are used routinely to follow concentrations of common non-metallic atoms, such as H, O, N, Cl, Br, S, and I. This approach has led to a method of studying the kinetics of atomic reactions, which has been extremely fruitful in leading to many hitherto unknown rate constants; for a review, see Baulch et al. (1980). Kinetic uses of atomic resonance fluorescence in the vacuum ultraviolet, and the closely related resonance absorption have been reviewed recently by Clyne and Nip (1979) and by Donovan and Gillespie (1975).

B. Linewidth Considerations

Factors influencing achievable signal-to-noise ratio are common to atoms and molecules, although additional factors are introduced in the consideration of molecular fluorescence. We consider initially the influence of linewidth on atomic fluorescence using radiation in the visible and ultraviolet range. At low pressures (~ 1 Torr or less) and at temperatures $T \geqslant 100$ K, the line profile of an absorbing atom is determined almost entirely by the Doppler width Δv_D, given by (Mitchell and Zemansky, 1934)

$$\Delta v_D = 2v_D 2R \ln 2 (T/M)^{1/2}/c. \tag{1.1}$$

(This will not necessarily be the case for infrared fluorescence, for which the magnitude of Δv_D is much less than for ultraviolet visible radiation. Thus, the lineshape for absorption of infrared radiation tends towards a Lorentzian whose linewidth is determined by the radiative lifetime and by pressure broadening even in the torr pressure range.)

The Doppler line appropriate to absorption of ultraviolet/visible radiation has a Gaussian form given by

$$k_v = k_0 \exp(-\omega^2). \tag{1.2}$$

In Eq. (1.2), k_v is the absorption coefficient at frequency v and k_0 is the absorption coefficient at the line centre; ω is a reduced-frequency function given by

$$\omega = 2(\ln 2)^{1/2}(v - v_0)/\Delta v_D. \tag{1.3}$$

Clearly, the absorption of energy is maximized when the linewidth of the incident radiation Δv_s is equal to or less than the magnitude of Δv_D. Use of an extremely narrow linewidth, for example a few megahertz such as is available from a stable single-mode dye laser, may not be favourable. It is possible to burn out the centre of the absorption line so that no population remains in the lower state.

The case where $\Delta v = \Delta v_D$ is simple; also, it frequently approximates to the practical linewidth of an etalon-narrowed, pulsed dye laser. We assume optically thin conditions, i.e. when $I_{abs}/I_0 \to 0$, and a Doppler profile for the exciting radiation. Both assumptions will usually be good approximations. The expression for the integrated energy absorbed in a Doppler line is given approximately (Mitchell and Zemansky, 1934; Bemand and Clyne, 1973)

$$I_{abs}/I_0 \simeq k_0 l/(1 + \alpha^2)^{1/2}, \tag{1.4}$$

where l is the absorbing pathlength and $\alpha = \Delta v_s/\Delta v_D$. For equal linewidths of emitter and absorber, $\alpha = 1$; and thus

$$I_{abs}/I_0 = 0.7 k_0 l. \tag{1.4a}$$

Eq. (1.4) implies a proportional dependence of I_{abs}/I_0 upon absorber concentration N (Bemand and Clyne, 1973) since, for atomic absorption in the Doppler model, k_0 is proportional to N through

$$k_0 = 2fN[(\ln 2)/\pi]^{1/2}(\pi e^2/mc)\Delta v_D. \tag{1.5}$$

In Eq. (1.5), f is the atomic oscillator strength of the relevant transition.

When $I_{abs}/I_0 \gg 0$, optically thin conditions do not prevail and the dependence of I_{abs}/I_0 upon N becomes nonlinear. Curves of growth, relating I_{abs}/I_0 to N for simple cases, have been given by Mitchell and Zemansky (1934), Braun and Carrington (1969), and Bemand and Clyne (1973).

C. Oscillator Strength and Lifetime of an Atomic Transition

Eq. (1.5) indicates that k_0, the absorption coefficient at the line centre, is directly proportional to the oscillator strength f of the relevant atomic transition. Thus, the fluorescence intensity varies in direct proportion with f, assuming optically thin fluorescence. A similar conclusion is also valid for molecular fluorescence.

A number of different quantities are used to describe the probability of an electronic transition. In atomic spectroscopy, the oscillator strength f is commonly used. In molecular spectroscopy, various quantities are used, of which we select the electric dipole moment $|R_e|^2$ as one of the more fundamental parameters describing a transition probability. The magnitudes of f and $|R_e|^2$ are related in a relatively simple manner to absorption intensity, as for example through Eqs. (1.4) and (1.5) for atomic absorption.

The radiative lifetime of the upper state τ_R is clearly related closely to emission phenomena, such as the measured excited-state lifetimes. The value of τ_R^{-1} measures the sum of Einstein coefficients for all spontaneously emitting processes out of the upper state. Thus, if the branching ratio out of the upper state is known, τ_R can be related to f or to $|R_e|^2$ for a particular transition. For an atomic transition such that g_1 and g_2 are the statistical weights of the ground and excited states, the relevant relationship (Mitchell and Zemansky, 1934) is

$$\tau_R^{-1} = (8\pi^2 e^2 v_0^2/mc^3)(g_1/g_2)f. \tag{1.6}$$

D. Electric Dipole Moment and Lifetime of a Molecular Transition

The radiative lifetime for a molecular electronic transition is related to the corresponding electric dipole moment through (Zaraga *et al.*, 1976; Okabe, 1978)

$$\tau_R^{-1} = (64\pi^4/3h)\sum_{v''} |R_e|^2 q_{v',v''} v^3. \tag{1.7}$$

In Eq. (1.7), $q_{v',v''}$ is the Franck–Condon density for a particular vibrational transition $v'-v''$ at frequency v. The summation in Eq. (1.7) represents the fact that emission out of the initial state v' is distributed amongst a number of ground-state v'' levels, according to the Franck–Condon densities.

Eq. (1.7) cannot be solved to give a single value of $|R_e|^2$ from a measured value of τ_R, since it allows for the fact that $|R_e|^2$ can vary with v'', and thus with the r-centroid $\bar{r}$ of the transition. Often, the variation of $|R_e|^2$ over a limited range of $\bar{r}$ is small. The Eq. (1.7) may therefore be simplified by assuming a constant value of $|R_e|^2$, as in Eq. (1.8) (Zaraga *et al.*, 1976; Okabe, 1978)

$$\tau_R^{-1} = (64\pi^4/3h)|R_e|^2(\bar{v})^3. \tag{1.8}$$

$(\bar{v})^3$ is defined as

$$(\bar{v})^3 = \sum_{v''} v^3 q_{v',v''}.$$

Substitution of the measured value for τ_R then leads directly to a mean value of $|R_e|^2$, averaged over the range of r-centroids of importance in the transition.

The accuracy of this approach is dependent upon the availability of reliable Franck–Condon densities. For many molecules, such as the halogens (Coxon, 1971), interhalogens (Clyne and McDermid, 1976a), OH (Crosley and Lengel, 1975; Chidsey and Crosley, 1980), CN (Rao and Lakshman, 1972), etc., accurate RKR potential energy curves and the derived Franck–Condon densities are now available over considerable ranges of v' and v''. However, for other molecules, the spectroscopic data base is insufficient, and relatively unreliable Franck–Condon densities based on Morse potential energy curves only are available. It is important, therefore, that acquisition of good-quality spectroscopic data by conventional and laser-based methods should advance more rapidly, if progress in this field is not to be held up.

E. Scope of this Review

Some important spectroscopic aspects of quantum-resolved dynamics experiments have been summarized above. In particular, the relationship of radiative lifetime to other measures of the probability of an electronic transition has been discussed briefly.

In the experimental section, techniques for obtaining tunable radiation from pulsed dye lasers are reviewed, with particular reference to narrow-band lasers. The importance of narrow-band lasers in the present context lies in the possibility of selectively exciting resolved ro-vibrational states of electronically excited molecules, using the lasers. Methods for obtaining tunable ultraviolet radiation are also discussed.

Following the experimental section, we present the results of selected studies of small molecules using laser-induced fluorescence. Both dynamical and spectroscopic results are considered.

II. EXPERIMENTAL ASPECTS: FREQUENCY AND BANDWIDTH CONTROL OF PULSED LASERS IN THE VISIBLE AND ULTRAVIOLET

A tunable laser makes possible the studies of the spectroscopy and kinetics of excited states described in this chapter. Although a number of different types of laser can be tuned, the wavelength range is usually very narrow (< 1 nm). The dye laser is unique in possessing broad-band, continuous tunability in the wavelength range from -300 nm to $-1\,\mu$m. The use of nonlinear techniques, such as second harmonic generation (SHG) and frequency mixing in suitable media, further extends this range of tunability to less than 200 nm in the ultraviolet. This section is concerned with the control of the frequency and the bandwidth in dye lasers.

A. Pulsed Tunable Dye Laser

The basic principles of dye laser operation were reviewed in detail by Schafer (1973) and a complete bibliography of all literature concerned with dye lasers published between 1966 and 1972 is also available (Magyar, 1974).

Since that time the development of dye lasers has proceeded very rapidly. The first really successful and widely used design for a pulsed dye laser was that of Hansch (1972) and is shown in Fig. 1. The essential features of this cavity are the high dispersion echelle grating wavelength selector and the inverted telescope for collimating and expanding the beam. Collimation increases the resolution of the grating since it reduces the angular spread of the fluorescence from the dye cell, and expansion increases the resolution since more lines of the grating are illuminated. The angular dispersion of a diffraction grating in a Littrow mount is given (Longhurst, 1973) by

$$\frac{d\theta}{d\lambda} = \frac{p}{d\cos\theta} = \frac{2\tan\theta}{\lambda}, \tag{2.1}$$

where p is the diffraction order, d is the groove spacing of the grating and θ is the angle of the diffracted beam. In these pulsed dye lasers, the excitation time is typically of the order of 10 ns, so that only a few light passes within the cavity are possible. As a result, the linewidth of the dye laser is not much less than the single-pass bandwidth of the cavity. The linewidth (FWHM) $\Delta\lambda$ of the spectral distribution of the output beam is given (Littman and Metcalf, 1978) by

$$\Delta\lambda/\lambda = 2(\pi l \sin\theta), \tag{2.2}$$

where λ is the wavelength, l is the width of the illuminated part of the grating and θ is the angle between the grating normal and the incident beam. For a typical dye laser at ~ 500 nm, the single-pass resolution is usually in the range 0·005–0·010 nm.

Although this design is very successful, the use of the telescope has some disadvantages. Exact alignment of the telescope on the cavity axis produces undesirable feedback caused by reflection at the lenses. The telescope must therefore be slightly misaligned (Hansch, 1972; Hnilo *et al.*, 1980) and the optimum position is difficult to find. Since beam expansion

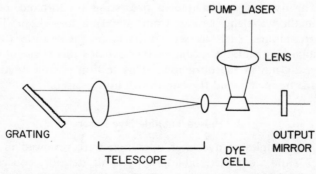

Fig. 1. Schematic diagram of a Hansch-type pulsed tunable dye laser.

is essential for narrow bandwidth operation of this type of dye laser cavity, a number of alternative types of beam expanders have been used. These include single- and multiple-prism devices (Myers, 1971; Stokes et al., 1972; Hanna et al., 1975; Nair, 1977; Klauminzer, 1977), reflective beam expanders (Eesley and Levenson, 1976; Beiting and Smith, 1979; Konig et al., 1980; Hnilo et al., 1980), and an arrangement whereby the expansion is realized directly by a grazing incidence upon the grating (Shoshan et al., 1977; Littman and Metcalf, 1978; Littman, 1978; Saikan, 1978; Shoshan and Oppenheim, 1978; Dinev et al., 1980; Godfrey et al., 1980; Nair and Dasgupta, 1980).

The advantages of a prism beam expander are that magnification occurs in only one dimension, thus making it easier to align the grating, and they also allow the laser cavity to be shorter than with a telescope expander. Fig. 2 shows two different designs for a cavity utilizing a prism beam expander. In the single-prism arrangement (Hanna et al., 1975; Stokes et al., 1972), beam expansion is achieved by placing the prism at high angle of incidence (see Fig. 2(a)). The main disadvantages of this design are that the single-prism expander is not achromatic and there are high losses at the prism face. In

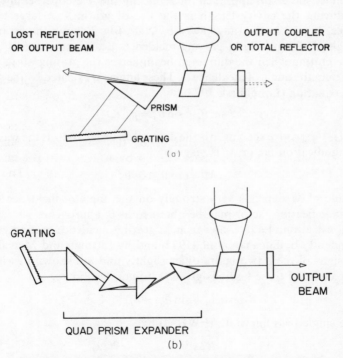

Fig. 2. Tunable dye laser cavities employing prism beam expanders.
(a) Single-prism arrangement (Hanna et al., 1975; Stokes et al., 1972).
(b) Quad-prism system (after Klauminzer, 1977).

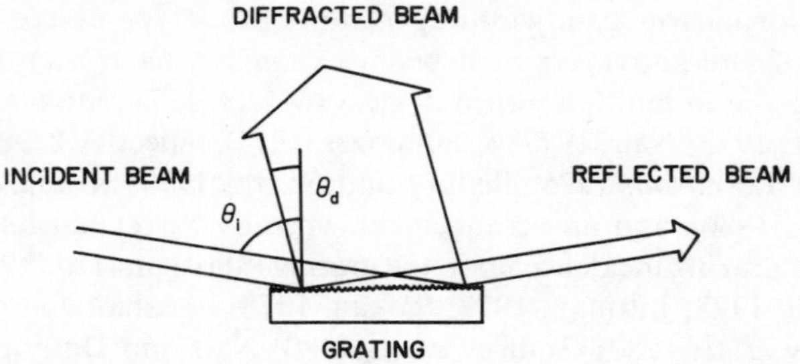

DIFFRACTED BEAM

Fig. 3. Configuration for using a grating to expand a beam.

the multiple-prism expander described by Klauminzer (1977), four prisms at low angles of incidence, approaching Brewster's angle, are configured in achromatic pairs. This quad-prism system produced one-dimensional beam magnification of 40 × with high efficiency. Its only disadvantage is that the initial alignment of the prisms is difficult and high-quality coatings are required to reduce losses.

The most successful approach for replacing the telescope beam expander and reducing the cavity length is the use of gratings at large angles of incidence, usually in the range $89° \leqslant \theta_i \leqslant 90°$. Fig. 3 shows the arrangement and angles for a grating at grazing incidence. Beam expansion is achieved with the grating when the diffracted beam leaves the grating at an angle θ_d smaller than the angle of incidence θ_i. These angles are related by the standard grating equation (Longhurst, 1973)

$$\sin \theta_i + \sin \theta_d = p\lambda/d, \tag{2.3}$$

where d is the groove spacing, p is the diffraction order and λ is the wavelength. The magnification factor M is given by

$$M = \cos \theta_d/\cos \theta_i. \tag{2.4}$$

The value of M depends very strongly on the angle of incidence θ_i and a large magnification factor may be obtained as θ_i approaches 90°.

Demonstration of a dye laser using a grazing-incidence grating was made independently by Shoshan et al. (1977) and by Littman and Metcalf (1978). The designs of these two lasers differ slightly and are shown in Fig. 4. The tunning curve of these lasers is given (Littman, 1978) by

$$\lambda = d/p(\sin \theta_i + \sin \theta_d) \simeq d/p(1 + \sin \theta_d), \tag{2.5}$$

and the single-pass linewidth (FWHM) by

$$\frac{\Delta\lambda}{\lambda} = \frac{2(2\lambda)^{1/2}}{\pi l(\sin \theta_i + \sin \theta_d)} = \frac{2(2\lambda)^{1/2}}{\pi l(1 + \sin \theta_d)} \tag{2.6}$$

These equations can be compared with those for the Hansch-type dye laser

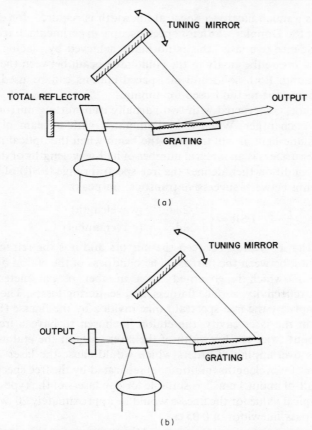

Fig. 4. Two configurations for a grazing-incidence grating dye laser.
(a) After Shoshan *et al.* (1977). (b) After Littman and Metcalf (1978).

with the grating in a Littrow mount (Eq. (2.1) and (2.2)). The expression
above predicts a linewidth about 30% less than for the Hansch laser.

Tuning of these lasers is controlled by the rotation of the tuning mirror.
Rotation of the grating varies the magnification factor, and once the cavity
is aligned the grating should remain fixed. The only limitation of this system
is the small amount of feedback from the grating. However, the dye laser
has sufficiently high gain to overcome this and, in principle, this method of
tuning is applicable to any high-gain laser system.

B. Narrow Spectral Bandwidth Output

The dye lasers described above, comprising a grating and a beam expansion
system, produce typical output linewidths of 0.2–$0.5\,\mathrm{cm}^{-1}$. For many

experiments a much narrower spectral linewidth is required, for example of the order of the Doppler width for spectroscopic experiments. Except for the grazing-incidence dye laser, this is normally achieved by placing a Fabry–Perot etalon inside the cavity in the collimated beam between the expander and the grating. Both solid and air-spaced etalons can be used, the only difference being the method used for tuning.

The etalon is constructed with two partially transmitting mirrors that are parallel to each other. When it is illuminated with a beam of coherent, monochromatic light, it will transmit the beam when the optical pathlength between the surfaces is an integral number of half-wavelengths of the incident light. This condition then defines the free spectral range (FSR) of the etalon as the spacing between successive transmission peaks

$$\text{FSR} = \begin{cases} \lambda^2/2nd & \text{(wavelength)} & (2.7a) \\ 1/2nd & \text{(wavenumber)} & (2.7b) \end{cases}$$

where d is the separation between the mirrors and n is the refractive index of the medium between the mirrors. The bandpass of the etalon depends on the finesse F, which is governed by a number of parameters: mirror parallelism, reflectivity, surface flatness, and scattering losses. The bandpass is given simply by the free spectral range divided by the finesse (FSR/F).

For use in the laser cavity, the etalon is chosen to have a free spectral range of about twice the linewidth of the laser without the etalon installed. This avoids overlapping of orders which would cause the laser to operate at a number of wavelengths (multimode) separated by the free spectral range. Thus an FSR of about $1\,\text{cm}^{-1}$ is suitable for dye lasers of the types discussed above. A typical value for the finesse would be approximately 20, which leads to a single-pass linewidth of $0.05\,\text{cm}^{-1}$.

The etalon transmission maximum and the maximum of the spectral profile from the grating must be synchronized before the system can be used. This can be achieved either by tilting the etalon, which shifts the position of the transmission maximum, or by tilting the grating to shift the position of the maximum of the grating profile. This synchronization must be maintained at all times and thus there are special requirements if the laser wavelength is to be scanned. The scanning techniques for air-spaced and solid etalons are different.

A solid etalon is mounted in an adjustable kinematic holder and wavelength tuning is achieved by changing the tilt angle θ. For a solid quartz etalon such as is commonly used, the wavelength change, for a given order, as the etalon is rotated through an angle $d\theta$ is given by

$$d\lambda = -\lambda_0\theta\,d\theta/2n_d, \qquad (2.8)$$

where n is the refractive index of quartz at wavelength λ_0, θ is the angle of rotation from the normal and λ_0 is the laser wavelength at $\theta = 0$. The dispersive index n_d is given by

$$n_d = n - \lambda_0 \left|\frac{dn}{d\lambda}\right|. \tag{2.9}$$

Thus there is no simple relationship between the scanning of the grating, a sine function, and the scanning of the etalon. In order to scan any significant range of wavelength, it is necessary to use a computer to control the angles of the grating and the etalon.

Scanning can be somewhat simplified by using an air-spaced etalon. Both the etalon and the grating are enclosed in a pressure vessel and the wavelength is changed by varying the refractive index by changing the pressure of the gas in the pressure vessel (Wallenstein and Hansch, 1974, 1975). The wavelengths selected by both the grating and the etalon vary linearly with n. From a knowledge of the refractive index under ambient conditions, which can readily be calculated from its value at STP, and also the dye laser wavelength under these conditions, the wavelength at any pressure can be calculated from

$$\lambda = \frac{\lambda_A}{n_A} l + (n_A - 1)\frac{P}{P_A} \tag{2.10}$$

where λ is the output wavelength, P_A is the local ambient pressure, n_A is the refractive index and λ_A the output wavelength at P_A, and P is the pressure in the etalon/grating chamber. For a pressure change of 1 atm, the change in output wavelength is given by

$$\Delta\lambda = \lambda(n - 1) \qquad \text{(wavelength)}, \tag{2.11a}$$

or

$$\Delta v = - v(n - 1) \qquad \text{(wavenumber)}. \tag{2.11b}$$

Thus for two typical scan gases, $N_2(n = 1 \cdot 000\,299)$ and $SF_6(n = 1 \cdot 000\,783)$, the scan range for a 1 atm change is $4 \cdot 98\,\mathrm{cm}^{-1}$ and $13 \cdot 05\,\mathrm{cm}^{-1}$ respectively, or at 600 nm, $0 \cdot 179$ nm and $0 \cdot 470$ nm. High-pressure systems have been built in which pressure scans of up to 5 nm can be achieved (P. Hargis, 1980, private communication).

Several methods have been used to narrow the output linewidth in grazing-incidence grating dye lasers. For example, Saikan (1978) has replaced the total reflector with a resonant reflector and used the zero-order reflection to couple out of the cavity. In a laser designed to operate narrow-band simultaneously at two wavelengths, Dinev et al. (1980) used multiple-element resonant reflectors in place of the tuning mirror. Perhaps the most interesting method (Shoshan and Oppenheim, 1978; Littman, 1978) is the use of a second grating in a Littrow mount in place of the tuning mirror. Two possible orientations of the second grating satisfy the Littrow condition and they are shown in Fig. 5. However, only one orientation (full line in Fig. 5) can be

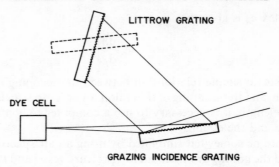

Fig. 5. Dual-grating, single-mode, pulsed tunable dye laser.

used since, for best operation, the dispersion of the grazing-incidence grating and the Littrow grating should be additive. The derivation of the tuning curves and single-pass linewidth of the double-grating system is complex and the reader is referred to the original papers of Shoshan and Oppeheim (1978) and Littman (1978). Since the wavelength is not a simple function of the grating rotation, the best method for continuous scanning is again pressure tuning. In operation, a laser of this design had a measured linewidth of about 750 MHz ($0.025 \, \text{cm}^{-1}$), but this was limited by jitter and the single-shot linewidth was estimated to be about 300 MHz ($0.01 \, \text{cm}^{-1}$) (Littman, 1978).

C. Oscillator–Amplifier Systems

The high-resolution dye laser oscillators described so far are not very efficient because of the losses associated with the intracavity dispersive optics. To obtain high output powers, a powerful pump laser can be used, but the risk of damage to optical elements in the cavity is increased. A more efficient method is to divide the pump laser beam and use some of its energy to drive a dye cell amplifier. Depending on the output power required, several stages of amplification may be employed (Wallenstein and Zacharias, 1980). Fig. 6 shows a diagram of an N_2-laser-pumped dye laser with two stages of amplification (Wallenstein and Hansch, 1975).

The influence of superfluorescence in oscillator–amplifier systems must be considered since it can compete with the external signal amplification, thus decreasing the overall efficiency. Several investigations of laser–dye amplifier systems have been published for both transversely pumped (Bonch-Breuvich et al., 1975a, b; Ganiel et al., 1975) and longitudinally pumped (Carlsten and McIlrath, 1973; Moriarty et al., 1976; Narovlyanskaya and Tihkonov, 1978) arrangements. It has been shown that superfluorescence restricts the efficiency of conversion of pump laser energy into a useful signal when the injected signal is small. When the external signal is increased, the influence of the

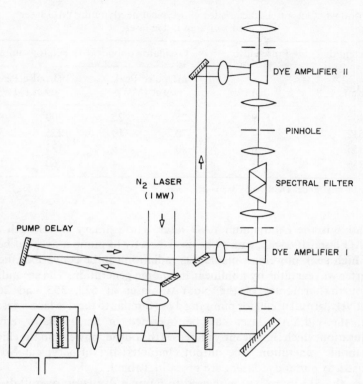

Fig. 6. Oscillator–amplifier dye laser system. Note the use of spectral and spatial filters between the amplifier stages to reduce the background of broadband spontaneous emission. After Wallenstein and Hansch (1975).

superfluorescence diminishes and the conversion efficiency increases. However, when the injected signal is very large, the efficiency is restricted as a result of saturation of the dye amplifier. Therefore the partitioning of the pump laser energy between oscillator and amplifier must be carefully judged.

D. Pump Sources

Little has been mentioned so far regarding suitable pump lasers for these dye lasers. Most of the lasers discussed above were originally designed to be pumped by nitrogen lasers at 337 nm in the ultraviolet. Even though the nitrogen laser proved to be a good pump laser, its peak power is limited to about 1 MW, or about 10 mJ per pulse maximum energy, which in turn limits the maximum energy available from the dye laser. Several other lasers have been found to be good dye laser pump sources. The most frequently used

TABLE I. Output characteristics of a typical neodymium:YAG laser[a]
used to pump a dye laser.

Wavelength	Pulsewidth	Oscillator only		Oscillator–amplifier	
(nm)	(ns)	mJ/pulse	Peak power (MW)	mJ/pulse	Peak power (MW)
1064	8–9	225	25	700	80
532	6–7	70	10	225	32
355	5–6	40	6	125	20
266	4–5	20	4	60	12

[a] Quanta-Ray model DCR Nd:YAG laser.

alternative is the neodymium:YAG laser. The primary wavelength of the Nd:YAG laser, 1064 nm, cannot be used directly to pump a dye, but because of the high peak powers available this radiation can be converted efficiently to shorter wavelengths by nonlinear harmonic generation. The second, third and fourth harmonics of the Nd:YAG laser at 532, 355 and 266 nm, respectively, are all useful for pumping dyes. Similar to the dye lasers described earlier, the Nd:YAG laser can be operated in an oscillator–amplifier configuration which significantly increases the output power and the efficiency of harmonic generation. The output characteristics of a typical Nd:YAG laser used to pump dye lasers are given in Table I.

When the 532 nm radiation is used to pump a dye laser, overall dye laser efficiencies of greater than about 30% can be achieved at wavelengths longer than 540 nm. When the UV harmonics are used, the output efficiencies fall to approximately 20%. The tuning ranges of laser dyes vary with the pump laser wavelength, and the efficiency is usually highest when the fluorescence wavelength is close to the excitation wavelength. There are several studies of the tuning ranges of various dyes when pumped by the different harmonics of the Nd:YAG laser. For example, Ziegler and Hudson (1980) have measured the power tuning curves for several dyes in the range 333–390 nm when pumped with 266 nm radiation and Guthals and Nibler (1979) have published similar curves for 355 nm pumped laser dyes in range 410–715 nm.

The excimer lasers, particularly KrF at 248 nm (Sutton and Capelle, 1976; Rulliere et al., 1977; Tomin et al., 1978, 1979) and XeCl at 308 nm (Corney et al., 1979; Uchino et al., 1979; Huffer et al., 1980) have also proved to be good pump sources. XeCl is perhaps the better of these sources since it has a long static fill lifetime; also its emission wavelength is more suited to pump a wide range of dyes thoughout the wavelength 340–710 nm. Typical output pulse energies of commercial excimer lasers are in the range 0·1–0·5 J and these units can operate at repetition rates up to 100 Hz. The peak powers available with excimer laser pumping are typically two orders of magnitude greater than with N_2-laser excitation. The availability of these short-wavelength sources has stimulated the search for better short-wavelength dyes and it is now possible to obtain lasing with good efficiency down to 320 nm.

E. Frequency Doubling and Mixing

The range of useful laser output of a tunable dye laser can be extended well into the ultraviolet through the nonlinear optical techniques of frequency doubling and frequency mixing. Efficient SHG frequency conversion requires phase matching between the fundamental wave and its second harmonic. This can be obtained by properly adjusting the angle between the optical axis of the crystal and the input beam (Bloembergen, 1965; Terhune and Maker, 1968). The UV frequency can be continuously tuned by varying this angle.

There are a number of crystals which have the necessary optical properties for these nonlinear techniques. The four most frequently used are ammonium dihydrogen phosphate (ADP) (Dunning et al., 1972), lithium formate monohydrate (LFM) (Dunning et al., 1973), potassium dihydrogen phosphate (KDP) (Johnson and Swagel, 1971), and potassium pentaborate (KPB) (Dewey et al., 1975; Dewey, 1976). The latter two crystals, KDP and KPB, can be used to cover the entire range from 217·3 to about 450 nm. Fig. 7 shows a series of tuning curves demonstrating how these crystals can be angle tuned to provide continuous output in this range. The efficiency of frequency doubling is dependent on beam quality, power, and wavelength. The highest efficiency (~ 10%) is obtained with KDP in the range 260–310 nm. The efficiency of KPB is considerably lower, 0·5–5% for the same input power.

In a scanning system it is necessary to maintain the correct phase-matched angle as the wavelength changes according to tuning curves, such as in Fig. 7. Several methods have been used to automate this procedure. Since the proper tilting angles of a grating and a crystal are different nonlinear functions of the wavelength, mechanical coupling of the two motions is difficult to achieve. One of the most elegant approaches was first developed by Kuhl and Spitschan (1975) who utilized the fact that, when the crystal is slightly misaligned, UV output can still be obtained but the direction of the UV beam is shifted. A similar but not identical technique has also been described by Bjorklund and Storz (1978). Two photodiodes were used to detect the spatial position of the UV beam. If the fundamental and harmonic beam spots are observed in a plane some distance from the SHG crystal, one would observe, as the crystal is rotated through the phase-matching angle, that the harmonic spot undergoes a displacement and grows in intensity as it approaches the fundamental spot and then fades again after passing through it. These diodes were then incorporated into a servo-feedback circuit so that

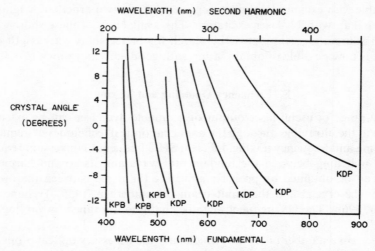

Fig. 7. Typical tuning curves, crystal angle versus wavelength, for correct phase matching of the fundamental and second harmonic beams in KDP and KPB.

any deviation of the UV output was automatically corrected by changing the tilt of the crystal.

Another method of automatic scanning can be used if a microprocessor is available to control the grating and crystal tilts. The tuning curves of the crystal can be determined experimentally and then fitted to some polynomical function. The coefficients of the fit can be stored in the computer and used to calculate the correct crystal angle at any wavelength. A motor and encoder can then be used to rotate the crystal to the correct angle.

Saikan (1976) and Saikan et al. (1979) have described a complicated system in which the frequency-doubling crystal remains fixed. The phase matching is achieved by changing the angle of the laser beam incident on the SHG crystal using dispersive components such as gratings and prisms. This system has been successfully used with KDP and LFM crystals.

Wavelengths shorter than 217 nm, which is the lowest limit for SHG, can be generated by sum frequency mixing (Dunning and Stickel, 1976; Kato, 1977a, b). Efficient sum frequency mixing of radiation at wavelengths λ_1 and λ_2, to produce radiation at wavelength λ_3, occurs when the phase-matching condition

$$\frac{n(\lambda_3)}{\lambda_3} = \frac{n(\lambda_1)}{\lambda_1} + \frac{n(\lambda_2)}{\lambda_2} \tag{2.12}$$

is satisfied, where

$$\frac{1}{\lambda_3} = \frac{1}{\lambda_1} + \frac{1}{\lambda_2} \tag{2.13}$$

and $n(\lambda)$ is the appropriate refractive index (Zernicke and Midwinter, 1973). By mixing, in KPB, the outputs of two tunable dye lasers, one operating in the infrared and one in the ultraviolet, Stickel and Dunning (1978) have obtained tunable radiation at wavelengths down to 185 nm.

In addition to providing shorter wavelengths than available from SHG, sum frequency mixing can also be used to generate higher-power radiation in the wavelength region where frequency doubling can be used. For example, megawatt power levels were generated by Stickel and Dunning (1978) in the range 240–248 nm by mixing the second harmonic of a ruby laser with the output of an infrared dye laser. The dye laser was pumped by the fundamental of the ruby laser. Similar schemes can be realized with the harmonics of a Nd:YAG laser.

The generation of tunable coherent vacuum ultraviolet (VUV) radiation has also been demonstrated (Mahon and Coopman, 1978; Wallenstein and Zacharias, 1980) using non-resonant frequency tripling in krypton gas. The rare gases exhibit only a small non-resonant third-order susceptibility and therefore efficient frequency tripling requires laser pulses of at least megawatt peak power. Using a pressure-tuned narrow-band dye laser with three amplifier stages and pumped by the second or third harmonics of a Nd:YAG laser, Wallenstein and Zacharias (1980) generated tunable light pulses at the Lyman-α wavelength (121·6 nm). With fundamental peak powers up to 9 MW and a bandwidth of more than 9×10^{-3} cm^{-1}, the dye laser output at 364·6 nm was frequency tripled to give radiation at 121·6 nm with peak powers above 2 W and a bandwidth of $5·2 \times 10^{-2}$ cm^{-1}. This method can be used for the generation of narrow-band VUV in several spectral regions between 105 nm, the transmission limit of LiF, and 147 nm, the lowest resonance line of xenon, in which the rare gases Kr and Xe exhibit negative dispersion.

F. Stimulated Raman Scattering

It is only very recently that stimulated Raman scattering (SRS) has found widespread use as a method of frequency conversion (Byer, 1980; Paisner and Hargrove, 1979). Efficient SRS is now possible because of the availability of high peak-power laser sources and the technique can be used to generate tunable radiation covering the electromagnetic spectrum from less than 200 nm to more than 1 μm.

In the SRS process, the pump laser pulse at frequency v_P is incident on the scattering molecule which is in an initial state i, and can promote it to a virtual state. If sufficient transitions to this virtual level are induced, a population inversion can be generated between this virtual level and a final level f. Using the hydrogen molecule as an example, the initial state is the ground vibrational level and the final state is the first vibrational level. Stimulated emission then occurs at longer wavelength corresponding to the frequency

$$v_P - (v_f - v_i) = v_P - v_R.$$

For hydrogen, the Raman frequency v_R, is simply the fundamental vibrational frequency which is equivalent to $4155\,cm^{-1}$. The pump laser light converted in this process produces the first Stokes radiation (S_1) output.

As in a conventional laser scheme, there is an energy threshold below which no S_1 radiation is generated. Once above threshold, S_1 radiation increases exponentially in intensity as the pump intensity increases. Thus two intense waves, the pump and the S_1, propagate in phase through the gas and these waves harmonically beat against one another to produce a high-frequency sideband on the light. This conversion of light to a higher frequency $(v_P + v_R)$ is called the first anti-Stokes radiation, AS_1. This process has no threshold once S_1 radiation appears. Higher-order anti-Stokes frequencies $(v_P + nv_R)$ are produced in an equivalent manner and parametric generation of higher-order Stokes frequencies also occurs.

The three most commonly used gases are H_2, D_2 and CH_4 for which the shift increments are $4155\,cm^{-1}$, $2987\,cm^{-1}$ and $2917\,cm^{-1}$ respectively. Using a Nd:YAG pumped tunable dye laser which produced 20 mJ, 5 ns pulses at 560 nm, Paisner and Hargrove (1979) reported observation of transitions AS_8(195·7 nm) to S_2(1048 nm). With 208 nm radiation, they observed AS_4 (191.7 nm) to S_6(927.9 nm). With high output powers available from commercial dye lasers, spectral coverage in the UV can be achieved to below 175 nm

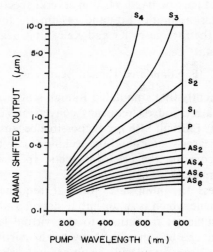

Fig. 8. Theoretical stimulated Raman scattering (SRS) wavelengths in H_2 (Quanta-Ray Corp).

using SRS with only a few dyes. In fact, this tuning range can be covered almost entirely using a rhodamine-6G dye laser alone.

To demonstrate the application of the technique of SRS, Paisner and Hargrove (1979) used the AS_7 component of the output of a rhodamine-6G dye laser to excite nitric oxide (NO). They obtained high-quality, high-resolution photoacoustic absorption spectra over the wavelength range 200–228 nm for the $A-X$ system of NO.

A commercial SRS wavelength conversion accessory is now available (Quanta-Ray). With this system, eight orders of anti-Stokes and five orders of Stokes wavelengths have been observed. By varying the gas pressure, typically in the range 100–300 psi, the output in a particular line can be optimized. Although the highest energies are produced in the first few Stokes and anti-Stokes components, useful energies are obtained for relatively high-order conversions. Fig. 8 shows the theoretical SRS tuning curves using H_2 as the scattering gas.

III. LASER-INDUCED FLUORESCENCE STUDIES OF THE DIATOMIC HALOGENS AND INTERHALOGENS

The low-lying electronic states of the diatomic halogens and interhalogens give rise to banded (except F_2) and continuous absorption spectra in the visible and ultraviolet regions. The spectroscopy of the transitions involved is well understood and was the subject of two useful reviews (Coxon, 1973; Child and Bernstein, 1973). The states giving rise to these transitions are the ground state $X^1\Sigma_g^+$ (the g, u symmetry property is relevant only for the homonuclear molecules); the first excited state $A^3\Pi(1_u)$, which also correlates with ground-state atoms; the $B^3\Pi(0_u^+)$ state, which lies at only slightly higher energy than the A state and which correlates with one spin–orbit excited atom and one ground-state atom. There are then two repulsive states, $^1\Pi(1_u)$ and $Y0_g^+$; the $^1\Pi(1_u)$ state gives rise to the continuous absorption in the homonuclear halogens and the $Y0^+$ state is responsible for the predissociation of the B states in the heteronuclear interhalogens. Because they have opposite symmetry properties, the $Y0_g^+$ state is allowed to cross the $B^3\Pi(0_u^+)$ state in the homonuclear halogens without causing predissociation. Although the A and B states lie at similar energies above the ground state, thus causing the $A-X$ and $B-X$ systems to overlap, the spectra are easily differentiated since the $A-X$ system contains one P, one Q and one R branch and the $B-X$ only one P and one R branch.

Potential energy curves (Morse curves) for the low-lying states are shown in Fig. 9. The A state has not been observed for some of the molecules, namely Cl_2, ClF, BrCl, and is dubious for BrF. The A states are shown in Fig. 9 only when they are well known. The exact forms of the $Y0^+$ states are unknown although the crossing point with the B state can be estimated from

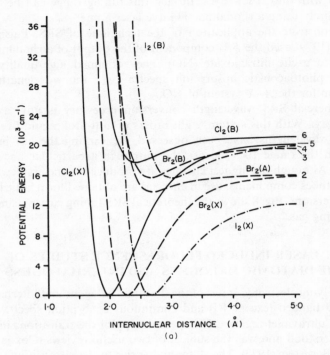

(a)

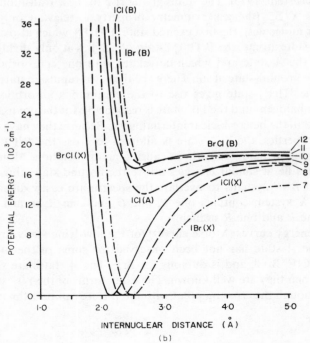

(b)

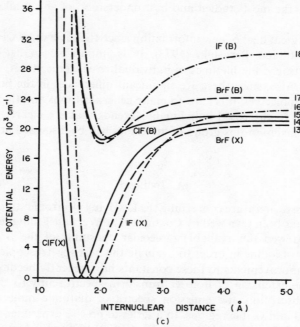

Fig. 9. Morse potential energy curves for the $B^3\Pi(0^+)$, $X^1\Sigma^+$, and, where known, the $A^3\Pi(1)$ states of the halogens and interhalogens. (a) I_2, Br_2, Cl_2. (b) IBr, ICl, BrCl. (c) IF, BrF, CIF. Dissociation products: 1, $I\,^2P_{3/2} + I\,^2P_{3/2}$; 2, $Br\,^2P_{3/2} + Br\,^2P_{3/2}$; 3, $Br\,^2P_{1/2} + Br\,^2P_{3/2}$; 4, $I\,^2P_{1/2} + I\,^2P_{3/2}$; 5, $Cl\,^2P_{3/2} + Cl\,^2P_{3/2}$; 6, $Cl\,^2P_{1/2} + Cl\,^2P_{3/2}$; 7, $I\,^2P_{3/2} + Br\,^2P_{3/2}$; 8, $I\,^2P_{3/2} + Cl\,^2P_{3/2}$; 9, $Br\,^2P_{3/2} + Cl\,^2P_{3/2}$; 10, $I\,^2P_{3/2} + Cl\,^2P_{1/2}$; 11, $I\,^2P_{3/2} + Br\,^2P_{1/2}$; 12, $Br\,^2P_{3/2} + Cl\,^2P_{1/2}$; 13, $Br\,^2P_{3/2} + F\,^2P_{3/2}$; 14, $Cl\,^2P_{3/2} + F\,^2P_{3/2}$; 15, $Cl\,^2P_{3/2} + F\,^2P_{1/2}$; 16, $I\,^2P_{3/2} + F\,^2P_{3/2}$; 17, $Br\,^2P_{1/2} + F\,^2P_{1/2}$; 18, $I\,^2P_{1/2} + F\,^2P_{3/2}$.

observations of the predissociation and it is known that they correlate with ground-state atoms.

Since they all have discrete absorptions in the visible spectrum, which can readily be obtained with tunable dye lasers, the halogens and interhalogens provide an interesting group of molecules for study with these techniques. As discussed in the introduction, the types of information available from an LIF study include radiative lifetimes, collisional energy transfer rates, predissociation rates and, sometimes, improvements to the spectroscopic constants. Since there are a number of curve crossings, detailed quantum-resolved measurements of the dynamics of the excited state can provide much information on the interesting effects of these crossings. As a group, the low-lying states of the homonuclear and heteronuclear diatomic interhalogens are probably the most studied and best understood of any similar group of molecules.

There is a great deal of new information since the reviews by Coxon (1973) and by Child and Bernstein (1973). It is impossible to list all of the measurements, e.g. the lifetimes of individual ro-vibronic states, in a review such as this. Also, since there are significant differences in the behaviour of the different molecules, it is hard to make generalizations and therefore each molecule will be discussed individually. Trends that are clearly due to the family relationship of this group of molecules will be discussed at the end of this section.

A. Iodine, I_2

The extensive literature concerning the low-lying electronic states of iodine, up to 1973, has been reviewed by Coxon (1973). Wei and Tellinghuisen (1974) critically reviewed the available molecular constants of the $B^3\Pi(0_u^+)$ and $X^1\Sigma_g^+$ states of iodine in order to evaluate the best spectroscopic constants. Significant improvements to these constants have recently become available following the extension of Fourier transform spectroscopy to the study of electronic absorption and emission spectra of diatomic molecules in the near-infrared and visible regions. In a series of precise experiments, Gerstenkorn et al. (1977), Gerstenkorn and Luc (1978, 1979) and Luc (1980) have recorded and assigned some 14 000 lines in 139 bands of the $B-X$ system of I_2. New molecular constants and Dunham expansion parameters have been calculated from these data (Luc, 1980) for bands with $0 \leqslant v'' \leqslant 9$ and $1 \leqslant v' \leqslant 62$. Thus, the spectroscopic constants of the $B-X$ system of I_2 are extremely well parametrized.

The resonance emissions of iodine, which can lead to spectroscopic information concerning the ground state, have long been the subject of many fluorescence studies (see, for example, Pringsheim, 1949). With the advent of lasers, the few transitions overlapping laser lines have been studied in more detail. Holzer et al. (1970a, b) observed fluorescence when iodine vapour was excited with the 514·5 and 501·7 nm lines of an Ar^+ laser, but they made no further analysis. The fluorescence from 514·5 nm excitation was investigated further by Ditman et al. (1975) and by Patterson et al. (1975). The experiments of these two groups were similar, and in both cases the fluorescence was excited by a narrow-bandwidth, single-frequency argon-ion laser and was resolved with a Fabry–Perot interferometer. Ditman et al. (1975) studied the emission line profile as the excitation wavelength was tuned through and away from resonance. They found that the intensity of the emission lines followed the absorption line profile whereas the frequency of the fluorescence was determined by the laser frequency. Patterson et al. (1975) studied the fluorescene spectrum when different modes of the argon-ion laser were used to excite the iodine. In addition to the previously reported transitions (Halldorsson and Menke, 1970), 43–0 $R(15)$, 43–0 $P(13)$ and 58–1 $R(99)$, they observed three new transitions in the wings of the laser gain curve, 55–1 $P(87)$, 60–1 $P(103)$ and 44–0 $P(48)$. In order to assign the vibrational levels associated with the above transitions and to confirm the J' values obtained from the spectra, calculations of the energies were made using Wei and Tellinghuisen's (1974) constants. For the higher J' transitions, these energies were found to be in error by up to $1\,\text{cm}^{-1}$, although the agreement for lower J' was good. Most recently, Bacis *et al.* (1980) have recorded the I_2 fluorescence excited by the 514·5 and 501·7 nm Ar$^+$ laser lines using a high-resolution Fourier transform spectrometer. Spectra were obtained with both multimode and single-mode argon-ion lasers. With single-mode excitation, the fluorescence spectra exhibited a reduced Doppler width and spectra were recorded for $v'' = 10$–100. Broadening of quasibound $X^1\Sigma_g^+$ rotational levels above the rotationless $X^1\Sigma_g^+$ dissociation limit was observed and perturbations of the ground state, $v'' \geqslant 92$, by two previously unobserved long-range I_2 molecular states was detected. These two states, assigned as 0_g^+ and 1_g, which both correlate with two ground-state $I^2P_{3/2}$ atoms, were also observed directly in the fluorescence spectrum.

The measurement of lifetimes and self-quenching cross sections of levels in the $B^3\Pi(O_u^+)$ state of I_2 has been facilitated by the development of lasers. The first systematic studies, using direct observation of the fluorescence decay, by Sakurai *et al.* (1971) and by Capelle and Broida (1973), were restricted to the measurements of lifetimes associated with incompletely resolved vibrational bands. Vibrational bands of the B state were excited by a broad-band, 0.3–0.7 nm, dye laser. The results of Capelle and Broida (1973) provided a survey of the lifetimes and quenching cross sections for different vibrational levels. The lifetimes were found to be strongly dependent on v', varying from less than 0·4 μs to greater than 7 μs. Self-quenching cross sections showed less v' dependence, varying between 47 and 90 Å^2. Three peaks in the value of $1/\tau_{\text{NR}}$ were found at different excitation wavelengths. Near the dissociation limit, this increase in $1/\tau_{\text{NR}}$ was attributed to an increase in the radiative decay rate due to infrared emission to repulsive states, the most likely being 0_g^+ (Le Roy, 1970). The other peaks, at 550 nm and $\geqslant 620$ nm, were explained as being due to spontaneous predissociation by the repulsive $^1\Pi(1_u)$ state.

The lifetime of I_2 B state levels excited by transitions coincident with Ar$^+$ and Kr$^+$ laser lines were measured by Paisner and Wallenstein (1974). Their results agreed well with previous measurements. Sakurai *et al.* (1976) have considered the effects of narrow-bandwidth detection as well as narrow-band excitation. Lifetimes of two single rotational levels of $I_2(B)$ were measured with both vibrationally resolved (1.0 nm) and rotationally resolved (0.015 nm) detection. Measured radiative lifetimes were consistent with previous measurements except that a 10% shorter lifetime was observed with single rotational line detection. The detector bandwidth dependence of the self-quenching cross sections were also measured by Sakurai *et al.* (1976). When all fluorescence is detected, the cross section refers to the average electronic quenching of the excited state and for initial excitation to $v' = 43$ the cross section $\sigma = 64\,\text{Å}^2$. When the fluorescence from the originally excited level and also the rotational levels populated by relaxation are detected, the measured cross section is due to both electronic and vibrational relaxation and $\sigma = 76\,\text{Å}^2$. For detection of just a single rotational transition, the cross section is due to the combined effects of electronic, vibrational, and rotational relaxation and $\sigma = 89\,\text{Å}^2$.

TABLE II. Summary of lifetime measurements of $I_2(B)$ as a function of v' and J'. The cross section σ for destruction of the excited-state population is also given. After Broyer *et al.* (1975)

Excited levels				
v''	J'	τ_0(ns)	$1/\tau_0(10^6\,\text{s}^{-1})$	$\sigma(\text{Å}^2)$
20	40	890 ± 50	$1\cdot12 \pm 0\cdot06$	$65\cdot5 \pm 4$
19	96	920 ± 40	$1\cdot09 \pm 0\cdot04$	$66\cdot5 \pm 3$
18	37	970 ± 35	$1\cdot03 \pm 0\cdot04$	$70\cdot5 \pm 3$
	58	955 ± 35	$1\cdot05 \pm 0\cdot04$	$69\cdot2 \pm 3$
	85	968 ± 35	$1\cdot03 \pm 0\cdot04$	68 ± 3
	104	977 ± 35	$1\cdot02 \pm 0\cdot04$	69 ± 3
17	27	1146 ± 40	$0\cdot87 \pm 0\cdot03$	66 ± 3
16	57	1227 ± 50	$0\cdot815 \pm 0\cdot03$	$61\cdot9 \pm 3$
15	63	1356 ± 60	$0\cdot737 \pm 0\cdot03$	$63\cdot7 \pm 3$
14	53	1314 ± 60	$0\cdot761 \pm 0\cdot03$	$63\cdot9 \pm 3$
13	11	1260 ± 60	$0\cdot794 \pm 0\cdot04$	$65\cdot2 \pm 3$
	73	1146 ± 50	$0\cdot873 \pm 0\cdot04$	$66\cdot3 \pm 3$
12	32	1090 ± 50	$0\cdot917 \pm 0\cdot04$	$66\cdot7 \pm 3$
	64	996 ± 40	$1\cdot004 \pm 0\cdot05$	$67\cdot7 \pm 3$
	97	797 ± 40	$1\cdot255 \pm 0\cdot06$	70 ± 3
11	8	920 ± 40	$1\cdot090 \pm 0\cdot05$	71 ± 3
	76	701 ± 30	$1\cdot43 \pm 0\cdot06$	72 ± 3
	90	605 ± 25	$1\cdot65 \pm 0\cdot07$	$71\cdot5 \pm 3$
	102	565 ± 25	$1\cdot77 \pm 0\cdot08$	$73\cdot5 \pm 3$
	112	478 ± 25	$2\cdot09 \pm 0\cdot1$	$72\cdot5 \pm 3$
	126	390 ± 30	$2\cdot56 \pm 0\cdot2$	68 ± 5
10	20	689 ± 30	$1\cdot45 \pm 0\cdot07$	$67\cdot5 \pm 3$
	70	531 ± 25	$1\cdot88 \pm 0\cdot09$	$71\cdot5 \pm 3$
	89	460 ± 25	$2\cdot17 \pm 0\cdot1$	$68\cdot5 \pm 3$
9	33	596 ± 30	$1\cdot68 \pm 0\cdot9$	$69\cdot5 \pm 3$
	39	569 ± 30	$1\cdot76 \pm 0\cdot9$	$72\cdot5 \pm 3$
	61	480 ± 30	$2\cdot08 \pm 0\cdot12$	$73\cdot5 \pm 3$
	84	380 ± 35	$2\cdot63 \pm 0\cdot25$	$76\cdot5 \pm 6$

A number of experiments, of very different types, have proved that the $B^3\Pi(0_u^+)$ state suffers a weak natural predissociation (see, for example, Chutjian, 1969, and references therein). Vigue et al. (1974, 1975) have considered the data on the magnetic and natural predissociations of the B state of iodine. They have concluded that the same electronic state, the 1_u state, is responsible for both observations. They have also shown, from a theoretical standpoint, that the natural predissociation should be dependent on the magnitude of $J'(J' + 1)$ according to

$$\Gamma_{pred} = k_{v'} J'(J' + 1). \tag{3.1}$$

(We recall that the lifetime, $1/\tau_0 = \Gamma_{rad} + \Gamma_{pred}$.) In order to confirm their predictions, Broyer et al. (1975) measured the lifetimes of about 30 individual ro-vibrational levels. Table II summarizes their lifetime measurements. For $9 \leqslant v' \leqslant 13$ the lifetimes show a clear dependence on J'. The dependence of $k_{v'}$ on the vibrational energy is shown in Fig 10.

Predissociation of $I_2(B)$ to unbound states other than the $^1\Pi(1_u)$ has recently been studied by Sullivan and Dows (1980). Low-pressure I_2 vapour was excited to a vibrational state v' and the broad-band fluorescence was monitored. A modulated electric field was then applied which induced crossing from the B state to a dissociative state, thus causing a reduction and a modulation of the fluorescence intensity. The electric field can couple the $B^3\Pi(0_u^+)$ state to predicted, unbound 0_g^+ and 1_g^+ states. The dependence of the induced predissociation on v' was studied and the results indicated that an unbound state, probably the 0_g^+, crosses the B state near $v = 3$–4 on the outer limb. The transition dipole moment between the B and the unbound state was 0·03–0·04 D.

High vibrational levels of the $B^3\Pi(0_u^+)$ state near its dissociation limit and also highly excited electronic states in the 5 eV region have been studied in a series of elegant two-photon experiments by Danyluk and King (1976a, b, 1977a, b) and King et al. (1980). In single-photon electronic absorption spectra of heavy molecules such as iodine, the very high density of ro-vibrational

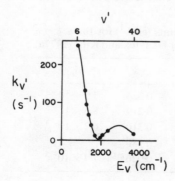

Fig. 10. $k_{v'}$ as a function of the vibrational energy E_v in the $B^3\Pi(0_u^+)$ state of I_2. The curve is just a rough interpolation between the experimental points. After Broyer et al. (1975).

states close to the dissociation energy produces a spectrum which is indistinguishable from the adjoining continuum. Single-photon laser excitation of I_2 $B-X$ gives a complex, partially resolved spectrum from which the dissociation energy can be estimated directly to only $\pm 12 \, \text{cm}^{-1}$ (Ornstein and Derr, 1976). Danyluk and King (1976a, 1977b) have used two-photon sequential absorption spectroscopy, in which the iodine molecule is first excited from the ground state to the B state and then to the higher E state. It was possible to excite selectively rotational band structure for the $B-X$ system up to within $0.5 \, \text{cm}^{-1}$ of the dissociation limit. The amount of rotational structure observed depended on the ratio of the bandwidth of one of the exciting dye lasers to the rotational constant in the E state. This ratio could be varied to change the observed rotational line density and complexity. From their study using this technique under very high resolution, Danyluk and King (1977b) obtained and improved value for the dissociation limit of the $B^3\Pi(0_u^+)$ state, $D_0^0 = (20\,043{\cdot}063 \pm 0{\cdot}20) \, \text{cm}^{-1}$. The higher levels studied also allowed the determination of values for the constants C_5, C_6, and C_8 in the long-range potential out to an internuclear separation of $15 \, \text{Å}$. Recently, King *et al.* (1980) have examined the analogous levels of $^{129}I_2$ by the same technique. Vibrational and rotational constants were obtained for B state levels with $v' = 71-79$ and the dissociation energy was measured to be $D_0^0(^{129}I_2) = 20\,043{\cdot}897 \, \text{cm}^{-1}$. The results from studies of both $^{127}I_2$ and $^{129}I_2$ were combined to give improved values of the long-range constants of the B state.

The ground $X^1\Sigma_g^+$ state of iodine has gerade symmetry. Thus, excited states of even parity are not accessible by single-photon absorption from the ground state and very little is known about them. Danyluk and King (1976b, 1977a), using a similar technique to that described above, have excited iodine molecules first to the ungerade B state and then into levels of gerade excited electronic states in the energy range of $5{\cdot}0-5{\cdot}5 \, \text{eV}$. Five vibrational progressions have been identified and analysed, and were labelled $\alpha, \beta, \gamma, \delta$, and ε. Vibrational and rotational constants were obtained for these five excited electronic states, which have ion-pair character. The symmetries of two of

TABLE III. Vibrational constants (cm^{-1}) for the $\alpha, \beta, \gamma, \delta$, and ε excited states of I_2. After Danyluk and King (1977a).

	T_e	ω_e	$\omega_e x_e$	$\omega_e y_e$
α	40281·8	104·83	0·324	$3{\cdot}9 \times 10^{-3}$
β	40925·1	104·07	0·219	$8{\cdot}6 \times 10^{-5}$
γ	41513·6	95·53	0·310	$6{\cdot}7 \times 10^{-3}$
δ	41682·9	100·41	0·180	—
ε	42493·7	95·75	0·0265	$-7{\cdot}4 \times 10^{-3}$

these states were identified as 0_g^+, and the other three as either 0_g^+ or 1_g^+. All five of the excited states have vibrational frequencies of the order of $100 \, \text{cm}^{-1}$ and are summarized in Table III. The equilibrium internuclear distances are $r_e = 3 \cdot 676$ and $3 \cdot 513 \, \text{Å}$ for the α and β states respectively.

B. Bromine, Br_2

The discrete region of the $B^3\Pi(0_u^+)-X^1\Sigma_g^+$ system of bromine extends from about 510 to 740 nm. Despite the simple P and R branch structure of these bands, the low magnitudes of the vibrational and rotational constants give rise to a strongly overlapped, dense spectrum. The complexity of the spectrum is enhanced by the natural occurrence of the three isotopic species $^{79}Br_2$, $^{79}Br^{81}Br$, $^{81}Br_2$ in the approximate ratio 1:2:1. The most comprehensive and accurate spectroscopic constants for the $B-X$ system were obtained from a high-resolution absorption study on separated isotopes by Barrow et al. (1974). High-resolution data for $v'' > 10$ do not exist, but low-resolution observation of the UV resonance fluorescence spectrum by Rao and Venkateswarlu (1964) provided constants ($\pm 1 \, \text{cm}^{-1}$) for v'' levels up to 36. Improved RKR curves for the B and X states were determined by Barrow et al. (1974) which confirmed the validity, for $v' \leqslant 19$, of the Franck–Condon factors calculated by Coxon (1972a).

Bromine is the only homonuclear interhalogen for which accurate data for the weaker $A^3\Pi(1_u)-X^1\Sigma_g^+$ transition have been obtained. This system extends from about 640 to 710 nm and is difficult to separate from the more intense, overlapping $B-X$ system. However, high-resolution absorption spectra were recorded by Horsley (1967) and by Coxon (1972b). RKR curves for the A state and Franck–Condon factors for the $A-X$ system were calculated by Coxon (1972c).

Laser-induced fluorescence on the $B-X$ system of Br_2 was first reported by Holzer et al. (1970a, b). A progression of fluorescence doublets was observed following excitation with the 514·5 nm line from an argon-ion laser. From a comparison of the measured vibrational constants with those from absorption spectra, the emitting isotope was identified as $^{81}Br_2$. The doublet spacing gave a J' value of 15, and, although the vibrational level could not be defined accurately, it was thought to be close to $v' = 39$. This suggestion was later confirmed by Coxon (1973), based on the agreement between the relative band intensities of the fluorescence progression and the calculated Franck–Condon factors. More recently, Hozack et al. (1980) have reported a new study of LIF in Br_2, excited by a Kr^+ laser.

Coxon (1973) calculated a value for the radiative lifetimes of the B and A states of Br_2 using values of the electronic transition moment R_e, obtained from absorption spectral data. His estimates were $\sim 25 \, \mu s$ for the B state lifetime and $\sim 2 \, \text{ms}$ for the A state lifetime. The first measurements of the radiative lifetime of the B state using laser-induced fluorescence techniques produced widely different values for τ_0. Capelle *et al.* (1971) used a dye laser with an output bandwidth of $1-8\,\text{Å}$ to excite vibrational levels between $v' = 1$ and 31. Lifetimes and self-quenching cross sections were measured by observing directly the decay of fluorescence as a function of pressure. Large variations, by a factor of 8, in the lifetime were observed, ranging from more than $1 \cdot 2\,\mu s$ near $v' = 27$ to less than $0 \cdot 15\,\mu s$ near $v' = 14$. This variation was interpreted as due to spontaneous predissociation of the B state through dissociative states. However, Coxon (1973) has pointed out the uncertainty with these measurements owing to possible contamination by molecular iodine.

Zaraga *et al.* (1976) used narrow-linewidth excitation near 558 nm to populate specific ro-vibrational (v', J') states in $^{81}Br_2(B)$. From absolute absorption measurements, which permitted an accurate calculation of the electronic transition dipole moment, an estimate of $20\,\mu s$ was made for the radiative lifetime of the B state. Actual fluorescence decay lifetimes at zero pressure were found to be much shorter than the radiative lifetime because of the effects of predissociation. Zero-pressure lifetimes of $0 \cdot 11$, $0 \cdot 31$ and $0 \cdot 5\,\mu s$ were measured for the levels (16,48), (19,40) and (23,46) respectively.

Several specific levels near the dissociation limit of $Br_2(B)$ were studied by McAfee and Hozack (1976). The lifetime was found to depend sharply on v' and J' and they reported lifetimes ranging from $3 \cdot 57$ to $1 \cdot 57\,\mu s$ for various levels with $40 \leqslant v' \leqslant 47$ and $16 \leqslant J' \leqslant 42$. The quenching cross sections were found to increase systematically, from $70\,\text{Å}^2$ for $v' = 40$ to $88\,\text{Å}^2$ for $v' = 46$ and 47, as the dissociation limit was approached. A more detailed study of this type has been reported recently by Luypaert *et al.* (1980). These workers report variations of lifetime with v' and J' that are consistent with heterogeneous predissociation of the B state (see below).

As for Cl_2 (see next section), the radiative lifetime of bromine trapped in solid argon has been measured (Bondybey *et al.*, 1976). In contrast with the chlorine case, excitation into the bound levels, or into the continuum of the $B^3\Pi(0_u^+)$ state of bromine isolated in a rare-gas matrix, resulted in emission from the $v' = 0$ level. Thus, there was fast vibrational relaxation but no radiationless transitions into other electronic states and no communication between the different components of the $^3\Pi$. The radiative lifetime was found to be $7 \cdot 6\,\mu s$ in argon, and $8 \cdot 6$ and $6 \cdot 4\,\mu s$ in neon and krypton respectively. Based on extrapolation, Bondybey *et al.* (1976) predicted a lifetime of $(12 \pm 2)\,\mu s$ for $v' = 0$ of the gas-phase molecules.

An extensive and systematic study of the laser excitation spectrum, radiative lifetimes, and quenching rate constants for both the $B^3\Pi(0_u^+)$ and $A^3\Pi(1_u)$ states of bromine has recently been made by Clyne and Heaven (1978) and Clyne *et al.* (1980b, c). Clyne and Heaven (1978) first reported the observation of the high-resolution laser excitation spectrum of the $B-X$ system. The

TABLE IV. Values of $k_{v'}$ and radiative lifetimes τ_∞ for various v' states of $Br_2 \, B^3\Pi(0_u^+)$. After Clyne and Heaven (1978).

v'	$k_{v'}(10^3 \, s^{-1})$	$1/\tau_\infty \, (10^5 \, s^{-1})$	$\tau_\infty(\mu s)$
11	7·3	1·95	$5·1^{+1·5}_{-0·9}$
14	6·7	1·35	$7·4^{+1·5}_{-1·0}$
19	4·9	0·88	$11·3^{+2·0}_{-1·5}$
20	3·7	1·65	$6·1^{+2·5}_{-1·4}$
23	3·9	0·64	$15·7^{+2·5}_{-1·9}$
24	2·7	0·98	$10·2^{+1·5}_{-1·2}$

observed rotational intensity distributions were strongly non-Boltzmann. The intensity maxima in all bands appeared at J'' values considerably lower than predicted by theory or observed in the absorption spectra. For example, the intensity maximum in the 14–2 band was near $J'' = 10$ rather than the expected value of $J'' = 35$ for a 298 K distribution. Anomalies were also observed in the relative intensities of the vibrational bands. For instance, the calculated relative intensities of the 5–3 and 3–2 bands are 210:1, but in the excitation spectrum they showed similar intensities. All of the abnormalities noted in the intensity distributions were explained on the basis of rotationally dependent predissociation as confirmed by their radiative lifetime measurements.

Measurements of collision-free lifetimes τ_0 of $Br_2(B)$ excited molecules were made over a range of rotational states J', for several vibrational levels in the range $23 \geqslant v' \geqslant 11$ (Clyne and Heaven, 1978), and for lower levels $v' > 2$ (Clyne and Heaven 1981a). A strong dependence of $1/\tau_0$ on J' was found and attributed to a heterogeneous predissociation of the emitting $B^3\Pi(0_u^+)$ state by the $^1\Pi(1_u)$ state. The measured lifetimes gave a good fit to the equation

$$1/\tau_0 = 1/\tau_\infty + k_{v'}J'(J' + 1).$$

The results are summarized (Table IV) in terms of values of τ_∞ and $k_{v'}$. The mean value of the extrapolated zero-pressure lifetime (τ_∞) was 8·1 μs for $24 \geqslant v' \geqslant 11$ (Clyne and Heaven, 1978). Examples of plots of $1/\tau_0$ against $J'(J' + 1)$ are given in Fig. 11 for $v' = 11$ and 14, showing the linear relationship of these parameters and the dependence of the predissociation on rotational quantum number.

In the later work by Clyne and Heaven (1980b), the radiative lifetime of $Br_2(B)$ was measured directly from fluorescence decay, using excitation of several rotational levels of the $v' = 2$ vibrational level, which is not subject to predissociation. The mean value found for τ_R was $(12·4 \pm 0·2) \, \mu s (1\sigma)$; this value is expected to be more reliable than the estimate, $\tau_\infty = 8·1 \, \mu s$, as found

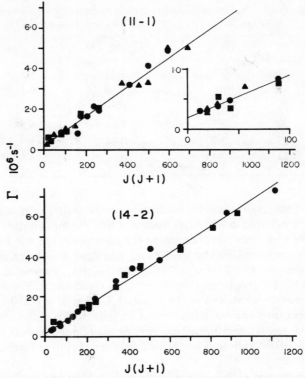

Fig. 11 Rotationally dependent predissociation in $Br_2(B)$: ▲,$^{79}Br_2$; ■,$^{81}Br_2$; ●,$^{79}Br^{81}Br$. After Clyne and Heaven (1978).

from extrapolation to $J' = 0$ of the data for predissociated levels.

Very few data are available in the literature for comparison with the present value of τ_R. Possibly the best comparison is with calculations from the work of Le Roy *et al.* (1976) on the absorption coefficient of Br_2 in the $B–X$ continuum. These calculations (see Clyne *et al.*, 1980c) give $\tau_R = (11.3 \pm 1.0)\,\mu s$ for $v' = 2$, in excellent agreement with the present work.

The vibrational level $v' = 7$ of $Br_2(B)$ in unique in that the lifetimes show a strong dependence upon isotopic composition. The $v' = 7$ rotational levels are relatively weakly predissociated, and unlike all other levels except $v' = 3$, $(\tau_0)^{-1}$ does not depend linearly on $J'(J' + 1)$. There is a periodic dependence of $(\tau_0)^{-1}$ upon $J'(J' + 1)$ which leads to some very interesting intensity anomalies in the excitation spectrum (Clyne *et al.*, 1980c),

These results for $v' = 7$, and the overall dependence of $k_{v'}$ upon v', can be rationalized in detail through calculations of the Franck–Condon overlap between the B-state and the $^1\Pi(1_u)$ state. The calculations give a well defined potential function for the $^1\Pi(1_u)$ state (Clyne and Heaven, 1981c).

Results for collisional deactivation of $Br_2(B)$ have also been reported. The analysis is complicated for predissociated levels, since rotational energy transfer changes the lifetime of the initially formed state. However, this problem does not arise for excitation of the $v' = 2$ levels, for which electronic quenching rate constants, as follows, were reported by Clyne et al. (1980c):

$$k_{Q,Br} = (5 \cdot 8 \pm 0 \cdot 3) \times 10^{-11} \, cm^3 s^{-1},$$

$$k_{Q,Ar} = (1 \cdot 7 \pm 0 \cdot 3) \times 10^{-11} \, cm^3 s^{-1}.$$

The rate constants for collisional predissociation for the $v' = 14$ level were found to be considerably larger in magnitude, namely $(4 \cdot 2 \pm 1 \cdot 3) \times 10^{-10} \, cm^3 s^{-1}$ for Br_2 as collision partner (Clyne et al., 1980a). As in the B state manifolds of other halogens and interhalogens, vibrational energy transfer is of significance (Clyne et al., 1980a).

Finally, note should be taken of a study of the $A^3\Pi(1_u)$ state of Br_2 (Clyne et al., 1980b). In this work, resolved ro-vibrational levels of $Br_2(A)$ were excited with a dye laser operating near 705 nm. No major problems of overlap with the B state were experienced in this range of excitation wavelengths. Identification of the $A-X$ ro-vibrational transitions was made through rotational combination differences and vibrational isotope splittings. Lifetimes of the F^- components of the Ω-doublet of the $v' = 12$, $J' = 23$ state were measured as a function of pressure of Br_2 and Ar, as bath gases.

The radiative lifetime of $Br_2(A)$ ($v' = 11$) was reported to be $(347 \pm 50) \, \mu s$ (1σ). This magnitude for τ_R is at the upper limit of lifetimes that can be measured using LIF. However, it is believed that systematic errors do not seriously affect the determination of τ_R for $Br_2(A)$ (Clyne et al., 1980b).

The collisional behaviour of the A-state manifold is dominated (for Br_2 and Ar as bath gases) by vibrational transfer, which for Br_2 bath gas occurs at virtually every collision. V–T transfer in $Br_2(A)$ with Ar is about one order of magnitude slower. On the other hand, electronic quenching of $Br_2(A)$ by Br_2 and Ar is slow.

It can be seen from the above summary of collision dynamics in Br_2 that LIF provides an extremely powerful method of examining the details of electronically excited molecules, including predissociation, radiation, and a multitude of collisional processes.

C. Chlorine, Cl_2

The $B^3\Pi(0_u^+)-X^1\Sigma_g^+$ visible band system of Cl_2 has been studied in much detail, providing accurate information on the lowest vibrational levels of the ground state and on levels from $v' = 5$ to almost the convergence limit of the excited state (Richards and Barrow, 1962; Douglas et al., 1963; Clyne and Coxon, 1970b). Further spectroscopic data for the ground state are available from a study of the resonance fluorescence spectrum in the vacuum ultraviolet by Douglas and Hoy (1975). Recently, Coxon and Shanker (1978) have reported the first rotational analysis of the afterglow emission spectrum of Cl_2 providing new information for B state levels, $v' \leqslant 6$. Accurate term values and rotational data are now available for all levels of the B state in the range $0 \leqslant v' \leqslant 31$. RKR curves and rotationally dependent Franck–Condon factors have been recalculated by Coxon (1980) in the light of the new data. Tentative evidence for weak emission from the $A^3\Pi(1_u)–X^1\Sigma_g^+$ system in the chlorine afterglow has been reported (Coxon, 1973).

The first observations of laser-induced fluorescence on the $B–X$ system of Cl_2 were by Holzer et al. (1970a, b). Raman and fluorescence spectra were excited with a multimode argon-ion laser at 514·5, 501·7, 496·5, 488·0, and 476·5 nm. Fluorescence was observed following excitation at 488·0 nm only. This fluorescence was very weak and even at 8 Torr the Raman band was 10 times more intense than the fluorescence lines. Hwang and Chang (1978) carried out similar experiments but, instead of a multimode laser, a single longitudinal mode argon-ion laser was used. The single mode was tuned through the lasing profiles of the argon-ion lines and five progressions of fluoresence lines were found. The excitation transitions were given as: $R37$ of the 22–0 band and $P13$ of the 18–0 band in $^{35}Cl_2$ at 488·0 nm; $P39$ of 16–0 in $^{35}Cl_2$ at 496·5 nm; $R23$ of 13–1 in $^{37}Cl_2$; and $R59$ of 18–1 in $^{35}Cl_2$ at 514·5 nm. The fluorescence intensities were very weak. It should be noted (see later) that all of these excited-state vibrational levels lie above the energy of dissociation of ground-state Cl_2 to two $Cl \, ^2P_{3/2}$ atoms. Also, these studies (Holzer et al., 1970a, b; Hwang and Chang, 1978), which were carried out using pressures of Cl_2 greater than about 10 Torr, showed little or no evidence for vibrational or rotational relaxation in the fluorescence spectra. This indicates that the excited B state levels with $v' \geqslant 13$ must be short-lived.

Laser-excited emission from chlorine trapped in rare-gas matrices has also been studied (Ault et al., 1975; Bondybey and Fletcher, 1976). Bondybey and Fletcher found that chlorine molecules initially excited either into the $B0_u^+$ or into the $^1\Pi(1)$ continuum underwent a fast radiationless transition and vibrational relaxation and emission was observed from the $v = 0$ level of a new low-lying electronic state. This new state had a vibrational spacing $\omega_e \simeq 280 \, cm^{-1}$ and was located $650 \, cm^{-1}$ below the $B^3\Pi(0_u^+)$ electronic state. The lifetime of this new state was found to be 76 ms in an argon matrix and it was assigned as the $^3\Pi(2_u)$ electronic state.

In the gas phase, emission is observed directly from the B state. The first study of the time-resolved $B–X$ laser-induced fluorescence of Cl_2 was by Huie et al. (1976). They used a broad-band ($\sim 16 \, cm^{-1}$) pulsed dye laser to excited pressures of Cl_2 between 140 and 1560 mTorr near 480 nm. Maximum fluorescence intensity was reported using 485 nm as the excitation wavelength and no fluorescence was seen for excitation wavelengths above 505 nm. Although the fluorescence signals were weak, Huie et al. (1976) reported rate constants for deactivating collisions of $Cl_2(B)$ with $Cl_2(X)$ and with Ar and ascribed these rate constants to collision-induced predissociation. They were unable to determine values for the radiative lifetime of $Cl_2(B)$, but found lifetimes as long as $5\,\mu s$ at pressures near 150 mTorr.

A detailed and extensive series of studies concerned with the spectroscopy and kinetics of the $B-X$ system of Cl_2 studied by laser-induced fluorescence technique has been reported by Clyne and McDermid (1978g, 1979b, c, d) and Clyne and Martinez (1980a, b). The first of these papers, Clyne and McDermid (1978g), reported the observation of the rotationally resolved excitation spectrum and the onset of predissociation in the $B-X$ system. For the $v'' = 0$ progression, the best Franck–Condon factors are for transitions to $14 \geqslant v' \geqslant 24$. Values for $\theta_{v'} q_{v',v''}$ exceed $2\cdot2 \times 10^{-4}$ for all these bands whose origins lie at wavelengths between $480\cdot8$ and $495\cdot6$ nm. Initial experiments were carried out with laser excitation wavelengths in this range but no fluorescence could be observed. However, when the scan was continued to longer wavelengths, above 501 nm, a relatively strong, structured, laser excitation spectrum was observed. The first band observed (i.e. shortest-wavelength band) was readily identified as the 12–0 $B-X$ band.

The laser excitation spectrum of the 12–0 band of Cl_2 is shown in Fig. 12. The P and R branches of $^{35}Cl_2$ are totally overlapped for all observed rotational lines. However, partial separation of the P and R lines in the $^{35}Cl^{37}Cl$ band can be seen for J' above that for the $P10 + R12$ blended lines.

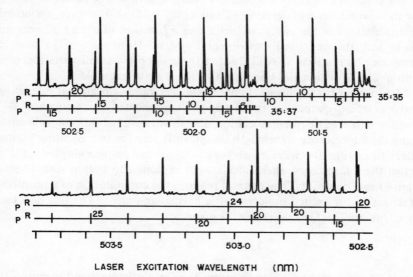

Fig. 12. Laser excitation spectrum of the 12–0 and of $Cl_2(B-X)$. Bandwidth, 1 pm. Note alternation of intensities with J' in $^{35}Cl_2$ band, and break-off in fluorescence at $J' \sim 22$ for $^{35}Cl_2$, $J' \sim 24$ for $^{35}Cl^{37}Cl$. After Clyne and McDermid (1978g)

Intensity alternation with odd–even J values was observed for all $^{35}Cl_2$ bands recorded, but not for the $^{35}Cl^{37}Cl$ bands. This is because the homonuclear $^{35}Cl^{35}Cl$ molecule exists in ortho- and para- forms, and the nuclear spin quantum number of ^{35}Cl is 3/2. Thus, a 5:3 intensity alternation with odd–even J values is expected and was observed (see Fig. 12). Well defined bands with $12 \geqslant v' \geqslant 7$ were readily identified in the excitation spectra recorded between 504 and 522 nm using 1–20 mTorr total pressure of Cl_2. Predissociation, as evidenced by a break-off in fluorescence in the excitation spectrum, was observed near (12, 22) for $^{35}Cl_2$ and near (12, 24) for $^{35}Cl^{37}Cl$. The predissociation will be discussed in more detail later.

The lifetimes of selected ro-vibrational states with $12 \geqslant v' \geqslant 7$ in the stable part of the B state manifold were initially determined by Clyne and McDermid (1979b) who found a mean collision-free lifetime $\tau_0 = (83 \pm 5)\,\mu s$. However, a subsequent redetermination of the radiative lifetime by Clyne and Martinez (1980a) showed that the results from this initial study were affected by diffusion of excited $Cl_2(B)$ molecules out of the fluorescence detection volume, thus leading to an underestimation of the lifetime. In the later study, a more energetic laser was used which made possible the observation of B state vibrational levels down to $v' = 5$. The $R15$ line of the 5–1 band of $^{35}Cl_2$ was selected by Clyne and Martinez (1980a) for lifetime and quenching studies. The $v' = 5$ vibrational level lies well below the predissociation in $Cl_2(B)$ at $v' = 12$; therefore, $Cl_2(B)$ excited molecules in $v' = 5$ are not very susceptible at low pressures to loss via collisional predissociation. The radiative lifetime for the $v' = 5$ level was determined to be $(305 \pm 15)\,\mu s$. However, systematic determinations of the radiative lifetime as a function of v' and J' were not made with the improved experimental system. However, on the basis of previous observations, it is likely that there is only a small variation of τ_0 with v' and J' within the stable part of the B state manifold.

Measurements of collisional energy transfer rates, vibrational relaxation, and electronic quenching in $Cl_2(B)$ are complicated since upward V–V transfer can lead to predissociation and downward vibrational relaxation shifts the fluorescence wavelength significantly into the red. By using various filters to change the wavelength sensitivity of the photomultiplier used to detect the fluorescence, these effects were exploited to obtain state-to-state transfer rate constants. The method requires the computation of a sensitivity of detection for excited molecules in a particular state v'. The function used by Clyne and McDermid (1979c) is given by

$$S_{v'} = \sum_{v''} v^3_{v',v''} q_{v',v''} \Phi_\lambda, \tag{3.2}$$

where $v^3_{v',v''} q_{v',v''}$ is the emission intensity of a particular band given by the cube of the centroid wavenumber of the band $v' - v''$ multiplied by the Franck–Kinz factor. The function Φ_λ describes the variation with wavelength of the quantum efficiency for the photomultiplier plus filter combination.

Table V summarizes the determinations of the total second-order rate constant k_M, for removal of $Cl_2(B)$ molecules as measured by Clyne and McDermid (1979c). No significant dependence of k_M upon initial rotational state J' was discerned over the range $40 \geqslant J' \geqslant 0$. However, the mean values of k_M for each vibrational state varied strongly with v'. Applying the concept of predissociation by vibrational transfer to the unstable levels of $Cl_2(B)$, k_M is equal to the sum of rate constants

$$\sum_{}^{\Delta v} k_{v,\Delta v}$$

for vibrational energy transfer in collisions out of the initial state v into all states $v + \Delta v$ that lie above the predissociation energy. Since multiple collisions were eliminated in the experiments by operating at low pressures, the large values of k_M found for levels $v' = 9$ and 10 indicate that vibrational energy transfer with $|\Delta v| > 1$ have an appreciable probability. Assuming that

TABLE V. (a) Measurements of collisional rate constants k_M for depletion of excited $^{35}Cl^{35}Cl(B, v', J'')$ molecules at 298 K.

Initial vibrational state v'	Range of J' values	No. of runs	k_M $(10^{-10}\,cm^3\,molecule^{-1}\,s^{-1})$
12	2–20	8	3.9 ± 0.3
11	6–36	30	2.4 ± 0.4
10	10–37	4	1.2 ± 0.3
9	10–40	4	$0.6_4 \pm 0.2_0$

TABLE V. (b) $k_{v,\Delta v}$ values and stabilization energies.

Initial vibrational state v'	Energy $\Delta U(cm^{-1})$ below predissociation[a]	$\exp(-\Delta U/kT)$ at 298 K	$k_{v,\Delta v}(10^{-10}\,cm^3\,molecule^{-1}\,s^{-1})$
12	0.0	1.000	$1.5\ (v = +1)$
11	131.0	0.656	$1.2\ (v = +2)$
10	272.9	0.415	$0.5_6\ (v = +3)$
9	425.7	0.254	
8	589.4	0.150	
7	764.1	0.086	

[a] Predissociation occurs at an energy of 19 999 cm^{-1} above the zero-point state of $Cl_2\,X^1\Sigma_g^+$. Energies tabulated are for $J' = 21$, near the Boltzmann maximum of J' at 298 K. After Clyne and McDermid (1979c).

the rate constants for vibrational transfer are independent of v, values for $k_{v,1}, k_{v,2}$, and $k_{v,3}$ can be deduced. These values are also listed in Table V.

The rate of electronic quenching of $Cl_2(B)$ has also been studied by Clyne and McDermid (1979c) and Clyne and Martinez (1980a) and the results of the two studies are in good agreement. The main conclusion of these studies is that electronic quenching of $Cl_2(B)$ by $Cl_2(X), O_2, N_2$, and Ar collision partners is uniformly inefficient. Values for k_Q range from $5 \times 10^{-12} \, cm^3$ molecule^{-1} s^{-1} or less for Ar, through $6 \cdot 4 \times 10^{-12} \, cm^3$ molecule^{-1} s^{-1} for Cl_2, up to $1 \cdot 0 \times 10^{-11} \, cm^3$ molecule^{-1} s^{-1} for O_2 and N_2 (Clyne and Martinez, 1980a).

Observation of predissociation in the laser excitation spectrum has already been noted. The exact rotational levels for the onset of predissociation were determined from lifetime measurements to be $J' = 21$ for $^{35}Cl_2$ and $J' = 24$ for $^{35}Cl^{37}Cl$ in the level $v' = 12$. The energies of these states can be calculated accurately, leading to new upper limits for the ground-state dissociation energy of Cl_2 (Clyne and McDermid, 1979b). Taking account of vibrational zero-point energies in both isotopic species, the corresponding energies referenced to the potential energy minimum of the ground state are $(12,21)$, $^{35}Cl_2 = 20\,278 \cdot 0 \, cm^{-1}$ and $(12,24)$, $^{35}Cl^{37}Cl = 20\,271 \cdot 9 \, cm^{-1}$. These data were based on the ground-state constants of Douglas and Hoy (1975).

Although all vibrational levels above $v' = 12$ are predissociated by inter-action of the $B^3\Pi(0_u^+)$ state with one or more states of 1_u character, Clyne and McDermid (1979d) were able to observe very weak fluorescence when Cl_2 was excited at wavelengths between $501 \cdot 3$ and $480 \cdot 0$ nm. These spectra involved bands of $Cl_2 B-X$ with $13 \leqslant v' \leqslant 25$. A heterogeneous predissociation of this type (with $\Delta\Omega = 1$) can be regarded as a Coriolis interaction (Hougen, 1970) having a matrix element which involves $[J(J + \frac{1}{2})]^{1/2}$. Thus the probability of predissociation is expected to show a dependence on $J'(J' + 1)$ according to

$$1/\tau_0 = 1/\tau_\infty + k_{v'}J'(J' + 1). \qquad (3.3)$$

Such behaviour of the data was indeed observed by Clyne and McDermid (1979d) and is shown in Fig. 13. This figure exemplifies the very strong dependence of the lifetime on initial rotational level. A summary of the $k_{v'}$ values for all vibrational levels studied is given in Table VI. These values of $k_{v'}$ for Cl_2 are much larger than the corresponding values for $Br_2(B)$ and $I_2(B)$ states. These measurements were confirmed in a study by Clyne and Martinez (1980b) with the additional observation that those states with zero rotation $(J' = 0)$ are not predissociated and have lifetimes equal to the radiative lifetime, i.e. $\sim 305 \, \mu s$.

The nature of the predissociation at levels $v' \geqslant 12$ is very interesting. Since $J' = 0$ states evidently are not predissociated, the interaction is heterogeneous; thus, the $B^3\Pi(0_u^+)$ state is predissociated by a 1_u-type state. This could be

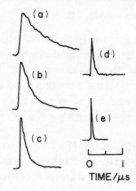

Fig. 13. Lifetime determinations of the $(13, J')$ rovibrational states in $^{35}Cl_2$ (B): fluorescence decay curves. Note shortening of lifetime as a function of J'. (a) $J' = 0$; (b) $J' = 1$; (c) $J' = 2$; (d) $J' = 4$; (e) $J' = 6$. After Clyne and McDermid (1979d)

TABLE VI. Rotationally dependent predissociation in $^{35}Cl^{35}Cl(B)$.

v'	$k_{v'}(10^5 \text{ s}^{-1})$		
	Clyne and Martinez (1980a)	Clyne and McDermid (1979d)	Recommended value
13	$8\cdot2 \pm 0\cdot4^a$	$7\cdot1 \pm 0\cdot4$	$7\cdot7 \pm 0\cdot4$
14		$5\cdot9 \pm 0\cdot4$	$5\cdot9 \pm 0\cdot4$
15		$5\cdot3 \pm 0\cdot4$	$5\cdot3 \pm 0\cdot4$
16		$4\cdot1 \pm 0\cdot5$	$4\cdot1 \pm 0\cdot5$
18		$3\cdot8 \pm 0\cdot5$	$3\cdot8 \pm 0\cdot5$
19		$4\cdot2 \pm 0\cdot7$	$4\cdot2 \pm 0\cdot7$
21		$2\cdot9 \pm 0\cdot5$	$2\cdot9 \pm 0\cdot5$
22		$2\cdot4 \pm 0\cdot4$	$2\cdot4 \pm 0\cdot4$
23		$1\cdot6 \pm 0\cdot5$	$1\cdot6 \pm 0\cdot5$
24	$1\cdot0 \pm 0\cdot2$	$1\cdot7 \pm 0\cdot5$	$1\cdot2 \pm 0\cdot3$
25	$1\cdot1 \pm 0\cdot2$	$1\cdot2 \pm 0\cdot4$	$1\cdot1 \pm 0\cdot3$

^{a} For $^{35}Cl^{37}Cl$, $k_3 = (6\cdot4 \pm 0\cdot3) \times 10^5 \text{ s}^{-1}$.

either the well known repulsive $^1\Pi(1_u)$ state, or the shallowly bound $A^3\Pi(1_u)$ state referred to above. By analogy with I_2 and Br_2, one would expect the $^1\Pi(1_u)$ state to be responsible. However, Clyne and Heaven (1981c) have found the Franck–Condon factors for the $B-^1\Pi(1_u)$ interaction to be very small, for all feasible configurations of the $^1\Pi(1_u)$ state. In addition, their calculations predict an oscillatory behaviour, similar to that found for I_2 and Br_2, for the variation of $k_{v'}$ with v'. This oscillatory behaviour is not observed in Cl_2. Clyne and Heaven found that the tentative potential energy function of Cl_2 $A^3\Pi(1_u)$ (given by Coxon) in fact gave a good prediction of the observed variation of $k_{v'}$ with v'. Thus, it appears that the $B-A$ predissociation is the dominant effect in Cl_2.

D. Iodine Monobromide, IBr

Three distinct band systems extending from the visible into the infrared spectral regions have been observed for IBr. These band systems were assigned by Brown (1932) as transitions from the $X^1\Sigma^+$ ground state to the $A^3\Pi(1)$, $B^3\Pi(0^+)$, and $B0^+$ states. All the systems were recorded in absorption with high resolution by Selin (1962a, b) and by Selin and Soderborg (1962). Some 1200 IBr lines were measured, thus providing a good set of spectroscopic constants for these electronic states.

The $B0^+ - X^1\Sigma^+$ system is well developed in IBr, more so than in any other of the interhalogens. The $B0^+$ state arises because the crossing between the bound $B^3\Pi(0^+)$ state and the repulsive 0^+ state is forbidden, so that a repulsive interaction occurs at the hypothetical crossing point. This is strong enough to form a potential minimum in the 0^+ curve, causing it to become a bound state containing a number of quantized energy levels. This is illustrated in Fig. 14. No levels which could unambiguously be assigned as $^3\Pi(0^+)$ have been seen above the crossing point, but it was postulated that the quantization of this state is not completely destroyed. Thus, while the $^3\Pi(0^+)$ levels are unstable and not sharply defined above the crossing, the rotational quantization still exists. Where these levels coincided with levels of bound 0^+ state with the same rotational quantum number, a transition to the latter is possible, resulting in a sharp spectral line. Theoretical models to describe the effects of the coupling of the $^3\Pi(0^+)$ and 0^+ states were

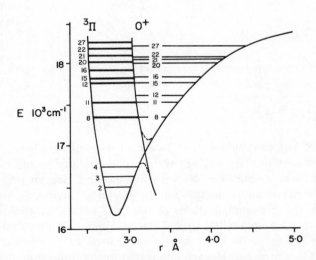

Fig. 14. Potential energy curves of the $B^3\Pi(0^+)$ and 0^+ states of IBr. The heavy lines indicate the positions of the vibrational levels to which transitions have been observed. After Selin and Soderborg (1962).

investigated by Child and Bandrauk (1970). Child (1976) concluded that the coupling must be of intermediate strength and extended the theoretical treatment developed by Child and Bandrauk (1970). The positions, widths, and intensities of the spectral lines in the $B0^+ - X^1\Sigma^+$ system were accurately predicted by this theory. Thus, Child (1976) was able to obtain spectroscopic constants and curve-crossing parameters for the $B0^+$ state.

In order to obtain information concerning the higher vibrational levels of the ground state, which cannot be observed in absorption at room temperature, Weinstock (1976) and Weinstock and Preston (1978) used a single-mode CW dye laser to excite fluorescence progressions of the $B0^+ - X^1\Sigma^+$ band. By resolving the fluorescence with a 1 m monochromator, they observed progressions corresponding to $19 \geqslant v'' \geqslant 0$. Assignment of the bands was made by comparison with Selin's absorption data. In the first of these studies, Weinstock (1976) recorded four fluorescence progressions for the $I^{79}Br$ molecule, and, in the second study, Weinstock and Preston (1978), four progressions were recorded for the $I^{81}Br$ molecule. The data were used in a nonlinear least-squares fitting routine to determine simultaneously the rotational and vibrational constants. These are probably the best constants available at this time for the $X^1\Sigma^+$ state of IBr.

Clyne and McDermid (1976b) attempted to record the laser excitation spectrum of the $B-X$ system in the region 620–657 nm. No fluorescence from IBr was seen because fluorescence from I_2 produced in the decomposition of IBr was sufficient to obscure totally any emission from IBr. The high sensitivity of their detection system was indicated by the observation of I_2 bands originating in $v'' = 5$ which has a thermal population of only 0.58% at 295 K. Failure to observe fluorescence in this study was not incompatible with the observations of Weinstock (1976) and Weinstock and Preston (1978).

The RKR turning points and Franck–Condon factors were originally calculated by Clyne and McDermid (1976a) for the $B-X$ system of IBr. Using the new molecular constants from Weinstock (1976) and Weinstock and Preston (1978), Clyne and Heaven (1980a) were able to obtain improved values for these parameters. The Franck–Condon factors for transitions to $v' \leqslant 5$, which originate from the lowest v'' ground-state levels, were very low. Thus, the transitions of $IBr(B-X)$ which should have the highest intensities in laser-induced fluorescence were hot bands, i.e. with $v'' > 0$.

Using the product of the Franck–Condon factor and the ground-state vibrational level population, Clyne and Heaven (1980a) made estimates of the intensities of a number of bands. In the light of these calculations, the region in which the most intense $B-X$ bands could be expected was found to be 630–645 nm. Preliminary experiments with IBr taken directly from a trap showed mainly fluorescence from I_2. In order to suppress iodine formation in the equilibrium,

$$IBr \rightleftharpoons \tfrac{1}{2}I_2 + \tfrac{1}{2}Br_2,$$

carefully metered amounts of Br_2 were distilled into the IBr. By adjustment of the amount of Br_2 added, it was possible to obtain IBr $(B-X)$ spectra which were free of I_2 $(B-X)$ lines and which showed only weak Br_2 $(B-X)$ lines.

Three bands of the $I^{79}Br$ and $I^{81}Br$ isotopic species were observed by Clyne and Heaven (1980a) at total IBr pressures in the range 2–5 mTorr. The bands were assigned as the 3–3, 2–2, and 2–3 transitions of IBr $(B-X)$. Rotational assignments of the 3–3 band were made by reference to Selin and Soderborg's (1962) high-resolution absorption data. The 2–2 and 2–3 bands had not been reported previously but rotational assignment up to $J'' = 52$ (2–2 band) and $J'' = 47$ (2–3 band) was possible by forming combination differences. The intensity factors for the 4–3 and 4–4 bands, which were not observed, were greater than those for the corresponding $v' = 3$ bands. Thus, it was concluded that predissociation in $v' = 4$ was at least 20 times stronger than in $v' = 3$.

Unblended rotational lines in both the 2–2 and 3–3 bands were selected for lifetime measurements. For several lines, measurements were made over a range of IBr pressures between 1 and 5 mTorr. No systematic trend of fluorescence decay rate upon pressure could be observed over this pressure interval, and therefore remaining lines were measured at a single pressure below 5 mTorr. Two major trends were noted from the lifetime measurements. First, the lifetime τ_0 showed a decrease as a function of increasing rotational energy for both $v' = 2$ and $v' = 3$. Secondly, the τ_0 values for $v' = 3$ were all

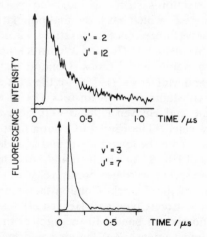

Fig. 15. Typical fluorescence decay curves of IBr (B). Decay curves are shown for the $(2, 12)$ and the $(3, 7)$ ro-vibrational states of I^8Br. Note the much shorter lifetime in the $v' = 3$ state. After Clyne and Heaven (1980a).

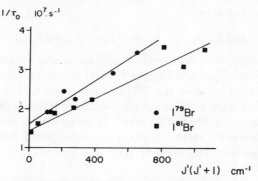

Fig. 16. Predissociated lifetimes in IBr (B). Plots of $1/\tau_0$ against $J'(J' + 1)$ for the $(3, J')$ ro-vibrational manifolds of I^{79}Br and I^{81}Br. After Clyne and Heaven (1980a).

much shorter than for $v' = 2$. The range of τ_0 for $v' = 2$ of I^{81}Br was from 290 ns for $J' = 7$ down to 70 ns for $J' = 34$. For $v' = 3$ of I^{79}Br, τ_0 varied from 72 ns for $J' = 3$ down to 28 ns for $J' = 32$.

The only other reported measurement of the radiative lifetime of IBr (B) was by Wright and Havey (1978). They measured the fluorescence decay rate following excitation of IBr with a high-energy flashlamp-pumped dye laser with a pulse duration of 150 ns. No excitation spectrum was recorded and therefore identification of the fluorescing species was impossible. It is therefore not surprising that their results bear no agreement with those from Clyne and Heaven (1980a).

As discussed above, Clyne and Heaven (1980a) found that the lifetimes of the $(3, J')$ and $(2, J')$ ro-vibrational states of IBr (B) showed strong dependences on the rotational quantum number J'. This is illustrated in Fig. 15 and 16. The rotational dependence showed that the predissociation was heterogeneous, i.e. $\Delta\Omega = 0$. For a heterogeneous predissociation, the first-order rate constant for predissociation Γ_p depends upon J' according to

$$\Gamma_p = k_{v'} J'(J' + 1). \tag{3.4}$$

Thus, the observed decay rate constant $1/\tau_0$ is given by

$$1/\tau_0 = 1/\tau_\infty + k_{v'} J'(J' + 1). \tag{3.5}$$

In cases where predissociation of rotation-free states $(J' = 0)$ is completely forbidden, the term τ_∞ may be equated to the radiative lifetime τ_R of the excited state. This was not found to be the case for IBr, and the predissociation of the $B^3\Pi(0^+)$ involved both rotationally dependent and rotation-free terms. The τ_∞ and $k_{v'}$ values measured by Clyne and Heaven (1980a) were

$$v' = 2 \text{ of } I^{81}\text{Br}: \qquad \tau_\infty = 441 \text{ ns}, \quad k_{v'} = 1\cdot1 \times 10^4 \text{ s}^{-1},$$

$$v' = 3 \text{ of } I^{81}Br: \qquad \tau_\infty = \ 70 \text{ ns}, \quad k_{v'} = 2 \cdot 1 \times 10^4 \, s^{-1},$$

$$v' = 3 \text{ of } I^{79}Br: \qquad \tau_\infty = \ 62 \text{ ns}, \quad k_{v'} = 2 \cdot 8 \times 10^4 \, s^{-1}.$$

Arguing by analogy with information that is available for the other halogens and interhalogens, Clyne and Heaven (1980a) concluded that it was probable that an interaction between the $B^3\Pi(0^+)$ state and a $^1\Pi(1)$ unbound state was responsible for the observed predissociation. If this is the case, there are two possible ways in which the predissociation could occur. First, heterogeneously via the 0^+ state or, secondly, a combined heterogeneous and homogenous predissociation via $^1\Pi(1)$.

E. Iodine Monochloride, ICl

Both the $A^3\Pi(1)$–$X^1\Sigma^+$ and $B^3\Pi(0^+)$–$X^1\Sigma^+$ systems are well known from the absorption experiments of Hulthen et al. (1958, 1960) and earlier workers. The early experiments to excite fluorescence on these transitions was hampered by interference from the much more intense fluorescence from I_2 which practically always contaminates the ICl. Disproportionation of ICl occurs according to

$$K = \frac{[Cl_2][I_2]}{[ICl]^2} = 4 \cdot 9 \times 10^{-6}$$

(Calder and Giauque, 1965). Since I_2 is soluble in ICl but Cl_2 is not, the vapour above the ICl is Cl_2 rich. ICl therefore, does not evaporate at constant composition, and it cannot be purified by distillation or by any other technique which involves pumping on it (Buckles and Bader, 1971). The reaction of ICl with water also produces I_2:

$$10ICl + 5H_2O \rightarrow I_2O_5 + 10HCl + 4I_2,$$

so care must also be taken to keep the ICl dry.

The first observation of fluorescence from ICl vapour was made by Holleman and Steinfeld (1971). They used a flashlamp-pumped tunable dye laser operating over the range 575–610 nm with a spectral bandwidth of 5–10 Å. Because of the large spectral bandwidth, identification of the fluorescing levels could not be made and the resulting fluorescence probably had significant components from impurity I_2 B–X, and ICl B–X together with the more intense fluorescence from ICl A–X which they were trying to measure. Stern–Volmer plots of the reciprocal lifetime versus ICl pressure were made in order to determine the collision-free lifetime and the rate constant for self-quenching of fluorescence. Pumping at 6068 Å gave a lifetime $\tau_0 = (110 \pm 20) \, \mu s$ and $\tau_0 = (76 \pm 15) \, \mu s$ when 5922 Å excitation was used. The corresponding self-quenching rates were $(7 \cdot 2 \pm 1 \cdot 1) \, \text{Torr}^{-1} \, \mu s^{-1}$ and $(8 \cdot 5 \pm 1 \cdot 3) \, \text{Torr}^{-1} \, s^{-1}$ respectively.

Bondybey and Brus (1975, 1976) have studied the radiative lifetimes of both the A and B state of ICl trapped in low-temperature matrices. For the A state ($v' = 0$) they reported a lifetime of 260–290 μs in the matrix and concluded that the gas-phase lifetime was probably 350–400 μs. For the B state, their ICl lifetime was roughly 2.5 μs, indicating a probable gas-phase lifetime of the order of 4 μs.

The most recent study of the A state lifetime (Harris et al., 1979) confirms the predictions of Bondybey and Brus (1976) by measuring lifetimes in the range (405 $\pm$ 40) μs to (460 $\pm$ 40) μs. Harris et al. (1979) were able to eliminate the influence of diffusion at low pressures which can allow excited molecules to move out of the observation area thus shortening the observed lifetime. Only a small variation of lifetime on excitation wavelength was observed over the range 589–669 nm. Fluorescence quenching rate constants for ICl(A) were also obtained for a number of foreign gases and for self-quenching. The self-quenching rates were $(3.1 \pm 0.5) \times 10^{-10}$ cm^3 molecule^{-1} s^{-1} (10.0 $\pm$ 1.5 Torr^{-1} s^{-1} at 604 nm, and $(2.5 \pm 0.4) \times 10^{-10}$ cm^3 molecule^{-1} s^{-1} (8.1 $\pm$ 1.2 Torr^{-1} s^{-1}) at 669 nm. These rate constants agree well with those of Holleman and Steinfeld (1971) and indicate that practically every collision of ICl(A) + ICl(X) results in electronic quenching. However, one should note the effect of vibrational relaxation as discussed for Br$_2$(A) by Clyne et al. (1980c). The lowest quenching rates were found for the inert gases, Xe, Ar, and Ne, for which the rate constants were 4.2×10^{-12}, 4.7×10^{-12}, and 4.5×10^{-12} cm^3 molecule^{-1} s^{-1} respectively.

In the wavelength region 600–620 nm, the $A^3\Pi(1)$–$X^1\Sigma^+$ absorption bands are intense and therefore it would be expected that fluorescence from the A state following laser excitation at these wavelengths would also be intense. Clyne and McDermid (1976b) recorded the laser excitation spectrum of unpurified ICl ($\sim$ 0.1 Torr in 2 Torr helium) in the range 600–620 nm and with a bandwidth of 0.01 nm. The appearance of the spectrum was complex and obviously due to many overlapped bands. Preliminary analysis showed that some of the bands were due to the A–X system of ICl and others were due to the B–X system of I$_2$, but several strong bands remained unassigned. In the course of making high-resolution (0.001 nm) measurements of these bands, it was noticed that the intensities of fluorescence from the A state of ICl and also somewhat from the B state of I$_2$ were sensitively decreased in ICl pressure, but that the yet-unassigned bands were barely affected at pressures up to 5 Torr. This immediately showed that the lifetime of the fluorescing state, later shown to be ICl $B^3\Pi(0^+)$, must be short, i.e. of the order of 1 μs or less.

High-resolution (0.001 nm) laser excitation spectra were then recorded in the region 568–650 nm at 5 Torr pressure of ICl (Clyne and McDermid, 1976b). The spectra showed a series of doublets, overlapped with a similar series of lower intensity due to the ^{37}Cl isotope. These sets of PR doublets and the absence of a Q branch confirmed the assignment of the transition as $B^3\Pi(0^+)-X^1\Sigma^+$. Only the vibrational levels $v' = 1, 2$, and 3 of the excited B state of ICl had been observed previously in absorption. Consideration of the Franck–Condon factors for the B–X system of ICl (Clyne and McDermid, 1976a) indicated that the best opportunity to observe fluorescence from $v' = 0$ was in the 0–2 band. This band was overlapped by the stronger 2–3 band and the predicted intensity of the 0–2 was about one-tenth of that of the 2–3 band. About 33 lines in all were picked out and positively assigned to the 0–2 band.

Measurements of the line positions of the 0–2 band were made by using the accurate measurements of the overlapping 2–3 band (Hulthen et al., 1960) as wavelength markers. Trial combination differences were formed, confirming that the ground state of this transition was $v'' = 2$ of $X^1\Sigma^+$ and the upper state was $v' = 0$ of $B^3\Pi(0^+)$. The measurements then allowed the first experimental determination of the spectroscopic constants for the $v' = 0$ level. Table VII was calculated using the data from Hulthen et al. (1960) and the new data on the 0–2 band from Clyne and McDermid (1976b). The bandhead wavelengths for a number of bands are shown and also indicated are those bands which were observed by laser-induced fluorescence.

As can be seen from Table VII, Clyne and McDermid (1976b) made a thorough search for fluorescence from $v' = 3$ but none was observed. In many cases, the Franck–Condon factors for transitions to $v' = 3$ are much more favourable than those for $v' < 3$ bands which were observed. It was concluded that the whole of the $v' = 3$ level was predissociated. The $v' = 2$ level was then examined carefully for signs of the onset of predissociation. It appeared that this occurred near $J' = 70$ but, because the Boltzmann population is very low for $J' = 70$ at 295 K, unequivocal identification of the precise J' for the onset of predissociation was not possible.

Radiative lifetimes for the B state of ICl are available from the laser experiments of Clyne and McDermid (1977) and R. G. Miller and J. R. McDonald (1978, private communication), and also from the absorption

TABLE VII. Wavelengths λ_{air}(nm) for ICl B–X.

v'	$v'' = 0$	1	2	3	4
0	578·373	591·420	604·956[a]	619·015	633·622
1	571·879	584·632[a]	597·856[a]	611·583[a]	625·835[a]
2	565·964	578·451	591·397[a]	604·825[a]	618·760[a]
3	560·777[b]	573·034[b]	585·733[b]	598·902[b]	612·565[b]

[a] Band observed.
[b] Band sought but not observed.

linewidth measurements of Olsen and Innes (1976) and Gordon and Innes (1979). Clyne and McDermid (1977) measured the lifetimes for initial excitation to several different rotational levels in the 2–3 and 1–3 bands. The radiative lifetimes were obtained from Stern–Volmer plots with a fairly long extrapolation to zero pressure. The lifetimes obtained were of the order of 450 ns for $v' = 2$ and 600 ns for $v' = 1$. Miller and McDonald were able to measure the lifetimes of two levels in $v' = 1$ and one in $v' = 2$. They used a Stern–Volmer plot to obtain τ_0, but since their range of pressures included measurements at below 1 mTorr the extrapolation required was very short. This should produce more reliable lifetimes than those obtained by Clyne and McDermid (1977). Miller and McDonald observed a constant fluorescence lifetime of 4·88 μs. The magnitudes of the self-quenching rate constants measured by both sets of authors were in good agreement, being $(2-3) \times 10^{-10}$ cm^3 molecule^{-1} s^{-1}. The only other fluorescence study of the B state lifetime was by Bondybey and Brus (1975) who observed fluorescence lifetimes of about 2·5 μs in rare-gas matrices. The best value for the gas-phase radiative lifetime of the stable levels of the B state of ICl, i.e. $v' \leqslant 2$, is probably that from Miller and McDonald, i.e. 4·88 μs.

No fluorescence has been observed from levels with $v' \geqslant 3$ and this has been assumed to be due to predissociation of the B state. This was confirmed by Olson and Innes (1976) who calculated the radiative lifetimes of rotational levels in $v' = 3$ from measurements of the absorption linewidth. Their lifetimes for the predissociated levels varied from about 0·2 ns for $J' < 20$ down to 0·03 ns for $J' = 41$. The radiative lifetime for the stable levels was measured to be 0·98 μs, which is in good agreement with the fluorescence studies. Gordon and Innes (1979) have recently made an in-depth examination of the nature of the predissociation in the B state from measurements of the absorption linewidth. Line profiles in the 3–1 and 2–0 bands of the B–X system, as well as several fragmentary P and R branches associated with an adiabatic B' state, were recorded using a Fabry–Perot interferometer. The potential curves of the B state and of the adiabatic B' state were found to form an avoided crossing near 3·6 Å and 18 140 cm^{-1}. The shapes of these curves were determined using the intermediate coupling model developed by Child (1976). Their measurements also provided evidence for a second curve crossing near 2·94 Å and 17 975 cm^{-1}, of the B state and an $\Omega = 1$ state. The potential curves for the $B^3\Pi(0^+)$ state and other low-lying states are summarized in Fig. 17.

Excitation to, and fluorescence from, the much higher-lying E state of ICl has also been observed using sequential two-photon laser excitation (Barnes et al., 1974). Two laser sources were used. The first laser was used to excite molecules to the A state through the A–X system and the second laser was used further to excite these molecules to the E state through the E–A system. Fluorescence was then observed on the E–A system. King and McFadden

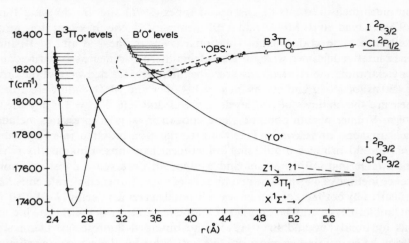

Fig. 17. Potential energy curves for the $B^3\Pi(0^+)$ state of ICl and other low-lying states of 0^+ or 1 (case (c)) symmetry. After Gordon and Innes (1979).

(1978) and King *et al.* (1979) have used this technique in a novel way to obtain information on the energy levels of the A state near its dissociation limit free from interference by other species. Two independently tunable dye lasers were pumped simultaneously by a nitrogen laser. The pump laser contained an intracavity etalon and could be pressure tuned with output bandwidths of about $0 \cdot 12 \, \text{cm}^{-1}$. The probe laser has much broader bandwidth $(1 \cdot 7 \, \text{cm}^{-1})$ and was tuned to the frequency of E–A vibrational transitions. The pump laser was then scanned through A–X transitions and fluorescence was observed on the E–A system. The X levels observed for $I^{35}Cl$ by Hulthen *et al.* (1960) were observed by this technique and a new observation of similar levels of the $I^{37}Cl$ isotope was reported. Depending on whether the pump or the probe laser is scanned, information can be obtained for both the A–X and E–A systems. King *et al.* (1979) have used both techniques and were able to determine molecular constants for the E state.

F. Bromine Monochloride, BrCl

Clyne and Coxon (1967) were the first to identify discrete bands due to BrCl lying in the red and near-infrared spectral regions. Initially, the BrCl band system was observed in emission from the chemiluminescent reaction of bromine with chlorine dioxide (ClO_2) (Clyne and Coxon, 1967; Hadley *et al.*, 1974). The recombination of ground-state bromine and chlorine atoms was also found to give emission on the same band system (Clyne and Coxon, 1967; Clyne and Smith, 1979). High-resolution absorption studies have been carried out (Clyne and Coxon, 1968, 1970a; Coxon, 1974) resulting in a complete analysis and confirmation of the assignment $B^3\Pi(0^+)-X^1\Sigma^+$.

The excited $B^3\Pi(0^+)-X^1\Sigma^+$ state correlates diabatically with one $^2P_{3/2}$ bromine atom and one spin–orbit excited $^2P_{1/2}$ chlorine atom. As with all of the heteronuclear interhalogens a strong interaction, leading to an avoided crossing, with a repulsive 0^+ state causes predissociation of the B state. No levels with $v' > 8$ have been observed in either the emission or the absorption spectra.

The Franck–Condon factors were computed by Coxon (1974) and they show that the transition probabilities for the $B–X$ system are low. Indeed, BrCl bands with $v'' = 0$ were not observed in Coxon's (1974) study and the continuum absorption was much stronger than the banded absorption at the shorter wavelengths. As a consequence, laser-induced fluorescence from BrCl(B) is weak and the only studies reported are those of Wright et al. (1977) and Clyne and McDermid (1978b, c, 1979a).

In their first study, Clyne and McDermid (1978b) recorded rotationally resolved laser excitation spectra of the $B^3\Pi(0^+)-X^1\Sigma^+$ transitions of both $^{79}Br^{35}Cl$ and $^{81}Br^{35}Cl$. For the laser-induced fluorescence studies, bromine monochloride was obtained as a product of the equilibrium between bromine and chlorine:

$$Br_2 + Cl_2 \rightleftharpoons 2BrCl, \qquad K_{eq}^{298} \simeq 7{\cdot}4.$$

In order to suppress fluorescence from Br_2, excess Cl_2 was used in the mixtures. A typical equilibrium composition of the mixture was 15% BrCl with 85% Cl_2. In spite of the unfavourable Franck–Condon factors for all vibrational bands accessible by absorption of radiation at 295 K, rotationally assigned and extensive bands with $7 \geqslant v' \geqslant 3$ were excited with a narrow-band pulsed dye laser. No fluorescence from $v' \geqslant 8$ was observed and the $v' = 7$ state of BrCl(B) was found to be extensively predissociated, based on the low photon yield of fluorescence in the 7–2 vibrational band.

The first measurements of the dynamics of the BrCl(B) state were made by Wright et al. (1977). Unfortunately, their study did not employ quantum-resolved excitation and the pressures were high, 2 Torr or more. Extrapolation of the Stern–Volmer plot to obtain the zero-pressure radiative lifetime gave a value of $\tau_0 = 18\ \mu s$ which agrees surprisingly well with the currently accepted value, $\tau_0 = (40{\cdot}2 \pm 1{\cdot}8)\ \mu s$, especially considering the long extrapolation required. Wright et al. (1977) also noted that electronic quenching of BrCl(B) by chlorine and bromine monochloride was inefficient.

This observation of inefficient electronic quenching by Cl_2 and BrCl(X) was confirmed by Clyne and McDermid (1978c). Although electronic quenching is inefficient, rapid vibrational energy transfer occurs during the lifetime of BrCl(B) molecules initially formed in a defined (v', J') quantum state. Thus fluorescence may be emitted not only from the initially formed state but also from other states populated by energy transfer. The trend is to shift the emission further into the red as the $BrCl(B)$ molecules are relaxed to lower v' levels. With the types of photomultiplier commonly used, the quantum efficiency of the photocathode falls to longer wavelength and thus the fluorescence detection efficiency is wavelength dependent. Clyne and McDermid (1978c) attempted to quantify these effects by recording the fluorescence with different filters in front of the photomultiplier.

In their first study of the state-selected kinetics of the excited $B^3\Pi(0^+)$ state of BrCl, Clyne and McDermid (1978c) used total pressures in the wide range, 30 mTorr–11 Torr. Resolved rotational levels in the vibrational states $7 \geqslant v' \geqslant 3$ were excited and the fluorescence lifetimes were measured as a function of J', v', and pressure. All rotational levels of the $v' = 7$ state were predissociated with lifetimes which shortened systematically with increase in rotational energy, from $0.91\,\mu s$ for $J' = 7$ down to $0.24\,\mu s$ for $J' = 16$. Because of the vibrational transfer effects discussed above, the lifetime measurements were carried out in two separate total pressure ranges. The low-pressure range was 30–200 mTorr and the high-pressure range was 3–11 Torr. At low pressures, a strong pressure dependence of the fluorescence decay rate constant Γ was found. This pressure-dependent rate was attributed almost entirely to rapid vibrational transfer, which depleted the population of emitters by upward V–V transfer into the unstable part of the energy level manifold above $v' = 6$. Values for the lifetime, extrapolated to zero pressure, for $v' = 3$ and 4 were determined to be $\tau_0 = 35^{+11}_{-9}\,\mu s$. At high pressures and after the first few microseconds, the fluorescence decay was exponential. In this regime Γ showed a much less marked dependence on total pressure than at low pressures. Vibrational transfer was essentially complete under these conditions and the remaining pressure dependence of Γ was attributed to quenching of a Boltzmann distribution of $BrCl(B)$ molecules, consisting of about 75% in $v' = 1$ and 25% in $v' = 2$. For a mixture of BrCl (15%) with Cl_2, the electronic quenching rate measured by Clyne and McDermid (1978c) was $k_M = 3.9 \times 10^{-13}\,cm^3$ molecule^{-1} s^{-1}. The agreement with the results of Wright et al. (1977) who obtained $k_M = 3.4 \times 10^{-13}\,cm^3$ molecule^{-1} s^{-1} is exceptional.

In the most recent laser-induced fluorescence study of the $B-X$ system of BrCl, Clyne and McDermid (1979a) used an improved experimental arrangement to remeasure the lifetimes and quenching rates. This new system permitted operation at total pressures as low as 0.15 mTorr, and thus it was possible to determine τ_0 under collision-free conditions. In addition, it was possible to determine directly the rate constants for vibrational energy transfer within $BrCl(B)$.

An example of the laser excitation spectrum from this study is shown in Fig 18. This spectrum, of part of the 4–1 and 5–1 bands, was recorded at a total pressure of 5 mTorr and shows the rotational assignment of all four isotopic bands. In previous work, only the more intense $^{79}Br^{35}Cl$ and

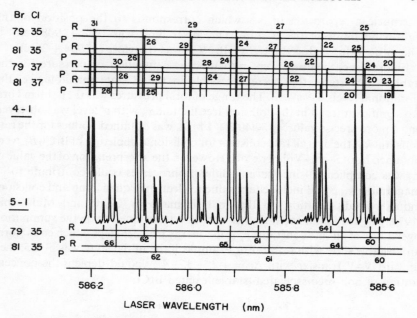

Fig. 18. Laser excitation spectrum of the 4–1 and 5–1 (high J) bands of BrCl $B–X$. After Clyne and McDermid (1979a).

$^{81}Br^{35}Cl$ bands could be identified. The higher rotational levels ($J' \geqslant 42$) in $v' = 6$ could not be observed in the laser excitation spectrum recorded at low pressures of BrCl. All levels with $v' = 7$ were also absent. If the pressure was increased, these levels could be observed because they were stabilized by relaxation. By increasing the pressure, it was thus possible to tune the laser to the exact frequency of the predissociated line in order to try to measure their lifetimes. In $v' = 6$ the lifetime of $J' = 41$ was $40·6\,\mu s$, falling to $8·8\,\mu s$ for $J' = 42$. This was the first observation of predissociation in $v' = 6$ of BrCl(B). Following this observation, a careful search was made of the high J levels of the $v' = 5$ level to see if predissociation could also be observed there (see Fig. 18 for example). A break-off in the excitation spectrum was not observed for J' levels up to 71. Higher levels were obscured by overlapping with the 3–1 bandhead. These results were used to show that the type of curve crossing causing the predissociation is Herzberg's case I(c) which is caused by an avoided crossing with a repulsive state. They were also used to determine a new upper limit for the dissociation energy of BrCl. Effective B values were calculated and used to correct the energy, in order to allow for the rotational energy barrier, of the known predissociation in the state (6, 42). The upper limit was thus calculated to be $D_0^0(^{81}Br^{35}Cl) \leqslant 17\,959·9$ cm^{-1}. A lower limit was also found by calculating an effective B for an internuclear separation of $3\,\text{Å}$ which corresponds to the position of the right-hand limits of the B state. The lower limit was $D_0^0 \geqslant 17\,908\cdot3\,\text{cm}^{-1}$. The mean value deduced for $D_0^0(^{81}\text{Br}^{35}\text{Cl})$ was therefore $(17\,934 \pm 26)\,\text{cm}^{-1}$.

Fluorescence lifetimes were measured as a function of pressure for the stable (v', J') states that were accessible in the B state manifold, i.e. the vibrational levels $6 \geqslant v' \geqslant 3$. The range of total pressures was $0\cdot11$–$1\cdot0\,\text{mTorr}$. No significant trend in the collision-free lifetime τ_0 with v' level was observed and a more precise value, $\tau_0 = (40\cdot2 \pm 1\cdot8)\,\mu\text{s}$, was obtained. Values for the rate constant k_M, the overall rate constant for collisional depletion of $\text{BrCl}(B)$, were obtained from Stern–Volmer plots. However, the interpretation of the value of k_M was complicated since several different processes could contribute to its magnitude: k_M could include terms due to electronic quenching and collision-induced predissociation. In the analysis made by Clyne and McDermid (1979a), k_M for initial excitation of state v' was assumed to be the summation over all Δv of all rate constants $k_{v,\Delta v}$ for energy transfer steps which could form an unstable state with $v' \geqslant 7$. For example, if the initial state was $v' = 5$, then upward V–V transfer with $\Delta v = +2, +3, +4, \ldots$ could deplete the concentration of non-predissociated (stable) excited BrCl:

$$k_{M,5} = k_{5,2} + k_{5,3} + k_{5,4} + \ldots.$$

The following equations were then deduced, assuming that $k_{v,\Delta v}$ was not a strong function of v:

$$k_{M,6} = k_{v,1} + k_{v,2} + k_{v,3} + k_{v,4} + \ldots,$$
$$k_{M,5} = \phantom{k_{v,1} + {}} k_{v,2} + k_{v,3} + k_{v,4} + \ldots,$$
$$k_{M,4} = \phantom{k_{v,1} + k_{v,2} + {}} k_{v,3} + k_{v,4} + \ldots.$$

Using the experimentally observed values of k_M, the following results were calculated for $\text{BrCl}(B, v')$ collisions with Cl_2.

$$k_{v,1} = k_{M,6} - k_{M,5} = (1\cdot1 \pm 0\cdot4) \times 10^{-10}\,\text{cm}^3\,\text{molecule}^{-1}\,\text{s}^{-1},$$

$$k_{v,2} = k_{M,5} - k_{M,4} = (3\cdot4 \pm 2\cdot0) \times 10^{-11}\,\text{cm}^3\,\text{molecule}^{-1}\,\text{s}^{-1},$$

$$k_{v,3} + k_{v,4} + k_{v,5} + \ldots = k_{M,4} = (6\cdot2 \pm 2\cdot3) \times 10^{-11}\,\text{cm}^3\,\text{molecule}^{-1}\,\text{s}^{-1}.$$

The corresponding rate constants for downward vibrational transfer were about a factor of 2 larger than these data for upward transfer.

G. Iodine Monofluoride, IF

Absorption spectra of the labile molecule IF have never been reported since this diatomic interhalogen is thermochemically and kinetically unstable. Emission spectra of the B–X system have been observed from several different sources. The initial discovery of this species was made by Durie (1950, 1951), who studied the emission from an $I_2 + F_2$ flame. Durie (1966) has also reported the most complete spectroscopic study of this emission. He observed bands with $11 \geqslant v' \geqslant 0$ and $12 \geqslant v'' \geqslant 0$ and was able to detect line broadening at and above the $(11, 45)$ state of IF(B) which allowed him to set an upper limit of 23 341 cm^{-1} for the ground-state dissociation energy. IF$(B–X)$ emission spectra have been observed also by Birks *et al.* (1975) in low-pressure systems containing $I_2 + F_2$; in addition, these authors reported the first observation of the emission spectrum of the $A^3\Pi(1)–X^1\Sigma^+$ system of IF.

The B state of IF is not efficiently populated from recombination of $I\,^2P_{3/2} + F\,^2P_{3/2}$ ground-state atoms. However, Clyne *et al.* (1972) have found that in the presence of metastable oxygen, $O_2\,^1\Delta_g$, IF(B) is efficiently formed in a stream of recombining ground-state atoms. It is assumed that $I\,^2P_{3/2}$ atoms are excited to $I\,^2P_{1/2}$ atoms on collision with $O_2\,^1\Delta_g$ molecules. Then, $I^2P_{1/2} + F^2P_{3/2}$ atoms can recombine directly to give IF(B). Chemiluminescence from the A and B states of IF has been observed, by Coombe and Horne (1979), in the reactions of I_2, ICl and HI with O_2F. Although emission was observed from the direct reaction of these reagents, a significant enhancement of the emission was observed when metastable oxygen molecules were present.

Ground-state iodine monofluoride can be produced in a flow system by the rapid reaction of $F\,^2P$ atoms with iodine monochloride:

$$F + ICl \rightarrow IF + Cl.$$

The rate constant k^{298} is equal to $(5 \pm 2) \times 10^{-10}$ cm^3 molecule^{-1} s^{-1} (Appelman and Clyne, 1975). The reaction of fluorine atoms with iodine,

$$F + I_2 \rightarrow IF + I,$$

is also fast, with k^{298} equal to $(4 \cdot 3 \pm 1 \cdot 1) \times 10^{-10}$ cm^3 molecule^{-1} s^{-1} (Appelman and Clyne, 1975). In the first study of the laser excitation spectrum of IF, Clyne and McDermid (1976b) used the first reaction to produce IF. Iodine monochloride usually contains significant amounts of I_2 unless rigorous purification is carried out. Since the second reaction is also rapid, any impurity I_2 in the ICl which might otherwise interfere in the excitation spectrum is also converted to IF. The alternative reaction of F atoms with ICl to form ClF,

$$F + ICl \rightarrow ClF + I,$$

has been shown by Appelman and Clyne (1975) to be a minor channel ($\sim 20\%$) and no evidence of ClF was found in the IF excitation spectra recorded by Clyne and McDermid (1976b, 1978a).

In their study (Clyne and McDermid, 1976b), IF was produced in a discharge flow system where the maximum reaction time of the IF before the fluorescence cell was about 5 ms and the total flow tube pressure was

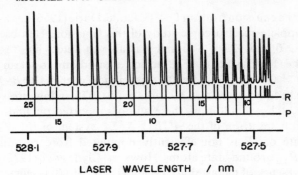

Fig. 19. Laser excitation spectrum of the 0–0 band of IF B–X with 1 pm resolution. This is the weakest of the IF bands with a Franck–Condon factor $q_{0,0} = 4.8 \times 10^{-3}$. After Clyne and McDermid (1978a).

roughly 1 Torr. The concentration of IF was estimated to be less than 10^{13} cm^{-3}. Initially, low-resolution laser wavelength scans were carried out between 430 and 500 nm. For most bands, the P and R branches were totally resolved, except very close to the bandhead, even at low resolution (0.01 nm). Since there are only single significant natural isotopes of I and F, the $B^3\Pi(0^+)$–$X^1\Sigma^+$ excitation spectrum comprises a single set of regular PR doublets. Although the initial studies carried out in the flow tube provided valuable new information on the spectroscopy and predissociation (Clyne and McDermid, 1976b) and estimates of the radiative lifetime (Clyne and McDermid, 1977), the more recent measurements by the same authors (Clyne and McDermid, 1978a) using narrow-band excitation and total pressures below 1 mTorr have provided superior data.

In this latter study (Clyne and McDermid, 1978a), high-resolution (1 pm resolution) laser excitation spectra were recorded throughout the region 440–530 nm. These cover bands with $10 \geqslant v' \geqslant 0$. Typically, rotational levels up to $J' = 65$ were observed. Transitions with $v'' = 0$ have large Franck–Condon factors (Clyne and McDermid, 1976a) and in most regions these were dominant bands. However, some bands with $v'' = 1$, though none with $v'' = 2$, were observed. The weakest transition is the 0–0 band and no observation of this band had previously been reported. Fig. 19 shows a part of this band exemplifying the high sensitivity of laser-induced fluorescence.

Fig. 20 shows part of the 9–0 band near the bandhead at 447·762 nm. The intensity contour of this spectrum recorded under collision-free conditions, below 1 mTorr total pressure, is very unusual. The lowest J' lines, i.e. P branch lines up to and including $P7$ and R branch lines up to and including $R5$, are anomalously intense in relation to the higher J' lines. This was a clear indication of the onset of predissociation, as was confirmed by measurements of the radiative lifetime as a function of J'. Levels up to and

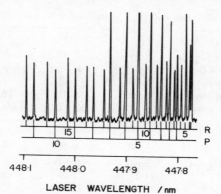

Fig. 20. Laser excitation spectrum of the 9–0 band of the IF B–X system using 1 pm laser excitation bandwidth. Note the anomalously weak rotational structure for transitions with $J' \geqslant 7$. After Clyne and McDermid (1978a).

including $J' = 6$ have normal lifetimes, $\tau_0 = (8 \cdot 8 \pm 0 \cdot 9)\,\mu s$. The subsequent rotational levels showed much shorter lifetimes ($\sim 1 \cdot 0\,\mu s$) which were virtually invariant up to at least the (9, 41) level.

Clyne and McDermid (1978a) also measured the radiative lifetime as a function of v' and J' for the entire manifold of B–X transitions observed by laser-induced fluorescence. These experiments were carried out at a total pressure between 0·5 and 1·0 mTorr. Under the conditions used, the frequencies of $IF(X) + IF(B)$ and $He + IF(B)$ collisions at 298 K were calculated to be $1 \cdot 80 \times 10^{-10}$ and $2 \cdot 40 \times 10^{-10}\,cm^3$ molecules$^{-1}\,s^{-1}$ respectively. Thus, the time $(1/e)$ between $IF + He$ collisions was $200\,\mu s$, which was equivalent to at least 20 lifetimes of $IF(B)$, and thus the experiments were essentially collision free.

Good, single-exponential, decay curves were obtained from the average of 1000 laser shots. No major irregularities in the lifetime values were found below the energy of the (v', J') level (8, 52). The mean values of the lifetimes τ_0, averaged for each vibrational state over all J' levels, are given in Table VIII. There is a slight, but significant, trend for τ_0 to increase with increasing v'.

From their observations of the excitation spectrum and the measurement of the fluorescence lifetimes, Clyne and McDermid (1978a) were able to determine that all levels with $v' = 10$ were affected by predissociation. In the 10,0 band the reciprocal lifetime varied almost linearly with rotational energy. Values of τ_0 varied from about $0 \cdot 8\,\mu s$ for low J', down to about $0 \cdot 3\,\mu s$ for $J' = 20$. The onset of this predissociation could clearly be seen at the levels (9,7) and (8,52). The energies of these levels can be determined quite accurately using the measured frequencies of transitions to these levels and

TABLE VIII. Mean lifetimes and $|R_e|^2$ values for vibronic transitions of IF(B).

v'	0	1	2	3	4
Range of J	5–45	3–49	3–48	3–50	3–59
$\bar{\tau}_0(\mu s)$	7.0 ± 0.5	6.7 ± 0.3	7.1 ± 0.6	6.9 ± 0.3	7.4 ± 0.6
$A_v(10^4\,s^{-1})$	1·44	1·49	1·42	1·45	1·34
$\bar{v}(cm^{-1})$	16 266	16 259	16 239	16 203	16 149
$|R_e|^2(10^{-1}D)$	1.07 ± 0.07	1.11 ± 0.05	1.06 ± 0.09	1.09 ± 0.04	1.02 ± 0.08

v'	5	6	7	8	8
Range of J	3–57	3–57	4–57	5–52	0–6
$\bar{\tau}_0(\mu s)$	8.1 ± 0.4	8.2 ± 0.4	8.6 ± 0.4	8.6 ± 0.5	8.8 ± 0.9
$A_v(10^4\,s^{-1})$	1·23	1·21	1·16	1·16	1·14
$\bar{v}(cm^{-1})$	16 081	15 997	15 900	15 785	15 618
$|R_e|^2(10^{-1}D)$	0.95 ± 0.05	0.95 ± 0.05	0.93 ± 0.04	0.95 ± 0.05	0.95 ± 0.10

the ground-state rotational constants (Durie, 1966). Thus, it was found that the energy difference $E(8,52) - E(9,7)$ was $225.3\,cm^{-1}$. This datum was used to determine an exact value for the dissociation energy of IF(X). An effective B value, and thus the internuclear distance corresponding to the position of the potential energy maximum in the predissociating state, could be calculated. The calculations gave $B = 0.0834\,cm^{-1}$ and $r = 3.50\,\text{Å}$. These results indicated that the predissociation in IF(B) belonged to Herzberg's case I(b), where the predissociating state is weakly bound.

There are a limited number of electronic states which can be expected to lie in the low-energy region around the predissociation, that is between the energy of two ground-state atoms and that of a ground-state iodine atom and a spin–orbit excited F atom. Clyne and McDermid (1978a) considered that curve crossings between the left-hand limb of the B state and a repulsive state were unlikely to be responsible for the observed nature of the predissociation. They proposed that the most likely predissociating state was the $^3\Pi(0^+)$, which they designated $C(0^+)$. Child and Bernstein (1973) had indicated that this should be at least weakly bound and should dissociate diabatically to $I\,^2P_{1/2} + F\,^2P_{3/2}$ atoms. These states and their estimated positions are shown in Fig. 21. It is believed that all observations could be accommodated with such a scheme.

Recently, the emission and excitation spectra of IF trapped in a solid argon matrix at 12 K have been recorded (Miller and Andrews, 1980). The 488·0 and 457·9 nm lines of an argon-ion laser were used as the excitation source for recording the fluorescence emission spectrum and a pulsed dye laser operating in the region 460–517 nm was used to record the excitation spectrum. The spectral resolution of these experiments was low and no new or improved data for IF was forthcoming from these experiments.

Laser-induced fluorescence of IF(B–X) has been used by Donovan *et al.*

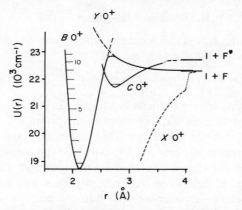

Fig. 21. Potential energy curves for the low-lying 0^+ states of IF. After Clyne and McDermid (1978a).

(1980) to determine the vibrational populations and rotational temperature of the IF product formed in the reaction

$$F + I_2 \rightarrow IF + I.$$

Their experiment employed a crossed-beam laser-induced fluorescence technique. Measurements of the relative populations in the $v = 0, 1$, and 2 levels showed a strong population inversion with an effective temperature of 3000 K, corresponding to a fraction of the total energy, $f \langle f_v \rangle = 0.6 \pm 0.1$, appearing in product IF vibration. In contrast, the product was rotationally cold with $\langle f_R \rangle \sim 2$.

Stein et al. (1980) have also used laser-induced fluorescence to study in detail the internal vibrational and rotational product distributions of IF formed in the reactions

$$F + CH_3I \rightarrow IF + CH_3$$

and

$$F + CF_3I \rightarrow IF + CF_3.$$

The mean fraction of E_{tot} entering product vibration and rotation was determined to be $\langle f_v' \rangle = 0.15 \pm 0.02$, $\langle f_R' \rangle = 0.14 \pm 0.03$ for $F + CH_3I$; and $\langle f_v' \rangle = 0.11 \pm 0.02$, $\langle f_R' \rangle = 0.10 \pm 0.02$ for $F + CF_3I$. From the comparison of the fluorescence intensities, the ratio of the total average cross-sections of the two reactions was determined to be

$$\sigma_{tot}(CF_3I)/\sigma_{tot}(CH_3I) = 2.2 \pm 0.5.$$

The vibrational and rotational product state distributions were essentially statistical, indicating that the reactions proceed through a long-lived complex.

H. Bromine Monofluoride, BrF

Bromine monofluoride is thermochemically unstable and at 300 K rapidly disproportionates via the reaction (cf. IF)

$$3BrF \rightarrow Br_2 + BrF_3.$$

Additional problems are experienced in handling BrF at pressures that would be sufficient to obtain a satisfactory absorption spectrum (> 1 Torr) because of its excellent fluorinating properties. Thus, although detailed investigations of the $B-X$ absorption systems of ClF (Stricker and Krauss, 1968) and IF (Durie, 1966) have been available for some time, similar studies of the $B-X$ absorption system of BrF have only recently been published (Coxon and Curran, 1979). Before the laser-induced fluorescence study of Clyne et al. (1976) and the high-resolution absorption study by Coxon and Curran (1979), the only data on the visible electronic spectrum of BrF were the absorption studies by Brodersen and Sicre (1955), and the emission studies in flames by Durie (1951) and in a 300 K flow system by Clyne et al. (1972). All of these studies included vibrational, but not rotational, analyses of the main $^3\Pi(0^+)-X^1\Sigma^+$ system of BrF. Additionally, Brodersen and Sicre (1955) presented evidence for a weaker $^3\Pi(1)-X^1\Sigma^+$ system. However, this assignment has been criticized by Coxon (1975) and the system could not be observed in the later high-resolution study (Coxon and Curran, 1979).

In order to study the laser excitation spectrum of BrF$(B-X)$, Clyne et al. (1976) produced concentrations of BrF between 1×10^{15} and 3×10^{15} cm^{-3} in a discharge flow system, through the rapid reaction

$$F + Br_2 \rightarrow BrF + Br.$$

An excess of F atoms was used to ensure that there was no residual Br$_2$ which could fluoresce.

Initial laser excitation scans were carried out near the 9–0 bandhead around 475–480 nm. No fluorescence was detected with excitation in this region, but, when the scan was continued to longer wavelengths, strong fluorescence from the 8–0 band was readily observed. Spectra were obtained for the $v'' = 0$ progression to excited-state vibrational levels with $8 \geqslant v' \geqslant 3$. Even though the excitation bandwidth gave a resolution of only about 0·01 nm, there was no difficulty in assigning the spectrum. The positions of the lines were measured and used to determine the rotational constants for the $B^3\Pi(0^+)$ state. This work also provided confirmation of Brodersen and Sicre's (1955) vibrational numbering. Using the new spectroscopic constants obtained from the laser excitation spectra, Clyne et al. (1976) calculated the RKR potential curves for the $X^1\Sigma^+$ and $B^3\Pi(0^+)$ states and also the Franck–Condon factors for the $B-X$ system.

Using a similar system, but with much narrower linewidth (1 pm) and greater sensitivity, Clyne and McDermid (1978d) were able to observe bands involving $v' = 0, 1$, and 2 which were not seen in the previous LIF study. Indeed, the level $v' = 0$ had not been observed at all, and so rotational and vibrational constants for this level were derived from the high-resolution laser excitation spectrum.

A striking break-off in fluorescence for excitation to levels with $v' \geqslant 9$ was first noted by Clyne et al. (1976). Although a careful examination of the line intensities in the 8–0 band was made, exact identification of the rotational level at which the onset of predissociation occurred was not possible. Clyne and McDermid (1978e, f) have made detailed studies of the dynamics of the excited state in the region of the predissociation.

In the first of these studies (Clyne and McDermid, 1978e), excitation at BrF (B) was carried out in a discharge flow system where the minimum operating pressure was of the order of 65 mTorr. Thus, the effects of collisional energy transfer were experimentally inseparable from the radiative lifetime in the measured fluorescence decay rate. Non-exponential decay curves were observed for some levels around the point of the onset of predissociation. In order to interpret these decay curves and also to determine the rotational level for which the onset of predissociation occurred, a kinetic model for R–T transfer was assembled and used to predict fluorescence decay rates in a variety of conditions. The spacing of rotational levels in BrF(B) (e.g. between $J' = 20$ and 21) is of the order of $10 \, \text{cm}^{-1}$, compared with the much larger magnitude of $\frac{3}{2}kT = 311 \, \text{cm}^{-1}$, which is the mean translational kinetic energy per molecule at 298 K. Thus, rotational energy transfer in BrF could be approximated to a classical model in which the collision efficiency with helium would be high, and multiquantum transitions would have a large probability. Such multiquantum transitions have been reported for collisions of rare gases with both heavy and relatively light molecules. Steinfeld and Klemperer (1965) reported R–T collisions with $\Delta J \geqslant + 10$ to have high collision probabilities in excited $I_2 \, B^3\Pi(0_u^+)$, whilst Polanyi and Woodall (1972) suggested $\Delta J = 1$ to 5 as being important for rotational relaxation of the light HCl molecules in collisions with H_2. Broida and Carrington (1963) found $\Delta J = \pm 1$ to ± 5 for rotational relaxation of NO $A^2\Sigma^+$ by several gases. On the basis of these arguments, Clyne and McDermid (1978e) constructed a simple, five-level model for upward and downward R–T transfer. The particle densities of the excited rotational levels were determined by sets of coupled differential rate equations. These were solved by computer and used to predict the fluorescence intensity at any particular time after the initial excitation to a particular rotational level.

From the model, Clyne and McDermid (1978e) concluded that the onset of predissociation was at $J' = 28$ in $v' = 7$. This was later confirmed in collision-free studies (Clyne and McDermid, 1978f), thus justifying the kinetic model. Other conclusions from this initial LIF study of BrF in a discharge flow system were that all rotational levels of $v' = 8$ were predissociated with lifetimes shortening systematically from $(1·74 \pm 0·15)\,\mu s$ for $J' = 1$ to 3 down to $(0·11 \pm 0·06)\,\mu s$ for $J' = 31$. The lower levels, $v' \leqslant 6$, were unperturbed with lifetimes up to $40\,\mu s$.

Improved lifetime measurements were made in a low-pressure system providing essentially collision-free conditions (Clyne and McDermid, 1978f). The lifetimes of resolved (v', J') levels throughout the B state manifold were measured at pressures below 1 mTorr. For all rotational states with $5 \geqslant v' \geqslant 0$ the lifetimes were regular with only small variations in the range $42–56\,\mu s$. The lower rotational levels of $v' = 6$ and 7 showed similar collision-free lifetimes. Predissociation in $v' = 6$ at $J' = 49$ was observed for the first time in this study. The observation of predissociation at (6,49) and (7,30) showed that the predissociation belongs to Herzberg's case I(b) with formation of a potential maximum at $r \simeq 5·3\,\text{Å}$. This is similar to the case for IF, and the consequences for the nature of the predissociating state are the same and are discussed in the section on IF. From the observations of the predissociation, the dissociation energy could be established within narrow bounds: $D_0^0(^{79}\text{BrF}) = (20\,622 \pm 20)\,\text{cm}^{-1}$.

The collision-free kinetics of the $v' = 8$ level were interesting as they were dissimilar to those of all the other levels. The entire state is considerably more stable than would be expected, considering that all its rotational levels lie at energies higher than the predissociation energy. For both isotopic species of BrF, the reciprocal lifetime was found to vary nearly with rotational energy for $J' = 4$ to 21. The lifetimes in this range were a good fit to

$$1/\tau_0 = 1/\tau_\infty + k_{v'}J'(J' + 1). \tag{3.6}$$

Fig. 22 shows plots of $1/\tau_0$ against $BJ'(J' + 1)$ for both ^{79}BrF and ^{81}BrF. Further information on the potential energy functions of the states of BrF is required before these results can be explained in detail.

Clyne and Liddy (1980) have recently redetermined the radiative lifetime of $\text{BrF}\,(B)$, and extended their study to investigate vibrational energy transfer and electronic quenching within the B state manifold. Their results gave improved values of τ_R for $7 \geqslant v' \geqslant 3$ in the range $64·2\,\mu s \geqslant \tau_R \geqslant 55·5\,\mu s$. These values are slightly greater (by $\sim 10\,\mu s$) than those reported by Clyne and McDermid (1978f), and this is explained in terms of the effects of diffusion of excited $\text{BrF}\,(B)$ molecules out of the fluorescence region in the earlier work.

Broad-band wavelength resolution of fluorescence and computer modelling were used by Clyne and Liddy (1980) to obtain rate constants at 293 K for electronic quenching and for vibrational energy transfer following collisions of $\text{BrF}\,(B)$ with the gases, O_2, Cl_2, Ar, $CHFCl_2$, and HCl. By using various combinations of bandpass and cut-off filters, the relative sensitivity of detection of fluorescence originating from different vibrational levels of the excited state could be varied. The extraction of rate constants for vibrational

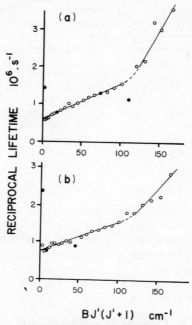

Fig. 22. Dependence of collision-free lifetime upon rotational energy in the $(8, J')$ manifold of BrF(B): rotationally dependent predissociation. (Full circles are data from overlaped lines.) (a) ^{79}BrF; (b) ^{81}BrF. After Clyne and McDermid (1978f).

energy transfer from the resulting decay curves required a computer model, and was somewhat indirect. However, since the fluorescence intensity was not sufficient to use a monochromator to isolate particular spectral regions, this was the only approach possible.

Various forms of the v' dependence of $k(v' \rightarrow v' \pm 1)$ were investigated. To explain the observed trends adequately, it was necessary that these rate constants for vibrational energy transfer increase with increasing v'. The exponential transition probability function of Troe (1977) was used to describe this dependence. A summary of the measured and modelled rate constants for energy transfer in BrF(B) with various collision partners is given in Table IX.

Electronic quenching was found to be inefficient with $k_{Q,M}$ varying from about 1.2×10^{-11} cm^3 molecule^{-1} s^{-1} for M = HCl and CHFCl$_2$, down to less than 1×10^{-13} cm^3 molecule^{-1} s^{-1} for M = Ar. For most of the bath gases used, vibrational energy transfer was more efficient than electronic quen-

TABLE IX. Summary of measured and modelled rate constants for energy transfer in BrF (B) with various collision partners.

Collision partner	Measured rate constant k_v for net removal of BrF$(B), v' = 3$ in collisions (10^{-11} cm^3)	Modelled rate constant $k_{v' \to v' \pm 1}$ for total removal of BrF$(B), v' = 3$ in collisions (10^{-11} cm^3)	Rate constant $k_{Q,M}$ for electronic quenching of BrF(B) (10^{-12} cm^3 molecule^{-1} s^{-1})	
			Measured	Modelled
O_2	3·6	2·7	2·6	2·2
Cl_2	0·5	0·85	<0·1	0·1
Ar	≤0·25	0·35	<0·1	0·05
$CHFCl_2$	2	4·1	6	6
HCl	5	5	12	12

ching. There were evidently some structural effects in the variation of $k_{v,M}$ with the nature of M. However, a wide variety of collision partners, O_2, Cl_2, $CHFCl_2$, and HCl, all gave $k_{v,M}$ values in the range 10^{-11} to 10^{-10} cm^3 molecule^{-1} s^{-1}.

I. Chlorine Monofluoride, ClF

Chlorine monofluoride is the only kinetically stable fluorine-containing diatomic interhalogen; BrF and particularly IF rapidly disproportionate to give BrF_3 and IF_5. It is surprising, therefore, that ClF is also the least well studied of all of the interhalogens. The only LIF study of ClF reported to date was by McDermid (1980). In that study, the laser wavelength was

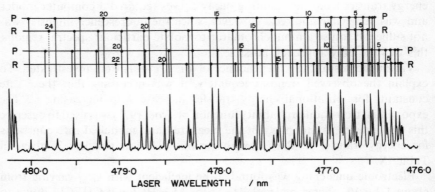

Fig. 23. Low-resolution excitation spectra of the 9–0 band in the $B-X$ system of ClF. Note the break-off in fluorescence at $J' = 23$ in ^{35}ClF and $J' = 25$ in ^{37}ClF. Upper assignments are for ^{37}ClF; lower for ^{35}ClF. After McDermid (1980)

scanned from 464 to 520 nm, which covers $B-X$ transitions with $17 \geqslant v' \geqslant 3$. No fluorescence was observed for excitation wavelengths shorter than 476 nm which corresponds to levels $v' \geqslant 10$. Although only low-resolution excitation spectra were recorded, at 0·01 nm resolution, the wide spacing of the rotational levels permitted virtually complete resolution of the P and R branches. Fig. 23 shows the entire low-resolution excitation spectrum of the 9–0, $B-X$ band.

Predissociation was observed in $v' = 9$ of the $B^3\Pi(0^+)$ state of ClF (McDermid, 1980). The predissociation is evidenced by a sudden break-off in the laser excitation spectrum (Fig. 23) at $J' = 23$ (first missing level) in ^{35}ClF and $J' = 25$ in ^{37}ClF. The intensities of the rotational lines leading up to the predissociation appeared to follow a normal Boltzmann distribution. In this respect, the predissociation differs from that observed in IF and BrF where the onset of predissociation did not appear sharp and the intensities of a number of rotational transitions showed a gradual reduction. Also, no levels at all have been observed above the apparently sharp onset of predissociation in $v' = 9$ of ClF, in contrast to the situation for BrF and IF (see earlier).

The observation of predissociation provides an upper limit for the ground-state dissociation energy, $D_0'' = (21\,126 \pm 6)\,\text{cm}^{-1}$ for ^{35}ClF and $D_0'' = (21\,136 \pm 6)\,\text{cm}^{-1}$ for ^{37}ClF. These energies were obtained from observation of the break-off of fluorescence in the excitation spectrum. These values should be confirmed by future measurements of the radiative lifetime as a function of v' and J'.

The question still arises as to whether the B state of ClF correlates with $F\,^2P_{1/2} + Cl\,^2P_{3/2}$ or with $F\,^2P_{3/2} + Cl\,^2P_{1/2}$. The problem arises because the energy difference between the $^2P_{1/2}$ and $^2P_{3/2}$ spin–orbit splittings or Cl and F is only 478 cm^{-1}. Using the convergence limit of the $B-X$ system gives two possible values for the ground-state dissociation energy, 20 632 and 21 110 cm^{-1}. McDermid (1980) has argued that, since the upper limit for D_0'' obtained from observation of predissociation is just slightly higher than this latter value, this favours the products $Cl\,^2P_{3/2} + F\,^2P_{1/2}$. Results from a study of the thermodynamics of the

$$F + Cl_2 \rightleftharpoons ClF + Cl$$

equilibrium (Nordine, 1974) strongly support this conclusion; however, results from the photoionization spectrum (Dibeler et al., 1970) favour the alternative products. Coxon (1975), in considering the dissociation energies of the diatomic halogen fluorides, concluded that an unequivocal choice between the two alternatives could not be made. This is still the case, even though an exact energy for the onset of predissociation is now available.

The three lowest energy levels of Cl + F atoms lie within a narrow energy range, 881 cm^{-1}. Therefore, the mixing of electronic states in this region is severe and, in addition to the predissociation, perturbations of the rotational

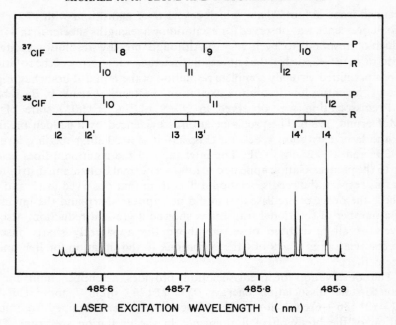

Fig. 24. Section of the high-resolution laser excitation spectrum of the 7–0 band of ClF $B-X$. Splitting of rotational lines in the R branch can clearly be observed. After McDermid and Laudenslager (1980).

structure of the $B-X$ system are also observed (McDermid, 1980 McDermid and Laudenslager, 1980). These perturbations have been observed in $v' = 7, 8$, and 9 and are manifest by a splitting and shifting of some of the rotational levels. Fig. 24 shows the high-resolution excitation spectrum of a section of the 7–0, $B-X$ band in the region of the perturbations. Although more data are required to say anything definite about the perturbing state, some calculations can be made which indicate that the internuclear separation of the Cl and F atoms is greater than for the B state.

No lifetime studies of ClF (B) have yet been reported. These are clearly needed and will be especially interesting in the regions of rotational perturbation. Additional interest in the low-lying states of ClF stems from the recent observation of lasing at 285 nm (Diegelmann *et al.*, 1979). The laser transition is not yet unambiguously identified but is thought to terminate on an electronic state at about the same energy as the B state and has tentatively been assigned as the ionic-to-covalent transition, $^3\Pi(2) \to {}^3\Pi(2)$, analogous to the homonuclear halogens (Diegelmann, 1980; Diegelmann *et al.*, 1979). Further experiments to determine the details of this transition would also be interesting.

J. Summary

The homonuclear halogens, I_2, Br_2, and Cl_2, and the heteronuclear halogens, IBr, ICl, BrCl, IF, BrF, and ClF, form two well defined series. The radiative lifetimes τ_R of the B states of the homonuclear halogens decrease with increasing molecular mass, as would be expected with gradual breakdown of the spin selection rule for the $B^3\Pi(0_u^+)$–$X^1\Sigma_g^+$ radiative transition. Insufficient information on the $A^3\Pi(1_u)$ state lifetimes is available to define any trends, although it is clear that the τ_R values are much longer than those for the corresponding B states (cf. Br_2).

The B states of I_2, Br_2, and Cl_2 all show heterogeneous predissociation by one or more 1_u-type states. The strength of the predissociation increases as one proceeds from I_2 to Cl_2. However, in the case of $Cl_2(B)$, the principal predissociating state is believed to be the bound $^3\Pi(1_u)$ state, whilst $Br_2(B)$ and $I_2(B)$ interact with the relevant $1\Pi(1_u)$ states that are unbound. These results can be interpreted in detail on the basis of Franck–Condon overlap of the B states and the predissociating states.

As in the homonuclear halogens, predissociation affects the manifolds of all the B states of the interhalogens. However, the predissociation is of a different type, and it varies considerably with the nature of the molecular species. IBr is a particularly well studied example of predissociation, thanks to the work of Child and others.

The radiative lifetimes of the B states of the interhalogens show the expected decrease with increasing molecular mass, and at least limited data are available for all members of the series except ClF.

Self-quenching rates and the rates of quenching by common molecular gases are generally found to have a low collisional efficiency, although collisional predissociation, as studied in $Br_2(B)$ and $I_2(B)$, has a collisional efficiency approaching unity. Vibrational relaxation in the B state manifolds of the halogens and interhalogens has a high collisional efficiency (10^{-2} to 1), even for monatomic gases as collision partners. These results mainly are consistent with the expected magnitudes of V–T transfer at 300 K, when the small sizes of the vibrational quanta ($0.3 kT$ to kT for the heavier members of the series) are considered.

Certain features of the collision dynamics of the halogens and interhalogens are favourable for possible lasing on the B–X transitions. These features include displaced potential energy curves, favouring inversion between the B and X states; radiative lifetimes in the microsecond time domain; and relatively low electronic quenching efficiencies for the excited B states. We note that optically pumped lasing has actually been demonstrated for the B–X transitions of I_2 and Br_2, and, more recently, that of IF. These lasing systems are in addition to a number of ultraviolet and visible lasers working on the D'–A' transitions of the halogens and interhalogens which do not involve the B state.

IV. LASER-INDUCED FLUORESCENCE STUDIES OF SELECTED FREE RADICALS

The electronic absorption spectra of free radicals and transient molecules provide direct and relatively simple data for the evaluation of the nature, spectroscopic constants, and lifetimes of their excited states. However, it is a basic limitation of absorption spectroscopy that a reasonably large optical depth of absorber is required, thus leading to the need for large absorber concentrations. For labile species, this requirement sometimes can be satisfied successfully by flash photolytic production and absorption spectroscopy in real time (Herzberg, 1971).

An alternative approach, which eliminates the need for high radical concentrations and is thus potentially very powerful, is to use a narrow-band dye laser to excite fluorescence. When undispersed fluorescence intensity is observed, as a function of laser frequency, the resulting laser excitation spectrum resembles the corresponding absorption spectrum, although the line intensities may be modified by perturbations or predissociations. Examples of laser excitation spectra, and their interpretation, are discussed in the accompanying sections on interhalogens and halogens.

Further examples are provided by the group of oxide free radicals XO, where $X = C, S, B, P$, and Sn, which have been studied recently in our laboratory. In addition to fundamental spectroscopic data, detailed information has become available on the quantum-resolved dynamics of the excited electronic states involved, including lifetimes.

A. CO $a^3\Pi$ Metastables

Fig. 25 shows the potential energy curves of a selection of excited electronic states of CO, after the data of Tilford and Simmons (1972). The allowed transitions of CO are either singlet–singlet or triplet–triplet in type; intercombination transitions are formally forbidden for light molecules. However, such transitions in CO, occurring weakly, either radiatively or non-radiatively have a profound effect, particularly on the dynamics of several triplet states of CO. The nesting of the potential energy curves indicated in Fig. 25, starting with $a^3\Pi$ and $A^1\Pi$, indicates that there are numerous curve crossings between $A^1\Pi$ and various triplet states including the $d^3\Delta$ state.

Several studies of the lifetimes and dynamics of triplet excited states of CO have been reported. Electron-impact excitation and the delayed coincidence method have been used, for example, for the study of $d^3\Delta$ lifetimes (van Sprang et al., 1977; Paske et al., 1980). The data analysis may be complicated by cascade from higher-energy states (Paske et al., 1980). Also, Slanger and Black (1973) have studied the dynamics of the $d^3\Delta, e^3\Sigma^-$, and $A^1\Pi$ states of CO indirectly, using steady-state photolysis.

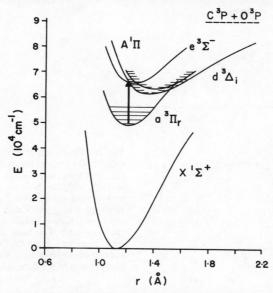

Fig. 25. Potential energy curves for CO. Note nesting of excited states and multiple crossings involving the $d^3\Delta$ state.

The work of Clyne and Heaven (1981a) was the first study of the dynamics of triplet states of CO by laser-induced fluorescence. Metastable CO $a^3\Pi$ molecules were produced by reaction of a trace of CO_2 with a stream of Ar* $^3P_{2,0}$ metastable atoms, formed by a low-power hollow-cathode discharge:

$$Ar^* \, ^3P_{2,0} + CO_2 \rightarrow CO \, a^3\Pi + O + Ar.$$

The concentration of CO $a^3\Pi$ molecules was estimated to be 10^8–$10^9 \, cm^{-3}$, and their lifetime is a few milliseconds (Taylor and Setser, 1973).

In Clyne and Heaven's (1981a) study, pulses from a 1 pm bandwidth dye laser were used to excite the $d^3\Delta$–$a^3\Pi$ 'Triplet' transition of CO, using radiation at wavelengths between 560 and 605nm. The laser excitation spectrum could be assigned to discrete ro-vibrational states of the d–a transition, which contains 27 Λ-doubled branches per band. Using the appropriate optical filters, it was possible to discriminate between fluorescence emanating from the vibrational levels $v' = 3, 4$, and 5 of the $d^3\Delta$ manifold.

The $v' = 4$ level was excited in the $\Omega = 1$ sub-band at the $P_{31}(10)$ line. The fluorescence intensity out of the $v' = 4$ level showed a satisfactory linear logarithmic decay plot, as shown in Fig. 26(a). There was no significant trend of fluorescence lifetime with argon pressure, over the rather limited range of 0·13 to 2·11 Torr. The mean value for the lifetime was $(5·75 \pm 0·70) \, \mu s$, which

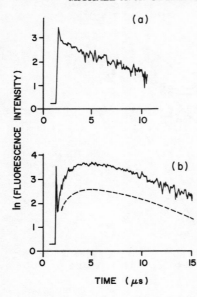

Fig. 26. Fluorescence decay kinetic of CO $d^3\Delta$. (a) Typical logarithmic decay of the $v = 4$ level; initial excitation of $\Omega = 1$ in the $P_{31}(10)$ line. Note good fit to single-exponential decay. (b) Typical logarithmic plot of growth and decay of the $v = 3$ level, following initial excitation of the $v = 4$ level. Note laser spike, marking zero time for the kinetics of $v = 3$. The broken curve, displaced for clarity, is a computer fit of the data. After Clyne and Heaven (1981a).

can be equated to the radiative lifetime τ_R of CO $d^3\Delta, v' = 4$. Agreement was obtained with the electron-impact excitation studies of van Sprang *et al.* (1977) and Paske *et al.* (1980). The data of Clyne and Heaven (1981a) involved initial excitation of a defined rotational and spin–orbit state of known parity. However, since argon pressures up to 2 Torr were used, extensive rotational or spin–orbit relaxation could be expected. Therefore, the values for τ_R should be closely comparable to the values of van Sprang *et al.* (1977) and Paske *et al.* (1980).

In a second experiment of Clyne and Heaven (1981a), the same $v' = 4$ ro-vibronic state was excited, but fluorescence intensity out of the next lower $v' = 3$ level was monitored. In this case, intensity (Fig. 26(b)) increased for the first few Microseconds, reached a maximum value, and then declined, reaching a single-exponential rate that was similar to that of the $v' = 4$ level described above. Thus, although non-collisional (radiative removal of excited states is a major pathway in depleting $d^3\Delta (v' = 4)$, particularly at pressures below 0·5 Torr, collisional relaxation also is an important channel.

Vibrational relaxation with Ar of $v' = 4$ molecules is a significant channel on the $5\,\mu s$ timescale set by the lifetime of $d^3\Delta$. A simple modelling scheme for energy transfer in CO $d^3\Delta (v' = 3, 4)$ was used by Clyne and Heaven (1981a) to obtain a value of $3\cdot2 \times 10^{-12}\,\mathrm{cm^3\,s^{-1}}$ for the rate constant k_v of the process

$$CO(d^3, v' = 4) + Ar \rightarrow CO(d^3, v' = 3) + Ar.$$

The high magnitude of the collisional efficiency for Ar in vibrational relaxation of CO $d^3\Delta(v' = 4)$ is inconsistent with the well known very low efficiencies of monatomic gases for V–T transfer in light molecules. However, recently, examples of electronically excited (A,v) molecules have been found in which rare gases are extremely efficient in causing vibrational transfer (Katayama *et al.*, 1980; Bondybey and Miller, 1978). The high rate constants are found to be confined to a few vibrational levels in the vicinity of a curve crossing with a second electronic state (B), which may be the ground state. Thus, vibrational relaxation of A,v occurs by a curve crossing to B followed by crossing back to $A,(v - 1)$. In this mechanism, state B should be long-lived, and it effectively constitutes a reservoir for A state population. Miller and coworkers (Katayama *et al.*, 1979, 1980; Bondybey and Miller, 1978) have described several examples of this behaviour, including $N_2^+ \, A^2\Pi_u$, CN $A^2\Pi$, and $CO^+ \, A^2\Pi$.

Rapid vibrational transfer by argon of $CO \, d^3\Delta(v' = 4)$, which has been reported by Clyne and Heaven (1981a) and by Slanger and Black (1973), evidently can be explained by a curve-crossing mechanism similar to that suggested by Miller and coworkers. The most probable process for vibrational transfer in CO would be (Clyne and Heaven, 1981a)

$$d^3\Delta(v' = 4) \overset{Ar}{\to} e^3\Sigma^-(v = 1) \to d^3\Delta(v = 3).$$

Fluorescence decay out of the $v' = 5$ level of CO $d^3\Delta$ was found by Clyne and Heaven (1981a) to show much shorter lifetimes than those for the $v' = 4$ and 3 levels. These results are consistent with the existence of a rotationally dependent, intense perturbation in the $\Omega = 1$ substate of the $v' = 5$ level. Initially stable ro-vibrational states of $d^3\Delta(v' = 5)$ can undergo collisional transfer into the unstable (perturbed) part of the manifold, leading to a reduction in lifetime.

B. SnO Transient Molecules

The structures of electronically excited states of SnO resemble in many respects those of CO which is the lightest member of the series of group IV monoxides. Thus, the lowest-energy bound singlet states are the ground $X^1\Sigma^+$ and the excited $A^1\Pi$ states. As for CO, there are a number of triplet excited states of SnO, which are nested with $A^1\Pi$. However, these states of SnO are less well studied than those of CO. Fig. 27 shows potential energy curves of several states of SnO, after Clyne and Heaven (1981b).

One difference between CO and SnO is that, whilst singlet–triplet transitions are weak for CO, these transitions become more allowed for the heavier monoxide radicals, such as SnO. There is a tendency towards Hund's case (c) coupling in the excited states of SnO, thus leading to radiative transitions between various Ω substates of $^3\Pi$ and the ground $X^1\Sigma^+$ state (see Fig. 27).

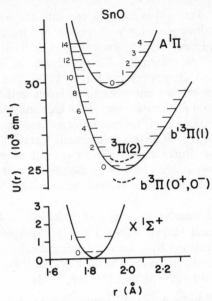

Fig. 27. Potential energy curves of SnO. Data shown are Morse functions for $X^1\Sigma^+$, $A^1\Pi$, and $b'^3\Pi(1)$; $^3\Pi(2)$ and $^3\Pi(0^+,0^-)$ minima are indicated. Note nesting of $A^1\Pi$ with the components of $^3\Pi_i$, which facilitates mixing of states and overlapping of spectra in certain wavelength regions.

Laser-induced fluorescence has been used by Clyne and Heaven (1981b) to study the nature and natural lifetimes of excited SnO. Ground-state SnO molecules were generated near 0·5 Torr in a discharge flow system, using the reaction of $Sn(CH_3)_4$ with a flow of $O\ ^3P$ atoms that were produced by a microwave discharge in O_2. Detection of SnO fluorescence was made in a flowing cell, using pulses from a frequency-doubled, narrow-band dye laser, between 310 and 340 nm.

The diversity of isotopic species in natural Sn leads to considerable complexity in the spectra of SnO. Therefore, a sample of pure $^{120}Sn\ (CH_3)_4$ was synthesized from isotopically pure ^{120}Sn. Most studies were carried out on ^{120}SnO.

Fig. 28 shows the laser excitation spectrum of the 1–0 band of the A–X transition of natural SnO. (This is the only band of the $v'' = 0$ progression that has been rotationally analysed previously (Lagerqvist et al., 1959).) For $J > 25$, the spectrum appeared as distinct clumps of lines, which could be assigned to the seven most abundant isotopic species of Sn. The expected single $P, Q,$ and R branches were readily identified for this $^1\Pi - ^1\Sigma^+$ transition.

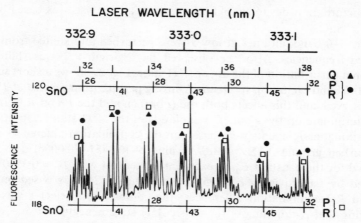

Fig. 28. Laser excitation spectrum of the 1–0 band of SnO$(A–X)$. Natural SnO. Note clumps of lines, with fine structure due to the small isotopic shifts. After Clyne and Heaven (1981b).

The isotopic shifts in the 1–0 band could readily be measured. Accurate dispersion for the SnO spectra was obtained by determining an excitation spectrum of BrCl$(B–X)$ under the same conditions. The ground-state combination differences for the BrCl spectrum were formed. Since the ground-state rotational constants $B_{v''}$ are known accurately from Coxon's (1974) work, the spectral dispersion is reliably defined.

As an example, the 120–118 isotopic shift δv^i was measured for six pairs of Q branch lines between $J = 12$ and $J = 20$, and determined to be $(0.443 \pm 0.017)\,\mathrm{cm}^{-1}(2\sigma)$. For low J values, most of the isotopic shift is due to the vibrational shift δG^i, which may be calculated from

$$\delta G^i = (\rho - 1)\omega'_e(v' + \tfrac{1}{2}) - (\rho^2 - 1)\omega'_e x'_e(v' + \tfrac{1}{2})^2$$
$$- (\rho - 1)\omega''_e(v'' + \tfrac{1}{2}) + (\rho^2 - 1)\omega''_e x''_e(v'' + \tfrac{1}{2})^2. \qquad (4.1)$$

Using data from Huber and Herzberg (1979), the magnitude of δG^i was calculated to be $0.437\,\mathrm{cm}^{-1}$. There is a small rotational isotope correction of $-0.023\,\mathrm{cm}^{-1}$ to be made, giving $0.414\,\mathrm{cm}^{-1}$ as the calculated value δv^i for the 1–0 band. Considering the small magnitude of δv^i (some 4.9 pm in wavelength), the agreement of observed and calculated values is quite satisfactory. The sign and magnitude of the isotopic shift confirms unequivocally the vibrational numbering of the $A^1\Pi$ state given by Lagerqvist et al. (1959).

Laser excitation spectra of the 4–0 and 3–0 bands of SnO$(A–X)$ showed heads near the expected wavelengths. These bands have been seen previously only at low resolution, in work carried out many years ago (Connelly, 1933). The present rotationally resolved spectra of natural SnO showed a complex line structure. In order to simplify the spectra, all analysis was based on the use of ^{120}SnO.

In the 4–0 band, all lines at low J were perturbed and shifted from their expected frequencies. Also, the observed band origin $v_{4,0}$ was shifted by rotational perturbation of the upper $A^1\Pi$ state. Fig. 29 shows a short section of the 4–0 band excitation spectrum. A strong perturbation centred at $J' = 18$ was observed, and this affects both the Q branch and the P and R branches, which terminate on the e and f Λ sublevels of the $A^1\Pi$ state, respectively.

Rotational assignments were based on the combination differences of the unperturbed ground $X^1\Sigma^+$ state. A value of (0.3457 ± 0.0006) cm^{-1} was obtained for the constant B_0'', in excellent agreement with $B_0'' = 0.3550$ cm^{-1} reported by Lagerqvist et al. (1959). Our value for B_0'' was based on both PR and RQ combination differences; the combination defect ε was zero, indicating that Λ-doubling of the upper $A^1\Pi$ state is negligible:

$$R(J) - Q(J + 1) = Q(J) - P(J + 1) + \varepsilon \simeq 2B_{v''}(J + 1). \qquad (4.2)$$

Several 'extra' lines, originating from transitions to the perturbing state, were observed; some of these are shown in Fig. 29.

In the long-wavelength tail of the 4–0 band of SnO A–X, the laser excitation spectrum showed a new band near 315.7 nm. It showed single P, Q, and R branches and the analysis was straightforward. The rotational numbering was fixed by combination differences formed on the lower state. The rotational constant thus obtained, namely $B'' = (0.3554 \pm 0.0006)$ cm^{-1}, identifies the lower state as $X^1\Sigma^+$, $v'' = 0$.

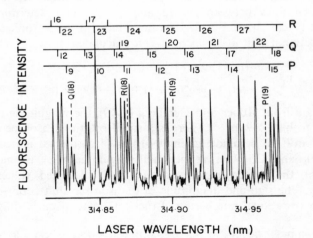

Fig. 29. Laser excitation spectrum of the 4–0 band of ^{120}SnO(A–X). Note strong perturbation centred at $J' = 18$, affecting all three branches. Extra lines due to the perturbing state are also shown. After Clyne and Heaven (1981b)

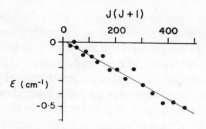

Fig. 30. Λ-doubling in the $b'^3\Pi(1)$ state, $v' = 14$, of SnO. The figure shows a plot of the combination defect ε against $J(J + 1)$. After Clyne and Heaven (1981b).

Combination differences formed on the upper state from the PR branches were regular and showed no evidence for perturbations, up to the highest J' level examined, namely $J' \leqslant 28$. The rotational constant B' for the appropriate Λ-doublet was determined to be $(0.2605 \pm 0.0008) \, \text{cm}^{-1}$, i.e. considerably less than B'_4 for the $A^1\Pi$ state. The band origin v_0 was determined to be $(31\,666.5 \pm 0.2) \, \text{cm}^{-1}$.

Combination relations involving the Q branch (see Eq. (4.2)) revealed the presence of significant Λ-doubling in the upper state. The data are shown in Fig. 30 in terms of a plot of ε against $J(J + 1)$, where ε is defined in Eq. (4.2). Within the scatter of the data, a linear correlation is obtained. Least-squares fitting to the equation

$$\varepsilon/2 = qJ(J + 1)$$

gave

$$q = B'_d - B'_e = 0.0012 \, \text{cm}^{-1}.$$

The observation of single $P, Q,$ and R branches, and the identification of the lower state as $X^1\Sigma^+$, imply that the upper level is a Π state. This conclusion is further supported by the relatively large magnitude of the Λ-doubling. SnO has only two low-lying Π states, namely $A^1\Pi$ and $^3\Pi_i$. Thus, the upper state must be one of the Ω-components of $^3\Pi_i$. $^3\Pi(2)$ and $^3\Pi(0^-)$ components are eliminated due to the selection rules governing radiative transitions. $^3\Pi(0^+)$ is ruled out, since a $^3\Pi(0^+)-X^1\Sigma^+$ band shows two branches and not three as observed. Therefore, we conclude that the upper state is $b'\,^3\Pi(1)$, in accordance with the development of three branches per band, and the form of the dependence of ε on J.

The excitation spectrum of the 3–0 band showed single $P, Q,$ and R branches, as expected, but the structure was extensively perturbed. In addition to the 3–0 band of SnO(A–X), a second band γ has been observed in the laser excitation spectrum; the bandhead of γ lies to shorter wavelengths of the 3–0 bandhead.

The new band γ was regular in form, and showed single $P, Q,$ and R lines, suggesting that it belonged to the same $b'^3\Pi(1)-X'\Sigma^+$ transition as the new band X seen in the tail of the 4–0 band. Other arguments were presented (Clyne and Heaven, 1981b), which suggest that X is the 14–0 band and γ is the 13–0 band of the $b'^3\Pi(1)-X^1\Sigma^+$ transition of SnO.

In addition to spectroscopic analysis, measurements of the lifetimes τ of the $A^1\Pi$ and $b'^3\Pi(1)$ excited states of SnO were determined in the torr pressure range. Initial excitation of several ro-vibrational states of ^{120}SnO was carried out. The resulting values for the extrapolated zero-pressure lifetimes τ_0, and for the quenching rate constant $k_M(M = O_2)$, were as follows:

$$v' = 1 \quad (A^1\Pi): \qquad \tau_0 = (160 \pm 20)\,\text{ns},$$
$$k_M = (2\cdot2 \pm 0\cdot6) \times 10^{-10}\,\text{cm}^3\,\text{s}^{-1};$$

$$v' = 3 \quad (A^1\Pi): \qquad \tau_0 = (140 \pm 10)\,\text{ns},$$
$$k_M = (6\cdot9 \pm 0\cdot6) \times 10^{-10}\,\text{cm}^3\,\text{s}^{-1};$$

$$v' = 4 \quad (A^1\Pi): \qquad \tau_0 = (130 \pm 20)\,\text{ns},$$
$$k_M \leqslant 1\cdot4 \times 10^{-10}\,\text{cm}^3\,\text{s}^{-1};$$

$$v' = 14 \quad (b'^3\Pi(1)): \qquad \tau_0 = (580 \pm 34)\,\text{ns},$$
$$k_M = (3\cdot6 \pm 1\cdot0) \times 10^{-11}\,\text{cm}^3\,\text{s}^{-1}.$$

The only other work on the lifetime of SnO states of which we are aware is a preliminary study of Capelle and Linton (1976), who used a nitrogen laser in broad-band excitation of fluorescence from the SnO + N_2O reaction. They reported two decay times, extrapolated to zero pressure: $(160 \pm 20)\,\text{ns}$ with a moderate pressure dependence, and $(130 \pm 60)\,\text{ns}$ with a strong pressure dependence. These magnitudes for τ_0 are broadly in agreement with the present work, although it is clear that several transitions were excited simultaneously in the work of Capelle and Linton (1976), thus making a detailed interpretation impossible.

As expected, the values of τ_0 for the $A^1\Pi$ state is considerably shorter ($\sim 1/4$) than that for the $b'^3\Pi(1)$ state, in their radiative transitions to the $X^1\Sigma^+$ ground state. However, the 580 ns lifetime of the $b'^3\Pi(1)$ state of SnO is much shorter than the values longer than $10^2\,\mu\text{s}$ found for the lifetimes of the much more metastable $A^3\Pi(1)$ states of Br_2 (Clyne et al., 1980c) and ICl (Harris et al., 1979). The $b'-X$ transition of SnO thus is an almost fully allowed radiative transition.

The considerable variation in the k_M values for the various initially formed states is interesting. It is noted that the maximum value of k_M was found for the most perturbed level ($v' = 3$) of $A^1\Pi$. This result would indicate that collisional, as well as non-collisional, curve crossing has a maximum probability for the $v' = 3$ state. Collisional transfer from the $A^1\Pi$ to the $^3\Pi(2)$

state would result in a decrease in lifetime, since this triplet state cannot radiate in allowed transitions to the ground $X^1\Sigma^+$ state.

Clyne and Heaven (1981b) discussed and analysed the perturbations in the $v' = 3$ and 4 levels of SnO $A^1\Pi$ using the simple theory given by Kovacs (1969) and others. The results for $v' = 4$ indicated that there is a strong perturbation extending down to $J' = 0$.

The selection rules for perturbations allow interactions characterised by $\Delta\Lambda = 0, \pm 1$ in case (a), or $\Delta\Omega = 0, \pm 1$ in case (c.) Homogeneous interactions (i.e. $\Delta\Lambda = 0$ or $\Delta\Omega = 0$) occur only through rotational coupling of the states; consequently, the strength of such an interaction is proportional to $J(J + 1)$. This behaviour was not exhibited by the data for the $v' = 4$ level of SnO $A^1\Pi$. The interaction evidently is homogeneous. Consideration of the electronic structure of SnO shows that the only states of suitable energy for the observed interaction are the components of the low-lying $^3\Pi_i$ state, namely $^3\Pi(2)$, $^3\Pi(1)$, $^3\Pi(0^+)$, and $^3\Pi(0^-)$, $b^3\Pi(0^+)$ and $^3\Pi(0^-)$ are eliminated since these states can only perturb one Λ-doublet component of $A^1\Pi$, whereas both Λ-doublets in fact are symmetrically perturbed. $b'^3\Pi(1)$ is improbable, since the $b'-X$ transition radiates strongly, whereas the perturbring state appears as a few 'extra' lines. Therefore, $^3\Pi(2)$ is the most probable perturbing states in case (a) approximation.

The perturbation in the $v' = 3$ level differs from that in the $v' = 4$ level. The $v' = 3$ level of $A^1\Pi$ mutually perturbs the $v' = 13$ level of $b'^3\Pi(1)$, as shown by the analysis of Clyne and Heaven (1981b). Thus, the $b'^3\Pi(1)$ is the perturbing state in the 3–0 band.

C. BO Radicals

Clyne and Heaven (1980b) reported quantum-resolved LIF of the BO free radical, and measurements of the radiative lifetime of the $A^2\Pi$ excited state of BO. The BO radical belongs to an important group of isoelectronic molecules containing also CN and CO^+, whose $X^2\Sigma^+$ ground states have analogous electronic structures. Also, the BO radical occupies an important place in the development of molecular spectroscopy. The α–$(A^2\Pi$–$X^2\Sigma^+)$ and β–$(B^2\Sigma^+$–$X^2\Sigma^+)$ band systems of BO in emission were first correctly identified by Mulliken (1925). He showed that the relatively large magnitudes of the ^{10}B–^{11}B isotopic splittings in these band systems of BO demonstrate the existence of half-integral vibrational quantum numbers, as had been suggested earlier for J state quantum numbers of atoms by Heisenberg (1922). Vibrational and rotational analyses of the band systems of BO were carried out by the early workers (Mulliken, 1925; Jenkins and McKellar, 1932; Scheib, 1930), with only one more recent study of significance on the spectroscopy of the A–X system (Dunn and Hanson, 1970). The A–X band system is clearly established as $A^2\Pi$–$X^2\Sigma^+$, with the excited $A^2\Pi$ state

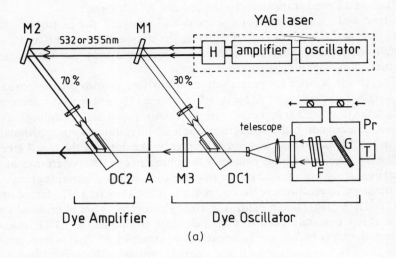

(a)

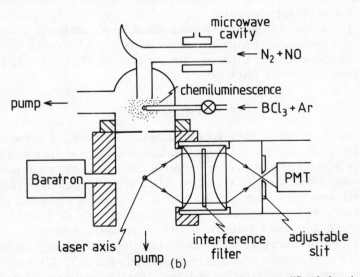

Fig. 31. (a) Schematic of dye laser based on an oscillator–amplifier design. A, aperture; DC1 and DC2, flowing dye cells; F, Fabry–Perot etalon; G, echelle grating; H, harmonic generator; L, cylindrical lens; M1, beam splitter; M2, total reflector; M3, dye oscillator front mirror; Pr, pressure chamber; L, linear transducer. (b) Laser fluorescence cell for studies of BO and BO_2. Note hole between reaction zone and fluorescence chamber: 6 mm diameter for BO studies, 2 mm diameter for BO_2 studies. After Clyne and Heaven (1980b)

belonging to Hund's case (a). However, the spectroscopic constants of both states, being based on early work, are of limited accuracy (Huber and Herzberg, 1979), and extend only up to the $v' = 4$ vibrational level of the $A^2\Pi$ state.

Future LIF studies of BO can provide a direct method of following the concentration of ground-state BO radicals for kinetic studies of the ground state. At the present time, virtually nothing is known about the reaction kinetics of BO radicals. This area is expected to be very fruitful for future study.

In Clyne and Heaven's (1980b) study, a narrow-band pulsed dye laser was used to excite specific ro-vibrational levels (v', J') of BO $A^2\Pi$. Excitation was carried out in a laser fluorescence cell that formed part of a flow system; it is shown schematically in Fig. 31. The laser crossed the flow system 2 cm downstream from the zone of reaction between BCl_3 and $O\,^3P$ atoms, at a total pressure near 1 Torr. Between the reaction zone and the fluorescence cell was a glass plate with either a 6 mm or a 2 mm hole, which served both to provide a pressure gradient and to reduce the observation of chemiluminescence from the reaction zone by the fluorescence detector.

When O^3P atoms were generated by a 2·45 GHz discharge in O_2 and mixed with a trace of BCl_3, green $BO_2\,(A–X)$ chemiluminescence was emitted from the reaction zone. Under these conditions, intense LIF of BO_2 radicals could be observed. However, LIF of BO radicals could be detected.

Therefore, O^3P atoms were generated by reaction of NO with a stream of ground state N^4S atoms formed by a 2·45 GHz discharge in N_2:

$$N\,^4S + NO \rightarrow N_2 + O\,^3P.$$

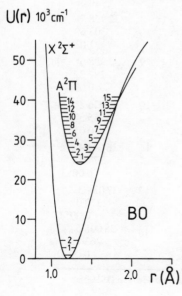

Fig. 32. Potential energy curves for low-lying states of BO. Morse functions are shown for the $X^2\Sigma^+$ and $A^2\Pi$ states

BO $(A-X)$ and BO $(B-X)$ chemiluminescence was emitted from the reaction between a trace of BCl_3 and $O\,^3P$ atoms under these conditions. Also, LIF of BO $(A-X)$ could be seen, indicating the presence of ground-state BO $X^2\Sigma^+$ radicals in the reaction zone. The intensity of LIF of BO was a maximum when approximately half the $N\,^4S$ atoms had been titrated by NO, i.e. when $[N^4S] = [O^3P]$, and appreciable free nitrogen atoms remained in the reacting gas.

Fig. 32 shows the Morse potential energy curves for the ground $X^2\Sigma^+$ and $A^2\Pi$ states of the BO radicals. The potential energy curves of the A and X states are displaced, and evidently cross in the vicinity of the excited state level $v' = 14$.

Table X shows a summary of the vibrational bands of interest, including Franck–Condon factors (Liszt and Smith, 1971), in the study of BO $(A-X)$ in laser-induced fluorescence by Clyne and Heaven (1980b). The tabulated wavenumbers and wavelengths are calculated for the origins of the higher-energy sub-bands ($\Omega = 1/2$) of the $A^2\Pi–X^2\Sigma^+$ transition (see below).

TABLE X. BO $A^2\Pi–X^2\Sigma^+$ bands expected to be observed in laser-induced fluorescence. Order of entries is: band origin ($\Omega = 1/2$) (cm^{-1}); air wavelength (nm); Franck–Condon factor $q_{v',v''}$ for $J' = J'' = 0$.

v'	$v'' = 0$	v'	$v'' = 0$
0	23 646·4[a]	7	31 846·5[a]
	422·90		314·01
	0·036		0·061
1	24 884·8	8	32 928·7
	401·85		303·69
	0·101		1·039
2	26 100·9	9	33 988·6[a]
	383·13		294·22
	0·155		0·024
3	27 294·6	10	35 026·2[a]
	366·37		285·50
	0·172		0·015
4	28 466·1[a]	11	36 041·4[a]
	351·30		277·46
	0·158		0·0087
5	29 615·2	12	37 034·3
	337·66		270·02
	0·126		0·0051
6	30 742·0	13	38 005·5
	325·29		263·12
	0·091		0·0030

[a] Band observed in LIF by Clyne and Heaven (1980b).

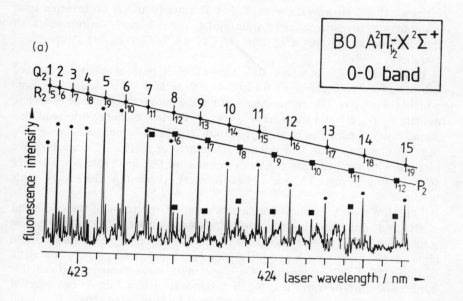

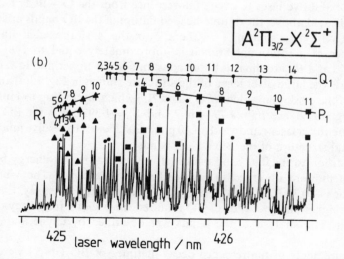

Fig. 33. Laser excitation spectrum of the 0–0 band of $BO(A-X)$ at 1 pm laser bandwidth. (a) The $A^2\Pi_{1/2}-X^2\Sigma^+$ sub-band, showing overlapped $Q_2 + R_2$ branches and P_2 branch. (b) The $A^2\Pi_{3/2}-X^2\Sigma^+$ sub-band, showing R_1 head and P_1, Q_1 branches. After Clyne and Heaven (1980b).

The ground $X^2\Sigma^+$ state of BO has a very small spin splitting (Dunn and Hanson, 1970). However, the excited $A^2\Pi$ state of BO is an inverted case (a) state with a relatively large splitting ($A_0 = 122\cdot26\,\text{cm}^{-1}$) compared with that for the analogous $A^2\Pi$ state of CN ($A_0 = 52\cdot64\,\text{cm}^{-1}$) (Huber and Herzberg, 1979).

Thus, each $A-X$ band of BO consists of a pair of partly separated sub-bands, corresponding to transitions to the $\Omega = 1/2$ and $\Omega = 3/2$ components of $A^2\Pi$(a). The higher-energy sub-band shows three intense branches, namely P_2, Q_2, R_2, and a weaker R_{21} branch. The lower-energy sub-band also shows four branches, namely P_1, Q_1, and R_1 with a somewhat weaker P_{12} branch. The remaining four branches expected for a $^2\Pi-^2\Sigma^+$ transition are blended with the principal branches, because of the negligible ground-state spin splitting. Figures 123 and 124 of Herzberg (1950) show the energy levels involved.

Fig. 33(a) and 33(b) show the laser excitation spectrum of the 0–0 band of BO at 1 and 10 pm laser bandwidths, using stilbene-420 as the dye. The double-headed 0–0 band from 422 to 428 nm (Fig. 33(a) and 33(b)) showed very open and highly divergent rotational structure, characteristic of the $A^2\Pi-X^2\Sigma^+$ system of BO. Wavelength measurements showed this band to be the 0–0 transition. No 'hot' bands from $v' = 1$ could be observed at longer wavelengths, as had been suggested by Huie. et al. (1978), who used a broad-band dye laser to excite fluorescence from the $O + BCl_3$ reaction.

As indicated above, the double-headed nature of the BO bands arises from the relatively large splitting ($\sim 122\,\text{cm}^{-1}$) of the $A^2\Pi$(a) excited state. The divergence of a rotational branch is approximately equal to $2(B_{v''} - B_{v'})$, where $B_{v''}$ and $B_{v'}$ are the rotational constants in the upper and lower states. Therefore, the large divergence of the structure shown in Fig. 33(a) and 33(b) is due to the dissimilar values of $B_{v''}$ and $B_{v'}$, which, according to Huber and Herzberg (1979), are $B_0'' = 1\cdot774\,\text{cm}^{-1}$ and $B_0' = 1\cdot392\,\text{cm}^{-1}$ for BO ($A-X$). Consequently, a laser bandwidth of 10 pm was sufficient to resolve much of the rotational structure of the BO ($A-X$) bands.

In addition to the 0–0 band of BO ($A-X$), several higher-energy bands of this system were observed in laser-induced fluorescence. The vibrational transitions 4–0, 7–0, 9–0, 10–0, and 11–0 were selected for study. Excitation of these higher-energy bands of BO required the use of a frequency-doubled dye laser. The analysis of the bands was described by Clyne and Heaven (1980b).

Measurements of fluorescence decay-lifetimes of BO ($A-X$) were made. Fig. 34 shows typical fluorescence decay curves in their exponential and logarithmic forms. In Fig. 34(a) broad-band excitation near the R_1 head of the 0–0 band was used. Fig. 34(b) refers to the same conditions as Fig. 34(a), except that narrow-bandwidth excitation in the same R_1 head was used. The decay curves were essentially the same.

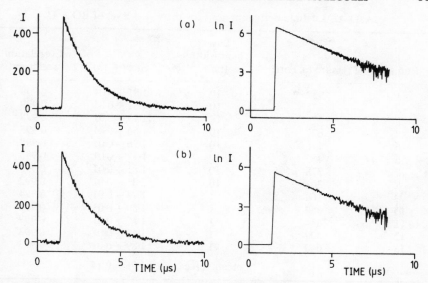

Fig. 34. Fluorescence decay of BO $A^2\Pi$. (a) 10 pm laser bandwidth; excitation near the R_1 bandhead of the 0–0 band of ^{11}BO. (b) 1 pm laser bandwidth; same excitation. Total pressure is 600 mTorr; transient recorder at 10 ns/channel. Left-hand curves show decay of fluorescence intensity; right-hand lines are the same data in logarithmic form. After Clyne and Heaven (1980b).

The noise on the baseline of the decay curves was due to a CW signal observed from the BO chemiluminescence emitted from the $O + BCl_3$ reaction. The signal-to-noise ratio was extremely high, however, and a maximum of 2000 laser pulses sufficed to form a good BO decay curve. The decay gave a good fit to single-exponential behaviour over two or more lifetime.

Least mean-squares analysis of decay curves was carried out over more than two lifetimes, giving correlation coefficients above 0·992 for all runs. As shown in Table XI (Clyne and Heaven, 1980b), there was no significant variation of lifetime τ with laser bandwidth, nor with total pressure (of N_2) over the range 300 to 1150 mTorr. The mean value was $\tau = (1·78 \pm 0·10)$ μs (1σ) for excitation of the $v' = 0$ level.

Similar results were obtained for excitation of the 4–0, 7–0, 9–0, 10–0, and 11–0 bands (table 3 of Clyne and Heaven, 1980b). There was little variation of τ with v', as shown by these data. As for excitation of the $v' = 0$ level, no significant pressure dependence was found. Therefore, τ could be equated to τ_R, the radiative lifetime of BO $A^2\Pi$, since predissociation of this state is not energetically feasible.

No significant dependence of τ upon rotational quantum number N', nor upon Ω ($\Omega 1/2$ or $3/2$), could be detected from the experiments where different

TABLE XI. Lifetime determinations for the $v' = 0$ level of BO $A^2\Pi$.

Run no.	Pressure (mTorr)	Laser bandwidth (pm)	$\tau(\mu s)(\pm 1\sigma)$	Correlation coefficient
1	932	10	1·81 ± 0·01	0·999
2	875	10	1·81 ± 0·01	0·999
3	778	10	1·79 ± 0·01	0·998
4	677	10	1·66 ± 0·01	0·995
5	576	10	1·76 ± 0·01	0·998
6	455	10	1·77 ± 0·01	0·995
7	378	10	1·81 ± 0·01	0·995
8	301	10	2·01 ± 0·02	0·992
9	1168	1	1·73 ± 0·01	0·994
10	1052	1	1·80 ± 0·01	0·994
11	634	1	1·87 ± 0·01	0·997
12	507	1	1·40 ± 0·01	0·993
13	640	1	1·89 ± 0·01	0·993
		mean value	1·78 ± 0·14	

Transient recorder at 10 ns/channel; 200 to 1000 laser shots per run; photomultiplier voltage 1·4–1·6 kV.

levels were excited. Any such dependence would not be observed easily at pressures near 500 mTorr, since rotational and spin–orbit relaxation in collisions would be appreciable during the $\sim 1\cdot8\,\mu s$ lifetime of BO $A^2\Pi$. Vibrational relaxation under the same conditions, however, is not expected to be significant.

As yet, the higher vibrational levels $v' = 14$ and 15 of BO $A^2\Pi$ have not been probed by LIF. As can be seen from Fig. 32, the $X^2\Sigma^+$ ground state of BO is expected to cross the $A^2\Pi$ excited state near $v' = 14$. This crossing would probably show as a rapid quenching rate constant for the $A–X$ fluorescence near $v' = 14$, as has been observed for crossing of the analogous A and X states of CO^+ (Bondybey and Miller, 1978).

Based on the failure to observe any trend of τ with N_2 pressure over the range employed, an upper limit of $k_N > 2 \times 10^{-11}\,cm^3\,molecule^{-1}\,s^{-1}$ can be placed on the rate constant of quenching of BO $A^2\Pi$ by N_2.

The value $\tau_R = (1\cdot76 \pm 0\cdot13)\,\mu s$ for the levels $11 \geqslant v' \geqslant 0$ or BO $A^2\Pi$ has been used to evaluate the electric dipole moment $|R_e|^2$ for the $A–X$ transition, and also the absorption coefficients of typical rotational lines in the spectrum (Clyne and Heaven, 1980b). It was found the $|R_e|^2 = (0\cdot26 \pm 0\cdot02)\,D^2$, and was virtually invariant for the vibrational levels $11 \geqslant v' \geqslant 0$. At 300 K, and for an intense transition (the $Q_2(7)$ line), the line absorption coefficients were estimated to be $2\cdot5 \times 10^{-17}\,cm^2$ for the 0–0 band and $1\cdot1 \times 10^{-16}\,cm^2$ for the 4–0 band (one of the most intense bands.).

It is of interest to consider possible laser action on the $B-X$ transition of BO, in the light of the favourable lifetime of the $A^2\Pi$ excited state. The 4–0 band of BO $A-X$ lies between 351 and 354 nm, and is thus favourable for optical pumping by the 353 nm line of the XeF excimer laser.

Assuming that an optical depth of BO radicals equal to 0·1 cm Torr can be achieved, considerable absorption of energy from the XeF laser can be expected in one pass. Probably, a pressure of 10^{-2} Torr of BO radicals can be produced in the O + BCl_3 reaction, and thus a 10 cm absorbing length would be required for an optical pumping experiment. Excitation at the $v' = 4$ level of BO $A^2\Pi$ is favourable, since Franck–Condon factors for several emission bands terminating on $v'' > 0$ are large. Examples of such bands are the 4–4 band near 473 nm, the 4–5 band near 516 nm, the 4–7 band near 628 nm, and the 4–8 band near 703 nm.

Factors which favour lasing by optical or chemical pumping of BO $A-X$ are the lifetime ($\tau_R = 1\cdot8\,\mu s$) and the slow rate constants for quenching of the excited state. In any future study of a chemically pumped BO laser, it will be necessary to investigate the kinetics of the O + BCl_3 precursor reaction, and to determine the quantum yield of emission of BO $A-X$ chemiluminescence from this reaction. The mechanism of formation of BO radicals in the O + BCl_3 reaction is still obscure. The work of Clyne and Heaven (1980b) shows that BO $X^2\Sigma^2$ radicals are not observed in the O^3P + BCl_3 reaction in the presence of O_2, although large concentrations of BO_2 radicals are generated under these conditions. Furthermore, the presence of N^4S atoms, as well as of O^3P atoms, appears to be necessary for the formation of BO radicals. Presumably the initial step is the stripping of a Cl atom:

$$O + BCl_3 \rightarrow BCl_2 + ClO.$$

Secondary steps could be of two types, namely

$$O + BCl_2 \rightarrow BOCl + Cl$$

or

$$O + BCl_2 + BCl + ClO.$$

BO radicals would then be formed by reactions such as

$$O + BOCl \rightarrow BO + ClO$$

or

$$O + BCl \rightarrow BO + Cl.$$

ClO radicals are converted to Cl atoms by further reaction with O atoms. Probably, at least one rate-controlling step in the above scheme involves an N atom, rather than an O atom as written. A systematic kinetic study of the O + BCl_3 reaction using atomic resonance fluorescence, mass spectrometry, and LIF is clearly required.

D. The PO Free Radical

The PO radical resembles the BO radical in that the spectroscopy of several transitions has been studied, whilst little or no information has been available regarding the dynamics of the ground- or excited-state species. The ground state of PO is $X^2\Pi_r$, i.e. the analogue of the stable free radical NO. However, PO $X^2\Pi$ is difficult to form and is short-lived. The differences in reactivity and in kinetic behaviour between PO and NO are attributable, in part, to the differences in thermochemistry between the relevant oxides of phosphorus and nitrogen. Most of the meagre information about the kinetic behaviour of PO $X^2\Pi$ comes from an informative (although mainly qualitative) study by Davies and Thrush (1968), using observations of chemiluminescence from the $O + P_4$ and $O + PH_3$ reactions in a discharge flow system.

The work of Clyne and Heaven (1981c) shows that LIF of PO, in the $B^2\Sigma^+-X^2\Pi$ transition, may be used as a sensitive means of following ground-state PO concentrations. Detection of PO $X^2\Pi$ was carried out using the reaction of PH_3 with a mixture of $N\,^4S$ and $O\,^3P$ atoms near 0.5 Torr as the source. Excitation of the so-called 'β-system' of PO was carried out in the 0–0 sub-band of the $B^2\Sigma^+-X^2\Pi_{1/2}$ transition. Rotationally resolved laser excitation spectra were reported by Clyne and Heaven (1981c), and the radiative lifetime of PO $B^2\Sigma^+$ has been determined for the first time.

A Nd:YAG-pumped narrow-band dye laser was used for these studies. Using ethanolic CVP to generate laser output near 645 nm, 5 mJ pulses of 10 ns duration were obtained at a repetition rate of 10 Hz. Frequency doubling was effected with an angle-tuned KDP crystal, giving 250 μJ/pulse at 325 nm and a bandwidth of 0.04 cm^{-1}.

It is useful to summarize the relevant aspects of the spectroscopy of PO before describing further the LIF studies of Clyne and Heaven (1981c). The $B-X$ transition, β-system, of PO has been observed in absorption by Verma and Singhal (1975), and in emission from a discharge by the same workers and by Singh (1959). Rotational analysis of several bands has been described (Verma and Singhal, 1975).

The rotational structure of bands of the $B-X$ system of PO is closely analogous to that of the $A^2\Sigma^+-X^2\Pi$ bands of NO ('γ-system'). The relevant potential energy curves for PO are shown in figure 1 of Clyne and Heaven (1981c). Since $r'_e < r''_e$ for both PO and NO, the 0–0 bands of both the $^2\Sigma^+-X^2\Pi$ transitions are violet-degraded. However, since the difference in r_e value for PO is not so large, the rotational branches are compact and diverge (or converge) very slowly from the band origin.

The regular $X^2\Pi$ ground state of PO belongs to Hund's case (a), with a relatively large spin splitting of 224 cm^{-1} for $v'' = 0$. On the other hand, the spin splitting of the $A^2\Sigma^+$ is very small (case (b)). Thus, each PO band consists of two sub-bands, namely, $B^2\Sigma^+-X^2\Pi_{1/2}$ and $B^2\Sigma^+-X^2\Pi_{3/2}$. The sub-bands show little mutual overlap of lines belonging to a 300 K rotational envelope. The structure of each sub-band is similar, consisting of six branches per sub-band. For example, the shorter-wavelength $B^2\Sigma^+ - A^2\Pi_{1/2}$ sub-band consists of six branches, $P_1, Q_1, R_1, {}^SR_{21}, {}^QP_{21}$, and ${}^RQ_{21}$. Only four branches, P_1, Q_1, R_1, and ${}^SR_{21}$, are fully resolved, since the small spin splitting in the $B^2\Sigma^+$ state leads to blending of the main Q_1 branch with the satellite ${}^QP_{21}$ branch, and to blending of the main R_1 branch with the satellite ${}^RQ_{21}$ branch.

The excitation spectrum of the PO radical was sought in Clyne and Heaven's (1981c) work, by recording undispersed fluorescence intensity as a function of laser wavelength near 324 nm. Linear wavelength scans, each of about 0·15 nm, were obtained by pressure tuning the dye oscillator enclosure from 0 to 2 atm pressure of nitrogen.

Fig. 35 shows the spectral region from 324·3 to 324·9 nm. The resolved rotational structure was readily identified as the 0–0 sub-band of PO $B^2\Sigma^+ - X^2\Pi_{1/2}$. The signal-to-noise ratio of the spectrum was high. Discrimination against CW background signal due to chemiluminescence was optimized using an input impedance of 1 kΩ and a gate width of 1 μs on the boxcar integrator.

Rotational assignments of the four principal branches are shown in Fig. 35.

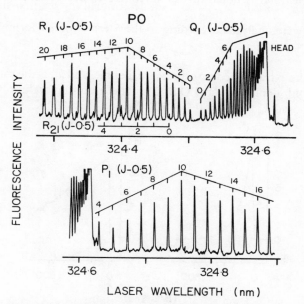

Fig. 35. Laser excitation spectrum of the 0–0 band of PO $B^2\Sigma^+ - X^2\Pi_{1/2}$. Laser bandwidth, 1 pm. Note compact rotational structure and splitting of R_1 branch above $J = 11\frac{1}{2}$. After Clyne and Heaven (1981c).

The Q_1 branch shows a well defined head near 324·6 nm, whilst the slowly converging P_1 branch does not reach a head. This is because the PO ground state radical is found to be in a rotational Boltzmann distribution for 300 K, which shows a maximum population near $J = 10\frac{1}{2}$ and little population above $J > 25\frac{1}{2}$ where the P_1 head is formed. This statement is illustrated by the rotational envelopes of the branches shown in Fig. 35.

The R_1 branch is split into two components above $J = 11\frac{1}{2}$, due to the resolution of the R_1 and $^RQ_{21}$ lines which are totally blended at lower J values. Splitting of the blended Q_1 and $^QP_{21}$ branches cannot be observed, because of the convergence of the Q_1 lines at high J.

The second sub-band of the 0–0 transition was not recorded at high resolution. A search was made for the 1–0 and 2–0 bands, but this was unsuccessful, probably because of unfavourable Franck–Condon densities.

Fluorescence decay measurements on PO $B^2\Sigma^+ (v' = 0)$ were made. Excitation was carried out at the Q_1 bandhead. The pressure of N_2 in the fluorescence cell was varied from 90 to 920 mTorr. Fig. 36 shows typical data for fluorescence decay, in this case at 795 mTorr using 1000 laser pulses. A good single-exponential decay was observed, up to at least four 1/e lifetimes (Fig. 36). Table 1 of Clyne and Heaven (1981c) shows a summary of the data, from which a trend is noted for the lifetime τ to decrease as a function of N_2 pressure.

The data were analysed according to the Stern–Volmer formulation

$$\tau^{-1} = \tau_0^{-1} + k_M[M], \tag{4.3}$$

where τ_0 is the extrapolated zero-pressure lifetime, and k_M is the rate constant for collisional depletion of PO $B^2\Sigma^+$. The resulting values were $\tau_0 = (250 \pm 10)$ ns and $k_{N_2} = (2·7 \pm 0·1) \times 10^{-10}$ cm^3 s^{-1} at 300 K.

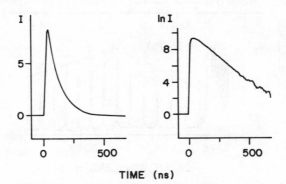

Fig. 36. Fluorescence decay curves for PO $B^2\Sigma^+$. Pressure of N_2 is 795 mTorr. Note good linearity of plot of ln I against time over at least four lifetimes. After Clyne and Heaven (1981c).

In the absence of any known perturbations or predissociations of PO $B^2\Sigma^+$, τ_0 may be identified with τ_R, the radiative lifetime of the $v' = 0$ level. k_{N_2} can be identified as the rate of electronic deactivation (quenching) of the $B^2\Sigma^+$ state of PO. The value of k_{N_2} is not affected by vibrational relaxation of the excited-state manifold, since $v' = 0$ was the initially excited state.

Electronic quenching by N_2 of PO $B^2\Sigma^+$ is found to occur at a rate which closely approaches the hard-sphere bimolecular collision frequency. Electronic energy transfer from PO $B^2\Sigma^+$ to $N_2 X^1\Sigma_g^+$ would appear to be ruled out as a quenching mechanism, since there are no energetically accessible states of N_2. One possible quenching mechanism for PO $B^2\Sigma^+$ is collision-induced curve crossing. The most likely such process would be transfer to the $a^4\Pi$ state, which would be a highly metastable state and the analogue of the well known low-lying $a^4\Pi$ state of NO. Although $B^2\Sigma^+$–$a^4\Pi$ collisional transfer in PO is spin-forbidden, an analogous process in NO, namely, $C^2\Pi$–$a^4\Pi$, has been suggested to be efficient due to strong spin-orbit coupling in the $a^4\Pi$ state (Callear and Pilling, 1970).

We note that quenching of the $A^2\Sigma^+$ state of NO by N_2 is inefficient. This process is the analogue of quenching of the $B^2\Sigma^+$ state of PO. Evidently, different mechanisms are applicable to the $^2\Sigma^+$ states of NO and PO.

The value for the radiative lifetime of PO $B^2\Sigma^+$ ($v' = 0$), $\tau_R = (250 \pm 10)$ ns, can be compared with that for the $A^2\Sigma^+$ ($v' = 0$) state of NO, namely (215 ± 5) ns from Zacharias et al. (1976), (175 ± 10) ns from Benoist d'Azy et al. (1975) and (205 ± 10) ns from Brzozowski et al. (1974). As expected, similar values for τ_R apply to these $^2\Sigma^+$–$X^2\Pi$ transitions of PO and NO.

E. The SO Free Radical

Ground-state SO $X^3\Sigma^-$ free radicals differ from BO and PO in that kinetic information on SO has become available as a result of several methods, including mass spectrometry and emission spectroscopy. Unlike BO and PO radicals, SO in its ground state is long-lived and is relatively unreactive species. Therefore, it is possible to generate and maintain reasonably large concentrations of SO $X^3\Sigma^-$ radicals, of the order of 10^{13}–10^{14} cm^{-3}.

Two studies on LIF of the SO radical have been described recently: these involved the $A^3\Pi$–$X^3\Sigma^-$ transition, and the relevant potential energy curves are shown in Fig. 37. In the first study, Clyne and McDermid (1979e) generated SO $X^3\Sigma^-$ from the reaction of O ^3P atoms with CS_2 in a flow system near 0.15 Torr. In order to achieve collision-free conditions and to minimize chemiluminescence, the SO radicals were sampled from the flow tube, via a pinhole, into a test chamber near 0.5 mTorr. Laser excitation was carried out in the test chamber at wavelengths between 259 and 262 nm, using frequency-doubled radiation from a 1 pm bandwidth dye laser pumped by a nitrogen laser. The generation of coherent UV radiation by the KDP

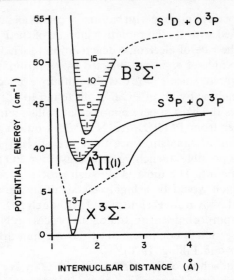

Fig. 37. Potential energy curves for the $X^3\Sigma^-$ ground-state and for the $A^3\Pi(1)$ and $B^3\Sigma^-$ excited states of SO. After Clyne and McDermid (1979e).

crystal was inefficient due to the relatively low fundamental energy of the dye laser ($\sim 10\,\mu J/$pulse).

However, laser excitation spectra of the 1–0 band of the SO $A^3\Pi - X^3\Sigma^-$ were observed successfully, as shown in figure 2 of Clyne and McDermid (1979e). Each vibrational band of this transition contains 24 strong branches. However, because of the large spin splitting of the case (a) $A^3\Pi$ upper state, each band divides into three almost non-overlapped sub-bands, namely $A^3\Pi(1) - X^3\Sigma^-$, $A^3\Pi(2) - X^3\Sigma^-$, and $A^3\Pi(0) - X^3\Sigma^-$. The $A^3\Pi(1)$ sub-band is the simplest in structure, and possesses six rotational branches, whilst the other two sub-bands consist of nine branches each. The analysis of the bands has been described by Colin (1969, 1980, and private communication) and by Clyne and McDermid (1979e). An interesting observation was the absence from the excitation spectrum of the $Q_{21}(15)$ line, and related lines involving the same upper-state rotational quantum number (Clyne and McDermid, 1979e). This result indicates a curve crossing from the fluorescing $A^3\Pi(1)$ state into a non-radiating state. Colin (1980) has observed a strong rotational perturbation in absorption at the same quantum state that was observed to show breaking-off in fluorescence. He assigned the perturbation to a crossing with the $\Omega = 1$ component of the $^3\Delta$ state, i.e. $^3\Delta(1)$. Clearly, selection rules forbid radiation of $^3\Delta(1)$ to the ground $X^3\Sigma^-$; and there are no low-lying electronic states to which radiation by $^3\Delta(1)$ could occur.

In Clyne and McDermid's (1979e) work, lifetimes of various rotational and Λ-doublet levels of $A^3\Pi(v' = 1)$ were determined. Essentially collision-free conditions were used. Apart from the $J' = 15$ state (discussed above), no significant variation of collision-free lifetime τ_0 was found. The mean value was $(16 \cdot 4 \pm 1 \cdot 4)\,\mu s$ for $v' = 1$; this quantity was identified with the radiative lifetime for the $A^3\Pi$ state.

The value of $\tau_R = 16 \cdot 4\,\mu s$ is surprisingly long for a fully allowed $^3\Pi - ^3\Sigma^-$ transition, and suggests that it may be possible to design an electronic transition laser operating on the $A-X$ transition of SO (Clyne and McDermid, 1979e). The reason for the long lifetime of SO $A^3\Pi$ is not completely clear. However, it is noted that both the upper and lower states of the $A-X$ transition dissociate to ground-state S + O atoms. Also, the bonding type in SO $A^3\Pi$ appears to change as a function of r-centroid (see below).

Limited quenching data on SO $A^3\Pi$ were reported by Clyne and McDermid (1979e). The most complete study was with O_2 as bath gas, giving $k_{O_2} = (6 \cdot 4 \pm 1 \cdot 6) \times 10^{-11}\,cm^3\,s^{-1}$. Other measurements on N_2, N_2O, and CS_2 were also described. However, experimental limitations restricted the range of bath gas pressures to 6·3 mTorr or less, thus giving a rather small variation of τ and limited accuracy for k_Q determinations.

Using a modified apparatus, Clyne and Liddy (1981) were able to carry out quenching studies over a wider pressure range, up to 300 mTorr, using excitation in the $v' = 5$ level of SO $A^3\Pi$. Their results for the bath gases CS_2, O_2, N_2, Ar, and SF_6 are summarized in Table XII.

Clyne and Liddy (1981) extended Clyne and McDermid's (1979e) work, by determination of τ_R as a function of vibrational energy in SO $A^3\Pi$, over the range $6 \geqslant v' \geqslant 0$. The extension to levels $v' \geqslant 2$ was made possible by using frequency-doubled dye laser radiation at wavelengths down to 246 nm. This was accomplished using more powerful pulses ($\sim 300\,\mu J$) of fundamental

TABLE XII. Rate constants for electronic quenching k_Q, of selected rotational levels of the $v' = 5$ vibrational level of SO $A^3\Pi(1)$.

Quenching gas	Rotational level N'	Λ-doublet component	$k_Q(cm^3\,molecule^{-1}\,s^{-1})$ $(\pm 2\sigma)$
CS_2	5 and 6[a]	c	$(5 \cdot 8 \pm 0 \cdot 4) \times 10^{-10}$
CS_2	8	d	$(4 \cdot 5 \pm 0 \cdot 3) \times 10^{-11}$
O_2	5 and 6[a]	c	$(1 \cdot 2 \pm 0 \cdot 2) \times 10^{-10}$
O_2	8	d	$(8 \cdot 2 \pm 0 \cdot 8) \times 10^{-11}$
N_2	5 and 6[a]	c	$(9 \cdot 0 \pm 1 \cdot 0) \times 10^{-12}$
Ar	5 and 6[a]	c	$(7 \cdot 5 \pm 1 \cdot 0) \times 10^{-12}$
SF_6	5 and 6[a]	c	$(2 \cdot 6 \pm 0 \cdot 5) \times 10^{-11}$

[a] Overlapped absorption lines ($R_{21}(4)$ and $R_{21}(5)$).

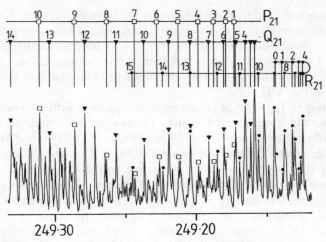

Fig. 38. Laser excitation spectrum of the 5–0 band of SO $A^3\Pi(1)$–$X^3\Sigma^-$. After Clyne and Liddy (1981)

dye laser radiation than in Clyne and McDermid's (1979e) work, and by frequency doubling with a relatively inefficient lithium formate monohydrate (LFM) crystal. Although the LFM very easily became burned, particularly at shorter wavelengths, this crystal was found to be better than an old sample of potassium pentaborate (KPB) which was almost blind in the wavelength region of interest.

Clyne and Liddy (1981) used the slow O + OCS reaction as a source of SO radicals, in preference to the fast O + CS_2 reaction. The O + OCS reaction had the advantage of not producing highly metastable CS radicals, which show an extremely intense and complex laser excitation spectrum that obliterates parts (e.g. the 2–0 band) of the weaker SO A–X system.

Fig. 38 shows the laser excitation spectrum of the 5–0 band of SO A–X (Clyne and Liddy, 1981). The structure is regular and no perturbations were noted.

Determinations of the collision-free lifetimes of SO $A^3\Pi$ were made, following excitation of the c- and d-components of the Λ-doublets. However, no study of τ_R as a function of J' was attempted for vibrational bands other than $v' = 1$, since there was no significant variation of τ_R within the 1–0 band, according to Clyne and McDermid (1979e).

Table XIII summarizes the lifetime results of Clyne and Liddy (1981) for $6 \geqslant v' \geqslant 0, \Omega = 1$; the $v' = 5, \Omega = 0, 2$ levels also were studied. The results for the $v' = 0$ level of SO $A^3\Pi$ were of lower reliability than those for the other levels, because of the weakness of the fluorescence out of this state and the consequent need for appreciable corrections due to SO_2 fluorescence.

The results show clearly that τ_R of SO $A^3\Pi$ decreased strongly as a function

TABLE XIII. Collision-free lifetime of SO $A^3\Pi$ as a function of v'.

Initial v'	Ω	No. of determinations	Mean lifetime $\tau_0(\mu s)$ ($\pm 1\sigma$)
6	1	5	9.8 ± 0.8
5	2	6	9.4 ± 0.8
	1	38	9.2 ± 0.9
	0	7	9.4 ± 1.4
4	1	6	10.4 ± 0.8
3	1	9	11.5 ± 1.3
2	1	12	14.1 ± 1.1
1	1	14	17.9 ± 1.4
0	1	4	35 ± 3

of increasing vibrational energy. This trend is not due to Franck–Condon effects, and represents an increase in the electric dipole moment of the $A–X$ transition with increasing vibrational energy. The effect is relatively large; for example, by contrast, the electric dipole moment of the $B–X$ transition of I_2 shows only a two-fold variation over a wide range of r-centroid values extending over most of the potential energy well.

The potential energy function of SO $A^3\Pi$ is known to be irregular above $v' = 4$, and it is possible that it shows a maximum. In addition, the higher vibrational levels of the potential energy curve for SO $A^3\Pi$ may interact with another bound state, possibly a $^3\Pi$ state (Colin, 1980), thus accounting for the irregular nature of the $A^3\Pi$ state, and the strong variation of electric dipole moment of the $A–X$ transition with v' in SO $A^3\Pi$.

V. STATE-TO-STATE REACTION DYNAMICS

Of fundamental importance to the understanding of the kinetics of a chemical reaction are the effects of the internal state distributions of the reactants on the state distribution of the products. For this purpose, it is necessary to perform kinetic studies under single-collision conditions so that collisional relaxation will not significantly change the product state distribution. A means for determining the population in specific quantum levels is also required. Pioneering work in this field was carried out by Zare and coworkers (Cruse et al., 1973; Zare and Dagdigian, 1974; Pruett and Zare, 1975, 1976) who utilized molecular beam methods to obtain single reactive collisions and laser-induced fluorescence to probe the internal states of the products.

In the first of a series of elegant experiments (Cruse et al., 1973), Zare and coworkers studied the series of reactions

$$Ba + HX \rightarrow BaX(X^2\Sigma^+) + H; \qquad X = F, Cl, Br, I$$

using laser-induced fluorescence of the $C^2\Pi-X^2\Sigma^+$ systems to detect the BaX products. The experimental arrangement, which consisted of a simple molecular beam apparatus, a pulsed tunable dye laser for exciting fluorescence, and a gated optical detection system, is shown in Fig. 39. In this study, the relative reaction rates producing BaX in all different vibrational states were determined. The specification of the reactant HF in the ground vibrational level occurred naturally from the Boltzmann population at room temperature. From this work it was concluded that the product internal states were not thermal and that an unusually small fraction of the total energy available appeared as product vibration or rotation.

In a subsequent study, Pruett and Zare (1976) studied the reaction of barium atoms with hydrogen fluoride with an additional step of selectively

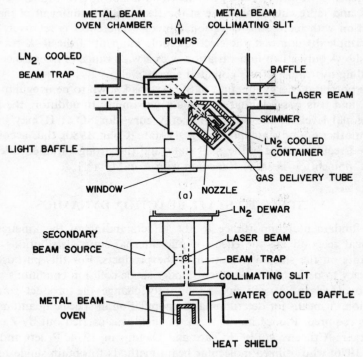

Fig. 39. Molecular beam apparatus: (a) top view; (b) side view. In (b) the laser beam is perpendicular to the plane of the paper and the secondary beam source has been rotated by 45° about the Ba beam axis for clarity. After Cruse *et al.* (1973).

preparing the HF molecule in the $v = 1$ state prior to collision:

$$Ba + HF(v = 1) \rightarrow BaF(v) + H.$$

In order to excite the HF in the reaction chamber, an infrared HF laser oscillating on the $P(2)$ line of the $(1-0)$ transition was used. This laser delivered $110\,\mu J$ in a $500\,ns$ pulse. The determination of the exact fraction of HF molecules excited to $v = 1$ was difficult and only an upper limit ($\leqslant 0.01 \pm 0.0005$) was obtained. The excitation spectrum of the BaF product could be taken with or without the HF laser firing. When the HF laser was firing, the tunable probe beam was delayed with respect to the HF pulse so that the HF $(v = 1)$ molecules had sufficient time to react with Ba atoms. The best spectra were obtained with a $50\,\mu s$ delay.

The laser excitation spectra were recorded for BaF produced by reaction with only HF $(v = 0)$ and also for BaF produced when the HF reactant was irradiated with the HF laser. The long-wavelength tails of these spectra, shown in Fig. 40, reveal the appearance of new bandheads of higher vibrational numbering. These bandheads indicate that higher vibrational states of BaF are populated, up to $v = 11$, in the reaction of Ba with HF $(v = 1)$. In order to make quantitative statements regarding the state-to-state reaction rates, it is necessary to obtain the excitation spectrum characteristic of the products of the reaction $Ba + HF(v = 1)$ only, i.e. in the absence of any contribution from the reaction $Ba + HF(v = 0)$. This is essentially the difference between the two spectra shown in Fig. 40. Because of the very large contribution from HF $(v = 0)$, the difference spectrum is degraded at shorter wavelengths (lower v) by all of the problems associated with

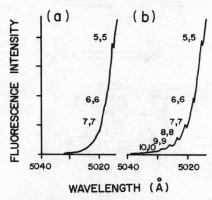

Fig. 40. Simple BaF excitation spectra (high sensitivity) for the long-wavelength tails: (a) HF laser off, HF$(v = 0)$ + Ba; (b) HF laser on, HF$(v = 0, 1)$ + Ba. After Pruett and Zare (1976).

subtracting two very large signals to form a small difference. The results of this study were that the BaF product was found to retain an average of 64% of the initial reactant vibrational energy and the distribution of product states for $Ba + HF(v = 1)$ had a broad maximum shifted to $v = 6$ from the value $v = 1$ found for $Ba + HF(v = 0)$.

Recognizing the problems caused by the domination of products formed by ground-state reagents, Torres-Filho and Pruett (1980) have developed a system which utilizes two simultaneously probed beam reactions and allows very sensitive and quantitative measurements of small differences in product state distributions in molecular beam reactions. Details of the apparatus are shown in Fig. 41. Because the two atom beams travel in different directions, a single laser pulse propagating in one direction in the plane of the two atom beams would be seen by an absorber as having two different frequencies separated by the effective Doppler shift between the two product zones. To eliminate this problem, counter-propagating laser pulses of equal intensity, produced by splitting the probe dye laser beam, were used to detect the products. When the two reaction zones and detection systems were properly balanced, any external effects on one of the reaction zones, such as vibrational excitation of one of the reagents, could be sensitively measured in the difference between the two excitation spectra.

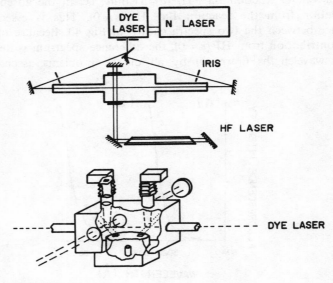

Fig. 41. Dual molecular beam apparatus showing the two reaction zones used in differential measurements. Also shown is the optical alignment of the intracavity HF laser and the counter-propagating probe laser beams. After Torres–Filho and Pruett (1980).

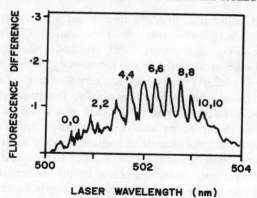

Fig. 42. Excitation difference spectrum of the BaF products $(^2\Sigma^+ - X^2\Pi_{1/2})$. Zero signal implies no difference in the state-to-state rate constant for the product level at that spectral region. After Torres–Filho and Pruett (1980).

When this system was used to study the reaction of $Ba + HF(v = 1)$, contrary to the earlier results (Pruett and Zare, 1976) it was found that low vibrational levels of BaF were populated predominantly by this reaction. The excitation difference spectrum of the BaF products observed by Torres-Filho and Pruett (1980) is shown in Fig. 42. This shows that the production of BaF $v = 0, 1$, and 2 has equivalent detailed rates from HF $v = 0$ or $v = 1$, since the difference in this region is near zero. Also, even though there are fewer HF $v = 0$ molecules present after the IR laser excitation, there are no negative contributions to the difference spectrum, implying that all state-to-state rate constants from $HF(v = 1)$ are as large as or larger than from $HF(v = 0)$. This then means that the overall reaction rate is increased and the reaction cross section for $Ba + HF(v = 1)$ was found to be 3 times greater than that for $Ba + HF(v = 0)$.

In another molecular beam study of this reaction, $Ba + HF$, Gupta *et al.* (1980) have studied the effect of reagent translational energy on the product state distribution. The relative collision energy was varied between 3 and 13 kcal mol^{-1} using a crossed-beam geometry in which a seeded HF beam intersected a thermal Ba beam. The vibrational and rotational distributions of the BaF product were determined from computer simulations of the excitation spectra. The reaction cross section was found to have a low threshold, ~ 1 kcal mol^{-1}. With increasing collision energy, the cross section increased to a maximum in the range 6–8 kcal mol^{-1} and an upper limit of 15Å^2 was placed on the absolute value of the reaction cross section. The fraction of energy appearing in translation, rotation or vibration of the products was roughly constant over the range of collision energies studied, with nearly half going into product translation and the remainder being nearly equally divided between rotation and vibration. The effects of reagent vibration (Torres-Filho and Pruett, 1980) and collisional energy (Gupta *et al.*, 1980) are similar with regard to the reaction dynamics.

Although laser-induced fluorescence is a very sensitive detector of product distributions, it cannot be applied universally. The products must have a strong electronic absorption band in a region which can be covered with a tunable laser. The spectroscopy and radiative properties of the upper state must be well known and the fluorescence quantum yield must be appreciable. These restrictions usually limit the usefulness of LIF to diatomic and some selected small polyatomic molecules (Zare, 1979). Molecular beam systems are complicated and can be difficult to set up. Possibly for these reasons, the number of state-to-state studies in the literature is limited.

Reactions other than $Ba + HF$ that have been studied by molecular beam–laser-induced fluorescence include $H + NO_2$ (Silver *et al.*, 1976; Mariella *et al.*, 1978), $H + ClO_2$ (Mariella *et al.*, 1978), and $O + CS_2$ (Clough and Johnston, 1980). Silver *et al.* (1976) studied the relative populations of rotational states in the $v = 0$ and $v = 1$ vibrational states of OH produced in the reaction

$$H + NO_2 \rightarrow NO + OH.$$

The excited vibrational state was found to be produced at $(1 \cdot 3 \pm 0 \cdot 3)$ times the rate of $v = 0$. High degrees of rotational excitation were also observed. Primary energy disposal in the OH product of the reactions

$$H + NO_2 \rightarrow NO + OH$$

and

$$H + ClO_2 \rightarrow ClO + OH$$

was studied by Mariella *et al.* (1978). In their study of the

$$O + CS_2 \rightarrow CS + SO$$

reaction, Clough and Johnston (1980) used a CW frequency-doubled dye laser to determine the initial vibrational distribution of the CS product. The $A^1\Pi - X^1\Sigma^+$ system was excited and rotationally resolved spectra were obtained. The vibrational distribution found for $v = 0, 1$, and 2 was $1 \cdot 0, 0 \cdot 27$, and $0 \cdot 11$, respectively, yielding a value of 6% for the fraction of reaction exoergicity entering vibration of the CS product. No evidence of high rotational excitation was detected. In an adaptation of a crossed molecular beam experiment, Allison *et al.* (1979) have studied the internal state distributions of N_2^+ ions formed from N_2 by electron-impact ionization. Analysis of the $B-X$ (0, 0) band showed that the rotational distribution could be characterized by a temperature which was found to increase slightly with decreasing electron energy (60–100 eV).

A different approach to the measurement of state-to-state reaction rates has been developed by Breckenridge *et al.* (1978, 1979), Breckenridge and Kim Malmin (1979), and Breckenridge and Oba (1980). They have described a laser 'pump-and-probe' technique which allows the determination of initial quantum-state energy distributions of single-collision events under bulb rather than beam conditions. The basis of the technique is to use a very fast observation technique rather than low particle densities to create the single-collision kinetic conditions. The time resolution is achieved by producing excited atoms with one short laser pulse, then probing the products of collisions of the excited atoms with a molecular substrate with another laser pulse, all within a time period of about 20 ns. A schematic diagram of the laser pump-and-probe apparatus is shown in Fig. 43. The system has been

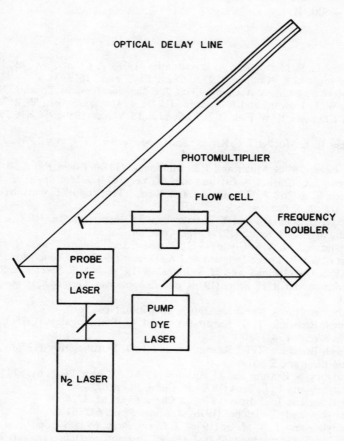

Fig. 43. Schematic diagram of the laser 'pump-and-probe' apparatus used by Breckenridge *et al.* (1978).

used to measure the initial distributions of $Cd(^3P_{0,1,2})$ electronic states in the E to V, R transfer process,

$$Cd(^1P_1) + M \rightarrow Cd(^3P_{0,1,2}) + M,$$

where M represents a variety of quenchers. For $M = N_2$, for example, the initial $^3P_0 : ^3P_1 : ^3P_2$ distribution was $0.05 : 0.27 : 0.68$ (Breckenridge, Malmin, Nikolai and Oba, 1978). The technique has also been used to determine the initial distribution of rotational quantum states of the diatomic product, CdH, of the chemical reaction,

$$Cd(^1P_1) + H_2 \rightarrow CdH(v, N) + H,$$

where $v = 0$ and N is the rotational quantum number. The rotational distribution was found to approximate closely to a Boltzmann distribution with $T_{eff} \sim 5000$ K.

References

N. A. Allison, G. R. Hinton and A. J. Bottomley (1979), *J. Chem. Phys.*, **69**, 1628.

J. G. Anderson and F. Kaufman (1972), *Chem. Phys. Lett.*, **16**, 375.

E. H. Appelman and M. A. A. Clyne (1975), *J. Chem. Soc. Faraday Trans. I*, **71**, 2072.

B. S. Ault, W. F. Howard and L. Andrews (1975), *J, Mol. Spectrosc.*, **55**, 217.

R. Bacis, S. Churassy, R. W. Field, J. B. Koffend and J. Verges (1980), *J. Chem. Phys.*, **72**, 34.

R. H. Barnes, C. E. Moeller, J. F. Kircher and C. M. Verber (1974), *Appl. Phys. Lett.*, **24**, 610.

A. P. Baronavski, R. G. Miller and J. R. McDonald (1978), *Chem. Phys.*, **30**, 119.

R. F. Barrow, T. C. Clark, J. A. Coxon and K. K. Yee (1974), *J. Mol. Spectrosc.*, **51**, 428.

D. L. Baulch, R. A. Cox, R. F. Hampson, J. A. Kerr, J. Troe and R. T. Watson (1980), *J. Phys. Chem. Ref. Data*, **9**, 295.

K. H. Becker, U. Schurath and T. Tatarczyk (1975), *Appl. Opt.*, **14**, 310.

E. J. Beiting and K. A. Smith (1979), *Opt. Commun.*, **28**, 355.

P. P. Bemand and M. A. A. Clyne (1973), *J. Chem. Soc Faraday Trans. II*, **69**, 1643.

O. Benoist d'Azy, R. Lopez-Delgado and A. Tramer (1975), *J. Chem. Phys.*, **9**, 327.

J. W. Birks, S. D. Gabelnick and H. S. Johnston (1975), *J. Mol. Spectrosc.*, **57**, 23.

G. C. Bjorklund and R. H. Storz (1978), *Bell Laboratories Technical Memo* TM-78-1313-23.

N. Bloembergen (1965), *Nonlinear Optics*, Benjamin, New York.

A. M. Bonch-Breuvich, T. K. Razumova and I. O. Starobogatov (1975a), *Sov. J. Quantum Electron.*, **4**, 1380.

A. M. Bonch-Breuvich, T. K. Razumova and I. O. Starobogatov (1975b), *Sov. J. Quantum Electron.*, **5**, 867.

V. E. Bondybey, S. Bearder and C. Fletcher (1976), *J. Chem. Phys.*, **64**, 5243.

F. E. Bondybey and L. E. Brus (1975), *J. Chem. Phys.*, **62**, 620.

V. E. Bondybey and L. E. Brus (1976), *J. Chem. Phys.*, **64**, 3724.

V. E. Bondybey and C. Fletcher (1976), *J. Chem. Phys.*, **64**, 3615.

V. E. Bondybey and T. A. Miller (1978), *J. Chem. Phys.*, **69**, 3597, 3602.

W. Braun and T. Carrington (1969), *J. Quant. Spectrosc. Radiat. Transfer*, **9**, 1133.

W. H. Breckenridge, R. J. Donovan and O. Kim Malmin (1979), *Chem. Phys. Lett.*, **62**, 608.

W. H. Breckenridge and O. Kim Malmin (1979), *Chem. Phys. Lett.*, **68**, 341.

W. H. Breckenridge, O. Kim Malmin, W. L. Nikolai and D. Oba (1978), *Chem. Phys. Lett.*, **59**, 38.

W. H. Breckenridge and D. Oba (1980), *Chem. Phys. Lett.*, **72**, 455.

P. H. Brodersen and J. E. Sicre (1955), *Z. Phys.*, **141**, 515.

H. P. Broida and T. Carrington (1963), *J. Chem. Phys.*, **38**, 136.

W. G. Brown (1932), *Phys. Rev.*, **42**, 355.

M. Broyer, J. Vigue and J. C. Lehmann (1975), *J. Chem. Phys.*, **63**, 5428.

J. Brzozowski, N. Elander and P. Erman (1974), *Phys. Scr.*, **9**, 99.

R. E. Buckles and J. J. M. Bader (1971), *Inorg. Synth.*, **9**, 130.

R. L. Byer (1980), *Electro-optics System Design*, **12**, 24.

R. L. Byer and R. L. Herbst (1978), *Laser Focus*, **14**, 48.

G. V. Calder and W. F. Giauque (1965), *J. Phys. Chem.*, **69**, 2443.

A. B. Callear and M. J. Pilling (1970), *Trans. Faraday Soc.*, **66**, 1618.

G. A. Capelle and H. P. Broida (1973), *J. Chem. Phys.*, **58**, 4212.

G. A. Capelle, and C. Linton (1976), *J. Chem. Phys.*, **65**, 5361

G. A. Capelle, K. Sakurai and H. P. Broida (1971), *J. Chem. Phys.*, **54**, 1728.

J. L. Carlsten and T. J. McIlrath (1973), *Opt. Commun.*, **8**, 52.

I. L. Chidsey and D. R. Crosley (1980), *J. Quant. Spectrosc. Radiat. Transfer*, **23**, 187.

M. S. Child (1976), *Mol. Phys.*, **32**, 1495.

M. S. Child and A. D. Bandrauk (1970), *Mol. Phys.*, **19**, 95.

M. S. Child and R. B. Bernstein (1973), *J. Chem. Phys.*, **59**, 5916.

A. Chutjian (1969), *J. Chem. Phys.*, **51**, 5414.

P. N. Clough and J. Johnston (1980), *Chem, Phys. Lett.*, **71**, 253.

M. A. A. Clyne and J. A. Coxon (1967), *Proc. R. Soc. A*, **298**, 428.

M. A. A. Clyne and J. A. Coxon (1968), *Nature*, **217**, 448.

M. A. A. Clyne and J. A. Coxon (1970a), *J. Phys. B*, **3**, L9.

M. A. A. Clyne and J. A. Coxon (1970b), *J. Mol. Spectrosc.*, **33**, 381.

M. A. A. Clyne, J. A. Coxon and L. W. Townsend (1972), *J. Chem. Soc. Faraday Trans. II*, **68**, 2134.

M. A. A. Clyne, A. H. Curran and J. A. Coxon (1976), *J. Mol. Spectr.*, **63**, 43.

M. A. A. Clyne and M. C. Heaven (1978), *J. Chem. Soc. Faraday Trans. II*, **78**, 1992.

M. A. A. Clyne and M. C. Heaven (1980a), *J. Chem. Soc. Faraday Trans. II*, **76**, 49.

M. A. A. Clyne and M. C. Heaven (1980b), *Chem. Phys.*, **51**, 299.

M. A. A. Clyne and M. C. Heaven (1981a), *J. Chem. Soc. Faraday Trans. II*, **77**, 1735.

M. A. A. Clyne and M. C, Heaven (1981b), *Faraday Discuss. Chem. Soc.*, no.71.

M. A. A. Clyne and M. C. Heaven (1981c), *Chem. Phys.*, **58**, 145.

M. A. A. Clyne, M. C. Heaven and S. J. Davis (1980a), *J. Chem. Soc. Faraday Trans. II*, **76**, 961.

M. A. A. Clyne, M. C. Heaven and E. Martinez (1980b), *J. Chem. Soc. Faraday Trans. II*, **76**, 177.

M. A. A. Clyne, M. C. Heaven and E. Martinez (1980c), *J. Chem. Soc. Faraday Trans. II*, **76**, 405.

M. A. A. Clyne and J. P. Liddy (1980), *J. Chem. Soc. Faraday Trans. II*, **76**, 1569.

M. A. A. Clyne and J. P. Liddy (1981), *J. Chem. Soc. Faraday Trans. II*, in press.

M. A. A. Clyne and I. S. McDermid (1976a), *J. Chem. Soc. Faraday Trans.* **72**, 2242.

M. A. A. Clyne and I. S. McDermid (1976b), *J. Chem. Soc. Faraday Trans.* **72**, 2252.

M. A. A. Clyne and I. S. McDermid (1977), *J. Chem. Soc. Faraday Trans.* **73**, 1094.

M. A. A. Clyne and I. S. McDermid (1978a), *J. Chem. Soc. Faraday Trans.* **74**, 1644.

M. A. A. Clyne and I. S. McDermid (1978b), *J. Chem. Soc. Faraday Trans.* **74**, 798.
M. A. A. Clyne and I. S. McDermid (1978c), *J. Chem. Soc. Faraday Trans.* **74**, 807.
M. A. A. Clyne and I. S. McDermid (1978d), *J. Chem. Soc. Faraday Trans.* **74**, 664.
M. A. A. Clyne and I. S. McDermid (1978e), *J. Chem. Soc. Faraday Trans.* **74**, 644.
M. A. A. Clyne and I. S. McDermid (1978f), *J. Chem. Soc. Faraday Trans.* **74**, 1376.
M. A. A. Clyne and I. S. McDermid (1978g), *J. Chem. Soc. Faraday Trans.* **74**, 1935.
M. A. A. Clyne and I. S. McDermid (1979a), *Faraday Discuss. Chem. Soc.* no. 67, 316.
M. A. A. Clyne and I. S. McDermid (1979b), *J. Chem. Soc. Faraday Trans.* **75**, 280.
M. A. A. Clyne and I. S. McDermid (1979c), *J. Chem. Soc. Faraday Trans.* **75**, 1313.
M. A. A. Clyne and I. S. McDermid (1979d), *J. Chem. Soc. Faraday Trans.* **75**, 1677.
M. A. A. Clyne and I. S. McDermid (1979e), *J. Chem. Soc. Faraday Trans.* **75**, 905.
M. A. A. Clyne and E. Martinez (1980a), *J. Chem. Soc. Faraday Trans.* **76**, 1275.
M. A. A. Clyne and E. Martinez (1980b), *J. Chem. Soc. Faraday Trans.* **76**, 1561.
M. A. A. Clyne and W. S. Nip (1979), *Reactive Intermediates in the Gas Phase*, ed. D. W. Setser, Academic Press, New York.
M. A. A. Clyne and D. J. Smith (1979), *J. Chem. Soc. Faraday Trans.* **75**, 704.
R. Colin (1969), *Can. J. Phys.*, **47**, 979.
R. Colin (1980), *J. Chem. Soc. Faraday II*, in press.
F. C. Connelly (1933), *Proc. Phys. Soc.*, **45**, 780.
R. D. Coombe and R. K. Horne (1979), *J. Phys. Chem.*, **83**, 2435.
A. Corney, J. Manners and C. E. Webb (1979), *Opt. Commun.*, **31**, 354.
J. A. Coxon (1971), *J. Quant. Spectrosc. Radiat. Transfer*, **11**, 443.
J. A. Coxon (1972a), *J. Quant. Spectrosc. Radiat. Transfer*, **12**, 639.
J. A. Coxon (1972b), *J. Mol. Spectrosc.*, **41**, 548.
J. A. Coxon (1972c), *J. Mol. Spectrosc.*, **41**, 566.
J. A. Coxon (1973), *Molecular Spectroscopy*, vol. 1, *Chem. Soc. Spec. Per. Rep.*, The Chemical Society, London.
J. A. Coxon (1974), *J. Mol. Spectrosc.*, **50**, 142.
J. A. Coxon (1975), *Chem. Phys. Lett.*, **33**, 136.
J. A. Coxon (1980), *J. Mol. Spectrosc.*, **82**, 264.
J. A. Coxon and A. H. Curran (1979), *J. Mol. Spectrosc.*, **75**, 270.
J. A. Coxon and R. Shanker (1978), *J, Mol. Spectrosc.*, **69**, 109.
D. R. Crosley and R. K. Lengel (1975), *J. Quant. Spectrosc. Radiat. Transfer*, **15**, 579.
H. W. Cruse, P. J. Dagdigian and R. N. Zare (1973), *Faraday Discuss. Chem. Soc.*, no. 55, 277.
M. D. Danyluk and G. W. King (1976a), *Chem. Phys. Lett.*, **43**, 1.
M. D. Danyluk and G. W. King (1976b), *Chem. Phys. Lett.*, **44**, 440.
M. D. Danyluk and G. W. King (1977a), *Chem. Phys.*, **22**, 59.
M. D. Danyluk and G. W. King (1977b), *Chem. Phys.*, **25**, 343.
P. B. Davies and B. A. Thrush (1968), *Proc. R. Soc. A*, **302**, 243.
D. D. Davis, W. S. Heaps, D. Philen, M. Rodgers, T. McGee, A. Nelson and A. J. Moriarty (1979), *Rev. Sci. Instrum.*, **50**, 1505.
C. F. Dewey, W. R. Cook, R. T. Hodgson and J. J. Wynne (1975), *Appl. Phys. Lett.*, **26**, 714.
H. J. Dewey (1976), *IEEE J. Quantum Electron.*, **QE-12**, 303.
V. H. Dibeler, J. A. Walker and K. E. McCulloh (1970), *J. Chem. Phys.*, **53**, 4414.
M. Diegelmann (1980), *PhD Thesis*, Munich.
M. Diegelmann, K. Hohla and K. L. Kompa (1979), *Opt. Commun.*, **29**, 334.
S. G. Dinev, I. G. Koprinkov, K. V. Stamenov, K. A. Stankov and C. Radzewicz (1980), *Opt. Commun.*, **32**, 313.
L. S. Ditman, R. W. Gammon and T. D. Wilkerson (1975), *Opt. Commun.*, **13**, 154.

R. J. Donovan D. P. Fernie, M. A. D. Fluendy, R. M. Glen, A. G. A. Rae and J. R. Wheeler (1980), *Chem. Phys. Lett.*, **69**, 472.

R. J. Donovan and H. Gillespie (1975), *Gas Kinetics and Energy Transfer*, vol. 1, *Chem. Soc. Spec. Per. Rep.*, The Chemical Society, London.

A. E. Douglas and A. R. Hoy (1975), *Can. J. Phys.*, **53**, 1965.

A. E. Douglas, C. K. Moller and B. P. Stoicheff (1963), *Can. J. Phys.*, **41**, 1174.

T. H. Dunn and L. Hanson (1970), *Can. J. Phys.*, **47**, 1657.

F. B. Dunning and R. E. Stickel (1976), *Appl. Opt.*, **15**, 3131.

F. B. Dunning, E. D. Stokes and R. F. Stebbings (1972), *Opt. Commun.*, **6**, 63.

F. B. Dunning, F. K. Tittel and R. F. Stebbings (1973), *Opt. Commun.*, **7**, 181.

R. A. Durie (1950), *Proc. Phys. Soc.*, **63**, 1292.

R. A. Durie (1951), *Proc. R. Soc. A.*, **207**, 388.

R. A. Durie (1966), *Can. J. Phys.*, **44**, 337.

G. L. Eesley and M. D. Levensen (1976), *IEEE J. Quantum Electron.*, **QE-12**, 440.

U. Ganiel, A. Hardy, G. Neumann and D. Treves (1975), *IEEE J. Quantum Electron.*, **QE-11**, 881.

K. R. German (1975a), *J. Chem. Phys.*, **62**, 2584.

K. R. German (1975b), *J. Chem. Phys.*, **63**, 5252.

K. R. German (1976), *J. Chem. Phys.*, **64**, 4065.

S. Gerstenkorn and P. Luc (1978), *Atlas du Spectre d'Absorption de la Molecule de l'Iode*, Editions du CNRS, Paris.

S. Gerstenkorn and P. Luc (1979), *J. Mol. Spectrosc.*, **77**, 310.

S. Gerstenkorn, P. Luc and A. Perrin (1977), *J. Mol. Spectrosc.*, **64**, 56.

L. A. Godfrey, W. C. Egbert and R. S. Meltzer (1980), *Opt. Commun.*, **34**, 108.

R. D. Gordon and K. K. Innes (1979), *J. Chem. Phys.*, **71**, 2824.

A. Gupta, D. S. Perry and R. N. Zare (1980), *J. Chem. Phys.*, **72**, 6237.

D. M. Guthals and J. W. Nibler (1979), *Opt. Commun.*, **29**, 322.

S. G. Hadley, M. J. Bina and G. D. Brabson (1974), *J. Phys. Chem.*, **78**, 1833.

T. Halldorsson and E. Menke (1970), *Z. Naturf.*, **259**, 1356.

D. O. Hanna, P. A. Kárkkáinen and R. Wyatt (1975), *Opt. Quantum Electron.*, **7**, 115.

T. W. Hansch (1972), *Appl. Opt.*, **11**, 895.

T. W. Hansch, S. A. Lee, R. Wallenstein and C. Wieman (1975), *Phys. Rev. Lett.*, **34**, 307.

S. J. Harris, W. C. Natzle and C. Bradley Moore (1979), *J. Chem. Phys.*, **70**, 4215.

W. Heisenberg (1922), *Z. Phys.*, **8**, 273.

G. Herzberg (1950), *Molecular Spectra and Molecular Structure*, part 1, *Spectra of Diatomic Molecules*, Van Nostrand, New York.

G. Herzberg (1971), *Spectra and Structure of Simple Free Radicals*, Cornell University Press, New York.

A. A. Hnilo, F. A. Manzano and A. H. Burgos (1980), *Opt. Commun.*, **33**, 311.

G. W. Holleman and J. I. Steinfeld (1971), *Chem. Phys. Lett.*, **12**, 431.

W. Holzer, W. F. Murphy and H. J. Bernstein (1970a), *J. Chem. Phys.*, **52**, 399.

W. Holzer, W. F. Murphy and H. J. Bernstein (1970b), *J. Chem. Phys.*, **52**, 469.

J. A. Horsley (1967), *J. Mol. Spectrosc.*, **22**, 469.

J. T. Hougen (1970), *Calculation of Rotational Energy Levels and Rotational Line Intensities in Diatomic Molecules*, *NBS Monograph* no. 115.

R. S. Hozack, A. P. Kennedy and K. B. McAfee (1980), *J. Mol. Spectrosc.*, **80**, 239.

K. P. Huber and G. Herzberg (1979), *Molecular Spectra and Molecular Structure*, vol. IV, *Constants of Diatomic Molecules*, Van Nostrand Reinhold, New York.

W. Huffer, R. Schieder, H. Telle, R. Raue and W. Brinkwerth (1980), *Opt. Commun.*, **33**, 85.

R. E. Huie, N. J. T. Long and B. A. Thrush (1976), *Chem. Phys. Lett.*, **44**, 608.

R. E. Huie, N. J. T. Long and B. A. Thrush (1978), *Chem. Phys. Lett.*, **55**, 404.

E. Hulthen, N. Jarlsater and L. Koffman (1960), *Ark. Fys*, **18**, 479.

E. Hulthen, N. Johansson and U. Pilsater (1958), *Ark, Fys.* **14**, 31.

D. M. Hwang and H. Chang (1978), *J. Mol. Spectrosc.*, **69**, 11.

F. A. Jenkins and A. McKellar (1932), *Phys. Rev.*, **42**, 464.

F. M. Johnson and M. W. Swagel (1971), *Appl. Opt.*, **10**, 1624.

D. H. Katayama, T. A. Miller and V. E. Bondybey (1979), *J. Chem. Phys.*, **71**, 1662.

D. H. Katayama, T. A. Miller and V. E. Bondybey (1980), *J. Chem. Phys.*, **72**, 5469.

K. Kato (1977a), *IEEE J. Quantum Electron.*, **QE-13**, 544.

K. Kato (1977b), *Appl., Phys. Lett.*, **30**, 583.

G. W. King, I. M. Littlewood, R. G. McFadden and J. R. Robins (1979), *Chem. Phys.*, **41**, 379.

G. W. King, I. M. Littlewood, J. R. Robins and N. T. Wijeratne (1980), *Chem. Phys.*, **50**, 291.

G. W. King, and R. G. McFadden (1978), *Chem. Phys. Lett.*, **58**, 119.

G. K. Klauminzer (1977), *IEEE J. Quantum Electron.*, **QE-13**, 920.

R. Konig, G. Minkwitz and B. Christov (1980), *Opt. Commun.*, **32**, 301.

I. Kovacs (1969), *Rotational Structure in the Spectra of Diatomic Molecules*, Hilger, London.

J. Kuhl and G. Marowsky (1971), *Opt. Commun.*, **4**, 125.

J. Kuhl and H. Spitschan (1975), *Opt. Commun.*, **13**, 6.

A. Lagerqvist, N. E. L. Nilsson and K. Wigartz (1959), *Ark. Fys.*, **15**, 521.

R. J. Le Roy (1970), *J. Chem. Phys.*, **52**, 2678.

R. J. Le Roy, R. G. McDonald and G. Burns (1976), *J. Chem. Phys.*, **65**, 1485.

H. S. Liszt and W. H. Smith (1971), *J. Quant. Spectrosc. Radiat. Transfer.*, **11**, 1043.

M. G. Littman (1978), *Opt. Lett.*, **3**, 138.

M. G. Littman and H. J. Metcalf (1978), *Appl. Opt.*, **17**, 2224.

R. S. Longhurst (1973), *Geometrical and Physical Optics*, 3rd edn., Longman, London, p. 267.

P. Luc (1980), *J. Mol. Spectrosc.*, **80**, 41.

P. Luypaert, G. De Vlieger and J. Van Craen (1980), *J. Chem. Phys.*, **72**, 6283.

K. B. McAfee and R. S. Hozack (1976), *J. Chem. Phys.*, **64**, 2491.

I. S. McDermid (1980), *J. Chem. Soc. Faraday Trans. II*, in press.

I. S. McDermid (1981), *Comprehensive Chemical Kinetics*, suppl. vol. 1, eds. C. H. Bamford and C. F. H. Tipper, Elsevier, Amsterdam.

I. S. McDermid and J. B. Laudenslager (1980), *Chem. Phys. Lett.*, in press.

G. Magyar (1974), *Appl. Opt.*, **13**, 25.

R. Mahon and D. W. Coopman (1978), *Appl. Phys. Letts.*, **33**, 305.

R. P. Mariella, B. Lantzsch, V. T. Maxson and A. C. Luntz (1978), *J. Chem. Phys.*, **69**, 5411.

J. C. Miller and L. Andrews (1980), *J. Mol. Spectrosc.*, **80**, 178.

A. C. G. Mitchell and M. W. Zemansky (1934), *Resonance Radiation and Excited Atoms*, Cambridge University Press, Cambridge.

A. Moriarty, W. Heaps and D. D. Davis (1976), *Opt. Commun.*, **16**, 324.

R. S. Mulliken (1925), *Phys. Rev.*, **25**, 259.

S. A. Myers (1971), *Opt. Commun.*, **4**, 187.

L. G. Nair (1977), *Opt. Commun.*, **23**, 273.

L. G. Nair and K. Dasgupta (1980), *IEEE J. Quantum Electron.*, **QE-16**, 111

N. M. Narovlyanskaya and E. A. Tihkonov (1978), *Sov. J. Quantum Electron.*, **8**, 173.

P. C. Nordine (1974), *J. Chem. Phys.*, **61**, 224.

H. Okabe (1978), *Photochemistry of Small Molecules*, Wiley-Interscience, New York.

C. D. Olson and K. K. Innes (1976), *J. Chem. Phys.*, **64**, 2405.

M. H. Ornstein and V. E. Derr (1976), *J. Opt. Soc. Am.*, **66**, 233.

J. Paisner and S. Hargrove (1979), *Lawrence Livermore Laboratory, Energy and Technology Review*, UCRL-52000-79-3.

J. A. Paisner and R. Wallenstein (1974), *J. Chem. Phys.*, **61**, 4317.

W. C. Paske, J. R. Twist, A. W. Garrett and D. E. Golden (1980), *J. Chem. Phys.*, **72**, 6134.

G. D. Patterson, S. H. Dworetsky and R. S. Hozack (1975), *J. Mol. Spectrosc.*, **55**, 175.

J. C. Polanyi and K. B. Woodall (1972), *J. Chem. Phys.*, **56**, 1563.

P. Pringsheim (1949), *Fluorescence and Phosphorescence*, Interscience, New York.

J. G. Pruett and R. N. Zare (1975), *J. Chem. Phys.*, **62**, 2050.

J. G. Pruett and R. N. Zare (1976), *J. Chem. Phys.*, **64**, 1774.

T. V. R. Rao and S. V. J. Lakshman (1972), *J. Quant. Spectrosc. Radiat. Transfer*, **12**, 1063.

Y. Rao and P. Venkateswarlu (1964), *J. Mol. Spectrosc.*, **13**, 288.

W. G. Richards and R. F. Barrow (1962), *Proc. Chem. Soc.*, **1962**, 297.

C. Rulliere, J. P. Morand and O. De Witte (1977), *Opt. Commun.*, **20**, 339.

S. Saikan (1976), *Opt. Commun.*, **18**, 439.

S. Saikan (1978), *Appl. Phys.*, **17**, 41.

S. Saikan, D. Ouw and F. P. Schafer (1979), *Appl. Opt.*, **18**, 193.

K. Sakurai, G. A. Capelle and H. P. Broida (1971), *J. Chem. Phys.*, **54**, 1220.

K. Sakurai, G. Taieb and H. P. Broida (1976), *Chem. Phys. Lett.*, **41**, 39.

F. P. Schafer (ed.) (1973), *Topics in Applied Physics*, vol. 1, *Dye Lasers*, Springer, Berlin, Heidelberg, New York.

W. Scheib (1930), *Z. Phys.*, **60**, 74.

L. E. Selin (1962a), *Ark. Fys.*, **21**, 479.

L. E. Selin (1962b), *Ark. Fys.*, **21**, 529.

L. E. Selin and B. Soderborg (1962), *Ark. Fys.*, **21**, 515.

K. Shimoda (ed.) (1976), *Topics in Applied Physics*, vol. 13, *High Resolution Laser Spectroscopy*, Springer, Berlin, Heidelberg, New York.

I. Shoshan, N. N. Danon and U. P. Oppenhein (1977), *J. Appl. Phys.*, **48**, 4495.

I. Shoshan and U. P. Oppenheim (1978), *Opt. Commun.*, **25**, 375.

J. A. Silver, W. L. Dimpfl, J. H. Brophy and J. L. Kinsey (1976), *J. Chem. Phys.*, **65**, 1811.

N. L. Singh (1959), *Can. J. Phys.*, **37**, 136.

T. G. Slanger and G. Black (1973), *J. Chem. Phys.*, **58**, 194, 3121, 4367.

H. A. van Sprang, G. R. Mohlmann and F. J. de Heer (1977), *Chem. Phys.*, **24**, 429.

L. Stein, J. Wanner and H. Walther (1980), *J. Chem. Phys.*, **72**, 1128.

J. I. Steinfeld and W. Klemperer (1965), *J. Chem. Phys.*, **42**, 3475.

R. E. Stickel and F. B. Dunning (1978), *Appl. Opt.*, **17**, 1313.

E. D. Stokes, F. B. Dunning, R. F. Stebbings, G. K. Walters and R. D. Rundel (1972), *Opt. Commun.*, **5**, 267.

W. Stricker and L. Krauss (1968), *Z. Naturf. A*, **23**, 1116.

B. J. Sullivan and D. A. Dows (1980), *Chem. Phys.*, **46**, 231.

D. G. Sutton and G. A. Capelle (1976), *Appl. Phys. Lett.*, **29**, 563.

G. W. Taylor and D. W. Setser (1973), *J. Chem. Phys.*, **58**, 4840.

R. W. Terhune and P. D. Maker (1968), *Non Linear Optics*, vol. II, *Lasers*, ed. A. K. Levine, Marcel Dekker, New York.

S. G. Tilford and J. D. Simmons (1972), *J. Phys. Chem. Ref. Data*, **1**, 147.

V. I. Tomin, A. J. Alcock, W. J. Sarjeant and K. E. Leopold (1978), *Opt. Commun.*, **26**, 396.

V. I. Tomin, A. J. Alcock, W. J. Sarjeant and K. E. Leopold (1979), *Opt. Commun.*, **28**, 336.

A. Torres-Filho and J. G. Pruett (1980), *J. Chem. Phys.*, **72**, 6736.

J. Troe (1977), *J. Chem. Phys.*, **66**, 4745.

O. Uchino, T. Mizunami, M. Maeda and Y. Miyazoe (1979), *Appl. Phys.*, **19**, 35.

R. D. Verma and S. R. Singhal (1975), *Can. J. Phys.*, **53**, 411.

J. Vigue, M. Broyer and J. C. Lehmann (1974), *J. Phys. B*, **7**, L158.

J. Vigue, M. Broyer and L. C. Lehmann (1975), *J. Chem. Phys.*, **62**, 4941.

R. Wallenstein and T. W. Hansch (1974), *Appl. Opt.*, **13**, 1625.

R. Wallenstein and T. W. Hansch (1975), *Opt. Commun.*, **14**, 353.

R. Wallenstein and H. Zacharias (1980), *Opt. Commun.*, **32**, 429.

H. Walther (ed.) (1976), *Topics in Applied Physics*, vol. 2, *Laser Spectroscopy of Atoms and Molecules*, Springer, Berlin, Heidelberg, New York.

C. C. Wang and L. I. Davis (1974), *Phys. Rev. Lett.*, **32**, 349.

J. W. Wei and J. Tellinghuisen (1974), *J. Mol. Spectrosc.*, **50**, 317.

E. M. Weinstock (1976), *J. Mol. Spectrosc.*, **61**, 395.

E. M. Weinstock and A. Preston (1978), *J. Mol. Spectrosc.*, **70**, 188. *Air Force Weapons Laboratory Tech. Rep.* no. AFWL-TR-67-30, vol. 1.

J. J. Wright and M. D. Havey (1978), *J. Chem. Phys.*, **68**, 864.

J. J. Wright, W. S. Spates and S. J. Davis (1977), *J. Chem. Phys.*, **66**, 1566.

H. Zacharias, J. B. Halpern and K. H. Welge (1976), *Chem. Phys. Lett.*, **43**, 41.

F. Zaraga, N. S. Nogar and C. Bradley Moore (1976), *J. Mol. Spectrosc.*, **63**, 564.

R. N. Zare (1979), *Faraday Discuss. Chem. Soc.* no. 67, 7 (Polanyi Memorial Lecture).

R. N. Zare and P. J. Dagdigian (1974), *Science*, **185**, 739.

F. Zernicke and J. E. Midwinter (1973), *Applied Nonlinear Optics*, ed. S. Ballard, Wiley, New York.

L. D. Ziegler and B. S. Hudson (1980), *Opt. Commun.*, **32**, 119.

Dynamics of the Excited State
Edited by K. P. Lawley
© 1982 John Wiley & Sons Ltd.

INFRARED MULTIPHOTON EXCITATION AND DISSOCIATION

DAVID S. KING

*National Bureau of Standards, Division of Molecular Spectroscopy,
Washington, DC 20234, USA*

CONTENTS

I. INTRODUCTION

Since the first reports of Isenor and Richardson,[1,2] Lyman *et al.*,[3,4] and Ambartzumian *et al.*[5] indicating the absorption of many infrared photons by virtually isolated molecules, there has been both vigorous research and growth in the still young field of infrared multiphoton processes (review articles are listed in refs. 6–15). Much of the early work was inspired by the hope of driving chemical reactions in either a bond-specific or isotopically selective fashion, with their respective commercial or tactical implications. Thus, many of the early workers were primarily concerned with the amount of energy deposited into a system and the chemical or isotopic assay of the stable end-products. Recently, it has been realized by many research groups that the ability to drive isolated molecules to dissociation along a ground-state potential surface in a 10^{-7}–10^{-9} s time frame opens up new possibilities in the study of molecular dynamics and the opportunity for developing new chemical physics. Specifically, it is becoming possible to probe the interactions of matter and intense, phase-coherent radiation fields for different regions of molecular excitation; to probe molecular physics as a function of the level of internal excitation in a controlled manner; and to study energy transfer processes and chemical reactivity of moderately-to-highly excited species.

Central to an understanding of the multiphoton dissociation process is the recognition that the vibrational energy level spectrum of a moderate-sized polyatomic molecule presents three very different dynamical regions, as depicted in Fig. 1. The first few infrared (IR) photons to be absorbed by a cold molecule excite transitions between separate, discrete vibrational states located in the low-lying region of the energy spectrum, the so-called region I. As a consequence, certain aspects of multiphoton excitation (MPE) and

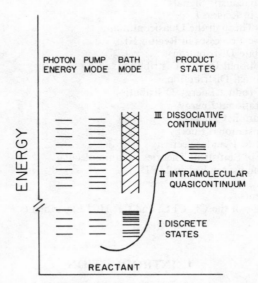

Fig. 1. Schematic diagram of vibrational level structure of a simple polyatomic. Three regions are distinguished: discrete levels (I), quasicontinuum (II), and true dissociative continuum (III).

dissociation (MPD), i.e. frequency response, isotope separation factors, and laser coherence effects, may depend strongly on the *detailed spectroscopy* of the low-lying levels. However, at higher levels of excitation, the spacing between individual vibrational states becomes increasingly smaller due to molecular anharmonicities. Thus it is impossible for transition into exact energy eigenstates to occur on any practical timescale due to rapid intramolecular relaxations. Excitation through this 'quasicontinuum' is often thought of as a sequence of incoherent transitions between homogeneously broadened states that are specified by total internal energy content alone.[16-22] For reasons still only partially understood at the microscopic level, many polyatomics can efficiently absorb the 20, 30 or more infrared photons required for excitation above the dissociation threshold. Radiative excitation continues in this true continuum (region III) in competition with one or several decomposition channels.

From even such a simple physical picture it becomes apparent that intensity-dependent processes will be occurring in regions I and III. For experiments designed to probe the details of laser excitation in these regions, the time-dependent *intensity* profile of the laser source must be well characterized. Equally important is the use of a laser output of well characterized spatial distribution (e.g. a TEM_{00} single longitudinal mode). On the other hand, a widely used approach to the study of MPD is to monitor total product yield or macroscopic dissociation rates[20] as a function of laser energy or wavelength. The results of this type of experiment give information on, and only on, the overall rate-limiting process. For many species, especially those with high activation energies, this rate-limiting step is the sequential absorption of those 10 to 30 photons required to traverse the quasicontinuum. In these cases the experimental results, that is to say some gross frequency response for dissociation or absorption cross section, will show a strong fluence dependence. Several key studies of the spectroscopy and dynamics of highly excited molecules have appeared in the last two years. Ideally, the source laser should exhibit a well behaved temporal output, ideally a TEM_{00} single longitudinal mode output modified by a programmable Pockell cell. Such a profile would then be a rectangular function of variable intensity and/or time duration. Subsequently, either the spatial beam profile must be cleaned and clipped or the probe in a pump–probe experiment must sample a region of constant uniform flux such as the centre of a Gaussian radial profile (see, for example, the Appendix). Only then will *all molecules in the probe region have been exposed to identical intensities/fluences.* There is, unfortunately, no one experiment that can follow in real time the evolution of an ensemble of molecules driven to dissociation by an infrared laser. Indeed, currently there are no spectroscopic formalisms to relate observed spectra to population distributions for highly excited polyatomics. Similarly, no single experiment is simple in interpretation. For example, an optoacoustic measurement of $\langle n \rangle$, the average number of photons absorbed per reactant, is quite insensitive to actual population distributions. At low laser intensities, this technique is useful in discerning overtone transition. However, at higher intensities, many molecules will be driven into the quasicontinuum and the measurement of $\langle n \rangle$ will reflect a strong convolution of the spectroscopies of region I and of region II. Experiments aimed specifically at determining molecular properties of systems in regions I to III are presented in Secs. III to V respectively.

One of the great promises of MPD is the potential, for the first time, stringently to test existing statistical theories of energy flow and unimolecular decay. Since one cannot directly probe the distribution of energy in the reactants just prior to decomposition, one is forced to consider the distribution of energy in the newly born fragments and their rates of formation (equal to the rates of reactant decay). Two exciting side issues arising from this approach concern the influence of the reaction sufrace on the partitioning of available excess energy between the separating fragments and over their internal degrees of freedom (i.e. dynamical effects in the exit channel), and the influence of the laser field on the microscopic *rates* of unimolecular decay. The results of measurements of product energy distributions are discussed in Sec. VI

One of the broader, more interesting studies of multiphoton excitation and dissociation from the viewpoint of gas-phase chemistry is the work on CF_2CFCl. Stephenson, Bialkowski, and King were successful in determining, in real time, both the *absolute macroscopic dissociation rates* (s^{-1}) and the *total product states distribution* under various conditions of irradiation and buffer gas pressure. Theoretically, Goodman Stone, and Thiele were successful[23] in separating the dynamical contributions from regions of different vibrational energy content including rotational hole filling and power broadening in region I and microscopic decay in region III and in determining the form and energy scaling of the key parameters (e.g. $T_2(E)$, the absorption cross section $\sigma(E)$ and a V–T transfer rate for modelling pumping in region II). The theoretical model of Goodman, Stone, and Thiele is presented in Sec. II and the experimental results of King and Stephenson are presented in the appropriate sections.

II. THEORETICAL CONSIDERATION

A. General Approach

As the experimental data from MPD experiments have become more refined, there have been significant changes in theoretical attitudes concerning the MPD of isolated molecules.[16-21,24-32] The early observed dissociation threshold dependence of SF_6 on laser fluence[33,34] led to the advocation[12,19] of a simple set of rate equations, justified on the basis of Fermi's golden rule, for transitions in the dense manifold (i.e. region II):

$$\frac{dN_i}{dt} = W_{i \leftarrow i-1} N_{i-1} + W_{i \leftarrow i+1} N_{i+1}$$
$$- (W_{i+1 \leftarrow i} + W_{i-1 \leftarrow i}) N_i - (2\Gamma_i/\hbar) N_i, \tag{1}$$

where N_i is the total population of vibrational states i laser quanta above the ground state, the $W_{i \leftarrow i \pm 1}$ and $W_{i \pm 1 \leftarrow i}$ are stimulated emission and absorption rates, and $2\Gamma_i/\hbar$ is the microscopic rate of decay from the ith level. Such a treatment implies a strong fluence dependence (intensity independence) for both the rates and threshold for decomposition. This condition would be satisfied if the rate-limited step in the MPD process were sequential absorption in the quasicontinuum. In relating an expected fluence dependence in N_i to an apparent fluence dependence in decomposition rate, this treatment utilizes the commonly accepted idea that it is sufficient to know the vibrational energy content of a molecule to determine its microscopic decay rates. However, in the absence of extensive information on the form and energy scaling of the radiative transition terms W, there is no way a priori to calculate the fraction of molecules having a given total vibrational energy without arbitrarily assuming some absorption cross section.[24,35]

The strict rate equation approach presumes explicitly that energy randomization occurs on a rapid timescale in comparison to radiative pumping. The effects of intramolecular energy randomization have recently been treated[36-38] using a set of optical Bloch equations

$$\frac{d\rho_{jj}}{dt} = \frac{i}{\hbar}[\rho, H]_{jj} + \sum_{k \neq j} [(1/T_1)_{j \leftarrow k} \rho_{kk} - (1/T_1)_{k \leftarrow j} \rho_{jj}],$$
$$\frac{d\rho_{jk}}{dt} = \frac{i}{\hbar}[\rho, H]_{jk} - (1/T_2)_{jk} \rho_{jk}, \qquad j \neq k, \tag{2}$$

containing both T_1 and T_2 relaxation times and the Hamiltonian

$$H = H_0 + (a^+ + a^-)\alpha_{01} A \sin \chi t$$

for the molecular pump mode[21] and its interaction with the laser field. The density matrix elements ρ_{ii} refer to the pump mode, $(1/T_1)_{i \leftarrow j}$ represents the rate of transition from level j to level i in the pump mode due to intramolecular relaxation, and $(1/T_2)_{ij}$ is the rate of dephasing between levels i and j.

Unfortunately, as written, Eqs. (2) cannot be used to calculate chemical decomposition rates which are functions of total vibrational energy or to determine how intramolecular relaxation affects the fluence dependence of decomposition. The inability to calculate decomposition rate results from the absence of any information concerning the overall level of molecular excitation. This is to say that the pump mode population contained in these equations group together all molecules with the same degree of pump mode excitation irrespective of the amount of energy in the remaining (heat bath) modes. In order to assess the effect of intramolecular relaxation on the fluence dependence of decomposition, or indeed to determine if dependence on laser intensity and not on fluence is occurring, a more general representation of intramolecular relaxation is required to calculate both pump mode populations n_i and total shell populations N_i under arbitrary relaxation and radiative pumping conditions.

Leaving aside certain controversies concerning the possible role of energy localization, [39,40] the MPD process is reasonably understood in a qualitative fashion. The time has now come to establish rigorous, quantitative means for interpreting experimental results in a comprehensive model with strong predictive capabilities. The rest of this section will be devoted to a more detailed description of the theoretical model developed by Goodman, Stone, and Thiele and applied to the experimental results on CF_2CFCl. The model describes pumping through both the discrete and quasicontinuum regions, as well as intramolecular T_2 dephasing and collisional effects in a unified way using Bloch equations. The dual purpose of this collaborative effort was to gain physical insight into the dynamics of competing processes simultaneously going on within a polyatomic molecule, e.g. absorption of intense, phase-coherent radiation, intra- and intermolecular relaxation, and unimolecular decay; and to define key experiments that will be capable of testing both the predictive capabilities of the model and the validity of currently accepted theories of chemical dynamics.

B. Experimental Counterpart

The theoretical model was applied to the buffer gas *pressure* and infrared laser *intensity* dependences in the microscopic *rates* of MPD of CF_2HCl[32] and CF_2CFCl.[23] The specific details of the experimental procedure and data acquisition are presented in the Appendix. The model calculations are based on a set of eight molecule-specific parameters, as shown in the following section.

Table I includes a listing of these parameters along with the appropriate values for CF_2CFCl. It is important to observe that much of the required information is obtainable from alternative spectroscopic and kinetic sources. The remaining few parameters must be extracted from experiment.

Any comprehensive theoretical treatment of MPD must be capable of predicting molecular dynamics occurring under collision-unimportant to collision-dominated conditions (i.e. the reactant very dilute in a non-absorbing bath gas such that neither bulk heating nor reactant–reactant collisions are important). Under collision-free conditions, the details of radiative excitation in regions 1 and III will be important as discussed in Secs. III and IV respectively.

Collisions with the cold buffer gas will contribute to the MPD process in the following ways:

(i) T_2 dephasing of any coherent pumping that might otherwise occur under collision-free conditions;

(ii) repopulation during the laser pulse of those low-lying rotational states depopulated because allowed transitions out of those states are closely resonant with the laser field, i.e. rotational hole filling;

(iii) collisional deactivation, usually by a V–T process, of molecules while they are being pumped up through the quasicontinuum;

(iv) collisionally induced relaxation of vibrational energy among the molecule's internal degrees of freedom, i.e. the elimination of any effects due to energy localization (a T_1 type process).

Collisional deactivation (of type (iii)) was treated using the approach involving the exponential dependence on energy transferred used by Tardy and Rabinovitch[41] and Marcoux and Setser[42] from chemical activation studies. This introduces into the theoretical analysis the parameter ΔE equal to the mean energy transferred per collision to the cold heat bath.

Collisional dephasing (i) and hole filling (ii) were taken into account by combining the effects of these collisions into a parameter f_p. Although the effects of these types of collisions might in principle be determined exactly from a knowledge of the detailed spectroscopy of the low-lying levels, these data are generally either unavailable or incomplete.[31,32]

C. Basic Theory

When T_1 energy scrambling processes among the molecule's internal degrees of freedom are sufficiently rapid, then a description dependent on only the total internal energy content is sufficient to specify the molecular state.[5] Under these conditions, the coarse-grained density matrix of the absorbing molecule evolves in time according to an equation of the form

$$\frac{d\rho}{dt} = \frac{d\rho}{dt}\bigg|_{\text{dynamical}} + \frac{1}{\tau}(\mathbf{K} - 1)\rho - \mathbf{K}_2\rho - \mathbf{K}_r\rho. \tag{3}$$

The terms on the right-hand side of Eq. (3) are (a) a formal term describing the dynamics and interaction with the light field for the absorbing molecule, for which explicit expressions can be given in two limiting cases,[21] (b) a term involving the operator $\mathbf{K}$ describing the effects (types (i) and (iii)) of collisions with inert buffer gas molecules at a frequency $1/\tau$, (c) a term involving the operator $\mathbf{K}_2$ describing intramolecular T_2 dephasing due to a homogeneous broadening of the pumped levels, and (d) a term involving the operator $\mathbf{K}_r$

describing unimolecular reaction from energy states above the activation energy threshold.

The collision operator $\mathbf{K}$ is defined by[43-45]

$$(\mathbf{K}\rho)_{ij} = 0, \qquad i \neq j,$$

and

$$(\mathbf{K}\rho)_{ii} = \sum_j P_{i\leftarrow j}\rho_{jj},$$

where

$$P_{j\leftarrow j} = 1 - \sum_{i \neq j} P_{i\leftarrow j}.$$

Here $P_{i\leftarrow j}$ is the probability that upon collision with a buffer gas molecule the absorber will make a transition from state j to state i; up-and-down transition probabilities are related by microscopic reversibility. The mean energy transferred upon collision to a cold heat bath (i.e. a heat bath which can only deactivate but not energize reactants) for a molecule in state j prior to collision is

$$(\Delta E)_j = \sum_{i=0}^{j} (\varepsilon_j - \varepsilon_i) P_{i\leftarrow j}, \qquad \varepsilon_i = i(\hbar\chi),$$

where $\hbar\chi$ is the energy of the laser photon. In the limit of large j,

$$\Delta E = \lim_{j \to \infty} (\Delta E)_j = \frac{\hbar\chi}{\exp(\beta\hbar\chi) - 1}.$$

The collisional transition probabilities are then specified in terms of a single parameter β, or equivalently ΔE. The exponential form for the transition probabilities surely overestimates the amount of energy transferred from low-lying states. However, *since the radiative pumping is fastest in the low-lying states,*[32] *most molecules are pumped through the low-lying states before suffering any collisions.* As a consequence, the exact form of the collisional transition probabilities for the lowest states is not critical.

The operator $\mathbf{K}_2$, describing intramolecular dephasing, is defined by[38]

$$(\mathbf{K}_2\rho)_{ij} = (1/T_2)_{ij}\rho_{ij}, \qquad i \neq j,$$

$$(\mathbf{K}_2\rho)_{ii} = 0,$$

$$(1/T_2)_{ij} = (1/T_2)_{ji},$$

By the above definition we have a loss of phase information (in the absence of collisions) for the off-diagonal density matrix elements with

$$\rho_{ij} \sim \exp\left[-(1/T_2)_{ij}t\right].$$

This dephasing is, in principle, observable as a homogeneous line broadening for the i to j transition and can be thought of as arising from a convolution of the homogeneous broadenings (w_i and w_j) of levels i and j, respectively, such that[14,32]

$$\left(\frac{1}{T_2}\right)_{ij} = \frac{w_i + w_j}{\hbar}.$$

Specification of line-broadening factors w_i, or more specifically their functional dependence on the energy ε_i, is in our approach the central problem in modelling the radiative pumping. Assuming the validity of the single quantum exchange theory,[46]

$$w_i = 2\pi H_{anh}^2 \rho^{sqe}(\varepsilon_i),$$

where ρ^{sqe} is a single-quantum exchange density of states (discussed later), and H_{anh} is a parameter related to the strength of the anharmonic interaction that spreads oscillator strength over many vibrational levels.

The operator $\mathbf{K}_r$, describing unimolecular reaction from levels above threshold, is defined as

$$(\mathbf{K}_r\rho)_{ii} = k_i\rho_{ii},$$

$$(\mathbf{K}_r\rho)_{ij} = 0, \qquad i \neq j.$$

Here k_i is the microscopic rate constant, calculated from RRK or RRKM theory,[47,49] for reaction from level i. For energies below the activation energy, $k_i = 0$.

From a Bloch equation point of view, the distinction between coherent pumping through low-lying discrete levels and incoherent pumping through the quasicontinuum is one of degree only, dependent on the ratio of an effective Rabi frequency Ω^{eff} to a total dephasing rate $1/T_2$. In the limiting case where

$$\Omega^{eff} > 1/T_2,$$

the pumping is at least partially coherent, and one has for the dynamical term in Eq. (3):[21]

$$\left.\frac{d\rho}{dt}\right|_{dynamical} = \frac{i}{\hbar}[\rho, H].$$

In this case, applicable to the pumping through the low-lying discrete levels, Eq. (3) has precisely the form of a Bloch equation.

In the other limiting case where

$$\Omega^{eff} \ll 1/T_2,$$

the pumping is incoherent, and one can in effect reduce Eq. (3) to an ordinary master equation[21,34,35,38] in which the dynamical term for diagonal elements is

$$\frac{d\rho_{ii}}{dt}\bigg|_{dynamical} = R_{i\leftarrow i+1}\rho_{i+1,i+1} + R_{i\leftarrow i-1}\rho_{i-1,i-1}$$
$$- R_{i-1\leftarrow i}\rho_{ii} - R_{i+1\leftarrow i}\rho_{ii}. \tag{4}$$

The radiative transition rate

$$R_{i\pm1\leftarrow i} = \frac{(\langle\alpha^2\rangle_{i\pm1,i}A^2/2\hbar^2)[1/T_2]_{i\pm1,i}}{[1/T_2]^2_{i\pm1,i} + [\chi - |\varepsilon_i - \varepsilon_{i\pm1}|/\hbar]^2}$$

contains a Lorentzian lineshape factor appropriate to homogeneously broadened transitions. Here, $\langle\alpha^2\rangle_{i\pm1,i}$ is an averaged square dipole coupling element, A and χ are the amplitude and frequency of the laser field, respectively, and the ε_i locate energy levels in the so-called pumped mode[21] spectrum. When Eq. (4) applies, i.e. for pumping through the quasicontinuum, Eq. (3) reduces to ordinary rate equations for the diagonal density matrix elements. These rate equations have the explicit form

$$\frac{d\rho_{ii}}{dt} = R_{i\leftarrow i+1}\rho_{i+1,i+1} + R_{i\leftarrow i-1}\rho_{i-1,i-1} - R_{i-1\leftarrow i}\rho_{ii} - R_{i+1\leftarrow i}\rho_{ii}$$
$$+ \frac{1}{\tau}\sum_{j\neq i}P_{i\leftarrow j}\rho_{jj} - \frac{1}{\tau}\sum_{j\neq i}P_{j\leftarrow i}\rho_{ii} - k_i\rho_{ii}. \tag{5}$$

Eq. (5), describing pumping through the quasicontinuum, can be interfaced in a natural way with a limited number of Bloch equations describing the pumping through the low-lying discrete levels.[14,47] However, if the effects of collisions with a buffer gas are of primary interest or when spectroscopic details about the low-lying levels are unknown, it is neither possible nor even desirable to attempt such a careful treatment of the discrete level pumping. One can assume that the rate equations describe pumping all the way up from the ground state. Then, by introducing a theoretically undetermined parameter f_p, one can account for the fact that only a fraction of ground-state molecules are readily pumped through the discrete levels up to the beginning of the quasicontinuum. Theoretical calculation under collisionless conditions would amount simply to solving the rate equations with the initial condition corresponding to a ground-state population

$$\rho_{00}(t=0) = f_p.$$

The real value of lumping all the discrete level information into the single parameter f_p emerges only when one considers the effects of collisions: rotational hole filling allows one to set $f_p = 1$ at high pressures (> 100 Torr) of the buffer gas. Also, at lower pressures, rotational hole filling can be included in the rate equations in a very simple way. Letting ρ_{00} and n_h be the ground-state populations of laser-pumpable and non-pumpable states, respectively, one has[32]

$$\frac{dn_h}{dt} = \frac{1}{\tau_{rot}}[(1 - f_p)\rho_{00} - f_p n_h]$$

$$\frac{d\rho_{00}}{dt} = (R_{0 \leftarrow 1}\rho_{11} - R_{1 \leftarrow 0}\rho_{00})$$

$$+ \left(\frac{1}{\tau}\sum_{j \neq 0} P_{0 \leftarrow j}\rho_{jj} - \frac{1}{\tau}\sum_{j \neq 0} P_{j \leftarrow 0}\rho_{00}\right)$$

$$+ \frac{1}{\tau_{rot}}[f_p n_h - (1 - f_p)\rho_{00}].$$

One further simplification of the rate equations is possible if the laser frequency χ is assumed to be always on resonance; then

$$R_{i \pm 1 \leftarrow i} = \frac{\langle \alpha^2 \rangle_{i \pm 1, i} A^2}{2\hbar^2}\left(\frac{1}{T_2}\right)^{-1}_{i \pm 1, i}.$$

D. Consequence and Prediction

1. Single-Quantum Exchange

To predict MPD results requires a detailed knowledge of appropriate radiative transition rates and energy and phase relaxation rates at every level of vibrational excitation. Assuming, in a simplified Fermi golden rule approach,[50] each state is coupled to a dense background of ζ_{total} states by interaction matrix elements all having the same value H_{anh}, then the interaction width w of a state at energy E can be expressed as

$$w(E) = 2\pi H^2_{anh}\zeta_{total}(E).$$

If one were to apply the degenerate rate equations using $\zeta_{total}(E)$ given by the Whitten–Rabinovitch formula,[51] the linewidths of even moderate-sized molecules would increase astronomically with increasing energy. Such broad linewidths lead to vanishingly small radiative pumping rates and the prediction that collision-free MPD should be impossible, in direct contradiction to experiment.

What is required to calculate the linewidth is not a total density of states but only a count of those states that are strongly coupled. Specifically, if direct transfer of large numbers of quanta from one set of vibrational states to another is negligible compared with exchanges involving a single or perhaps two quanta,[46] then, for a given highly excited zero-order state of the molecule, even though there will be many other zero-order states nearby in energy, most of these states will differ so strongly from the given state in their internal energy distribution that direct coupling will be extremely weak and certainly not of the same order as the average coupling that prevails at much lower energies.

In the derivation of the *single-quantum exchange theory*,[46] Stone, Goodman, and Thiele assert the following:

(i) The mean squared coupling strength between nearby zero-order states depends strongly on the number of quanta that must be exchanged to go between the states but is, for a given quantum exchange number m, approximately independent of total energy, exhibiting at most a slight increase with increasing energy.

(ii) The coupling between states related by m-quantum exchange, designated $H_{anh}(m)$, is a strongly decreasing function of m.

(iii) Within each subgroup of coupling elements (defined by a particular quantum exchange number), the actual values of the coupling elements are randomly distributed with a mean value of zero.

If the details of the energy distribution within the heat bath degrees of freedom can be ignored, the width of a state with j quanta in the pump mode and $n - j$ quanta in the heat bath becomes

$$w_{j,n-j} = 2\pi \sum_m [H_{anh}(m)]^2 \left(\frac{1}{f_{n,j}} \sum_{\{n_i\}} \xi(\{n_i, n_p = j\} ; m) \right) \qquad (6)$$

Here n_p denotes the number of quanta in the pump mode and $f_{n,j}$ is the number of states with n quanta, j of which are in the pump mode.

Such interaction widths are sufficient to determine the optical linewidths and T_2 rates. If we want to calculate the intramolecular energy transfer rates T_1 into or out of the pump mode, the density of states term in the RHS of Eq. (6) must be evaluated in greater detail to isolate contributions from states when the pump mode excitation has various final values j' not equal to the initial value j.

Calculating the density of states that can be reached by an m-quantum exchange with the restrictions that in the final state the pump mode contains j' quanta, T_1 relaxation rates are determined from

$$\left(\frac{1}{T_1} \right)_{j' \leftarrow j, n-j} = 2\pi \sum_m [H_{anh}(m)]^2 \left[\left(\frac{1}{f_{n,j}} \right) \sum_{\{n_i\}} \xi(\{n_i, n_p = j\} ; m, n_p = j') \right]. \qquad (7a)$$

Now, the interaction width can be expressed in terms of T_1 rates and a proper dephasing rate $\hat{w}_{j,n-j}$

$$w_{j,n-j} = \hat{w}_{j,n-j} + \sum_{j' \neq j} (1/T_1)_{j' \leftarrow j, n-j}. \qquad (7b)$$

The rapid fall-off of intramolecular coupling $H_{anh}(m)$ with quantum exchange number provides conditions where the sums over m in Eqs. (6) and (7a) may be truncated after $m = 1$, or some other small value of m.

Explicit evaluation of $\hat{w}$ and T_1 gives the simple, useful result for an s-fold degenerate oscillator:

$$w_{j,n-j} = 2\pi H_{anh}^2(m = 1)\langle \xi(m = 1)\rangle_{Av, np = j}, \qquad 0 < j < n$$

where

$$\langle \xi(m = 1)\rangle_{Av, np = j} = \frac{1}{\hbar\omega_0}\frac{2(n-j)(s-1) + (n-j+1)(s-1)(s-2)}{n-j+s-2}$$

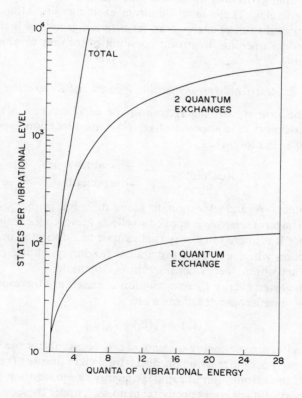

Fig. 2. Energy dependence of vibrational densities of states. The total density of states calculated by the formalism of Whitten and Rabinovitch[51] and the number of states coupled by one- or two-quantum exchanges[46] are plotted against energy for a molecule with 15 degenerate degrees of freedom, i.e. SF_6. (After J. Stone, E. Thiele, and M. G. Goodman.[46] Reproduced by permission of North-Holland Publishing Company, Amsterdam.

and

$$\left(\frac{1}{T_1}\right)_{j'\leftarrow j,n-j} = \frac{2\pi H_{anh}^2(m=1)}{\hbar\omega_0}\left(\frac{(n-j)(s-1)}{n-j+s-2}\delta_{j',j+1} + (s-1)\delta_{j,j-1}\right).$$

Most significant, the dependence of $w_{j,n-j}$ on total energy n exhibits a levelling off of the relaxation rates with energy, instead of the continued increase which would be predicted if $w_{j,n-j}$ were taken proportional to the total density of states (see Figs. 2 and 16).

It must be noted that the large spread in normal mode frequencies in real polyatomic will impose additional restrictions on single- or few-quantum exchange couplings leading to similar linewidth saturation effects at high vibrational energies. These same quantum exchange limitations may also result in restricted rates of energy transfer between groups of high- and low-frequency modes opening up strong possibilities to observe mode-selective laser chemistry.[52-54]

2. Restricted Intramolecular Vibrational Relaxation

One of the primary driving factors in the early days of MPD was the (realized) possibility of isotope-selective and (hoped-for) bond-specific chemistry. Consider the following

$$\text{Reactant} \left| \begin{array}{l} \hbar\chi_A \rightarrow \text{product A} \\ \hbar\chi_B \rightarrow \text{product B} \end{array} \right.$$

where products A and B originate from different primary dissociation processes. The term *selective* is applied to effects in which the ratio of product A to product B is altered in a way that cannot be duplicated in a thermally driven reaction where energy is distributed randomly amongst the various bonds and activation energy considerations dominate.

If, at some given energy E_i, two reaction channels are thermodynamically allowed, the microscopic decay rates are

$$k_r(E_i) = k_A(E_i) + k_B(E_i),$$

where k_A and k_B are the energy-dependent rate constants for the formation of each of the two possible reaction products. Implicit in the use of rate equations for MPD is the assumption of rapid T_1 energy randomization precluding *a priori* any possibility of laser selectivity: in no way, under these conditions, can any property of the laser (e.g. frequency) alter the fraction of products A and B formed from reactant molecules excited to a given energy. A type of 'pseudoselectivity'[40] is possible for the case where increased laser intensity is sufficient to raise the radiative pumping rates to be competitive with the lower-lying channel dissociation rates. In this manner, higher activation energy products may be formed in primary processes. This is not a new concept. In fact, pseudoselectivity has been observed in experiments on ethylvinyl ether,[55,56] CF_2Cl_2,[57-59] CF_2Br_2,[59] and CF_2CFCl (see Sec. VII).

The concept of pseudoselectivity requires a critical analysis of the effects of laser pulse shape on MPD. Thiele, Goodman, and Stone[60] present the results of model-based calculations for two limiting pulse profiles. Both pulses are of a rectangular temporal shape; in the short-pulse case a δ-function is assumed, whereas in the long-pulse case the intensity is so low that the radiative pumping rates are negligibly small compared to the decay rates. Notably, in the short-pulse case no reaction occurs during the laser pulse, while in the long pulse case all reaction occurs from the first level above the dissociation threshold. Although the total product *yield* appears to be relatively insensitive to 'intrinisic' pulse shape effects, the relative *distributions* of products in a multichannel reaction are intrinsically sensitive to the pulse shape. If the reaction to yield B has an activation threshold of only a few quanta above that to for A, we might obtain the results shown in Fig. 3. In the long-pulse limit all

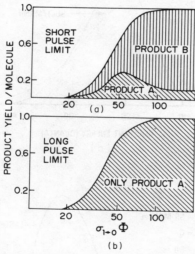

Fig. 3. An example of the pseudoselective effect. In a two-channel reaction the distribution of products can be intrinsically quite sensitive to actual pulse shape, i.e. instantaneous intensity. In these calculations, the threshold for formation of product B is at $\sigma\Phi = 26$ photons absorbed, two quanta above that for A formation; the ratio of pre-exponential factors was 40, favouring the high-energy channel. After E. Thiele, M. F. Goodman, and J. Stone.[60] Reproduced by permission of North-Holland Publishing Company, Amsterdam

reaction occurs from the lowest reactive level giving product A only. In the short-pulse limit the laser pulse establishes a distribution of population over a number of reactive levels. A considerable amount of B may then be observed, especially in those cases where the high pressure pre-exponential factor for the higher-energy channel is larger than for the lower channel (as is the case for ethylvinyl ether).

The overall product yield is principally determined by the flux across the dissociation threshold into the continuum and is relatively insensitive to pulse shape. However, an intrinsic pulse shape effect based on a rate equation

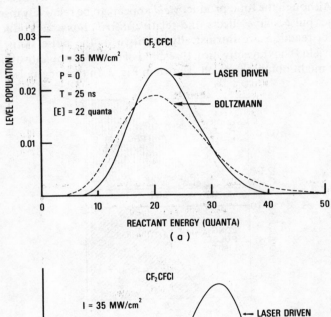

(a)

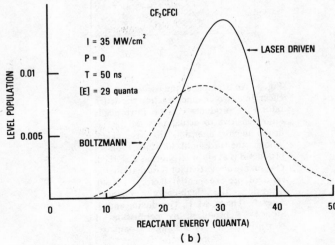

(b)

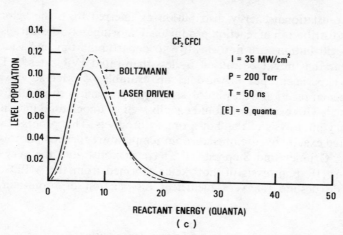

Fig. 4. Level population distributions calculated for the MPE of CF$_2$CFCl under the conditions listed in the figures. (Calculations by J. Stone, M. F. Goodman, and E. Thiele.)

analysis[60] would predict the product yield to be smaller for equi-fluence pulses of shorter duration. When the product yield is observed to increase for shorter pulses, the occurrence of some effect outside the province of rate equations (such as coherent pumping through low-lying levels or dynamic Stark broadening) can be correctly inferred (as observed for SF$_6$,[33] CF$_2$HCl,[61,62] and CF$_3$I[63]).

But let us come back to the more basic, and more stimulating, possibility of actual bond-selective MPD. A logical extension of the single-quantum exchange analysis is to provide a barrier, or at least a long tortuous path, for the coupling of isoenergetic states of widely differing vibrational identity. As such, energy localization could occur when radiative pumping and unimolecular decay were competitive with (the proposed) restricted intramolecular vibrational relaxation (IVR). In general, the existence of a critical energy at which the IVR rate and selective reaction rate become comparable is a *necessary* requirement for selective bond breaking even with lasers of arbitrarily high intensity. Selectivity in vibrational mode excitation does *not* guarantee selective bond breaking unless the reaction rate competes favourably with the IVR rate at the given level of excitation.

The database relating to restricted IVR and the possibility for mode-selective multiphoton chemistry is too small for a critical evaluation. For example, the elegant crossed laser–molecular beam experiments of Lee and coworkers have been taken as indicating very rapid IVR rates because of an agreement between RRKM predictions for the translational energy distributions (for simple bond-rupture reactions) and experimental measurements of the translational energy distribution as inferred from the velocity and angular distribution of reaction product(s). Unfortunately, these translational energy distributions are not sensitive to departures from intrinsic RRKM behaviour that might be caused by less-than-rapid IVR rates.[46] The only current experiment put forward as an example of laser selectivity in a unimolecular reaction is the photolysis of cyclopropane at either 1050 cm^{-1} or 3100 cm^{-1}. This experiment could equally well be understood in terms of the secondary photolysis of the 'hot' primary products. This process cannot be duplicated exactly by attempts at room-temperature thermal photolysis. For instance, Grimley and Stephenson[64] have demonstrated the very efficient secondary IR photolysis of 'hot' ketene formed from the MPD of acetic anhydride despite the very inefficient dissociation of room-temperature ketene.

3. *Reactant Energy Distribution*

From the theoretical analysis of intensity and pressure dependences in the MPD *rate* for CF_2CFCl,[23] it is possible to predict the evolution of population distributions in time during laser excitation. Such a prediction takes into account the appropriate energy scalings of T_2 rates, absorption cross section, and V–T energy transfer processes. Three such computer simulations are presented in Fig. 4 for CF_2CFCl. The distributions obtained under collision-unimportant conditions are never Boltzmann thermal. They are, as was also found for CF_2HCl,[32] somewhat narrower and peaked at a higher value of photons absorbed than the corresponding thermal distribution. In the calculations for the addition of buffer gas, collision-induced T_1 and T_2 processes and especially rotational hole filling contribute to the magnitude of the respective level population while V–T losses control the overall degree of excitation.

It is also possible, in principle, to predict the distribution of energy in the reactants that are driven into levels sufficiently energetic to dissociate.

4. *Optoacoustic Signals*

The energy absorbed from the light field is an observable quantity that can be readily computed from the theoretical model. As shown earlier and in ref. 60, when collisional effects compete with radiative transitions the excitation process becomes fluence dependent, implying both that $f_p \sim 1$ and that a rate equation approach to the system dynamics will be valid. The rate at which energy is absorbed from the light field is readily calculated in terms of the radiative rates

$$\frac{d\varepsilon_{abs}}{dt} = \hbar\chi \sum_{all\ i} (R_{i+1 \leftarrow i} - R_{i-1 \leftarrow i})\rho_{ii}.$$

When no reaction occurs, all of the energy absorbed from the light field is available for optoacoustic detection at some later time, typically after the laser pulse has terminated. When reaction does occur, that portion of the absorbed energy which goes into bond breaking will not be detected. Defining $\varepsilon^*_{\mathrm{abs}}$ to be that portion of the absorbed energy available to optoacoustic detection, one has, assuming $E_{\mathrm{act}} = \Delta H^0_0$ (which is generally a good approximation for simple bond breaking),

$$\frac{d\varepsilon^*_{\mathrm{abs}}}{dt} = \frac{d\varepsilon_{\mathrm{abs}}}{dt} - E_{\mathrm{act}} \sum_{\mathrm{all}\ i} k_i \rho_{ii}.$$

It follows that

$$\varepsilon^*_{\mathrm{abs}}(t) = \hbar\chi \sum_{\mathrm{all}\ i} \left(R_{i+1 \leftarrow i} - R_{i-1 \leftarrow i} - \frac{E_{\mathrm{act}}}{\hbar\chi} k_i \right) \int_0^t \rho_{ii}(t')dt', \qquad (8)$$

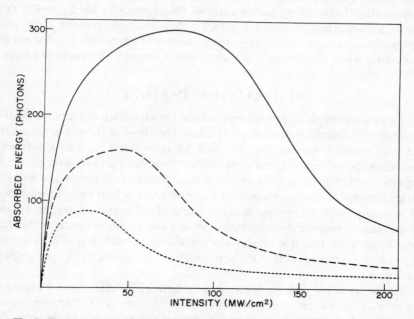

Fig. 5. The quantity of energy (per absorbing CF_2CFCl molecule) absorbed from the light field and available to optoacoustic detection. Curves are plotted for three buffer gas pressures: $P_{\mathrm{Ar}} = 100$ Torr (---), $P_{\mathrm{Ar}} = 200$ Torr (----), and $P_{\mathrm{Ar}} = 400$ Torr (——). These curves are a theoretical prediction of the pressure dependence of the multiphoton absorption. The dependence on peak laser intensity is calculated assuming square-wave pulses of 120 ns duration and varying intensity. The pronounced fall-off with increasing laser intensity occurs when reactions begin to deplete absorbers and also make that fraction of absorbed energy that goes into chemical bond breaking unavailable to the optoacoustic detector. After J. Stone, E. Thiele, M. F. Goodman, J. C. Stephenson, and D. S. King.[23]

taking account of the fact that some molecules, excited above reaction threshold at the end of the pulse, will eventually react while others will be collisionally deactivated before they can react.

Since an optoacoustic detector gathers energy on the timescale of several microseconds or longer, the observable quantity is

$$\varepsilon_{abs}^* = \lim_{t \to \infty} \varepsilon_{abs}^*(t).$$

We have used Eq. (8) to calculate[23] ε_{abs}^* for square-wave pulses of 120 ns duration and various intensities. Results of these calculations are presented in Fig. 5. Curves, for constant buffer gas pressure, all show a peak in energy absorption at those values of laser intensity for which chemical reaction begins to become appreciable (i.e. about 10%). The decline in energy absorbed that accompanies chemical reaction is due both to the loss of absorbers and to the loss of that fraction of absorbed energy that goes into bond breaking. It seems noteworthy that so many photons, about 300 photons for the $P_{Ar} = 400$ Torr curve, are absorbed (by each CF_2CFCl molecule initially present) at the peak in the absorption curve. Clearly a large number of deactivating collisions are occurring while the molecule is being excited through the quasicontinuum.

III. EXCITATION IN REGION I

There is no simple or general way to treat the excitation of simple molecules through the discrete states region up to the threshold of the quasicontinuum. Several molecule-specific factors must be considered. In many low-yield experiments, for instance, the contribution from hot band transitions can be significant. At high intensities, one must also consider in detail the potential for true multiphoton transitions and for the occurrence of field-induced wavefunction mixing and Rabi cycling. Specific details of the excitation process are just beginning to become apparent through the use of optoacoustic techniques which make the detection of overtone transitions possible and of laser-excited fluorescence techniques that allow population evolutions to be monitored in real time.

The onset of the quasicontinuum is thought of in terms of that region of molecular excitation where intramolecular relaxation (T_2) becomes competitive with the excitation process. This level can be estimated theoretically, based on certain assumptions, but there has previously been no firm experimental evidence to test these assumptions. Concomitant with increased excitation is the increase in the rate of intramolecular energy flow $(1/T_1)$. Several experiments have been aimed at bracketting the level of internal excitation required for facile energy randomization. The experiments in this section deal with these questions of population evolution and intramolecular relaxations. Spectroscopically, overtone transitions have been observed to play an important role in MPE even at modest levels of irradiation, T_2 type level broadening becomes apparent at vibrational states densities as low as $100/cm^{-1}$, and intramolecular energy flow has been observed below level densities of $400/cm^{-1}$. It should not be surprising that most of these studies were performed with small hydrogen-containing molecules. Here, where rotational spacings are relatively large and the vibrational density of states low, only a small fraction of the initial thermal ensemble would have an allowed one-photon transition resonant with the exciting laser frequency. Under these conditions, dynamical Stark effects, resulting in spectral shifts and wavefunction mixing, and coherent processes will be most significant. In addition, if the activation energy is sufficiently low that sequential excitation through the quasicontinuum is not the rate-limiting process for dissociation, then the intensity dependences of excitation in region I will be preserved in measurements of dissociation yields. Such a prototype species and one that has received considerable attention[32,35,59,61,62,65-70] is CF_2HCl.

Intensity-dependent results in collision-unimportant MPD were first reported by King and Stephenson[61] for CF_2HCl. A single-mode CO_2 laser was used to dissociate CF_2HCl at very low pressures. The CF_2 products were subsequently probed in real time by laser-excited fluorescence (LEF) techniques, probing a region of constant uniform fluence. Using a calibrated detection system, the observed LEF signals were directly related to absolute yields, allowing valid estimates of the actual importance of collisional processes. In analysing the MPD results, the authors emphasized the importance of plotting the yield obtained between time $t = 0$ and $t = \tau$ (defined relative to the leading edge of the CO_2 pump pulse) against the integral of the instantaneous intensity experienced in the probe region. Thus the effects of a long-duration, low-intensity pulse can be directly compared with those of a short, intense pulse of comparable shape and equal fluence. Comparisons were also made between single-mode excitation and irradiation by mode-locked pulse trains.

The results of these studies were quite dramatic. In the collision-unimportant regime the yields from long, weak pulses were much less than for short, intense pulses. This effect was largest for lower fluences. In the range of $1\ J\ cm^{-2}$, a six-fold increase in average intensity resulted in a 400-fold increase in yield (see Fig. 6). Similar, although less striking, effects were seen in comparison of smooth single-mode pulses to mode-locked pulse trains. In experiments dominated by $Ar-CF_2HCl$ collisions (e.g. 2 mTorr CF_2HCl in 300 Torr Ar), all intensity-dependent effects were washed out. These results are clearly contrary to the intrinsic pulse shape expectations described by Thiele et al.[60] for a model where only fluence was considered, i.e. for those situations where the rate-limiting process would be excitation through the quasicontinuum. If fluence were the critical interaction, then one should obtain larger net yields for longer pulses as compared with shorter pulses of equal fluence.

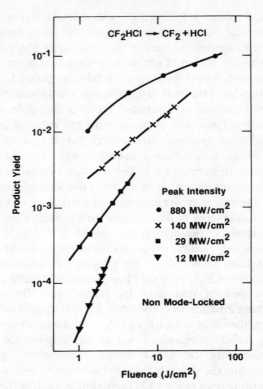

Fig. 6. Intensity dependence in the MPD of CF_2HCl. Absolute product yields from the dissociation of CF_2HCl (9.3 μm $R(32)$ CO_2 laser transition at 1086 cm^{-1}) for non-mode-locked laser pulses of four different total energies. The pulses all exhibited identical temporal envelopes and are labelled according to the peak intensity. The data given for each different pulse represent the yield of CF_2 observed at various times τ_D during the IR laser pulse plotted against the fluence $F(\tau_D)$ for that time and laser pulse. The total pulse energies were 2, 5, 24, and 150 mJ, respectively, for the 12, 29, 140, and 880 MW cm^{-2} pulses. The values of τ_D (ns) corresponding to the data points ranged from 25 to 200 ns. For the most intense pulse, there is 400 times more product formed during the first 25 ns than during 155 ns irradiation by the weakest pulse, although both subject the reactant to the same fluence. The absence of data at fluences below 1 J cm^{-2} for the weaker pulses is due to an induction period; there seems to be a critical fluence requirement of about 0·8 J cm^{-2} before any significant collision-free dissociation of the CF_2HCl can occur at this IR wavelength. After D. S. King and J. C. Stephenson.[61]

Fig. 6 clearly demonstrates the inapplicability of strict fluence considerations for the MPD of CF_2HCl. The high-pressure buffer gas results imply that the effect of increasing intensity is to increase f_p, the fraction of molecules in laser-pumpable states. The high Ar pressures achieve this same end by collision-induced rotational relaxation to repopulate those states depleted by laser excitation. The higher intensities can overcome this rotational bottlenecking by power broadening and by promoting coherent processes such as Rabi cycling and multiphoton transitions.

There are only a few direct spectroscopic measurements of the population evolution due to exposure to intense, phase-coherent radiation from an IR laser. Optoacoustic (OA) techniques have been used extensively[71-76] towards this end. However, while OA techniques are powerful probes of weak spectral transitions (used also in trace impurity analysis[77] and overtone spectroscopy[78]), there is no information on actual final states distributions, and nor can these measurements give time-resolved information during the pumping process.

In an OA study of the 10 μm MPE of ethylene,[71] Fukumi reported two new vibrational bands absent in the low signal IR absorption spectrum of ethylene. Fukumi's experiment involved irradiation of moderate pressures of ethylene, i.e. 0.5 Torr, by an unfocused multimode TEA laser pulse. Two interesting results were obtained:

(1) For the two new bands, the slope of log(OA signal) versus log fluence was 2, at very low fluence, while for the corresponding one-photon (IR) band this slope was 1.
(2) The new band positions correlated very well to frequencies *expected* to arise from Fermi mixing of the nearby isoenergetic overtones $2v_7$ and $2v_8$.

These two facts strongly support the contention[71] that this is evidence of a true two-photon transition. The threshold for this two-photon process in ethylene appears to be well below an *average* energy fluence of 2 mJ cm^{-2} and, under the reported experimental conditions, seemed to saturate above about 5 mJ cm^{-2}. In order to obtain valid estimates of threshold conditions and a physical understanding of the exact nature of the apparent saturation, one would require experiments designed to probe regions of uniform, well known intensity.

Several aspects of MPD at 3·3 μm are distinctly different than at 10 μm. In ethylchloride, for instance, the vibrational level density[79] is 70/cm^{-1} near 6000 cm^{-1} and 1300/cm^{-1} at 9000 cm^{-1}. The use of 3·3 μm (i.e. 3000 cm^{-1}) radiation assures that the quasicontinuum is reached following the absorption of two or three photons. Thus only modest intensities are required for excitation and sharp spectral features characteristic of discrete state transitions may be preserved. In contrast, at 10 μm several step-by-step resonances are needed to drive the molecule through region I into the quasicontinuum. The MPD yield wavelength dependence will then be the convolution of these several steps. Dai, Kung, and Moore reported[79] just such sharp resonances in the 3·3 μm MPD of ethylchloride.

Dai et al. used a YAG-pumped LiNbO$_3$ optical parametric oscillator to provide continuously tunable pump pulses of 0·15 cm^{-1} spectral width and 10 ns duration. The MPD yields were determined by gas chromatographic analysis for ethylene, after several thousand photolysis shots. Dissociation was observed throughout the 2850 to 3050 cm^{-1} spectral range associated with the five CH stretching fundamentals. The dissociation yield spectrum followed, approximately, the linear absorption spectrum. In particular, the band centre frequency and band contour of the sharp, 0·4 cm^{-1} FWHM peak in dissociation yield observed at 2943·8 cm^{-1} matched the fundamental Q branch almost exactly. The additional sharp feature (2·5 cm^{-1} FWHM) at 2913 cm^{-1} did not correspond to any feature of the $v = 1 \leftarrow 0$ spectrum. It did coincide with half of the frequency of a strong sharp Q branch in the first overtone spectrum. Far in the red end of the MPD yield spectrum there appeared a broader resonance (about 20 cm^{-1}) that corresponded to one-third of the frequency of a strong resonance in the second overtone region. Based on this evidence, for C_2H_5Cl the levels $v_{CH} = 1$ appear to be discrete, $v_{CH} = 2$ show evidence of broadening, and $v_{CH} = 3$ are well within the quasicontinuum; at level densities as low as 70/cm^{-1} there is direct evidence for the onset of behaviour attributed to states within the quasicontinuum. The importance of resonance in the $2 \leftarrow 1$ transition is clearly demonstrated by the strong enhancement in yield for excitation at 2913 cm^{-1}. This is the one case where strong, sharp resonances in MPD yield spectra have been correlated exactly with known overtone transitions.

Time-resolved spectroscopic diagnostics will, in principle, offer more complete information about the *dynamics* of pumping through region I than measurements of energy deposition or gross dissociation yields. Several types of spectroscopic pump–probe or double-resonance experiments have been aimed at studying MPE in the low-lying discrete states. Kowk and Yablonovitch[80] tried picosecond IR–IR absorption to study intramolecular vibrational relaxation in SF_6; IR–IR double-resonance experiments were used by Alimpiev et al.[81] to study rotational hole burning in SF_6; IR–UV double-resonance experiments were used by Ambartzumian et al.[82] and by Sudbo et al.[83] to study gross effects of pumping through region I. Similar results have been obtained in our laboratory by studying the appearance rate of product species.[23,32]

Bagratashvili et al.[84] report the initial attempts at time-resolved Raman studies of moderate-pressure (0·5 Torr) gases undergoing MPE. Experimentally, this is one of the best conceived and executed studies of MPE performed to date. The pump laser was a truncated CO_2 TEA laser of 10 ns FWHM duration. The Raman signal was generated in a multipass geometry using a frequency-doubled ruby laser (20 ns FWHM) and a triple spectrograph with multiplex (optical multichannel analyser or photomultiplier tube array) detection. *The experimental geometry was such as to ensure initial sampling of species all experiencing identical fluence.*

Analysis of the resulting, time-resolved Raman signals depends on the understanding of two fundamental molecular properties: (1) spectral shifts and line broadenings for highly vibrationally excited molecules, and (2) the dependence of the Raman scattering cross section on the distribution of vibrational energy within a molecule (i.e. specific anharmonic couplings). Initial experiments were aimed at determining the fraction f_p of molecules directly interacting with the IR pump laser under collision-unimportant conditions. Using Xe as a buffer gas to induce rotational hole filling, they obtained the ratio of zero-pressure Raman signal to maximum signal, for SF_6, to be 0·43 for a 10 ns pulse of 1·2 J cm^{-2} fluence.

A more exciting result involved the time dependence of the Raman spectrum during and immediately after the IR pump pulse. In principle, if collisions are unimportant and if energy is initially localized in one or at most a few vibrational modes of the molecule, then there should be some change in the Raman spectrum due to collision-free intramolecular (T_1) vibrational relaxation. Under collision conditions, there is also the possibility for collisional processes to remove excited-state population causing a decrease in the signal. Within their experimental resolution, at 0·5 Torr SF_6 or CF_3I, they observed no significant changes in the anti-Stokes spectrum for $v_1(SF_6)$ or $v_5(CF_3I)$ over the time period 10^{-8} to 10^{-6} s. It appears from these results that energy randomization has occurred on the timescale of their measurements, i.e. 20 ns. Using OA estimates of energy absorption, an upper limit can be placed on the energy barrier for rapid (i.e. 5×10^7 s^{-1}) stochastization of energy. These estimates place the onset region for rapid energy flow within these molecules at state densities below 10^4/cm^{-1} for CF_3I and below 7×10^3/cm^{-1} for SF_6. The authors claim to be in the process of extending the sensitivity of this technique to allow studies at 1–2 orders of magnitude lower laser fluence. When these conditions are met, the laser-excited species may lie below the quasicontinuum and direct time-resolved spectroscopic measurements of intra- and intermolecular T_1 and T_2 processes should be possible.

The observation of fast apparent energy randomizations at the energies obtained by Bagratashvili *et al.* is not surprising in view of current theories of intramolecular relaxation dynamics. Hudgens and McDonald[85] place a much lower limit on the state density required for ergodic behaviour, at 400/cm^{-1} for C_2F_5Cl. Their experiment[85] involves the spectral observation of infrared emission from very low-pressure C_2F_5Cl following TEA laser excitation. Their spectra were obtained using a cryogenic IR detector and circular variable interference filters and a reagent pressure of 10^{-4} Torr at a total (He) pressure of 10^{-3} Torr. The detector response was gated to observe spontaneous emission for $42\,\mu s$ starting some few microseconds after the laser pulse. The laser pulse was unfocused, exhibiting a trapezoidal profile that made observing regions of constant fluence impossible. In addition, spectra of thermal gas samples could be obtained from C_2F_5Cl effusing from an oven source for spectral (e.g. thermal) comparison. A spectrum obtained from each type of experiment is displayed in Fig. 7. The 1145 K thermal spectrum clearly shows five strong bands which have been previously assigned in absorption.[86,87] The band maxima are red-shifted compared with the 300 K spectrum, as might be expected in terms of molecular cross-anharmonicity, but are still sharp and clearly resolvable.

Fig. 7 also shows the IR emission spectrum following $967\,cm^{-1}$ excitation of the ν_4 mode of a 220 K beam of C_2F_5Cl. The sharp emission feature at $982\,cm^{-1}$ is presumably $\nu_4 = 1 \rightarrow 0$ fluorescence from those C_2F_5Cl species having absorbed exactly one IR laser photon.[85] The broad, red-shifted emission in the 950–$1000\,cm^{-1}$ range is strong evidence for direct radiative excitation into higher-lying levels of the ν_4 pump mode. Strong emission is also observed from modes not directly populated by radiative transitions. Emission from non-laser-pumped states can only arise from collisional or intramolecular vibrational energy flow (these experiments being performed at pressures sufficiently low to exclude within reason V–T exchange processes). The data strongly suggest that C_2F_5Cl experiences rapid intramolecular

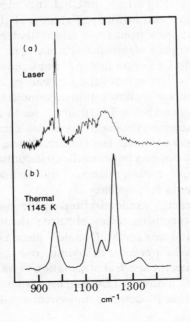

Fig. 7. Emission spectra of C_2F_5Cl produced by laser irradiation at $967{\cdot}7\,cm^{-1}$ (a) and by thermal heating at 1145 K (b). After J. W. Hudgens and J. D. McDonald.[85]

vibrational redistribution following excitation to levels equal to or in excess of $2v_4$, at $1950\,\text{cm}^{-1}$. This conclusion is based on the similarities[85] between the laser-prepared and 1125 K thermal emission spectra and implies that rapid energy flow can proceed, in C_2F_5Cl, at vibrational state densities of about $400/\text{cm}^{-1}$.

The sharp emission feature at $982\,\text{cm}^{-1}$ exhibited by the laser-prepared ensemble is intriguing. It may, as the authors suggest,[85] represent those species that are never excited beyond the $v_4 = 1$ level due to radiation bottlenecking effects, or may just be due to a large percentage of sampled species having been subjected to the weaker intensity wings of the laser profile, i.e. an artifact due to non-uniform excitation. Molecules of similar size and complexity, specifically C_2F_3Cl, have not shown significant resonance or bottlenecking effects[23] when probed using other techniques. The IR emission studies of Hudgens and McDonald[85] are certainly much more sensitive to this issue and future experiments of this type should add to the understanding of excitation and energy flow in regions of low levels of excitation.

In many respects, the most exciting and comprehensive experiments on excitation within region I are the IR–visible double-resonance experiments performed under molecular beam conditions by Brenner, Brezinsky, and Curtis.[88,89] A grating-tuned multimode CO_2 TEA laser was used to excite propynal, $HC\equiv C-CHO$. Changes in the initial population distributions due to IR wavelength and energy and the population decays due to intramolecular relaxation processes were monitored by time-resolved laser-excited fluorescence (LEF) techniques.

The molecular beam apparatus contained a glass-nozzled effusive source 15 mm in diameter. The dye laser probe was timed for a 225 ± 25 ns delay with respect to the CO_2 TEA pump laser (to probe just after the intense central spike) and aligned in a crossed laser geometry. With background pressures of 2×10^{-5} to 3×10^{-4} Torr, collisional effects were presumed to be unimportant. Unfortunately, due to the crossed laser geometry, these experiments sample intensity-averaged effects, the probe laser interrogating molecules experiencing widely differing instantaneous intensities. This type of geometry does, however, allow for good comparison with optoacoustic measurements under conditions of similar irradiance. The CO_2 laser can excite two fundamental propynal transitions, the v_{10} CH wag ($981\,\text{cm}^{-1}$) and the v_6 CC stretch ($944\,\text{cm}^{-1}$).

Dramatic frequency-dependent effects were observed, using the LEF technique, for excitation across just a $14\,\text{cm}^{-1}$ region when pumping the v_6 fundamental (see Fig. 8). The ro-vibronic populations of v_6 and $2v_6$ were observed to vary in a non-uniform fashion under excitation conditions which for higher-pressure samples (600 mTorr) gave relatively uniform OA signals for all CO_2 laser lines. The authors estimated[88] a typical power broadening of $0.45\,\text{cm}^{-1}$. This should have been sufficient for the laser excitation to overlap

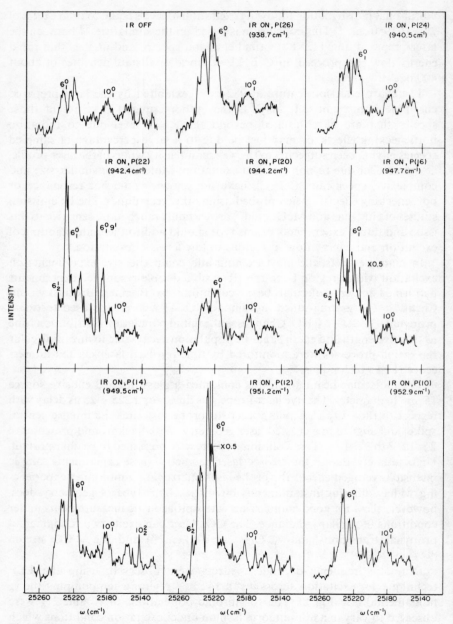

Fig. 8. Laser-induced fluorescence excitation spectra of propynal in the frequency region of v_6 at room temperature (IR off) and at higher vibrational temperatures (IR on), produced by CO_2 laser pumping at specified frequencies. Background pressure is $(2-3) \times 10^{-4}$ Torr; laser fluence around $750 \, mJ \, cm^{-2}$; cross-beam geometry. The time delay between pump and probe laser is (225 ± 25) ns. After D. M. Brenner, K. Brezinski, and P. M. Curtis.[89] Reproduced by permission of North-Holland Publishing Company, Amsterdam.

with some 8 to 20 individual ro-vibrational transitions. However, the data in Fig. 8 indicate the actual number of excited ro-vibrational levels to be much larger. Nevertheless, the changes observed in the LEF spectrum showed clear evidence of resonances and a dramatic difference from the OA results, implying the OA results to be influenced by collisional processes (such as rotational hole filling).

The state $2v_{10}$ can be populated directly by pumping the v_{10} manifold at $977 \cdot 2 \, \text{cm}^{-1}$. No evidence for collision-free relaxation into the isoenergetic $v_4 + v_{12}$ level was observed, although LEF signals could be observed on the 4^0_1–12^1_1 transition under the collisional conditions of 10 mTorr and a 200 ns delay ($P\tau = 2 \times 10^{-9}$ s Torr). At a level of internal excitation of $1954 \, \text{cm}^{-1}$, the propynal molecule does not show any significant degree of intramolecular relaxation.

This work suggests very strongly that *one must consider explicitly changes in populations, and not just average absorption properties.* Indeed, in comparing the time-resolved spectroscopic measurements with OA measurements, it appears as though near-resonant collision-induced vibrational relaxation processes will occur at a rate $(1/P\tau)$ of about $5 \times 10^8 \, \text{s}^{-1} \, \text{Torr}^{-1}$. Clear evidence is given for the participation of hot band transitions, for instance excitation $v_6 + v_9 \leftarrow v_9$, and, although direct evidence is given for radiative excitation to states such as $2v_6$, no suggestion is given as to whether this state is reached by two successive steps or a single two-photon transition.

When probing population distributions in levels where there are no strong interactions to randomize state populations, it is important to consider the influence of the intense laser field on the preparation of the initial ensemble of excited states. Duperrex and van den Bergh[65] have attempted to address this question in much greater detail in the MPD of CF_2HCl in an external magnetic field.

Excitation by the output of an unfocused multimode TEA laser was performed in a small Pyrex cell located between the poles of a 14 kG electromagnet. The magnetic field homogeneity was stated at better than ± 3% over the total cell volume. MPD analysis was performed, as a function of laser fluence or gas pressure, by UV absorption of CF_2 products. The influence of the magnetic field on the MPD yield from CF_2HCl is quite dramatic even at pressures as high as 100 mTorr (where, based on pressures, yields, and time frames, it is hard to claim to be in a collision-free regime). At the lowest fluence levels studied, magnetic field enhancements of about three-fold were reported. The degree of enhancement decreased sharply as the laser fluence was increased, becoming negligible above 3 J cm^{-2}. The pressure dependence of both the high-field and zero-field results were obtained. Both showed no change in yield up to some critical pressure, presumably that pressure where collisional effects began to be manifested. In the absence of any external field, this pressure was 0·6 Torr. In the presence of the 14 kG magnetic field, collisional effects were not apparent below 1·5 Torr. The reason for this unexpected behaviour is unclear. Very recent results by Gozel and van den Bergh[90] on effects of 5·4 kV cm^{-1} electric fields on the CF_2HCl MPD show enhancements at high field, low pressure, and low laser fluence similar to the magnetic field effects. The zero-field and high electric field MPD yields, however, follow the same pressure dependence. Van den Bergh advocates that enhanced yields at low pressure and low laser fluence result from a breakdown in angular momentum selection rules in the strong external fields, resulting in an increased effective density of states. The pressure (of buffer gas) and fluence dependence of this enhancement parallel the laser intensity dependences of King and Stephenson[61] discussed earlier.

IV. EXCITATION THROUGH THE QUASICONTINUUM

In many cases, the rate-limiting step in multiphoton dissociation is the sequential absorption of photons in the quasicontinuum. This will become particularly apparent if one measures total yields or macroscopic dissociation rates.[20] There will be a strong experimental appearance of fluence dependence and a rate equation model will be suitable. The experiments presented in this section fall into this category. There are several points regarding the dynamical processes occurring within highly excited species that these experiments attempt to address. These include the potential for energy localization to allow for bond-specific or non-equilibrium chemistry and a development of a sounder theoretical basis for predicting the energy scalings of T_1 and T_2 type intramolecular processes and for collisional energy transfers. These are some of the facets arising from the MPD experiments that will be of interest to a much broader group of chemists than were initially interested in laser isotope separation.

During the initial period of interest in MPD, two-colour experiments were generating a high degree of interest and of controversy. In principle, the two-colour MPD experiment[91-93] offers time-resolved information on pumping through either region I or region II, or through both. The experiment involves two independently line-tunable IR lasers that can be fired synchronously in time with the second laser fired some variable delay time after the first. Thus, if the second laser has sufficient fluence to drive all molecules reaching the quasicontinuum to dissociation, variations in the frequency of the first laser will give explicit information regarding pumping in the discrete levels, i.e. frequency response and intensity dependence. Conversely, for fixed initial pump conditions, variation in the intensity, frequency, and time delay of the second laser give information regarding the level of excitation achieved by the first pulse f_p, the gross spectral features of the quasicontinuum, and the relaxation mechanisms of highly excited molecules. Of particular interest to earlier investigations was the significantly higher yield obtainable using such two-colour techniques concomitant with much greater initial (isotopic) selectivity.

In two recent papers, Ambartzumian et al.[91,92] use two-colour IR probing to study the MPD dynamics of OsO_4. They show evidence for sharp structure on the Q branch transitions $v_3 = 1 \leftarrow 0$ and $v_3 = 2 \leftarrow 1$; no dissociation was observed when excitation was in the OsO_4 v_3 R branch. Some effort was expended to attempt to obtain wavelength-dependent absorption cross sections for pumping through the quasicontinuum and to relate these to molecular properties. The lack of success in this, other than in a general fashion, arises from some of the same aspects that caused a waning interest in similar two-laser experiments. Due to the nature of the MPD process, the initial pulse creates a broad distribution of reactant energies which are then probed indiscriminately. In addition, the second pulse may, itself, contribute to excitation through the higher levels of region I into region II and finally dissociation. It appears as though, with the new generation of short-pulse, single-mode IR lasers being developed, that two-laser IR MPD experiments will give valuable information on the coherence and intensity requirements for pumping through the low-lying discrete states and on intramolecular relaxation processes in the region of the interface between regions I and II.

A new type of two-colour experiment designed to probe specific regions of the quasicontinuum was reported by Heller and West.[94] In addition, they claim it 'provides a new tool for exploring intramolecular interactions in strong and weak fields and for studying the mechanism of MPA.' The technique employs electronically induced non-resonant multiphoton absorption (EINMA); the test species was chromylchloride, CrO_2Cl_2. An initial visible-frequency photon was used to excite the CrO_2Cl_2 electronically, followed by spontaneous intramolecular non-radiative decay. This radiationless decay produces molecules in the vibrationally hot quasicontinuum of the ground electronic state with a uniform amount of energy equal to the photon energy. By this technique, large concentrations of such highly excited isoenergetic species can be prepared. These molecules directly access region II (see Fig. 9); subsequently, an off-resonance (with respect to any fundamental vibrational transition) infrared pulse can produce further MPE leading to higher-energy molecular processes such as dissociation or emission. The experiment was performed in a flowing gas cell. The visible laser was a frequency-doubled YAG (532 nm, 8 ns FWHM, saturation level below $50 \, mJ \, cm^{-2}$), the second pump source was a CO_2 TEA laser ($9\text{–}10 \, \mu m$, 40 ns + tail, 0·1 J/pulse) synchronized in time to ± 5 ns. Absorption was monitored using an optoacoustic detector, and visible luminescence (resulting from multiphoton up-pumping) was monitored using a filtered photomultiplier tube. A linear pressure dependence in the EINMA was reported over the range 0·02 to 0·5 Torr (lowest pressure corresponds to 0·1 hard-sphere collisions per reactant molecule, on average).

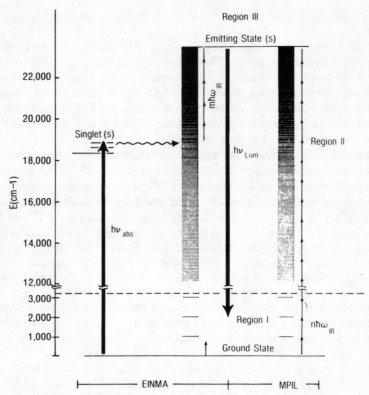

Fig. 9. Schematic energy level diagram for electronically induced non-resonant multiphoton absorption (EINMA) and for normal multiphoton excitation (MPIL) in CrO_2Cl_2, both of which result in visible luminescence. After D. F. Heller and G. A. West.[94] Reproduced by permission of North-Holland Publishing Company, Amsterdam.

    In the EINMA experiments, off-resonance excitation was competitive with on-resonance pumping. Also, dissociation was observed at much lower fluences for EINMA excitation. Although, as the authors point out, the concept of a threshold is hard to justify under the focused conditions of these experiments, the observed yield *versus* fluence curves showed behaviour characteristic of a threshold. No significant enhancement was observed for excitation to the blue of the one-photon signal. In this respect, their results parallel the thermal heating results of Tsay, Riley, and Ham.[95]

    It is to be hoped that techniques such as EINMA will prove useful in preparing an ensemble of reactants at a sharp, well defined level of internal excitation within the quasicontinuum. Thus, absorption cross sections and collisional energy transfer probabilities could be obtained as a uniformly varying function of internal energy content.[96,97]

Coggiola, Cosley, and Peterson[98] discuss experiments sensitive to the interface region between the quasicontinuum and dissociative states. They report on the strength and frequency dependence of IR absorption of highly vibrationally excited CF_3X^+ ions (X = I, Br, Cl). These results may be compared to low-power (CW) IR laser photolysis of large ions such as $C_3F_6^+$ or $(Et_2O)_2H^+$ performed by Beauchamp, Woodin, and Bomse[99–101] in ion cyclotron resonance (ICR) spectrometers and by Rosenfeld, Jasinski, and Brauman[102] and von Hellfeld et al.[103] using high-intensity CO_2 TEA lasers in crossed-beam experiments.

The interpretation of the data of Coggiola et al. requires the active dissociation mechanism to be a single-photon process. This is supported in their data by the linear power dependence of the dissociation signal as well as

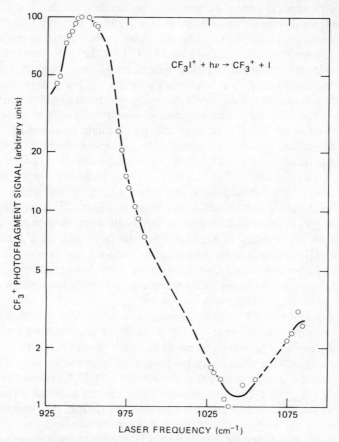

Fig. 10. Frequency dependence on the CF_3^+ fragment yield from CF_3I^+. After M. J. Coggiola, P. C. Cosby, and J. R. Peterson.[98] Reproduced by permission of the American Institute of Physics.

by careful consideration of the laser flux and interaction times. The experiment utilizes a laser–ion coaxial beam spectrometer. Parent ions are formed by electron impact at 100 eV and then accelerated to a kinetic energy of 2·5 keV. The laser output was 2–3 W CW. Although the laser output was TEM_{00}, the experimental geometry does not allow one to probe the effects of regions of known, constant intensity or fluence. Phase-sensitive detection was required to distinguish the photodissociation products from collision-induced dissociation (CID) products.

The data clearly show strong maxima of photodissociation probability with wavelength. As shown in Fig. 10 this maximum occurs at 947 cm^{-1} for CF_3I^+. A similar feature was observed at 953 cm^{-1} for CF_3Br^+. For CF_3Cl^+, there was no apparent wavelength dependence on photodissociation cross section; however, the apparent cross section was considerably smaller. For CF_3I^+, an additional, weaker peak was observed at $v > 1080$ cm^{-1}.

Spectroscopically, in the CF_3I neutral, the mode closest to the observed 947 cm^{-1} absorption of the highly excited CF_3I^+ is the C–F symmetric stretch (v_1) at 1073 cm^{-1}. This correlation would imply that ionization and vibrational excitation lead to a weakening of the C–F bond. Support for this behaviour is found in the far-UV absorption spectrum of CF_3I where the v_1 mode in a Rydberg state was assigned at about 950 cm^{-1} by Sutcliffe and Walsh.[104] Likewise, the CF_3Br v_1 at 1082 cm^{-1} might be expected to show a similar red shifting for the ionic species into the observed 950 cm^{-1} region. In similar experiments, single-photon dissociations were observed for CH_3F^+, CH_2F^+, $C_2F_5I^+$, and $C_3F_7I^+$. CH_3I^+, which lacks any appropriate vibrational modes near the frequency range of the CO_2 laser, showed no evidence of dissociation, even under irradiations of up to 40 W cm^{-2}. It appears, despite the high degree of ion excitation and the large density of states present, that the oscillator strength tends to peak around the frequency of a fundamental molecular vibrational mode. Although the absorption process leading to dissociation shows vibrational structure, at these low irradiance levels the observed dissociation channel always represented the lowest-energy pathway irrespective of which mode was excited.

In measurements of the kinetic energy of the product species by time of flight, it was determined that for CF_3I^+ and CF_3Br^+ the fragment translational energy $E_{trans} \ll h\chi$ (E_{trans} being 4–5 meV compared with the 0·12 eV photon energy). This result may be compared with the focused TEA laser results of Sudbo et al.[70] where the dissociation of CF_3X resulted in product species with $E_{trans} \approx 1·2 h\chi$, the dissociation occurring after the absorption of 4–5 excess photons. Under the more intense irradiation, the reactant species are driven to higher-lying levels of the dissociative continuum.

Two intriguing examples of excitation in region II have come out of the Exxon Research and Engineering Laboratories. They both make claims at demonstrating non-statistical behaviour in the quasicontinuum. The first of these involves the pumping of cyclopropane by IR pulses at either $9.5\,\mu m$ to excite the low-frequency CH_2 wag or at $3.22\,\mu m$ to excite high-frequency CH asymmetric stretches.[105] The second involves the laser photolysis of large uranyl prototype molecules where the thermal energy of the molecule places it effectively in the quasicontinuum.[106]

The question raised by the cyclopropane experiment is one of statistical behaviour. Apparent, non-statistical behaviour in observed product branching ratios has been reported by Brenner[55] for ethylvinyl ether and by Hudgens,[107] Lee et al.,[56,59] and Morrison and Grant[58] for CF_2Cl_2. These experiments, however, all involved excitation in the $9-11\,\mu m$ range, and the results could possibly be explained in terms of pseudoselectivity (see Sec. IID) arising simply from the competition of radiative up-pumping and microscopic decay, consistent with RRKM theory. The work of Hall and Kaldor[105] is one of the first experiments aimed at elucidating non-statistical behaviour using two very different excitation wavelengths. For cyclopropane there are two energetically competitive channels, isomerization

$$\text{cyclopropane} \rightarrow \text{propylene} \qquad k_\infty = 10^{15.2} \exp(-65.0\,\text{kcal}/RT)$$
$$\Delta H = -7\,\text{kcal}\,\text{mol}^{-1},$$

and fragmentation

$$\text{cyclopropane} \rightarrow H_2C{=}CH_2 + :CH_2, \qquad k_\infty \sim 10^{15} \exp(-E_a/RT)$$
$$E_a > 100\,\text{kcal}\,\text{mol}^{-1},$$

The experiments (unfortunately) were performed under focused geometries in static cells with two distinctly different lasers: an HF laser at $3.22\,\mu m$ producing fluences of $2\,\text{J}\,\text{cm}^{-2}$ for a 40 ns FWHM pulse to excite the CH asymmetric stretch at $3100\,\text{cm}^{-1}$; and a CO_2 laser producing fluences of $60\,\text{J}\,\text{cm}^{-2}$ in a 450 ns FWHM pulse to excite the $9.5\,\mu m$ CH wag at $1050\,\text{cm}^{-1}$. In both cases the product yields were very low, of the order of 1–4% for 20 000– 40 000 pulses. The results reported can be summarized as follows:

(A) *High-fluence CH stretch excitation*
  (i) The high-frequency CH excitation resulted in the observation of *only* the low-energy channel isomerization reaction products at pressures of about 1 Torr neat cyclopropane.
  (ii) Addition of an inert buffer gas (Ar) served to increase the yields in the range 0–10 Torr due to rotational hole filling; and to enhance the relative yields from the high-energy fragmentation process compared to isomerization in the range 100 to 500 Torr.

(B) *Low-fluence CH$_2$ excitation*
  (iii) For the low-frequency CH_2 excitation both reaction channels were seen to be competitive.

(iv) Addition of Ar (1 to 20 Torr) was seen to favour dissociation via the low-energy channel as would be expected on the basis of simple V–T transfer.

Hall and Caldor[105] claim that both because the branching ratio of reaction depends on which mode is excited and because buffer gas collisions show distinctly different effects these experiments demonstrate non-statistical or mode-selective behaviour. It is their argument that if the molecule were behaving statistically the only effect of a buffer gas collision during excitation would be to remove energy from the reactant, thereby favouring the lower-energy dissociation channel.

Although the reported results are consistent with this very exciting interpretation, there is a second, equally valid, albeit more mundane, explanation for the observed results. Specifically, for the 9·5 μm excitation the molecules are driven by long-duration, high-fluence pulses. For cyclopropane, the pre-exponential A_∞ factor is not exceedingly high and the molecule is sufficiently large that the RRKM macroscopic decay rates should vary slowly with increasing energy. Thus, there will be competition between continued radiative up-pumping, unimolecular decay, and (with the addition of Ar) collisional V–T transfer. The relative yields of the high-energy versus low-energy channels should show pseudoselectivity as discussed previously. The added argon then allows collisional V–T transfer to compete with the microscopic decay through both channels. That the yields begin falling off in the 10–20 Torr range indicates that the average rate for radiative excitation under the excitation conditions reported was moderately low, in the 10^5 to 10^6 s^{-1} range.

To understand the 3·22 μm results, it should also be borne in mind that the yields are very small. Thus, pumping through the quasicontinuum is very inefficient under these excitation conditions. The shorter, less powerful 3·2 μm pump is either not strong enough or long enough to pump the ground-state cyclopropane sufficiently high into the continuum region to effect a significant yield of dissociation products. In the presence of Ar, rotational hole filling of the *hot* isomerization products formed during the initial portion of the pulse may allow a macroscopic fraction of these nascent isomerization products to be photolysed to yield fragmentation (i.e. simple secondary photolysis). This analysis would only require that there be some type of 'low-lying' bottleneck to the MPD of cyclopropane. The large enhancement in yields for the addition of 10 Torr Ar is consistent with rotational hole filling overcoming such a bottleneck. In this reaction, the nascent isomerization product, propylene, will be born with a great excess of internal energy (i.e. $E_a - \Delta H \gtrsim 72$ kcal mol^{-1}), i.e., in its quasicontinuum. Although the concept of energy localization either in one single chemical bond or between groups of low- and high-frequency modes is exciting (and even observed for low levels of molecular excitation,[108] experiments involving more careful excitation and time-resolved probes will be required to demonstrate this point conclusively.

In their experiments on the prototype $UO_2(hfacac)_2THF$ (*bis*-hexafluoroacetylacetonate uranyl tetrahydrofuran), Cox and Horsley[106] explore the role of thermal energy on MPD processes in very large molecules. $UO_2(hfacac)_2THF$ contains 44 atoms. Its average thermal energy is comparable with the activation energy for dissociation $(24-32\,kcal\,mol^{-1})$. This thermal energy acts to place the molecule in a region of high state density

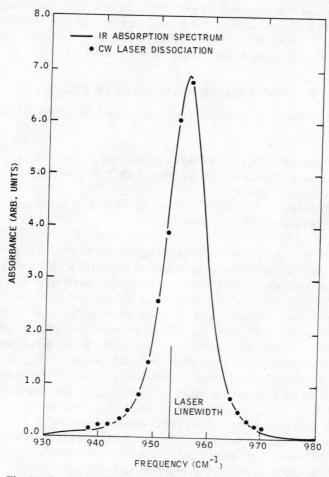

Fig. 11. Comparison between the infrared absorption spectrum of $UO_2(hfacac)_2THF$ and the relative photodissociation spectrum obtained with a CW CO_2 laser. After D. M. Cox and J. A. Horsley.[106] Reproduced by permission of the American Institute of Physics.

where sequential absorption will predominate and to provide a significant fraction of the energy required for decomposition. The observed wavelength dependence of photodissociation is exactly equal to the infrared absorption spectrum (see Fig. 11); surprisingly, no shifts or broadenings are apparent. Together with the linear fluence dependence and the temperature dependence of the dissociation yields, their data indicate that the MPA process in $UO_2(hfacac)_2THF$ can be considered to be a $1 \leftarrow 0$ excitation in the pump mode followed by rapid intramolecular relaxation (T_1) of this vibrational quantum into the strongly coupled heat bath modes. In the case of rapid T_1 relative to radiative pumping the excitation can be satisfactorily described by rate equations. In a simplified model calculation, Cox and Horsley[106] only require $T_1 < 1$ ns for the 5–10 photons to be absorbed from the incident laser field to give 10% dissociation during the laser pulse.

V. COMPETITIVE PROCESSES IN REGION III

Several competitive processes can be vizualized for region III. These will include:

(i)   continued radiative excitation, depending on laser intensity;
(ii) collision-induced energy transfer, both V–T loss and loss of energy localization effects;
(iii) unimolecular dissociation along one or several pathways;
(iv) collision-free curve crossings.

In terms of our current understanding, continued radiative excitation in region III can be treated by a rate equation approach with a nearly constant net absorption crosssection obtained from the single-quantum exchange theory.[46] Thus, the rate of excitation is directly proportional to the instantaneous laser intensity and is independent of the level of excitation. Collisions with a cold bath gas leading to deactivation and collision-free unimolecular decay will increase in efficiency with increasing level of excitation as predicted by chemical activation studies[41,42] and statistical reaction theories.[48,49] Of great experimental interest is the possibility of demonstrating a controllable influence over competitive chemical reactions (i.e. involving energy localization) or relaxation phenomena such as the controversial collision-free inverse inter-system crossing.[109] Two questions important to an understanding of the competitive MPD processes in region III are the validity of statistical theories for unimolecular decay and the (presumed) independence of processes (ii)–(iv) on the laser field.

The two systems most extensively studied in an attempt to show energy localization effects or non-statistical chemical decay are ethylvinyl ether (EVE) and dichlorodifluoromethane (CF_2Cl_2). In addition, one should consider the work of Kaldor and Hall[105] on cyclopropane (discussed in Sec. IV) and the CF_2Br_2 system (which is similar[59] to CF_2Cl_2).

CF_2Cl_2 has been studied by several different groups.[57-59,107,110-112] The two competitive reaction channels are

$$CF_2Cl_2 \rightarrow CF_2Cl + Cl \qquad \Delta H = 78\text{--}82 \, kcal \, mol^{-1} \qquad (9)$$

and

$$CF_2Cl_2 \rightarrow CF_2 + Cl_2 \qquad \Delta H \sim 71 \, kcal \, mol^{-1}. \qquad (10)$$

The presence of CF_2Cl has been confirmed in the molecular beam experiments of Sudbo et al.[59] while CF_2 has been observed in situ by King and Stephenson[57,110] using time-resolved spectroscopic techniques. Unfortunately, any detection scheme sensitive only to CF_2 and not to either Cl, Cl_2 or CF_2Cl cannot readily distinguish reaction (10) from subsequent

$$CF_2Cl(hot) \rightarrow CF_2 + Cl \qquad \Delta H = 54 \, kcal \, mol^{-1}. \qquad (11)$$

Cl_2 has been observed in the beam-sampled, low-pressure experiments of Hudgens.[107]

Morrison and Grant[58] reported the apparent branching ratio for channels (9) and (10). The experiment involved scavenging of CF_2Cl_2 products by Br_2 in a 10-fold excess at a total pressure of 1 Torr. Following irradiation by 50–3000 pulses of uniform beam profile, but not uniform intensity or fluence, GC analysis was made for CF_2Cl_2 (remaining parent), CF_2Br_2 (from CF_2), and CF_2ClBr (from CF_2Cl). Unfortunately the effects of rephotolysis, collisional effects, and the fact that these experiments show an effective integration over the spatial beam profile including the high intensity in the centre and the low intensity wings cannot be ignored. Therefore, quantitative conclusions drawn from the data are subject to scrutiny. However, the results show an interesting qualitative behaviour (see Fig. 12).

It is seen that, as the laser energy increases, the relative amount of CF_2 (recaptured as CF_2Br_2) goes through a maximum and then decreases. Although channel (10) has a lower endothermicity, there should be a $10 \, kcal \, mol^{-1}$ activation barrier for the reverse reaction, thus giving the two competitive channels comparable activation energies. If, as Lee suggests,[59] the atomic elimination involves a loose complex and the three-centre molecular elimination involves a much tighter complex with significantly higher vibrational frequencies in the critical configuration, then the macroscopic rate constants for chemical decay will be divergent, with the macroscopic rate for direct production of Cl atoms growing much faster with increasing internal energy. Under exactly these conditions one would expect to observe pseudoselectivity.

Where the results of CF_2Cl_2 MPD are subject to scrutiny due to the possibility of secondary photolysis, the results of Brenner and more recently of

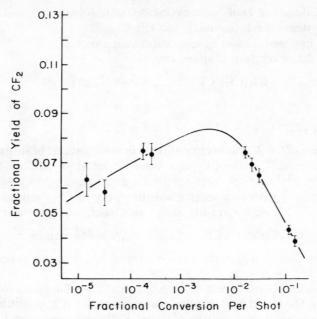

Fig. 12. Fractional yield of CF_2, measured as CF_2Br_2, as a function of laser energy in the MPD of CF_2Cl_2 in an excess of Br_2 scavenger. After R. J. S. Morrison and E. R. Grant.[58] Reproduced by permission of the American Institute of Physics.

Lee and coworkers[56] on EVE clearly show pseudoselectivity. For EVE the lower-energy channel has a pre-exponential factor of $10^{11.6}$ and an enthalpy of $44 \, \text{kcal mol}^{-1}$. The higher-energy channel, at $65 \, \text{kcal mol}^{-1}$, has a much higher Arrhenius factor of 10^{15}.

In Brenner's work, two different laser pulses of equal fluence were used to photolyse 0·4 Torr of neat EVE, one of 0·2 μs duration the other of about 2 μs duration. The stable end-products were assayed later by GC–mass spectrometry. In the fluence range 0·5 to 1 J cm^{-2}, there was no change in relative branching ratio for short-pulse excitation. The higher-energy channel was always slightly favoured. However, upon long-pulse photolysis only low-energy-channel products were apparent. Examining these results[55] in terms of a quantum RRK formulation (taking into account the full 33 vibrational degrees of freedom of EVE) the observed product yield ratio in the short-pulse regime only required an absorption and subsequent statistical distribution of 35 photons (104 kcal mol^{-1}). This is equivalent to excitation to a level with a corresponding microscopic decay rate of about $10^7 \, \text{s}^{-1}$. This behaviour is quite reasonable in terms of microscopic rates attributed to species (see, for example, Sudbo *et al.*[59]) and is completely compatible with current theories of statistical reaction energetics. No information is available on the absorption cross section in this energy region for EVE. However, this apparently low decay rate provides an attractive basis for further studies of competitive excitation and decay processes.

Danen, Koster, and Zitter[112] have proposed an alternative probe for non-statistical behaviour. Many classes of concerted reactions are known for which the products resulting from thermal (i.e. ground-state) or photochemical (i.e. excited-state) processes differ greatly. These differences are accountable on the basis of the well established Woodward–Hoffmann rules of conservation of orbital symmetry.[113] The MPD of cis-3, 4-dichlorocyclobutene involves two energetically competitive reactions:

$$E_a \approx 30 \, \text{kcal mol}^{-1}$$
$$\log A_\infty \approx 13\cdot 4$$
(12)

$$E_a \approx 45 \, \text{kcal mol}^{-1}$$
$$\log A_\infty \approx 14$$
(13)

Only (12) is produced by thermal heating in agreement with Woodward and Hoffmanns[113] predictions. Under modest irradiation conditions, i.e. $I_{max} = 20 \, \text{MW cm}^{-2}$, only (12) was observed. This was always true up to the point where more complex product arrays appeared due to secondary photolysis. *Experiments of this type offer the possibility to study the effects of the intense field created by the laser on removing symmetry restraints on reaction.* The results here are strongly contrasted to Brenner's work on EVE[55] where the higher channel competed successfully under high fluence conditions. However, for EVE, the high-energy channel had a higher Arrhenius factor A_∞ by almost 10^4. For the case of the dichlorocyclobutene, the A_∞ factors are approximately equal.

Von Hellfeld et al.[103] report to have measured *directly* the distribution of lifetimes of highly MPE SF_5^+ species using time-of-flight ionbeam techniques. As pointed out by Bunker and Hase,[114] the true test of statistical theories of intramolecular energy flow in unimolecular decomposition is the measurement of the distribution of lifetimes of a set of *monoenergized molecules*. Unfortunately, von Hellfeld's experiment was designed around a crossed laser–ion beam geometry. It is not possible to describe the intensity experienced by each dissociating SF_5^+. In addition, the internal energy

DAVID S. KING

distribution resulting even from uniform intensity MPE will be broad. Consequently, the distributions measured did not correspond to a 'monoenergized ensemble' of reactants. Still, the experimental results are very informative with regard to the distribution on excited reactants. In the experiment, SF_5^+ (produced from SF_6 by MPD) was exposed to the output of a CO_2 laser. The highly excited SF_5^+ was then accelerated by a variable electric field (2–8 KV cm^{-1}) into an ion detector, and the SF^+_5 ion current monitored as a function of transit time from the photolysis zone to the detector. The time-of-flight data are shown in Fig. 13.

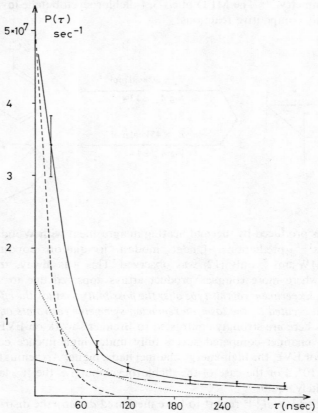

Fig. 13. Experimental measured distribution of lifetimes of infrared-pumped SF_5^+ ions(——). Other curves are theoretical RRKM lifetime distributions for three values of the unimolecular rate constant k_a: $k_a = 5 \times 10^7 \, \text{s}^{-1}$ (---); $k_a = 1\cdot6 \times 10^7 \, \text{s}^{-1}$ (.......); $k_a = 5 \times 10^6 \, \text{s}^{-1}$ (-.-.-). For the case of excited SF_5^+ ions, these rate constants correspond to three levels of internal excess energy above dissociation barrier: 6, 5, and 4 photons (1 photon = 0·12 eV). After A. von Hellfeld, D. Feldman, K. Welge, and A. P. Fourier.[103] Reproduced by permission of the American Institute of Physics.

In their analysis von Hellfeld *et al.* do not use these experimental data to derive the distribution of lifetimes of MPE species; rather, the authors assume statistical (i.e. RRKM) theory to hold and estimate the distribution of excitation in the dissociating SF_5^+ species. Dividing the excited SF_5^+ into three groups, containing ions with 0–4, 4–6, and 6–8 excess photons, with respective RRKM decay constants of 5×10^6, $1 \cdot 6 \times 10^7$, and $5 \times 10^7 \, s^{-1}$, the distribution shown in Fig. 13 was reproduced under the following conditions: 10% of the population in those levels 0 to 4 photons above the dissociation threshold, 60% in those levels between 4 and 6 excess photons, and 30% in the levels between 6 to 8 excess photons. Although this distribution is non-thermal in nature, it is consistent with the calculations of Goodman and Stone[23,32] and others (see also Sec. II). The results of von Hellfeld *et al.* clearly show that MPE processes are not really suitable for preparing a monoenergized ensemble of reactants. A better test of statistical decay theories will be made using a single high-energy photon to excite molecular overtone transitions,[115,54] staying on the ground-state potential surface.

Although excitation rates, and phase and energy relaxation times (T_2 and T_1) have been theoretically calculated for laser-driven reactions,[14] direct measurement of such rates is generally not feasible because of the difficulties in doing measurements on a fast (e.g. 10^{-12} s) timescale, and in understanding the

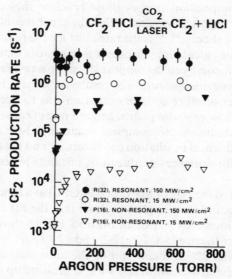

Fig. 14. Macroscopic rates of CF_2 production during the dissociation of CF_2HCl by CO_2 laser pulses. The rates were determined (see text) for the CF_2HCl very dilute in Ar at the indicated pressures.

spectroscopy of polyatomic molecules which are highly vibrationally excited. Fortunately, readily performed measurements of the macroscopic rates of reaction are sensitive to assumptions concerning reactant excitation and relaxation rates. The results of such experiments (see Appendix for experimental details) on the intensity and buffer gas pressure dependences in the rates of CF_2HCl (Fig. 14)[32] and CF_2CFCl (Fig. 15)[23] are presented below. There is one marked difference in the pressure dependence of the dissociation rates for these two reactants. For CF_2CFCl–Ar there is a dramatic decrease in decomposition rate with increasing buffer gas pressure above 100 Torr. This is not observed for CF_2HCl–Ar. This effect can readily be ascribed to a competition between radiative pumping and collisional deactivation in the quasicontinuum. By modelling the MPE process are can derive absorption cross sections for this region of high vibrational excitation. Although we did not observe[32] this high-pressure rate fall-off for CF_2HCl, we note that Duperrex and van den Bergh[66] did observe this effect pumping somewhat further to the blue.

There are a variety of parameters which may influence the rate of MPD: laser intensity, wavelength, and gas pressure are experimental variables most practicable to study. In all instances the rates of CF_2 production closely followed the rise time of the laser pulse (see Fig. 24). Thus, the initial slope giving the rise of CF_2 density with time cannot provide a measure for the unimolecular decomposition rate of those reactant molecules which have absorbed enough energy to dissociate. What the experiment is sensitive to, and what is predicted by theory,[23,32] is the rate at which the reactant absorbs CO_2 laser radiation. Our calculations suggest (as expected) that those molecules which have enough energy to decompose do so at a rate far in excess of the observed average *macroscopic rate*. The results on CF_2HCl (Fig. 14) indicate that the main effect of off-resonance excitation was to decrease both f_p, as determined by the low-pressure points, and the rate of radiative excitation, as reflected under conditions of complete rotational hole filling at higher pressures. Under identical irradiation conditions, f_p and the rates of excitation are lower for the off-resonance line, due to significantly lower absorption cross sections.

We have measured the rate of MPD of CF_2CFCl as a function of CO_2 laser fluence and Ar pressure for a single CO_2 laser line, the $P(14)$, which is exactly resonant with an intense CF_2CFCl absorption. Fig. 15 shows results for four values of CO_2 laser fluence (5·8, 7·2, 12·2, and 37·1 J cm^{-2}, corresponding to $I_{max} = 35$, 47, 73, and 220 MW cm^{-2}). For the most intense pulse, there is a barely significant initial increase in rate with Ar pressure up to about 100 Torr, followed by a gradual decrease in rate by a factor of 3 as the pressure is increased to 700 Torr. For the lower intensity data, there is less initial increase in reaction rate with Ar pressure, while the decrease in rate at high Ar pressures is dramatic. In the low-pressure (collision-unimportant) limit, the CF_2CFCl

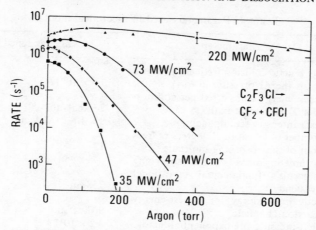

Fig. 15. Rates of production of CF_2 during the dissociation of CF_2CFCl by CO_2 laser pulses. The rates were determined for the CF_2CFCl very dilute in Ar. The pressure dependence was determined at four values of peak laser intensity, as indicated, with the laser operating on the $P(14)$ transition at $1052\,cm^{-1}$. At pressures above 100 Torr, there occurs a pronounced decrease in production rate due to energy loss to the buffer gas in collisions. The full curves are theoretical fits to the experimental data obtained using the model described in Sec. II. An energy loss per collision of $\Delta E = 2{\cdot}6\,kcal\,mol^{-1}$ was obtained from this fit.

reaction rate decreases only by a factor of 5 as the I_{max} drops from 220 to $35\,MW\,cm^{-2}$. At a pressure of 200 Torr, however, the $I_{max} = 35\,MW\,cm^{-2}$ rate is more than 10^4 times less than that for $I_{max} = 220\,MW\,cm^{-2}$.

The full curves in Fig. 15 are the best theoretical fit to the CF_2CFCl rate data obtained using the theory[23] discussed in Sec. II and the parametric values included in Table I. Of the eight parameters listed in Table I, three (s, $\langle w \rangle$, and χ) are known *a priori* from the properties of the absorbing molecule and the laser frequency. The two parameters E_{act} and A_∞ are, in principle, independently obtainable from thermal or shock-wave excitation experiments. The problem of determining 'best-fit values' for the parameters δ and ΔE is enormously simplified by the fact that, at Ar pressures above 100 Torr or so, the rotational hole filling is essentially complete, making the theoretically calculated rates essentially independent of f_p. We therefore assume $f_p = 1$ and adjust the two parameters δ and ΔE to obtain the best fit possible for the high-pressure ($>$ 100 Torr) experimental data. For these pressures, the dominant collisional effect is deactivation of highly excited molecules by V–T transfer of energy to the heat bath. The fact that the parameter δ is a scaling factor for the intensity I_{max} also simplifies the fitting problem. In effect, to make the high-pressure fit, one needs a database of theoretically calculated rates for various values of just three parameters, δI_{max}, ΔE, and P_{Ar}.

TABLE I. Parameters used in theoretical model.

Parameter		Value assigned
$1/\tau$	gas kinetic collision frequency (where P_{Ar} is the Ar pressure in torr)	$7\cdot5 \times 10^6 (P_{Ar}) \, \text{s}^{-1} \, \text{Torr}^{-1}$
ΔE	mean energy transferred per collision to a $T = 0\,\text{K}$ heat bath	$2\cdot6 \, \text{kcal mol}^{-1}$
δ	laser intensity scaling factor	$8 \times 10^{-6} (\text{MW cm}^{-2})^{-1}$
s	number of vibrational degrees of freedom of the absorbing molecule	12
$\langle \omega \rangle$	arithmetic mean of the absorbing molecule's fundamental vibrational frequency	$758 \, \text{cm}^{-1}$
E_{act}	activation energy for lowest-energy reaction channel	$96 \, \text{kcal mol}^{-1}$
A_∞	high-pressure pre-exponential factor	$3\cdot2 \times 10^{16} \, \text{s}^{-1}$
χ	laser frequency	$1052 \, \text{cm}^{-1}$

In the absence of collisions (i.e. $P_{Ar} = 0$) the theoretically calculated rates are proportional to f_p. Therefore, using the parameters ΔE and δ that are obtained from the high-pressure fit, one extracts, for each intensity, a value

$$f_p = \frac{\text{experimental rate}}{\text{theoretical rate } (f_p = 1)} \bigg|_{P_{Ar} = 0}.$$

In this way the f_p's were obtained. Theoretical values of the rate at intermediate pressures $0 < P_{Ar} < 100 \, \text{Torr}$ are then obtained by solving the full set of rate equations (Eq. (8)), using an f_p as determined from the zero-pressure results. As described earlier, f_p characterizes in one parameter the pumping through the low-lying discrete levels. We regard the extraction of f_p values for various molecules as a potentially very useful outcome of this type of analysis of pressure-dependent experiments. The experimentally observed intensity dependence of f_p for CF_2CFCl can be approximated as

$$f_p \sim \sqrt{I_{max}},$$

consistent with the notion that f_p is dependent on the Rabi frequency (Ω) and closely related power-broadening phenomena, where

$$\Omega \sim \sqrt{I_{max}}.$$

An important aspect of the MPD rate fit is the sensitivity of the measured rate to the net microscopic absorption cross section for radiative pumping. The extensive study of the buffer gas pressure and laser intensity dependences in the CF_2CFCl MPD (the results of which are shown in Fig. 15) uniquely determines the magnitude and energy scaling of the radiative pumping rates (see Fig. 16). Knowing these pumping rates exactly, one can calculate both

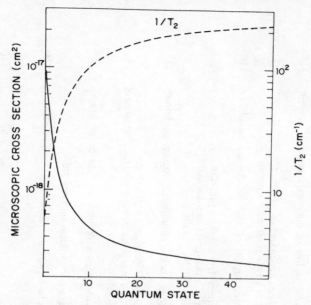

Fig. 16. Relative pumping cross sections and $1/T_2$ dephasing rates for CF_2CFCl MPD as a function of quantum number (for values of the model parameters listed in Table I). (The calculations were carried out by J. Stone, M. F. Goodman, and E. Thiele.)

energy deposition into a system and the distribution of this absorbed energy (Section IIE).

VI. PRODUCT STATE DISTRIBUTIONS

The most dramatic point to be made regarding detailed studies of excitation in region III is the dynamics of the competition between radiative excitation and microscopic decay. Although there is still no clear evidence for energy localization effects, pseudoselectivity is readily observed and understood on the basis of accepted energetics. Multiphoton excitation produces, in a short time frame, a distribution of energized reactants. From detailed studies of the distributions of energy in the nascent products, it will be possible both to infer the dynamical effects of the exit channel potential surface and, to a certain extent, to test the applicability of currently accepted theories of unimolecular reaction to the MPD process occurring on a nanosecond timescale in the absence of collisions. There have been three distinct types of measurements of energy distributions in MPD products (see Table II to V). These include molecular beam measurements of product velocity distributions, spontaneous infrared luminescence determinations of nascent vibrational distributions,

TABLE II. Energy distribution in products of MPD.

Reactant	Product	Measured distribution	Comments	Reference
	Laser-excited fluorescence			
CH_3NC	$\tilde{X}^2\Sigma^+$ CN	$T_R = 912 \pm 45$ K $T_v = 1350 \pm 100$ K	Did not probe region of uniform fluence or known intensity.[a] Product energies dependent on pressure.[b] T_v the result of a two-point measurement.	116
CH_3CN	$\tilde{X}^2\Sigma^+$ CN	$T_R = 585 \pm 62$ K $T_v = 840$ K	Did not probe region of uniform fluence or known intensity. Product energies dependent on pressure. T_v the result of a two-point measurement. Compare with CH_3NC results.	116
	$\tilde{X}^2\Sigma^+$ CN	$T_R(v=0) = 630\text{–}730$ K $T_R(v=1) = 809$ K $T_v = 793 \pm 200$ K	Did not probe region of uniform fluence or known intensity. Product energies dependent on pressure. T_v the result of a two-point measurement (at a pressure of 100 mTorr).	117
	$\tilde{X}^2\Pi$ CH	$T_R(v=0) = 622 \pm 113$ K	Did not probe region of uniform fluence or known intensity. Product energies dependent on pressure.	117
CH_3OH	$\tilde{X}^2\Pi$ OH	$T_R = 1250$ K at 100 ns $T_R = 400$ K at 1600 ns $T_T \sim 300$ K	Did not probe region of uniform or known intensity.	116
CH_3NH_2,[e]	$\tilde{X}^2B_1$ NH$_2$	$T_R(v=0,1) \simeq 500$ K $T_T(v=0) = 760 \pm 150$ K $T_T(v=1) = 960 \pm 110$ K	Cross laser-molecular beam; cannot define laser intensity. T_T measured by time-of-flight mass spectrometry.	118, 119
	$\tilde{X}^2B_1$ NH$_2$	$T_R(v=0) = 400$ K	Did not probe region of uniform fluence or known laser intensity.	120

Molecule	Species	Temperatures	Comments	Ref.
C_2H_4	$\tilde{a}^3\Pi_u\ C_2$	$T_R \sim 1000$ K $T_v \sim 1100$ K	Did not probe region of uniform fluence or known intensity. Product energies dependent on pressure. T_v the result of three measurements.	121
$C_2H_3CN^e$	$\tilde{X}^2\Sigma^+\ CN$	$T_R = 970$ K at 160 ns $T_R = 435$ K at 3 μs	Did not probe region of uniform fluence or known intensity. At 10 mTorr probably not due to collisional effects. Time dependence arising from either (i) competition between radiative excitation and decay (intensity-dependent) or (ii) RRKM lifetime effect.	122
	$\tilde{X}^2\Sigma^+\ CN$	$T_R = 682$ K at 560 ns $T_R = 435$ K at 2·85 μs	Crossed laser–molecular beam geometry; cannot define laser intensity. Time dependence arising from either (i) competition between radiative excitation and decay (intensity-dependent) or (ii) RRKM lifetime effect.	123
	$\tilde{X}^2\Sigma^+\ CN$	$T_R = 1000 \pm 200$ K $T_v \lesssim 1000$ K $E_T \sim 2\cdot3$ kcal mol^{-1}	Did not probe region of uniform fluence or known intensity. T_R observed to be constant with time delay contrary to refs. 122 and 123. No $\tilde{a}$ CH$_2$, $\tilde{X}$ C$_3$ or $\tilde{X}$ CH observed.	124
	$\tilde{a}^3\Pi_u\ C_2$	$T_R = 700 \pm 200$ K	Did not probe region of uniform fluence or known intensity. C_2 source presumed $C_2CN \rightarrow C_2 + CN$. Total excess energy in C_2CN estimated at 8·2 kcal mol^{-1}.	124
	$\tilde{a}^3\Pi_u\ C_2$	$T_T(\text{c.m.}) \sim 360$ K	Did not probe region of uniform fluence or known intensity. Optically detected time-of-flight technique. Recoil energy of 1·1 kcal mol^{-1}	125

TABLE II. (Contd.)

Reactant	Product	Measured distribution	Comments	Reference
$CF_2Cl_2{}^e$	$\tilde{X}^1A_1\,CF_2$	$T_R = 550 \pm 50$ K $T_v = 1050 \pm 100$ K $T_T = 510$ K	Probing region of uniform fluence in collision-unimportant pressure regime. T_v measurement of many levels. T_R and T_T independent or vibrational state. Recoil energy $1.5\,\text{kcal mol}^{-1}$.	57, 68
$CF_2Br_2{}^e$	$\tilde{X}^1A_1\,CF_2$	$T_R = 450 \pm 25$ K $T_v = 790 \pm 70$ K $T_T = 1130$ K	As above for CF_2Cl_2. Recoil energy $3.5\,\text{kcal mol}^{-1}$.	57, 68, 126
CF_2HCl^e	$\tilde{X}^1A_1\,CF_2$	$T_R \sim 2000$ K $T_v = 1160 \pm 100$ K $T_T = 2300$ K	As above for CF_2Cl_2. Bending (v_2) and stretching (v_1, v_3) manifolds equilibrated in the absence of collisions (calibrated detection apparatus). Recoil energy $3.5\,\text{kcal mol}^{-1}$.	57, 68
CF_2CFCl^f	$\tilde{X}^1A_1\,CF_2$	$T_R = 1550 \pm 150$ K $T_v(v_2) = 1860 \pm 250$ K $T_v(v_1,v_3) \sim 1100$ K $E_T = 0.4\,\text{kcal mol}^{-1}$	As above for CF_2Cl_2. $T_v(v_2)$ measurement of 9 levels giving Boltzmann distributions. $T_v(v_1,v_3)$ determined by knowing $T_v(v_2)$ and vibrational partition function. Bending (v_2) and stretching (v_1,v_3) modes *not* equilibrated!	126
	$\tilde{X}^1A_1\,CFCl$	$T_R \geqslant 400$ K $T_v(v_2) = 1550 \pm 300$ K $T_v(v_1,v_3) \sim 900$ K $E_T = 0.3\,\text{kcal mol}^{-1}$	Probing region of uniform fluence in collision-unimportant pressure regime. $T_v(v_2)$ result of only two-point measurement. T_R experimentally $\geqslant 400$ K; assumed to be about 1550 K due to similarities between CF_2 and CFCl. Greater degree of vibrational excitation in CF_2CFCl fragment.	126
	CF_2CFCl	$\langle E_{excess}\rangle_{ave} \sim 19.2\,\text{kcal mol}^{-1}$	Total mean excess energy absorbed by the CF_2CFCl, available for product excitation.	

			Comments	Ref.
NO_2	$\tilde{X}^2\pi NO$	$T_R \sim 300$ K $T_v \lesssim 240$ K	Did not probe region of uniform excitation; one visible plus two or three infrared photons. T_R determined by visual inspection; T_v by comparison with impurity NO in the NO_2.	127

Time-of-flight mass spectrometry[c]

Performed in molecular beam, collision-free, does not allow probing regions of uniform excitation.
No real-time information.
Cannot, in general, quantitatively relate signal strengths from different chemical fragments.
Effects of fragment internal excitation on ionization cross section unknown.
Can, however, distinguish primary and secondary photolysis processes.

			Comments	Ref.
CF_3Cl	Cl, CF_3	$\langle E_T \rangle = 1 \cdot 1\,\text{kcal mol}^{-1}$	Detected as Cl^+, CF_2^+, and CF^+; only minor CF_3^+ peak.	70
CF_3Br	CF_3	$\langle E_T \rangle = 1 \cdot 2\,\text{kcal mol}^{-1}$	Detected as CF_2^+, CF^+.	70
CF_3I	CF_3	$\langle E_T \rangle = 1 \cdot 1\,\text{kcal mol}^{-1}$	Detected as CF_2^+, CF^+.	70
CF_2Cl_2	CF_2Cl, Cl	$\langle E_T \rangle = 2 \cdot 0\,\text{kcal mol}^{-1}$	Detected as CF_2Cl^+, CF_2^+, and CF^+ and Cl^+. Did not see any Cl_2^+ above background; places 10% limit on Cl_2 formation.	70
CF_2Br_2	CF_2Br, Br	$\langle E_T \rangle = 1 \cdot 6\,\text{kcal mol}^{-1}$	Detected as CF_2Br^+, CF_2^+, and CF^+ and Br^+. Br_2 less than 10% level of detection limit.	70
$CFCl_3$	$CFCl_2, Cl$	$\langle E_T \rangle = 1 \cdot 2\,\text{kcal mol}^{-1}$	Detected as $CFCl_2^+$, $CFCl^+$, and CF^+ and Cl^+. No primary Cl_2^+ observed. Secondary photolysis occurring above $10\,\text{J cm}^{-2}$.	70
N_2F_4	NF_2	$\langle E_T \rangle = 0 \cdot 4\,\text{kcal mol}^{-1}$	Detected as NF_2^+, NF^+, and F^+; no $N_2F_3^+$ or $N_2F_2^+$.	70
SF_6	SF_5	$\langle E_T \rangle < 1 \cdot 5\,\text{kcal mol}^{-1}$	Detected as SF_4^+.	128

TABLE II. (*Contd.*)

Reactant	Product	Measured distribution	Comments	Reference
CHF_2Cl	CF_2, HCl	$\langle E_T \rangle = 8 \pm 3\,\text{kcal mol}^{-1}$	Detected as CF^+ and as Cl^+ and HCl^+. Three-centre elimination. Velocity distribution peaked at non-zero value.	59
$CHFCl_2$	CF_2, HCl	$\langle E_T \rangle = 8 \pm 3\,\text{kcal mol}^{-1}$	Detected as CF^+ and as HCl^+ and Cl^+; no $CHFCl^+$, $CFCl^+$ or HF^+ observed. Velocity distribution peaked at non-zero value. Three-centre elimination.	59
CH_3CCl_3	CH_2CCl, HCl	$\langle E_T \rangle = 8 \pm 4\,\text{kcal mol}^{-1}$	Detected as CH_2CCl^+ and as HCl^+. No CH_3^+, $CH_3CCl_2^+$, CH_3CCl^+ or $CH_2CCl_2^+$ observed. Four-centre elimination.	59
CH_3CF_2Cl	HCl	$\langle E_T \rangle = 12 \pm 4\,\text{kcal mol}^{-1}$	Detected as HCl^+. Velocity distribution peaked at non-zero value. Four-centre elimination.	59
$CHClCF_2$	HCl, C_2F_2	$\langle E_T \rangle = 1\text{-}1{\cdot}5\,\text{kcal mol}^{-1}$	Detected as HCl^+ and Cl^+ and as $C_2F_2^+$ and CF^+. Velocity distribution peaked at zero unlike other three- or four-centre eliminations. Three-centre elimination.	59
Infrared luminescence[d]			Spontaneous luminescence studies do not allow the observation of regions of uniform fluence, nor the determination of real-time response. No information can be gained regarding the relative population of the product species vibrational ground state.	
C_2H_3F	HF	$HF(1) > HF(2) > HF(3)$ $\langle E_v \rangle_{ave} \leqslant 10\,\text{kcal mol}^{-1}$	Non-Boltzmann vibrational distribution, no population inversions. Vibrational distribution obtained with addition of 12 Torr He to effect rotational equilibration.	129, 130

Reactant	Product		Observations	Ref.
		No C_2H_2 or C_2H_3F luminescence observed. Average HF vibrational energy determined in prescence of He buffer.		
C_2H_5F	HF	$HF(1) > HF(2) > HF(3)$ $T_R = 800$ K neat $\langle E_v \rangle_{ave} \leqslant 12\,\text{kcal mol}^{-1}$	Non-Boltzmann vibrational distribution, no population inversions. Vibrational distribution obtained with addition of 12 Torr He to effect rotational equilibration. No apparent change in $\langle E_v \rangle_{ave}$ between 0.2 Torr neat and with addition of 12 Torr He.	130
CF_2HCH_3	HF	$HF(1) > HF(2) > HF(3)$ $T_R = 1000$ K neat $\langle E_v \rangle_{ave} \leqslant 11\,\text{kcal mol}^{-1}$	Non-Boltzmann vibrational distribution, no population inversions. Vibrational distribution obtained with addition of 12 Torr He to effect rotational equilibration.	130
C_2H_3F	HF	$HF(1) > HF(2) > HF(3)$ $\langle E_v \rangle_{ave} \leqslant 10\,\text{kcal mol}^{-1}$	Non-Boltzmann vibrational distribution, no population inversions. Vibrational distribution obtained with addition of 12 Torr He to effect rotational equilibration.	130
$1,1\text{-}C_2H_2F_2$	HF	$HF(1) > HF(2) > HF(3)$ $\langle E_v \rangle_{ave} \leqslant 9\,\text{kcal mol}^{-1}$	Non-Boltzmann vibrational distribution, no population inversions. Vibrational distribution obtained with addition of 12 Torr He to effect rotational equilibration.	130

[a] It is inherent in MPD that the degree of reactant excitation, and therefore the amount of energy available to product excitation, will be critically dependent on laser intensity.

[b] In experiments giving pressure-dependent results, the reported 'temperature' generally is observed to increase with increasing pressure. The results tabulated here refer to the lowest-pressure results reported.

[c] See also Table III.

[d] See also Table IV.

[e] Many experiments have utilized the very sensitive LEP probing of product state distributions. Most of these, probe regions of ill-defined laser fluence and intensity, and often at pressures too high to claim the observations to be characteristic of collision-unimportant excitation and dissociation processes. Those few experiments performed under suitably controlled conditions that should give meaningful insight into MPE and unimolecular decay are indicated by symbol[e].

[f] Only this one experiment has fully analysed the total energy distributions of both separating fragments in an MPD experiment.

and visible–ultraviolet laser-excited fluorescence (LEF) determinations of electronic, vibrational, and rotational states distributions and average kinetic energy.

There is currently no theoretical basis for predicting *a priori* how the available energy in a reactant will be partitioned into the products' degrees of freedom. In fact *the distribution of energy among the various vibrational modes and rotational and translational degrees of freedom is not particularly sensitive to the details of laser excitation or relaxation processes in the reactant, but is exceedingly sensitive to the potential energy surfaces along which the reaction occurs.*[136] For the case of simple bimolecular reactions, it has been possible to understand why energy preferentially goes into particular product quantum states or degrees of freedom.[136] A similar understanding should be possible for these laser-induced unimolecular reactions of larger molecules, for which such data are becoming increasingly available (see Table II).

A. Translational Energy Distributions

The molecular beam experiments can provide detailed information on both velocity and angular distributions. This information allows the distinction between primary and secondary photolysis processes and the determination of the actual distribution of translational energy. It also, in principle, provides the basis for deducing the dynamics of the MPD process. Lee and his coworkers have examined two types of unimolecular decay in great detail: (1) simple bond rupture[70] and (2) three- and four-centre eliminations.[59] It is their claim throughout[70] this work that 'the dissociation of all these molecules proceeds in accordance with a statistical unimolecular dissociation model.' This simply implies that T_1 times are very short for these highly excited reactants. For cases of simple bond rupture, they examined the MPD of CF_3Cl, CF_3Br, CF_3I, CF_2Cl_2, CF_2Br_2, $CFCl_3$, and N_2F_4 under molecular beam conditions. The experimental results of this work are summarized in Table III. Briefly, all seven species were readily dissociated by moderate *fluence* pulses of $1-30$ J cm^{-2}, showing a significant preference for simple bond rupture. Only in the two cases of CF_2Br_2 and CF_2Cl_2 would the molecular elimination be expected to compete, based on endothermicity, and in these two cases Sudbo *et al.*[70] admit to not having sufficient sensitivity to place a precise value on the branching ratio for these two potentially competitive channels, their detection limit being at about the 10% level.

In these seven instances of simple unimolecular bond rupture, the product species showed uniformly low (i.e. a few kilocalories per mole) average translational energies with the actual distribution being of an isotropic nature and peaking near zero excess energy. In applying their simplified RRKM scheme, *assuming equilibration prior to dissociation*, the authors calculated both the approximate amount of excess energy required to give the observed

TABLE III. Molecular beam IR MPD results.[59,70]

Parent	Reaction	Experimental results				Best RRKM fit		Energy barrier in back-reaction (kcal mol^{-1})
		Fluence (J cm^{-2})	Ave. trans. energy (kcal mol^{-1})	FWHM spread (kcal mol^{-1})	Peak near zero	Ave. excess energy (kcal mol^{-1})	τ (s)	
CF_3Cl	$CF_3 + Cl$	5	1·1	<1	yes	4	5×10^{-9}	no
CF_3Br	$CF_3 + Br$	12	1·2	<1	yes	5	2×10^{-9}	no
CF_3I	$CF_3 + I$	30	1·1	<1	yes	4	1×10^{-9}	no
CF_2Cl_2	$CF_2Cl + Cl$	4	2·0	<1	yes	10	5×10^{-9}	no
CF_2Br_2	$CF_2Br + Br$	8	1·6	<1	yes	7	5×10^{-9}	no
$CFCl_3$	$CFCl_2 + Cl$	5	1·2	<1	yes	5	12×10^{-9}	no
N_2F_4	$2NF_2$	4	0·4	<1	yes	2	1×10^{-9}	no
CHF_2Cl	$CF_2 + HCl$	—	8	6	no	—	—	6
$CHFCl_2$	$CFCl + HCl$	—	—	—	no	—	—	~0
CF_2CHCl	$C_2F_2 + HCl$	—	1	1	yes	—	—	55
CH_3CF_2Cl	$CH_2CF_2 + HCl$	—	12	8	no	—	—	55
CH_3CCl_3	$CH_2CCl_2 + HCl$	—	8	8	no	—	—	42

time-of-flight results and the corresponding RRKM lifetime for this average state. In these studies, for fluences in the range 4–30 J cm^{-2} the average RRKM rates calculated were in the range $(0.8 - 10) \times 10^8 5^{-1}$. For CF_2Cl_2, CF_2Br_2, and N_2F_4 the authors looked at product translational distributions for different fluence levels. No significant changes were observed for changes in exciting fluence of up to a factor of 4. This is surprising in view of the earlier arguments regarding competitive processes in the dissociative region (see Sec. V). However, two points must be borne in mind: (1) these results were obtained under experimental conditions that integrated laser intensities over all space and time, thereby weighting the observations against detecting significant intensity-dependent results, and (2) it is the authors' contention that the translational energy is determined primarily by the exit channel potential surface and any barrier to the reverse association reaction and is not very sensitive to the amount of energy absorbed by the molecule over and above the minimum requirement for dissociation. Since the observed velocity distributions are peaked near zero, there must be small exit channel barriers and presumably very little rotational energy in the activated complex. Otherwise the centrifugal effect would suppress the low-energy portion of the translational energy distribution. In certain instances, the RRKM calculations[70] require a significant amount of excess energy to be absorbed by the reactant to give a microscopic decay rate in the range of 10^8 s^{-1}. This excess energy would then be available for internal excitation of the dissociation products. Specifically, for CF_2Cl_2 and CF_2Br_2 there is expected to be about 8 and 5.4 kcal mol^{-1} for excitation of the fragments CF_2X (X = Cl or Br respectively). It is exactly for these two cases that secondary photolysis of the hot, nascent fragments has been observed (see Sec. V).

The results from three- and four-centre elimination reactions[59] are significantly different from those for simple bond rupture. Most notable is the large amount of translational energy observed in the fragments (see Fig. 17). This translational energy is derived, in part, from the exit channel barrier. In both cases of four-centre elimination (CH_3CF_2Cl and CH_3CCl_3) an amount of energy equal to 20% of the exit channel barrier was found in product kinetic energy. Not only are the translational distributions for the multicentre reactions peaked at non-zero energies, but they are fairly broad (6 to 8 kcal mol^{-1} FWHM compared with 0.5 to 1 kcal mol^{-1} for the simple bond case). Under these conditions, it is impossible to use simple RRKM formulations even to attempt a derivation of the average excess energy absorbed. Nonetheless, in these three- and four-centre elimination reactions the translational information derived from these types of experiments will, eventually, yield information on the nature of the potential surface of the unimolecular reaction. The actual physical presence of the potential energy barrier to back-reaction implies that there is considerable interaction between the fragments even after the critical configuration is passed. Thus the

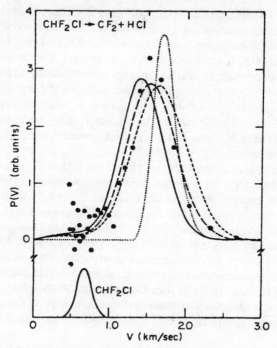

Fig. 17. Speed distribution of 10° of HCl fragments from CF$_2$HCl: experiment (●); RRKM distribution (...) calculated for a 1·5 ns dissociation lifetime; other curves for statistical distributions with average translational energies of 6 (——), 8 (— — —), and 10 (- - -) kcal mol^{-1}. After Aa. S. Sudbo, P. A. Schultz, Y. R. Shen and Y. T. Lee.[59]

partitioning between vibrational, rotational, and translational degrees of freedom of the energy available to the fragments cannot be predicted without further modelling of the exit channel potential energy surface.

This point seems particularly poignant for the CF$_2$HCl case. Lee's RRKM calculations (assuming no exit channel barrier) place the amount of translational energy associated with the reaction coordinates in the critical configuration at 2–4 kcal mol^{-1}. Experimentally, the average amount of energy released to translational energy was 8 kcal mol^{-1}. Clearly, in this case, the bulk of the potential energy of the back-reaction barrier (6 kcal mole^{-1}) must have been converted into translational energy. This measurement must reflect the character of the potential surface along which the fragments separate rather than the intramolecular dynamics of the excited parent molecule. The main effect of this barrier appears to be to provide a repulsive impulse to the two separating fragments. The LEF studies of

Stephenson and King[68] on CF_2HCl MPD gave the same 3·5 kcal mol^{-1} value for the CF_2 average translational energy (for a non-zero peaked distribution) and characteristic vibrational temperature of 1160 K and rotational temperature of about 2000 K for the nascent CF_2 fragments. To the extent that this non-statistical partioning of available energy reflects effects of the exit channel, one might simply say that the repulsive force between the separating CF_2 and HCl was non-central in nature thus giving a large degree of angular momentum to the separating fragments, and that while in the critical configuration the CF_2 and HCl are not greatly distorted from their equilibrium conditions.

For the four-centre elimination of HCl from CH_3CCl_3 a simple RRKM calculation[59] based on an average decomposition rate of 10^8 s^{-1} gave an average excess energy of about 22 kcal mol^{-1}. When added to the 42 kcal mol^{-1} barrier it appears that the measured average translational energy (i.e. 8 kcal mol^{-1}) only represents 10–15% of the total excess, and 20% of the exit channel barrier. This is somewhat lower than the figure of 30% based on the work by Setser[42] on four-centre eliminations. This *rough* calculation indicates that approximately 85% of the total energy available to the products in these unimolecular reactions transforms efficiently into internal excitation of the molecular fragments. Experimental measurements of the partitioning of this available energy are now becoming available for validation of model predictions.

As beautifully demonstrated by Lee's group, the molecular beam method of study for MPD gives fragment translational energy distributions. With the help of RRKM theory, estimates can be made of the excess energy released in the dissociation, provided of course that there is no significant degree of energy localization. Unfortunately, for reactions more complicated than simple two-centre atomic elimination, the translational energy distribution alone is not sufficient to characterize the dynamics of the reaction. The mass spectrometric experiment cannot give any information regarding internal energy content nor distribution in the fragments. This is particularly limiting for reactions exhibiting back-reaction barriers. In order fully to understand the dissociation dynamics, time-of-flight and angular distributions obtained by molecular beam–mass spectrometric techniques must be complemented by techniques sensitive to internal excitation such as spontaneous IR luminescence absorption or laser-excited fluorescence.

B. Vibrational Distributions

Quick and Wittig[129–131] report the results of a series of measurements of energy partitioning in HF fragments formed by MPD using spontaneous IR luminescence. Typical infrared radiative lifetimes are a few to tens of milliseconds. Thus, although it is straightforward to have collision-free MPD, it is

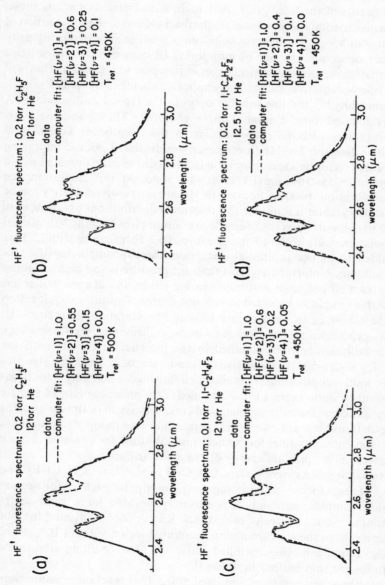

Fig. 18. HF spontaneous fluorescence following MPD under the conditions listed in the figure. The laser fluence was 40 J cm^{-2}; the computer fits assume the rotational temperature, T_{rot} to be equivalent for all vibrational levels. After C. R. Quick, Jr, and C. Wittig.[130]

(a) HF* fluorescence spectrum: 0.2 torr C_2H_3F
12 torr He
—— data
- - - computer fit: $[HF(\nu=1)] = 1.0$
$[HF(\nu=2)] = 0.55$
$[HF(\nu=3)] = 0.15$
$[HF(\nu=4)] = 0.0$
$T_{rot} = 500K$

(b) HF* fluorescence spectrum: 0.2 torr C_2H_5F
12 torr He
—— data
- - - computer fit: $[HF(\nu=1)] = 1.0$
$[HF(\nu=2)] = 0.6$
$[HF(\nu=3)] = 0.25$
$[HF(\nu=4)] = 0.1$
$T_{rot} = 450K$

(c) HF* fluorescence spectrum: 0.1 torr I_1-$C_2H_4F_2$
12 torr He
—— data
- - - computer fit: $[HF(\nu=1)] = 1.0$
$[HF(\nu=2)] = 0.6$
$[HF(\nu=3)] = 0.2$
$[HF(\nu=4)] = 0.05$
$T_{rot} = 450K$

(d) HF* fluorescence spectrum: 0.2 torr I_1-$C_2H_2F_2$
12.5 torr He
—— data
- - - computer fit: $[HF(\nu=1)] = 1.0$
$[HF(\nu=2)] = 0.4$
$[HF(\nu=3)] = 0.1$
$[HF(\nu=4)] = 0.0$
$T_{rot} = 450K$

much harder to work at pressures sufficiently low as to ensure monitoring of the products prior to any collisions. Unfortunately measurements of spontaneous luminescence integrate intensity effects occurring over the entire spatial and temporal profile of the laser pulse. Although, measurements such as these, being sensitive to total energy content in the reactant and to the competition of excitation with V − T relaxation in collisions, should show a strong intensity dependence, observations made by monitoring IR luminescence will not give any information on intensity dependences in pumping in region III. They will however offer qualitative insight into applicable chemical dynamics.

Experimentally,[130] the line-tunable output of a typical multimode TEA laser was focused into a large sample chamber. The spontaneous IR luminescence was detected perpendicular to the laser beam with a InSb photovoltaic detector. The HF fluorescence was dispersed prior to impinging on the photodetector by a low-resolution IR monochromator, operated with a resolution of 0.035–0.040 μm (FWHM) at 2.5 μm. At this resolution, the individual vibration–rotation lines of the HF are not resolved, as they can be at significantly greater resolution. The rotational distributions obtained were restricted to computer best fits of envelope contours (see Fig. 18). Vibrational level populations were obtained at a pressure of 0.2 Torr reagent with 12 Torr of He added to induce rotational relaxation, thereby simplifying the vibrational analysis. Unfortunately, this reactant pressure is too high to ignore reactant–reactant collisions and bulk heating effects and at these buffer gas pressures there might be expected to be some degree of vibrational relaxation during the HF 10 μs IR fluorescence lifetime (the vibrationally excited HF fragments experiencing some 300 gas kinetic collisions before fluorescing). Indeed, rotational contours obtained in the presence of the He bath gas showed $T_{rot} \sim 450$–500 K, indicative of moderate bulk heating. The vibrational level populations obtained following essentially collision-unimportant excitation versus those obtained in the presence of the 12 torr He were very similar, sufficiently so, that valid conclusions can be drawn from the level populations obtained under these conditions. Due to low signal-to-noise ratio it was not possible for Quick and Witting to examine fluence dependences on the product energy or energy distribution.

Four reactant species were studies, CH_2CHF, CH_3CH_2F, CH_3CHF_2, and CF_2CH_2. These species all decompose *primarily* through a four-centre elimination complex, although it is certainly possible for a three-centre elimination to occur in three of these cases. Kim et al.[132] indicated that for 1, 1-difluoroethane the α,α elimination accounted for only about 10% of the total dissociation following chemical activation. The resulting vibrational distributions are summarized in Table IV.

The rotational 'temperatures' obtained at 0.2 Torr reactant pressure were uniformly in the range 800 to 1000 K. Upon addition of 12 Torr He the rotational distribution was cooled to 450–500 K and the vibrational

TABLE IV. HF level populations obtained by IR luminescence.

Reactant	Excitation method	Relative HF(v) population					Reference
		$v=0$	1	2	3	4	
CH_2CHF	IR MPD	—	1·0	0·55 ± 0·1	0·15 ± 0·05	< 0·05	130
CH_2CF_2	IR MPD	—	1·0	0·40 ± 0·1	0·10 ± 0·05	< 0·05	130
	Triplet Hg	—	1·0	0·40 ± 0·02	0·21 ± 0·01	0·072 ± 0·011	134
CH_3CH_2F	IR MPD	—	1·0	0·60 ± 0·1	0·25 ± 0·05	< 0·15	130
CH_3CHF_2	IR MPD	—	1·0	0·60 ± 0·1	0·20 ± 0·05	< 0·10	130
CH_3CF_3	Chemical activation	—	1·0[a]	0·43 ± 0·02	0·13 ± 0·007	0·033 ± 0·007	134
	Chemical activation	1·25 ± 0·1	1·0[a]	0·55 ± 0·08			135
	Chemical activation	1·43 ± 0·2	1·0[b]	0·55 ± 0·08			135

[a] From $CH_3 + CF_3$
[b] From $H + CH_2CF_3$

distribution analysed. In all cases the vibrational level populations decreased with increasing value of vibrational quantum number; however, *the vibrational level populations could not be characterized by a simple Boltzmann distribution*. In most cases the ratio $HF(v = 2)/HF(v = 1)$ was 0·4–0·6. No information on the $v = 0$ level can be obtained by this technique. In the case of CH_2CF_2 the MPD results are very similar to results obtained by triplet Hg sensitized decomposition (for CH_2CF_2, triplet sensitization and MPD provide the reactant with approximately the same amount of energy, 112 kcal mol^{-1}, which is 28 kcal mol^{-1} above the minimum amount required for dissociation).

Berry[133] has shown that a statistical treatment for the description of the HF or HCl level populations resulting from four-centre elimination from halogenated ethanes and ethylenes excited by chemical activation, triplet Hg sensitization, and UV photolysis is inadequate. The excess energy[134], which on the basis of RRKM theory is statistically distributed within the transition state, is simply not sufficient to populate the high vibration levels of HF which are observed. The relative level population obtained by Quick and Witting[130] in MPD are believed, by the authors, to represent the composite result of partitioning of the *excess* energy (i.e. that amount of energy in excess of the activation energy) according to some probability function, and partitioning of the *localized* energy (i.e. $E_a - \Delta H$) associated with the transition state according to some other probability function. Assuming the partitioning of the excess and localized energies as independent processes and a statistical treatment to be valid for the partitioning of the excess energy (a similar assumption was made by Lee *et al.*[70] in modelling their molecular beam results when no barrier was present to the back-reaction, i.e. when the localized energy was very small), then the partitioning of the 40–60 kcal mol^{-1} of localized energy in these four-centre elimination reactions can be inferred.

For the systems studied, a statistical partitioning of the about 20–35 kcal mol^{-1} of excess energy expected for decomposition rates of about 10^8 s^{-1} would only represent a small fraction (i.e. $< 10\%$) of the excited HF level population.[133] Under the conditions that these assumptions hold and that the apparent level populations are not an artifact of laser profile and hot spots, the excitation within the HF fragments originates primarily from the localized energy,[130] i.e. the barrier to the back-reaction must couple strongly into the internal degrees of freedom of the HF pre-product. To estimate the average vibrational energy of the HF products, Quick and Witting assumed that the ratio $HF(v = 1)/HF(v = 0) = 0.8$ as was found[135] for CH_3CF_3 by chemical activation. The results thus obtained indicate that the HF fragment carries away some 15–30% of the localized energy as vibrational excitation. This is a molecule-specific factor and depends on the specific precursor. Lee *et al.*[59] in beam studies of four-centre elimination in HCl analogues found that about 20% of the localized energy was converted into translational energy, leaving some 40–60% of the localized energy and most of the excess energy available for rotational excitation of the HCl or HF and internal excitation of the larger polyatomic fragment. Such an expectation would be consistent with the energy distribution obtained in the CF_2 formed from MPD of CF_2HCl by King and Stephenson.[32]

C. Total Product Energy Distribution

For the MPD of CF_2CFCl it is possible to determine the energy distributions of both separating fragments, giving, for the first time, the total amount of energy in the products (E_{tot} relative to the vibrational ground state). If E_{tot} is measured, then the energy which the average reactant had when it dissociated, E_{react}, is known, since $E_{react} = E_{tot} + \Delta H_0$, where the enthalpy ΔH_0 is the thermodynamic minimum energy necessary for reaction. The distribution of energy in the reactant above the dissociation limit is determined by the ratio of the laser excitation rate, characterized by an energy-dependent absorption cross section $\sigma_{abs}(E)$, to the microscopic unimolecular reaction rate $\Gamma(E)$, and to the collisional vibrational relaxation rate described by a deactivation cross section $\sigma_{vt}(E)$. Measurements of E_{tot} as a function of laser intensity and buffer gas pressure probe the relative size of σ_{vt}, Γ, and σ_{abs} and their dependence on the level of excitation (E). Any consistent theory of IR MPD must explain not just product yields but also product formation rates and the energy content of the products as a function of pressure and laser intensity.

1. Vibrational Energy

A detailed discussion is given in ref. 57 of the relation between the population N_v in vibrational level v, and the integrated intensity $S^{v'' \leftarrow v' \leftarrow v}(V_{v' \leftarrow v''})$ of the LEF signal resulting from UV laser excitation of the molecules initially formed in $\tilde{X} CF_2(v)$ to $\tilde{A} CF_2(v')$ followed by detection of fluorescence on the $\tilde{A} CF_2(v') \rightarrow \tilde{X} CF_2(v'')$ vibronic transition. The relative populations in the different vibrational levels were determined by LEF techniques both under conditions where (a) the observed CF_2 fragments had not suffered any collisions (neat reactant at a pressure of 10 mTorr, a delay time τ_D of 0.2 μs) and, therefore, retained their nascent rotational energy distribution, and (b) rotational, but not vibrational, relaxation had occurred (1 Torr He added and $\tau_D = 1 \mu$s). The measured distribution of vibrational energy in the products was the same for these two different conditions. For both reactant species, the population distributions within the CF_2 product bending mode (v_2) followed the statistical expression

$$P(E_{v_2}) = \exp[-E_{v_2}/kT_v(v_2)],$$

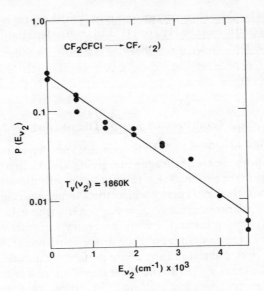

Fig. 19. Probability of CF_2 product molecules being created with energy E_{v_2} in the bending vibration in the MPD of CF_2FCl. Measurements were performed by laser-excited fluorescence at a pressure of 0·01 Torr at a delay time of 300 ns. A least-squares fit of the data gives a vibrational temperature of
$$T_v = (1860 \pm 125)K.$$

defining a vibrational temperature for the $CF_2 v_2$ manifold. Populations[137-140] in the levels $v_2 = 0$ to $v_2 = 7$ were measured. For the CF_2HCl precursor $T_v(v_2) = (1160 \pm 100)K$; for CF_2CFCl, $T_v(v_2) = (1860 \pm 250)K$, for laser pulses of $50\,J\,cm^{-2}$ fluence (see Fig. 19).

We used a vibrational relaxation method[137] to estimate the amount of vibrational energy in the two stretching modes. In this method, the $CF_2 (0,0,0)$ state is probed as a function of time after the IR photolysis pulse in mixtures very dilute (10^{-5} mole fraction) in Ar. The signal S_0 at early time (e.g. $2\,\mu s$), after rotational relaxation is complete but before vibrational relaxation has occurred, is due solely to molecules initially formed in the ground vibrational level. For CF_2 in Ar, the rate of rotational relaxation is 10^5 times the rate of vibrational relaxation,[138] and product formation ceases promptly when the laser pulse terminates ($< 1\,\mu s$). At longer times, the 2_0^2 LEF signal increases due to relaxation of the vibrationally excited CF_2 species back to the vibrational ground state, until it achieves its long-time value S_∞, characteristic of $T_v = 298$ K. The details of this procedure and the vibrational relaxation rate of $\tilde{X} CF_2$ in Ar ($k_{vt} = 2 \times 10^{-15}\,cm^3\,s^{-1}\,molecule^{-1}$) are given in ref. 138. The ratio S_0/S_∞ equals the fraction f_0 of CF_2 molecules initially formed in the vibrational ground state divided by f_∞, the fraction in that state when vibrational relaxation is complete. That is

$$S_0/S_\infty = f_0/f_\infty = f_0 Q_v(T = 298\,\text{K}),$$

where $Q_v(298) = 1.04$ is the vibrational partition function of room-temperature CF_2. If the form of the initial distribution of vibrational energy in CF_2 is known, then this measurement gives the vibrational energy content of the molecule.

Specifically, if it is a Boltzmann distribution, then the initial vibrational temperature T_v is easily deduced from the relation

$$f_0^{-1} = Q_0(T_v) = Q(298)S_\infty/S_0.$$

For CF_2HCl MPD $f_0 = 0.33 \pm 0.02$, consistent with a Boltzmann vibrational distribution with $T_v = (1160 \pm 100)$ K for the nascent CF_2 as was measured for the v_2 manifold. For CF_2CFCl, if the energy in v_1 and v_3 is characterized by the same vibrational temperature as v_2, i.e. $T_v(v_2) = 1860$ K, then the result should be $f_0(1860) = 0.13$. However, for CF_2CFCl, $f_0 = 0.24 \pm 0.03$. This value corresponds to an average vibrational temperature of $T_v = (1400 \pm 100)$ K. These results suggest very strongly that *for the CF_2 from CF_2HCl all three vibrational modes are in equilibrium, whereas for the CF_2 from CF_2CFCl vibrational energy goes preferentially into the bending rather than stretching modes.*

Similar measurements were done to determine the distribution of vibrational energy in the newly formed $\tilde{X}$ CFCl product. The LEF spectrum of CFCl[140] is substantially more complicated than that of CF_2 because of the lower values of the vibrational frequencies and because transitions involving all three vibrational modes ($v_1 = 1158\,\text{cm}^{-1}$, $v_2 = 448\,\text{cm}^{-1}$, $v_3 = 750\,\text{cm}^{-1}$) are allowed by symmetry and by favourable Franck–Condon factors (resulting in the overlapping of many vibronic transitions in the excitation and fluorescence spectra of CFCl when it is vibrationally and rotationally excited). We directly measured the relative populations of CFCl$(0,0,0)$ and CFCl$(0,1,0)$ levels following the collision-free dissociation of CF_2CFCl, by exciting the 2_0^2 and 2_1^2 vibronic transitions. The results was CFCl$(v_2 = 1)/$CFCl$(v = 0) = 0.66 \pm 0.06$ at a fluence of $50\,\text{J cm}^{-2}$. This corresponds to a vibrational temperature $T_{v_2} = (1550 \pm 300)$ K. However, measurements of only two states do not prove that the energy in the bending mode of CFCl has a Boltzman distribution.

The vibrational relaxation method was used to determine $f_0 = 0.19 \pm 0.02$; if all three modes had the same vibrational temperature, this values of f_0 would correspond to $T_v = (1150 \pm 100)$ K. *This suggests that for CFCl, as for CF_2, vibrational energy preferentially goes into the bending rather than stretching modes.*

2. *Rotational Energy*

The rotational energy content of the CF_2 formed from both CF_2HCl and CF_2CFCl was sufficiently high that direct assignment of rotational quantum numbers and hence a direct measurement of the collision-free distribution of energy was not quite possible with our $1\,cm^{-1}$ spectral resolution. As described in ref. 68 we used a rotational relaxation method to determine E_{rot}. By adding a small amount of He diluent, such that the IR multiphoton dissociation process remained unaffected by collisions, and by going to successively longer delay times τ_D, we were able to observe the effects of rotational relaxation on the nascent $\tilde{X}\,CF_2$ fragments. Extrapolation back to $\tau_D = 0$ gave a rotational energy content compatable with $T_R \sim (2000 \pm 150)\,K$ for CF_2 molecules formed in the vibrational ground state from CF_2HCl and $T_R = (1550 \pm 150)\,K$ from CF_2CFCl. Analysis of the rotational contour of the $CF_2\,2_0^2$ LEF band obtained under collision-free CF_2CFCl MPD showed that there were no differences between the actual LEF spectrum and a simulated spectrum derived for a $T_R = 1550\,K$ thermal distribution. Similar rotational contours were obtained for the $CF_2\,v_2 = 0, 1$, and 5 vibrational states, showing the rotational distribution function to be independent of vibrational state.

The complexity of the CFCl LEF spectrum makes impossible the assignment of rotational quantum numbers (with our experimental resolution) even for a room-temperature thermal distribution of rotational and vibrational states. However, band contour analysis of the spectra of nascent CFCl formed from the collision-free dissociation of CF_2CFCl clearly shows that the rotational energy content of the CFCl is substantially higher than room temperature. One might assume, based on the similarity of molecular size and rotational constants, that both CF_2 and CFCl products to have similar rotational distributions following the MPD of CF_2CFCl.

3. *Translational Energy*

The measurements of the laboratory velocity of the CF_2 and CFCl fragments used the simple method described in ref. 68. At very low pressures, the LEF signals observed in our collinear geometry decrease with increasing time delay after the IR laser pulse due to collision-free gas expansion of the newly formed cylindrical cloud of CF_2 or CFCl radicals from the narrow region probed by the UV–visible laser. If the initial distribution of products is given by the source function

$$G(r) = N_0 \exp(-r^2/L^2)$$

and is probed by a collinear beam of profile

$$I = I_0 \exp(-r^2/R_p^2),$$

then the LEF signal probed at the delayed time t decays as:[137]

$$S(t) = \frac{S_{max}(L^2 + R_p^2)}{(L^2 + R_p^2 + c^2 t^2)},\tag{14}$$

where S_{max} is the maximum signal ($t \sim 0$), L is experimentally measured (L is approximately the radius of the CO_2 laser beam), and $c = (2kT/m)^{1/2}$ is the most probable velocity. Our measurements of E_T do not give any details about the specific shape of the velocity distribution function. However, since our LEF measurements are quantum-state specific, we can determine $E_T(v, J, K)$ for molecules in known internal energy states. Such state-specific data have not yet been obtained in beam studies in IR MPD.

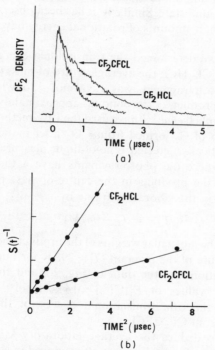

Fig. 20. (a) Laser-excited fluorescence signals from the ground-state CF_2 fragments formed in the collision-free MPD of CF_2HCl and CF_2CFCl as a function of delay between photolysis and probe pulses. The measurements were performed at a pressure of 3 m Torr, neat. (b) The inverse of the LEF signal in (a) plotted against delay time squared (see ref. 126). The slopes of the resulting lines are proportional to the laboratory kinetic energy of the probed fragments.

Fig. 20 shows translational decay curves for vibrational ground-state CF_2 formed from CF_2CFCl, and from CF_2HCl obtained under identical experimental conditions. The LEF signal (CF_2 density) decays faster for the CF_2 from the CF_2HCl than from the CF_2CFCl. According to Eq. (14), a plot of $S^{-1}(t)$ against t^2 should be linear with a slope proportional to c^2, the square of the most probable (laboratory) velocity (i.e. the slope is proportional to the laboratory temperature T_L). Fig. 20(b) shows the data plotted in this way. The slope for CF_2HCl is five times the slope for the CF_2CFCl. The same relative translation decay curves were obtained when the $v_2 = 5$ state of CF_2 formed from these reactants was probed. This indicates that CF_2 ($v_2 = 5$) molecules from CF_2CFCl and CF_2HCl, containing $3320\,cm^{-1}$ of excess vibrational energy, have the same amount of translational energy as those products formed in the vibrational ground state. Similarly, it has been observed that molecules born with low and high amounts of rotational energy have the same average translational energy.[68,126]

Sudbo et al.[59] have determined the translational energy distribution for CF_2 produced from CF_2HCl; the average centre-of-mass (c.m.) translational energy released to both fragments was $8\,kcal\,mol^{-1}$, of which $3.5\,kcal$ would belong to the CF_2 product, corresponding approximately to a temperature $T_{c.m.} = 1130\,K$. Campbell et al.[125] have used the LEF method to determine the velocity distribution of C_2 formed from CO_2 laser-induced decomposition of C_2H_3CN via optically detected time-of-flight measurements. They show that for the case where the photofragments have a thermal distribution, the temperature of the products in the centre-of-mass frame is related to the temperature T_L in the laboratory frame by

$$T_{c.m.} = T_L - (m/M)T_i,$$

where m and M are the molecular weights of the product and reactant, and T_i is the initial temperature of the reactant ($T_i = 298\,K$ for our experiments). We have used this relationship, the data of Fig. 20, and the molecular beam results[59] to derive values of $E_T(CF_2) = 0.4\,kcal\,mol^{-1}$ and $E_T(CFCl) = 0.3\,kcal\,mol^{-1}$ for the c.m. translational energy of the CF_2 and $CFCl$ fragments formed from CF_2CFCl.

This approach to deduce the average laboratory kinetic energy of the products assumes that the products are all produced in a short time, $t < L/c$. This may be a better approximation for CF_2HCl, where microscopic dissociation rates increase rapidly with energy, than for CF_2CFCl, where the distribution of lifetimes may be broader. If a significant fraction of the CF_2CFCl dissociated during the $0.5–3\,\mu s$ period shown in Fig. 20, the result would be an apparently slower product decay.

A summary of the results for CF_2CFCl MPD is given in Table V. Direct measurement in the collision-free regime showed that the energy in the

TABLE V. Nascent energy distributions[a] from the MPD of CF_2HCl and CF_2CFCl.

Parent =	CF_2HCl^{32}	CF_2CFCl^{23}	
Product =	CF_2	CF_2	CFCl
$T_v(v_2)$ (K)	1160	1860	1550
$T_v(v_1, v_3)$ (K)	1160	1100	900
E_v (kcal mol^{-1})	3·6	4·7	4·1
T_R (K)	2000	1550	>400
E_R (kcal mol^{-1})	6·0	4·7	4·7[b]
E_T (kcal mol^{-1})	3·5[c]	0·4	0·3
E_{total} (kcal mol^{-1})	13·1[c]	9·8	9·1[b]
E_{excess} (kcal mol^{-1})	—	~ 19	

[a] Following resonant IR laser excitation at a fluence of 50 J cm^{-2}.
[b] Assuming T_R(CFCl) to be comparable with T_R(CF$_2$).
[c] Assuming a δ-function velocity distribution.

bending mode of CF_2 was distributed according to the Boltzmann law:

$$P(E_{v_2}) = \exp[- E_{v_2}/kT(v_2)].$$

An estimate of the amount of vibrational energy in CF_2 is desirable in order to estimate the energy of the reactant when it dissociated. We have not deduced a distribution function for the energy in v_1 or v_3. However, since all available evidence shows Boltzmann-like distributions in the products of IR MPD (see Table II and references therein), we have assumed that the stretching modes ($v_1 = 1186$ cm^{-1}, $v_3 = 1112$ cm^{-1}) have thermal distributions characterized by the same value of T_v. We have calculated that value from

$$f_0^{-1} = Q_1(T_{v_1})Q_2(1860)Q_3(T_{v_3}),$$

to give an effective $T_v = 1100$ K for v_1 and v_3. If the preceding discussion for CF_2 is applied to CFCl, the measured value of $f_0 = 0.19$ is consistent with a vibrational temperature $T_v = (1150 \pm 100)$ K in all three modes. If the value $T_v(v_2) = 1550$ K, based on the spectroscopic measurement of the CFCl(0, 1, 0)/CFCl(0, 0, 0) ratio, is assumed for v_2, then f_0 is consistent with $T_v(v_1 \text{ and } v_3) = 900$ K. The value of the total vibrational energy in CFCl in Table V of $E_v = 4.1$ kcal mol^{-1} is based on this latter assumption. However, this same value of E_v is obtained under the assumption that $T_v = 1150$ K for all three modes.

The values of E_T in Table 1 are approximate, being based on the assumption of a Maxwell–Boltzman velocity distribution in the products. In the CF_2HCl decomposition, there is an energy barrier to the decomposition which appears as an activation energy of 6–$10\,\text{kcal mol}^{-1}$ to the reverse reaction

$$CF_2 + HCl \rightarrow CF_2HCl,$$

and which results[59] in higher kinetic energy release in the decomposition. Although thermal reaction rate data on CF_2CFCl are not available, the reaction is probably analogous to the

$$CF_2CF_2 \rightleftharpoons 2CF_2$$

system.[141] For the recombination of two CF_2 fragments, there is little activation energy ($E_a \simeq 0.4\,\text{kcal mol}^{-1}$), and hence one might expect less kinetic energy and a thermal distribution of the products. We can only hope the crude measurements of E_T in Table V will be superseded by molecular beam measurements on this reactant.

Our LEF measurements are adequate for assessing the dependence of E_T (as shown in the laboratory translational decay rate) on the degree of vibrational and rotational excitation in the product. For CF_2 formed from both CF_2HCl and from CF_2CFCl, products with no vibrational excitation had the same E_T (i.e. translational decay rate) as molecules formed vibrationally hot ($v_2 = 5, E_v = 3320^{-1}$), and molecules born with little rotational excitation ($E_R \simeq 40\,\text{cm}^{-1}$) had the same E_T as products created with substantial rotational excitation ($E_R \simeq 240\,\text{cm}^{-1}$). Hence, the rotational energy distribution and the average kinetic energy seem to be independent of the vibrational energy content of the product.

Owing to the presence of the competition between radiative excitation and unimolecular decay, the actual results obtained for product state distributions must show some dependence on instantaneous laser intensity. This is in fact observed for CF_2CFCl and is quite dramatic. No significant change in the CF_2 energy distribution with laser intensity is observed for CF_2HCl MPD, presumably owing to the much steeper rate of increase in decay rate with excess energy for the CF_2HCl (the results presented in Table V were obtained with pump fluences of $50\,\text{J cm}^{-2}$).

D. A Case of Pseudoselectivity

The experiments reported in this part demonstrate the competition between continued radiative excitation of highly vibrationally excited CF_2CFCl molecules in the continuum and unimolecular reaction. In brief, the nascent vibrational energy content of the CF_2 fragments from C_2F_3Cl was observed to depend on IR laser intensity. Similar intensity dependences in product species energy content were reported in the MPD of C_2H_3CN by Ashfold et al.,[120,122] Miller and Zare,[123] and Yu, Levy, and Wittig[124] and of CH_2CO (ketene) by Grimley and Stephenson.[64] These effects were notably absent in parallel studies on CF_2HCl, and represent a clear example of MPD pseudoselectivity.

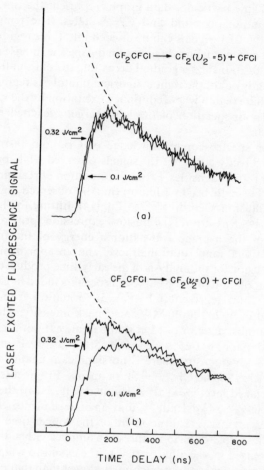

Fig. 21. Real-time laser-excited fluorescence signals observed for CF_2 products formed in either the (a) $v = 0$ or (b)$v_2 = 5$ level following collision-unimportant MPD of CF_2CFCl. Two different fluences were used. The broken curve represents diffusion of the nascent products out of the cylindrical photolysis region (see ref. 137). All vibrationally excited products (i.e. $CF_2 v_2 = 5$) are formed during the intense portion of the CO_2 laser, from 0 to 200 ns. There is *substantial* formation of the less energetic $CF_2 v = 0$ species continuing out to much later times for weak pulse photolysis.

The data graphs depicted in Fig. 21 represent CF_2 evolution from CF_2CFCl at about 7 mTorr pressure for IR pulses of 200 (strong) or 30 MW cm^{-2} (weak) peak intensity. These particular data graphs represent the time evolution of CF_2 in the vibrational ground and $v_2 = 5$ states. The temporal profiles exhibited by all the LEF signals can be divided into three regions: (1) a sharp rise due to rapid excitation and dissociation during the intense portion of the CO_2 laser pulse, 0–200 ns; (2) a gradual decay due to collision-free diffusion of the photofragments out of the tight cylindrical photolysis region, 1–6 μs; and (3) that intermediate region where continued excitation by the very weak laser tail, unimolecular dissociation of the less energetic, excited reactants, and diffusion compete.

Based on our spectroscopic knowledge of CF_2 and instrumental calibration, we can directly convert LEF signals observed while probing the vth level of CF_2 to product densities. From the cessation of reactant dissociation until collisions begin to become important, the observed signal should be controlled by diffusion according[137] to Eq. (14). Fitting the long-time LEF signals for $2 \leqslant t \leqslant 6 \mu$s to this diffusion expression provides us with a determination of the average translational energy of the fragments in a particular vibrational (and rotational) level and an approximate correction for fragment diffusion for the analysis of the early-time product evolution. This diffusional extrapolation (broken curve) has been included in Fig. 21. The two graphs in Fig. 21(a) represent $CF_2(v_2 = 5)$ formation from C_2F_3Cl dissociation by pulses of 200 or 30 MW cm^{-2} peak intensity. Although the net yield drops by almost a factor of 15, it is apparent that both traces follow the diffusional relation for $t > 1\cdot25 \mu$s, i.e. simply $S(t)^{-1} \propto c^2t^2$. Even for this factor of 7 change in laser intensity, there is no significant change in the average translational energy of the $CF_2(v_2 = 5)$ fragments. Fragments formed in the vibrationally relaxed level $v = 0$ also exhibited (Fig. 21(b)) the same average translational energy, $0\cdot4$ kcal mol^{-1}, and an independence of translational energy on laser intensity. The good fit of the longer-time behaviour, i.e. $t > 1\cdot25 \mu$s, to a simple diffusion expression implies either that all product formation has ceased by this time or that to the extent to which new product evolution continues it happens at rates much slower than that observed during the laser pulse. In the absence of any further experimental evidence, we shall assume collision-free product formation to be complete (i.e. $> 95\%$) by $t = 1\cdot25 \mu$s under these experimental conditions.

Nearly all, i.e. more than 90%, of the observed $v_2 = 5$ product is formed during the intense 200 ns laser spike (see Fig. 24). This is true for all laser pulse intensities studied, even when the total dissociation yield is only a few per cent. Probing the $v = 0$ products, this is only true for the highest intensity pulses; for the weaker pulses, significant $CF_2(v = 0)$ formation, up to almost 50%, occurs during that period 200–800 ns containing less than 10% of the integrated pulse energy and where the intensity is less than one-twentieth of the maximum value. For the IR MPD of CF_2HCl, under similar conditions, prompt formation is the rule for both $CF_2(v = 0)$ and $CF_2(v_2 = 5)$.

Taking the LEF signals observed at a delay time of $1.25\,\mu s$ to be representative of the total photolysis yield, it becomes apparent that CF_2 populations in $v_2 = 5$ or $v = 0$ exhibit different intensity dependences. These results are summarized in Fig. 22. As the laser intensity is reduced, the ratio of vibrationally hot fragments to vibrationally cold fragments decreases. Such behaviour is *not* observed for CF_2HCl photolysed under similar conditions. A reduction in laser intensity of a factor of 7 resulted, in the collision-free C_2F_3Cl IR MPD, in a six-fold reduction in the total $CF_2(v_2 = 5)$/total $CF_2(v = 0)$ ratio. The increased yield of $CF_2 v_2 = 5$ with increased laser intensity is evidence of the successful competition of continued radiative up-pumping in the continuum (region III) against the microscopic rates of unimolecular dissociation. This type of effect was not observed for CF_2HCl, reflecting significant differences in the energy scalings of microscopic decay rates and absorption cross sections. Over the intensity range studied, the value derived for $T_v(v_2)$ increased from about 1030 K to 1860 K; there was no *significant*

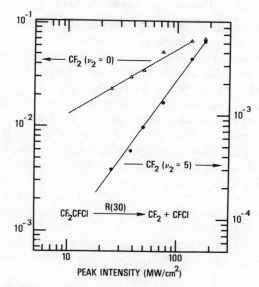

Fig. 22. Absolute yields of CF_2 products in the $v = 0$ and $v_2 = 5$ levels following collision-unimportant MPD of CF_2CFCl. The horizontal scale ('peak intensity') refers to pulses of the general type shown in Fig. 24. As the peak laser intensity increases, a larger fraction of reactants are excited into those higher-lying levels capable of producing the highly vibrationally excited $CF_2 v_2 = 5$ products.

change in either the rotational or translational energy content of the CF_2 fragments.

These experiments point out two important facets of laser intensity dependence in MPD, observed[64,120,122-124] using time-resolved laser-excited fluorescence techniques:

(1) The average energy content of the products, and thereby of the CF_2CFCl reactant, depends on the laser intensity.

(2) Those products formed with a large amount of excess energy (i.e. $CF_2(v_2 = 5)$) are formed instantaneously and only during the most intense portion of the IR pump pulse, while those products formed with a low amount of excess energy (i.e. $CF_2(v = 0)$) may also be formed from those reactants in low-lying energy levels with significantly lower microscopic decay rates.

Neither point was apparent in parallel studies on CF_2HCl performed under identical experimental conditions.

E. A Test of Statistical Theories of Unimolecular Reaction as Applied to MPD

No measurement sensitive only to total MPD yield or rate of dissociation or even to product velocity distributions can verify or disprove the adequacy of any statistical theory that might be used to model microscopic decay constants. Only by characterizing the *total* energy distribution of the multiphoton excited reactant will it be possible to obtain, directly, information on the magnitude of the relavent microscopic decay constants.

Information obtained from a product energy distribution measurement will complement that obtained from the more usual MPD experiment in which total product yield, produced during and after the laser pulse, is determined as a function of laser fluence and/or intensity. It can be shown[142] that the total yield is essentially independent of the unimolecular decay rates and pumping rates for states above the reaction threshold, a fact initially overlooked.[19] Effects connected with pumping through the low-lying discrete levels are known to play a role in determining total product yield.[143] The distribution of total energy in the reaction products is determined mainly by the microscopic unimolecular decay rates and radiative pumping rates *between reactive states only*.

Since, for polyatomics, absorption is more probable than stimulated emission, most molecules which arrive at the first reactive level ($i = m$) go on to react without crossing back into a non-reactive state. We can then view the total dissociation yield as

$$Y = \sum_{i=m} Y_i = \int_{-\infty}^{\infty} S(t)\, dt, \qquad (15)$$

where Y_i is the yield from each level i above threshold and $S(t)$ is some source function defined by the flux across the reaction threshold at level m:

$$S(t) = R_{m \leftarrow m-1} N_{m-1} - R_{m-1 \leftarrow m} N_m.$$

$S(t)$ is only slightly affected by the details of pumping through the reactive states, and therefore the yield is insensitive to the details of the rates of unimolecular decay and radiative pumping in the reactive levels.

In the continuum, it is the competition between radiative pumping (dependent on $w = 2\pi^2 H_{anh}^2 p^{sqe}(E)$) and microscopic decay (given by quantum RRK theory as $k_i = A\rho(E_i - E_{a_1})\rho(E_i)$) that determines the energy content in the products. Conversely, the overall dependence of yields or macroscopic dissociation rates are quite insensitive to the actual magnitude of A or the scaling of k_i with energy.

The average energy content of the CF_2CFCl MPD product was calculated from the data of Table V. The average vibrational energy is $1600 \, cm^{-1}$ in CF_2, $1400 \, cm^{-1}$ for CFCl, and the average rotational energy is $1600 \, cm^{-1}$ in each fragment. The average translational energy $E_T(v, J)$ found in the products was very small, only about $200 \, cm^{-1}$, and was found to be independent of v and J. The distribution function for E_T was taken to be thermal. Since E_T is much smaller than the products, internal energy, any errors in the kinetic energy distribution function would not significantly affect the overall probability function for a molecule to be formed with a particular total energy. Since the rotational, vibrational, and translational energy distributions within a product were independent, the probability of a product having an energy content $E = E_v + E_R + E_T$ is proportional to

$$P(E) = P(E_v)P(E_R)P(E_T),$$

where the individual probability functions are known. The average product energy derived from these distributions is about $3400 \, cm^{-1}$ for CFCl and $3700 \, cm^{-1}$ for CF_2. The distribution of total energy in the products shown in Fig. 23 is obtained assuming that the energies of the CF_2 and CFCl fragments are uncorrelated. This latter assumption, however, does not affect the mean of the distribution.

Reliable estimates of radiative pumping rates are available for CF_2CFCl. These come from our previously reported[23] analysis of the time-resolved rates of MPD determined as a function of laser intensity and added inert buffer gas pressure (see Sec. VII B).

In our original calculations[23] of the pressure and intensity dependence of the macroscopic MPD rates (see Fig. 15), we treated δ and ΔE as adjustable, making a priori assignments to the other parameters based on data other than MPD. We have redone[142] these calculations treating δ, ΔE, and A_∞ as adjustable and adding the further constraint that the calculated mean total

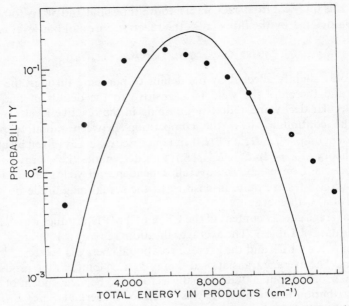

Fig. 23. Distribution of total energy deposited in both products, CF_2 and CFCl, during the MPD of CF_2CFCl as measured under conditions where theoretical calculations indicate that reaction is essentially complete during the time of the pulse. The CO_2 laser delivered $50\,J\,cm^{-2}$ with a peak intensity of $300\,MW\,cm^{-2}$. The full circles are experimental determinations (see Table V) of that fraction of products whose total energy lies in a $1000\,cm^{-1}$ window centred on the dot. The theoretical curve, calculated on the assumption of a rectangular pulse profile, has been fitted to agree with the experimental mean of $7050\,cm^{-1}$.

energy in products should agree with the measured value of $7050\,cm^{-1}$. The following points summarize the results:

(i) The calculated fit for the macroscopic dissociation rates are insensitive to A_∞, large variations in A_∞ being compensated by quite small variations in δ.

(ii) Only the fit of pressure and intensity dependence in the macroscopic rates is required to determine ΔE uniquely. The value found, $\Delta E = 2\cdot7\,kcal\,mol^{-1}$, lies within the range $1 < \Delta E < 3\,kcal\,mol^{-1}$ expected on the basis of chemical activation studies.[41]

(iii) The calculated mean total energy in products, $7050\,cm^{-1}$, is quite sensitive to A_∞ and relatively insensitive to small uncertainties in δ. The parameter ΔE does not enter into this calculation. The value found to give the best fit for mean excess energy is $A_\infty = 3\cdot2 \times 10^{16}\,s^{-1}$.

This value, $A_\infty \sim 3 \times 10^{16}\,\mathrm{s}^{-1}$, is the first to be determined from MPD measurements.

It is possible, using Eq. (15), to calculate the probability distribution for total energy in products. These results are also shown in Fig. 23. *That the theoretical curve is somewhat narrower than the experimental one is most likely due to the assumption of a square pulse in the theoretical calculation.* With a pulse that is not flat, different molecules see different laser intensities as they are being pumped between reactive states. As a result, more molecules react with energy at the low- and high-energy tails of the distribution when compared with the flat pulse case. However, this should not result in a significant shift in the mean energy.

This work offers the first quantitative test of the applicability of statistical theories to MPD because reliable estimates of the radiative pumping rates are known from separate studies of the MPD rates and the level of excitation in the reactant species is known directly. At every level, there is the potential competition between radiative pumping and unimolecular decay. The maximum flux is expected[70] from the level for which these rates are about equal. Quantum RRK and RRKM calculations using a transition state constrained to give $A = 3 \times 10^{16}\,\mathrm{s}^{-1}$ show equality at about 7.5 and 8.5 excess quanta respectively (1 quanta $= 1052\,\mathrm{cm}^{-1}$). Here the rates are about $1.5 \times 10^{10}\,\mathrm{s}^{-1}$. This is in remarkable agreement with the observed mean excess energy of 6.8 quanta.

Suppose energy were not randomized. Then the *microscopic rates* would be drastically different than those calculated from statistical theories. If one assumes, for instance, that rapid energy flow is limited to only six of the 12 molecular vibrational degrees of freedom of CF_2CFCl, one obtains $A = 2.3 \times 10^{13}\,\mathrm{s}^{-1}$, which is outside the range of 10^{15} to 10^{17} expected for this type of reaction.[49] These results support the conclusion that in this experiment energy is fairly well randomized before reaction occurs.

VII. SUMMARY

Many recent advances have been made in experimental techniques. It has become possible to ask detailed questions about the molecular physics involved in multiphoton processes and in evolving theoretical models capable of predicting experimentally verifiable dynamics. It has been shown that:

(i) excitation in the discrete region can be a coherent process—attention must be paid to laser intensity effects and the detailed states evolution, i.e. measurements of the optoacoustic type are not sufficient;

(ii) pumping through the quasicontinuum region can be described quite well with a rate equation formalism letting the single-quantum density-of-states formalism determine the magnitude and energy scaling of $T_2(E)$;

(iii) competition in the dissociative continuum region occurs between radiative up-pumping and microscopic decay, in agreement with statistical theories of unimolecular reaction, there being no firm evidence for mode selective or non-statistical (i.e. energy localization) effects.

In addition to a fuller understanding of the MPD processes, these recent advances in MPD experiment and theory are generating insight into molecular dynamics at all levels of excitation. Strong, sharp overtone transitions have been observed and the effects of intense fields on wavefunction mixing and collisional cross sections explored. Newly developed lines of research are providing quantitative information on the energy dependence of intramolecular (both T_1 and T_2 type) and collision-induced relaxation processes. The single-quantum exchange formalism, developed to follow apparent energy dependences in radiative absorption cross sections, are now allowing predictions to be made for fundamental optoacoustic and macroscopic rate measurements of MPD. The results of the product state distribution measurements provide a basis for the development of theoretical models for describing potential surface interactions in ground-state unimolecular reaction and for critically examining the applicability of currently accepted models of unimolecular decay to the MPD process.

Acknowledgements

I wish to express my appreciation to Drs John C. Stephenson, Myron F. Goodman, James Stone, and Everett Thiele for their contributions to this work and for many stimulating discussions.

Appendix: Details of the CF_2CFCl and CF_2HCl Experiments

These experiments involved the real-time *in situ* spectroscopic detection of the $\tilde{X}\,CF_2\,(v,J)$ and $\tilde{X}\,CFCl(v,J)$ fragments formed in the MPD reactions:

$$CF_2CFCl \rightarrow \tilde{X}\,CF_2(v,J) + \tilde{X}\,CFCl(v,J) + E_T(v,J), \qquad \Delta H = 96\,\text{kcal mol}^{-1} \tag{16}$$

and

$$CF_2HCl \rightarrow \tilde{X}\,CF_2(v,J) + HCl + E_T(v,J), \qquad \Delta H = 49\,\text{kcal mol}^{-1} \tag{17}$$

where v and J represent vibrational and rotational quantum numbers.

The reactant flowed through a $400\,\text{cm}^3$ cell ($T = 297\,\text{K}$) at low pressures, typically 1–10 mTorr, either neat or diluted to a mole fraction of 10^{-5} in Ar at a total pressure 5–760 Torr. Pulses from a thyratron-fired CO_2 TEA laser operating on a single line dissociated a fraction of the reactant species and the concentration of the CF_2 or CFCl product was determined by laser-excited fluorescence (LEF). For instance, a frequency-doubled N_2-pumped dye laser (3 ns FWHM, at 261·7 nm) probed the $CF_2(v=0, \text{low } J)$ density at a time τ_D after the start of the CO_2 laser pulse. The 261·7 nm wavelength excited the $\tilde{A} CF_2(0, 2, 0) \leftarrow \tilde{X} CF_2(0, 0, 0)$ vibronic band. The single vibronic level (SVL) fluorescence from the laser-excited $\tilde{A} CF_2(0, 2, 0)$ state was spectrally resolved by a $f/1·6$ UV monochromator to discriminate against laser scatter and fluorescence from vibrational interferences, detected with a photomultiplier tube and gated electronics, and recorded digitally using a signal averager. Similar excitation and fluorescence schemes were used to probe other vibrational levels of the CF_2 and of the CFCl.

The nascent fragments were initially generated by the CO_2 laser pulse in a well defined cylindrical geometry. The concentration of either CF_2 or CFCl fragment at any later time τ_D within this cylinder was probed by UV or visible laser pulses. The IR and probe beams propagated collinearly, but in opposing directions, through the cell. Both beams were loosely focused (30 cm focal length lenses) into the cell, and the beam waists of both lasers were constant in diameter over a distance of 2 cm at the focus. In this 2 cm long focal region, the radius of the IR beam was $4·7 \times 10^{-2}$ cm, while that of the probe beam was 5×10^{-3} cm. The probe beam had a radius so much smaller than that of the photolysis pulse that only the central region of constant IR fluence was probed. The monochromator entrance slit was oriented parallel to the laser beams, but was apertured so as to view only fluorescence from the central 8 mm of the 2 cm long region of constant beam profile. *The important point is that we examine regions of constant, known, CO_2 laser fluence.*

The timing τ_D between the CO_2 photolysis pulse and the tunable probe pulse was controlled by a programmable digital delay generator. Although the time delay could be scanned in 1 ns increments, the jitter between the pump and probe laser pulses was approximately 5 ns, which is the practical limit of the experimental time resolution. Only those fragments in the path of the probe beam in the appropriate quantum states (chosen by the probe laser frequency) at the time τ_D contribute to the LEF signal. By scanning the digital delay generator in time and recording LEF signal strengths, we observed directly the increase in product density due to reactant dissociation. Such data are shown in Fig. 24, where the production of $\tilde{X} CF_2(0, 0, 0)$ is shown during the CO_2 laser pulse, for CF_2CFCl photolysed at very low pressure (i.e. about 2 mTorr). After the laser pulse and on a longer timescale (e.g. 5 μs), the CF_2 density in the spatial region probed by the UV laser decreases due to collision-free gas expansion out of the narrow photolysis zone into the large cell.[137] This decrease has been used to derive the average kinetic energy of the newly formed fragments in the collision-free region.[68] Also shown is a trace of CO_2

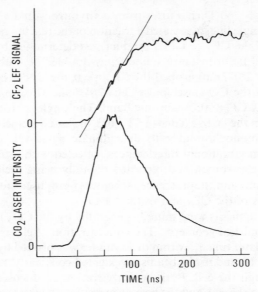

Fig. 24. Laser pulse shape and concentration of $\tilde{X}\,CF_2\,v = 0$ as a function of time delay after the start of the laser pulse.

laser intensity as a function of time, recorded by an Au–Ge detector with a 4 ns response time. This is the sum of 100 pulses digitized on a Tektronix R7912 transient digitizer and summed in a minicomputer. Approximately 80% of the energy in the CO_2 pulse is contained in the central 100 ns portion. Both probe and photolysis laser pulses were recorded with the same Au–Ge detector, so that τ_D was accurately known and is correctly shown on the horizontal axis of Fig. 24. The IR laser pulse energy was measured with a Laser Precision RK 3230 energy meter. From the beam waist[68] and pulse energy, the intensity of the CO_2 laser pulse was calculated for the temporal shape shown in Fig. 24. It is this peak value of intensity which labels the data traces discussed herein.

As described in ref. 110, we calibrated our detection system so that we could relate the PMT signal to the absolute number density of $\tilde{X}\,CF_2$ fragments in the central region of the photolysis zone. Since the initial CF_2HCl and CF_2CFCl pressure and the time-dependent $\tilde{X}\,CF_2$ concentration are both known, one may define at any time τ_D a macroscopic production rate $\kappa(\tau_D)$ by

$$\frac{d[CF_2]}{dt} = -\frac{d[\text{reactant}]}{dt} = \kappa(\tau_D)\,[\text{reactant}].$$

In our experiments only a fraction of the reactant in the centre of the photolysis region was dissociated, so [reactant] may be treated essentially as a constant. In Fig. 24, the rate of CF_2 production is approximately constant during the time $20 \leqslant \tau_D \leqslant 60$ ns. We drew a straight line through this initial rise in CF_2, and have identified that as the rate constant (s^{-1}) of CF_2 production for this set of experimental conditions. This approximation of fitting the initial CF_2 increase to a constant κ over this range of τ_D was made, with an equally good fit, for the rest of the data used to deduce dissociation rates as a function of laser intensity and argon pressure. The experiments to determine the effect on κ of added Ar pressure were performed and analysed in a similar manner.

The interpretation of some MPD experiments has been difficult because of gas heating or reactant–reactant collisions. Neither of these effects was important in the experiments reported here. To prove that gas heating was unimportant, LEF excitation spectra of the $\tilde{X}$ $CF_2 2_0^2$ transition were taken at various values of τ_D in the range of $0–2 \mu s$, for a mixture of CF_2CFCl dilute in Ar (i.e. a mole fraction $X_{CF_2CFCl} = 1 \cdot 1 \times 10^{-5}$) at total pressures of 30, 100, 200, and 650 Torr. Even for the highest fluences used $(37 J cm^{-2})$, for which gas heating would be most important, the measured CF_2 rotational temperature was $T_R = (300 \pm 30)$ K. However, when a more concentrated sample $(X_{CF_2CFCl} = 10^{-3})$ was photolysed, there was a significant increase in T_R, demonstrating that gas heating can be important effect at concentrations as dilute as $X = 10^{-3}$. To demonstrate that reactant–reactant collisions are unimportant, we studied rates as a function of reactant concentration in the range $5 \times 10^{-6} \leqslant X_{CF_2CFCl} \leqslant 5 \times 10^{-5}$. Even at high pressure (650 Torr) and high fluence conditions, there was no significant change in the normalized CF_2 production rate with CF_2CFCl concentration.

References

1. N. R. Isenor and M. C. Richardson (1971), *Appl. Phys. Lett.*, **18**, 224.
2. N. R. Isenor, V. Merchant, R. S. Hallsworth and M. C. Richardson (1973), *Can. J. Phys.*, **51**, 1281.
3. J. L. Lyman and R. J. Jensen (1972), *Chem. Phys. Lett.*, **13**, 421.
4. J. L. Lyman, R. J. Jensen, J. P. Rink, C. P. Robinson and S. D. Rockwood (1975), *Appl. Phys. Lett.*, **27**, 87.
5. R. V. Ambartzumian, V. S. Letokhov, E. A. Ryabov and N. V. Chekalin (1974), *JETP Lett.*, **20**, 273.
6. V. Letokhov and C. B. Moore (1976), *Sov. J. Quantum Electron.*, **6**, 259.
7. R. Ambartzumian and V. Letokhov (1977), in *Chemical and Biochemical Applications of Lasers*, vol. 3, ed. C. B. Moore, Academic Press, New York.
8. R. Ambartzumian and V. Letokhov (1977), *Acc. Chem. Res.*, **10**, 61.
9. J. L. Lyman, S. D. Rockwood and S. M. Freund (1977), *J. Chem. Phys.*, **67**, 4545.
10. S. Kimel and S. Speiser (1977), *Chem. Rev.*, **72**, 437.
11. C. D. Cantrell, S. M. Freund and J. L. Lyman (1978), in *Laser Handbook*, vol. 3, ed. M. Sitch, North-Holland, Amsterdam.
12. N. Bloembergen and E. Yablonovitch (1978), *Phys. Today*, **31**, 32.
13. P. A. Schultz, Aa. S. Sudbo, D. J. Krajnovich, H. S. Kwok, Y. R. Shen and Y. T. Lee (1979), *Annu. Rev. Phys. Chem.* **30**, 379.

14. M. F. Goodman, J. Stone and E. Thiele (1980), in *Multiple-Photon Excitation and Dissociation of Polyatomic Molecules*, ed. C. D. Cantrell, Springer, Berlin.
15. J. L. Lyman, G. P. Quigley and O. P. Judd (1980), in *Multiple-Photon Excitation and Dissociation of Polyatomic Molecules*, ed. C. D. Cantrell, Springer, Berlin.
16. J. L. Lyman (1977), *J. Chem. Phys.*, **67**, 1868.
17. E. R. Grant, P. A. Schultz, Aa. S. Sudbo, Y. R. Shen and Y. T. Lee (1978), *Phys. Rev. Lett.*, **40**, 115.
18. W. Fuss (1979), *Chem. Phys.*, **36**, 135.
19. J. G. Black, E. Yablonovitch, N. Bloembergen and S. Mukamel (1977), *Phys. Rev. Lett.*, **38**, 1131.
20. M. Quack (1978), *J. Chem. Phys.*, **69**, 1282.
21. J. Stone and M. F. Goodman (1979), *J. Chem. Phys.*, **71**, 408.
22. J. G. Black, P. Kolodner, M. J. Schultz, E. Yablonovitch and N. Bloembergen (1979), *Phys. Rev.*, **A19**, 704.
23. J. Stone, E. Thiele, M. F. Goodman, J. C. Stephenson and D. S. King (1980), *J. Chem. Phys.*, **73**, 2259.
24. M. Quack (1979), *J. Chem. Phys.*, **70**, 1069.
25. D. M. Larsen and N. Bloembergen (1976), *Opt. Commun.*, **17**, 254.
26. D. M. Larsen (1976), *Opt. Commun.*, **19**, 404.
27. N. Bloembergen (1975), *Opt. Commun.*, **15**, 416.
28. S. Mukamel and J. Jortner (1976), *J. Chem. Phys.*, **65**, 5204.
29. S. Mukamel and J. Jortner (1976), *Chem. Phys. Lett.*, **40**, 150.
30. M. Tamir and R. D. Levine (1977), *Chem. Phys. Lett.*, **46**, 208.
31. O. P. Judd (1979), *J. Chem. Phys.*, **71**, 4515.
32. J. C. Stephenson, D. S. King, M. F. Goodman and J. Stone (1979), *J. Chem. Phys.*, **70**, 4496.
33. P. Kolodner, C. Winterfeld and E. Yablonovitch (1977), *Opt. Commun.*, **20**, 119.
34. J. L. Lyman, J. W. Hudson and S. M. Freund (1977), *Opt. Commun.*, **21**, 112.
35. A. Baldwin, J. R. Barker, D. M. Goldin, R. Duperrex and H. van der Bergh (1979), *Chem. Phys. Lett.*, **62**, 178.
36. D. P. Hodgkinson and J. S. Briggs (1976), *Chem. Phys. Lett.*, **43**, 451.
37. E. Yablonovitch (1977), *Opt. Lett.*, **1**, 87.
38. J. Stone and M. F. Goodman (1978), *Phys. Rev.*, **A18**, 2618.
39. I. Oref and B. S. Rabinovitch (1979), *Acc. Chem. Res.*, **12**, 166.
40. E. Thiele, M. F. Goodman and J. Stone (1980), *Opt. Eng.*, **19**, 10.
41. D. C. Tardy and B. S. Rabinovitch (1977), *Chem. Rev.*, **77**, 369.
42. P. J. Marcoux and D. W. Sester (1978), *J. Phys. Chem.*, **82**, 97.
43. J. Stone, E. Thiele and M. F. Goodman (1973), *J. Chem. Phys.*, **59**, 2909.
44. M. F. Goodman, J. Stone and E. Thiele (1973), *J. Chem. Phys.*, **59**, 2919.
45. M. F. Goodman and E. Thiele (1972), *phys. Rev.*, **A5**, 1535.
46. J. Stone, E. Thiele and M. F. Goodman (1980), *Chem. Phys. Lett.*, **71**, 177.
47. J. A. Horsley, J. Stone, M. F. Goodman and D. A. Dows (1979), *Chem. Phys. Lett.*, **66**, 461.
48. P. J. Robinson and K. A. Holbrook (1972), *Unimolecular Reactions*, Wiley–Interscience, New York.
49. W. Forst (1973), *Theory of Unimolecular Reactions*, Academic Press, New York.
50. M. Bixon and J. Jortner (1968), *J. Chem. Phys.*, **48**, 725.
51. G. Z. Whitten and B. S. Rabinovitch (1963), *J. Chem. Phys.*, **38**, 2466.
52. D. M. Cox, R. B. Hall, J. A. Horsley, G. M. Kramer, P. Rabinowitz and A. Kaldor (1979), *Science*, **205**, 390.
53. R. B. Hall and A. Kaldor (1979), *J. Chem. Phys.*, **70**, 4027.

54. K. V. Reddy and M. J. Berry (1979), *Chem. Phys. Lett.*, **66**, 223.
55. D. M. Brenner (1978), *Chem. Phys. Lett.*, **57**, 357.
56. Y. T. Lee, (1980), private communication.
57. D. S. King and J. C. Stephenson (1977), *Chem. Phys. Lett.*, **51**, 48.
58. R. J. S. Morrison and E. R. Grant (1979), *J. Chem. Phys.*, **71**, 3537.
59. Aa. S. Sudbo, P. A. Schulz, Y. R. Shen and Y. T. Lee (1978), *J. Chem. Phys.*, **69**, 2312.
60. E. Thiele, M. F. Goodman and J. Stone (1980), *Chem. Phys. Lett.*, **72**, 34.
61. D. S. King and J. C. Stephenson (1979), *Chem. Phys. Lett.*, **66**, 33.
62. R. Duperrex and H. van den Bergh (1980), *J. Mol. Struct.*, **61**, 291.
63. I. Hermann and J. Marling (1980), *J. Chem. Phys.*, **72**, 516.
64. A. J. Grimley and J. C. Stephenson (1981), *J. Chem. Phys.*, **74**, 447.
65. R. Duperrex and H. van den Bergh (1980), *J. Chem. Phys.*, **73**, 585.
66. R. Duperrex and H. van den Bergh (1979), *J. Chem. Phys.*, **71**, 3613.
67. M. Quack, P. Humbert and H. van den Bergh (1980), *J. Chem. Phys.*, **73**, 247.
68. J. C. Stephenson and D. S. King (1978), *J. Chem. Phys.*, **69**, 1485.
69. E. Wurzberg, L. J. Kovalenko and P. L. Houston (1978), *Chem. Phys.*, **35**, 317.
70. Aa. S. Sudbo, P. A. Schultz, E. R. Grant, Y. R. Shen and Y. T. Lee (1979), *J. Chem. Phys.*, **70**, 912.
71. T. Fukumi (1979), *Opt. Commun.*, **30**, 351.
72. V. N. Bagratashvili, I. N. Knyazev, V. S. Letokhov and V. V. Lobko (1976), *Opt. Commun.*, **18**, 525.
73. J. G. Black, E. Yablonovitch and N. Bloembergen (1977), *Phys. Rev. Lett.*, **38**, 1131.
74. T. F. Deutch (1977), *Opt. Lett.*, **1**, 25.
75. D. M. Cox (1978), *Opt. Commun.*, **24**, 336.
76. G. P. Quigley (1979), *Opt. Lett.*, **4**, 84.
77. C. K. N. Patel (1979), *Am. Chem. Soc. Symp. Ser. A*, **94**, 177.
78. See, for example: K. V. Reddy and M. J. Berry (1977), *Chem. Phys. Lett.*, **52**, 111 and (1979), *Faraday Discuss. Chem. Soc.*, **67**, 222; B. R. Henry and W. R. A. Greenlay (1980), *J. Chem. Phys.*, **72**, 5516.
79. H.-L. Dai, A. H. Kung and C. B. Moore (1980), *Phys. Rev. Letts.*, **43**, 761 (1979)., **72**, 5525.
80. H. S. Kwok and E. Yablonovitch (1978), *Phys. Rev. Lett.*, **41**, 745.
81. S. S. Alimpiev, V. N. Bagratashvili, N. V. Karlov, V. S. Letokhov, V. V. Lobko, A. A. Makarov, B. G. Sartakov and E. M. Khokhlov (1977), *JETP Lett.*, **25**, 547.
82. R. V. Ambartzumian, G. N. Makarov and A. A. Puretzky (1978), *Opt. Commun.*, **27**, 79.
83. Aa. S. Sudbo, P. A. Schultz, D. J. Krajnovich, Y. T. Lee and Y. R. Shen (1979), *Opt. Lett.*, **4**, 219.
84. V. N. Bagratashvili, Yu. G. Vainer, V. S. Doljikov, S. F. Koliakov, A. A. Makarov, L. P. Malyavkin, E. A. Ryabov, E. G. Silkis and V. D. Titov (1980), *Appl. Phys.*, **22**, 101.
85. J. W. Hudgens and J. D. McDonald (1981), *J. Chem. Phys.*, **74**, 1510.
86. F. B. Brown, A. D. H. Clague, N. D. Heitkamp, D. F. Foster and D. Danti (1967), *J. Mol. Spectrosc.*, **24**, 163.
87. J. R. Nielsen, C. Y. Liang, R. M. Smith and D. C. Smith (1953), *J. Chem. Phys.*, **21**, 383.
88. D. M. Brenner and K. Brezinsky (1979), *Chem. Phys. Lett.*, **67**, 36.
89. D. M. Brenner, K. Brezinsky and P. M. Curtis (1980), *Chem. Phys. Lett.*, **72**, 202.
90. P. Gozel and H. van den Bergh (1981), *J. Chem. Phys.*, **74**, 1724.

91. R. V. Ambartzumian, G. N. Makarov and A. A. Puretzky (1978), *Opt. Lett.*, **3**, 103.
92. R. V. Ambartzumian, V. S. Letokhov, G. N. Makarov and A. A. Puretzky (1978), *Opt. Commun.*, **25**, 69.
93. R. V. Ambartzumian, G. N. Makarov and A. A. Puretzky (1978), *Opt. Commun.*, **27**, 79.
94. D. F. Heller and G. A. West (1980), *Chem. Phys. Lett.*, **69**, 419.
95. W. Tsay, C. Riley and D. O. Ham (1979), *J. Chem. Phys.*, **70**, 3558.
96. P. Avouris, W. M. Gelbart and M. A. El-Sayed (1977), *Chem. Rev.*, **77**, 793.
97. K. F. Freed (1976), in *Topics in Applied Physics*, vol. 15, ed. F. K. Fong, Springer, Berlin.
98. M. J. Coggiola, P. C. Cosby and J. R. Peterson (1980), *J. Chem. Phys.*, **72**, 6507.
99. R. L. Woodin, D. S. Bomse and J. L. Beauchamp (1979), in *Chemical and Biochemical Applications of Lasers*, vol. 3, ed. C. B. Moore, Academic Press, New York.
100. D. S. Bomse, R. L. Woodin and J. L. Beauchamp (1979), *J. Am. Chem. Soc.*, **101**, 5503.
101. R. L. Woodin, D. S. Bomse and J. L. Beauchamp (1978), *J. Am. Chem. Soc.*, **100**, 3248 and (1979), *Chem. Phys. Lett.*, **63**, 630.
102. R. N. Rosenfeld, J. M. Jasinski and J. I. Brauman (1979), *J. Am. Chem. Soc.*, **101**, 3999.
103. A. von Hellfeld, D. Feldmann, K. H. Welge, and A. P. Fournier (1979), *Opt. Commun.*, **30**, 193.
104. L. H. Sutcliffe and A. D. Walsh (1961), *Trans. Faraday Soc.*, **57**, 873.
105. R. B. Hall and A. Kaldor (1979), *J. Chem. Phys.*, **70**, 4027.
106. D. M. Cox and J. A. Horsley (1980), *J. Chem. Phys.*, **72**, 874.
107. J. W. Hudgens (1978), *J. Phys. Chem.*, **68**, 777.
108. See, for instance: E. Weitz and G. W. Flynn (1979), in *Annual Review of Physical Chemistry*, ed. H. Eyring, Annual Reviews, Palo Alto, and references therein.
109. A recent review of this subject is: J. Nieman and A. M. Ronn (1980), *Opt. Eng.*, **19**, 39.
110. D. S. King and J. C. Stephenson (1978), *J. Am. Chem. Soc.*, **100**, 7151.
111. G. Folcher and W. Braun (1978), *J. Photochem.*, **8**, 341.
112. W. C. Danen, D. F. Koster and R. N. Zitter (1979), *J. Am. Chem. Soc.*, **101**, 4281.
113. R. B. Woodward and R. Hoffmann (1970), *The Conservation of Orbital Symmetry*, Academic Press, New York.
114. D. L. Bunker and W. L. Hase (1973), *J. Chem. Phys.*, **59**, 4621.
115. B. D. Cannon and F. F. Crim (1980), *J. Chem. Phys.*, **73**, 3013.
116. K. W. Hicks, M. L. Lesiecki and W. A. Guillory (1979), *J. Chem. Phys.*, **83**, 1936.
117. M. L. Lesiecki and W. A. Guillory (1978), *J. Chem. Phys.*, **69**, 4572.
118. R. Schmiedl, R. Boettner, H. Zacharias, U. Meier and K. H. Welge (1980), *J. Mol. Struct.*, **61**, 271.
119. R. Schmiedl, R. Boettner, H. Zacharias, U. Meier and K. H. Welge (1980), *Opt. Commun.*, **31**, 329.
120. M. N. R. Ashfold, G. Hancock and G. Ketley (1979), *Faraday Discuss. Chem. Soc.*, **67**, 204.
121. J. H. Hall, Jr, M. L. Lesiecki and W. A. Guillory (1978), *J. Chem. Phys.*, **68**, 2247.
122. M. N. R. Ashfold, G. Hancock and M. L. Hardaker (1980), *J. Photochem.*, **14**, 85.
123. C. M. Miller and R. N. Zare (1980), *Chem. Phys. Lett.*, **71**, 376.
124. M. H. Yu, M. R. Levy and C. Wittig (1980), *J. Chem. Phys.*, **72**, 3789.
125. J. D. Campbell, M. H. Yu, M. Mangir and C. Wittig (1978), *J. Chem. Phys.*, **69**, 3854.

126. J. C. Stephenson, S. E. Bialkowski and D. S. King (1980), *J. Chem. Phys.*, **72**, 1161.
127. D. Feldmann, H. Zacharias and K. H. Welge (1980), *Chem. Phys. Lett.*, **69**, 466.
128. M. J. Coggiola, P. A. Schultz, Y. T. Lee and Y. R. Shen (1977), *Phys. Rev. Lett.*, **38**, 17.
129. C. R. Quick, Jr, J. J. Tiee, T. A. Fischer and C. Wittig (1979), *Chem. Phys. Lett.*, **62**, 435.
130. C. R. Quick, Jr, and C. Wittig (1980), *J. Chem. Phys.*, **72**, 1694.
131. C. R. Quick, Jr, and C. Wittig (1978), *Chem. Phys.*, **32**, 75.
132. K. C. Kim, D. N. Setser and B. E. Holmes (1973), *J. Phys. Chem.*, **77**, 725.
133. M. J. Berry (1974), *J. Chem. Phys.*, **61**, 3114.
134. P. N. Clough, J. C. Polanyi and R. J. Taguchi (1970), *Can. J. Chem.*, **48**, 2912.
135. E. R. Sirkin and M. J. Berry (1979), *IEEE J. Quantum Electron.*, **QE-10**, 701.
136. J. C. Polanyi and J. L. Schreiber (1977), *Discuss. Faraday Soc.*, **62**, 267.
137. S. E. Bialkowski, D. S. King and J. C. Stephenson (1980), *J. Chem. Phys.*, **72**, 1156.
138. D. L. Akins, D. S. King and J. C. Stephenson (1979), *Chem. Phys. Lett.*, **65**, 257.
139. D. S. King, P. K. Schenck and J. C. Stephenson (1979), *J. Mol. Spectrosc.*, **78**, 1.
140. S. E. Bialkowski, D. S. King and J. C. Stephenson (1979), *J. Chem. Phys.*, **71**, 4010.
141. D. S. Y. Hsu, M. E. Umstead and M. C. Lin (1978), *Am. Chem. Soc. Symp. Ser.*, **66**.
142. J. C. Stephenson, S. E. Bialkowski, D. S. King, E. Thiele, J. Stone and M. F. Goodman (1981), *J. Chem. Phys.*, **74**, 3905.
143. J. L. Lyman, W. C. Danen, A. C. Nilsson and A. V. Nowak (1979), *J. Chem. Phys.*, **71**, 1206.

Note

Dynamics of the Excited State
Edited by K. P. Lawley
© 1982 John Wiley & Sons Ltd.

THE PHOTON-AS-CATALYST EFFECT
IN LASER-INDUCED PREDISSOCIATION
AND AUTOIONIZATION

ALBERT M. F. LAU

Corporate Research Science Laboratories,
Exxon Research and Engineering Company,
PO Box 45, Linden, NJ 07036, USA

CONTENTS

I. INTRODUCTION

In this chapter, we consider the atomic or molecular decompositions due to a laser-induced multiphoton process known as the photon-as-catalyst effect (PCE). This effect was first proposed and analysed by Lau and Rhodes (1977a, b). Specifically, the bound–free decomposition processes addressed in this chapter include (1) photon-catalysed (PC) predissociation of molecules AB,

$$AB + N\hbar\omega \rightarrow A + B + N\hbar\omega, \tag{1.1}$$

and (2) PC autoionization of atoms and molecules A,

$$A + N\hbar\omega \rightarrow A^+ + e^- + N\hbar\omega; \tag{1.2}$$

where the reactions conserve the total number N of photons in a given mode of the electromagnetic field (EMF), such as that provided by single-mode lasers. As illustrated in Fig. 1, the PCE consists of a parent atom or molecule (hereafter 'molecules' is sometimes used to include atoms as well) initially in a bound state $|i\rangle$ being induced to make transition via the intermediate states $|m\rangle$ to an 'equal-energy' continuum $|f\rangle$ leading to the fragments by the virtual or real absorption of one (or more) photons of a given EMF mode and the stimulated emission of an equal number of laser photons of the same mode (i.e. the same frequency ω, polarization $\hat{\varepsilon}$ and propagation vector $\hat{k}$). Since the rate constants of these unimolecular reactions are increased while the number of photons N is conserved, the photons are acting as catalysts. When the laser

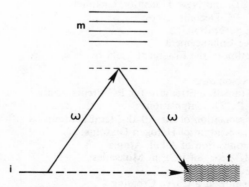

Fig. 1. The photon-catalysed decompositions (autoionization or predissociation) consist of an atom or a molecule initially in a bound state $|i\rangle$ making a transition to the continuum of $|f\rangle$ via the intermediate states $|m\rangle$ by the absorption of one laser photon and the stimulated emission of another photon of the same EM mode. In some molecular states, natural decomposition from $|i\rangle$ to $|f\rangle$ may also be present (dashed arrow).

frequency ω is non-resonant with the transitions $|i\rangle \to |m\rangle$, the virtual absorption and stimulated emission are simultaneous (i.e. being a multiphoton process). In the case in which ω is resonant with the $|i\rangle \to |m\rangle$ transition, the real absorption and stimulated emission may be simultaneous or stepwise (i.e. delayed) events. The definitive characteristic of this photoeffect is the conservation of photons in the overall atomic or molecular transformation. Since net absorption or stimulated emission of laser photons is zero, the total energy (i.e. including the relative kinetic energy of the fragments) of the final state is the same as the initial bound state, whereas the sum of *internal* energies of the fragments is usually lower.

The initial bound states for PC decomposition are in many cases electronically excited states; or in the case of polyatomic molecules, these may sometimes be excited rotational–vibrational states in the ground electronic states. Population in these excited states can be the result of one of many excitation processes: thermal population, charged- or neutral-particle collisional excitations, photoexcitations, chemical reactions, intramolecular relaxation from other states, and so on.

For polyatmoic molecules, however, the initial states are not necessarily highly excited states. They may be states thermally populated at ambient temperature or even the ground state (Lau, 1981a). Figs. 2(a) and (b) illustrate two such situations. The figures show the potential surfaces along the decomposition coordinate. The initial states of the molecule AB are stable due to the activation energy barrier. In Fig. 2(a), the sum of internal energies E_∞ of the decomposed products A and B is lower than the energies E_i of some thermally populated states $|i\rangle$ of the molecule AB, but is higher than its ground-state energy E_g. The population in such states $|i\rangle$ can be decomposed

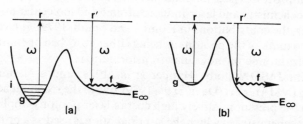

Fig. 2. Plots of potential surfaces along decomposition coordinate showing some examples of photon-catalysed predissociation of a polyatomic molecule in the ground electronic states. The initial states of the molecule are stable due to the activation barrier. The same intermediate state $|r\rangle$ may be involved in both the absorption and the stimulated emission, or after excitation into $|r\rangle$ the molecule may relax into another state $|r'\rangle$ from which stimulated emission to the dissociative continuum $|f\rangle$ occurs. (a) Energy E_g of the ground state is less than E_∞ of the products, while the thermally populated E_i is greater than E_∞. (b) E_g is greater than E_∞.

by the PCE via real absorption into an excited state and followed by stimulated emission from the same excited state, or from another excited state into which the excited molecule AB* has relaxed. In Fig. 2(b), the energy E_∞ of the products is less than the ground-state energy E_g of AB, so that *all* population of the bound molecules may be decomposed photocatalytically in a similar fashion. In both cases, if the high activation energy barrier forbids any tunnelling, the PCE in effect opens up a new dissociation channel by circumventing the potential barrier. This is in addition to the possibility of opening up new channels due to new selection rules (Lau and Rhodes, 1977b). Having the molecules in the thermally populated ground states as initial states for PCE is advantageous because no additional energy of any form is needed to prepare the initial-state population; and because the PCE can act over the long lifetimes of the ground states. This would probably reduce the required laser power (see below). Although one may call the situations described in this paragraph photon-catalysed dissociation, we shall include it under PC predissociation. The reason is that the term 'predissociation further conveys the notion that the dissociation does not require the net absorption of an amount of external energy in order to overcome the dissociation barrier.

A. General Remarks

Our original motivation of pursuing the study of PCE is to show that laser radiation can be used to modify the dynamics of atomic and molecular processes and to initiate new reactions without consuming any photons (Lau and Rhodes, 1977a, b; Lau, 1978). This concept of photons as catalysts leading to atomic and molecular transformations was novel to the studies of multiphoton processes (see, for example, the review by Lambropoulos, 1976) and to photochemistry and laser-induced atomic and molecular processes (see, for examples, the contributions in Kompa and Smith, 1979). It has been more customary to think of the photons as being either absorbed or emitted in order to initiate the atomic or molecular transformations.

In a recent JASON study (Happer *et al.*, 1979) and a National Science Foundation (NSF) report (Davis *et al.*, 1979), one of the recommendations was that, due to the present relatively high cost of laser photons, applied research in laser photochemistry in which the laser radiation is used as a primary source of energy should put more emphasis on 'high leverage' areas where laser photons are either used highly non-stoichiometrically or used to produce high-cost chemicals. In the light of this recommendation, it may be interesting to point out that the PCE is a highly non-stoichiometric use of laser photons to do laser photochemistry. As indicated above, the PCE consumes in principle zero photons per product molecule for this particular reaction channel. Of course, there could be other processes causing the loss of laser photons. Raman and Rayleigh light scatterings by the molecules converting some of the laser photons into photons of other electromagnetic field modes are always present. There could also be undesired single-photon or multiphoton absorption by the molecules. Depending on the molecule and the laser radiation under consideration, some of these undesirable losses may be minimized or avoided. As discussed below, *the magnitude of the rate 'constant' for the PCE depends directly on the laser intensity and frequency*. Therefore, the PCE may be made to dominate over other loss processes by tuning up the laser intensity. When the PCE is the dominant process, it is expected that the attenuation of the laser beam by the irradiated medium is considerably less than a corresponding experiment requiring actual single-photon or multiphoton absorption. An experimental set-up to optimize the use of laser photons with this PCE is to have the molecular medium situated between a pair of reflecting mirrors so that the laser radiation can be 'used over again' to photon-catalyse more molecules.

The PCE was first analysed theoretically for the idealized cases of no intermediate states (Lau and Rhodes, 1977a) and for the more common cases with intermediate states (Lau and Rhodes, 1977b). They adopted an alternative but equivalent perspective that the effect was due to the mixing of the initial state with the final states induced by the applied electromagnetic field. Their analysis, being non-perturbative in the radiative interaction, includes all higher-order contributions (such as the absorption of two laser photons and the stimulated emission of two laser photons) as well as the second-order process of one-photon absorption and one-photon stimulated emission. It should be emphasized that the lowest non-vanishing order of this mixing of equal-energy states is the second order, not the first order. This fact can be seen from the radiation–matter interaction Hamiltonian, i.e. that any *actual* transition from one molecular state to another resulting from the first-order radiative interaction necessarily involves the absorption or emission of one photon (or $\omega_{fi} \pm \omega = 0$ if the electromagnetic field is treated classically). Therefore in the present situation where the energy difference $\hbar\omega_{fi}$ between the initial and final states is zero, actual transitions between these two states by the first-order radiative interaction *cannot* occur according to the principle of energy conservation. The same conclusion also holds when ω_{fi} is very small compared to ω of an EMF. Thus for radiative transitions between equal-energy states induced by a 'high-frequency' EMF, there is no analogy with the familiar *first-order* mixing of these states by an electrostatic or Stark field. A closer analogy with PCE is the second-order. Stark mixing of states (Lau, 1981b). But even here, the two are distinguished by a frequency (ω) dependence in PCE, which becomes very prominent with near-resonant intermediate molecular states. To convey both the apparent 'electrostatic-field-like' nature of PCE (namely, causing transitions between *equal-energy* states) and its difference in frequency dependence with the second-order Stark-field mixing, one may also think of the PCE as *second-order* or *nonlinear* (to include higher-orders) *electromagnetic mixing* of molecular states.

It was noted (Lau and Rhodes, 1977b) that PC bound–free decompositions such as predissociation require, generally speaking, lower laser power to induce a given probability of occurrence than PCE-modified dissociations and collisions. In these latter processes, the molecule (or quasimolecule) is already in the continuum of internuclear motion when PCE induces transition to an equal-energy continuum belonging to a different channel. The physical explanation is that the radiative interaction can act over the lifetime of the initial bound states in PC decompositions. These lifetimes are usually much longer than the transit time of about 10^{-12} s or less for the fragments to fly apart in PCE-modified dissociation or to fly by each other in collision. Therefore, a small transition probability rate per unit time induced by a weaker laser intensity can accumulate to a significant probability over the relatively longer lifetimes in PC decompositions. Since lower laser intensity is preferable in experiments, the subsequent theoretical works by the present author and his colleagues (Lau, 1978, 1979a, b, c, 1980, 1981a, b; Lau et al., 1981; S. N. Dixit, 1980, private communication) tend to emphasize the topic to PC decompositions. On the other hand, Weiner (1980) has applied the theory of Lau and Rhodes to examine the PCE-modified dissociation of the alkali halides. To the author's knowledge, the above are all the works done so far that fall under the present subject of photon-catalysed unimolecular decompositions, and their results form the core of material covered in this chapter.

The realtionships among some of these works are briefly summarized as follows. In view of the fact that many intense lasers are not tunable in frequency, the PCE in molecules was initially analysed without emphasis on near-resonance of the laser frequency with the transition from the initial state to the intermediate discrete vibrational state (Lau and Rhodes, 1977b). The PCE was then called 'the non-resonant effect' to convey the idea that this process does not necessarily require resonant absorption of the laser photons. The theoretical analysis treated the vibrational motion semiclassically and neglected the rotational structure. All subsequent analyses on PC predissociation treated both the vibrational and rotational motions quantum mechanically (Lau, 1978, 1979a, b). Resonance enhancement was then proposed to reduce the laser power requirement of the non-resonant PCE (Lau, 1979c). At near-resonance, coherent saturation of the initial–intermediate transition was shown to be important but could also be exploited to broaden the class of applicable molecules (Lau, 1980). Recently, a more general theory containing the above features and the dependence on lifetimes has been given (Lau, 1981a). The theoretical formalism and analytic results in these works are applicable to both PC predissociation and PC autoionization. Although examples of specific systems with semiquantitative evaluations were suggested for possible experimental observation of PCE, detailed calculations are now in progress for I_2 (Lau et al., 1981) and other systems. Results of the dependence of PCE on the radiation polarizations (Lau, 1981b) and on the bandwidth of the laser radiation (Lau, 1979c; S. N. Dixit, 1980, Private Communication) have also been obtained.

The PCE has not yet been observed, so all the works reviewed here are theoretical. Being among the simplest chemical reactions, unimolecular processes such as PC predissociation and PC autoionization (especially those that produce fluorescing or charged fragments) may be easier for experimental observation. Our theoretical analyses have yielded a set of general criteria for choosing the molecular systems to optimize the effect (Lau, 1981a; and see the Sec. VI below). For the choice of specific systems, certainly small atoms and simple molecules with fewer channels of decomposition and more precise spectroscopic and calculated information are preferable for unambiguous demonstration. These considerations have guided us in our selection of specific systems for calculations to predict the PCE.

Although the PCE has not yet been observed, it is reasonable to expect that it should occur. Besides the more detailed analyses reported later in this chapter, the expectation is also based on consideration of general principles. We note that the PCE, in its lowest order, is just a kind of two-photon process: the closest analogy to PC predissociation is two-photon dissociation, and to PC autoionization is two-photon ionization. In these latter processes, virtual (or real) absorption of one photon excites the molecule into a virtual (or real) intermediate state, from which absorption of another photon brings the molecules into the continuum. This is illustrated by the broken arrows in Fig. 3. In PC decompositions, the first 'step' is identical whereas the second 'step' in contrast is a stimulated emission into the continuum instead of further

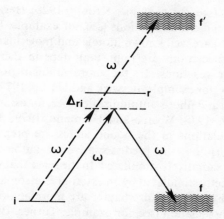

Fig. 3. Comparison of the photon-catalysed decomposition (solid arrows), a kind of two-photon process, with the familiar two-photon ionization or dissociation (dashed arrows).

absorption. It is well known that if the matrix element for the absorption is equal in magnitude to that for stimulated emission, the probabilities of occurrence for the two processes are equal. This is analogous to the statement for a bound–bound transition that the Einstein B coefficient for absorption is the same as that of stimulated emission between the same pair of discrete states. Thus, generally speaking one can conclude that PC decompositions are not any less likely to occur than the two-photon dissociation or ionizations from the same bound states. Now, two-photon dissociation (see, for example, Sorokin and Lankard, 1971; Wang and Davis, 1975) and two-photon ionization (see, for example, Granneman et al., 1976; Morellec et al., 1980)— and even higher-order bound–free processes (see, for example, Chin et al., 1969; Mainfray et al., 1972; Johnson et al., 1975; Tai and Dalby, 1978; Baravian et al., 1979)—of many different atoms and molecules have been observed. Therefore with the corresponding laser intensities about the same orders of magnitude as those of two-photon ionization (or dissociation) and with the appropriate choice of molecular systems, the PCE in autoionization (or predissociation) should occur with significant probability to render it observable. It has been pointed out (Lau, 1980) that the PCE may be occurring along with other multiphoton dissociation and/or ionization processes, thus providing another channel of decomposition of atoms or molecules. It is possible that the PCE was not detected along with these multiphoton processes in some cases simply because no effort was made to observe such a channel. We may also point out that our analytic results and discussions in Sec. II should be directly applicable to two-photon dissociation and ionization.

Predissociations (see, for example, Kronig, 1928; Herzberg, 1950, 1966; Lehmann, 1978b) and autoionizations (see, for example, Fano and Cooper, 1968; Garton, 1966) are well known atomic and molecular processes. Case I and case II predissociations are important steps in thermal unimolecular decompositions or reactions. In the more familiar processes of photo-predissociation (see, for example, Beswick and Durup, 1979; Riley et al., 1974; Gough et al., 1979) and photo-autoionization (see, for example, Bradley et al., 1973; Cooke et al., 1978; Wynne and Hermann, 1979; Sugar et al., 1979), although the populations in the bound states are prepared by single- or multiphoton absorption, the predissociative or autoionizing bound–free transitions are not radiatively stimulated. In the sense that the PC bound–free transitions are directly stimulated by an external electromagnetic field, the PC decompositions discussed in this chapter are more analogous to magnetic-field-induced predissociation (see, for example, Turner, 1930; Chapman and Bunker, 1972), electric-field-induced predissociation (see, for example, Zener, 1933; Sullivan and Dows, 1980), the recently proposed laser-induced pre-dissociation (Lau, 1979a), and laser-induced autoionization (Lambropoulos and Zoller, 1981) involving net absorption or stimulated emission of laser photons.

B. Organization of this Chapter

In Sec. II, we discuss our current understanding and several important aspects of PC decompositions from a unifying and rather simple theory. With respect to the degree of resonance of the laser frequency with the molecular transition between the initial state and the discrete intermediate state, the situations can be separated into three categories: non-resonant, resonant, and far off-resonant. The last two situations are the two opposite extremes of the more general first category. The resonant situation is important in the sense that resonance enhancement can increase the signals by orders of magnitude or equivalently reduce the required laser power needed for a given signal size. At near-resonance, coherent saturation between the initial and discrete intermediate states becomes probable and the important consequence is that the probability rate of PCE, being ordinarily a second-order process, becomes the same order as a single-photon bound–free process. Thus, under coherent saturation conditions, the PCE is as likely to occur as a single-photon bound–free process and, under some optimization of transition moments for a particular molecule, it may even dominate over a competing single-photon bound–free process from the same initial states. It is also possible to increase the yield of PC fragments by appropriate choice of the lifetimes of the initial state and the intermediate state. The selection rules for allowed photon-catalysed transitions in diatomic and polyatomic molecules are given in Sec. 11 B. They are, of course the same regardless of the degree of resonance. Of particular importance are the new transition channels which are the new states coupled by PCE but forbidden in the absence of the external applied laser field.

In Sec. III, the rotational transition strengths for PC predissociation of symmetric top molecules (including diatoms) by a linearly polarized laser are given. Application of these analytic results to individual states requires only simple substitution of the appropriate rotational quantum numbers.

In Sec. IV, proposed systems for PC predissociation and PC autoionization are described, utilizing many results and features of the theory in the previous two sections. The lowest triplet states of the alkali-metal diatoms are examples of those systems where the PCE does not have to compete with the lower-order single-photon dissociation process. However, the lack of more precise information especially among the heavier alkalis prevents us from saying definitely at present if these systems are really favourable for experimentation. The other class of systems, the $B(0_u^+)$ states of the diatomic halogens, represents those molecular systems where PCE has to compete with single-photon dissociation. The features of resonance and coherent situation are exploited to make PCE dominant over the single-photon dissociation. Some preliminary calculated results in I_2 indicate that PC predissociation in I_2 may be favourable for experimentation. Finally PC autoionization of the alkali atoms in the metastable quartet states is discussed.

In Sec. V, we give the semiclassical theory of PC predissociation where the rotational and vibrational motions are treated quasiclassically (the Wentzel–Kramers–Brillowin (WKB) approximation) or classically. The theory is formulated in a much simpler way than that presented in Lau and Rhodes but recovers some of their more important results for weak to moderate radiative interaction strength. The important features of PC transitions at curve crossings and of field-induced avoided crossings are discussed. For PC predissociation, this is a more limited theory than the quantum theory of Sec. II. Unlike the results of Secs. II and III, the theory here can be applied directly to PCE-modified dissociation. Examples of PCE-modified dissociation branching ratios of alkali halides due to Weiner are given.

In the presentation of this chapter, we tend to emphasize more our current understanding and results on the topic. In fact, most of the results presented here are either recently or will shortly be published. Some of the results given here are new. A summary of the results of this chapter is given in Sec. VI.

II. QUANTUM THEORY OF PHOTON-CATALYSED (PC) DECOMPOSITIONS

For the PC decompositions of Eqs. (1.1) and (1.2), the yield of one of the fragments collected over the time of an experiment is given by Y, where

$$Y = N_m P. \tag{2.1}$$

N_m is the total number of molecules (or atoms) in all the initial states $|i\rangle$ irradiated by the PC laser. P is the probability of PC decomposition from all the states $|i\rangle$ averaged over the initial probability distribution of population g_i in the states $|i\rangle$, i.e.

$$P = \sum_i g_i P_i, \tag{2.2}$$

where g_i is normalized according to $\Sigma_i g_i = 1$ and P_i is the probability of PC decomposition from a particular state $|i\rangle$. In this section, we give the theory for the calculation of the P_i and the rate constant γ, first for the non-resonant situation in Sec. II A, then for the resonant situation in Sec. II C, and then we shall discuss several interesting aspects of these results in Secs. II D to F. The selection rules for PCE are given in Sec. II B. The far off resonant situation is presented in Sec. II G.

The total Hamiltonian H for the physical system of a molecule irradiated by the PC laser consists of the Hamiltonian H_m for the molecule, the Hamiltonian

H_r for the electromagnetic field mode of the PC laser, the interaction Hamiltonian H' between the molecule and the laser radiation, and the intramolecular interaction Hamiltonian H'', which is the difference between the exact molecular Hamiltonian and H_m. We shall leave H_m unspecified on purpose for it depends on the level of approximation in the molecular states in a particular calculation. In the case of predissociation, H'' can be the non-adiabatic interaction terms (see, for example, Born and Huang, 1956; Kolos and Wolniewicz, 1963). In the case of autoionization, H'' is the configuration interaction (see, for example, Condon and Shortley, 1967). By including H'' in our theory, we can treat PC decomposition in the presence of the corresponding natural decomposition and their interferences (Lau, 1978; and see below).

The eigenvalues and eigenstates of H_m and H_r are defined by the following relations,

$$H_m|m\rangle = E_m|m\rangle \equiv \hbar\omega_m|m\rangle, \tag{2.3}$$

and

$$H_r|n\rangle = \hbar\omega a^+ a|n\rangle = n\hbar\omega|n\rangle, \tag{2.4}$$

where a^+ and a are the usual creation and destruction operators of photons (Heitler, 1954). Although higher multipole moments may readily be included (for example, as shown in Lau, 1979a), it is sufficient for most purposes to consider the electric dipole interaction only. In this case, the electric dipole interaction Hamiltonian H' is given by

$$H' = -\sum_l q_l \mathbf{r}_l \cdot \mathbf{E}, \tag{2.5}$$

where the electric field operator is given by

$$\mathbf{E} = i\left(\frac{2\pi\omega\hbar}{V}\right)^{1/2} (a - a^+)\hat{\varepsilon}, \tag{2.6}$$

where ω and $\hat{\varepsilon}$ are the angular frequency and polarization vector, respectively, of the electromagnetic field mode of the laser field, and V is the quantization volume which does not appear in the final results (Heitler, 1954). Expanding the total wavefunction Ψ for the system of molecule and the interaction field in terms of the basis states $|m'n'\rangle \equiv |m'\rangle|n'\rangle$

$$\Psi(t) = \sum_{m'n'} b_{m'n'}(t) e^{-i\omega_{m'n'}t}|m'n'\rangle, \tag{2.7}$$

one obtains from the Schrödinger equation,

$$i\hbar\frac{db_{mn}(t)}{dt} = \sum_{m'n'} H'_{mnm'n'} e^{-i\omega_{m'n'mn}t} b_{m'n'}(t)$$

$$+ \sum_{m'} H''_{mm'} e^{-i\omega_{mm'}t} b_{m'n}(t), \tag{2.8}$$

where the molecular states $|m\rangle$ and $|m'\rangle$ refer to both continua as well as discrete states. The interaction matrix element $H'_{mnm'n'} \equiv \langle mn|H'|m'n'\rangle$ will be expressed in terms of the electric dipole transition moment

$$D_{mm'} \equiv \langle m|\sum_l q_l r_l|m'\rangle \cdot \hat{\varepsilon}, \qquad (2.9)$$

or the Rabi flopping frequency $\chi_{mm'}$ (Rabi, 1937) by the following relations,

$$H'_{mnm'n'} = \pm i\left(\frac{2\pi}{c}I\right)^{1/2} D_{mm'}\delta_{n,n'+1} = \pm i\frac{\bar{h}}{2}\chi_{mm'}\delta_{n,n'\pm1}, \qquad (2.10)$$

where the upper (lower) sign corresponds to emission (absorption) of a laser photon in going from state $|m'\rangle$ to $|m\rangle$. The laser intensity I in Eq. (2.10) is the spectrally integrated laser intensity and is related to the mean number N of photons in the electromagnetic field mode by the relation,

$$I = N\hbar\omega c/V \simeq (N \pm 1)\hbar\omega c/V. \qquad (2.11)$$

The last equality holds because N is very much larger than unity. The frequency difference $\omega_{m'n'mn}$ used in Eq. (2.8) is defined by

$$\omega_{m'n'mn} \equiv \omega_{m'} - \omega_m + (n' - n)\omega. \qquad (2.12)$$

Since the intramolecular interaction Hamiltonian does not operate on the states $|n\rangle$ of the radiation field, it couples states $|mn\rangle$ of different molecular states $|m'\rangle$ but of the same photon number n. Hence,

$$H''_{mnm'n'} = H_{mm'}\delta_{nn'}. \qquad (2.13)$$

Therefore, in Eq. (2.8), the sum over n' has been evaluated and the frequency differences

$$\omega_{m'm} = \omega_{m'} - \omega_m \qquad (2.14)$$

appear instead of $\omega_{m'nmn}$.

The composite non-interacting system of molecule and laser field is initially in the state $|iN\rangle \equiv |i\rangle|N\rangle$, i.e. the molecule in state $|i\rangle$ and the field in the photon-number state $|N\rangle$. As seen from Eqs. (2.8) and (2.10), this state can be coupled to the intermediate states $|mn\rangle$ (where n may be $N+1$ or $N-1$) through the radiative interaction H'_{iNmn}. Similarly, the states $|mn\rangle$ are coupled to the state $|fN\rangle$ which is the composite state of the final continuum $|f\rangle$ and the field state $|N\rangle$ with the number of photons N conserved. In addition, the initial state $|i\rangle$ (and $|m\rangle$) may be coupled by H''_{if} (and H''_{mf}) to the continuum $|f\rangle$ leading to natural decompositions. The solutions for the probability amplitudes b_{iN}, b_{mn}, and b_{fN} of these principal states in Eq. (2.8) are somewhat different depending on whether or not the laser frequency ω is near-resonant with any molecular transition frequency ω_{mi} between the initial state $|i\rangle$ and some discrete molecular state $|m\rangle$. We shall discuss the solutions and results of the two cases separately in Secs. II A and C.

A. Non-Resonant PC Decomposition and Interference with Natural Decomposition

For the radiative transition between states $|i\rangle$ and $|m\rangle$, the transition is called non-resonant if the frequency detuning,

$$\Delta_{mi}(\pm\omega) = \omega_m - \omega_i \pm \omega, \tag{2.15}$$

between the transition frequency ω_{mi} and the laser photon frequency ω is much greater than the molecular linewidth γ_{mi}, i.e.

$$\Delta_{mi}^2 \gg \tfrac{1}{4}\gamma_{mi}^2. \tag{2.16}$$

The same transition is said to be coherently unsaturated if its Rabi flopping frequency χ_{mi} is much less than the frequency detuning and its linewidth, i.e.

$$\Delta_{mi}^2 + \tfrac{1}{4}\gamma_{mi}^2 \gg \chi_{mi}^2. \tag{2.17}$$

The non-resonant PC decomposition considered in this subsection is characterized by the above two conditions. Since none of the intermediate states $|m\rangle$ is resonant, radiative transitions to or from the intermediate states are virtual. Although this case has been considered before (Lau and Rhodes, 1977b; Lau, 1978, 1979a), some results in this subsection are given for the first time.

Suppose a molecule is in state $|i\rangle$ at time $t = 0$ when the radiative interaction of approximately constant intensity is switched on. In Appendix A, we solve the Schrödinger equation (2.8) and show that during subsequent time the population in state $|i\rangle$ decays as

$$|b_{iN}(t)|^2 = e^{-\gamma_i' t}, \tag{2.18}$$

where γ_i', the total decay rate of the state $|i\rangle$, is a sum of the natural decay rate γ_i (including natural predissociation, intramolecular relaxation, radiative decay, collisional quenching, etc.) and any laser-stimulated decay $\gamma_{f'i}$ such as single-photon dissociation (see Eqs. (A.9) and (A.11)). In Appendix A, it is also shown that the probability at time t of the molecule in the state $|i\rangle$ being decomposed by both the PCE and the intramolecular interaction is given by

$$P_i(t) = \Gamma_i(1 - e^{-\gamma_i' t})/\gamma_i', \tag{2.19}$$

where the decomposition probability rate per unit time Γ_i is given by Eqs. (A.19) and (2.10),

$$\Gamma_i = \frac{2\pi}{\hbar}\sum_f{}'\left|\frac{4\pi}{\hbar c}I\sum_m\frac{\omega_{im}D_{fm}D_{mi}}{\omega_{im}^2 - \omega^2} + H_{fi}''\right|^2 \rho_f, \tag{2.20}$$

where $\rho(E_f)$ is the density of the continuum of the final state evaluated at $E_f = E_i$, the energy conservation condition. The density ρ is defined in the normalization of the continuum states $\langle E_f | E_f' \rangle = \rho^{-1} \delta(E_f - E_f')$.

Eqs. (2.19) and (2.20) are especially useful to calculate $P_i(\tau_p)$ for the case of irradiation by short laser pulses whose duration τ_p is shorter than the effective lifetime $\tau_i \equiv 1/\gamma_i'$ of the state $|i\rangle$ and whose intensity can be approximated as fairly constant over τ_p. For a continuous-wave or long laser pulse $(\tau_p \gg \tau_i)$ with fairly constant intensity, one obtains from Eq. (2.19) the simple result,

$$P_i(\infty) = \Gamma_i \tau_i. \tag{2.21}$$

For a laser pulse of considerable but smooth variation in intensity over τ_p or τ_i, one may calculate $P_i(t)$ more accurately by

$$P_i(t) = \int_0^t dt' \, \Gamma_i(t') |b_{iN}(t')|^2, \tag{2.22}$$

where $\Gamma_i(t')$ is given by Eq. (2.20) with its laser intensity $I(t')$ describing the temporal laser pulse shape (see Eqs. (A.21)–(A.22)).

We see from the expression for the rate constant Γ_i in Eq. (2.20) or Eq. (A.19) that the rate constant for *non-resonant* PCE is proportional to the square of the laser intensity I, to the square of the product of electric dipole transition moments D_{fm} and D_{mi}, and inversely proportional to the square of the frequency detuning Δ_{mi}. The rate constant Γ_i depends on the laser polarization through the transition moments D_{fm} and D_{mi}, and on the laser frequency ω through the frequency detuning in the denominator. There could be additional laser dependence in the probability P_i through γ_i' if the laser-stimulated decays $\gamma_{f'i}$ in γ_i' are significant. Note that unlike the natural decomposition, the PCE does not depend on the overlap of the wavefunctions of the states $|i\rangle$ and $|f\rangle$, but rather it depends on those between $|i\rangle$ and $|m\rangle$, and between $|m\rangle$ and $|f\rangle$. We shall see in the next subsection that there are molecular states with symmetries such that both terms in Eq. (2.20) for the PC decomposition and the natural decomposition are allowed. Then Eq. (2.20) shows that interference between the two processes will arise.

As mentioned before, population in some initial states $|i\rangle$ may, depending on the molecule, be photodecomposed by absorption (or stimulated emission) of single photon into the continuum $|f'\rangle$ of another or the same channel as $|f\rangle$. The rate constant $\gamma_{f'i}$ for such a process is given by Eq. (A.11). The energy difference between $E_{f'}$ and E_f is one photon energy. For order-of-magnitude comparisons between the two rate constants, Γ_i may be written for a typical term as

$$\Gamma_i \sim \frac{4\pi^2}{\hbar c} I |D_{fm}|^2 \rho \frac{\frac{1}{4}\chi_{mi}^2}{\Delta_{im}^2}, \tag{2.23}$$

where we have substituted in the Rabi flopping frequency according to Eq.

(2.10). If $|D_{fm}|^2 \rho(E_f = E_i)$ is about the same order as $|D_{f'i}|^2 \rho(E_{f'} = E_i \pm \hbar\omega)$, then the ratio

$$\frac{\Gamma_i}{\gamma_{f'i}} \sim \frac{\frac{1}{4}\chi_{mi}^2}{\Delta_{im}^2} \tag{2.24}$$

is less than 1 if $\frac{1}{4}\chi_{mi}^2 < \Delta_{im}^2$. A numerical relation to calculate χ_{mi} (in cm^{-1}) is given by

$$\chi_{mi} = 4\cdot6 \times 10^{-4} I^{1/2} D_{mi}, \tag{2.25}$$

where the laser intensity I is in watts per square centimetre and D_{mi} is in debyes. So if the laser intensity is not so high that the coherently unsaturated condition (2.17) holds, then the rate constant Γ_i for PCE, being a second-order process, is smaller than that of the single-photon decomposition (Lau, 1979a). Therefore, when there is the possibility of single-photon decomposition from the same initial states, one has to be careful in the evaluation of its magnitude and try to discriminate its effect from the PCE. However, at near-resonance when Δ_{im} is small, then χ_{im} can become comparable with Δ_{im}, and Γ_i becomes comparable with $\gamma_{f'i}$ (Lau, 1980). This will be elaborated further in Sec. II D on coherent saturation and examples in iodine diatoms will be given in Sec. IV.

B. Selection Rules for the Photon-as-Catalyst Effect (PCE) and New Transition Channels

The PCE between states $|i\rangle$ and $|f\rangle$ is said to be allowed if the expression $D_{fm}D_{mi}$ in Eq. (2.20) is non-zero for some intermediate state $|m\rangle$. If both the transition moments are electric dipole. then two successive applications of the known electric dipole selection rules for the molecule of interest will give its selection rules for the PCE (Lau, 1978, 1979a, b). The electric dipole transition moment D_{mi} is non-vanishing if the symmetries of the two states $|m\rangle$ and $|i\rangle$ are such that the product of these states and the electric dipole moment operator is totally symmetric with respect to the symmetry operations common to the two states (Herzberg, 1966). The above general rules also apply to PC autoionization of atoms and molecules.

As specific examples, Table I compares the selection rules of single-photon transitions (Garstang, 1962), those of non-adiabatic transitions (Kronig, 1928; Herzberg, 1950), and those of PC predissociation in diatomic molecules. It is noted that some allow both non-adiabatic and photon-catalysed transitions. But there are important differences: for the electronic selection rules (6)–(8) in the table, $\Delta\Lambda = \pm 2$ and $\Sigma^+ \leftrightarrow \Sigma^-$ for the Hund's case (a), $\Delta K = \pm 2$ for Hund's case (b), and $\Delta\Omega = \pm 2$ for Hund's case (c) are allowed for PCE but not for non-adiabatic transitions. Thus, these electronic transition channels would be opened up by the applied laser—the so-called *new channels* in Lau and Rhodes (1977b). It would be most interesting to look for the manifestation of

TABLE I. Selection rules for the single-photon electric dipole transition, the photon-as-catalyst effect, and the non-adiabatic transition in diatomic molecules. The notation (a), (b) or (c) refers to the Hund's case (a), (b) or (c), respectively.

	Single-photon Transition	Photon-as-catalyst effect	Non-adiabatic transition
1	$\Delta J = 0, \pm 1$	$\Delta J = 0, \pm 1, \pm 2$	$\Delta J = 0$
2	$\Delta M = 0, \pm 1$	$\Delta M = 0$	$\Delta M = 0$
3	$+ \leftrightarrow -$	$+ \nleftrightarrow -$	$+ \nleftrightarrow -$
4	$s \nleftrightarrow a$	$s \nleftrightarrow a$	$s \nleftrightarrow a$
5	$g \leftrightarrow u$	$g \nleftrightarrow u$	$g \nleftrightarrow u$
6 (a),(b)	$\Delta\Lambda = 0, \pm 1$	$\Delta\Lambda = 0, \pm 1, \pm 2$	$\Delta\Lambda = 0, \pm 1$
	$\Sigma^+ \nleftrightarrow \Sigma^-$	$\Sigma^+ \leftrightarrow \Sigma^-$	$\Sigma^+ \nleftrightarrow \Sigma^-$
7 (a),(b)	$\Delta S = 0$	$\Delta S = 0$	$\Delta S = 0$
(a)	$\Delta\Sigma = 0$	$\Delta\Sigma = 0$	$\Delta\Sigma = 0$
(b)	$\Delta K = 0, \pm 1$	$\Delta K = 0, \pm 1, \pm 2$	$\Delta K = 0$
8 (c)	$\Delta\Omega = 0, \pm 1$	$\Delta\Omega = 0, \pm 1, \pm 2$	$\Delta\Omega = 0, \pm 1$
	$0^+ \nleftrightarrow 0^-$	$0^+ \leftrightarrow 0^-$	

PCE in such new channels (such as going from a Σ to a Δ electronic state). Also it may require less laser power for an observable signal because there are no natural predissociated fragments as background noise in such an experiment.

Even when the electronic selection rules allow for both PCE and non-adiabatic transition, we notice from Table I that the total angular momentum selection rules for PCE are $\Delta J = 0, \pm 1$ and ± 2 and $\Delta M = 0$ whereas those for non-adiabatic transitions are $\Delta J = 0$ and $\Delta M = 0$. This implies then that according to Eq. (2.20), when both PCE and non-adiabatic transitions are allowed, interference effects between these two processes will occur only between states with $J_f = J_i$ and not for states with $J_f = J_i \pm 1, J_i \pm 2$, because the non-adiabatic term H''_{fi} vanishes for the latter cases.

It has been noted (Lau, 1979a) that any states $|i\rangle$ and $|f\rangle$ that allow PCE also allow two-photon absorption and two-photon stimulated emission as far as symmetries are concerned. However, the energies of the final continua are quite different: $E_f = E_i$ for PCE, $E_f = E_i + 2\hbar\omega$ for two-photon absorption, and $E_f = E_i - 2\hbar\omega$ for two-photon stimulated emission. This means that the fragments generated by these different processes can be distinguished by their large difference in relative kinetic energies. Furthermore, because the final vibrational continua (and the significant intermediate states in some cases) are different, the magnitude of the transition matrix elements are quite different (e.g. due to the Franck–Condon factors) so that only one of these three processes may be dominant.

C. Resonant PC Decomposition

It was noted in Sec. II A that the PC decomposition is inversely proportional to $\Delta^2_{mi}(\pm \omega) = (\omega_m - \omega_i \pm \omega)^2$. Therefore, as this frequency detuning becomes smaller either by tuning the laser frequency ω, if possible, or by finding the right combination of initial and intermediate states, the PCE can be enhanced very significantly (Lau, 1979c). However, as Δ_{mi} approaches zero, Eq. (2.20) is no longer accurate. Near resonance with a discrete intermediate state $|r\rangle$, it is necessary to take into account the decay FWHM γ_{ri} and the coherent saturation of the $|i\rangle \rightarrow |r\rangle$ transition resulting in power broadening of the line. An accurate description requires a separate solution for this near-resonant situation. In this subsection, we shall not consider explicitly the interference with natural decomposition. The effect of the latter is included in the natural decay of the discrete states, as shown in Appendix A, and the total decomposition rate constant will be the sum of the resonant PC decomposition rate constant Γ_{iR} given below and that for natural decomposition.

We shall assume that $|r\rangle$ is the only discrete state that is near-resonant with a given initial state $|i\rangle$ and that all other discrete states are non-resonant. Then it is evident for example from the expression (2.20) that the state $|rN'\rangle$ contributes predominantly to the sum over intermediate states. Unlike the case of non-resonance where both $|r, N' = N + 1\rangle$ and $|r, N' = N - 1\rangle$ may contribute significantly, only $|r, N' = N - 1\rangle$ for the scheme (a) and only $|r, N' = N + 1\rangle$ for the scheme (b) in Fig. 4 will contribute significantly for the case of near-resonance. Scheme (a) uses for resonant intermediate state the excited states higher in energy than the initial state, whereas scheme (b) uses lower-lying states, such as those unpopulated rotational–vibrational states in the ground electronic states. Quite often our knowledge of these lower-lying states is much more accurate than that of the higher excited states. Sometimes loss processes competing with PC decomposition are also different for the two schemes.

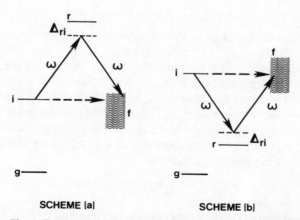

SCHEME (a)   **SCHEME (b)**

Fig. 4. Resonant photon-catalysed decompositions with the near-resonant state $|r\rangle$ lying higher in energy—the scheme (a); or lying lower in energy, such as those unpopulated rotational–vibrational states in the ground electronic state for molecules–the scheme (b).

In the lowest order of coupling, the Schrödinger equations governing the probability amplitudes of the principle states are thus,

$$\frac{d}{dt}b_{iN} = -\frac{\gamma_i}{2}b_{iN} - \frac{i}{\hbar}H'_{iNrN'}b_{rN'}e^{-i\omega_{rN'iN}t}, \tag{2.26}$$

$$\frac{d}{dt}b_{rN'} = -\frac{\gamma_r}{2}b_{rN'} - \frac{i}{\hbar}H'_{rN'iN}b_{iN}e^{-i\omega_{iNrN'}t}, \tag{2.27}$$

and

$$\frac{d}{dt}b_{fN} = -\frac{i}{\hbar}H'_{fNrN'}b_{rN'}e^{-i\omega_{rN'fN}t}, \tag{2.28}$$

where the γ_m are decay rates of the discrete states $|m\rangle$ in the absence of the laser. In these equations here, we neglect for simplicity the radiative interactions with other states $|mn\rangle$ and shall show in Appendix B that they can be included in the theory to give rise to optical Stark shifts of the discrete energies and the laser-stimulated decays that further broaden the lines. In particular, the interaction term proportional to $H'_{rN'fN}b_{fN}$ has been left out on the right-hand side of Eq. (2.27) and will be included in Appendix B to yield the necessarily present laser-stimulated single-photon decay γ_{fr} from $|rN'\rangle$ to $|fN\rangle$. The fact that we neglect it here in Eq. (2.27) and that we include it only as decay channel of the state $|rN'\rangle$ in the treatment in Appendix B means that we neglect the coherent Rabi population oscillation from the continuum $|f\rangle$ back to the discrete states $|r\rangle$. This assumption is intuitively plausible, since molecules once in the region of small interfragment distance of the continuum will fly apart rapidly into the large-distance region where there is negligible transition to the bound wavefunctions.

Eq. (2.26)–(2.28) are to be solved with the initial-value conditions

$$b_{iN}(0) = 1, \; b_{rN'}(0) = b_{fN}(0) = 0. \tag{2.29}$$

This solution and results given in this subsection follows closely that in Lau (1981a). The factor $\omega_{rN'fN}$ can be rewritten as $\omega_{rN'iN} - \omega_{fi}$. Then consider a time interval δt such that it is small compared to the duration τ_p of the slow and smooth time-varying laser pulse, to the Rabi flopping period χ_{ir}^{-1}, to the inverse of the frequency detuning $\omega_{rN'iN}^{-1}$, and to the lifetimes γ_r^{-1} and γ_i^{-1}. Then over the short time domain, $\omega_i^{-1}, \omega_f^{-1} \ll t < \delta t$, the laser intensity $I(t)$ in $H'_{fNrN'}$ and $b_{rN'}e^{-i\omega_{rN'iN}t}$ may be assumed to be essentially constant and may be taken out of the integral in the time integration of Eq. (2.28) over this domain. The results after taking absolute square is

$$|b_{fN}(t)|^2 = \frac{4\pi^2}{\hbar^2 c}I|D_{fr}|^2|b_{rN'}|^2 t\delta(\omega_{fi}), \tag{2.30a}$$

where the indentity (Heitler, 1954) for $t \gg \omega_f^{-1}$ and ω_i^{-1},

$$\frac{1 - \cos\omega_{fi}t}{\omega_{fi}^2} = \pi t\delta(\omega_{fi}), \tag{2.30b}$$

and Eq. (2.10) have been used. Summing over all the final states (including the integration over the continuum energy), one obtains a total probability of PC decomposition at time t of the form $\tilde{\gamma}_i(t)t$, where $\tilde{\gamma}_i(t)$ is an 'instantaneous' (at coarse-grained time interval δt) probability rate given by

$$\tilde{\gamma}_i(t) = \gamma_{fr}(t)|b_{rN'}(t)|^2, \tag{2.31}$$

where

$$\gamma_{fr}(t) \equiv \frac{4\pi^2}{hc} I(t) \sum_f{}' |D_{fr'}|^2 \rho(E_f = E_i), \tag{2.32}$$

is the 'instantaneous' probability rate of transition from $|r\rangle$ to $|f\rangle$. The summation $\sum_f'$ is carried out over the discrete quantum numbers of the final states. Eq. (2.31) says that the photon-catalysed probability rate is given by the probability $|b_{rN'}(t)|^2$ of finding the molecule in the near-resonant state $|r\rangle$ multiplied by the probability rate of transition from $|r\rangle$ to $|f\rangle$. In an experiment where the fragments are collected, the probability P_i of PC decomposition in Eq. (2.2) is obtained by integrating $\tilde{\gamma}_i(t)$ over the time from $t = 0$ when the radiative interaction is turned on to the time τ, which is the shorter of the interaction time (laser pulse duration) or the collection time:

$$P_{iR} = \int_0^\tau dt\, \gamma_{fr}(t)|b_{rN'}(t)|^2. \tag{2.33}$$

For a smooth laser pulse of significant time variation over the pulse duration or over the lifetimes of the states $|i\rangle$ and $|r\rangle$ but of negligible variation over the coarse-grained time interval δt, γ_{fr} can be calculated and $b_{rN'}$ can be solved at successive fixed values of the laser intensity $I(t)$ over the temporal pulse profile and then P_{iR} is calculated according to Eq. (2.33). However, for some special laser pulse shapes, analytic results for γ_{fr}, $b_{rN'}$, and P_{iR} may be obtained. One such case is given below.

Consider a laser pulse with constant intensity over a time interval τ, whose value may range from short to very long compared to the lifetimes γ_i^{-1} and γ_r^{-1}. Then during this time, $H'_{iN rN'}$ in Eqs. (2.26) and (2.27) can be treated as constant. Exact analytic solutions for b_{iN} and $b_{rN'}$ can be obtained (Rabi, 1937; see also, for example, Sargent et al., 1974) by eliminating $b_{rN'}$ and $db_{rN'}/dt$ in Eq. (2.27) using Eq. (2.26) to obtain a second-order differential equation in b_{iN} with constant coefficients. This is then solved by standard methods subject to the initial-value condition (2.29) to obtain

$$b_{iN}(t) = \left(\cos\left(\tfrac{1}{2}\Omega t\right) + i\frac{\Delta_{ri} + i\tfrac{1}{2}(\gamma_i - \gamma_r)}{\Omega}\sin\left(\tfrac{1}{2}\Omega t\right) \right)\exp\left[-i\tfrac{1}{2}\left(\Delta_{ri} - i\frac{\gamma_i + \gamma_r}{2}\right)t \right],$$
$$\tag{2.34}$$

and

$$b_{rN'}(t) = \mp\frac{\chi_{ri}}{\Omega}(\sin\tfrac{1}{2}\Omega t)\exp\left[i\tfrac{1}{2}\left(\Delta_{ri} + i\frac{\gamma_i + \gamma_r}{2}\right)t \right], \tag{2.35}$$

where χ_{ri} is the Rabi frequency in Eq. (2.10) and

$$\Omega \equiv \{[\Delta_{ri} + i\tfrac{1}{2}(\gamma_i - \gamma_r)]^2 + |\chi_{ir}|^2\}^{1/2}, \qquad (2.36)$$

and the frequency detuning,

$$\Delta_{ri} \equiv \omega_r - \omega_i \mp \omega. \qquad (2.37)$$

The upper and lower signs in Eqs. (2.38) and (2.40) correspond to $N' = N - 1$ for the scheme (a) and $N' = N + 1$ for the scheme (b), respectively.

Substituting the solution for $b_{rN'}(t)$ into Eq. (2.33) and performing the time integration, one obtains the probability for resonant PC decomposition at time τ (Lau, 1981a),

$$P_{iR}(\tau) = A_i + B_i(\tau), \qquad (2.38)$$

where A_i is a time-independent term,

$$A_i = \frac{\gamma_{fr}|\chi_{ri}|^2\gamma_{ri}}{4\gamma_i\gamma_r\Delta_{ri}^2 + \gamma_{ri}^2|\chi_{ri}|^2 + \gamma_i\gamma_r\gamma_{ri}^2}, \qquad (2.39)$$

and $B_i(\tau)$ is a time-dependent term,

$$B_i(\tau) = -\frac{\gamma_{fr}|\chi_{ri}|^2}{|\Omega|^2}e^{-\frac{1}{2}\gamma_{ri}\tau}\left(\frac{2\Omega_i\sinh(\Omega_i\tau) + \gamma_{ri}\cosh(\Omega_i\tau)}{\gamma_{ri}^2 - 4\Omega_i^2}\right.$$
$$\left. + \frac{2\Omega_r\sin(\Omega_r\tau) - \gamma_{ri}\cos(\Omega_r\tau)}{\gamma_{ri}^2 + 4\Omega_r^2}\right), \qquad (2.40)$$

with the definitions: (a) γ_{ri} is the total decay rate of the $|i\rangle \leftrightarrow |r\rangle$ transition due to all the incoherent depopulation processes,

$$\gamma_{ri} = \gamma_r + \gamma_i; \qquad (2.41)$$

and (b) Ω_r and Ω_i are, respectively, the real and imaginary parts of Ω defined in Eq. (2.36),

$$\Omega_r = \{\tfrac{1}{2}[\Delta_{ri}^2 + |\chi_{ri}|^2 - \tfrac{1}{4}(\gamma_r - \gamma_i)^2]$$
$$+ \tfrac{1}{2}[(\Delta_{ri}^2 + |\chi_{ri}|^2 - \tfrac{1}{4}(\gamma_r - \gamma_i)^2)^2$$
$$+ (\gamma_r - \gamma_i)^2\Delta_{ri}^2]^{1/2}\}^{1/2}, \qquad (2.42)$$

and

$$\Omega_i = \frac{-(\gamma_r - \gamma_i)\Delta_{ri}}{2\Omega_r}. \qquad (2.43)$$

From its definition Eq. (2.33), we see that $P_{iR}(\tau) \geqslant 0$ and $P_{iR}(\tau \neq \infty) \leqslant P_{iR}(\tau = \infty)$. We can show using the result $P_{iR}(\infty) = A_i$ from Eq. (2.38) that $P_{iR}(\infty) \leqslant 1$ even under very strong coherent saturated pumping (see Eq. (2.73) below). To summarize the conditions of validity of the result (2.38), it is derived by assuming (a) that there is only one discrete state $|r\rangle$ near-resonant with the initial state $|i\rangle$, and (b) that the interaction strength (i.e. laser intensity) is constant during the interaction time or the observation time τ.

Two separate limits of Eqs. (2.38)–(2.40) are of particular interest. In one case, when the lifetime of the initial state is infinite (i.e. $\gamma_i = 0$), such as the case of PC predissociation from the ground states shown in Fig. 2,

$$P_{iR}(\tau) = \frac{\gamma_{fr}}{\gamma_r} + B_i(\tau), \qquad (2.44)$$

where $B_i(\tau)$ is formally the same as Eq. (2.40) but with γ_i set equal to zero. In connection with studying the second case, the long-time limit, we note first that the magnitude of $B_i(\tau)$ is almost equal to A_i for very small τ, i.e. $|B_i(\tau)| \to A_i$ as $\tau \to 0$. The term $B_i(\tau)$ consists of summands, each of which contains an exponential time decay factor. In the limit of long interaction time (due to a long pulse or continuous wave) and long collection time, i.e. when

$$\tau \gg \frac{2}{\gamma_r'(1 - |\Delta_{ri}/\Omega_r|)}, \qquad (2.45a)$$

or when

$$\tau \gg \frac{2}{\gamma_i(1 - |\Delta_{ri}/\Omega_r|)}, \quad \text{if } \gamma_i \neq 0, \qquad (2.45b)$$

it can be proved by using the property $|\Delta_{ri}/\Omega_r| = 1$ for $|\chi_{ri}| \neq 0$ that $B_i(\tau)$ becomes negligible compared to A_i. In the case $\gamma_i = 0$, only condition (2.45a) is needed to determine the $B_i(\tau) \simeq 0$ result. Then P_{iR} of Eq. (2.38) becomes

$$P_{iR} = \begin{cases} \dfrac{\gamma_{fr}\beta_{ri}|\chi_{ri}|^2\tau_{ri}}{\Delta_{ri}^2 + \beta_{ri}|\chi_{ri}|^2 + \frac{1}{4}\gamma_{ri}^2}, & \text{if } \gamma_i, \gamma_r \neq 0; \qquad (2.46) \\[4mm] \dfrac{\gamma_{fr}}{\gamma_r}, & \text{if } \gamma_i = 0; \qquad (2.47) \end{cases}$$

where the dimensionless saturation enhancement factor,

$$\beta_{ri} \equiv (\gamma_i + \gamma_r)^2/(4\gamma_i\gamma_r), \qquad (2.48)$$

and the lifetime,

$$\tau_{ri} \equiv \gamma_{ri}^{-1}. \qquad (2.49)$$

The right-hand side of Eq. (2.46) is essentially the A_i term with the numerator and denominator divided by $4\gamma_i\gamma_r'$. According to the result Eq. (2.47), the probability of PC decomposition for the case $\gamma_i = 0$ in the long-time limit is simply the branching ratio of stimulated $|r\rangle \to |f\rangle$ transition rate to the total decay rate from $|r\rangle$, as it should be.

A time-averaged probability rate per unit time Γ_{iR} for a molecule in a given initial state $|i\rangle$ to be decomposed can be extracted from Eq. (2.46) by noting that $P_{iR} = 2\Gamma_{iR}\tau_{ri}$, where

$$\Gamma_{iR} \equiv \tfrac{1}{2}\gamma_{fr} \frac{\beta_{ri}|\chi_{ri}|^2}{\Delta_{ri}^2 + \beta_{ri}|\chi_{ri}|^2 + \tfrac{1}{4}\gamma_{ri}^2}. \tag{2.50}$$

with the restriction of time t short compared to all τ_{ri}, an overall probability rate γ may be defined by

$$\gamma = \sum_i g_i \Gamma_{iR}, \tag{2.51}$$

which can be useful in rate equations and in considerations of competition with other processes. Note that without such restriction, an overall probability rate γ for all initial states cannot be defined from Eq. (2.46) in conjunction with Eq. (2.1) because the τ_{ri} for different initial states $|i\rangle$ could be quite different in magnitude. When $\beta_{ri} = 1$ (i.e. $\gamma_i = \gamma_r'$), Eqs. (2.50) and (2.51) are the results used in Lau (1980). The results given here are therefore more general. The probability rate Γ_{iR} can be interpreted quite physically as the probability rate γ_{fr} of stimulated transition from the intermediate state to the continuum multiplied by the time-averaged probability of finding the molecule in the intermediate state. Similarly P_{iR} is simply this probability rate Γ_{iR} multiplying an effective time duration of $2\tau_{ri}$. We note from the denominator of P_{iR} in Eq. (2.46) or Γ_{iR} in Eq. (2.50) that the $i \rightarrow r$ transition now has a power broadened FWHM, $2(\beta|\chi_{ri}|^2 + \tfrac{1}{4}\gamma_{ri}^2)^{1/2}$, due to the radiative interaction.

So far in this subsection, we have not considered optical Stark shifts and further line broadening induced by the applied laser. In Appendix B, it is shown that the results in this subsection can still be used when these effects are important, provided the replacement of the following quantities by the corresponding perturbed quantities are made:

$$\omega_i \rightarrow \omega_i + \delta_i, \quad \omega_r \rightarrow \omega_r + \delta_r, \tag{2.52}$$
$$\gamma_i \rightarrow \gamma_i', \quad \gamma_r \rightarrow \gamma_r',$$

where δ_i and δ_r are the optical Stark shifts and γ_i' and γ_r' are the new effective decays of state $|i\rangle$ and $|r\rangle$ defined in Appendix B. In particular, $\gamma_r' = \gamma_r + \Sigma_{f'}\gamma_{f'r}$, includes the necessarily present probability rate γ_{fr} corresponding to the bound–free transition in the PC decomposition process (see Eqs. (2.31)–(2.33) or (2.38)). This turns out to be important to achieve the correct limiting behaviour for P_{iR}, as discussed later. Therefore, γ_r' should always be used instead of γ_r. In all applications, we shall replace γ_r by γ_r' in the results of this subsection. We note also that any laser-stimulated broadening contribution to γ_i' is particularly important when γ_i is negligibly small. When $\gamma_i' \neq 0$ even though $\gamma_i = 0$, we would not have the case $\gamma_i = 0$ described in Eqs. (2.44) and (2.47).

While there are useful applications and other simplifications of the results

Eqs. (2.38)–(2.44) under various conditions, we shall defer their discussion to future publications on specific systems. In the rest of the paper, we shall concentrate our study on the long-time limit, Eqs. (2.46) and (2.50).

D. Coherent Saturation

It has been shown that the bound–bound $|i\rangle \rightarrow |r\rangle$ transition can saturate at rather weak laser power under the near-resonance condition (Lau, 1980). The probability P_{iR} (or its probability rate Γ_{iR}) for this second-order process PCE becomes in magnitude that of the single-photon decomposition from the intermediate state $|r\rangle$ to the continuum state $|f\rangle$. Therefore, the PC decomposition rate can be comparable to or sometimes even larger than the rate of any single-photon decomposition (from the initial state $|i\rangle$ to a continuum state $|f'\rangle$) that may be present in some molecular systems. This leads to the important conclusion that even for molecular states that allow single-photon decomposition the PC decomposition under the coherent saturation condition may still be a significant or sometimes the dominant process. This certainly broadens the scope of applicable molecules (Lau, 1980). In this subsection, we shall demonstrate these points using the results in previous subsections. These results are more general than that used in Lau (1980), leading to some new features.

When the $|i\rangle \rightarrow |r\rangle$ transition is coherently saturated, i.e. when

$$\beta_{ri}|\chi_{ri}|^2 \gg \Delta_{ri}^2 + \tfrac{1}{4}\gamma_{ri}, \tag{2.53}$$

the probability rate Γ_{iR} becomes

$$\Gamma_{iR} = \tfrac{1}{2}\gamma_{fr}, \tag{2.54}$$

which may be interpreted as the probability of the molecule in the intermediate state $|r\rangle$ being $\tfrac{1}{2}$, multiplied by the single-photon decomposition probability rate γ_{fr} from $|r\rangle$ to $|f\rangle$. Comparing the expressions of γ_{fr} in Eq. (2.32) and that of $\gamma_{f'i}$ for single-photon decomposition from $|i\rangle$ to $|f'\rangle$, one sees that both are first-order processes and generally speaking can be of equal order of magnitude. Therefore, PCE can compete on an equal basis with any single-photon decomposition originating from the same states. In fact it can be argued further (Lau, 1980) that $\Gamma_{iR} \propto \gamma_{fr}$ can even be larger than $\gamma_{f'i}$ when $|D_{fr}|^2 \rho(E_f = E_i)$ is larger than $|D_{f'i}|^2 \rho(E_{f'} = E_i \pm \omega h)$. This condition may be satisfied when the Franck–Condon factor and/or electronic transition moment in $|D_{fr}|^2$ are larger than those in $|D_{f'i}|^2$. In PC predissociation, an example is the situation in which the classical turning points of the vibrational states of $|r\rangle$ *and* $|f\rangle$ are at about equal internuclear distance whereas those of $|i\rangle$ and $|f'\rangle$ are far apart.

The bound-state to bound-state transitions are usually saturated coherently at rather low laser power. According to the above result, Γ_{iR} is linear in laser intensity I after coherent saturation. In the coherently unsaturated regime when

$$\Delta_{ri}^2 + \tfrac{1}{4}\gamma_{ri}^2 \gg \beta_{ri}\chi_{ri}^2, \tag{2.55}$$

Γ_{iR} (and Γ_i) is quadratic in laser intensity (apart from possible further intensity dependence in γ_r' and γ_i'). This change in intensity dependence from the unsaturated regime is illustrated in Fig. 5. It is a plot of the intensity dependence factor,

$$F_{ri}(I) = \tfrac{1}{4}I\chi_{ri}^2/(\Delta_{ri}^2 + \chi_{ri}^2 + \tfrac{1}{4}\gamma_{ri}^2), \tag{2.56}$$

in Γ_{iR} ($\propto F_{ri}$) for a pair of resonant ($\Delta_{ri} = 0$) levels i and r with the special values $\beta_{ri} = 1$, $\gamma_{ri} = 0.01\ \mathrm{cm}^{-1}$, and $D_{ri} = 2.3\ \mathrm{D}$. It is seen that the coherent saturation occurs at a relatively weak laser intensity ($\sim 10^2\ \mathrm{W\,cm}^{-2}$). Since $\chi_{ri} \propto I_s^{1/2}D_{ri}$,

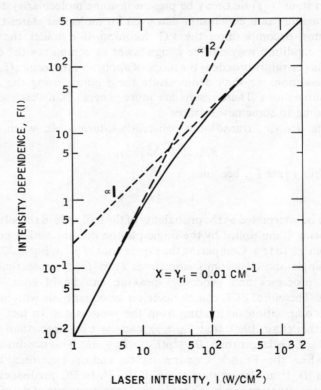

Fig. 5. Plot of the intensity dependence factor $F(I)$ in the probability rate of photon-catalysed decomposition, showing the transition from the quadratic intensity dependence of a second-order rate to the linear dependence of a first-order rate. From Lau (1980). Reproduced by permission of the American Physical Society.

this saturation intensity scales as $|D_{ri}|^{-2}$ for a given value of $\chi_{ri} \sim \gamma_{ri}$. So for more typical transition moment $D_{ri} \sim 0.23$ D in diatoms, $I_s \sim 10^4$ W cm^{-2} for the above example. It is also seen from Fig. 5 that the magnitude of the transition probability would be orders of magnitude too large if coherent saturation is not taken into account in the theory.

The above discussion shows that in optimizing the resonant PC decomposition through the transition moments D_{fr} and D_{ri}, one would choose the resonant state $|r\rangle$ to have the largest possible value of D_{fr} so that $\Gamma_{iR} \propto \gamma_{fr}$ is large. But its D_{ri} does not have to be optimum because the bound–bound transition may already be coherently saturated at the laser power necessitated from other considerations. For then the magnitude of D_{ri} does not contribute any more in the magnitude of $\Gamma_{iR} \propto \gamma_{fr}$.

The coherent saturation behaviour (and power broadening) depends on the enhancement factor β_{ri} multiplying the square of the Rabi flopping frequency χ_{ri}. According to its definition (2.48), this factor can be rewritten as

$$\beta_{ri} = \tfrac{1}{4}(2 + R + R^{-1}), \tag{2.57}$$

where R is the ratio of the rates γ_i and γ'_r of the two discrete states,

$$R \equiv \gamma_i/\gamma'_r, \ \text{ or } \ R \equiv \gamma'_r/\gamma_i. \tag{2.58}$$

The plot of β_{ri} as a function of R in Fig. 6 shows that it has a minimum value $\beta_{ri} = 1$ at $R = 1$ (i.e. $\gamma_i = \gamma'_r$). For any $R \neq 1$, β_{ri} is greater than unity. This means that by having unequal total decay rates (or unequal lifetimes), coherent saturation occurs at a lower laser intensity than that calculated only from the

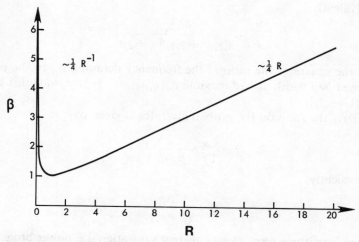

Fig. 6. Plot of the enhancement factor β (Eq. 2.57) in coherent saturation as a function of the ratio R of the lifetimes of the initial and intermediate states. From Lau (1981a).

Rabi flopping frequency χ_{ri}. This enhancement of the coherent saturation behaviour is large when this ratio R (or R^{-1}) becomes larger, as seen from the limiting behaviour of $\beta_{ri} \approx \frac{1}{4}R$ for $R \gg 1$; and $\beta_{ri} \approx \frac{1}{4}R^{-1}$ for $R \ll 1$ (Lau, 1981a).

E. Resonance Enhancement

It has been shown that the probability rate for the PCE can be increased by orders of magnitude due to resonance enhancement (Lau, 1979c). In this subsection, we show this using the more general results presented.

For non-resonant PC decomposition, the probability P_i can be rewritten from Eq. (2.21) for a typical term as

$$P_i \sim \tfrac{1}{4}\gamma_{fr}\chi_{ri}^2\tau_i/\Delta_{ri}^2, \tag{2.59}$$

whereas for resonant PC decomposition, i.e.

$$\Delta_{ri}^2 \ll \beta_{ri}|\chi_{ri}|^2 + \tfrac{1}{4}\gamma_{ri}^2, \tag{2.60}$$

it is

$$P_{iR} = \frac{\gamma_{fr}\beta_{ri}|\chi_{ri}|^2\tau_{ir}}{\beta_{ri}|\chi_{ri}|^2 + \tfrac{1}{4}\gamma_{ri}^2}. \tag{2.61}$$

For the same laser intensity in both cases, the enhancement ratio is given by

$$\frac{P_{iR}}{P_i} \sim \frac{\Delta_{ri}^2}{\beta_{ri}|\chi_{ri}|^2 + \tfrac{1}{4}\gamma_{ri}^2}\left(1 + \frac{\gamma_i}{\gamma_r'}\right), \tag{2.62}$$

or equivalently,

$$\frac{P_{iR}}{P_i} \sim \frac{4\Delta_{ri}^2\gamma_i}{(|\chi_{ri}|^2 + \gamma_r'\gamma_i)(\gamma_r' + \gamma_i)}. \tag{2.63}$$

This is the square of the ratio of the frequency detuning Δ_{ri} to the power-broadened half width at half maximum, $(\beta_{ri}|\chi_{ri}|^2 + \tfrac{1}{4}\gamma_{ri}^2)^{1/2}$, modified by the factor $(1 + \gamma_i/\gamma_r')$.

Similarly the ratio on the probability rates is given by

$$Q \equiv \frac{\Gamma_{iR}}{\Gamma_i} \sim \frac{2\Delta_{ri}^2\beta_{ri}}{\beta_{ri}\chi_{ri}^2 + \tfrac{1}{4}\gamma_{ri}^2}, \tag{2.64}$$

or equivalently,

$$Q \sim \frac{2\Delta_{ri}^2}{\chi_{ri}^2 + \gamma_r'\gamma_i}. \tag{5.65}$$

It is seen from these results that coherent saturation (i.e. power broadening) reduces the resonance enhancement from what it would be without power broadening. To get an idea of the huge resonance enhancement possible, we consider some typical numbers: for a laser intensity $I = 10^7$ W cm^{-2} and a transition moment $D = 0.1$ D, the Rabi flopping frequency $\chi \sim 0.14$ cm^{-1}. Even for rather short-lived states ($\gamma_i = \gamma_r' = 0.1$ cm^{-1}), and the typical range of frequency detuning of a few cm^{-1} to 10^4 cm^{-1} (about 1 eV), the enhancement ratio Γ_{iR}/Γ_i is in the range 10^3 to 10^{10}! If the $i \rightarrow r$ transition in the resonant PCE is not coherently saturated, the enhancement ratio Q is independent of laser intensity, i.e. $Q \sim 2\Delta_{ri}^2/\gamma_r'\gamma_i$. But if that transition is already coherently saturated, then $Q \sim 2\Delta_{ri}^2/\chi_{ri}^2$ is inversely proportional to the laser intensity.

The large resonance enhancement also means that in order to achieve a given probability rate, the required laser power will be greatly reduced if resonance can be achieved. This reduction is desired in experiments to avoid the high intensities at which other nonlinear processes such as multiphoton dissociation, ionization, and gas breakdown may become significant. To achieve the same probability rates $\Gamma_{iR} = \Gamma_i$, the ratio in the reduced laser intensity I_R for the resonance case to I_N for the non-resonance case goes as the inverse of the square root of the enhancement ratio Q, i.e.

$$\frac{I_R}{I_N} \sim Q^{-1/2}, \tag{2.66}$$

for the coherently unsaturated resonant PCE. But for the coherently saturated case, this ratio goes as the inverse of Q,

$$\frac{I_R}{I_N} \sim Q^{-1}. \tag{2.67}$$

F. Optimization of the Fragment Yield by Lifetimes

For non-resonant PC decomposition, the probability $P_i(\infty)$ is proportional to the initial-state effective lifetime τ_i and is not dependent on the lifetimes of the intermediate states (see Eq. (2.21)). For resonant PC decomposition, P_{iR} depends on the effective lifetimes of both the initial and the near-resonant states,

$$P_{iR} = \frac{\frac{1}{4}\gamma_{fr}|\chi_{ri}|^2(\tau_i + \tau_r)}{\Delta_{ri}^2 + \beta_{ri}|\chi_{ri}|^2 + \frac{1}{4}\gamma_{ri}^2}, \tag{2.68}$$

where β_{ri} and γ_{ri} are, of course, also functions of the lifetimes

$$\tau_i \equiv \gamma_i^{-1}, \text{ and } \tau_r \equiv 1/\gamma_r'. \tag{2.69}$$

This result, under the coherently unsaturated condition (2.55), has several interesting limits (Lau, 1981a):

(a) if $\gamma_i = \gamma_r' \equiv \tau_i^{-1}$,

$$P_{iR} = \left(\frac{\frac{1}{2}\gamma_{fr}|\chi_{ri}|^2}{\Delta_{ri}^2 + \frac{1}{4}\gamma_{ri}^2}\right)\tau_i, \tag{2.70}$$

(b) If $\gamma'_r \gg \gamma_i$,

$$P_{iR} = \left(\frac{\frac{1}{4}\gamma_{fr}|\chi_{ri}|^2}{\Delta_{ri}^2 + \frac{1}{4}\gamma_r'^2} \right)\tau_i, \tag{2.71}$$

(c) If $\gamma_i \gg \gamma'_r$,

$$P_{iR} = \left(\frac{\frac{1}{4}\gamma_{fr}|\chi_{ri}|^2}{\Delta_{ri}^2 + \frac{1}{4}\gamma_i^2} \right)\tau_r, \tag{2.72}$$

so that in each case it is a product of a probability rate, the quantity in parentheses, and a lifetime. In each of the cases (b) and (c), the lifetime involved is the *longer* of the two lifetimes. This seems reasonable if in case (b), P_{iR} is considered as the probability rate of a second-order process multiplying the effective lifetime of the initial state. But in case (c), since the initial population decays away rapidly, the final probability P_{iR} of going into the continuum $|f\rangle$ consists of the probability, $\frac{1}{4}\chi_{ri}|^2/(\Delta_{ri}^2 + \frac{1}{4}\gamma_i^2)$, of the molecule being in the state $|r\rangle$, multiplied by the probability of stimulated transition at a rate γ_{fr} from $|r\rangle$ to $|f\rangle$ acting over the effective lifetime τ_r. It is seen from these equations that without coherent saturation, a choice of τ_r or τ_i being very long would enhance the yield of this effect.

However, when the coherent saturation condition (2.53) is satisfied, P_{iR} becomes

$$P_{iR} = \gamma_{fr}\tau_{ri}, \tag{2.73}$$

which has the limits:

(a) for $\gamma'_r = \gamma_i$,

$$P_{iR} = \frac{1}{2}\gamma_{fr}\tau_i, \tag{2.74}$$

(b) for $\gamma_i \ll \gamma'_r$,

$$P_{iR} = \gamma_{fr}\tau_r, \tag{2.75}$$

(c) for $\gamma'_r \ll \gamma_i$,

$$P_{iR} = \gamma_{fr}\tau_i. \tag{2.76}$$

These results state that after coherent saturation of the $|i\rangle \rightarrow |r\rangle$ transition, the probability P_{iR} is reduced simply to the probability rate γ_{fr} of stimulated transition from the intermediate state $|r\rangle$ to the final continuum $|f\rangle$ times the effective lifetime. In each of the cases (b) and (c), the effective lifetime is the *shorter* of the two. This can be understood since, after coherent saturation, the rapid Rabi cycling of population among the two bound states makes the faster decay of the short-lived state the dominant loss of total population. From these equations, it is seen that after coherent saturation, *both* τ_i and τ_r have to be long in order to have an enhancement of yield. Since γ'_r includes γ_{fr}, the probability $P_{iR} \leqslant 1$, as it should be.

G. Far Off-Resonance

In Secs. II C–F, we examined the case of near-resonance. We turn now to consider the other extreme: when all the intermediate states are far away from any resonance. We shall illustrate the simplification in calculation of Eq. (2.20) with PC decompositions of *molecules*.

If we neglect writing the H''_{fi} term in Eq. (2.20) explicitly for the present discussion, the remaining term can be written equivalently as

$$\Gamma_i = \frac{2\pi}{\hbar} \sum_f{}' \left| \frac{2\pi}{\hbar c} I \sum_m D_{fm} D_{mi} (\Delta_{mi}^{-1}(\omega) + \Delta_{mi}^{-1}(-\omega)) \right|^2 \rho_f, \qquad (2.77)$$

where the energy denominators $\Delta_{mi}(\pm\omega)$ are defined in Eq. (2.15). For the non-resonant situation, no particular intermediate states contribute to the sum predominantly, so that the actual computation sometimes requires carrying out the sum over many of the rotational–vibrational states of each electronic intermediate state. For far off-resonance situations to be described below, this tedious procedure can be avoided so that only the sum over the electronic intermediate states needs to be carried out (Lau, 1979a, Sec. III). Let us define the frequency difference

$$\Delta_m(R) \equiv \omega_m - w_m(R), \qquad (2.78)$$

between the actual molecular frequency ω_m and the electronic potential curve $w_m(R)$ (in angular frequency units), and similarly $\Delta_i(R) \equiv \omega_i - w_i(R)$. The $|i\rangle \to |m\rangle$ transition is said to be far off-resonant if the condition

$$\frac{\Delta_m(R) - \Delta_i(R)}{w_m(R) - w_i(R) \pm \omega} \ll 1 \qquad (2.79)$$

is satisfied over the region of R with important Franck–Condon overlap between $|i\rangle$ and $|m\rangle$. This says that the difference in the rotational–vibrational energies is much smaller than the energy off-resonance in the electronic energy. Condition (2.73) can quite often be decided from inspecting the potential curves. If this condition is satisfied, $\Delta_{mi}(\pm\omega)$ in Eq. (2.77) can be approximated as

$$\Delta_{mi}(\pm\omega) \approx w_m(R) - w_i(R) \pm \omega. \qquad (2.80)$$

The above condition needs to be satisfied only by those m states with significant magnitude of $D_{fm} D_{mi}$. For the other m states, we can make the approximation (2.80) without satisfying condition (2.79) and yet do not commit any significant error. After the above manipulations, the sums over the rotational and vibrational states in Σ_m can be carried out trivially to give unity using the completeness relations, since the denominators are now independent of these states. Then a new expression for Γ_i, valid for far off-resonance, is

$$\Gamma_i = \frac{2\pi}{\hbar} \sum_f{}' |\langle E_f J_f | \hat{H}_{fi} | v_i J_i \rangle|^2 \rho_f, \qquad (2.81)$$

where the second-order effective interaction matrix element,

$$\hat{H}_{fi} \equiv \frac{4\pi}{\hbar c} I \sum_m \mu_{fm} \mu_{mi} w_{mi} / (w_{mi}^2 - \omega^2), \qquad (2.82)$$

consists of the electronic transition moments $\mu(R)$ evaluated between the electronic wavefunctions, the electronic frequency differences $w_{mi}(R) \equiv w_m(R) - w_i(R)$, and where the $|v_i J_i\rangle$ and $|E_f J_f\rangle$ are the rotational–vibrational parts of the initial state $|i\rangle$ and final state $|f\rangle$ respectively. The above result is essentially the same as those given in Eqs. (3.8) and (3.9) of Lau (1979a) and Eqs. (1.7)–(1.10) of Lau (1978). The effective interaction matrix element $\hat{H}_{fi}$, is still a function of the internuclear separations R because the μ and w_{mi} are functions of R. The integration over R must be carried out in evaluating the final matrix element $\langle E_f J_f | \hat{H}_{fi} | v_i J_i \rangle$. Using the simplified expression (2.81), one can derive many of the selection rules in Sec. II B and the rotational transition strengths for far off-resonance situations (Lau, 1978).

The same problem of simplifying the sum over intermediate molecular states also occurs in the calculation of the polarizability for Raman light scattering (Schulman et al., 1972; Davydkin and Rapoport, 1975). Explicit evaluation of the full summation over the intermediate states in Eq. (2.77) and the simplified sum over only the electronic parts in Eq. (2.82) shows that, when condition (2.79) is satisfied, their difference is less than a few per cent (Davydkin and Rapoport, 1975). Therefore, when condition (2.79) is valid, the results (2.81)–(2.82) offer the advantage of much simplified calculations.

III. ROTATIONAL TRANSITION STRENGTHS FOR PC PREDISSOCIATION

According to the results of the last section, evaluation of the transition matrix elements D_{fm} and D_{mi} is needed to calculate the probability or rate constant of photon-catalysed predissociation. If the Born–Oppenheimer approximation is taken, each of these transition moments is a product of the electronic transition moment, the square root of the Franck–Condon factor, and a rotational factor. Unlike the electronic transition moments and the Franck–Condon factors, which in almost all cases must be calculated numerically or deduced from experiments for the individual molecule of interest, the rotational transition strengths for this PCE in symmetric top molecules (including diatoms) can be given in analytic forms. Application to specific states in a molecule then requires only the simple substitution of the appropriate rotational quantum numbers. The rotational transition strength for non-resonant PCE induced by a linearly polarized laser in a symmetric top molecule have been given (Lau, 1978, Appendix). In this section, the results for resonant PCE induced by linearly polarized laser radiation are given. These results are adapted from Lau (1981a).

The wavefunctions (Rademacher and Reiche, 1927; Kronig, 1928) of a symmetric top molecule can be written as $|s\alpha v J K M\rangle$, i.e.

$$|\pm \alpha v J K M\rangle = \frac{1}{\sqrt{2}}(|\alpha v J K M\rangle \pm |\alpha v J, -K, M\rangle), \quad \text{for } K \neq 0,$$

and

$$|\pm \alpha vJ0M\rangle = |\alpha vJ0^{\pm}M\rangle, \text{ for } K = 0, \tag{3.1}$$

where s stands for either $+$ or $-$ sign, α is the electronic state label, v stands for all labels $(v_1, v_2, \ldots)$ for the vibrational states, J is the quantum number of the total angular momentum $\mathbf{J}$, M is the quantum number of its component along the space-fixed quantization axis, and K is the quantum number of the component $\mathbf{K}$ of the angular momentum $\mathbf{J}$ along the symmetric top axis. By convention, K is always positive. In general, for symmetric top polyatoms, $\mathbf{K}$ may have contributions from rotational, vibrational, and electronic angular momenta (Herzberg, 1966). In diatoms, $\hbar K$ is the magnitude of the electronic orbital angular momentum along the internuclear axis and is usually denoted by $\hbar\Lambda$. It is well known that the states $|s\alpha vJKM\rangle$ for different M values are $(2J + 1)$-fold degenerate. In addition, for $K \neq 0$, the states with K and $-K$ values are doubly degenerate. The question, therefore, arises whether the analytic formulae for resonant PC predissociation in Sec. IIC–F are applicable, since it has been assumed that for each initial state there is only one resonant intermediate state.

Since the laser bandwidth is usually narrow compared to the rotational energy separation between the P, Q, and R transitions from a given initial state, one can select a particular J' level as near-resonant. Then for a given initial state $|s\alpha vJKM\rangle$ and a given $\alpha'v'J'$ level, it can be shown that the initial state is coupled to only one state out of the $2(2J' + 1)$-fold degenerate states $|s'\alpha'v'J'K'M'\rangle$ for $K' \neq 0$, or one out of the $(2J' + 1)$-fold degenerate states for $K' = 0$. To see this, we note first that the selection rule for an electric dipole transition induced by linearly polarized light is $M' - M = 0$, if the space-fixed quantization axis is chosen along the polarization direction $\hat{\varepsilon}$. This means that for $K' = 0$, only one intermediate state $M' = M$ is coupled; and that for $K' \neq 0$, only two $M' = M$ states corresponding to $s' = s$ and $s' = -s$ are coupled. For the latter case, only one state has the opposite parity to the initial state and is therefore allowed by electric dipole selection rules; the other state has the same parity and is forbidden. This can be seen explicitly using the property under the spatial inversion $\mathbf{r}_l \to -\mathbf{r}_l$,

$$|s\alpha vJKM\rangle \to (-1)^{\delta_{s-}}(-1)^{K+J}|s\alpha vJKM\rangle, \tag{3.2a}$$

one obtains with spatial inversion invariance for the electric dipole matrix element,

$$\langle s'\alpha'v'J'K'M'|\sum_l q_l \mathbf{r}_l|s\alpha vJKM\rangle$$
$$= (-1)^{\delta_{s's}+J'+K'+J+K}\langle s'\alpha'v'J'K'M'|\sum_l q_l \mathbf{r}_l|s\alpha vJKM\rangle, \tag{3.2b}$$

where $\delta_{s's}$ equals 1 if s' and s are of the same sign and equals 0 if s' and s are of opposite sign. It is, therefore, seen that for non-vanishing matrix elements, a given state $|s\alpha vJKM\rangle$ and given quantum numbers $\alpha'v'J'K'M'$, only one value s' is possible.

From Eq. (2.1) with substitutions from Eqs. (2.46), (2.32), and (2.10), the probability P of the PC predissociation in the long-time case is

$$P = \frac{2^5\pi^3}{\hbar^3 c^2}I^2\sum_i g_i\sum_f{}' \rho_f\frac{\beta_{ri}\tau_{ri}|D_{fr}|^2|D_{ri}|^2}{\Delta_{ri}^2 + \frac{1}{4}\gamma_{ri}^2 + (8\pi I\beta_{ri}|D_{ri}|^2/\hbar^2 c)}, \quad (3.3)$$

where i stands for $s\alpha vJKM$, r for $s'\alpha'v'J'K'M$, and f for $s''\alpha''v''J''K''M$. The calculation becomes simpler when the transition matrix elements D are expressed in terms of $|\alpha vJKM\rangle$ only (i.e. free of the $|\alpha vJ, -K, M\rangle$ component in Eq. (3.1) for $K\neq 0$). This is accomplished by using the relationship

$$\langle s'\alpha'v'J'K'M|\sum_l q_l\mathbf{r}_l'|s\alpha vJKM\rangle$$

$$= l_{K'K}(1 + (-1)^{\delta_{s's} + J' + K' + J + K})\langle\alpha'v'J'K'M|\sum_l q_l\mathbf{r}_l'|\alpha vJKM\rangle, \quad (3.4)$$

where

$$l_{K'K} = \begin{cases} 1/\sqrt{2}, \text{if } (K'=0, K=1) \text{ or } (K'=1, K=0), \\ 1/2, \quad \text{in all other cases.} \end{cases}$$

Each of the wavefunctions $|\alpha vJKM\rangle$ can be written as a product of an electronic and vibrational part $|\alpha v\rangle$ and a rotational wavefunction $|JKM\rangle$,

$$|\alpha vJKM\rangle = |\alpha KvJ\rangle|JKM\rangle. \quad (3.5)$$

Thus, the matrix elements $\langle\alpha'v'J'K'M|\Sigma q_l\mathbf{r}_l'|\alpha vJKM\rangle\cdot\hat{\varepsilon}$ can be written as

$$\langle\alpha'v'J'K'M|\sum_l q_l\mathbf{r}_l'|\alpha vJKM\rangle\cdot\hat{\varepsilon} = \mu_{r'i'}A_{J'K'MJKM}, \quad (3.6)$$

where $\mu_{r'i'}$ are electronic–vibrational electric dipole matrix elements defined in Table II and A is a rotational factor. These angular factors A are known (Mizushima, 1975) and can be expressed as a product of three factors,

$$A_{J'K'MJKM} = N_{J'J}C_{J'MJM}B_{J'K'JK}, \quad (3.7)$$

TABLE II. Definitions of the electronic–vibrational electric dipole matrix element $\mu_{\alpha'K'v'J'\alpha KvJ}$ in Eq. (3.6). The evaluation of these matrix elements involves integration over the coordinates $\mathbf{r}_l$ in the molecular body-fixed coordinate system.

K'	$\mu_{\alpha'K'v'J'\alpha KvJ}$
$K+1$	$\langle\alpha', K+1, v'J'|\sum_l q_l(x_l + iy_l)|\alpha KvJ\rangle$
K	$\langle\alpha'Kv'J'|\sum_l q_l z_l|\alpha KvJ\rangle$
$K-1$	$\langle\alpha', K-1, v'J'|\sum_l q_l(x_l - iy_l)|\alpha KvJ\rangle$

TABLE III. Expressions of the factors $N_{J'J}$, $C_{J'MJM}$, and $B_{J'K'JK}$ in Eq. (3.7). The $C_{J'MJM}$ are valid only for linear polarization $\hat{\varepsilon}$ chosen as the rotational quantization space-fixed axis $\hat{z}$. Other than notational differences and factors of $\frac{1}{2}$ in the definition of the B, these factors are the same as those in Mizushima (1975).

	$J' = J+1$	$J' = J$	$J' = J-1$
$N_{J'J}$	$(J+1)^{-1}[(2J+1)(2J+3)]^{-1/2}$	$[J(J+1)]^{-1}$	$J^{-1}(4J^2-1)^{-1/2}$
$C_{J'MJM}$	$[(J+1)^2 - M^2]^{1/2}$	M	$[J^2 - M^2]^{1/2}$
$B_{J',K+1,JK}$	$-\frac{1}{2}[(J+K+1)(J+K+2)]^{1/2}$	$\frac{1}{2}[J(J+1)-K(K+1)]^{1/2}$	$\frac{1}{2}[(J-K)(J-K-1)]^{1/2}$
$B_{J'KJK}$	$[(J+1)^2 - K^2]^{1/2}$	K	$[J^2 - K^2]^{1/2}$
$B_{J',K-1,JK}$	$\frac{1}{2}[(J-K+1)(J-K+2)]^{1/2}$	$\frac{1}{2}[J(J+1)-K(K-1)]^{1/2}$	$-\frac{1}{2}[(J+K)(J+K-1)]^{1/2}$

where N, C, and B are given in Table III for convenience of reference. Substitution of Eqs. (3.4) and (3.6) into Eq. (3.3), gives

$$P = \left(\frac{8\pi}{\hbar}\right)^3 \left(\frac{I}{c}\right)^2 \sum_i g_i \sum_f{}' \rho_f \beta_{ri} \tau_{ri} l^2_{K''K} l^2_{K'K}$$

$$\times \frac{|\mu_{f'r'} \mu_{r'i'}|^2 A^2_{J''K''MJ'K'M} A^2_{J'K'MJKM}}{\Delta^2_{ri} + \frac{1}{4}\gamma^2_{ri} + (2^5 \pi I \beta_{ri} l^2_{K'K} |\mu_{r'i'}|^2 A^2_{J'K'MJKM}/\hbar^2 c)}, \tag{3.8}$$

where i, r, and f now stand for $\alpha v J K M$, $\alpha' v' J' K' M$, and $\alpha'' v'' E_f J'' K'' M$, respectively; i', r', and f' stand for $\alpha K v J$, $\alpha' K' v' J'$, and $\alpha'' K'' v'' J''$, respectively; and J' may take either one of the values $J-1$, J, $J+1$ while the summation over J'' includes all three values $J'-1$, J', $J'+1$. Eq. (3.8) is the general result of this section. It differs from Eq. (3.3) in that the sums over s, s', and s'' have been performed, and it exhibits the rotational transition factors A given by Eq. (3.7) and Table III.

The expression for P can be simplified further for the unsaturated regime and the coherently saturated regime. Since the Rabi flopping frequency χ_{ri} (or the third term in the denominator in Eq. (3.8)) differs for different pairs of $|i\rangle$ and $|r\rangle$ states, the condition (2.55) for non-saturation and the condition (2.53) for coherent saturation must be satisfied for all the significantly populated states $|i\rangle$. The following results in Eq. (3.11) and (3.12) for the two regimes are also further simplified in that the summation over M in Σ_i has been performed. In order to perform this summation, two simplifying assumptions have been made. The first one is that γ'_r (also Δ'_{ri} and γ'_i if Eq. (2.52) is used) are independent of M. The second one is that the initial population distribution is equal in states of different M but of the same J, i.e.

$$q_{\alpha v J K M} = \frac{1}{2J+1} g_{\alpha v J K}, \tag{3.9}$$

for all M. Another approximation, but an excellent one, is that the densities of final states $\rho_{J''}$ and transition matrix element $\mu_{f'r'}$ for *adjacent* rotational states are equal, i.e.

$$\rho_{J'} |\mu_{J'J'}|^2 = \rho_{J'-1} |\mu_{J'-1,J'}|^2 = \rho_{J'+1} |\mu_{J'+1,J'}|^2. \tag{3.10}$$

Under these assumptions, the result for P for near-resonant intermediate transition but before coherent saturation is

$$P = \left(\frac{8\pi}{\hbar}\right)^3 \left(\frac{I}{c}\right)^2 \sum_{i'} g_{i'} \sum_{\alpha''K''v''} \rho_{\alpha''K''v''J'} \beta_{r'i'} \tau_{r'i'} l^2_{K''K} l^2_{K'K}$$

$$\times \frac{|\mu_{\alpha''K''v''J',r'} \mu_{r'i'}|^2}{\Delta^2_{r'i'} + \frac{1}{4}\gamma^2_{r'i'}} R_{K''J'K'JK}, \tag{3.11}$$

TABLE IV. The $G_{J''J'J}$ factors defined in Eq. (3.13) and used in Eq. (3.12) for the calculation of rotational transition strengths of the PCE in the non-saturated case.

	$J'' = J'+1$	$J'' = J'$	$J'' = J'-1$
$J' = J+1$	$\dfrac{2}{15(J+1)(J+2)(2J+1)(2J+3)}$	$\dfrac{J}{15(J+1)^3(J+2)(2J+1)}$	$\dfrac{4(J+1)^2+1}{15(J+1)^3(2J+1)^2(2J+3)}$
$J' = J$	$\dfrac{J+2}{15J(J+1)^3(2J+1)}$	$\dfrac{3J(J+1)-1}{15J^3(J+1)^3}$	$\dfrac{J-1}{15J^3(J+1)(2J+1)}$
$J' = J-1$	$\dfrac{4J^2+1}{15J^3(2J-1)(2J+1)^2}$	$\dfrac{J+1}{15J^3(J-1)(2J+1)}$	$\dfrac{2}{15J(J-1)(4J^2-1)}$

where the rotational transition strength R is given by

$$
\begin{aligned}
R_{K''J'K'JK} = & B^2_{J'+1,K''J'K'}B^2_{J'K'JK}G_{J'+1,J'J} \\
& + B^2_{J'K''J'K'}B^2_{J'K'JK}G_{J'J'J} \\
& + B^2_{J'-1,K''J'K'}B^2_{J'K'JK}G_{J'-1,J'J},
\end{aligned}
\tag{3.12}
$$

and the G are defined by

$$
G_{J''J'J} \equiv \frac{N^2_{J''J'}N^2_{J'J}}{(2J+1)} \sum_{M=-J_m}^{J_m} C^2_{J''MJ'M}C^2_{J'MJM},
\tag{3.13}
$$

where J_m is the smallest J'', J', and J. These G are evaluated, and the results are tabulated in Table IV.

Under the same assumptions, the resulting P for near-resonance with intermediate states after coherent saturation is

$$
\begin{aligned}
P = & \frac{2^4\pi^2}{hc}I\sum_{i'}g_{i'}\sum_{\alpha''K''v''}\rho_{\alpha''K''v''J'}\tau_{r'i'}l^2_{K''K'} \\
& \times |\mu_{\alpha''K''v''J'r'}|^2 R_{K''J'K'},
\end{aligned}
\tag{3.14}
$$

where

$$
\begin{aligned}
R_{K''J'K'} = & B^2_{J'+1,K''J'K'}G_{J'+1,J'} + B^2_{J'K''J'K'}G_{J'J'} \\
& + B^2_{J'-1,K''J'K'}G_{J'-1,J'},
\end{aligned}
\tag{3.15}
$$

and the $G_{J''J'}$ are defined by

$$
G_{J''J'} = \frac{N^2_{J''J'}}{2J+1}\sum_{M=-J_m}^{J_m} C^2_{J''MJ'M}.
\tag{3.16}
$$

The results for $G_{J''J'}$ are tabulated in Table V.

With the analytic results given in this section, one can now calculate the rotational transition strengths of the PCE in symmetric top molecules by the simple substitution of the rotational quantum numbers for the states of

TABLE V. The $G_{J''J'}$ factors defined in Eq. (3.16) and used in Eq. (3.15) for the calculation of the rotational transition strengths of the PCE in the coherently saturated case.

	$J'' = J' + 1$	$J'' = J'$	$J'' = J' - 1$
$J'' = J + 1$	$\dfrac{J+4}{3(J+2)^2(2J+5)}$	$\dfrac{J}{3(J+1)(J+2)^2}$	$\dfrac{1}{3(2J+1)(J+1)}$
$J' = J$	$\dfrac{1}{3(J+1)(2J+1)}$	$\dfrac{1}{3J(J+1)}$	$\dfrac{1}{3J(2J+1)}$
$J' = J - 1$	$\dfrac{1}{3J(2J+1)}$	$\dfrac{2J-1}{3J(J-1)(2J+1)}$	$\dfrac{1}{3(J-1)(2J+1)}$

interest. These results are applicable to linearly polarized laser radiation, which is the most commonly encountered case. The analytic results for left and right circularly polarized radiation have also been derived and will be published elsewhere (Lau, 1981b).

IV. EXAMPLES OF PC DECOMPOSITIONS

In this section, we discuss some examples of photon-catalysed decompositions applying the results of the last two sections. Some photon-loss processes competing with the PCE are also discussed in the context of these examples.

A. PC Predissociation of the Alkali-Metal Diatoms

The lowest attractive $^3\Pi_u$ state of the alkali-metal diatoms is an example of those molecular states from which single-photon dissociation is not possible over a range of infrared laser frequencies. The size of this range depends on the initial rotational–vibrational levels. The reason for this is that, according to *ab initio* calculations (for Li_2, see Konowalow and Olson, 1979, and their earlier works cited therein; for Na_2, see Konowalow *et al.*, 1980), the nearest repulsive triplet gerade potential curve is the $^3\Pi_g$ state and this lies above the dissociation limit of the $^3\Pi_u$ state. But PC predissociation of the molecular population in the rotational–vibrational states of $^3\Pi_u$ to the adjacent continuum of the repulsive $^3\Sigma_u^+$ is allowed by symmetries, with the discrete states of a nearby attractive $^3\Sigma_g^+$ acting as intermediate states (Lau, 1979c). Na_2 was used as an example of the alkalis for discussion because much more information is known about it. But the relative position of the potential curves of $^3\Sigma_g^+$ and of $^3\Sigma_u^+$, according to the results of Konowalow *et al.* (1980), indicates that the bound–free Franck–Condon density in γ_{fr} may be quite small. The corresponding potential curves of Li_2 (Konowalow and Olson, 1979) indicate that the bound–free Franck–Condon densities for Li_2 are more favourable. On the other hand, Li_2 has other properties that makes it less desirable for experiment. One component set of states in the $^3\Pi_u$ is predissociated quite strongly (P. S. Julienne, 1980, Private communication). Some more detailed quantitative calculations of this PCE in these two systems are being planned. For the heavier alkalis, there are at present no calculated results of the relevant triplet states. For experimental information, some absorption spectra from the $^3\Sigma_u^+$ state to the $^3\Sigma_g^+$ state of K_2, Cs_2, and Na_2 are beginning to be recorded (Bhaskar *et al.*, 1979; Zouboulis *et al.*, 1980; Vasilakis *et al.*, 1980). Clearly there is a need to extend these works to study smaller internuclear distances and also to obtain discrete spectra of the $^3\Pi_u$–$^3\Sigma_g^+$ transitions before we can say if the PCE in the heavier alkalis is more favourable.

B. PC Predissociation of Halogen Diatoms: I_2

Preliminary detailed calculations of PC predissociation of the iodine molecule in the metastable $B(0_u^+)vJ$ states indicate that this process occurs with significant probability using available laser systems (Lau *et al.*, 1981). Some of these results will be discussed below in this section. They confirm the earlier qualitative assessment of the process in this molecular system (Lau, 1980). The $B(0_u^+)vJ$ states of the halogen diatoms (see, for example, Coxon, 1973) represent the class of molecular states from which both PC predissociation and single-photon stimulated dissociation can occur. Thus, in order to make the PC predissociation the dominant process, we exploit the feature of the PCE becoming a first-order process under coherent saturation, as discussed in Sec. II D.

We shall use the iodine molecule as an example because it is probably the

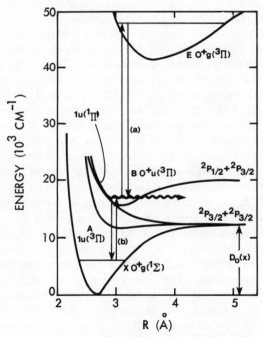

Fig. 7. Photon-catalysed predissociation of I_2 from the $B(0_u^+)vJ$ discrete level to the vibrational continuum (wavy line) of the $1_u(^1\Pi)$ and that of the $A\ 1_u(^3\Pi)$ states, producing two iodine atoms. In scheme (a) the resonant intermediate level is a $E(0_g^+)v'J'$ state, whereas in scheme (b) it is an unpopulated excited rotational–vibrational level in the ground electronic state. From Lau (1980). Reproduced by permission of the American Physical Society.

best studied halogen molecule and much information is available (see, for example, Mulliken, 1971; Lehmann, 1978a, b; and references cited below). The natural predissociation of the $B(0_u^+)vJ$ states of I_2 by the continua of the $1_u(^1\Pi)$ is known to occur (Tellinghuisen, 1972; Broyer et al., 1975, 1976; and references therein). Their predissociation by the $A\ 1_u(^3\Pi)$ continuum is also allowed by symmetry. When an external laser irradiates molecules excited in the $B(0_u^+)vJ$ states, these two ungerade $1_u(^1\Pi)$ and $A\ 1_u(^3\Pi)$ continua can act as the final states of the photon-catalysed predissociation (Lau and Rhodes, 1977b; Lau, 1980). This is illustrated in Fig. 7.

However, the electronic potential curve of the $B\ 0_u^+$ states is also crossed by two repulsive gerade states 1_g and 0_g^+. The existence of these two gerade states in I_2 was predicted by Mulliken (1971), inferred qualitatively from experiments of collisional predissociation (Steinfeld, 1972) and of Stark-field-induced predissociation (Sullivan and Dows, 1980), and calculated by R. L. Jaffe (1980, private communication). As shown in Fig. 8, populations in the

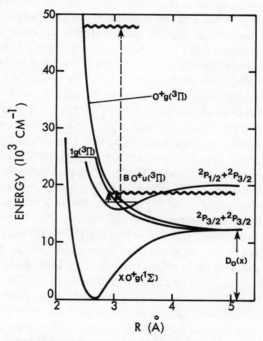

Fig. 8. Single-photon predissociation of I_2 in the $B(0_u^+)vJ$ state induced by lasers of far-infrared or lower frequency is probable (small solid arrows) but photodissociation by lasers of high frequency (dashed arrow) is highly improbable because of the Franck–Condon principle.

$B(0_u^+)vJ$ states of I_2, under irradiation by an infrared (such as CO_2) or low-frequency laser, would be predissociated into these two gerade states by single-photon absorption or stimulated emission (Lau, 1979a). For such low-frequency laser irradiation, one expects that this single-photon predissociation, being first order in laser intensity, would dominate over the second-order process of PCE. Therefore, the $B 0_u^+$ states of I_2 and other halogens are examples of the class of molecular systems in which PCE has to compete with the single-photon predissociation or dissociation process. (Incidentally, it may be worth noting that being a homonuclear diatom, I_2 in the $B vJ$ states does not photodissociate without transition to another electronic state because its permanent dipole moment is zero.)

Nevertheless, under the conditions of near-resonance and coherent saturation of the transition from the initial state to the intermediate state, the PCE possesses a first-order rate constant and can compete on an equal basis with or even dominate over the single-photon dissociation process (Lau, 1980). This statement does not contradict those of the previous paragraph: the difference in molecular response is due to the different laser frequencies needed in the two situations. Since the initial and the final states are ungerade, the intermediate states for PCE must be gerade for electric dipole interaction. Two known nearby gerade states are the X 0_g^+ ground state and the $E 0_g^+$ state (Wieland et al., 1972; Rousseau, 1975; Tellinghuisen, 1975). In order to satisfy the near-resonance condition, the laser frequency has to be at least $4 \times 10^3 \, \text{cm}^{-1}$ and $21 \times 10^3 \, \text{cm}^{-1}$, respectively. At such a laser frequency, the Franck–Condon density of the bound–free transition to the 0_g^+ or 1_g state is expected to be very small according to the Franck–Condon principle, for the needed change of relative kinetic energy from the small value of the bound state to that of close to $\hbar\omega$ is too large (see Fig. 8, dashed arrow). In contrast, the Franck–Condon density from the intermediate state $E(0_g^+)v'J'$ to the continuum of the $1_u(^1\Pi)$ in the bound–free step of PC predissociation is expected to be significant because, as seen from scheme (a) of Fig. 7, the change of relative kinetic energy is very small. If the above two expectations are true, then after coherent saturation, the probability rate $\Gamma_{iR} = \frac{1}{2}\gamma_{fr}$ for PC predissociation should be much greater than that of single-photon dissociation $\gamma_{f'i}$, as discussed below Eq. (2.48). It is possible to calculate fairly reliable Franck–Condon densities between $E(0_g^+)v'J'$ to $1_u(^1\Pi)$ and it is found that they are indeed favourable (Lau et al., 1981). Note that it is not yet possible to do any reliable calculation of Franck–Condon density for the single-photon dissociation from $B(0_u^+)vJ$ to the 0_g^+ or 1_g state because the potential curves of these two gerade states are rather poorly known. In fact, trying to deduce these potential curves by laser-induced single-photon predissociation (Lau, 1979a) or by Stark-field-induced predissociation (Sullivan and Dows, 1980) is an interesting problem in molecular spectroscopy.

Owing to the above considerations, it has been proposed (Lau, 1980) that one could observe PC predissociation of I_2 in the $B(0_u^+)vJ$ states,

$$I_2(BvJ) + N\hbar\omega \rightarrow 2I(^2P_{3/2}) + N\hbar\omega, \tag{4.1}$$

using as the resonant intermediate levels the $E(0_g^+)v'J'$ states in scheme (a) or the $X(0_g^+)v'J'$ states in scheme (b). One can populate selectively a particular rotational–vibrational level in the $B\,0_u^+$ state using a dye laser, as was done for example by Broyer *et al.* (1975). The excited I_2 population is known to fluoresce to the ground $X^1\Sigma_g^+$ state at very well known frequencies (Gerstenkorn and Luc, 1978; Luc, 1980; Bacis *et al.*, 1980; Tellinghuisen *et al.*, 1980 and references cited therein). It also undergoes natural predissociation. The ratios of radiative decay rate to natural predissociation rate are known for some states (Broyer *et al.*, 1975). We assume that the role of this first laser is merely to provide the population in the initial state $B\,vJ$. It can be turned off just before the second PC laser pulse arrives. The experimental observation can monitor a spectrally resolved fluorescence line from the populated $B\,vJ$ level and more importantly detect the iodine atoms. If the PC predissociation occurs when the PC laser irradiates the excited I_2 population, the monitored fluorescence should be depleted compared to its natural decay and, as a more definitive signal, the atomic fragments should be increased.

In scheme (a), the PC laser can be a dye laser, as described for example in Tai and Dalby (1978), in order to have the needed frequency range and intensity. When the laser frequency is resonant with an $E(0_g^+)v'J'$ level, resonant light (Rayleigh and Raman) scattering by the population in the BvJ level occurs with an estimated rate γ_s of about 10^7 s^{-1} by assuming that the $B\,vJ \rightarrow E v'J'$ transition is already saturated coherently. This light scattering is a competing loss mechanism of photons. The fraction with final continuum states will yield atomic iodine, whose distribution in kinetic energy is quite diffuse. But the PCE can compete and dominate over the light scattering. The above value of γ_s is already its maximum value and is constant even at further increase of laser intensity because of coherent saturation of the $|i> \rightarrow |r>$ step. But the PC predissociation $\Gamma_{iR} = \frac{1}{2}\gamma_{fr} \propto I$ still increases linearly with laser intensity. Therefore, there is always a laser intensity beyond which $\frac{1}{2}\gamma_{fr} \geqslant \gamma_s$ and fortunately, for favourable Franck–Condon density, this intensity is not too high–about 10^8 W cm^{-2} for $\gamma_s \sim 10^7$ s^{-1}. As discussed around Eq. (A.17) and Eq. (2.30), the distribution in kinetic energy of the $I(^2P_{3/2})$ fragments resulting from PC predissociation peaks sharply at the energy $\frac{1}{2}(E_{BvJ} - D_0(X))$, where $D_0(X)$ is the dissociation energy of the ground X state. This kinetic energy is in the range of 0·2 to 0·5 eV depending on E_{BvJ}. In a beam experiment, the location and the sharp feature of the distribution can be resolved to distinguish the PC predissociation from those resulting from light scattering.

In scheme (b), the unpopulated excited rotational–vibrational states are used as near-resonant intermediate states. A YAG laser of fixed frequency or a dye laser may be used for the PC laser. If a dye laser is used, its photon energy

$\hbar\omega$ should be less than $D_0(X)$ in order to avoid photodissociating any thermal population in the X state to another repulsive electronic state, unless the experiment can resolve in kinetic energy these photodissociated atoms from the PC atoms. The advantages of scheme (b) in I_2 is that the X state is much better known than the E state in scheme (a) and therefore calculated results are much more reliable. Furthermore, the usual light scattering (with an incident photon absorbed) of the PC laser radiation is far off-resonant with any discrete intermediate states. Also, the resonant light scattering (with the X $v'J'$ level resonant) is negligible because the permanent dipole moment is zero (and if any, the scattered emission would be in the far-infrared). Therefore, we expect that in scheme (b), light scattering is not as important as the depopulation of $I_{\frac{*}{2}}$ in the $|BvJ\rangle$ initial state by spontaneous radiative decay (typical rate $\sim 10^6$ s^{-1}). Again this competing process can be overcome at moderate laser power, as we see in the following specific calculation.

In both schemes (a) and (b), if there were any significant single-photon or multiphoton dissociation through electronic states not yet discussed, their dissociated I atoms should possess large kinetic energy difference from the PC atoms due to the large photon energy ($\hbar\omega \sim 1\,\mathrm{eV}$), and one or both of the atoms should be in an excited state which would fluoresce.

As a specific example, the PC predissociation of $I_{\frac{*}{2}}$ in the initial state $|i\rangle \equiv |B(0_u^+)v = 12, J = 144\rangle$.

$$I_2(B, 12, 144) + N\hbar\omega \rightarrow 2I(^2P_{3/2}) + N\hbar\omega, \qquad (4.2)$$

is calculated with the information of a YAG laser of frequency $9394{\cdot}6\,\mathrm{cm}^{-1}$, pulse duration $\tau_p \sim 15\,\mathrm{ns}$ provided by private discussions with R. Vetter, P. Cahuzac, J. L. Picque (1980). Fig. 9 shows the key energy levels. The intermediate state $|r\rangle \equiv |X, v = 41, J = 143\rangle$ is only about $\Delta = 0{\cdot}3\,\mathrm{cm}^{-1}$ away from the laser frequency. To consider the saturation behaviour, we need χ_{ri}^2

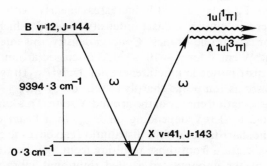

Fig. 9. Energy scheme for the specific example of PC predissociation of I_2 excited in the $(Bv = 12 J = 144)$ states given in the text.

and β_{ri}. Since $J \gg K$ and 1, Eq. (2.25) and the results of Sec. III give

$$\chi_{ri}^2 = 2 \cdot 1 \times 10^{-7} F_{ri} \mu_{ri}^2 \frac{J^2 - M^2}{4J^2} I, \tag{4.3}$$

where F_{ri} is the Franck–Condon factor, μ_{ri} is the electronic electric dipole transition moment (in debyes), and I is the laser intensity (in $\mathrm{W\,cm^{-2}}$). For these states here F_{ri} is calculated to be 0·044, and $\mu_{ri}^2 = 1 \cdot 4 \ \mathrm{D}^2$ is read off from experimental values in Koffend et al. (1979). For low pressures, we assume $\gamma_r \approx 0$ since there is no radiative decay nor predissociation allowed. Then $\beta_{ri} = (\gamma_i + \gamma_{fr})^2/(4\gamma_i\gamma_{fr})$ where $\gamma_i = \gamma_{\mathrm{rad}} + \gamma_{\mathrm{pred}} = 1 \cdot 67 \times 10^6 \ \mathrm{s}^{-1}$ according to measured values of Broyer et al. (1975); and where for large J,

$$\gamma_{fr} = 1 \cdot 57 \times 10^4 \tilde{F}_{fr} \mu_{fr}^2 \frac{(J^2 + M^2)}{J^2} I, \tag{4.4}$$

where $\tilde{F}_{fr}$ is the Franck–Condon density (in units of per cm^{-1}), μ_{fr} is the electronic electric dipole transition moment to the final continuum $|f\rangle$ which can be $1_u(^1\Pi)$ or $A\,1_u(^3\Pi)$. The chart below shows that transition to the $1_u(^1\Pi)$ channel is far more important then $A\,1_u(^3\Pi)$ for the given $|i\rangle$ and $|r\rangle$ states:

	$\tilde{F}_{fr}(\mathrm{per\,cm^{-1}})$	$\mu_{fr}^2(\mathrm{D}^2)$
$\mathrm{r} \to 1_u(^1\Pi)$	$8 \cdot 8 \times 10^{-5}$	0.15
$\mathrm{r} \to A\,1_u(^3\Pi)$	$1 \cdot 4 \times 10^{-5}$	0·044

Here the electronic transition moments μ_{fr} are taken from the experimental values of Tellinghuisen (1973). We may neglect the $A\,1_u(^3\Pi)$ channel from here on. Therefore, substituting in the various values above in Eq. (4.4) for $|f\rangle = |1_u, E\rangle$, one obtains $\gamma_{fr} = 0 \cdot 2I(J^2 + M^2)/J^2$. For $I \gtrsim 3 \times 10^7 \ \mathrm{W\,cm^{-2}}$, $\gamma_{fr} \gg \gamma_i$ and one obtains from Eq. (4.3) the expression,

$$\beta_{ri}\chi_{ri}^2 \approx 1 \times 10^{-16} I^2 (J^4 - M^4)/J^4, \tag{4.5}$$

in units of $(\mathrm{cm}^{-1})^2$. Therefore, the intensity I_c at which coherent saturation sets in is given by $\beta_{ri}\chi_{ri}^2 = (0 \cdot 3)^2$. It has the value of $I_c = 3 \times 10^7 \ \mathrm{W\,cm^{-2}}$ for most of the M states with $(J^4 - M^4)/J^4 \approx 1$. For those M states with values of M close to J, they would not be saturated but their transition strengths are also weaker (equal 0 for $M = J$). Thus, for an intensity range higher than $3 \times 10^7 \ \mathrm{W\,cm^{-2}}$, the bound–bound transition is already coherently saturated. Therefore, the photon-catalysed probability rate is simply

$$\Gamma_{i\mathrm{R}} = \tfrac{1}{2}\gamma_{fr} = 0 \cdot 1I. \tag{4.6}$$

For the pulse duration $\tau_p = 15\ \mathrm{ns}$, the probability of PC predissociation by the pulse is $P \approx \Gamma_{i\mathrm{R}}\tau_p$. The needed laser intensity I_1 such that $P = 1$ is $I_1 = 4 \cdot 8 \times$

10^8 W cm^{-2}. At such intensity or even slightly lower, Γ_{iR} is much larger than the radiative decay rate, 0.86×10^6 s^{-1}, for the initial state. Therefore, all the fraction of I$_2^*$ population normally decaying radiatively to the ground state of I$_2$ would now PC predissociate to yield iodine atoms. Therefore, the enhancement of iodine atom production is given by the ratio of radiative decay rate to the natural predissociation rate ($= 0.814 \times 10^6$ s^{-1}). This ratio is 1.06, i.e. the iodine atom yield is increased by more than 100%.

In the above example, we have not chosen those initial states with a large ratio of radiative decay rate to natural predissociation rate. There are some states with ratios close to the value of 100 (Broyer et al., 1975). Also, the needed laser intensity I_1 for $P = 1$ in the above example would be lower if we have longer pulses. In fact, for a laser pulse long compared to the lifetime of $|B, v = 12, J = 144\rangle$, the laser intensity I_1 so that $P = 1$ will only be 5×10^7 W cm^{-2}. The above results are approximate, given to illustrate the needed laser frequency, intensity, and molecular information. We have results based on the formula Eq. (2.38) which is more accurate for short pulses. We have also performed calculations for other molecular states of I$_2$ and for tunable dye lasers. These results will be published elsewhere (Lau et al., 1981).

C. PC Autoionization of Alkali Atoms

Autoionizing states have been proposed as possible candidates for experimental demonstration of PCE (Lau, 1981a). They have the advantage of producing charged fragments, which can be detected quite readily by such techniques as ejected-electron spectroscopy (see, for example, Ziem et al., 1975; Rassi et al., 1977). Those peaks in the electron spectrum due to PC autoionization should be narrow, equal in energy to the initial states, and possess laser parameter dependences described in Sec. II. To minimize required laser power, long-lived autoionizing states are preferable. Employing autoionizing states with very weak natural autoionization would minimize the noise background if the fragments are detected. Such states can be found in long-lived quartet states lying between the first and second ionization potentials of alkali atoms, known to be metastable against both autoionization and radiative decay (Feldman and Novick, 1967; Levitt et al., 1971; Bunge and Bunge, 1978). They are, therefore, possible good candidates for PC autoionization (Lau, 1981a). For intermediate states in the PC transition, one may use the metastable doubly excited even-parity ^2P states (see, for example, Bunge, 1979; Buchet et al., 1973; Nicolaides and Beck, 1978). An example for Li in the lowest quartet state PC autoionized via the $(1s2p^2 \, ^2P)$ state is illustrated in Fig. 10:

$$\text{Li}(1s2s2p \, ^4P^o) + N\hbar\omega \rightarrow \text{Li}^+ \, (1s^2 \, ^1S) + \text{e}^- + N\bar{\hbar}\omega. \tag{4.7}$$

The initial-state population can be prepared by electron or ion bombardment

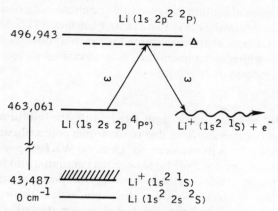

Fig. 10. Proposed photon-catalysed autoionization of Li($1s2s2p\,^4p^\circ$). From Lau (1981a).

of the neutral atoms in the ground state (see, for example, Feldman and Novick, 1967; Ziem *et al.*, 1975; Rassi *et al.*, 1977). For quantitative assessment of this example, the quartet-to-doublet bound–bound transition moments are available (Nussbaumer, 1980) while the doublet–continuum transition moment densities have yet to be calculated. Similar doublet–continuum transition densities have been calculated for autoionization by spontaneous radiation by Nicolaides and Beck (1978) but probably not those at energy equal to the initial quartet state as needed for PC autoionization. Certainly their method can be used for such calculation, and we plan to investigate quantitatively this PC autoionization of alkali atoms. This process may also be of interest in relation to the recent proposal to use the same quartet and doublet states in alkali atoms for the storage and the upper lasing states, respectively, for XUV lasers (Harris, 1980; Rothenberg and Harris, 1981). Since the transition from these doublet states to the final continuum states are electric-dipole-allowed, both PC autoionization and also two-photon ionization from the quartet states via the ($1s2p^2\,^2P$) should contribute in principle as population loss mechanisms when the proposed high-power transfer laser irradiates the quartet-state population. It would be interesting to determine quantitatively their relative importance to the laser proposal either by further calculation as indicated above or by experimental measurement.

V. SEMICLASSICAL THEORY OF PCE IN MOLECULES

The theory in Secs. II and III treats the rotational, vibrational, and electronic motion of the molecule quantum mechanically. When the quantum nature (such as discrete energy structures) of the rotational and vibrational motions is unimportant, these motions may be approximated by the WKB method or by classical motion governed by the electronic potential surfaces (see, for example, Messiah, 1961; Landau and Lifshitz, 1977). For the WKB method, the quasiclassical approximation for the rotational motion is valid if the corresponding angular momentum J is large compared to unity; and for the vibrational motion, it is valid if

$$\mu \hbar F/P^3 \ll 1, \tag{5.1}$$

where μ is the reduced mass associated with the vibrational motion, $F = -\,\mathrm{d}U/\mathrm{d}R$ is the classical force due to electronic potential curve $U(R)$, and $P = [2\mu(E - U(R))]^{1/2}$ is its momentum. Thus the WKB approximation and the classical motion are good for highly excited rotational and highly excited bound or continuum vibrational states (except near classical turning points).

In this section, we shall consider the semiclassical theory of PCE in molecules with the rotational and vibrational motion treated quasiclassically or classically while the electronic motion is treated quantum mechanically (Lau and Rhodes, 1977a, b; Lau, 1978). (Incidentally, for the present problem, one may treat the electromagnetic field classically or quantum mechanically, as we do here, and shall arrive at the same set of equations and results (Lau, 1976).) The advantage of the semiclassical formulation is that its results, unlike those of the quantum formulation in Sec. II, can be applied to PCE-modified dissociation, as well as PC predissociation. We shall, therefore, also discuss PCE-modified dissociation in this section. We give only a simple theory and the lowest-order results here. This is adequate for most applications. The more rigorous theory and more exact solutions are given in the referred original works.

A. A Simple Theory

The interaction of the electronic states $|e\rangle$ of the molecule with the laser radiation is described by the Schrödinger equation with the Hamiltonian $H = H_e + H_r + H'$, where H_r and H' are the same as defined in Eqs. (2.4) and (2.5) respectively, and H_e is the electronic adiabatic Hamiltonian.

$$H_e|e\rangle = \hbar w_e(R)|e\rangle, \tag{5.2}$$

where $w_e(R)$ is the electronic potential surface in angular frequency units. The problem can be solved perturbatively (Lau, 1978). If $|1\rangle, |2\rangle$, and $|e\rangle$ denote the electronic wavefunctions of the initial state $|i\rangle$, the final state $|f\rangle$, and the intermediate states $|m\rangle$ respectively, the effective interaction matrix element between the $|1\rangle$ and $|2\rangle$ is precisely $\hat{H}_{21}$ given in Eq. (2.82) but now with the subscript notations changed for simplicity to:

$$\hat{H}_{21} = \frac{4\pi}{\hbar}I \sum_e \mu_{2e}\mu_{e1} w_{e1}/(w_{e1}^2 - \omega^2). \tag{5.3}$$

This perturbative result agrees with the weak-field limit (Eqs. (3.64) and (3.57)) of the non-perturbative solution in Lau and Rhodes (1977b). With this result, we can proceed to use the quasiclassical approximate wavefunctions for the rotational and vibrational states, and evaluate the PC transition probability Γ_i given in Eq. (2.81).

Or we can proceed with approximating the rotational and vibrational motions by the classical orbits $R(t)$ governed by the electronic potential surfaces with the rotational contribution included. In order to calculate the transition probability for PCE, the time-dependent Schrödinger equations,

$$i\hbar\frac{db_1}{dt} = \hbar w_1 b_1 + (\hat{H}_{12} + H''_{12})b_2, \tag{5.4}$$

$$i\hbar\frac{db_2}{dt} = \hbar w_2 b_2 + (\hat{H}_{21} + H''_{21})b_1, \tag{5.5}$$

are to be solved with the initial-value conditions

$$b_1(t_0) = 1, \quad b_2(t_0) = 0, \tag{5.6}$$

at time $t = t_0$. The H''_{12} in these equations is the intramolecular interaction Hamiltonian H'' evaluated between the electronic states $|1\rangle$ and $|2\rangle$. In general, the $w, \hat{H}_{12}$, and H''_{12} are not simple functions of internuclear coordinates R and Eqs. (5.4) and (5.5) need to be integrated numerically to find the PC transition probability $|b_2(t)|^2$. However, sometimes it is possible to approximate these functions to obtain analytic solutions. One such case is given in the next subsection.

B. PC Transition at a Curve Crossing

It is well known that in the absence of any external field, inelastic transition between electronic states occurs with significant probability when their energy curves are nearly degenerate in a certain region of R (for a recent example, see Melius and Goddard, 1974). A particularly important configuration is the crossing of two potential curves 1 and 2 as shown in Fig. 11(a). With the addition of laser radiative interaction, the above statements remain true. We shall give a simple analytic formula for the PC transition probability at a curve crossing.

With reference to Fig. 11(a), suppose the molecular system is excited at time $t = 0$ to region A of electronic potential curve 1 to the left of the curve crossing C. We want to find the inelastic transition probability that the system will be in the electronic state 2 after the crossing C is traversed from the small R region to the large R region. If we assume (1) that the potential energy difference $w_2 - w_1$ can be represented by the function

$$w_2 - w_1 = -\alpha\tau \tag{5.7}$$

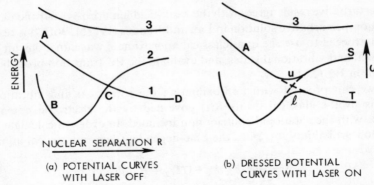

Fig. 11. (a) Potential curves showing a curve crossing between curves 1 and 2 in the absence of external laser radiation. (b) With the laser field on, an effective interaction is induced between states 1 and 2 via resonant intermediate state 3 by the photon-as-catalyst effect. This leads to photon-catalysed transitions between states 1 and 2. The new eigenenergies shows a field-induced avoided crossing.

where $\alpha \equiv d(w_2 - w_1)/dt$ is a constant and $\tau \equiv t - t_c$ is the time measured from the time t_c at which the crossing C is traversed and (2) that

$$\hat{H}_{12} + H''_{12} = \text{constant}, \tag{5.8}$$

we can solve the coupled equations (5.4) and (5.5) analytically for $b_2(t)$ and $b_1(t)$ (Zener, 1932). The particular solution satisfying the initial-value conditions (5.6) gives the asymptotic formulae

$$|b_2(\tau_1)|^2 = 1 - e^{-2\pi p}, \tag{5.9}$$

$$|b_1(\tau_1)|^2 = e^{-2\pi p}, \tag{5.10}$$

where

$$p \equiv |\hat{H}_{12} + H''_{12}|^2/(|\alpha|\hbar^2), \tag{5.11}$$

provided the approximations of Eqs. (5.7) and (5.8) remain good during the entire time interval $\tau_0 \equiv t_0 - t_c$ to τ_1 such that $\tau_0^2 \gg 1/\alpha$ and $\tau_1^2 \gg 1/\alpha$. This time interval defines a corresponding 'region of transition' on the energy surfaces. For small values of $2\pi p \ll 1$, we have the weak transition limit,

$$|b_2(\tau_1)|^2 \approx 2\pi p, \tag{5.12}$$

$$|b_1(\tau_1)|^2 \approx 1 - 2\pi p; \tag{5.13}$$

but for large values of $2\pi p \gg 1$, we have the adiabatic limit,

$$|b_2(\tau_1)|^2 \approx 1, \tag{5.14}$$

$$|b_1(\tau_1)|^2 \approx 0, \tag{5.15}$$

so that the transition probability is unity. A formula with the same formal mathematics as Eqs. (5.7)–(5.10) has been derived by Zener (1932) to consider non-adiabatic transitions. Landau (1932) using the WKB approximation for the vibrational wavefunctions and the method of stationary phase also arrived at the weak transition limit, Eq. (5.12). While the formal appearance is the same as the usual Landau–Zener formula, Eqs. (5.7)–(5.10) contain the physically new quantity $\hat{H}_{12}$ which describes the photon-as-catalyst effect induced by the applied laser radiation. When this quantity is zero (such as when the laser intensity I approaches zero), the above formula recovers the usual field-free Landau–Zener formula. We called this formula the modified Landau–Zener formula. It was first given for PCE in Lau and Rhodes (1977b) and Lau (1978).

The usual Landau–Zener formula has found much use in simple calculations and as a convenient basis of discussion on non-adiabatic transitions (see, for example, Baede, 1975). There have been extensions of (see, for example, Stückelberg, 1932) and criticisms on (see, for example, Bates, 1960) this formula mainly due to the restrictions imposed by the assumptions of Eqs. (5.7) and (5.8) and of classical motion. Baede (1975) gives more detailed discussions on the above points. We note that when these validity conditions are satisfied, the Landau–Zener formula is a rigorous mathematical result (Zener, 1932). Therefore, before applying its results, we should check if its validity conditions are satisfied by the system of interest. When these conditions are good approximations, then we expect that the formula should give at least semiquantitative accuracy.

C. Field-Induced Avoided Crossing

With the laser-induced radiative interaction, the new potential energies of the molecular system are given by solving the stationary eigenvalue problem of Eqs. (5.4) and (5.5) at fixed values of R,

$$Eb_1 = \hbar w_1 b_1 + (\hat{H}_{12} + H''_{12})b_2, \tag{5.16}$$

$$Eb_2 = \hbar w_2 b_2 + (\hat{H}_{12} + H''_{21})b_1. \tag{5.17}$$

The eigenvalues are given simply by

$$E_u(R) = \tfrac{1}{2}\hbar(w_2 + w_1) + \tfrac{1}{2}[\hbar^2(w_2 - w_1)^2 + 4|\hat{H}_{12} + H''_{12}|^2]^{1/2}, \tag{5.18}$$

$$E_l(R) = \tfrac{1}{2}\hbar(w_2 + w_1) - \tfrac{1}{2}[\hbar^2(w_2 - w_1)^2 + 4|\hat{H}_{12} + H''_{12}|^2]^{1/2}. \tag{5.19}$$

The relation of E_u and E_l to the original potentials $\hbar w_1$ and $\hbar w_2$ is depicted in Fig. 11(b). These new potential surfaces are dependent on the parameters $(I, \omega, \hat{\varepsilon})$ of the laser field and have been called field-dressed potential surfaces (Lau and Rhodes, 1977a, b). The energy separation between E_u and E_l is given

by

$$\hbar\Delta(R) \equiv E_{\mathrm{u}} - E_{\mathrm{l}} = [\hbar^2(w_2 - w_1)^2 + 4|\hat{H}_{12} + H''_{12}|^2]^{1/2}, \qquad (5.20)$$

which has a minimum Δ_{m} at the crossing point C where $w_2(R_{\mathrm{c}}) - w_1(R_{\mathrm{c}}) = 0$. This Δ_{m} is twice the interaction strength between w_1 and w_2,

$$\Delta_{\mathrm{m}} = 2|\hat{H}_{12} + H''_{12}|/\hbar. \qquad (5.21)$$

We see that the Landau–Zener formula can also be expressed in terms of this minimum energy separation of the field-induced avoided crossing, instead of $\hat{H}_{12} + H''_{12}$. The field-induced contribution $\hat{H}_{12}$ to this Δ_{m} can be varied by changing the laser field parameters and goes to zero as the laser intensity I goes to zero. The symmetry rules governing whether a field-induced avoided crossing is allowed or not are given by the electronic selection rules discussed in Sec. II B. For example, we expected that a field-induced contribution $\hat{H}_{12} \neq 0$ between two gerade or two ungerade electronic potential curves can occur, but not between a gerade and an ungerade curve. Another example of allowed field-induced avoided crossing is between a Σ and a Δ electronic potential curve with Π states as intermediate states. For a Σ to Δ crossing, the non-adiabatic coupling H''_{12} is zero and therefore the field-induced avoided crossing is purely radiative. A field-free true crossing will then change into an avoided crossing (Lau and Rhodes 1977a, b). This means there is new coupling between the Σ and Δ states by the PCE, the so-called new transition channel in Sec. II B. The field-induced avoided crossing between electron–field states $|1N\rangle$ and $|2N\rangle$ of equal photon number N is analogous to those between electron–field states $|1N\rangle$ and $|2N'\rangle$ of different photon numbers N and N' that were first introduced by Kroll and Watson (1976) and by Fedorov et al. (1975) for studying actual photon absorption involving continuum states and bound states in a molecular system, respectively. However, it should be noted that in the latter situations, the original potential curves w_1 and w_2 usually do not cross each other at all.

D. PCE-Modified Dissociation: Alkali Halides

The result (5.9) gives, within the Landau–Zener model, the probability of the total (photon-catalysed and non-adiabatic) transition from one electronic curve to another crossing electronic curve. For a large signal-to-noise ratio, it is clearly desirable to have H''_{12} as small as possible. For a large PCE signal, it is desirable to have a large value of $\hat{H}_{12}$ and a small $|\alpha|$. For a large $\hat{H}_{12}$ value, it is desirable to have, apart from the laser intensity I, large electronic transition moments μ_{2e} and μ_{e1}, and w_{e1} close to ω (while keeping in mind that Eq. (5.3) does not hold for near-resonance $w_{e1} \approx \omega$). The quantity α defined in Eq. (5.7) can be written for a one-dimensional system as

$$\alpha = \frac{d(w_2 - w_1)}{dR} \frac{dR}{dt}, \qquad (5.22)$$

where $d(w_2 - w_1)/dR$ is a force difference and dR/dt is the velocity at the crossing point. For a small α value, both of these quantities have to be small. Small force differences can occur at large internuclear separation R where the crossing potential curves both have small and nearly equal slopes. Low velocity is associated with thermal collision but it should not be so low to violate the validity of classical-motion approximation.

Weiner (1980) has examined the possible modification by the PCE of the branching ratio of ion production vs neutral production in the dissociation of the alkali halides. If the dissociation of an alkali halide is diabatic (following curve 2 in Fig. 11(a) from B to large R), the dissociated products are both ions. But if the dissociation is adiabatic (following the curve from B to C and then to D), the dissociated products are both neutrals. There are also intermediate cases. A quantitative criterion covering the whole range of adiabaticity ($\beta \gg 1$) to diabaticity ($\beta \approx 0$) is the value of the branching ratio β given by Berry (1957) and Ewing et al. (1971). This branching ratio is proportional to the energy gap Δ_m^4 at the crossing point. Therefore, if the addition of the laser field induces an additional component ($2\hat{H}_{12}$ in Eq. (5.21)) to the avoided crossing gap Δ_m, then the branching ratio would be modified. Weiner gave two examples whose results are tabulated below:

	Laser $I = 10^{10}\,\mathrm{W\,cm^{-2}}$	β	β'	β'/β
NaCl	Ruby	1.5×10^{-2}	0.12	8
KI	Nd:YAG	1.4×10^{-2}	0.39	28

Here β' and β are the branching ratios with and without the laser field respectively. Therefore, a decrease in production of ions and a significant increase in neutral fragments should be possible with these lasers. Because of various approximations, it was emphasized that these results were meant to be qualitative.

E. A Semiclassical Formula for PC Predissociation

In the approximation of quasiclassical or classical motion for the vibrational motion, the vibrating molecule oscillates classically v_i times per unit time in a given bound vibrational state $|i\rangle$. Owing to these large numbers v_i of oscillations, even a very small PC predissociation probability per oscillation can lead to a significant PC predissociation rate per unit time. Therefore, only moderate field intensity is usually sufficient. Thus, the result (5.12) for small transition probability is adequate. The PC transition probability per oscillation is given by $4\pi p$, which contains in the transition moments μ in $\hat{H}_{12}$ the dependence on the angles between the molecular moments and the polari-

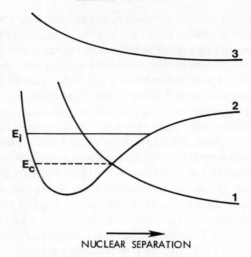

Fig. 12. Semiclassical description of photon-catalysed predissociation of molecules in level E_i.

zation $\hat{\varepsilon}$ of the laser radiation. If the molecules are initially randomly oriented, they will remain so under irradiation by the laser light because a rapidly oscillating field does not preferentially orient the molecules with respect to its polarization. Therefore, the angle-averaged probability rate per unit time for PC predissociation is given by

$$\Gamma_i = \frac{v_i}{h} \int_0^{2\pi} d\psi \int_{-1}^{1} d(\text{Cos }\theta) |\hat{H}_{12} + H''_{12}|^2 \left(\frac{d(w_2 - w_1)}{dR} \frac{dR}{dt}\right)^{-1}, \quad (5.23)$$

where dR/dt is related to the energy E_i of state $|i\rangle$ and the energy at the crossing point (see Fig. 12) by

$$dR/dt = [2(E_i - E_c)/\mu]^{1/2}$$

where μ is the reduced mass of the vibrating molecule. This semiclassical formula is valid only if the assumptions of the Landau–Zener model are a good approximation. If the expression of $\hat{H}_{12}$ in Eq. (5.3) is used, it is also restricted to far off-resonance from any intermediate states. Finally, it is clearly valid if we have a crossing away from any classical turning points and $E_i - E_c$ large. There are many other cases in predissociation where the curves do not cross at all (see Mulliken, 1960, Fig. 1). For the above reasons, the quantum results in Sec. II are preferred to this semiclassical expression.

F. Other Aspects

There are several relevant aspects that we do not cover in this chapter. The solution of the interaction of the radiation with the molecule given in $\hat{H}_{12}$ is perturbative. A non-perturbative solution is given in Lau and Rhodes (1977a, Sec. III: 1977b, Sec. III). But it was shown (Lau, 1978, Sec. II) that the weak-field limit agrees with the perturbative results given here. Only when strong field intensity is applied may we need this non-perturbative solution. The formulation of the theory there for calculating transition probabilities started from these non-perturbative solutions, the field-dressed states (see also Lau, 1977). But again, it was reduced to the equivalent formulation in Sec. V A here in the weak-field limit.

In the works of Lau and Rhodes (1977a, b), the PCE was also applied to examine PC transitions during atomic and molecular collisions. For similar reasons as in PCE-modified dissociation where only one or two crossings are involved in one event, the laser intensities used in PCE-modified collisions (see also Lau, 1979d; DeVries et al., 1978; Light and Altenberger-Siczek, 1979) were higher than those needed for PC predissociation. However, the laser intensity requirement should be lower with more optimization such as avoiding competition with field-free processes known to have large cross sections (Lau and Rhodes, 1977b). For example, the high laser intensity shown to be necessary in DeVries et al. (1978) to induce a noticeable difference from the field-free collisions is due to an already large cross section of the latter.

The effect of the laser linewidth in PC decompositions is unimportant for the non-resonant case but should be important for the near-resonant case (Lau, 1979c). S. N. Dixit (1980, private communication) has obtained results that include the laser linewidth.

VI. SUMMARY

In this chapter, we have presented our current understanding of photon-catalysed predissociation and autoionization. The theoretical results enable us to calculate the rate constants and fragment yield for these processes. For this purpose, the key results are: (a) for non-resonant PC predissociation, Eqs. (2.19)–(2.20); (b) for resonant PC predissociation, Eqs. (2.33) (2.38), and (246); and (c) for far off-resonant PC predissociation, Eqs. (2.81)–(2.82). We have applied some of these results and the results of rotational transition strengths in Sec. III to examples of molecular systems to show that the $B(0_u^+)$ states of iodine appear to be favourable for observing PC predissociation. The key result for photon-catalysed transition during dissociation are Eqs. (5.9)–(5.10).

The selection rules in Sec. II B enable us to determine if PCE is allowed or not. From our discussions of various features (resonance enhancement, coherent saturation, lifetime dependence, etc.), a set of criteria for optimizing the photon-catalysed bound–free process become clear. These are useful in selecting the most favourable set of molecular states for calculation or observation. The set of criteria for PC decompositions are as follows.

(1) The initial states are preferred to be long-lived for population storage

and enhancement of yield, as discussed in Sec. II F. The states that are thermally populated or in the ground states as illustrated in Fig. 2 are most desirable. For a given laser frequency range, it is preferred that single-photon dissociation from such initial states is forbidden either by symmetry or by energetics. For a large signal-to-noise ratio, it is desirable not to have a natural transition from the initial state to the same final state as the PC decomposition.

(2) For the choice of the intermediate discrete states, the most important criterion is to be as near resonance with the laser frequency as possible (Sec. II E). This can be achieved either by having a laser tunable over the molecular transitions or by choosing the right combination of initial–intermediate states if the laser frequency is fixed. The next most important criterion for the intermediate states is that the transition moment to the final continuum states is as large as possible. It is also desirable to have a large transition moment between the initial and intermediate states for the coherently unsaturated case. But once coherent saturation is achieved for the given laser intensity, its magnitude does not enhance the overall process any more (see Sec. II D). Whether the lifetime of the intermediate state enhances the yield or not depends on various cases. For the non-resonant situation or if the laser-stimulated decay rate from it is larger than its natural decay rate, then its natural lifetime is not important at all. If both of the above cases are not true, then for the near-resonant situation a longer lifetime enhances the yield (a) for the coherently unsaturated case if the lifetime is longer than the lifetime of the initial state, and (b) for the coherently saturated case if the lifetime is shorter than the lifetime of the initial state (see Sec. II F).

(3) For the choice of the final states, it has been mentioned in item (1) above that it is desirable to have a zero or very small intramolecular coupling with the initial states. Another desirable feature is for its decomposed fragments to be readily detected, such as excited fragments giving fluorescence or ionic fragments such as those in PC autoionization.

The reader of this chapter will realize that the topic of the photon-as-catalyst effect in various processes is wide open for more research. We expect that there will be more activity in this area. Certainly experimental observation of this effect will be most interesting. We have used the general principle (Sec. I A) and specific calculations (Sec. IV B) to show that its observation should be feasible with presently available equipment. The effect is novel and adds a new dimension to photochemistry and laser-induced processes in atoms and molecules. Its feature of non-stoichiometric use of laser photons is attractive for investigation into possible applications.

Acknowledgements

I am very grateful to Drs R. L. Jaffe of NASA Ames Research Center, P. S. Julienne of NBS, and S. N. Dixit of the University of Southern California, and to Professors R. Bacis, D. A. Dows, R. W. Field, W. Happer, D. D. Konowalow, and J. Weiner for sending me their research results before publication. I appreciate a helpful discussion with my colleague Dr M. S. Chou during the writing of this chapter.

Part of the work in this chapter was performed while I held the appointment of Professeur Associé at the Université de Paris-Sud, Orsay, France, in the autumn of 1980. It is a pleasure to acknowledge stimulating discussions with Drs R. Vetter, J. L. Picque, P. Cahuzac, C. Brechignac, P. Luc, and S. Gerstenkorn of Laboratoire Aimé Cotton and to thank its director, Professor S. Feneuille, and its staff for their helpfulness.

Over the past few years during which much of the reported research was performed, I have benefited from many informative discussions with Professors W. C. Stwalley of the University of Iowa and D. Herschbach of Harvard University, and from the encouragement and support of Professor C. K. Rhodes of the University of Illinois, and Drs F. R. Gamble and R. W. Cohen of Exxon.

Appendix A

In this appendix, the results (2.18), (2.19)–(2.20), and (2.22) in the text for the non-resonant PC decomposition with the presence of natural decomposition are derived. As discussed before, the principal states are the initial state $|iN\rangle$. the intermediate states $|mn\rangle$ where $n = N + 1$ or $N - 1$, and the final states $|fN\rangle$. Besides the laser-induced radiative interaction H' and intramolecular interaction H'' that are explicitly taken into account, there are other interactions such as spontaneous radiative decay, intramolecular interactions to states other than $|f\rangle$, and collisions that broaden the transitions between discrete states. The effects of these interactions are included in the theory by assigning a decay rate γ_{0m} for each discrete state $|m\rangle$. The equations (2.8) for the probability amplitudes of the principal states with the γ_{0m} inserted are

$$\frac{d}{dt}b_{iN} = -\frac{\gamma_{0i}}{2}b_{iN} - \frac{i}{\hbar}\sum_{mn} H'_{iNmn}b_{mn}e^{-i\omega_{mniN}t} - \frac{i}{\hbar}\sum_{f} H''_{if}b_{fN}e^{-i\omega_{fi}t}, \qquad (A.1)$$

$$\frac{d}{dt}b_{mn} = -\frac{\gamma_{0m}}{2}b_{mn} - \frac{i}{\hbar}\sum_{m'n'} H'_{mnm'n'}b_{m'n'}e^{-i\omega_{m'n'mn}t} - \frac{i}{\hbar}\sum_{f} H''_{mf}b_{fn}e^{-i\omega_{fm}t}, \quad (A.2)$$

$$\frac{d}{dt}b_{fN} = \frac{-i}{\hbar}\sum_{mn} H''_{fNmn}b_{mn}e^{-i\omega_{mnfN}t} - \frac{i}{\hbar}\sum_{m} H''_{fm}b_{mN}e^{-i\omega_{mf}t}. \qquad (A.3)$$

A formal solution to Eq. (A.2) for the probability amplitude b_{mn} of the intermediate states is

$$b_{mn} = \frac{1}{\hbar}\sum_{m'n'} \frac{H'_{mnm'n'}b_{m'n'}}{\omega_{m'n'mn} + i\frac{1}{2}\gamma_{0m}}e^{-i\omega_{m'n'mn}t} + \frac{1}{\hbar}\sum_{f} \frac{H''_{mf}b_{fn}}{\omega_{mf} + i\frac{1}{2}\gamma_{0m}}e^{-i\omega_{fm}t}, \qquad (A.4)$$

where it has been assumed that the fractional changes in time of $H'_{mnm'n'}b_{m'n'}$ and of $H''_{mf}b_{fn}$ over the time periods of $\omega_{m'n'mn}^{-1}$ and of ω_{fm}^{-1}, respectively, are small compared to unity. Similarly, the formal solution to Eq. (A.3) is

$$b_{fN} = \frac{1}{\hbar}\sum_{mn}\frac{H'_{fNmn}b_{mn}}{\omega_{mnfN} + i\frac{1}{2}\gamma_f}e^{-i\omega_{mnfN}t} + \frac{1}{\hbar}\sum_{m}\frac{H''_{fm}b_{mN}}{\omega_{mf} + i\frac{1}{2}\gamma_f}e^{-i\omega_{mf}t}, \qquad (A.5)$$

where a fictitious width γ_f for the continuum state $|f\rangle$ is introduced but will not appear in the final results. Substituting these into Eq. (A.1) and keeping only the lowest-order transition from state $|iN\rangle$, one obtains

$$\frac{d}{dt}b_{iN} = \left(-\frac{\gamma_{0i}}{2} - \frac{i}{\hbar^2}\sum_{mn}\frac{|H'_{iNmn}|^2}{\omega_{iNmn} + i\frac{1}{2}\gamma_m} - \frac{i}{\hbar^2}\sum_{f}\frac{|H''_{if}|^2}{\omega_{if} + i\frac{1}{2}\gamma_f}\right)b_{iN}. \qquad (A.6)$$

Using the identity (Heitler, 1954),

$$\lim_{\gamma \to 0}\frac{1}{x + i\gamma} = P\frac{1}{x} - i\pi\delta(x), \qquad (A.7)$$

where P stands for the principal value of integration, one obtains from Eq. (A.6),

$$\frac{d}{dt}b_{iN} = (-\tfrac{1}{2}\gamma'_i - i\delta_i)b_{iN}, \qquad (A.8)$$

where

$$\gamma'_i = \gamma_i + \sum_{f'}\gamma_{f'i}, \qquad (A.9)$$

is the total width of the state $|i\rangle$ in the presence of the laser. It is the sum of the natural width γ_i and any additional broadening $\gamma_{f'i}$ due to laser-stimulated dissociation or ionization from the state $|i\rangle$ to the continuum $|f'\rangle$. Here γ_i is given by

$$\gamma_i = \gamma_{0i} + \sum_{f}{}'\frac{2\pi}{\hbar}|H''_{if}|^2\rho_f \qquad (A.10)$$

with the density of the vibrational continuum $\rho(E_f)$ evaluated at $E_f = E_i$ and the sum carried over different continua $|f\rangle$. In application to specific molecular levels, the value of γ_i can be taken from the experimentally measured FWHM of the state $|i\rangle$ in the absence of the laser. The laser-stimulated width $\gamma_{f'i}$ in Eq. (A.9) is given by

$$\gamma_{f'i} = \frac{4\pi^2}{hc}I|D_{if'}|^2\rho(E_{f'} = E_i \pm \hbar\omega). \qquad (A.11)$$

In Eq. (A.8), the optical Stark shift $\hbar\delta_i$ of the energy level of the state $|i\rangle$ is given

by

$$\delta_i = \frac{4\pi}{\hbar c} I \sum_{d \neq i} \frac{(E_i - E_d)|D_{id}|^2}{(E_i - E_d)^2 - (\hbar\omega)^2}$$

$$+ \frac{2\pi}{\hbar c} I \sum_c \left(P \int dE_c \rho_c \frac{|D_{ci}|^2}{E_i - E_c - \hbar\omega} + P \int dE_c \rho_c \frac{|D_{ci}|^2}{E_i - E_c + \hbar\omega} \right), \quad (A.12)$$

where d stands for discrete states, c for all the continua, and P for the principal value. Eq. (A.12) is valid if the condition

$$\frac{8\pi}{c} I |D_{dj}|^2 \ll (|E_d - E_j| - \hbar\omega)^2 \quad (A.13)$$

holds. In Eq. (A.8), there is also an energy shift of the state $|i\rangle$ due to the H'' interaction. But this shift is already included in the molecular energy E_i. The solution to Eq. (A.8) is obviously.

$$b_{iN}(t) = \exp\left[-(\tfrac{1}{2}\gamma_i' + i\delta_i)t\right], \quad (A.14)$$

where the initial-value condition, $b_{iN}(0) = 1$, has been used.

Similarly, by substituting Eq. (A.4) into Eq. (A.3) and keeping only the terms with the non-zero b_{iN} while neglecting higher-order interactions, one obtains

$$\frac{d}{dt} b_{fN} = \left(\frac{-i}{\hbar^2} \sum_{mn} \frac{H'_{fNmn} H'_{mniN}}{\omega_{iNmn} + i\tfrac{1}{2}\gamma_m} - \frac{i}{\hbar} H''_{fi} \right) b_{iN} e^{-i\omega_{if}t}. \quad (A.15)$$

We now assume that the laser intensity is fairly constant over the lifetime $\tau_i = 1/\gamma_i'$ of the initial state $|i\rangle$ so that, when we integrate Eq. (A.15) with $b_{iN}(t)$ given by Eq. (A.14), the matrix elements $H'_{fNm'n'} H'_{m'n'iN}$ can be approximated as constant. With the initial-value condition $b_{iN}(0) = 1$, the solution to Eq. (A.15) is

$$|b_{fN}(t)|^2 = \frac{1}{\hbar} \left| \frac{1}{\hbar} \sum_{mn} \frac{H'_{fNmn} H'_{mniN}}{\omega_{iNmn} + i\tfrac{1}{2}\gamma_m} + H''_{fi} \right|^2$$

$$\times \frac{e^{-\gamma_i t}(1 - 2e^{\frac{1}{2}\gamma_i t} \cos\omega_{if}t) + 1}{\omega_{if}'^2 + \tfrac{1}{4}\gamma_i'^2}. \quad (A.16)$$

The energy spectrum of the fragments is given by

$$|b_{fN}(\infty)|^2 \rho = \frac{1}{\hbar^2} \left| \frac{1}{\hbar} \sum_{mn} \frac{H'_{fNmn} H'_{mniN}}{\omega_{iNmn} + i\tfrac{1}{2}\gamma_m} + H''_{fi} \right|^2 \frac{\rho(E_f)}{\omega_{if}'^2 + \tfrac{1}{4}\gamma_i'^2}, \quad (A.17)$$

where $\rho(E_f)$ is the density of final continuum states $|f\rangle$. If ρ is fairly smooth function, then the above spectrum has a sharp peak shifted from $\omega_f = \omega_i$ to the new position $\omega_f = \omega_i + \delta_i$ and a narrow FWHM $\gamma_i' \ll \omega_i$, ω_f.

The probability of total decomposition from the state $|i\rangle$ at time t is given by summing the contribution from all the continuum states,

$$P_i(t) = \int_0^\infty dE_f \rho(E_f) |b_{fN}(t)|^2 = \Gamma_i(1 - e^{-\gamma_i t})/\gamma_i', \qquad (A.18)$$

where

$$\Gamma_i = \frac{2\pi}{\hbar} \sum_f{}' \left| \frac{1}{\hbar} \sum_{mn} \frac{H'_{fNmn} H'_{mniN}}{\omega_{iNmn} + i\frac{1}{2}\gamma_m} + H''_{fi} \right|^2 \rho(E_f = E_i'), \qquad (A.19)$$

is the decomposition probability rate per unit time per molecule by the PCE and intramolecular interaction. In evaluating the integral in Eq. (A.18), the sharp energy spectrum has been used and leads to the energy conservation condition,

$$E_f = E_i + \hbar\delta_i \qquad (A.20)$$

in the expression for Γ_i. Thus we have derived Eqs. (2.19)–(2.20).

In contrast to the above assumption of approximately constant laser intensity, we now consider those cases where the incident laser pulse has significant variation in intensity over the time τ_p of the pulse or the lifetime τ_i of the initial state. Then we approach the problem by first considering a time interval δt such that

$$\omega^{-1}, \omega_{mi}^{-1} \ll \delta t \ll \tau_i \text{ or } \tau_p. \qquad (A.21)$$

If the laser intensity $I(t)$ does not change significantly over δt, the same expression for $\Gamma_i(t)$ as Eq. (A.19) can be obtained by second-order perturbation theory (Lau, 1979, Sec. III) for a given value of laser intensity $I(t)$ (see similar analysis from Eqs. (2.30)–(2.35)). Then the probability of total decomposition is given by

$$P_i(t) = \int_0^t dt' \Gamma_i(t') |b_{iN}(t')|^2, \qquad (A.22)$$

where $|b_{iN}(t)|^2$ is given by Eq. (A.14) with its laser-stimulated decay $\gamma_{f'i}$, if important, also parametrized by $I(t)$. This is the result Eq. (2.22).

Appendix B

The results in Sec. II C are derived from Eqs. (2.26)–(2.28) where the radiative interactions between the principal states with other states are neglected. In a real system, these interactions will give rise to optical Stark shifts of the discrete energies of the molecules and to laser-stimulated transitions from the discrete states resulting in further broadening of their spectral lines. In this appendix, adapted from Lau (1981a), we show how these effects can be included. It turns out that the expressions of all the results in Sec.

II C can still be used if the frequencies and linewidths occurring in them are generalized to include the optical Stark shifts and laser-stimulated decays. The inclusion of the laser-stimulated decay γ_{fr} of the near-resonant state $|r\rangle$ is important for achieving the correct limiting behaviour of the results.

The extensions of Eqs. (2.26) and (2.27) to include the radiative interaction with other states are similar and, therefore, only Eq. (2.27) is treated explicitly. We start with

$$\frac{\mathrm{d}}{\mathrm{d}t}b_{rN'} = -\frac{\gamma_r}{2}b_{rN'} - \frac{i}{\hbar}H'_{rN'iN}b_{iN}\mathrm{e}^{-i\omega_{iNrN'}t}$$

$$-\frac{i}{\hbar}\sum_{mn\neq iN}H'_{rN'mn}b_{mn}\mathrm{e}^{-i\omega_{mnrN'}t}, \tag{B.1}$$

and substitute for b_{mn} with the solution

$$b_{mn} = \sum_{m'n'}\frac{H'_{mnm'n'}b_{m'n'}}{\hbar(\omega_{m'n'mn}+i\frac{1}{2}\gamma_m)}\mathrm{e}^{-i\omega_{m'n'mn}t}, \tag{B.2}$$

obtained in a similar fashion as Eq. (A.4). By neglecting those terms corresponding to higher-order transitions from $|rN'\rangle$ (i.e. terms with $m'n' \neq rN'$ in Eq. (B.2)) and using identity Eq. (A.7), one obtains the reduced equation

$$\frac{\mathrm{d}b_{rN'}}{\mathrm{d}t} = -\frac{\gamma'_r}{2}b_{rN'} - i\delta_r b_{rN'} - \frac{i}{\hbar}H'_{rN'iN}b_{iN}\mathrm{e}^{-i\omega_{iNrN'}t}; \tag{B.3}$$

and similarly for b_{iN},

$$\frac{\mathrm{d}b_{iN}}{\mathrm{d}t} = -\frac{\gamma'_i}{2}b_{iN} - i\delta_i b_{iN} - \frac{i}{\hbar}H'_{iNrN'}b_{rN'}\mathrm{e}^{-i\omega_{rN'iN}t}. \tag{B.4}$$

The decay rates γ'_i and γ'_r in these equations are defined as in Eqs. (A.9) and (A.11) to include any laser-induced single-photon dissociation or ionization. We note in particular that in

$$\gamma'_r = \gamma_r + \sum_{f'}\gamma_{f'r}, \tag{B.5}$$

the component γ_{fr} is the probability rate for transition from $|r\rangle$ to the final state $|f\rangle$ which is necessarily present because it is part of the photon-as-catalyst transition (see Eqs (2.31) and (2.38)). The optical Stark shifts δ_i and δ_r are defined in the same way as Eq. (A.12) but in each case with both the $d = i$ and the $d = r$ terms excluded from the sum over the discrete states $|d\rangle$. This is due to the fact that the radiative interaction between the states $|i\rangle$ and $|r\rangle$ has been separated out in Eqs. (B.3) and (B.4) and it remains to be solved.

Letting $b_{iN} = b'_{iN}\mathrm{e}^{-i\delta_i t}$ and $b_{rN'} = b'_{rN'}\mathrm{e}^{-i\delta_r t}$ in Eqs. (B.3) and (B.4), one obtains the coupled equations describing the interaction between the perturbed (shifted and broadened) states $|i'\rangle$ and $|r'\rangle$,

$$\frac{d}{dt}b'_{iN} = -\frac{\gamma'_i}{2}b'_{iN} - \frac{i}{\hbar}H'_{iNrN'}b'_{rN'}e^{-i\omega'_{rN'iN}t},$$ (B.6)

and

$$\frac{d}{dt}b'_{rN'} = -\frac{\gamma'_r}{2}b'_{rN'} - \frac{i}{\hbar}H'_{rN'iN}b'_{iN}e^{-i\omega'_{iNrN'}t},$$ (B.7)

where the shifted frequency detuning $\omega'_{rN'iN}$ includes the optical Stark shifts δ_i and δ_r,

$$\omega'_{rN'iN} = \omega_r + \delta_r - \omega_i - \delta_i + (N' - N)\omega.$$ (B.8)

It may happen that the near-resonant ($\omega_{rN'iN} \neq 0$) transition is optically Stark shifted into resonance ($\omega'_{rN'iN} = 0$) (Lau, 1976; Loy, 1976).

Similarly, substitution of $b_{rN'} = b'_{rN'}e^{-i\delta_r t}$ into Eq. (2.28) gives

$$\frac{d}{dt}b_{fN} = \frac{-i}{\hbar}H'_{fNrN'}b'_{rN'}e^{-i\omega'_{rN'fN}t},$$ (B.9)

where

$$\omega'_{rN'fN} \equiv \omega_r + \delta_r - \omega_f + (N' - N)\omega.$$ (B.10)

Comparison of Eqs. (B.6), (B.7), and (B.9) with Eqs. (2.26)–(2.28) shows that they are formally identical and therefore the results in Sec. II C can also be used when the optical Stark shifts and laser-stimulated decays are significant, provided that the appropriate quantities are replaced according to Eq. (2.52) in the main text.

References

R. Bacis, S. Churassy, R. W. Field, J. B. Koffend and J. Verges (1980), *J. Chem. Phys.*, **72**, 34.

A. P. M. Baede (1975), in *Molecular Scattering: Physical and Chemical Applications*, ed., K. P. Lawley, Wiley, New York, p. 463.

G. Baravian, J. Godart and G. Sulton (1979), *Appl. Phys. Lett.*, **34**, 190.

D. R. Bates (1960), *Proc. R. Soc.*, **A257**, 22.

R. S. Berry (1957), *J. Chem. Phys.*, **27**, 1288.

J. A. Beswick and J. Durup (1979), In *Proceedings of the Summer School on Chemical Photophysics, Les Houches*, eds. P. Glorieux, D. Lecler, and R. Vetter, Editions du CNRS, Paris, pp. F1–F75.

N. D. Bhaskar, E. Zouboulis, T. McClelland and W. Happer (1979), *Phys. Rev. Lett.*, **42**, 640.

M. Born and K. Huang (1956), *Dynamical Theory of Crystal Lattices*, Oxford University Press, New York.

D. J. Bradley, P. Ewart, J. V. Nicholas, J. R. D. Shaw and D. G. Thompson (1973), *Phys. Rev. Lett.*, **31**, 263.

M. Broyer, J. Vigue and J. C. Lehmann (1975), *J. Chem. Phys.*, **63**, 5428.

M. Broyer, J. Vigue and J. C. Lehmann (1976), *J. Chem. Phys.*, **64**, 4793.

J. P. Buchet, M. C. Buchet-Poulizac, H. G. Berry and G. W. F. Drake (1973), *Phys. Rev.*, **A7**, 922.

C. F. Bunge (1979), *Phys. Rev.*, **A19**, 936 and references therein.

C. F. Bunge and A. V. Bunge (1978), *Phys. Rev.*, **A17**, 816 and 822.

G. D. Chapman and P. R. Bunker (1972), *J. Chem. Phys.*, **57**, 2951 and references therein.

S. L. Chin, N. R. Isenor and M. Young (1969), *Phys. Rev.*, **188**, 7.

E. U. Condon and G. H. Shortley (1967), *The Theory of Atomic Spectra*, Cambridge University Press, Cambridge, Chap. 15.

W. E. Cooke, T. F. Gallagher, S. A. Edelstein and R. M. Hill (1978), *Phys. Rev. Lett.*, **40**, 178.

J. A. Coxon (1973), in *Molecular Spectroscopy*, vol. 1, ed. R. F. Barrow, *Chem. Soc. Spec. Per. Rep.* The Chemical Society, London, p. 177.

J. Davis, M. Feld, C. P. Robinson, J. I. Steinfeld, N. Turro, W. S. Watt and J. T. Yardley (1979), *Laser Photochemistry and Diagnostics: Recent Advances and Future Prospects*, National Science Foundation, Washington, DC.

V. A. Davydkin and L. P. Rapoport (1975), *Sov. J. Quantum Electron.*, **4**, 1123.

P. L. DeVries, M. S. Mahlab and T. P. George (1978), *Phys. Rev.*, **A17**, 546.

J. J. Ewing, R. Milstein and R. S. Berry (1971), *J. Chem. Phys.*, **54**, 1752.

U. Fano and J. W. Cooper (1968), *Rev. Mod. Phys.*, **40**, 441.

M. V. Fedorov, O. V. Kudrevatova, V. P. Markarov and A. A. Samokhin (1975), *Opt. Commun.*, **13**, 299.

P. Feldman and R. Novick (1967), *Phys. Rev.*, **160**, 143.

R. H. Garstang (1962), in *Atomic and Molecular Processes*, ed. D. R. Bates, Academic Press, New York.

W. R. S. Garton (1966), *Adv. Atom. Mol. Phys.*, **2**, 93.

S. Gerstenkorn and P. Luc (1978), *Atlas du Spectre d' Absorption de la Molecule d' Iode*, Editions du CNRS, Paris.

T. E. Gough, R. E. Miller and G. Scoles (1979), in *Laser Induced Processes in Molecules*, eds. K. L. Kompa and S. D. Smith, Springer, New York, p. 433.

E. H. A. Granneman, M. Klewer, K. J. Nygaard and M. J. Van der Wiel (1976), *J. Phys. B*, **9**, 865.

W. Happer, J. Chamberlain, H. Foley, N. Forston, J. Katz, R. Novick, M. Ruderman and K. M. Watson (1979), *Laser Induced Photochemistry*, SRI International, Arlington, VA.

S. E. Harris (1980), Opt. Lett., **5**, 1.

W. Heitler (1954), *Quantum Theory of Radiation*, 3rd edn, Clarendon Press, Oxford, Chap. 2.

G. Herzberg (1950), *Spectra of Diatomic Molecules*, Van Nostrand Reinhold, New York.

G. Herzberg (1966), *Electronic Spectra and Electronic Structure of Polyatomic Molecules*, Van Nostrand Reinhold, New York, pp. 455–482.

P. M. Johnson, M. R. Berman and D. Zakheim (1975), *J. Chem. Phys.*, **62**, 2500.

J. B. Koffend, R. Bacis and R. W. Field (1979), *J. Chem. Phys.*, **70**, 2366.

W. Kolos and L. Wolniewicz (1963), *Rev. Mod. Phys.*, **35**, 473.

K. L. Kompa and S. D. Smith (eds.) (1979), *Laser Induced Processes in Molecules*, Springer Ser. Chem. Phys. 6, Springer, New York.

D. D. Konowalow and M. L. Olson (1979), *J. Chem. Phys.*, **71**, 450.

D. D. Konawalow, M. E. Rosenkrantz and M. L. Olson, (1980), *J. Chem. Phys.*, **72**, 2612.

N. M. Kroll and K. M. Watson (1976), *Phys. Rev.*, **A13**, 1018.

R. de L. Kronig (1928), *Z. Phys.*, **50**, 347.

P. Lambropoulos (1976), *Adv. Atom. Mol. Phys.*, **12**, 87.

P. Lambropoulos and P. Zoller (1981), *Phys. Rev.*, **A24**, 379. We note that PC autoionization was not studied by these authors.

L. D. Landau (1932), *Phys. Z. Sowjetunion*, **1**, 88, and **2**, 46.

L. D. Landau and E. M. Lifshitz (1977), *Quantum Mechanics*, Pergamon Press, New York, Chap. 7.

A. M. F. Lau (1976), *Phys. Rev.*, **A14**, 279.

A. M. F. Lau (1977), *Phys. Rev.*, **A16**, 1535.

A. M. F. Lau (1978), *Phys. Rev.*, **A18**, 172.

A. M. F. Lau (1979a), *Phys. Rev.*, **A19**, 1117.

A. M. F. Lau (1979b), in *Laser Induced Processes in Molecules*, eds. K. L. Kompa and S. D. Smith, Springer, New York, p. 167.

A. M. F. Lau (1979c), *Phys. Rev. Lett.*, **43**, 1009.

A. M. F. Lau (1979d), in *Laser Induced Processes in Molecules*, eds. K. L. Kompa and S. D. Smith, *Springer Ser. Chem. Phys.* 6, Springer, New York, p. 163.

A. M. F. Lau (1980), *Phys. Rev.*, **A22**, 614.

A. M. F. Lau (1981a), Resonant photon-catalyzed predissociation and autoionization fragment yield, rate constant and rotational line strengths, to be published.

A. M. F. Lau (1981b), to be published.

A. M. F. Lau, S. N. Dixit and J. Tellinghuisen (1981), The photon-as-catalyst effect in laser-induced predissociation of iodine diatom, to be published.

A. M. F. Lau and C. K. Rhodes (1977a), *Phys. Rev.*, **A15**, 1570.

A. M. F. Lau and C. K. Rhodes (1977b), *Phys. Rev.*, **A16**, 2392.

J. C. Lehmann (1978a), *Rep. Prog. Phys.*, **41**, 1609.

J. C. Lehmann (1978b), *Contemp. Phys.*, **19**, 449.

M. Levitt, R. Novick and P. D. Feldman (1971), *Phys. Rev.*, **A3**, 130.

J. C. Light and A. Altenberger-Siczek (1979), *J. Chem. Phys.*, **70**, 4108.

M. M. T. Loy (1976), *Phys. Rev. Lett.*, **36**, 1454.

P. Luc (1980), *J. Mol. Spectrosc.*, **80**, 41.

G. Mainfray, C. Manus and I. Tugov (1972), *Zh Eksp. Teor. Fiz. Piz'ma Red.*, **16**, 19 (*JETP Lett.*, **16**, 12).

C. F. Melius and W. A. Goddard, III (1974), *Phys. Rev.*, **A10**, 1541.

A. Messiah (1961), *Quantum Mechanics*, vol. 1, North-Holland, Amsterdam, Chap. 6.

M. Mizushima (1975), *The Theory of Rotating Diatomic Molecules*, Wiley, New York.

J. Morellec, D. Normand, G. Mainfray and C. Manus (1980), *Phys. Rev. Lett.*, **44**, 1394.

R. S. Mulliken (1960), *J. Chem. Phys.*, **33**, 247.

R. S. Mulliken (1971), *J. Chem. Phys.*, **55**, 288.

C. A. Nicolaides and D. R. Beck (1978), *Phys. Rev.*, **A17**, 2116.

H. Nussbaumer (1980), *Opt. Lett.*, **5**, 222.

I. I. Rabi (1937), *Phys. Rev.*, **51**, 652.

H. Rademacher and F. Reiche (1927), *Z. Phys.*, **41**, 453.

R. Rassi, V. Pejcev and K. J. Ross (1977), *J. Phys.* B, **10**, 3535.

S. J. Riley, R. K. Sander and K. R. Wilson (1974), in *Laser Spectroscopy*, eds. R. Brewer and A. Mooradian, Plenum, New York, p. 597.

J. E. Rothenberg and S. E. Harris (1981), *IEEE J. Quantum Electron.*, **QE-17**, 418.

D. L. Rousseau (1975), *J. Mol. Spectrosc.*, **58**, 481.

M. Sargent, III, M. O. Scully and W. E. Lamb, Jr (1974), *Laser Physics*, Addison-Wesley, Reading, MA.

J. M. Schulman, R. Detrano and J. I. Musher (1972), *Phys. Rev.*, **A5**, 1125.

P. P. Sorokin and J. R. Lankard (1971), *J. Chem. Phys.*, **54**, 2184.

J. I. Steinfeld (1972), *Faraday Discuss. Chem. Soc.*, **53**, 155 and references therein.

E. C. G. Stückelberg (1932), *Helv. Phys. Acta*, **5**, 370.

J. Sugar, T. B. Lucatorto, T. J. McIlrath and W. A. Weiss (1979), *Opt. Lett.*, **4**, 109.

B. J. Sullivan and D. A. Dows (1980), *Chem. Phys.*, **46**, 231 and references therein.

C. Tai and F. W. Dalby (1978), *Can. J. Phys.*, **56**, 183.

J. Tellinghuisen (1972), *J. Chem. Phys.*, **57**, 2397.

J. Tellinghuisen (1973), *J. Chem. Phys.*, **58**, 2821.

J. Tellinghuisen (1975), *Phys. Rev. Lett.*, **34**, 1137.

J. Tellinghuisen, M. R. McKeever and A. Sur (1980), *J. Mol. Spectrosc.*, **82**, 225.

L. A. Turner (1930), *Z. Phys.*, **65**, 464.

A. Vasilakis, N. D. Bhaskar and W. Happer (1980), *J. Chem. Phys.*, **73**, 1490.

C. C. Wang and L. I. Davis, Jr (1975), *J. Chem. Phys.*, **62**, 53.

J. Weiner (1980), *Chem. Phys. Lett.*, **76**, 241.

K. Wieland, J. B. Tellinghuisen and A. Nobs (1972), *J. Mol. Spectrosc.*, **41**, 69.

J. J. Wynne and J. P. Hermann (1979), *Opt. Lett.*, **4**, 106.

C. Zener (1932), *Proc. R. Soc. London*, **A137**, 1696.

C. Zener (1933), *Proc. R. Soc. London*, **A140**, 660.

P. Ziem, P. Bruch and N. Stolterfoht (1975), *J. Phys. B*, **8**, L480.

E. Zouboulis, N. D. Bhaskar, A. Vasilakis and W. Happer (1980), *J. Chem. Phys.*, **72**, 2356.

Dynamics of the Excited State
Edited by K. P. Lawley
©1982 John Wiley & Sons Ltd.

PHOTOFRAGMENT DYNAMICS

STEPHEN R. LEONE

Joint Institute for Laboratory Astrophysics, National Bureau of Standards and University of Colorado; and Department of Chemistry, University of Colorado, Boulder, CO 80309, USA

CONTENTS

I. INTRODUCTION

Molecular photodissociation has been studied by a wide variety of techniques for many years. Photofragment dynamics represents a subfield of photodissociation with special emphasis on fragmentation details, such as final state distributions, dissociation lifetimes, product angular distributions, fluorescence polarization of fragments, and translational energy distributions. The field of photofragment dynamics is still remarkably young. Experimental tools have become available only very recently to explore the photofragmentation process in such great detail. At least for simple molecules, it is possible that their photofragment dynamics can accurately be described theoretically.

In the last decade there have been intense experimental and theoretical investigations of photofragmentation processes. Our state of understanding has made rapid and dramatic progress. In 1977 there were several outstanding reviews of the fields. Simons[1] gives a complete account of the small-molecule photofragment processes up to this time. Gelbart[2] considers also predissociation phenomena and radiationless processes of larger molecules. Freed and Band[3] discuss in depth the theoretical treatments of small-molecule dissociation dynamics. An excellent review of all small-molecule photodissociation processes is given in the recent book by Okabe.[4] Ashfold, Macpherson, and Simons[5] recently reviewed the entire field of vacuum ultraviolet photochemistry of simple polyatomic molecules. They considered in great depth the photofragment dynamics of $H_2O, D_2O, H_2S, H_2Se, H_2Te, NH_3, ND_3,$ $PH_3, PD_3, HCN, DCN, CO_2, OCS, OCSe, CS_2, CSe_2, N_2O, ICN, BrCN,$ $ClCN, C_2N_2, CH_3CN, CH_3NC,$ and CF_3CN.

This article will consider predominantly the general field of photofragment dynamics and the advances since 1977, but with occasional excursions back in time to provide a complete discussion when necessary. The focus will be on small-molecule photofragment dynamics where in-depth experimental and theoretical work is available. Much less emphasis will be given to molecules where radiationless processes play a major role. Primarily visible and ultraviolet, single-photon, gas-phase dissociation processes will be considered. There will be no discussion of infrared multiple-photon dissociation processes or the photofragmentation of molecular ions.

In the last few years all of the fields of molecular dynamics have witnessed phenomenal improvements in experimental and theoretical capabilities. Many of the developments in photofragmentation studies have been motivated by quests for new lasers and laser isotope separation, or by problems relating to the atmosphere or combustion. This has had the effect of kindling an intrinsic interest and a tremendous new capability in studying photofragmentation processes. A quick glimpse at these capabilities is valuable. Experimentally, new methods have been developed for both preparing initial states and probing final products. Powerful new laser sources are commercially available, including tunable dye lasers, frequency-doubled dye lasers, and rare-gas halide excimer lasers. The high pulsed energies, wide tuning ranges, and short wavelengths from these lasers make many more experiments possible than ever before. Vacuum ultraviolet (VUV) photodissociation is vastly improved with high-energy, monochromatic, pulsed synchrotron radiation sources. Laser-induced fluorescence probing is used routinely to obtain complete vibrational and rotational state information on molecular dissociation fragments. The time resolution can be set to probe essentially collision-free dynamics, a capability formerly reserved only for sophisticated molecular beam experiments. Narrow-bandwidth, tunable lasers are used to interrogate Doppler velocity profiles to measure translational energy distributions. New, laser photofragment, time-of-flight (TOF) spectrometers have been constructed to obtain high-resolution angular and translational energy distributions. With such machines, excited-state lifetimes as short as 10^{-14} s have been observed for dissociative states. Accurate bond dissociation energies are obtained. The TOF techniques have been applied to three-body dissociation dynamics with spectacular success. Supersonic nozzle beams are used to prepare cooled molecular systems with well characterized initial states and to synthesize novel van der Waals molecular complexes for dissociation studies. Infrared lasers have been used to deposit energy into specific molecular vibrations to enable the study of photofragment dynamics of vibrationally excited molecules. Investigations delve into larger molecular systems with much less trepidation than before. Direct and sequential multiphoton dissociation processes abound, offering new vistas for the imaginative researcher.

The theories of photoejection dynamics have been extended from linear molecules to more general polyatomics and to predict complete angular and polarization effects. A theory of two-photon photoejection dynamics has been presented. Full quantum-mechanical calculations are being carried out to describe the detailed state distributions of photofragments. They take into account multidimensional Franck–Condon factors, final state recoil interactions, good potential surfaces, and correct excited-state geometries. It is now possible in many cases to predict the full effect of Franck–Condon factors and recoil interactions on the final state distributions in the photofragmentation process. The behaviour of mode-specific vibrational excitation on the absorption cross section and the quantum yield of excited product formation can be predicted. The effect of predissociation and specific Rydberg state excitations on the dynamics of dissociation and quantum yields is being discovered. As a result, we have a better understanding than ever before of simple, direct dissociation processes.

The remainder of this review considers in turn the developments in experimental technique (not intended to be exhaustive), the theoretical advances in describing photofragment dynamics, and finally the results for a number of specific systems.

II. EXPERIMENTAL TECHNIQUES

The *classic* techniques for studying photofragmentation have been vastly improved in the last few years, offering greater sensitivity, resolution, and detail. In addition, there are several major innovations which allow probing of the finest details of photofragment dynamics: individual vibration–rotation states, precise translational and angular distributions, lifetimes of the dissociative state, and excited-state symmetries, geometries, and transition moments. Most of the experiments are carried out under collision-free conditions, yielding high-quality data unobscured by the effects of subsequent relaxation.

A. UV Dissociation, Fragment Fluorescence

One of the simplest experiments, which also provides some of the most detailed information on photofragmentation, involves UV or VUV dissociation followed by monitoring fluorescence from excited atomic and molecular fragments. These experiments can now be carried out at low enough pressures with good signal-to-noise ratio to obtain initial, collision-free product distributions. Many of the improvements in capability are made possible by the availability of new light sources of high intensity. Tunable dye lasers, excimer lasers, and synchrotron radiation sources are used for many of the experiments. Better collection optics, monochromators, and phototubes are also available. The experimentalist can now pay closer attention to details of the molecular processes than to the apparatus. The fragment fluorescence experiments provide some of the highest-quality information available on excited atomic, vibrational, and rotational states.

Some of the most extensive studies involve VUV dissociation of various CN containing compounds with detection of vibrationally and rotationally resolved $CN(B^2\Sigma^+)$ emission. The technique has been applied to HCN,[6-9] BrCN,[7,10-12] ClCN,[7,12] ICN,[7,12] CH_3CN,[13] CF_3CN,[13] CH_3NC,[13] and deuterated analogues. These results have provided some excellent data for theoretical comparison. Experimentation on the production of atomic excited states has also been vigorous. Measurements have been made on the generation of $Se(^1S)$ from OCSe,[14-16] $O(^1S)$ from OCS,[17] and $S(^1S)$ from OCS.[18] All of these experiments involve detection of highly metastable electronic states of the atoms, often necessitating the use of collision-induced emission to enhance the signal strength. Several of the experiments[16,17] make use of the high pulsed energy of an ArF excimer laser at 193 nm. Synchrotron radiation sources have been used to accomplish VUV dissociation in very short ($\sim$ 17–130 nm) wavelength regimes. One of the earliest uses of synchrotron radiation was in 1975 with the VUV photodissociation of N_2O,[19] and more recently, for example, a synchrotron source was used to measure $OH(A^2\Sigma^+)$ emission in the dissociation of H_2O.[20] Fluorescence from the $CS(a^3\Pi)$ state in the photodissociation of CS_2[21] and from a variety of atomic and diatomic fragments in the dissociation of metal halide compounds, TlI, InI, HgI_2, $HgBr_2$, CdI_2, and ZnI_2,[22] has been reported recently. Fluorescence has also been studied from a number of polyatomic fragments, $NH_2(\tilde{A}^2A_1)$ from NH_3 dissociation,[23,24] $PH_2(\tilde{A}^2A_1)$ from PH_3,[25,26] and $CF_2(^1B_1)$ from C_2F_4[27] and CF_2Br_2.[28] $NO(A^2\Sigma^+)$ emission has been recorded from the VUV photofragmentation of CF_3NO and RONO compounds.[29,30] It is now possible to study these moderately sized polyatomic systems in great detail.

B. Molecular Beam Photofragmentation: Time-of-Flight (TOF) and Angular Distributions

One of the most elegant ways to study photofragment dynamics is by crossing a light source with a molecular beam and measuring the angular distributions and time-of-flight arrival of the fragments with a mass spectrometer (Fig. 1). Impressive new strides have been made in this technology by the introduction of lasers to replace arc lamp sources. The methods have been amply discussed and reviewed.[1,31–33] Retention of the angular polarization of the incident light beam in the fragments can be related to the lifetime of the excited state and to the orientation of the transition moment with respect to the molecular axis. For a direct dissociation process, such as in CH_3I, the lifetimes observed are 10^{-13} s or less.[34] For a predissociation process or a dissociation which involves some form of intersystem crossing, as in the aryl halides, lifetimes of 10^{-10} to 10^{-12} s are inferred.[34] For those longer lifetimes, the portion of the angular anisotropy which contains the lifetime information is lost. The theory of the basic photoejection dynamics and angular anisotropy has been described in detail for linear molecules by Zare[35] and for more general polyatomics by Yang and Bersohn.[36] From the translational energy distributions of the fragments, a tremendous amount of information is also obtained on the internal energy content of the fragments.

A classic and very powerful machine was constructed by Wilson and coworkers.[37] With this device, extensive studies of the photofragment dynamics were made on the halogens,[38,39] HI,[40] ICN,[38,41] NO_2,[42,43] NOCl,[44] alkyl iodides,[32] and C_2H_5ONO.[45] A new technique was also devised to study the photofragmentation of electronically excited states by adding a second, delayed laser pulse after the first exciting pulse.[46] This method was applied to a study on the $B^3\Pi_{0^+u}$ state of I_2.

In recent years, several other groups have made extensive use of the TOF

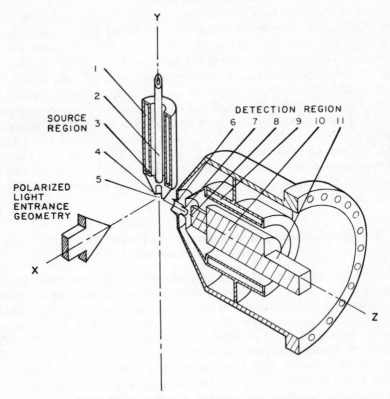

Fig. 1. Molecular beam photofragmentation apparatus, cut-away view: 1, molecular beam reaction chamber liquid-nitrogen trap; 2, molecular beam source introduction tube; 3, effusive array; 4, beam skimmer; 5, interaction zone; 6, buffer chamber orifice; 7, detector chamber orifice; 8, high-efficiency electron bombardment axial ionizer; 9, detector chamber liquid-nitrogen cold trap; 10, mass quadrupole filter; 11, particle multiplier. From Dzvonik *et al.*[34] Reproduced by permission of the American Institute of Physics.

and/or angular distribution techniques to explore bigger and more interesting molecules. Riley and coworkers[47] used the machine of Wilson to explore the dissociation of CH_2I_2 by one and two (sequential) photons:

$$CH_2I_2 \rightarrow CH_2I + I, \; CH_2I \rightarrow CH_2 + I.$$

A high degree of vibrational and rotational excitation is observed in the CH_2I radical. The bond dissociation energies, excited-state symmetries, and yields of electronically excited halogen atoms were obtained for a large collection of alkali iodides,[48] bromides,[49] and chlorides.[50] The primary dissociation process of

$$UF_6 \rightarrow UF_5 + F$$

has also been studied by the molecular beam method.[51] Three-body dissociation processes have now been successfully carried out in the molecular beam on CH_3COI[52] and CF_3COI[53]

$$CH_3COI(CF_3COI) \rightarrow CH_3(CF_3) + CO + I.$$

Bersohn and coworkers have studied in depth the photodissociation dynamics of alkyl and aryl iodides and bromides.[1,34] In the aryl halides, intersystem crossing pathways are important since the initial excitation is in the aromatic ring system and not in the carbon–halogen bond which ultimately breaks. Other studies have been reported on CS_2,[54] CdI_2,[55] TlI,[56,57] N_2O,[58] O_2,[58] O_3,[59] and KI.[60]

A new generation of higher-resolution TOF apparatus has recently been constructed by Lee and coworkers.[61] Most of the previous machines have mutually perpendicular directions between the light source, molecular beam, and fragment detector. Fragments recoiling slowly from the fast molecular beam may not be 'seen' by the detector. They might be detected by moving the mass spectrometer nearer to the beam axis. In addition, higher resolution can be obtained by having longer flight paths and sometimes by detecting backward-scattered fragments, taking advantage of the effect of subtraction of the molecular beam velocity. Such a machine, incorporating a 170° angular scan range from the beam axis, a 34·1 cm flight path, and a supersonic

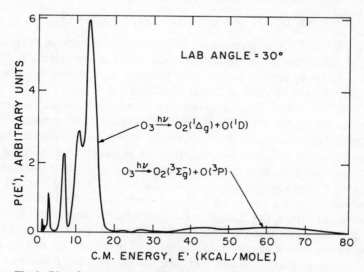

Fig. 2. Photofragment time-of-flight spectrum for O_2 fragments produced by dissociation of O_3 at 266 nm, converted into centre-of-mass translational energy space. The peaks in the $O_2(^1\Delta_g)$ correspond to $v = 3, 2, 1$, and 0 from left to right. From Sparks et al.[62] Reproduced by permission of the American Institute of Physics.

expansion beam, has been developed.[61] The cooled supersonic beam provides a more precise internal energy for the parent molecule when analysing the energetics. With this apparatus, high-resolution analysis of the classic CH_3I dissociation has obtained more detailed information on the vibrational excitation in the CH_3 'umbrella' out-of-plane bending motion.[61] The photofragmentation of ozone at 266 nm yields the extent of $O(^1D)$ and $O(^3P)$ formation, as well as fully resolved vibrational state populations in the $O_2(^1\Delta_g)$ product (Fig. 2).[62] The precise energy of the $O_2(^1\Delta_g, v = 0)$ peak also allows a separate estimate of the extent of rotational excitation in the fragmentation process. The TOF molecular beam method remains one of the most elegant and detailed forms of investigating photofragment dynamics.

C. Quantum Yield Measurements

Classic techniques of quantum yield measurement continue to be important in elucidating the details of electronically excited states and photofragment dynamics. There has been substantial activity in the measurements of both absolute and relative quantum yields as a function of wavelength.

The $O(^1D)$ quantum yield from the photolysis of O_3 is important because of its relevance in the chemistry of the atmosphere. This yield has been extensively studied in the fall-off region, 295–325 nm, using laser photolysis and NO_2 chemiluminescence from the reaction sequence:

$$O(^1D) + N_2O \rightarrow 2NO, NO + O_3 \rightarrow NO_2^* + O_2$$

(see Fig. 14).[63–66] Although there are still discrepancies between the measurements, possibly because of the complex chemistry of the system, it appears that reasonable agreement is now being reached. A surprising discovery was made that the $O(^1D)$ yield at 266 nm and 248 nm is not unity, but rather ~ 0.9.[59,62,67,68] There is now agreement between resonance fluorescence measurements on $O(^3P)$[67,68] and molecular beam TOF results.[59,62] The older results of Amimoto et al.[69] have been shown to be incorrect because of time resolution problems in measuring the initial $O(^3P)$ concentration.[68]

Vacuum ultraviolet absorption and resonance fluorescence methods have been used to measure many of the quantum yields for excited atom production. In some cases, both the ground and excited atoms can be monitored. In others, the change in the ground-state population is detected upon quenching of the excited atoms. This assumes, sometimes incorrectly, that all of the excited atoms are quenched and that none are reacted or lost from the system. These measurements are also fraught with difficulties owing to lineshape parameters of the atomic lamp and the extremely high absorption strength of the atomic resonance transitions in the VUV. As a result, there are sometimes more than the usual number of discrepancies in the reported results. Such methods have been applied to the quantum yields of $I^*(^2P_{1/2})$

from I_2[70] and alkyl iodides,[71-73] to the yields of $Br^*(^2P_{1/2})$ from Br_2 (see Fig. 8)[74,75] and $HCCBr$,[76] and to detect $Cl^*(^2P_{1/2})$ from CCl_4.[77] The detection of I^* by direct infrared fluorescence has been employed to map the I^* yield from ICN (see Fig. 12)[78] and HgI_2 (see Fig. 7)[79] as a function of tunable laser wavelength and at discrete wavelengths for CH_3I and CH_2I_2.[80] The results of these relative yields are normalized at one wavelength to the I^* signal from a molecule such as C_3F_7I, whose I^* yield is reported to be unity.[72] Similarly, infrared fluorescence has been observed directly from Br^* upon photodissociation of Br_2,[81] or by monitoring CO_2 vibrational emission from the electronic-to-vibrational transfer pathway,[82]

$$Br^* + CO_2 \rightarrow Br + CO_2(0,0,1).$$

Relative Br^* quantum yields have been measured in this way for Br_2 and IBr with a tunable laser (see Fig. 8).[82] The results are in good agreement with the experiments carried out by VUV detection.[74,75] Resonance fluorescence of O atoms has been used as the product monitor to investigate the quantum yields of CO_2 dissociation.[83] Similarly, the quantum yield of CO formation from single vibronic state excitation of glyoxal has been probed by resonance emission in the VUV.[84,85]

A variety of other quantum yield measurements have been carried out by direct fluorescence or subsequent chemiluminescence. Maya has reviewed the quantum efficiencies of fluorescence for a large number of metal halide vapours.[22,86] Quantum yields have been reported for $O(^1S)$ from VUV dissociation of CO_2,[87] for the $PH_2(\tilde{A})$ excited state upon 193 nm dissociation of PH_3,[88] for the $CN(A^2\Pi)$ and $CN(B^2\Sigma^+)$ states in the synchrotron VUV dissociation of HCN,[89] and for $OH(A^2\Sigma^+)$ by VUV dissociation of H_2O.[90] Chemiluminescence from excited HNO was used to determine H atom quantum yields from single vibronic level dissociation of formaldehyde.[91] The quantum yield of chlorine nitrate ($ClONO_2$) photolysis has been studied by direct mass spectrometric detection of the products from a 'bulb' type of apparatus rather than a molecular beam.[92] Photoacoustic measurements have been carried out on the fragmentation of CH_3I[93] and UF_6.[94] There is a real difficulty in interpreting these results, although the technique may offer promise for future quantum yield measurements.

Chemical trapping, followed by conventional gas chromatography–mass spectrometry analysis of stable products also continues to offer valuable solutions to the quantum yield problem. Dissociation of

$$CF_3NO \rightarrow CF_3 + NO$$

was demonstrated by trapping the CF_3 radical by reaction with Cl_2.[95] The yields of Cl atoms from $CFCl_3$ and CF_2Cl_2 were shown to be unity at wavelengths greater than 214 nm and nearly 2 at 147 nm.[96] In the fragmentation of $Fe(CO)_5$ at 248 nm it has been shown that $Fe(CO)_4$, $Fe(CO)_3$, and $Fe(CO)_2$ can be trapped quantitatively with PF_3 to obtain the precise fragment distributions.[97,98] The threshold energies for production of $CH_2(^3B_1)$ and $(^1A_1)$ states in the photolysis of ketene have been determined by conventional analysis of CO yields as a function of wavelength.[99] One of the most dramatic applications of chemical trapping of photolysis products involves the photodissociation and isotope separation of CS_2.[100] With a narrow-bandwidth ArF laser at 193 nm, isotopically enriched solid products were collected off the walls. In a precision laser deposition experiment, $2\,\mu m$ thick films of metals have been generated by photodissociation of organometallic compounds such a $Al(CH_3)_3, Cd(CH_3)_2$, and $Sn(CH_3)_4$.[101] It is postulated that free metal atoms are generated which then condense directly on the substrate.

D. Laser-Induced Fluorescence

Laser-induced fluorescence (LIF) is one of the most powerful new methods for obtaining detailed state information on photofragmentation processes. The method can be applied to probe vibration–rotation levels of ground electronic states, and it has also been used between two electronically excited states. Fig. 3 shows the basic experimental set-up for laser-induced fluorescence. A laser or other flashlamp source is used to photolyse the parent

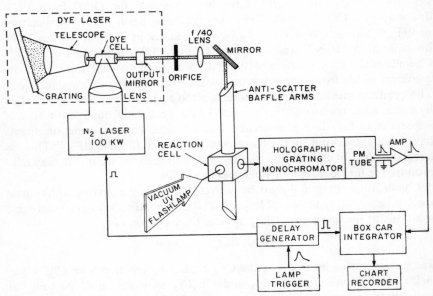

Fig. 3. Experimental apparatus for laser-induced fluorescence analysis of photofragmentation processes. From Cody et al.[102] Reproduced by permission of the American Institute of Physics.

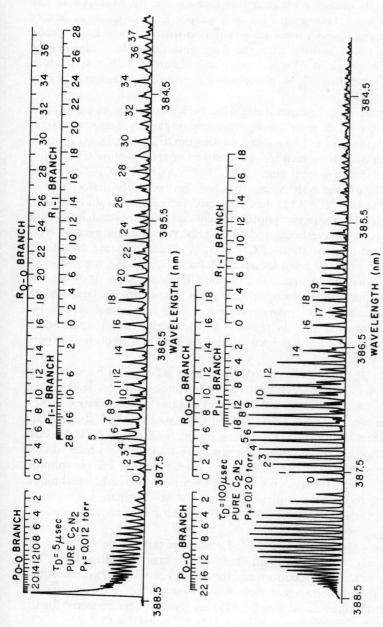

Fig. 4. Laser-induced fluorescence spectra of $CN(X^2\Sigma^+ \leftrightarrow B^2\Sigma^+)$ from C_2N_2 photolysis, measured under two different experimental conditions and plotted as a function of the laser wavelength. Lower spectrum: This spectrum was measured under conditions in which the rotational levels of the $CN(X)$ radicals were collisionally equilibrated before detection. Therefore, each band has a rotational temperature of about 300 K. The (0,0) and (1,1) bands account for most of the intensity in the spectrum. Upper spectrum: This spectrum was measured under conditions in which the $CN(X)$ radicals were unrelaxed rotationally. Approximately 95% of the total fluorescence signal is accounted for by the (0,0) and (1,1) band systems. From Cody et al.[102] Reproduced by permission of the American Institute of Physics.

molecule. Then a tunable dye laser is scanned through spectral transitions of the fragments to obtain resonance fluorescence spectra. Analysis of the fluorescence spectra taking into account proper Franck–Condon factors provides vibrational and rotational population distributions. If the probe laser pulse is timed shortly after the fragmentation pulse, collision-free dynamics can be obtained. This is dramatically illustrated in Fig. 4, in which the $CN(X^2\Sigma^+)$ state from C_2N_2 photolysis was probed both before and after rotational relaxation.[102]

The laser-induced fluorescence method has been used for a growing number of elegant studies on diatomic and polyatomic fragments. The most extensively studied is the $CN(X^2\Sigma^+)$ state by exciting the $B^2\Sigma^+ \leftarrow X^2\Sigma^+$ transition. In this way the vibration–rotation information in ground-state $CN(X)$ has been obtained for fragmentation of C_2N_2,[102] C_4N_2,[103] $ClCN$,[104] and ICN.[105,106] The excited $NH(^1\Delta)$ state has been probed in the dissociation of HN_3[107,108] and $HNCO$.[109] LIF has been used successfully to study OH,[110] CH,[111] and NO[112] diatomic photofragments as well. Several important polyatomic systems have been studied. Highly vibrationally excited NO_2 was probed in the dissociation of CH_3NO_2.[113] The results are difficult to interpret because of the complex spectral features and the high level of excitation. LIF has been carried out on the $CH_2(\tilde{a}^1A_1)$ state to obtain the threshold for appearance of singlet methylene in the tunable laser photo-dissociation of ketene.[114–116] An upper limit of $9.8 \pm 1.5\,kcal\,mol^{-1}$ ($1\,kcal = 4.18\,kJ$) was obtained for the $\tilde{a}^1A_1 - \tilde{X}^3B_1$ singlet–triplet splitting.[116] CH_3O radical production has been studied by LIF in a photolysis experiment on CH_3ONO.[117]

A novel two-photon, laser-induced fluorescence scheme has been reported for the detection of the $NO(X^2\Pi_{1/2,3/2})$ vibration–rotation states from CF_3NO.[118,119] A red dye laser (600–670 nm) was used to dissociate CF_3NO, followed by a blue dye laser (450–500 nm) to excite the $A^2\Sigma^+$ state of NO via a two-photon transition. UV emission on the $A–X$ transition is detected with a solar blind photomultiplier. The same LIF transition has been studied also by single-photon excitation in the range 224–250 nm.[112] The two-photon experiment illustrates the tremendously high sensitivity of the LIF technique. It was possible to obtain complete vibrational state information, rotational temperatures, the lifetime for predissociation, and a qualitative analysis of the $^2\Pi_{1/2}/^2\Pi_{3/2}$ populations.[118,119]

With narrow-bandwidth lasers, the LIF technique can also be used to probe Doppler velocity widths to obtain translational energy distributions. The link between the Doppler spectrum and the velocity profile has been described in detail for the photofragmentation of a simple diatomic, NaI.[120] Translationally hot Doppler linewidths ($\sim 5000\,cm^{-1}$) have been reported for the $NH(^1\Delta)$ fragments in the dissociation of HN_3[107] and $HNCO$.[109]

Laser absorption probing of photofragments also offers a high degree of sensitivity and selectivity, especially if the absorption can be carried out intracavity. A number of crucial experiments have been achieved where conventional techniques were lacking. Infrared CO probe laser absorption has been used to obtain the appearance rate and vibrational distribution of the CO product from H_2CO photolysis.[121] The rate of CO production was found to be several orders of magnitude slower than the fluorescence lifetime of the formaldehyde excited singlet state, requiring the existence of an unknown intermediate state in the decomposition. In a similar fashion, the vibrational distribution of CO from CH_3CHCO fragmentation has been obtained.[122] Intracavity laser magnetic resonance has been used to probe the production of $Cl^*(^2P_{1/2})$ atoms in the dissociation of Cl_2[123] and ICl[124] using a $^{13}CO_2$ infrared laser. In the case of ICl gain was actually observed, indicating a population inversion between the $Cl^2P_{1/2}/^2P_{3/2}$ levels.[124]

E. Polarized Photofluorescence Spectroscopy

Polarized fluorescence from the fragments of a dissociation process provides another way to obtain the symmetry and lifetime of the excited state. This method is entirely complementary to molecular beam angular distribution techniques. The polarized fluorescence method measures the degree of alignment or orientation of the separating fragments with respect to the initial polarized light beam. It can be applied when the products of the fragmentation process are left in fluorescing excited states. A complete theory which describes the expected degree of polarization for direct dissociation of a general triatomic molecule has recently been given by Macpherson. Simons, and Zare.[125] Earlier theoretical work on specific molecules has also been discussed by Simons and coworkers.[126-129]

Experimentally there have already been numerous observations of polarized fluorescence from the fragments of dissociation. Oddly enough, until recently all measurements were on molecules which are known to undergo predissociation, rather than direct dissociation. The exact degree of polarization is sensitively dependent on the lifetime of the dissociated state, and so these molecules do not serve as a precise check on the theoretical predictions. Polarized fluorescence has been observed from $CN(B^2\Sigma^+)$ in the predissociation of HCN,[126,127] BrCN,[126,128] ICN,[128] and ClCN[128] and from the $OH(A^2\Sigma^+)$ state in predissociation of $H_2O(D_2O)$.[129] Qualitatively, increased lifetimes for predissociation have been observed through decreased polarization at longer wavelengths,[127] and the symmetries of the excited states have been assigned in several cases.[126,128,129] Polarized photofluorescence spectroscopy has also been carried out on ICN with a synchrotron radiation source.[130] Time-resolved decays were recorded for both parallel and perpendicular polarization analysis to obtain the degree of polarization. For ClCN, BrCN, and ICN, the polarization parameter varies significantly depending on the photolysis wavelength,[128,130] possibly indicating a substantial number of different Rydberg states and excitation to various P, Q, R branch transitions, which affect the expected degree of polarization. The first observation of polarized fluorescence from fragments of a directly dissociated state have been reported for the $HgBr(B^2\Sigma^+)$ product of $HgBr_2$ with a 193 nm ArF laser.[131] The measured degree of polarization (11·9%) is in close agreement with the predicted theoretical value of 14·3%. It was found that single collisions with Kr rare gas cause almost complete loss of polarization.

Early attempts to observe polarized fluorescence from atomic products of Rb_2 dissociation were unsuccessful.[132] Recently, polarized $Na(^2P_{3/2})$ emission from the dissociation of Na_2 has been observed.[133] This represents the first reported polarization of atomic fluorescence in a photofragment process, an effect first predicted by Van Brunt and Zare in 1968.[134] In order to observe such an effect, there must be an angular anisotropy in the dissociation products, and also a preferential population of the m_J sublevels of the excited state.[132] Both these conditions appear to be met in the Na_2 case.

F. Photofragmentation of Vibrationally Excited Molecules

In the last few years, there has been a tremendous surge of interest in studying the effect of vibrational excitation on photofragment processes. This is partly motivated by ideas involving sequential two-photon (IR + UV) photodissociation for isotope separation schemes and also by a need to know photodissociation cross sections for vibrationally excited molecules of atmospheric interest and relevant to lasers. Vibrational excitation can result in one or more effects:[17] the ultraviolet absorption threshold and spectrum can shift to the red due to the additional vibrational energy; the vibrational excitation can cause large alterations in the Franck–Condon factors for absorption, resulting in dramatic changes in the ultraviolet spectrum; excitation in bending vibrations can lower the symmetry of a polyatomic molecule, making new electronic transitions allowed for the distorted geometry. Experimentally, two types of measurements can be carried out: (1) the absorption cross section for the vibrationally excited molecule and (2) the quantum yield of excited product formation. The latter experiment provides a delicate probe for changes in the fractional absorption of different electronic states due to the vibrational excitation. With the availability of pulsed and CW infrared lasers, experiments are possible to probe the dynamics of mode-specific vibrational excitation.

One of the simplest examples, which in principle should be possible to calculate, involves the dissociation of a diatomic. Zittel and Little have photolysed HBr in the long-wavelength tail of its directly dissociated absorption continuum with a doubled dye laser at 258·9 nm.[135] An HBr chemical laser is used to excite the HBr to $v = 1$. Using on-line mass spectrometric detection of photolysis products, they measure a factor of 34 increase in the dissociation cross section for the vibrationally excited HBr. The change in absorption cross section was estimated to be a factor of 90·6. In considering these cross sections, only single potential surfaces for the bound and repulsive state were used. The poor agreement suggests that it may be necessary to include accurate values for the electronic dipole matrix element as a function of internuclear separation as well as a complete set of repulsive potential energy surfaces. The variation of the electronic transition dipole moment was included in a study of the photodissociation of vibrationally excited Na_2.[136] The agreement between experiment and theory is good for the relative variation of absorption cross section versus wavelength. However, the absolute values obtained by experiment and theory differ significantly. The discrepancy is attributed to inaccuracies in the potential surfaces used. There is clearly much work that needs to be done to calculate repulsive potential surfaces accurately in order to account for even the simplest details of the effects of vibrational excitation.

Given the state of understanding in vibrationally excited diatomic dissociation, it is an even more difficult challenge to understand what will happen in polyatomics. There have already been a number of excellent and intriguing experimental results. The two-photon dissociation of

$$OCS \rightarrow CO + S(^1S)$$

has been carried out at 193 nm with and without excitation of the $2v_2$ bending vibration by a CO_2 laser.[17] The $S(^1S)$ yield was enhanced by 100%, and the entire increased yield was attributed to the enhancement of the total absorption cross section by the vibrational excitation.[17] A 25% enhancement of the $Se(^1S)$ yield was observed in a similar experiment on OCSe.[16] Once again, the entire increase was attributed to the enhancement of the absorption cross section. A contrasting result was obtained in the VUV dissociation of heated OCS, where a 190% increase in $S(^2S)$ yield was observed at 170 nm but without an accompanying change in the absorption cross section.[18] It is possible that the fractional absorption to different excited electronic states is altered by the vibrational excitation.

The dissociation of vibrationally excited ozone to yield $O(^1D)$ is important for atmospheric considerations in the long-wavelength threshold limit. Using a CO_2 laser to excite the v_3 mode of O_3, the $O(^1D)$ yield at 314·5 nm is enhanced by about 70, while the absorption cross section appears to increase by only about 8·4 (lower limit).[137] The effect of vibrational excitation was observed to decrease towards shorter wavelengths (see Fig. 15). Earlier work on the temperature-dependent $O(^1D)$ yield suggested that increased rotational energy was effective in promoting the fragmentation of O_3 for wavelengths below threshold.[138] Arguments involving promotion by the availability of rotational energy have also been made for NO_2.[139] This cannot be a factor

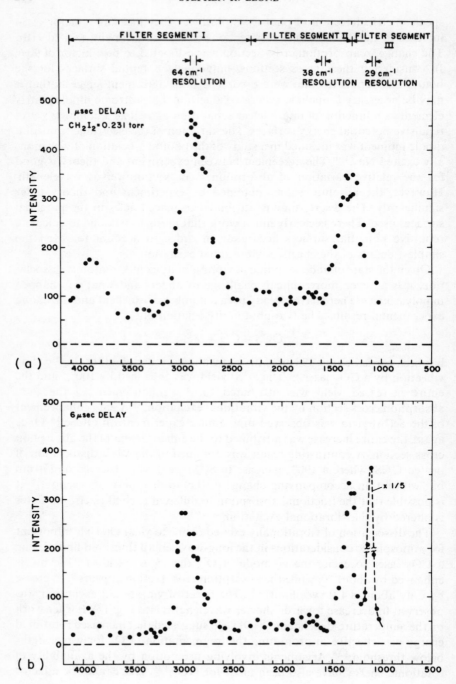

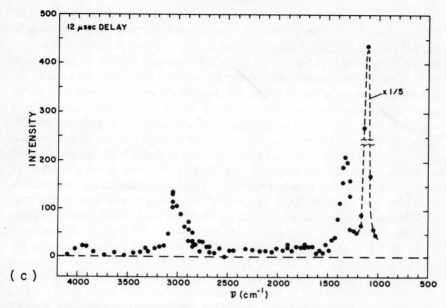

Fig. 5. Infrared emission spectrum of the CH_2I radical as a function of time after photolysis of CH_2I_2 at 248 nm: (a) $1\,\mu s$ after the photolysis pulse, (b) $6\,\mu s$ after the pulse, (c) $12\,\mu s$ after the pulse. The resolution of the interference filters is given as the full width at half-maximum at the top of the figure. The intensity scale is in arbitrary units. The dashed curve is assigned to a vibrational mode of the CH_2I_2 parent molecule excited by $V \rightarrow V$ energy transfer. From Baughcum and Leone.[80] Reproduced by permission of the American Institute of Physics.

in the CO_2 laser experiments since the temperature rise is negligible.[137] Therefore in those experiments the entire effect must be due to the vibrational excitation. The dissociation dynamics of vibrationally excited molecules remains an exciting field open to further exploration.

G. Spectrally Resolved Infrared Fluorescence

Infrared fluorescence and chemiluminescence have frequently been used to study product states of chemical reactions and energy transfer. As already discussed, infrared fluorescence has been used to study photofragmentation processes yielding the excited halogen atoms, $I^*(^2P_{1/2})$ and $Br^*(^2P_{1/2})$. Until recently, there were only a few selected reports of using the method to detect other species. For example, $NO(v = 1)$ was observed by infrared fluorescence in the dissociation of CF_3NO.[95] $CO(v \geqslant 1)$ was directly detected by infrared emission in the photodissociation of formaldehyde.[121] Until recently, there have not been any extensive, spectrally resolved infrared fluorescence studies of photofragments.

Baughcum and Leone first observed infrared emission from the CH_2I

radicals produced by photofragmentation of CH_2I_2 with a 248 nm KrF excimer laser.[140] Recently they reported a complete time- and wavelength-resolved study of the infrared emission from CH_2I in the 1000–4000 cm^{-1} region.[80] In these experiments, a liquid-helium-cooled Ge:Cu detector equipped with scanning, continuous-wavelength interference filters was used to detect the radical emission. The time response of the detector was about 150 ns and wavelength resolution was about 29–64 cm^{-1}. At each wavelength, a complete time decay of the total emission was taken. The radical emission could be distinguished by the time resolution from parent molecule emission excited by energy transfer. An entire spectrum of the radical was constructed at different time delays after the photolysis pulse to monitor the effect of energy removal. The results provide excellent complementary information to the molecular beam TOF spectrum,[47] in which it was determined that the CH_2I acquires a substantial fraction of the excess available photon energy as internal excitation. Fig. 5 shows the time-resolved infrared emission spectra from the CH_2I radicals. Features are observed at the correct positions for the CH_2 bend (~ 1300 cm^{-1}), the CH stretches (~ 2900 cm^{-1}), and a combination band (~ 3900 cm^{-1}). It is evident from the time-resolved spectra that the initial radical is formed extremely 'hot' vibrationally. The CH stretching region at ~ 2900 cm^{-1} has been fitted approximately with a series of statistically populated anharmonic bands.[80] There is a strong non-zero emission at all wavelengths due to the high density of states at about 15 000–20 000 cm^{-1} of internal energy. When the radical is collisionally deactivated, the spectral features sharpen and shift as the molecules tumble down anharmonic ladders of states to populate only the first few vibrational levels of each mode.

In a similar series of experiments, the CH_3 out-of-plane bending vibration was detected in the photofragmentation of CH_3I.[80] Recently this 'umbrella' mode emission has also been partially spectrally resolved.[141] At 266 nm the CH_3 from CH_3I is found to be highly excited in the out-of-plane bend. The CH_3 radical produced by photolysis of $Hg(CH_3)_2$ at 248 nm has also been observed with excitation in the C–H stretch and out-of-plane bend.[142] Excited CH_3 radicals from $Hg(CH_3)_2$ were previously observed by VUV absorption.[143] From the spectrally resolved infrared emission of the CH stretch, the radical is found to have very high initial rotational excitation as well.[142] Although the spectral resolution is still quite low, it is evident that many exciting and potentially powerful new results can be obtained by time- and wavelength-resolved infrared emission techniques.

H. Multiphoton Dissociation

Discoveries of visible and UV multiphoton and sequential multiple-photon fragmentation processes are rapidly increasing due to the availability of high-intensity excimer and dye lasers. Many of the processes are complex and there is difficulty in determining the specific electronic states involved in the fragmentation. Nevertheless, the field already provides significant new understanding of photofragment dynamics. A theory of the angular distribution of photofragments arising from resonant, sequential two-photon absorption has recently been given.[144] An earlier general discussion of the subject of multiphoton ultraviolet dissociation and its application to C_2N_2 has been published.[145,146] Sequential multiphoton absorption can be used in principle to probe the lifetimes of excited states by making the timescale for the absorption of the second photon comparable with the timescale for predissociation or direct dissociation. With laser fluxes higher than 10^{27} photons $cm^{-2} s^{-1}$, it is possible to achieve absorption rates which begin to be competitive with the rates of dissociative processes.

As early as 1969, a two-photon dissociation of I_2 was observed in a molecular beam TOF apparatus.[31] The two-photon process at 532 nm yielded two excited I* atoms per molecule, indicating absorption to a higher-lying excited state. Kroger, Demou, and Riley explored the sequential two-photon absorption in CH_2I_2 at 266 nm, yielding $CH_2 + 2I$.[47] An approximate cross section for fragmentation of the vibrationally excited CH_2I intermediate was obtained, and from the angular distribution of the CH_2, and estimate of the rotational excitation in the CH_2I fragment was made.

Since that time there has been an exponential growth of reported multiphoton processes. UV emission has been observed from excited I atoms produced by the multiple absorption of at least five visible laser photons in I_2.[147] Excited Hg atoms are observed following the sequential three-photon dissociation of $HgCl_2$, $HgBr_2$, and HgI_2 at 193 nm.[148] It is postulated that a single-photon dissociation of HgX_2 to HgX is followed by a two-photon dissociation of the HgX to yield the excited mercury atoms. Two 193 nm photons were shown to be absorbed *directly* to produce $O(^1S)$ from OCS.[17] The ArF laser at 193 nm has been used to dissociate NH_3 to $NH_2(\tilde{A}^2A_1$ and $\tilde{X}^2B_1)$ and $NH(A^3\Pi$ and $b^1\Sigma^+)$ sequentially,[24] to fragment C_2H_2 to $CH(A^2\Delta)$ and $C_2(A^1\Pi_u)$ via sequential photon absorptions,[149] and to generate $PH(A^3\Pi)$ and $PH_2(\tilde{A}^2A_1)$ from PH_3.[88] In many cases, the fragments are left in electronically excited states and can be detected by their emission. Sequential photoabsorption effects have also been reported in CF_2Br_2,[150,151] $CHBr_3$,[111] and CBr_4,[151] yielding ground-state and excited radicals, such as CF_2 and CH. Finally, whole new classes of metal-vapour photodissociation lasers have been obtained by multiple-photon dissociation processes in Hg_2[152] and metal triiodides.[153] The multiphoton process not only strips the molecule down to the bare metal atom when necessary, but also leaves the atom in a multitude of highly excited electronic states, which then lase. These processes have the potential to probe many of the higher-lying dissociative states of the molecules, about which almost nothing is presently known.

Although multiphoton dissociation dynamics is a very young field and may appear stultifying at times, clear patterns are emerging. If the first photon produces a fragment which is likely to have an absorption at the same wavelength and if the laser flux is high enough, a second photon will probably be absorbed. Successful studies of single-photon dynamics can always be carried out at light fluxes low enough to prevent the multiphoton process.[80] It is possible to strip atoms or fragments off the molecules in rapid succession, often ending only at fragments which no longer have large absorption rates at the particular exciting wavelengths and flux and/or density. In climbing up the ladders of molecular states, the process will often arrive at electronically excited fragments, just as is observed for single-photon dissociation. Vibrationally or electronically excited molecules often play a role in the intermediate steps. These excited fragments can enhance or alter the subsequent absorption process because of dramatic changes in Franck–Condon factors or symmetries. Multiphoton dissociation dynamics is certain to be a rewarding way to study high-lying excited states of molecules.

I. Photofragmentation Lasers

It was recognized more than 15 years ago that photodissociation processes can generate specific population inversions in the excited states of atomic and molecular fragments. A classic paper by Zare and Herschbach outlined the concept of photofragmentation lasers in detail.[154] That paper included a lengthy list of potential laser candidate molecules and their photofragment excited-state products. Simultaneously with these predictions, the first photofragment laser was demonstrated on the $I(^2P_{3/2} \leftarrow {}^2P_{1/2})$ transition by broad-band photolysis of CH_3I.[155] Since that time, an enormous number of photofragmentation atomic and molecular lasers have been discoverd. Many of the predicted laser candidates of Zare and Herschbach[154] have successfully lased. Many of the recent data are reviewed by Mikheev[156] and Maya.[86] Photofragmentation lasers can be used as a technique to study the excited states produced in a photodissociation process. A single important piece of information is always obtained when a photodissociation laser is demonstrated, i.e. a population inversion must exist between the upper and lower laser states, setting a limit on their populations. It is difficult to quantify precisely the yield of fragments in the different possible excited states, but often a limit can be placed on the quantum yields by laser energy and gain studies.

Over the years, more experimental tools have become available to produce photofragmentation lasers, especially optical pumps consisting of powerful pulsed lasers at new wavelengths. From TOF data, Hancock and Wilson predicted that a favourable branching ratio exists in I_2 and that an I* laser could be made by photolysis just above the dissociation limit for producing one I* and one I atom.[38] Recently, a tunable, high-energy dye laser was used to lase I* from I_2 in this manner.[157] Photodissociation of NaI with the fifth harmonic of a Nd:YAG laser at 2128 Å produces an inversion of the $3p\ ^2P$ level of Na with respect to the ground state, resulting in laser action.[158] Higher excited states of both In [159] and Tl [160] have lased by photodissociation of InI and TlI with a 193 nm ArF laser pump. Lasing has been observed on two forbidden transitions of Se, $(^1S_0-^3P_1)$ and $(^1S_0-^1D_2)$,[161] and on S($^1S_0-$ $^1D_2$)[162] by photodissociation of OCSe with 172 nm Xe_2^* excimer emission and dissociation of OCS with 146 nm Kr_2^* emission respectively. Photodissociation of the iodides of Na, K, Rb, and Cs at 193 nm results in dozens of superfluorescent laser lines in the alkali atoms.[163] The strongest lasing lines are from atomic states lying several thousands of cm^{-1} below the ArF photon energy. From the sidelight fluorescence, relative yields into the various electronically excited states were obtained for CsI.[163] Studies of this type can provide a tremendous amount of new information on the highly excited molecular dissociative states of these molecules. Metal-atom lasers have been discovered in the multiphoton UV dissociation of GaI_3,[153,164] InI_3,[153] AlI_3,[153] BiI_3,[153] and Hg_2.[152]

Numerous molecular fragment lasers have also been reported. Vibrationally excited hydrogen halide lasers have frequently been demonstrated by flash photolysis elimination from halogenated ethylenes.[165] Observation of these vibrational lasers has played a significant role in elucidating the photofragment dynamics of these molecules. Stimulated emission has been observed on CN vibrational levels and in some cases on the $A^2\Pi \to X^2\Sigma^+$ transition of CN in the photolysis of BrCN, C_2N_2, ICN, and CH_3NC.[166-168] Lasing has been reported on the $B^2\Sigma^+ \to X^2\Sigma^+$ transition of HgBr by photodissociation of $HgBr_2$ at 193 nm,[169] and on the $B_{1/2}-X^2\Sigma^+$ [170] and $C_{3/2}-A_{3/2}$ [171] transitions of XeF in the photolysis of XeF_2. In each case, only the major features of the fragmentation process are observed, but it can be determined that the dissociation of a certain molecular band preferentially populates a specific excited lasing state. Often this level of understanding provides important assignments of molecular absorption bands and direct information on the dissociation dynamics.

III. THEORETICAL DESCRIPTION OF THE PHOTOFRAGMENT PROCESS

State-of-the-art theoretical interpretations of photodissociation have made impressive strides in the past few years.[1-3] The earliest primitive models included only repulsive recoil interactions, where the departing fragments give each other a 'kick', imparting translational, rotational, or vibrational excitation to the products. While these models are often exceedingly helpful in describing qualitative effects, they do not provide precise interpretation of details. Photodissociation might be thought of as simple to calculate in principle. Yet even the best calculations have difficulty in reproducing experimental details. More often than not, this is due to a lack of accurate potential surfaces and not to a misunderstanding of the important aspects of the problem. Direct dissociation processes can now be described by full quantum-mechanical calculations. Multidimensional Franck–Condon factors are included in the photoabsorption step which prepares the excited-state geometry. Final-state (recoil) interactions allow for a full description of the effects of the repulsive release on the fragments. Unfortunately, there is still much less ability to describe systems which are predissociative or involve Rydberg state excitation. However, theorists are ambitiously tackling bigger and more interesting systems all the time, and they are predicting new effects to challenge the experimenter's skills.

A. Photoejection Dynamics

Photoejection dynamics provides the necessary general link between experimental measurements of fragment angular anisotropy and the details of molecular behaviour, e.g. excited-state lifetime, orientation of molecular transition moment, orientation of fragment recoil, and excited-state geometry. This single most important parameter is the lifetime of the excited molecule compared with its period of rotation. Whereas time-of-flight measurements and many other fluorescence, laser-induced fluorescene, and absorption probing experiments give details about fragment product states, angular anisotropy measurements provide information about the parent excited state before and during fragmentation. The two main methods of measurement are molecular beam angular orientation and fluorescence polarization spectroscopy. Theories have been developed to address to results of each.

The classic formulation of photoejection dynamics for diatomic and linear triatomic molecules was discussed by Zare,[35] and was considered in depth with the molecular beam method for the bent triatomic, NO_2, by Busch and Wilson.[43] These concepts were extended to treat several general categories of large polyatomic molecules.[36] All of this work has been extensively reviewed.[1,2,33] The basic idea is that the angular distribution of fragments is given by:

$$F(\theta) = \frac{1}{4\pi}[1 + \beta P_2(\cos\theta)]$$

where θ is the angle between the polarization of the light and the direction of the detected fragment, P_2 is the second Legendre polynomial,

$$P_2(\cos\theta) = \frac{3\cos^2\theta - 1}{2}$$

and β is the symmetry parameter[35] which is related to the orientation of the transition dipole μ in the molecule and lifetime τ of the excited dissociative state. If the molecule dissociates rapidly compared with the rotation period, β can be shown to be given by $2P_2(\cos\chi)$, where χ is the angle between the transition moment of the molecule and the direction of the fragments on leaving the molecule. For a diatomic molecule, χ can only be 0 (for a parallel transition, along the internuclear axis) or 90° (for a transition moment perpendicular to the internuclear axis). For a polyatomic molecule, there can be a distribution of χ values due to a distribution of excited-state geometries and vibrational and/or rotational motions. The value of β decreases by a factor of 4 if the rotational period is short compared with the excited-state lifetime.[36] Between these two limits, the lifetime of the excited state can often be extracted by angular orientation studies.

An important new contribution to the theory of photoejection dynamics is discussed by Band, Morse, and Freed.[172] They consider, for the specific case of ICN, the changes in the anisotropy parameter β which will occur when (a) the CN rotational angular momentum differs from the initial total angular momentum of the ICN and (b) the initial ICN bending motion is excited. Dramatic changes in β are predicted for specific angular momentum cases involving dissociation of ICN containing two quanta of bending vibration. However, these special cases of large angular momentum differences between the triatomic and diatomic fragments are unlikely to make a significant enough contribution to alter the classical value of β observed overall.[172]

A theory of angular distribution photoejection dynamics has been

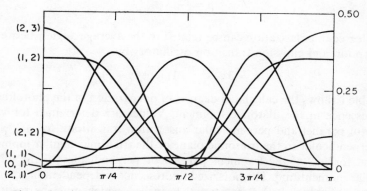

Fig. 6. Angular distributions of photofragments in the two-photon dissociation of a diatomic molecule which undergoes a $(\pi \leftarrow \pi \leftarrow \Sigma)$ transition sequence. The two excitation fields are linearly polarized. Numbers in parentheses indicate initial and intermediate rotational state quantum numbers. From Chen and Yeung.[144] Reproduced by permission of the American Institute of Physics.

discussed for two-photon photodissociation involving a long-lived resonant intermediate state.[144] In the two-photon process, the angular orientation of fragments is predicted to depend critically on the initial and intermediate rotational states (Fig. 6). This effect occurs for both linearly or circularly polarized light. In a theory for the photodissociation of optically active molecules, it has been shown that there is one term in the final expression for the angular distribution of photofragments which is different for the dextro-rotatory and laevo-rotatory forms.[173] No estimate is made on the magnitude of the effect, but it is suggested that this 'anomalous' term could be comparable in magnitude to the other ordinary terms. It may be possible to observe a small effect with either linearly or circularly polarized light.

The second major method of measuring angular distributions involves the detection of polarized fluorescence from the fragments of the dissociation process. Here the theory has recently received a major contribution. A number of specific molecular cases have been considered in the literature.[126-130] Recently Macpherson, Simons, and Zare provided a complete summary of all the possible cases applicable to triatomic molecules.[125] The incident polarized light field interacts with the transition moment μ_{abs} of the molecule. If the molecule dissociates in a time short compared with rotation, the diatomic fragments that are produced will have their angular momentum vector $\mathbf{j}$ oriented with respect to the incident polarization of the exciting light. Electronically excited diatomic fragments will emit preferentially polarized radiation along their transition moment μ_{em}. The degree of polarization is defined in the usual sense as:

$$p = \frac{I_{\parallel} - I_{\perp}}{I_{\parallel} + I_{\perp}}.$$

The degree of polarization can be related to the average angle γ between the absorption and emission transition oscillators by

$$p = (3\cos^2 \gamma - 1)/(\cos^2 \gamma + 3).$$

Table I shows the calculated degrees of polarization of the photofragment fluorescence in the diatomic fragment. Values are determined for various cases of parallel and perpendicular absorption transitions, for μ_{abs} parallel or perpendicular to the triatomic plane or the triatomic angular momentum $\mathbf{J}$, and for μ_{em} of the fragment parallel or perpendicular to the diatomic angular momentum $\mathbf{j}$. The polarized fluorescence is dependent in some cases on the individual P, Q, R rotational branches, which must be resolved in emission to obtain the full effect. In principle, measurements of polarized fluorescence can be used to assign the symmetry of the electronically excited state involved in the dissociation. The values of p listed in the table are maximum effects, which will be expected to decrease markedly if the lifetime

TABLE I. Calculated degrees of polarization p and angle γ between absorption and emission oscillators, for the fluorescence photofragments (AB)* produced through the photodissociation of triatomic ABC molecules by linearly polarized light. From Macpherson et al.[125] Reproduced by permission of Taylor & Francis, Ltd.

Case	ABC transition	(AB)* transition	γ (deg)	$\cos^2\gamma$	p
		Nonlinear parent molecule or linear → bent transition			
I	μ_{abs} in (ABC)* plane	∥-type; $\mu_{em}\perp\mathbf{j}$	45	$\frac{1}{2}$	$\frac{1}{7}$
IIa	μ_{abs} in (ABC)* plane	⊥-type; $\mu_{em}\parallel\mathbf{j}(Q)$	90	0	$-\frac{1}{3}$
IIb		$\mu_{em}\perp\mathbf{j}(P,R)$	45	$\frac{1}{2}$	$\frac{1}{7}$
III	$\mu_{abs}\perp$(ABC)* plane	∥-type; $\mu_{em}\perp\mathbf{j}$	90	0	$-\frac{1}{3}$
IVa	$\mu_{abs}\perp$(ABC)* plane	⊥-type; $\mu_{em}\parallel\mathbf{j}(Q)$	0	1	$\frac{1}{2}$
IVb		$\mu_{em}\perp\mathbf{j}(P,R)$	90	0	$-\frac{1}{3}$
		Linear → linear transition			
V	∥-type; $\mu_{abs}\perp\mathbf{J}$	∥-type; $\mu_{em}\perp\mathbf{j}$	45	$\frac{1}{2}$	$\frac{1}{7}$
VIa	∥-type; $\mu_{abs}\perp\mathbf{J}$	⊥-type; $\mu_{em}\parallel\mathbf{j}(Q)$	90	0	$-\frac{1}{3}$
VIb		$\mu_{em}\perp\mathbf{j}(P,R)$	45	$\frac{1}{2}$	$\frac{1}{7}$
VIIa	⊥-type; $\mu_{abs}\parallel\mathbf{J}(Q)$	∥-type; $\mu_{em}\perp\mathbf{j}$	90	0	$-\frac{1}{3}$
VIIb	$\mu_{abs}\perp\mathbf{J}(P,R)$		45	$\frac{1}{2}$	$\frac{1}{7}$
VIIa	⊥-type; $\mu_{abs}\parallel\mathbf{J}(Q)$	⊥-type; $\mu_{em}\perp\mathbf{j}(Q)$	0	1	$\frac{1}{2}$
VIIIb	$\mu_{abs}\perp\mathbf{J}(P,R)$		90	0	$-\frac{1}{3}$
IXa	⊥-type; $\mu_{abs}\parallel\mathbf{J}(Q)$	⊥-type; $\mu_{em}\perp\mathbf{j}(P,R)$	90	0	$-\frac{1}{3}$
IXb	$\mu_{abs}\perp\mathbf{J}(P,R)$		45	$\frac{1}{2}$	$\frac{1}{7}$

of the excited state is not short compared with the rotational period. In some cases, the transition moment in absorption may have mixed character, which can alter the expected value of p by a weighted combination of cases. Thus far, most of the experimental tests involved longer-lived predissociative states,[126-130] which are not expected to give the exact results in the table. However, a recent study on $HgBr_2$, which is directly dissociated, gives good agreement with prediction.[131] It should be possible to take full advantage of the fluorescence polarization effect to learn about the character of many excited dissociative states in the future.

B. Dynamical Theories of Photofragmentation

The basic ideas of what to include in the theory of dissociation dynamics have been around since the early 1970s.[1] There are two major considerations: (1) the initial Franck–Condon excitation, where changes in the geometry of the photoexcited state are taken into account (sometimes called intrafragment) and (2) the final-state interactions, where the impulsive recoil of the separating fragments causes an alteration of the final translational, vibrational, and rotational states (sometimes called interfragment). Many different methods have been proposed to calculate the two effects, both by simple models and by more rigorous quantum-mechanical formulations. These methods have been reviewed thoroughly from several different points of view.[1-3]

The earliest model proposed has become known as the quasidiatomic model, and in its simplest form it considered only the half-collision, or repulsive recoil mechanism for exciting the fragments.[32,174] In 1973 Simons and Tasker[175] proposed a model which incorporated the half-collision concept together with the possibility of intrafragment photoexcitation changes. For a more detailed review of the historical development of these theories and credit to other investigators, the reader is referred to the review of Simons.[1] The theory of Simons and Tasker established the more complete picture of the photofragmentation process as we know it today. Berry proposed a 'golden rule' model of photodissociation, which basically takes into account only the Franck–Condon changes in the photoabsorption step.[165,176,177] This assumes that the geometry changes in the excited state are the most important, and that the recoil, final-state interactions only slightly modify the Franck–Condon result. Simultaneously, Band and Freed reported a more proper treatment of the normal mode changes occurring in the photoexcitaion step, in which they also included separately the final-state interactions by adjusting the repulsive potential to reproduce experimental results.[178-180] Their earlier predicted isotope ratios for HCN/DCN dissociation were found to be enormously dependent on the choice of the repulsive potential, which in turn could not be precisely determined by the available experimental data.[181] They showed, however, that their theory was more complete than Berry's and reduced to his result in a one-dimensional limit.[182] Morse, Freed, and Band have considered the further possibilities of bent excited-state geometries and the effect that this has on rotational state distributions through the final-state interactions.[172,183-186]

Mukamel and Jortner[187,188] and Atabeck and coworkers[189-191] provided a unified quantum-mechanical approach which attempts to calculate both intrafragment and interfragment effects simultaneously. They found that the fragment product state distributions are not only sensitively dependent on initial state excitation, but also on the shape and steepness of the repulsive potential used. A more exact quantum-mechanical treatment of just the quasidiatomic model has been described[192,193] and more recently extended to the collinear dissociation of vibrationally excited N_2O[194] and to HCN/DCN.[195]

Shapiro and Bersohn have computed a two-coordinate potential surface (r_{C-I} and r_{C-H_3}) and used it to predict the vibrational excitation in the CH_3

fragment of CH_3I dissociation.[196] This calculation includes both the Franck–Condon distortion of the CH_3 radical in the initial excitation and the final-state recoil interactions in one complete quantum-mechanical event. They point out that the dissociation process cannot be thought of as two 'incoherent' events, i.e. excitation followed by T–V transfer on the repulsive surface, because there is substantial interference by various intermediate wavefunction amplitudes. The calculations of Shapiro and Bersohn[196] reproduce moderately well the distribution of excited states obtained qualitatively[61] for the CH_3 out-of-plane bend. However, the experiments surprisingly seem to indicate a more narrow distribution than the theory. Shapiro and Bersohn predict shifts of the distribution to higher vibrational levels with shorter excitation wavelengths, a result which can be checked experimentally.

A number of other notable predictions have come out of the recent work on linear triatomics by Morse, Freed, and Band.[172,184–186] They find considerable structure in the product rotational distributions arising from photodissociation of excited bending states.[184,185] The effect is most dramatic for $J = 0$ states. With cooled supersonic beams, it may be possible to prepare a small group of rotational states consisting mainly of $J = 0$ molecules to test this prediction. They predict generally how the rotational state distributions for ICN dissociation change with wavelength.[186] The results of the rotational distributions are remarkably sensitive to details of the unbound potential surface, more so than the vibrational and electronic state distributions.

For molecules that do not undergo direct dissociation processes, a variety of ideas on the statistical break-up of excited complexes has been proposed. Riley and Wilson showed that a simple statistical partitioning of energy in CH_3I would deliver far too much energy into the CH_3 radical and that a better picture of the dissociation process must be the impulsive recoil interaction.[32] However, for longer-lived excited states, statistical approaches can give a better picture of the fragmentation process. In the photodissociation of CH_3CN and other analogues, the vibrational distributions in the CN(B) fragment approximate a Boltzmann distribution.[13] The results are well described by an RRK formulation assuming a long-lived intermediate state and a statistical partitioning of the energy between the fragments. Several statistical models have been applied to the fragmentation of $Cd(CH_3)_2$, all of which show a significant degree of asymmetrical dissociation.[197] In the three-body dissociation of CH_3COI, the direct dissociation to $CH_3CO + I$ occurs promptly, followed by a longer-lived unimolecular fragmentation of CH_3CO to $CH_3 + CO$.[52] The translational TOF of the $CH_3 + CO$ fragments was successfully modelled by statistical RRKM ideas.

There is, however, a sharp dividing line between the successful application of statistical ideas and the more selective nature of dissociation processes involving predissociation via electronic curve crossing. Predissociation and

Rydberg state excitations are phenomena which are extremely difficult to treat theoretically. Many of the current theories have had to assume direct dissociation processes in order to calculate results for molecules which are in fact predissociative. This is unfortunate because of the enormous number of molecules which undergo photofragmentation through predissociated or Rydberg state transitions. Gelbart has reviewed many of the important cases of molecular predissociation in depth.[2] There are notable attempts to take into account simultaneous electronic relaxation effects and dissociation to fragments, for example, in H_2CO.[198] It will be important to have theoretical work in the future on simple models of Rydberg state decomposition and predissociation for small molecules such as O_2, H_2O, HCN, and NH_3. Extensive experimental data exist for these molecules excited to a variety of predissociated states and Rydberg levels.

C. Potential Surfaces

There have been substantial accomplishments in calculating molecular potential energy surfaces and relating these to results of photofragment dynamics *Ab initio* self-consistent field–configuration interaction (SCF–CI) calculations have been used to predict the low-lying singlet and triplet states of HOCl.[199] The computed transition strengths and absorption spectrum predict a single absorption peak at 220 nm. Various experiments are not consistent, but they seem to indicate an additional absorption feature at 320 nm as well.[199] A similar calculation was carried out on HOO, where $O(^3P) + OH(X^2\Pi)$ are the predicted dissociation products.[200] The electronic structures of $HgCl_2$ and $HgBr_2$ have been calculated, and an important prediction was made that the first electronically excited state should lead directly to three atomic products, $Hg + Cl + Cl$.[201] Experimentally the first long-wavelength absorption bands of both HgI_2 [79] and CdI_2 [54] are found to be composed of two states, one leading to I* and one producing an I ground-state product (Fig. 7). In both cases it appears that the products of the dissociation are $HgI + I(I^*)$ and $CdI + I(I^*)$. It remains to be verified whether the predictions of the theory are correct for $HgCl_2$ and $HgBr_2$. The theory does not attempt to treat the difficult spin–orbit splitting problem, which is easily detected experimentally. A detailed *ab initio* study on NH_3 has fully characterized the electronic states responsible for the dissociation and their products.[202] Large-scale calculations have been carried out to analyse the transition state geometries and energies for the dissociation dynamics of H_2CO,[203] C_3O_2,[204] cyclopropanone,[205] and ketene.[206]

One interesting possibility is ultimately to relate the observed dynamics of the dissociation process to potential surfaces and in turn to obtain comuputed absorption spectra or linewidths. Such an approach has been taken to calculate the HCN lineshape[191] and DCN/HCN absorption ratio.[190,195]

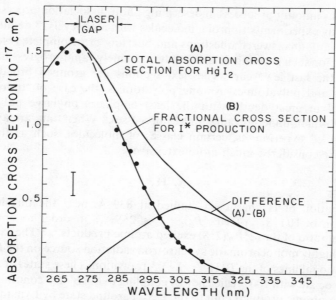

Fig. 7. Total absorption cross section of HgI_2 and fractional components leading to excited $I(5^2P_{1/2})$ and ground-state $I(5^2P_{3/2})$ atoms upon photodissociation. Measured by tunable laser, infrared fluorescence technique. From Hofmann and Leone.[79] Reproduced by permission of the American Institute of Physics.

The contributions of different vibrational product-state channels to the total absorption spectrum has been calculated for ICN.[207] The absorption spectrum of N_2O was predicted for excitation out of different vibrational levels, (000), (001), and (100), and these were related to predicted N_2 product vibrational distributions.[194] Shapiro and Bersohn have predicted the fractional contributions of the different CH_3 out-of-plane bending vibrational levels to the total absorption cross section of CH_3I.[196] Heller has proposed a time-dependent formulation of the total photodissociation cross sections to obtain an explanation of the lineshapes and absorption envelopes of symmetric triatomic XY_2 molecules.[208, 209] Finally, Pack has shown that polyatomic dissociative upper states can have considerable diffuse vibrational structure due to Franck–Condon overlaps with levels that are bound in other coordinates.[210] These results will certainly be important in interpreting the origins of spectra associated with dissociative states.

IV. SPECIFIC SYSTEMS

A review of specific molecular systems reveals many rewarding new developments in photofragment dynamics. It is clear that there has been a tremendous amount of success in describing photodissociation processes and in designing experiments to probe molecules with high levels of detail. There are also many unanswered questions and puzzling or incomplete results. By and large, for such a young field of study, extremely rapid progress has been made in the last few years. The next sections are grouped broadly into diatomics and polyatomics. Among polyatomics, the case of triatomics is considered in most depth, with the least emphasis on large polyatomic molecules which undergo radiationless processes.[2] When data are available from detailed experiments, certain classes of molecules, such as the alkyl halides, are considered much more extensively.

A. H_2

Dissociation of H_2 has been studied at 839 Å, near the threshold for production of $H(1^2S_{1/2}) + H(2^2P_{3/2,1/2}$ or $2^2S_{1/2})$, in order to obtain the branching ratio of the $(2^2P)/(2^2S)$ excited atomic products.[211] The excitation is with a highly monochromatic synchrotron radiation source on the $R(0)$ and $R(1)$ rotational lines of the $D^1\Pi_u^+ \leftarrow X^1\Sigma_g^+$ transition. These states are known to predissociate with 100% efficiency by coupling to the $B'^1\Sigma_u^+$ continuum. The $2^2P_{1/2,3/2}$ atomic products emit to the $1^2S_{1/2}$ ground state by Lyman α(1216) Å) emission with a lifetime of $1\cdot6 \times 10^{-9}$ s. The $2^2S_{1/2}$ state is metastable with a radiative lifetime of $0\cdot12$ s, but can be induced to emit the same Lyman α wavelength by applying an electric field to mix the ^2P and ^2S wavefunctions. It was found that 57% of the excited atoms are in the $2^2S_{1/2}$ state and 43% in the $2^2P_{3/2,1/2}$ states. If all the products resulted from adiabatic dissociation along the $B'^1\Sigma_u^+$, only $H(1^2S)$ and $H(2^2S)$ atoms would be produced. The substantial fraction of $H(2^2P_{3/2,1/2})$ production indicates that there must be a mixing of several other states. The observed branching ratio cannot be explained by simply taking the statistical weights of the atomic products or by allowing for equal production of the three relevant molecular states. Komarov and Ostrovsky have proposed a theory based on the interaction of the covalent and ionic $(H^+ + H^-)$ electronic states of H_2 at large internuclear separations.[212,213] They use a Landau–Zener formalism to calculate relative yields of $2^2P(60\%)$ and $2^2S(40\%)$, in good agreement with the experiment. It may be possible to measure the fluorescence polarization of the atomic products if a preferential population of the m_J sublevels is obtained in the dissociation.[133] In addition, it would be desirable to measure the angular distribution of atomic products with a molecular beam apparatus to obtain the lifetime of the dissociation, and to relate this to the timescale for ionic curve crossing.

B. Halogens and Interhalogens

The halogens have long been an attractive subject for the study of photofragment processes because of their readily detected atomic states,

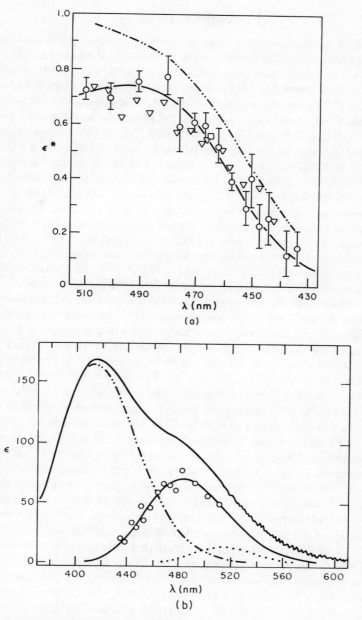

Fig. 8 (a) Plot of $\varepsilon^* =$ quantum yield of Br*($^2P_{1/2}$) formation for Br$_2$, comparing the results of Lindemann and Wiesenfeld[74,75] ($\bigcirc$) with those of Petersen and Smith[82] (∇) scaled to match, Busch *et al.*[37] ($\square$), and an unpublished theoretical prediction of R. J. LeRoy[215] (— .. —). The full curve is the best fit to the Lindemann and Wiesenfeld data. (b) Absorption spectrum of Br$_2$ decomposed into three curves with the experimental values of $\varepsilon(B \leftarrow X)$ from the work of Lindemann and Wiesenfeld[74,75] (∇). Peaks for the $^1\Pi_{1u} \leftarrow X$ (— .—) and $B \leftarrow X$ (—) are well characterized, while the weak $A \leftarrow X$ (.......) is not. From Lindemann and Wiesenfeld.[75] Reproduced by permission of the American Institute of Physics.

well-known molecular states, and interesting predissociation and curve-crossing phenomena. Photodissociation processes in the halogens have been extensively reviewed.[1] Child and Bernstein summarized the low-lying electronically excited states and dissociation products of the halogens and interhalogens.[214] Interestingly enough, in the interhalogens, electronic states are observed which correlate to only one or the other electronically excited $(^2P_{1/2})$ atom (e.g. Br* from IBr, Cl* from ICl, Cl* from BrCl). Thus far, bound or repulsive states which correlate to the I* limit in IBr or ICl, or to the Br* limit in BrCl, have not been observed. The halogens were some of the earliest molecules to be studied by time-of-flight/angular distribution molecular beam techniques.[1,31,38,39] I_2 was one of the first molecules in which a two-photon dissociation was observed.[31] It was also used for a double laser pulsed technique to probe the excited B state dissociation dynamics.[46]

Recent experiments have now explored the excited atom products, I* and Br*, as a function of tunable laser wavelength. Some of the earliest experiments of this type obtained the quantum yield of I* $(^2P_{1/2})$ in the photodissociation of I_2 near the I* + I dissociation limit.[70] Substantial fractions of I* for wavelengths below the dissociation threshold confirmed the mechanism of 'collisional release'. Lindemann and Wiesenfeld[74,75] and Petersen and Smith[82] recently measured the quantum yields of Br* atom production from Br_2 throughout the continuum absorption above the Br* + Br threshold. The measurements using transient absorption of resonance VUV radiation[74,75] give absolute yields of Br*/Br (Fig. 8). The data taken by the infrared fluorescence method[82] were not normalized to a standard, so only relative values were obtained. Petersen and Smith favour quantum yield values which match the calculations of LeRoy[215] (0·97 at 510 nm). The absolute quantum yields of Lindemann and Wiesenfeld are lower (0·72 at 510 nm). However, the Lindemann–Wiesenfeld results do agree with the molecular beam photofragmentation experiment at 466·2 nm[39] and a recent measurement of the $A–X$ continuum.[216] Consideration of some other data indicates that the results of Lindemann and Wiesenfeld are probably correct: Petersen and Smith also measured the relative Br* yields from IBr.[82] The ratio of magnitudes of the Br* yields from IBr and Br_2 was 0·9 at 503 nm. A tunable laser, time-of-flight experiment has been performed on the continuum of IBr as well.[217] By selecting the orientations of the light polarization and the mass spectrometer to detect a parallel transition, only the dissociation products of the $^3\Pi_{0^+}$ state are observed. At 500 nm, these TOF measurements give a yield of Br* of about 0·7. If this value is used to scale the data of Petersen and Smith, their quantum yields for Br* from Br_2 are in excellent agreement with the results of Lindemann and Wiesenfeld. Thus it would appear that the data of Lindemann and Wiesenfeld are the correct values to use but that the contributions of the $A–X$ and $B–X$ bands of Br_2 in the long-wavelength region need further theoretical study.[216]

The Petersen and Smith data[82] show only a small discontinuity in the Br* yield from IBr in the region of the avoided crossing of the $^3\Pi_{0^+}$ and 0^+ states at 545·2 nm (see Fig. 9 for the potential curves). A substantial fraction of the molecules excited below the I + Br* dissociation limit appear to produce Br*, indicating that they may live long enough to undergo collisional release rather than predissociation to I + Br. Child discussed the IBr molecule as a case of intermediate coupling strength, in which molecules excited to the $^3\Pi_{0^+}$ state above the I + Br* dissociation limit will tend to remain on the diabatic curve when passing through the crossing region, leading to the Br* product.[218] This has been tested experimentally by DeVries et al.,[217] who showed that the Br* yields were in rather good agreement with this Landau–Zener curve-crossing result using the potential surface parameters and coupling strengths of Child.[218]

A search for states which might lead to the I* product upon photodissociation of IBr and to Br* from BrCl was made using a tunable laser, infrared fluorescence apparatus.[219] These states are expected to exist, but may be much weaker owing to a displacement of the potential curves to larger internuclear separation with respect to the ground state[214] (Fig. 9). Even with many orders of magnitude single-to noise advantage,[219] no I* nor Br* was detected, proving again the elusive nature of these 'missing' states. Either the Franck–Condon factors for absorption of states correlating to Br–I* and Cl–Br* are very weak, or the states themselves undergo curve-crossing mechanisms which always lead to ground-state atoms or to the Br* and Cl* dissociation products which are observed. Child has suggested that it may be possible to access these states at larger internuclear separation with two sequential photons, the first to excite one of the sharper, long-lived states in the $^3\Pi_{0^+}$ manifold, and the second to reach the dissociative continuum from an outer turning point.[220]

In a unique experiment, Krasnoperov and Panfilov have reported the first detection of a population inversion between $Cl(^2P_{1/2})$ and $Cl(^2P_{3/2})$ when dissociating ICl with 532 nm light.[124] They detected gain on the Cl* → Cl transition with an intracavity laser magnetic resonance absorption technique using a $^{13}CO_2$ laser. There are no quantitative Cl* yield data as yet from any molecules and this method offers exceptional promise for such measurements in the future.

Laser-induced fluorescence techniques have provided a tremendous amount of new information on predissociation processes in the B states of the halogens. Clyne and Heaven have studied Br_2[221,222] and IBr,[223] and Clyne and McDermid have made an in-depth examination of Cl_2.[224,225] It was found that virtually all of the banded absorption features in the $Br_2(B^3\Pi_{0^+u})$ state lead to two $Br\,^2P_{3/2}$ atoms via predissociation. The predissociation has a strong rotational dependence, and the quantum yield of fluorescence is only 0·05. In contrast, the $A^3\Pi(1_u)$ state is not quenched

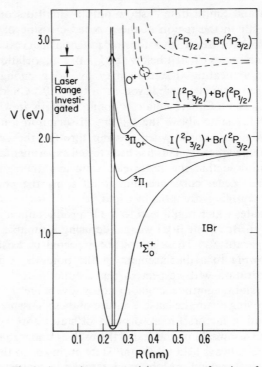

Fig. 9. Approximate potential curves as a function of internuclear separation for IBr, showing the three lowest possible dissociation limits, I + Br, I + Br*, and I* + Br, and the region of the $^3\Pi_{0^+}$ and 0^+ avoided crossing. The broken curves are unobserved states in IBr, leading to the I* product. The inability to observe these states may be due to either (a) poor Franck–Condon factors, schematically indicated by large displacement of the I* + Br states to larger internuclear separation than the vertical transition, or (b) by fortuitous curve crossings, indicated by the circled region in the figure, which produces Br* or ground-state atomic products whenever the I* + Br states are accessed. From Baughcum et al.[219] Reproduced by permission of the Royal Society of Chemistry.

efficiently and predissociation cannot occur below the dissociation limit, so long-lived behaviour is observed. Fluorescence from the $B^3\Pi_{0^+}$ state of IBr indicates that predissociation is moderately strong in the $v' = 2$ and 3 manifolds. The variations in linewidths are interpreted in terms of coupling the B and 0^+ states. The quantum yield of fluorescence for the $Cl_2(B^3\Pi_{0^+u})$ state was found to be only 3×10^{-5}, indicating very strong predissociation. It is clear that narrow-band laser techniques can be applied to explore photofragmentation processes of the halogens and interhalogens in even more depth in the future. It should be possible, for example, to study systematically the individual vibration–rotation states in the region of the $^3\Pi_{0^+}$ and 0^+ crossing of IBr. The quantum yields of excited atom formation from other molecules can be obtained, and more thorough searches for the 'missing' excited atom states can be made.

C. O_2

Vacuum UV photolysis has been used to study the quantum yields of excited atomic products, predissociation phenomena, and Rydberg states of O_2. TOF/angular distribution measurements on O_2 using wavelengths 120 nm, 124 nm, 147 nm, and 149 nm give a lifetime for predissociation of the $B^3\Sigma_u^-$ states as 2×10^{-13} s, and obtain $O(3P) + O(^1D)$ as the major products.[58] The formation of $O(^1D)$ and energy transfer with O_2 was measured with a novel H_2 laser excitation source at 160 nm.[226] More extensive tunable quantum yield measurements for $O(^1D)$ production were obtained in the range 116–177 nm with hydrogen lamp excitation.[227] The quantum yield is unity for wavelengths greater than 139 nm, but has a sharp cut-off at 175 nm. Below 139 nm the $O(^1D)$ quantum yields are strongly dependent on which absorption feature is excited (Fig. 10). The total O atom yield is unity in the entire region, but each peak in the spectrum has been identified as

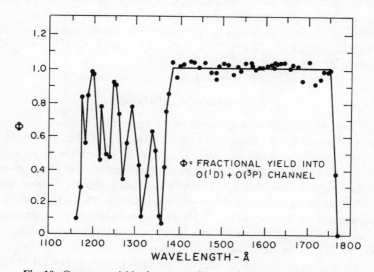

Fig. 10. Quantum yields for the $O(^1D) + O(^3P)$ produced by photo-dissociation of O_2 at 116–177 nm. The data are connected with straight lines to show the structure. From Lee *et al.*[227] Reproduced by permission of the American Institute of Physics.

producing either $^3P + {}^3P$ or $^1D + {}^3P$ channels.[227] Several of the peaks are identified as the $v = 0, 1, 2$ levels for the $2^3\Sigma_u^-$ Rydberg state, which is probably predissociated by the $B^3\Sigma_u^-$. There is still a great deal of theoretical work which can be done to characterize fully the locations and couplings of the various Rydberg states which give rise to the remarkable structure in the excited products of O_2.

D. Na_2, K_2, Rb_2

Several recent experiments have probed the population of excited $^2P_{3/2}$ and $^2P_{1/2}$ fine-structure states in alkali dimer dissociation. Feldman and Zare first reported preferential population of the $^2P_{3/2}$ component in Rb upon predissociation of Rb_2 at 4765Å.[132] It was necessary to use a supersonic nozzle beam expansion to investigate isolated molecule behaviour. Rothe, Krause, and Düren have measured the $^2P_{3/2}/P_{1/2}$ ratio for direct dissociation of the $Na_2(B)$ state and for predissociation of the $Rb_2(C)$ state at several wavelengths.[228] In each case there are wavelengths where the $^2P_{3/2}$ component is formed exclusively, indicating a correlation with the $B^1\Pi_u$ state either directly or by a curve crossing. In the case of Rb_2 they attribute the predissociation of the $C^1\Pi_u$ state entirely to the $B^1\Pi_u$ state, which correlates to the $^2S_{1/2} + {}^2P_{3/2}$ atomic products. They have also been able to observe polarization of the $^2P_{3/2}$ fluorescence in the case of Na_2.[133] In order to observe polarized atomic fluorescence, it is necessary that the m_J atomic sublevels are also preferentially populated at least partially. Predominant population of the $^2P_{3/2}$ excited-state component has now been demonstrated for $Na_2(B^1\Pi_u)$, $K_2(B^1\Pi_u)$, $Cs_2(C^1\Pi_u)$, $K_2(C^1\Pi_u)$, and $Rb(C^1\Pi_u)$ photofragmentation.[229]

E. Diatomic Metal Halides

There has been a tremendous amount of research on photodissociation processes of metal halide species in order to develop atomic and molecular lasers. For example, calculations have been carried out on the cross sections for absorption of the $X-A$ bound-to-continuum transition in HgCl, in which it was determined that this transition will absorb the $B-X$ laser line of HgCl.[230] Cool et al. reported a two-photon dissociation of HgX molecules (X = halogen) with 193 nm light.[148] Photofragment lasers have successfully operated on the higher excited states of Tl and In by photodissociation of TlI[160] and InI.[159] A long-standing problem concerns the Tl*($^2P_{3/2}$) and Tl($^2P_{1/2}$) branching ratios in the thallium halides. In the case of TlI, the spin-orbit splittings of the thallium and iodine atoms are nearly identical (Tl*($^2P_{3/2}$) = 7793 cm^{-1}, I*($^2P_{1/2}$) = 7603 cm^{-1}). Molecular beam TOF, angular distribution experiments have been carried out on TlI by several groups.[56,57] It is found that at 302·5 nm and 380 nm the TlI absorptions have both mixed parallel and perpendicular symmetries.[56] At 266 nm, there is also mixed parallel and perpendicular character.[57] The molecule seems to undergo direct dissociation. At 266 nm, the TOF spectrum indicates that the products must always include one excited atom, Tl* + I or Tl + I*, or a combination of both channels.[57] However, the resolution is not sufficient to determine the fractional yield of each excited state. It should be possible to investigate these atomic yields as a complete function of wavelength with either VUV resonance absorption or infrared fluorescence techniques. These results should provide important new information on the role of ionic $Tl^+ + I^-$ states in the photofragmentation. In a recent optical experiment on TlBr at 266 nm, it was found that the ground state $Tl(^2P_{1/2})$ is produced 99% of the time.[231] These experiments were carried out in a heated cell arrangement, and no estimates were discussed for possible quenching of the Tl*.

There has been extensive molecular beam photofragment work on the alkali halides. Some of the earliest work was carried out by Herm and coworkers[232,233] to measure the fluorescence efficiencies of the excited alkali atom. In that work, they obtained the translational velocity expression for the alkali atom. John and Dahler have provided a more complete description of the Doppler spectrum expected from a fluorescing alkali atom produced by photodissociation.[120] Bersohn has reviewed the entire field and considered the possible applications of these 'superalkali' atoms.[234] Population inversion and laser action on the Na resonance line has been achieved by dissociation of NaI at 2128 Å with the fifth harmonic of a Nd:YAG laser.[158]

The extent of parallel and perpendicular transitions in the photodissociation of alkali halides is of major importance. The ground states of these molecules are primarily ionic. The absorption transitions occur to predominantly covalent excited states which lead to excited and ground-state atomic products. Such a transition can be approximated by a one-electron charge transfer model,[235] which gives mainly perpendicular character. Zare and Herschbach showed that only 5% covalent character in the ionic ground state can give rise to a 50% contribution of parallel character to the transition.[235] This effect has been systematically investigated for NaI, KI, RbI, CsI, LiI,[48] NaBr, KBr, RbBr, CsBr, LiBr,[49] NaCl, KCl, RbCl, CsCl,[50] and in a separate experiment on NaI.[236] In the iodides there is a strong correlation between the covalent character of the ground state and the amount of parallel character in the transition. Such a trend was not as obvious for the bromides or chlorides. More detailed electronic state calculations are needed to consider these results. From the TOF data it was also possible to obtain the I/I* and Br/Br* branching ratios, and to detect that both the Cl and Cl* were being formed, although the Cl/Cl* ratio could be resolved because of the small splitting. A separate TOF experiment to measure the I/I* ratio as a function of tunable laser wavelength was carried out on KI.[60] Finally, bound dissociation energies were obtained for all of the alkali halides listed above.[48-50]

The alkali halides have complex UV absorption spectra that consist of numerous continua and diffuse bands. They represent the simplest systems to test theories of the interaction between ionic and covalent states. Tunable laser photofragment experiments along with better electronic structure calculations should provide important new insights into the spectroscopy and dynamics of these molecules. It should also be valuable to find ways to prepare the alkali halides cooled to their lowest states, so that investigations can probe a smaller Franck–Condon range of transitions than is presently achieved with heated effusive beams.

F. HCN, XCN, and Other CN-Containing Compounds

The group of polyatomic moleclues containing CN are some of the most extensively investigated both experimentally and theoretically. Photo-dissociation with VUV radiation produces the $CN(B^2\Sigma^+)$ state which emits directly and has a well assigned fluorescence spectrum. In addition, laser-induced fluorescence is readily carried out on ground-state $(X^2\Sigma^+)$ fragments using the same B–X transition.

Vibrational state distributions have been obtained for the $CN(B)$ state upon photodissociation of HCN and DCN at 121·6 nm,[8] 123·6 nm,[6,9] 129·5 nm,[6] and 130·4 nm.[6] The vibrational distributions show a monotonic decrease from $v = 0$ to $v = 4$ (Fig. 11(a)). In all cases the DCN has a greater relative population in the $v = 1, 2, 3$, and 4 levels, often by a factor of about 2. There have been many widely varying predictions for the vibrational distributions and deuterium isotope effect in HCN/DCN.[177,178,179,181,188-191,195] Many of these discussions model the dissociation as a direct process, even though HCN is a predissociated case. The interpretations of Ashfold et al.[6] appear to offer the most realistic conclusion to the theoretical description of this dissociation. They suggest that the disposal of vibrational energy is primarily dependent on the Frank–Condon factors, and is relatively in-sensitive to the final-state interactions.[6] This would be the case if the repulsive surface is relatively flat with a gentle slope, which appears to be confirmed by the observed rotational distributions.[6] The data show an almost exact mapping of J parent into j fragment (Fig. 11(b)), even for predissociated states involving several quanta of bending excitation. This is exactly the behaviour expected when the relatively light H or D atom breaks away from the heavy CN without any strong recoil, or final-state interaction, between the separating fragments.

The cyanogen halides, ClCN, BrCN, and ICN, have also been the subject of extensive theoretical and experimental study. ICN was one of the first triatomic molecules to be photodissociated in a molecular beam.[41] The

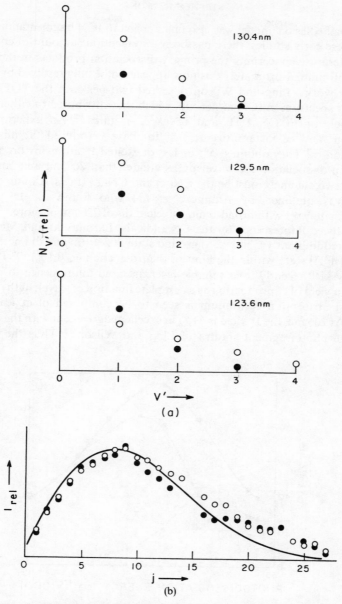

Fig. 11. (a) Relative vibrational populations in CN(B) from the photodissociation of HCN (●) and DCN (○) at three wavelengths. (b) CN(B, v = 0) rotational distributions from photodissociation of 0·2 Torr HCN at 147·0 nm (●) and 141·1 nm (○). 147 nm corresponds to populating the C(0, 3, 0) state and 141·1 nm the C(0, 6, 0) state. The full curve represents the distribution over rotational states in the parent HCN molecule at 298 K. From Ashfold et al.[6] Reproduced by permission of North-Holland Publishing Company, Ansterdam.

simple 'half-collision' model was also first applied to ICN fragmentation,[174,175] Since these early studies, there has been a tremendous amount of confusion in the literature concerning the primary dissociation products in the first $\tilde{A}$ state continuum. Fortunately this now appears to be fully resolved by several excellent works. Ling and Wilson observed two peaks in the TOF photofragment spectrum of ICN at 266 nm, which they attributed to either I* and I channels or to $CN(A^2\Pi)$ and $CN(X^2\Sigma^+)$ states.[41] An experiment by Donovan and Konstantatos seems to have produced a misleading conclusion.[237] They obtained 5% or less of excited I* atoms by broad-band flashlamp dissociation for wavelengths greater than 200 nm and concluded that the two channels must be the $CN(A)$ and $CN(X)$ states. A quite different result was obtained by Amimoto et al., who found an I* yield of approximately 0·5 with broad-band wavelengths of 220 nm or more.[238] Using laser-induced fluorescence on the CN radicals, Baronavski and McDonald found predominantly $CN(X^2\Sigma^+)$ ground state at 266 nm and did not observe the $CN(A^2\Pi)$ state within the limit of their detection ($\lesssim 0·1\%$).[166] Pitts and Baronavski[78] recently used tunable laser, infrared fluorescence to map the quantum yield of I* production as a complete function of wavelength (Fig. 12). The first absorption continuum is seen to be composed of at least three bands. At 266 nm, the I* yield is 61%, in excellent agreement with the estimate from the photofragment spectra of Ling and Wilson.[41] Thus the primary

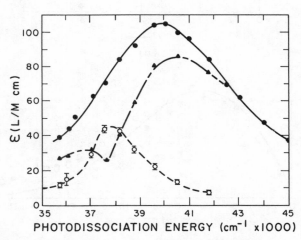

Fig. 12. Experimental results for the production of $I(^2P_{1/2})$ from ICN. The full curve and full circles is the absorption spectrum of ICN. The (short) dashed curve and open circles is $\varphi_{I(^2P_{1/2})}\varepsilon_{ICN}$. The (long) dashed curve and triangles is $\varepsilon_{ICN} - \varphi_{I(^2P_{1/2})}\varepsilon_{ICN}$. From Pitts and Baronavski.[78] Reproduced by permission of North-Holland Publishing Company, Amsterdam.

dissociation pathways are firmly established to be:

$$ICN(\tilde{A}) \rightarrow I^* + CN(X)$$
$$\rightarrow I + CN(X)$$

The results of Pitts and Baronavski also show why the flash photolysis experiments could obtain such drastically different results. The spectral features leading to I* comprises only a small part of the total integrated absorption. Therefore, the I* yield will depend sensitively on the spectral distribution of the broad-band flashlamp.

Laser-induced fluorescence studies on ICN have also provided vibrational and rotational information. At 266 nm, the $CN(X^2\Sigma^+)$ state is predominantly in the $v = 0$ state, with $N_{v=1}/N_{v=0} = 0.012$, and $N_{v=2}/N_{v=0} = 6 \times 10^{-4}$.[106] The rotational distribution has a temperature greater than or equal to 3000 K. With flashlamp dissociation of ICN for wavelengths of 145 nm or more, the vibrational distribution is bimodal: $0.38:0.28:0.25:0.32:1.0:0.17$ for $v = 0:v = 1:v = 2:v = 3:v = 4:v = 5$.[105] There are strong indications that the $CN(A^2\Pi)$ state is a partial product at the shorter wavelengths, and that it populates the $v = 4$ and 5 levels of the X state by collision-induced intersystem crossing.[105,106] For $\lambda \geqslant 220$ nm, only the $CN(X^2\Sigma^+)$ state is produced, with little vibrational excitation,[105] as was observed in the experiments at 266 nm.[106] The rotational energy disposal in ICN has been systematically investigated with 185 nm light.[7,12] The $CN(B)$ state acquires only very low rotational excitation. The predissociated upper state does not contain any bending excitation, so the dissociative state is mainly linear. In this case, the rotational excitation of the fragment will largely map that of the parent. Several polarized fluorescence experiments have been carried out on the $CN(B)$ fragment of ICN.[128,130] Both experiments show a significant amount of structure in the degree of polarization as a function of wavelength, owing to the excitation of different P, Q, and R branches and a variety of different Rydberg electronic transitions. In addition, a polarization is observed which can be attributed to underlying continuum excitation.[128]

Similarly VUV photolysis experiments have been carried out on BrCN.[7,10–12,126,128] Emission is again observed from the $CN(B)$ state, and in addition, excitation of the $CN(A)$ state is inferred.[10] In the absence of collisions, the B state has an inverted vibrational distribution,[10] in contrast to the results of Tatematsu et al.[11] They report that the vibrational distributions are unchanging below 0.1 Torr;[11] however, Ashfold and Simons find a marked change in the $v = 0$ populations below 0.1 Torr.[10] They attribute this to a repopulation of the $v = 0$ level of the B state by collision-induced intersystem crossing back from the $v = 10$ level of the long-lived A state.

The rotational distribution of the $CN(B)$ fragment from BrCN at 158 nm is a good fit to a Boltzmann distribution at 1700 K;[12] however, dissociation of ClCN at 147 nm produces a bimodal distribution in the CN.[12,104] The bimodal distribution in ClCN may be due to two overlapping electronic states. The degree of rotational excitation is greater in BrCN than in ICN, perhaps indicating a greater bending contribution in the initial dissociating state and a steeper repulsive potential, which would enhance the final-state 'recoil' interaction. In all cases the extent of rotational excitation increases with increasing photon energy. The remarkable fit of the rotational populations to Boltzmann distributions indicates that the predissociations must proceed, however, from nearly linear excited states.

The theory of photodissociation of the cyanogen halides has had a chequered history. Although the basic concepts and methods of calculation are now well understood, carrying out the calculations in difficult in practice. Good potential surfaces are always a major problem. Two publication were incorrect because of the early interpretation that dissociation of the $ICN(\tilde{A})$ state leads to $1 + CN(A)$, or $CN(X)$.[207,239] Another set of publications was able to fit the approximate $CN(B)$ state vibrational distributions, but then obtained results for CHN/DCN which were exactly opposite to experiment.[187,189] They predicted that DCN would produce less vibrational excitation in the $CN(B)$ fragment than HCN, whereas experimentally it actually increases substantially. It was shown that the early quasidiatomic model could be extended by a classical treatment incorporating a Franck–Condon transition using a Wigner distribution function for the initial vibrational conditions.[240] Reasonable vibrational distributions were obtained for ICN.[240] Other works attempted to fit the repulsive potential final interaction along with Franck–Condon calculations to reproduce experimental data on vibrational distributions.[178,179] However, the resultant potential surfaces were too steep and predicted physically unrealistic HCN/DCN isotope effects. Later they showed that the available experimental data in the case of HCN were insufficient to characterize completely the repulsive potential parameters.[181] There has only been moderate success at predicting experimentally observed rotational distribution in $ICN(\tilde{A})$ dissociation.[186] Recently, theoretical work has predicted interesting structure in the distribution of rotational product states in ICN,[172,184] which might be observed if single rotational levels of ICN can be prepared experimentally.

The varied success of predicting the experimental results of the cyanogen halides and HCN tells us that the electronic states and potential surfaces of these triatomics are extremely difficult to treat. Nevertheless, many rewarding insights have been obtained, and we certainly understand more about the dynamics of these molecules than is reflected in a close comparison of experiment and theory. The theorists' task has been doubly difficult in this case, since all of these molecules involve states which predissociate. A few results are now appearing in the literature for directly dissociated linear triatomics,[55,79,131] which might be more amenable to theory. We have learned the dangers in qualitative, broad-band flash experiments. Tunable, single-frequency laser and synchrotron radiation experiments should help to avoid confusion and erroneous results in the future.

Photofragmentation of several other CN-containing polyatomics has been investigated. Ashfold and Simons studied VUV photolysis of CH_3CN, CD_3CN, CF_3CN, and CH_3NC.[13] The vibrational distributions in the $CN(B^2\Sigma^+)$ fragment approximate Boltzmann behaviour for all except CH_3NC. The fraction of energy delivered into vibration increases slightly in the series $CH_3CN < CD_3CN < CF_3CN$. These results can be explained by a competition between two factors: (1) in the RRK statistical picture there will be increased vibrational excitation in the CX_3 fragment upon substitution of heavier X atoms due to an increase in the density of vibrational states in the CX_3, and (2) increased vibrational excitation of the CN will result from the recoil of heavier CX_3 fragments. If the RRK picture is correct, the data would

TABLE II. Distribution of excess energy in photodissociation of various OH- and CN- containing compounds. From Okabe.[243] Reproduced by permission of the American Institute of Physics.

Molecule	λ_{exc} (Å)	$E_{avl}{}^a$ (cm^{-1})	Process^b	Ground-state bond angle	Conversion of excess energy into diatomic fragments		Ref.
					Vibration	Rotation	
				(a) Bent molecules			
H_2O	1236	7200	H + OH(A)	H–O–H angle^c	0·10	0·61	243
	1216	8540		105°	0·08	0·68	244
H_2O_2	1236	31 100	OH + OH(A)	H–O–O angle^d	<0·002	0·23	245
				98°			
$HONO_2$	1236	31 800	NO$_2$ + OH(A)	H–O–N angle^e	0·02	0·07	243
				103°			
				(b) Linear molecules			
HCN	1236	14 000	H + CN(B)	180°^{f}	0·12	0·07	246
	1216	15 300	H + CN(B)		0·12	0·06	8
ClCN	1236	21 200	Cl + CN(B)	180°^{f}	0·28	0·21	246
	1470	8 300	Cl + CN(B)	^{g}	0·08	0·12	7
	1700	25 000	Cl + CN(X)	^{g}	0·05	0·05	104
BrCN	1236	26 100	Br + CN(B)	180°^{f}	0·21	0·07	246
	1216	27 400	Br + CN(B)		0·15	0·05	11
ICN	1236	29 600	I + CN(B)	180°^{f}	0·23	0·10	246
C_2N_2	1600	8 600	CN + CN(A)	linear^c	0·18	0·11	102

^{a} E_{avl} = excess energy available for excitation of vibration and rotation.

^{b} Unspecified products are those formed in the ground state.

^{c} From G. Herzberg (1966), *Molecular Spectra and Molecular Structure*, vol. III, *Electronic Spectra and Electronic Structure of Polyatomic Molecules*, Van Nostrand, Princeton, NJ.

^{d} From P. A. Giguere and T. K. K. Srinivasan (1977), *J. Mol. Spectrosc.*, **66**, 168.

^{e} From A. P. Cox and J. M. Riveros (1965), *J. Chem. Phys.*, **42**, 3206.

^{f} From D. R. Stull and H. Prophet (1971), JANAF thermochemical tables, 2nd edn., *Natl. Bur. Stand. Ref. Data Ser.*, 37.

^{g} The upper state may be bent to give more rotation than expected from linear configuration.

suggest that the second effect is slightly larger than the first. The rotational distributions have many complex features, indicating excitation to more than one electronic state, perhaps one of bent and one of linear configuration. Knudtson and Berry have shown that flash photolysis ($\lambda \gtrsim 155$ nm) in CH_3NC results in predominantly the $CN(A^2\Pi)$ state.[167]

Using laser-induced fluorescence, the $CN(X)$ state produced from C_2N_2[102,241,242] and C_4N_2[103] has been investigated. In C_2N_2, both the vibrational and rotational distributions give good fits to Boltzmann temperatures. At 160 nm, for example $T_{vib} = 2750$ K ($v = 0 : v = 1$ is $0.74 : 0.26$) and $T_{rot, v=0} = 1400$ K. Characteristically very little energy is deposited into rotation and vibration of the $CN(X)$ fragment, whereas a substantial amount of energy ends up in translation. Similarly, low vibrational and rotational excitation of the CN fragment is observed for C_4N_2. The Boltzmann rotational distributions can be explained qualitatively by the idea of a statistical predissociation.[103,242] Flash photolysis of ONCN at 147 nm results in $CN(B)$ state emission with evidence of considerable rotational excitation.[29] A qualitative comparison of some of the results for dissociation of CN-containing compounds is given in Table II, taken from the summary of Okabe.[243] It is apparent that only a small fraction of the available energy goes into vibration and rotation of the CN fragment. There is a dramatic difference in the extent of rotational excitation for the CN- and OH-containing compounds. All of the OH compounds are highly bent, favouring a larger disposal of energy into rotation than for the linear CN-containing molecules.

G. H_2O, H_2S, and Other OH-Containing Compounds

Molecules containing OH offer novel possibilities for fragmentation dynamics because of large changes in bending geometries upon excitation, the large rotational constant of the OH fragment, and the possibility of forming two electronic states, $OH(X^2\Pi)$ and $OH(A^2\Sigma^+)$. In several cases, very light atoms recoil from the heavy OH fragment, making for some interesting and predictable geometries. $OH(A^2\Sigma^+)$ emission has been studied in the VUV photolysis of H_2O.[20,90,243-245] Polarized fluorescence techniques have been used to make symmetry assignments of the excited states.[129] The vibrational distribution in $OH(A)$ with 121.6 nm and 123.6 nm excitation is found to be $1.0 : 0.3 : < 0.1$ for $v = 1 : v = 2 : v = 3$.[243,244] The rotational states are excited to a highly non-Boltzman distribution and actually contain the major fraction of the available energy (Fig. 13 and Table II).[243,247] Higher temperatures appear to give greater rotational excitation.[248] The vibrational distribution has been successfully modelled with a Franck–Condon theory,[249] since it is unlikely that the recoil of the rather, light H atom against the heavy O atom could provide much vibrational excitation. Using a simple recoil model, a good fit of the rotational distribution is obtained as well.[250] There is almost a complete

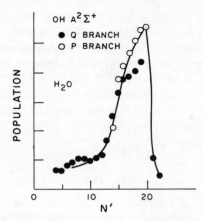

Fig. 13. Rotational population distribution of $OH(A^2\Sigma^+)$, $v' = 0$ state in the photodissociation of 0·2 Torr H_2O by the Kr lamp (mainly 123·6 nm): ●, Q branch; ○, P branch; N' is the rotational quantum number of the upper state. From Okabe.[243] Reproduced by permission of the American Institute of Physics.

separation of the dynamics of the vibrational and rotational effects because of the light recoiling mass and the highly bent dissociation geometry.

Quite different results are observed for H_2S. Hawkins and Houston used 193 nm excitation and laser-induced fluorescence detection of the ground $X^2\Pi_{1/2, 3/2}$ state of SH.[251] They find no vibrational excitation in either the SH or DH fragments. The rotational distribution of SH radicals is nearly Boltzmann with a temperature of 375 K for the $^2\Pi_{3/2}$ component and 220 K for the $^2\Pi_{1/2}$ component. The $^2\Pi_{3/2}/^2\Pi_{1/2}$ ratio is 3·75. The HS bond length is nearly identical to the H–S distance in H_2S, and apparently there is also no change in the excited-state geometry, since the SH product is vibrationally cold. The recoil interaction leading to rotational excitation is much less in H_2S than in H_2O. This can be attributed to the nearly orthogonal bond angle (92°) which must not change in the excited state. Thus 98% of the available energy appears in translational recoil of the fragments.

VUV photolysis of H_2O_2 at 123·6 nm also produces significant rotational excitation of the $OH(A^2\Sigma^+)$ state, up to N' levels as high as 28, but with little or no vibrational excitation.[245] For $HONO_2$ photolysis at 123·6 nm, relative $OH(A^2\Sigma^+)$ populations $v = 0:v = 1$ are found to be 1·0:0·28.[243] The rotational distribution for $HONO_2$ is a broad peak with a maximum at $N' = 10$, which probably results because of nearly similar bond angles in the ground and excited states.[243] The laser-induced fluorescence spectrum has been obtained for $OH(X)$ produced by 307–309 nm photodissociation of

HOCl.[110] The OH is found to have very little vibrational or rotational excitation,[110] again because of nearly similar geometries in the ground and excited states.[199] Complete *ab initio* SCF–CI calculations have been carried out to predict the geometries, absorption strengths, absorption spectra, and dissociation products of both HOCl[199] and HO$_2$.[200] The first long-wave-length dissociation processes are predicted to lead to OH + Cl and OH + O respectively. The theory for HOCl also predicts that there should be predominantly translational excitation of the products, which is now confirmed by the experiments.

H. O$_3$

The quantum yield of O(^1D) formation from ozone in the wavelength region 250–325 nm is important in order to model the production of O(^1D),

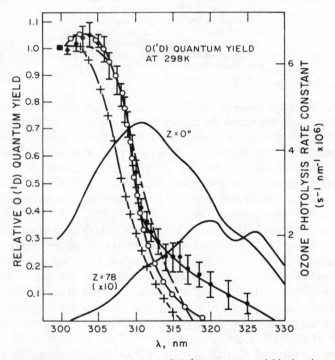

Fig. 14. Wavelength variation of O(^1D) quantum yield showing differences among various studies. The full curves are the wavelength variation of the ozone photolysis rate constant at two different zeniths. The quantum yield measurements are: +, ref. 64; O, ref. 63; ———, ref. 253; ●, ref. 66. ■, normalization point. From Brock and Watson.[66] Reproduced by permission of North-Holland Publishing Company, Amsterdam.

which subsequently reacts with H_2O to produce OH in the upper atmosphere. There have been recent extensive investigations of the $O(^1D)$ quantum yield, especially in the wavelength region where the O_3 absorption is decreasing (300–320 nm).[63–66,252,253] Most of the recent studies use frequency-doubled, tunable dye lasers as the photodissociation source. Detection is by chemical reaction with N_2O, ultimately to obtain NO_2 chemiluminescence from reaction of the NO product with O_3. The $O(^3P)$ yields have also been measured by absorption of the VUV resonance line. The most recent measurements on the $O(^1D)$ quantum yield are summarized in Fig. 14. The results appear to be approaching the stages of final agreement, with the exception of the long-wavelength 'tail' recently observed by Brock and Watson.[66] In addition, there is strong evidence that the quantum yield at 255 nm is not unity, as had been generally thought,[254] but is really more like 0·9. Two separate molecular beam TOF measurements obtain between 5 and 10% $O(^3P)$ production at 266 nm.[59,62] Recent VUV absorption measurements also give the $O(^3P)$ quantum yield at 266 nm as 0·1[67] and 0·15 at 248 nm.[68] The earlier results of Amimoto et al.[69] were shown to be incorrect due to inadequate time resolution in measuring the initial $O(^3P)$ concentration.[68] It appears that a 10–15% contribution of the $O(^3P)$ channel definitely occurs in these shorter-wavelength regions.

The most detailed dynamical results on O_3 were obtained by the molecular beam method. The TOF spectra at 266 nm distinctly show a broad feature due to the $O(^3P) + O_2(X^3\Sigma_g^-)$ products, as well as the major $O_2(^1\Delta_g) + O(^1D)$ channel.[59,62] Sparks et al. were also able to resolve fully the vibrational peaks due to the $v = 0$–3 levels of $O_2(^1\Delta_g)$ (see Fig. 2).[62] They obtain values of 0·57:0·24:0·12:0·07 for the $v = 0:v = 1:v = 2:v = 3$ distribution. In addition, they estimate that 17% of the available energy is deposited into rotation for the $O_2(^1\Delta_g)$ $v = 0$ channel. It is estimated that the $O(^3P)$ channel produces the $O_2(^3\Sigma_g^-)$ ground state in vibrational levels $v = 0$–10.[59] Experiments at 600 nm give only $O_2(X^3\Sigma_g^-) + O(^3P)$, with the O_2 excited only to $v = 0$ and 1.[59]

Because of the sensitivity of the $O(^1D)$ production with temperature in the fall-off region, there have been a number of investigations of the temperature dependence.[138,252] Kajimoto and Cvetanović discuss the possibility that enhanced rotational energy with increasing temperature can serve to promote the ozone over the $O(^1D)$ threshold.[138] They do not feel that vibrational excitation of the (010) state could account for the magnitude of the effect they observed. Zittel and Little,[137] however, have recently shown at 314·5 nm that the $O(^1D)$ yield is increased by nearly two orders of magnitude upon CO_2 laser excitation of the v_3 vibration (Fig. 15). The v_1 mode rapidly equilibrates with the v_3 mode, but the rotational temperature does not change significantly. Thus the entire effect is ascribed to vibrational enhancement. Mode-specific experiments of this kind should provide a detailed understanding of Franck–

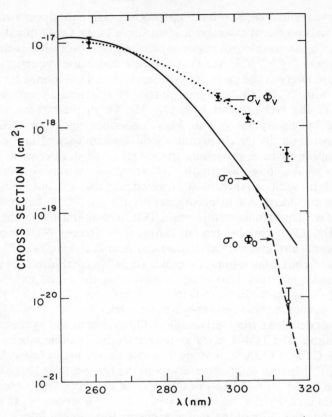

Fig. 15. Cross sections for photodissociative production of $O(^1D)$ from vibrationally excited and room-temperature ozone: $\bullet$, experimental results for ozone excited with a CO_2 laser on ν_3 band; ——, cross section for $O(^1D)$ production at room temperature; ——, cross section for absorption at room temperature; $\bigcirc$, measurement of the room-temperature $O(^1D)$ cross section at 314·5 nm from the present study. From Zittel and Little.[137] Reproduced by permission of the American Institute of Physics.

Condon overlaps and geometry considerations involved in the photofragment dynamics of specific electronically excited states.

I. CO_2, CS_2, OCS, and $OCSe$

CO_2 has numerous high-lying electronic states and Rydberg transitions. Several significant new results have been obtained in the past few years. Careful reinvestigation of the dissociation has proved that the yield of $CO + O$ is exactly unity at both 147 nm and 130·2–130·6 nm,[83] in contrast to earlier reports that the yield was 0·85 or less throughout the entire 120–170 nm range.[255] In the region 106–117·5 nm, the $O(^1S)$ yield is found to be near unity, but with an abrupt dip to less than 0·15 at 108·9 nm.[87] There is a strong Rydberg transition at this wavelength, which must correlate to different products. The quantum yield of $O(^1D)$ production at this wavelength was found to be 0·65.[87] At even shorter wavelengths, 70–90 nm, dissociation of CO_2 leads to dominant emission from the $CO(A^1\Pi \rightarrow X^1\Sigma^+)$ spectrum.[256] The vibrational population in the $CO(A^1\Pi)$ state follows a Poisson distribution. The onset of $CO(A^1\Pi) + O(^1D)$ production is detected at about 80 nm by an increase in the $CO(A^1\Pi, v = 0)$ population.[256] Between 42 and 83 nm, the emission contains many atomic carbon lines[256,257] and some atomic oxygen lines,[258] especially at the shorter wavelengths. The theories of Band and Freed[180] and Berry[176] have been applied to CO_2. The vibrational distributions in the electronically excited states of the CO product cannot by fitted assuming a direct dissociation, but can be obtained assuming predissociation. The final-state interactions are found to make a non-negligible contribution in CO_2, whereas in HCN and ICN they were of minor importance.[180]

Early studies on CS_2 from 69 nm to 133·7 nm found only the $CS(A^1\Pi)$ excited-state product. The presumed major products, $CS(A^1\Pi) + S(^3P)$, would be in violation of simple spin conservation ideas. More recently, the long-lived $CS(A^3\Pi)$ state has been detected for wavelengths below 160 nm,[21] and this state now appears to be the real major product. The maximum $CS(A^1\Pi)$ yield is thought to be only 7%, probably resulting from a weaker triplet state absorption in CS_2.[21] In the wavelength range 290–380 nm, no dissociation takes place at all. The molecule either radiates or internally converts to dissipate the energy into vibration and rotation. From 185 to 230 nm, dissociation to $S(^3P) + CS(X^1\Sigma)$ is also detected.[259] Yang et al. have studied the photofragment TOF spectrum and the laser-induced fluorescence spectrum of the $CS(X^1\Sigma)$ fragments from CS_2 in a molecular beam.[54] They observe channels leading to both $S(^1D)$ and $S(^3P)$ atoms, corresponding to excitation to a singlet and triplet state respectively. The dissociative state is found to be long-lived ($\tau \sim 1$ ps) and predissociated.[54] The vibrational distribution in the $CS(X^1\Sigma)$ fragment is strongly inverted, with excitation up to $v = 7$, where the LIF method ceases to be sensitive enough to detect any further. The highly inverted vibrational distribution indicates that Franck–Condon factors in the excitation geometry probably play a decisive role.

Photodissociation of OCS with 220–260 nm light produces $S(^1D)$ atoms in substantial yields.[260] Single-photon dissociation at 142–160 nm,[18] and two-photon dissociation at 193 nm,[17] produce $S(^1S)$ atoms with nearly unity quantum yield. Both direct heating (with 170 nm dissociation)[18] and vibrational excitation with a CO_2 laser (with two-photon 193 nm dissociation)[17] enhance the $S(^1S)$ yield by a factor of two. In the case of the CO_2 laser experiments, the entire enhancement is attributed to the increased total cross section for absorption. In the temperature experiments, no increase in the total absorption cross section is obtained. It is speculated that changes in the specific cross section leading to the $S(^1S)$ electronic state component are occurring.

A similar set of experiments has been carried out on OCSe. For 164–180 nm dissociation, the Se(^1S) yield is greater than 75%.[14] Single-photon dissociation at 193 nm gives a 30% Se(^1S) yield.[14-16] An enhancement of 25% is observed with CO_2 laser excitation of the vibrations.[16] The enhancement is attributed to the increased absorption cross section of the vibrationally excited molecule. Unfortunately, there are no data on the CO product state vibrational distributions. It would be especially interesting to determine whether the vibrational energy in the CO product might also change with initial vibrational excitation of the OCS and OCSe triatomic.

J. SO_2, N_2O, NO_2, NOCl, and Other NO-Containing Compounds

TOF fragment translational distributions have been obtained[261] at 193 nm for the process

$$SO_2 \rightarrow SO + O(^3P).$$

SO_2 is known to be predissociated, with a lifetime of 10 ps at 196 nm. The angle and bond lengths change drastically from the ground to excited state.[261] The bond length is 1·431 Å in the ground state and 1·56 Å in the excited state, and the bond angle goes from 119·3° to 104·3°. The bond length of the SO radical is 1·48 Å. Less than half of the available excess energy appears in translation of the fragments.[261] The remaining energy must be deposited into vibration and rotation of the SO product. Because of the large geometry changes, it is very likely that the Franck–Condon structural changes in the excitation dominate the dissociation dynamics.

N_2O has been studied by VUV excitation followed by fluorescence[19] and by molecular beam photofragmentation methods.[58,262] Using synchrotron radiation from 175 to 750 Å, a variety of N, O, and N_2 electronically excited states are observed in emission.[19] At 130 and 147 nm, peaks in the TOF spectrum are presumed to due to $O(^1D)$ and (^1S) excited states.[58] The rather low-resolution TOF spectrum can then be modelled to show that there is substantial vibrational excitation in the N_2 fragment. Shapiro has calculated the production of vibrationally excited N_2 as a function of initial N_2O vibrational state dissociation.[194]

Some of the earliest triatomic molecule TOF spectra were taken on NO_2.[42,43] A lifetime of less than 2×10^{-13} s indicated that the molecule is directly dissociated at 347 nm. The quantum yield of NO + O formation has been measured from 375 to 420 nm.[263] The quantum yield takes on values between 0·6 and 0·85 from 375 to 400 nm, where it rapidly decreases to zero by 420 nm. Perhaps the most elegant and extensive investigations have been carried out by Zacharias et al.[112] Using a frequency-doubled dye laser to probe the NO fragment of NO_2 dissociated at 337·1 nm, they obtain complete vibrational and rotational state distributions. The ratio of $v = 0 : v = 1 : v = 2$ is highly inverted 2:3:5, in contrast to the results with 347 nm excitation.[42,43] The exquisitely detailed rotational distributions show a truncation of the higher levels as less energy is available in the higher vibrational channels (Fig. 16).[112]

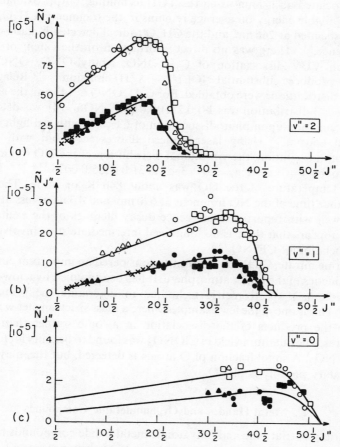

Fig. 16. Population distribution for the rotational–vibrational states of the NO product from dissociation of NO_2 at 337·1 nm; open symbols, $^2\Pi_{1/2}$ substate; filled symbols, $^2\Pi_{3/2}$ substate. The different symbols refer to different rotational branches. The ordinate scales should be divided by 30 to obtain absolute numbers. From Zacharias et al.[112] Reproduced by permission of Springer-Verlag.

Busch and Wilson first studied NOCl with a doubled ruby laser at 347 nm.[44] The molecule directly dissociates with approximately 70% of the excess energy appearing in translation. Recently, Grimley and Houston used time-resolved infrared emission from the vibrationally excited NO to examine the extent of NO excitation in the dissociation of NOCl and NOBr at different wavelengths.[264] For wavelengths greater than 480 nm, no $NO(v)$ was produced in the primary dissociation process. At 355 nm, substantial NO vibrational excitation is observed, in good agreement with the molecular beam results.[44] Photodissociation of a number of alkyl nitrites has been studied. C_2H_5ONO was dissociated at 347 nm using the TOF technique.[45] Approximately 64% of the available energy on average remains in the fragments. CH_3ONO has been dissociated at 266 nm, and the CH_3O radical detected by laser-induced fluorescence.[117] There was no direct information on the extent of internal excitation VUV dissociation of C_2H_5ONO, t-butyl-ONO, ONCN, and CF_3NO produces substantial $NO(A^2\Sigma^+ - X^2\Pi)$ emission.[29,30] Relative vibrational distributions were obtained. For C_2H_5ONO at 147 nm, the $v = 0:v = 1:v = 2:v = 3$ distribution was $1\cdot0:0\cdot2:0\cdot06:0\cdot03$.[30] $NO(v = 1)$ was detected by infrared emission upon photodissociation of CF_3NO with red light between 580 and 660 nm.[95] Using laser-induced fluorescence, but with a novel two-photon excitation, the vibrational distribution of the NO fragment was found to be $v = 0:v = 1:v = 2:v = 3 \equiv 1:0\cdot31:0\cdot0085:0\cdot0028$.[118,119] The rotational temperature of the NO was about 840 K or more. The 18·5 ns appearance time of the NO fragment at 670 nm, and 4·3 ns or less at 600 nm, agreed well with reported fluorescence decay lifetimes of the excited state. Thus it appears that there is no long-lived intermediate state involved in the predissociation of CF_3NO.[118]

Chlorine nitrate, $ClONO_2$, is recently considered an important constituent in the kinetics of the upper atmosphere. It has been studied by a low-pressure photolysis/mass spectrometric technique to determine the primary products.[92] This is one of the few examples where a mass spectrometer was used to analyse the products of the dissociation in a 'bulb' rather than a beam apparatus. The quantum yield of $ClONO_2$ was found to be unity for formation of $Cl + NO_3$. A small fraction of O atoms is detected, but this may be from a secondary photolysis of NO_3.

K. Metal Halides and Organometallic Compounds

A variety of triatomic and polyatomic metal halide compounds have been the subject of intensive investigation because of their prospects for atomic and molecular discharge lasers. Many of the properties of these molecules are best studied first by photofragmentation. Schimitschek et al., for example, studied lasing action on the HgBr $(B-X)$ transition produced by photofragmentation of $HgBr_2$ at 193 nm.[169] Eden and Waynant used dissociation of

HgBr$_2$ and HgI$_2$ at 195–200 nm and 230 nm respectively to prepare HgBr(B) and HgI(B) states for selective quenching studies.[265] A similar method in XeF$_2$ at 175 nm was used to prepare the XeF(B) state,[266] a molecule which is unbound in its ground state. Maya has remeasured the ultraviolet absorption cross sections of the mercury and tin halides[267] and has measured[22] and extensively reviewed[86] many of the fluorescence quantum yields of the fragments in the photodissociation of metal halides.

A molecular beam TOF experiment has been carried out on CdI$_2$ in its first long-wavelength absorption band at 278 and 300 nm.[55] Of the many possible channels (Cd + I + I, CdI + I, CdI + I*($^2P_{1/2}$)), it was conclusively shown that the major products are CdI + I and CdI + I*. The first absorption band was found from angular anisotropy measurements to be composed of at least two different excited states of parallel and perpendicular character. The TOF spectrum could be resolved into two different peaks identified as the I and I* components. Approximately equal amounts of energy are released into translation and internal excitation; however, the vibrational states of CdI could not be resolved. Tunable laser, infrared fluorescence quantum yield measurements have been performed on HgI$_2$ in the analogous long-wavelength absorption band.[79] The quantum yield of I* shows that the first absorption band is composed of two distinct components, leading to I and I* channels (see Fig. 7). The longest-wavelength portion of the band may even contain another component, since the residual part of the spectrum does not seem to fit an expected Gaussian form. Wadt has calculated the electronic structure of HgCl$_2$ and HgBr$_2$ in detail, although spin–orbit coupling has not been included as yet.[201] A major prediction of those calculations is that the first absorption bands of HgCl$_2$ and HgBr$_2$ should lead to three atoms for the products, since the first excited state of HgCl$_2$ correlates with the repulsive HgCl(A) state, not to a bound state. These results seem to be in contrast to the experiments on CdI$_2$ and HgI$_2$, where it appears that the lowest excited states produce a diatomic fragment. It should be extremely interesting to investigate the long-wavelength absorption features of HgCl$_2$, HgBr$_2$, and HgI$_2$, in more detail to obtain the major dissociation products.

Polarized photofragment fluorescence experiments have been carried out on the excited HgBr(B) state by dissociation of HgBr$_2$ at 193 nm.[131] This is the first example of such an experiment on a directly dissociated state, where the results should check closely with the theory.[125] The measured degree of polarization (11·9 $\pm$ 1·5%) is in good agreement with the predicted theoretical value of 14·3%.

More complex polyatomic molecules are now finding their way into molecular beam machines, often providing a crucial test of even just primary dissociation products. Kawasaki and Bersohn studied the TOF photofragmentation of AsI$_3$ between 280 and 300 nm.[268] Single-photon dissociation produces $AsI_2 + I$. It appears that I^* is formed predominantly with very little vibrational excitation of the AsI_2 fragment. Andreeva et al. showed by photofragmentation laser experiments that the long-wavelength absorption (~ 290 nm) of $(CF_3)_2$ AsI produces I^*, and that wavelengths less than 200 nm form I ground-state atoms.[269] This seems to be a somewhat general phenomenon for many of the metal iodide compounds, such as CdI_2, HgI_2, and AsI_3. Sequential multiple-photon absorption can take place in the polyatomic metal halides. Numerous examples have been given where atoms are stripped off one by one, often leaving an excited metal atom which emits or lases.[148,153,164]

Because of interest in isotope separation with lasers, the photophysics of gaseous UF_6 has received much attention.[270] Kroger et al. studied the TOF photofragment spectrum at 266 nm, which produces $UF_5 + F$.[51] They did not observe any sequential multiple-photon processes. A bimodal translational distribution was observed, which indicates that the UF_5 fragments must have two distinct internal energy distributions, possibly due to two different electronic states or a curve crossing. The rather large change in geometry on going from UF_6 to UF_5 will undoubtedly leave the UF_5 fragment with high vibrational excitation. The near-UV photophysics of UF_6 has been studied also in 'bulb' experiments from 340 to 410 nm.[271,272] The quantum yield of fluorescence is found to be between 10^{-6} and 5×10^{-4} on going from shorter to longer wavelengths.[271] The lifetimes have a distinct double-exponential decay at the shorter wavelengths, indicating complex interactions between the electronically excited states.[272] Classic chemical trapping experiments with H_2,[271,273] as well as an optoacoustic experiment,[94] showed that the quantum yield of UF_5 production is essentially unity.

Classic angular anisotropy experiments were first carried out in a bulb by observing the removal of metal mirrors from the walls by action of the radical reactions.[1] In the case of $Cd(CH_3)_2$, which is one of the early examples of a three-body dissociation, the Cd itself deposits a mirror on the walls. The anisotropic distribution of Cd could only be explained if one methyl–Cd bond breaks first and the second bond breaks only a short time later. Tamir et al. provided three models of the photofragmentation of $Cd(CH_3)_2$, all of which predicted asymmetrical dissociation patterns.[197] Recently, Deutsch et al. showed that the metal mirror deposition method is quite general and can also be used for precision deposits of metal atoms.[101] An ArF laser was used to photofragment $Al(CH_3)_3$, $Cd(CH_3)_2$, and $Sn(CH_3)_4$ to deposit small ($<2 \mu m$) spots of the metal. Presumably single- or multiple-photon absorptions strip the molecule down to the free metal atoms, which then condense on a substrate. Baughcum and Leone have studied wavelength-resolved infrared fluorescence from CH_3 radicals produced by 248 nm dissociation of $Hg(CH_3)_2$.[142] They observe excitation in both the CH stretch and the out-of-plane bend, with a concurrent rotationally excited envelope on the CH stretching band.

Several new experiments have been performed to elucidate the gas-phase primary dissociation dynamics of metal carbonyls. Single-photon dissociation of $Fe(CO)_5$ at 248 nm, followed by chemical trapping of the products with PF_3, reveals that the primary dissociation process yields $Fe(CO)_4$, $Fe(CO)_3$, and $Fe(CO)_2$.[97,98] The product ratios are 0·10;0·35;0·55 respectively. It is postulated that the initial vibrationally excited $Fe(CO)_4$ fragment undergoes unimolecular decomposition to $Fe(CO)_3$ and $Fe(CO)_2$. In condensed phases, only $Fe(CO)_4$, is observed, probably because of collisional stabilization. Focused 248 nm dissociation of $Fe(CO)_5$ produces stimulated emission, probably from atomic Fe states, by an unknown series of decomposition pathways.[274] Photodissociation of $(CO)_5ReRe(CO)_5$ and $(CO)_5MnMn(CO)_5$ in a molecular beam at 300 nm results in cleavage of the metal–metal bond.[275] No cleavage of the metal–CO bonds was observed. Two-thirds of the available energy is found as internal excitation in the fragments.

L. Alkyl and Aryl Halides and Fluorinated Analogues

The first absorption bands of the alkyl iodides and bromides involve $n-\sigma^*$ electronic transitions which results in directly dissociative states. The lifetimes for molecules like CH_3I are some of the shortest observed by molecular beam angular anisotropy experiments. The basic photofragment processes and previous results have been thoroughly reviewed.[1,32,33] Photodissociation can lead to I or I* products with increasing amounts of internal energy in the remaining alkyl fragment as the size of the fragment goes up. Photodissociation in the VUV at 140–170 nm results is breaking a C–H bond, rather than the C–I bond.[276].

Extensive I* and Br* quantum yield results have been presented using broad-band flash photolysis in the alkyl iodides,[71–73,93] fluorinated iodides,[71,72,277–282] alkyl bromides,[76,282] and fluorinated bromides.[282] In one case, the change in the ultraviolet absorption profile has been measured upon vibrational excitation of CF_3I with a CO_2 laser.[280] There is no simple, consistent picture which can be used to explain all of these results. Tunable laser measurements now show that many of the quantum yields are much more complex than previously thought. The results reveal numerous additional electronic states or, perhaps more likely, curve crossings, which must be involved in the dissociation.[80,283,284] In the simplest of these molecules, CH_3I, for example, it is not possible to fit the known absorption band components to the measured I* contributions without postulating a curve-crossing mechanism which alters the I* yield at certain wavelengths.[61,196] Using tunable laser, infrared fluorescence techniques, it should now be possible to map many of these yields in order to describe the dynamics of the dissociation more precisely.

A high-resolution TOF apparatus has been used to study the photo-fragmentation of CH_3I at 266 nm.[61] The data show the two major channels,

I and I*, with vibrational excitation of the CH_3 radical superimposed on each threshold. From symmetry considerations,[61] it is likely that only the out-of-plane bending vibration can be excited in the dissociation process. Baughcum and Leone also observed infrared emission only from the out-of-plane bend in this same process.[80] Recently Hermann and Leone have partially resolved the out-of-plane infrared emission.[141] From the observed peaks corresponding to the different vibrational levels, it is apparent that this 'umbrella' mode is highly excited.[141] Assuming that all of the radical excitation is in this $CH_3 v_2$ mode, the molecular beam results provide qualitative information that the CH_3 is excited from $v = 0$ to $v \simeq 4$ in the I* channel and from $v = 0$ to $v \simeq 10$ in the I channel.[61] However, the anharmonicity and rotational distribution were not yet taken into account in these estimates. Both Franck–Condon factors geometry changes and final-state 'recoil' interactions are expected to play an important role in the decomposition process. Shapiro and Bersohn have attempted a full quantum-mechanical calculation on a good potential surface for the CH_3I dissociation.[196] They analyse the CH_3 'umbrella' mode distribution explicitly for the I* channel only. Their results predict the approximate peak of the distribution, but not the exact width. The experiments appear to give a slightly more narrow distribution than the theory. Shapiro and Bersohn also predict that the distribution will shift to higher vibrational levels with shorter dissociating wavelengths. It should be possible to confirm this experimentally.

The photofragment dynamics of CH_2I_2 have also been studied in detail. Kawasaki et al. first reported single-photon dissociation in a molecular beam.[285] The dissociation takes place in a time short compared with rotation. A substantial fraction of the excess photon energy is delivered into internal excitation of the CH_2I fragment.[47,285] Kroger et al. also observed a sequential two-photon process which further dissociates CH_2I to $CH_2 + I$.[47] They were able to infer that the CH_2I fragment has a high degree of rotational excitation, as would be predicted from simple geometry considerations. Recently Baughcum and Leone have taken a complete time- and wavelength-resolved infrared emission spectrum of the highly excited CH_2I radical produced by the dissociation of CH_2I_2 (see Fig. 5).[80] There is evidence for excitation energies in the radical as high as $15\,000$–$20\,000\,\mathrm{cm}^{-1}$ from the 248 nm dissociation. Emission is observed from the CH stretches, the CH_2 bend, a combination band of these two modes, and a broad-band infrared background attributed to the high density of states. The CH stretches can be fitted approximately by a statistical distribution which shows excitation up to as high as $v = 5$. The CH_2 bending mode shows a symmetrical collapse from both the long- and short-wavelength edges, characteristic of initial high rotational excitation. The vibrational modes are probably populated statistically at the initial high levels of excitation; at least there was no evidence to the contrary.

The UV laser photochemistry of brominated hydrocarbons is now a popular subject. Multiphoton dissociation processes in molecules like $CHBr_3$ can sequentially strip off Br atoms, leaving CH radicals for further study.[111] It is also possible[1] that single-photon dissociation can lead to processes like

$$CHBr_3 \rightarrow CHBr_2^\dagger + Br, \ CHBr_2^\dagger \rightarrow CBr + HBr.$$

Sequential photon absorption of an excimer laser in CBr_4[151] and CF_2Br_2[150,151] leads to various atomic and radical (CF_2) emissions. VUV dissociation of CF_2Br_2 also produces electronically excited CF_2.[28] One of the most interesting questions involves the nature of the single-photon primary dissociation processes for such molecules as CF_2Br_2, where channels exist to make $CF_2 + Br_2$ or $CF_2Br + Br$ directly. The exact yield of these channels with wavelength and the nature of the electronically excited states which produce the molecular products is of fundamental importance. There have not yet been any detailed studies. The assignment and description of the dissociative electronic states is extremely complex, even for just the doubly substituted halocarbons.[80,285] A simple exciton model has been given to describe the lower-lying electronic states of CH_2I_2.[285] However, it is clear that a rigorous assignment of the symmetries and details of the dissociative electronic states of these molecules represents a challenging and difficult theoretical problem.

The aryl halides make an interesting case for study because the initial electronic excitation is in the aromatic ring system ($\pi \rightarrow \pi^*$), while the final dissociative state is localized on the C–halogen bond (σ^*). Bersohn and coworkers have extensively studied singly and doubly halogenated benzenes, naphthalenes, biphenyls, etc., by TOF techniques.[34,286,287] The halogen atoms are produced primarily in their ground electronic state. The lifetimes for the bromides can be several orders of magnitude longer than the iodides, indicating a slowing of the internal conversion process necessary to produce the dissociation. A variety of theoretical arguments has been offered concerning the mechanisms of the electronic changes which must occur before dissociation,[34,286–288] and there is not complete agreement on the nature of these processes. However a number of dramatically different translational TOF distributions are observed. From these, it is possible to describe whether the molecules undergo fragmentation by an electronic conversion directly from the initially excited state to the C–I bond or whether the molecules must convert all of their energy into vibrational excitation of the ground state, and then undergo a simple unimolecular decomposition. The latter effect occurs for $C_6H_5CH_2X$ molecules, where the translational energy distributions are observed to peak at zero energy.[286]

Broad-band flash photolysis experiments on C_6H_5I and C_6F_5I confirm that the quantum yield of I* is small.[289] Measurements on a variety of aryl halides using single wavelengths from an excimer laser show that the quantum yields actually vary substantially from band to band, but are typically very small for the strongest transitions (e.g. $\sim 9\%$ for C_6H_5I at 193 nm).[283] There is still a substantial amount of work that can be done both experimentally and theoretically to elucidate the complex photofragment dynamics that occur in the aryl halides. There is a need for better understanding of the electronic states and their lifetimes, the quantum yields of excited halogen production, and the mechanisms of electronic excitation transfer.

There are many recent studies on the photofragmentation processes of chlorinated and fluorinated hydrocarbons. Early studies were concerned primarily with quantum yields for forming atomic products. For example, it was shown that $CFCl_3$ and CF_2Cl_2 produce Cl atoms with a yield of essentially unity at 213.9 nm, but increasing to a yield of nearly 2 at much shorter wavelengths (147 nm).[96] No F atoms were observed. In ethyl chloride, 185 nm photons produce primarily Cl atoms, but at 124 nm elimination to form HCl and H_2 occurs.[290] Elimination of HCl has frequently been observed in the chloroethylenes.[165] HF elimination was recently reported for CF_3CCH.[291] Fletcher and Husain demonstrated a new capability to detect the $Cl^*(^2P_{1/2})$ excited atom by VUV absorption.[77] They observe Cl^* upon flash photolysis of CCl_4, although, unfortunately, no quantum yield information was obtained. Koda has studied the dissociation of C_2F_4 in the VUV, which produces CF_2 in the $^1B_1$ state.[27] The extent of vibrational excitation in the CF_2 can be simulated assuming a Boltzmann distribution at 1400 K. From the energetics, it appears that the products are $CF_2(^1A_1) + CF_2(^1B_1)$.

There continues to be an unusually keen interest in studying further the photochemistry of fluorocarbons because of their effect on the atmospheric ozone layer. In this regard, there have been extensive measurements as a function of temperature on the ultraviolet absorption cross sections of halogenated methanes and ethanes, the so-called freons.[292] It will be valuable to obtain data on the photofragment dynamics of these molecules as well. They offer some intriguing possibilities for more complex dissociation behaviour to test theory against experiment.

M. NH_3, PH_3, and Other NH-Containing Compounds

Photodissociation of NH_3 typically produces $NH_2 + H$, but at higher incident energies can also form $NH + H_2$.[293] Runau et al. have done an ab initio calculation on the electronic states of ammonia and its dissociation products.[202] They discuss in detail the electronically excited states which predissociate to the various dissociation products: $NH_2(\tilde{X}^2B_1) + H$, $NH_2(\tilde{A}^2A_1) + H$, and $NH + H_2$. Back and Koda studied the translational excitation of the H atoms produced by VUV photolysis of NH_3 using chemical reactions which compete for the hot and slow atoms.[294] The H atoms were found to be produced with high translational velocity, but their energy was not quantified. The $NH_3 \tilde{A}' A_2''$ excited state was found to predissociate into the excited $NH_2(\tilde{A}^2 A_1)$ state, which emits.[23] Photolysis of NH_3 at 193 nm produces both $NH_2(\tilde{A}^2 A_1)$ and $NH_2(X^2 B_1)$.[24] The $\tilde{A}$ state was observed directly from 620–1100 nm and the $\tilde{X}$ state was probed by dye-laser-induced fluorescence. The $\tilde{A}^2 A_1$ state population is only 2.5% of the $\tilde{X}^2 B_1$ state and is formed with an average vibrational energy of 1000 cm^{-1}. Observed fluorescence from $NH(A^3\Pi \to X^3\Sigma^-)$ and $NH(b^1\Sigma^+ \to X^3\Sigma^-)$ was attributed to a resonant two-photon dissociation process.[24] Similar emission from NH_2 and NH electronically excited states were reported upon dissociation of NH_3, N_2H_4, and $(CH_3)_4N_2$ in the region 86–164 nm.[295] Dissociation of HNCO [109] and HN_3 [107,108] by 193 nm and 266 nm respectively produces $NH(a^1\Delta)$ and the corresponding $CO(X^1\Sigma^+)$ and $N_2(X^1\Sigma^+)$. The $(a^1\Delta)$ state can be studied by laser-induced fluorescence to a higher $(C^1\Pi)$ state. The rotational temperatures of the NH in both cases are near 1000 K, but there is essentially no vibrational excitation. The translational energies, as measured by the laser-induced fluorescence Doppler linewidths, are substantial (~ 5000 cm^{-1} for HN_3 and ~ 2000 cm^{-1} for HNCO).[107–109] A large amount of the energy is still unaccounted for, and may end up in the N_2 and CO fragments.

PH_3 has also been studied by VUV[25,26] and single- and multiple-photon ArF laser dissociation.[88] $PH_2(\tilde{A}^2 A_1)$, $PH(A^3\Pi)$, and $PH(b^1\Sigma^+)$ emissions are observed by VUV photolysis. With the laser at 193 nm, $PH_2(\tilde{A}^2 A_1)$ is observed, but with a quantum yield of only 0.014. The main channel must be to form the $PH_2(\tilde{X})$ state. A two-photon dissociation produces the $PH(A^3\Pi)$ state[88] in an analogous fashion to that obtained for NH_3.[24]

N. Three-Body Dissociation: CH$_3$COI, CF$_3$COI, and s-Tetrazine

Elegant molecular beam experiments have been performed on several molecules which undergo three-body dissociation processes:

$$CH_3COI \to CH_3 + CO + I, {}^{52}$$

$$CF_3COI \to CF_3 + CO + I, {}^{53}$$

$$C_2N_4H_2 \to HCN + HCN + N_2. {}^{296}$$

In CH_3COI at 266 nm there is evidence for a two-step dissociation process, first to $CH_3CO + I$ and then in 10^{-11} s via a further decomposition of the CH_3CO to $CH_3 + CO$.[52] The second unimolecular process was successfully modelled by statistical RRKM ideas. In contrast, the fragmentation of CF_3COI appears to go to $CF_3 + CO + I$ in a single rapid step (10^{-13} s).[53] Classical trajectory calculations on the direct three-fragment decomposition were qualitatively consistent with the experimental observations. Possibly a non-adiabatic curve crossing which is necessary for the decay of the CF_3CO fragment is facilitated by the heavier F atom substitution.

Photofragmentation of s-tetrazine at 266 nm leads to one translationally hot HCN, one translationally cold HCN, and translationally hot N_2.[296]

$$\text{s-tetrazine} \xrightarrow{266\,\text{nm}} \text{HCN} + \text{HCN} + N_2.$$

The process occurs rapidly with respect to the timescale of molecular rotation. A large amount of the additional available energy must go into the HCN bending vibration and rotation. This would especially make sense because of the tremendous geometry changes necessary in going from the bent to linear HCN molecules. The only other experiment on s-tetrazine was carried out with 492 nm excitation and infrared fluorescence.[297] Only 1% of the HCN product molecules were found excited in their CH stretch. By a vibrational energy transfer experiment, it was ascertained that 10% of the N_2 molecules, or perhaps the v_1 stretch of the HCN molecules, are excited with a vibrational quantum. Again it appears that the rather large exothermicity ($\sim 400\,\text{kcal mol}^{-1}$, $1\,\text{kcal} = 4.18\,\text{kJ}$) in the decomposition process must be completely channelled into translation and the bending vibrations of the HCN.

O. Formaldehyde, Ketene, Glyoxal, and Others

The dissociation dynamics of H_2CO have been thoroughly reviewed.[2] A most notable photofragmentation experiment at the time of that review was the direct measurement of the appearance rate of the CO fragment by Houston and Moore.[121] They found that the CO production was much slower than the fluorescence lifetime of H_2CO^*, indicating the presence of some long-lived intermediate state or perhaps rotationally excited CO. Since that time, several theoretical works calculated the energies, transition states, and fragmentation pathways in formaldehyde.[198,203] There is still no conclusion to this problem, where explanations range from a mechanism involving population of the S_0 ground state to rearrangement to an HCOH intermediate. Quantum yields of H atom formation have been measured for single vibronic level excitation.[91] The yield varies smoothly from 0.4 to 0.8 as the excess energy increases in the $\tilde{A}^1 A_2$ state. Using intracavity dye laser absorption probing, the HCO radical was detected upon photolysis of H_2CO with a tunable UV laser.[298] The energy threshold for HCO production was confirmed to be $86 \pm 1\,\text{kcal mol}^{-1}$.

HCO was observed in its ground vibrational state, and also partly with one quantum of excitation in one of its vibrations. The quantum yield of HCO production was measured at several wavelengths. For example, at 294·2 nm it is 0·5.[299]

Atkinson *et al.* measured single vibronic level quantum yields for CO production from glyoxal, $C_2H_2O_2$, using resonance VUV emission excitation in CO at 147 nm.[84,85]

Ketene, CH_2CO, has been the subject of extensive investigations because it is a source of the $CH_2(\tilde{X}^3B_1)$ and $(\tilde{a}^1A_1)$ states. The energy splitting of these well known states has been an object of great controversy.[114–116,300] Laser-induced fluorescence has been used to probe the tunable laser photolysis production of the $CH_2(\tilde{a}^1A_1)$ state from ketene between 290 and 355 nm.[114–116] The two different groups established upper limits of 8–10 kcal mol^{-1} for the singlet–triplet splitting. Yamabe and Morokuma have carried out extensive calculations on the decomposition pathways, excited states, and transition geometries of ketene.[206]

Several other photofragmentation results on large molecules should be mentioned briefly and in rapid succession. A CO probe laser was used to measure CO product distributions in the photolysis of CH_3CHCO and CH_2CHCHO.[122] The vibrational populations are approximated by a Boltzmann distribution with a temperature of 3800 K. An intracavity dye laser absorption probe was used to measure HCO formation in the photolysis of CH_3CHO.[301] Both the relative yield and the rate of appearance are strong functions of the exciting wavelength. Large-scale molecular-orbital (MO) calculations have been carried out on the transition states and decomposition pathways of carbon suboxide, C_3O_2.[204] Photodissociation of thiophosgene, Cl_2CS, at 253·7 nm produces Cl atoms with unity quantum yield.[302] Dissociation of acetylene, C_2H_2, with a 193 nm ArF laser produces $CH(A^2\Delta \to X^2\Pi)$ emission due to a sequential two-photon dissociation.[149] VUV photolysis of CH_4 at 121·6 nm produces emission from $CH_2(\tilde{b}^1B_1 \to \tilde{a}^1A_1)$ and photolysis at 103·2 nm produces emission from $CH(B^2\Sigma \to X^2\Pi)$ and $CH(A^2\Delta \to X^2\Pi)$.[303] Dissociation of CH_3NO_2 and 2-$C_3H_7NO_2$ produces highly vibrationally excited NO_2, which has been probed by dye laser-induced fluorescence.[113] The n–π^* absorption is localized on the NO_2, so the fragmentation process must take place via an indirect predissociation. Photolysis of N_2O_4 at about 300 nm produces electronic emission from one of the NO_2 products.[304] In contrast to earlier proposals that the decomposition of N_2O_4 occurs by two paths, $NO_2 + NO_2$ and $NO + NO_2 + O$, Inoue *et al.* conclude that the path $NO_2 + NO_2$ has a quantum yield of unity.[304]

Given the complexities of the systems listed above, the reader should be provided with some encouragement for the future. Researchers have been able to tackle bigger and more difficult problems each year. If the rapidity with which the photofragmentation field has grown is any indicator, at the time of the next review of this subject, many of these systems will probably have been explored in much greater depth and with even more sophisticated techniques.

P. Van der Waals' Molecules

The first results of the photofragmentation of van der Waals' molecules, I_2He_n, have been reported.[305] The van der Waals' complex is excited on the B–X transition of I_2. In all cases, the first observed dissociation channel (removal of all the He atoms) involves the loss of n quanta of vibrational excitation from the I_2 stretch. Thus the cross section for dissociation of

$$I_2He_2 \rightarrow I_2 + He + He$$

is very small via a one-quantum change, but large for the loss of two vibrational quanta, even though the removal of both He atoms would be energetically allowed for the loss of only one vibrational quantum. The process in viewed as a two-step 'shedding' of He atoms, with the loss of each vibrational quantum used to kick off a He.

Such a beautiful result leaves one with the feeling that in some cases the photofragment dynamics can almost be seen in action. It is evident that the field of photofragmentation, very young only a few years ago, has dramatically matured. The workers who have contributed so many amazing experiments and excellent theories are to be congratulated. It is also apparent that there is still a lot of hard work to be done, since so few systems are really fully understood. Another generation of refinements and advances are in order, and most certainly they are on the way, given the vigour and resourcefulness of the scientists working in the field.

Acknowledgements

The author would like to thank his students, whose talents far exceed the bounds of expectation and imagination. He gratefully acknowledges the support of the US National Bureau of Standards, National Science Foundation, Department of Energy, and Air Force Office of Scientific Research. Acknowledgement is made to the Donors of the Petroleum Research Fund, administered by the American Chemical Society, for partial support of this research.

References

1. J. P. Simons (1977), in *Gas Kinetics and Energy Transfer*, Vol. 2, senior reporters P. G. Ashmore and R. J. Donovan, The Chemical Society, London, p. 58.
2. W. M. Gelbart (1977), *Annu. Rev. Phys. Chem.*, **28**, 323.
3. K. F. Freed and Y. B. Band (1977), in *Excited States*, vol. 3, ed. E. C. Lim, Academic Press, New York, p. 110.

4. H. Okabe (1978), *Photochemistry of Small Molecules*, Wiley, New York.
5. M. N. R. Ashfold, M. T. Macpherson and J. P. Simons (1979), *Top. Curr. Chem.*, **86**, 1.
6. M. N. R. Ashfold, M. T. Macpherson and J. P. Simons (1978), *Chem. Phys. Lett.*, **55**, 84.
7. M. N. R. Ashfold and J. P. Simons (1978), *J. Chem. Soc. Faraday Trans. 2*, **74**, 280.
8. S. Tatematsu and K. Kuchitsu (1977), *Bull. Chem. Soc. Jpn.*, **50**, 2896.
9. I. Stein and A. Gedanken (1978), *J. Chem. Phys.*, **68**, 2982.
10. M. N. R. Ashfold and J. P. Simons (1977), *Chem. Phys. Lett.*, **47**, 65.
11. S. Tatematsu, T. Kondow, T. Nakagawa and K. Kuchitsu (1977), *Bull. Chem. Soc. Jpn.*, **50**, 1056.
12. M. N. R. Ashfold and J. P. Simons (1977), *J. Chem. Soc. Faraday Trans. 2*, **73**, 858.
13. M. N. R. Ashfold and J. P. Simons (1978), *J. Chem. Soc. Faraday Trans. 2*, **74**, 1263.
14. G. Black, R. L. Sharpless and T. G. Slanger (1976), *J. Chem. Phys.*, **64**, 3985.
15. W. K. Bischel, G. Black, R. T. Hawkins, D. J. Kligler and C. K. Rhodes (1979), *J. Chem. Phys.*, **70**, 5589.
16. W. K. Bischel, J. Bokor, J. Dallarosa and C. K. Rhodes (1979), *J. Chem. Phys.*, **70**, 5593.
17. D. J. kligler, H. Pummer, W. K. Bischel and C. K. Rhodes (1978), *J. Chem. Phys.*, **69**, 4652.
18. G. Black and R. L. Sharpless (1979), *J. Chem. Phys.*, **70**, 5567.
19. L. C. Lee, R. W. Carlson, D. L. Judge and M. Ogawa (1975), *J. Phys. B*, **8**, 977.
20. L. C. Lee (1980), *J. Chem. Phys.*, **72**, 4334.
21. G. Black, R. L. Sharpless and T. G. Slanger (1977), *J. Chem. Phys.*, **66**, 2113.
22. J. Maya (1978), *Appl. Phys. Lett.*, **32**, 484.
23. G. Di Stefano, M. Lenzi, A. Margani and C. N. Xuan (1977), *J. Chem. Phys.*, **67**, 3832.
24. V. M. Donnelly, A. P. Baronavski and J. R. McDonald (1979), *Chem. Phys.*, **43**, 271.
25. G. Di Stefano, M. Lenzi, Z. Margani, A. Mele and C. N. Xuan (1977), *J. Photochem.*, **7**, 335.
26. G. Di Stefano, M. Lenzi, A. Margani and C. N. Xuan (1978), *J. Chem. Phys.*, **68**, 959.
27. S. Koda (1980), *Chem. Lett.*, **1**, 57.
28. C. A. F. Johnson and H. J. Ross (1978), *J. Chem. Soc. Faraday Trans. 1*, **74**, 2930.
29. C. A. F. Johnson, V. Freestone and J. Giovanacci (1978), *J. Chem. Soc. Perkin Trans. 2*, **74**, 584.
30. C. A. F. Johnson, V. Freestone and J. Giovanacci (1977), *Ber. Bunsenges.*, **81**, 218.
31. K. R. Wilson (1970), in *Excited State Chemistry*, ed. J. N. Pitts Jr, Gordon and Breach, New York.
32. S. J. Riley and K. R. Wilson (1972), *Faraday Discuss. Chem. Soc.*, **53**, 132.
33. R. Bersohn (1975), *Isr. J. Chem.*, **14**, 111.
34. M. Dzvonik, S. Yang and R. Bersohn (1974), *J. Chem Phys.*, **61**, 4408.
35. R. N. Zare (1972), *Mol. Photochem.*, **4**, 1.
36. S.-C. Yang and R. Bersohn (1974), *J. Chem. Phys.*, **61**, 4400.
37. G. E. Busch, J. F. Cornelius, R. T. Mahoney, R. I. Morse, D. W. Schlosser and K. R. Wilson (1970), *Rev. Sci. Instrum.*, **41**, 1066.
38. G. A. Hancock and K. R. Wilson (1973), in *Fundamental and Applied Laser Physics, Proceedings of the Esfahan Symposium*, eds. M. Feld, A. Javan and N. Kurnit, Wiley, New York.

39. R. J. Oldman, R. K. Sander and K. R. Wilson (1975), *J. Chem. Phys.*, **63**, 4252.
40. R. D. Clear, S. J. Riley and K. R. Wilson (1975), *J. Chem. Phys.*, **63**, 1340.
41. J. H. Ling and K. R. Wilson (1975), *J. Chem. Phys.*, **63**, 101.
42. G. E. Busch and K. R. Wilson (1972), *J. Chem. Phys.*, **56**, 3626.
43. G. E. Busch and K. R. Wilson (1972), *J. Chem. Phys.*, **56**, 3638.
44. G. E. Busch and K. R. Wilson (1972), *J. Chem. Phys.*, **56**, 3655.
45. A. F. Tuck (1977), *J. Chem. Soc. Faraday Trans, 2*, **73**, 689.
46. R. K. Sander and K. R. Wilson (1975), *J. Chem. Phys.*, **63**, 4242.
47. P. M. Kroger, P. C. Demou and S. J. Riley (1976), *J. Chem. Phys.*, **65**, 1823.
48. T. -M. R. Su and S. J. Riley (1979), *J. Chem. Phys.*, **71**, 3194.
49. T. -M. R. Su and S. J. Riley (1980), *J. Chem. Phys.*, **72**, 1614.
50. T. -M. R. Su and S. J. Riley (1980), *J. Chem. Phys.*, **72**, 6632.
51. P. M. Kroger, S. J. Riley and G. H. Kwei (1978), *J. Chem. Phys.*, **68**, 4195.
52. P. M. Kroger and S. J. Riley (1977), *J. Chem. Phys.*, **67**, 4483.
53. P. M. Kroger and S. J. Riley (1979), *J. Chem. Phys.*, **70**, 3863.
54. S. C. Yang, A. Freedman, M. Kawasaki and R. Bersohn (1980), *J. Chem. Phys.*, **72**, 4058.
55. M. Kawasaki, S. J. Lee and R. Bersohn (1979), *J. Chem. Phys.*, **71**, 1235.
56. M. Kawasaki, H. Litvak and R. Bersohn (1977), *J. Chem. Phys.*, **66**, 1434.
57. M. P. Sinha, R. T. Su and S. J. Riley (1979), *J. Chem. Phys.*, **70**, 4431.
58. E. J. Stone, G. M. Lawrence and C. E. Fairchild (1976), *J. Chem. Phys.*, **65**, 5083.
59. C. E. Fairchild, E. J. Stone and G. M. Lawrence (1978), *J. Chem. Phys.*, **69**, 3632.
60. N. J. A. Van Veen, M. S. DeVries and A. E. DeVries (1979), *Chem. Phys. Lett.*, **60**, 184.
61. R. K. Sparks, K. Shobatake, L. R. Carlson and Y. T. Lee (1981), *J. Chem. Phys.*, **75**, 3838.
62. R. K. Sparks, L. R. Carlson, K. Shobatake, M. L. Kowalczyk and Y. T. Lee (1980), *J. Chem. Phys.*, **72**, 1401.
63. I. Arnold, F. J. Comes and G. K. Moortgat (1977), *Chem. Phys.*, **24**, 211.
64. D. L. Philen, R. T. Watson and D. D. Davis (1977), *J. Chem. Phys.*, **67**, 3316.
65. P. W. Fairchild and E. K. C. Lee (1978), *Chem. Phys. Lett.*, **60**, 36.
66. J. C. Brock and R. T. Watson (1980), *Chem. Phys.*, **46**, 477.
67. J. C. Brock and R. T. Watson (1980), *Chem. Phys. Lett.*, **71**, 371.
68. S. T. Amimoto, A. P. Force, J. R. Wiesenfeld and R. H. Young (1980), *J. Chem. Phys.*, **73**, 1244.
69. S. T. Amimoto, A. P. Force and J. R. Wiesenfeld (1978), *Chem. Phys. Lett.*, **60**, 40.
70. D. H. Burde, R. A. McFarlane and J. R. Wiesenfeld (1974), *Phys. Rev.*, **A10**, 1917.
71. T. Donohue and J. R. Wiesenfeld (1975), *Chem. Phys. Lett.*, **33**, 176.
72. T. Donohue and J. R. Wiesenfeld (1975), *J. Chem. Phys.*, **63**, 3130.
73. T. D. Padrick and R. E. Palmer (1976), *J. Chem. Phys.*, **64**, 2051.
74. T. G. Lindeman and J. R. Wiesenfeld (1977), *Chem. Phys. Lett.*, **50**, 364.
75. T. G. Lindeman and J. R. Wiesenfeld (1979), *J. Chem. Phys.*, **70**, 2882.
76. A. H. Laufer (1979), *J. Phys. Chem.*, **83**, 2683.
77. I. S. Fletcher and D. Husain (1977), *Chem. Phys. Lett.*, **49**, 516.
78. W. M. Pitts and A. P. Baronavski (1980), *Chem. Phys. Lett.*, **71**, 395.
79. H. Hofmann and S. R. Leone (1978), *J. Chem. Phys.*, **69**, 3819.
80. S. L. Baughcum and S. R. Leone (1980), *J. Chem. Phys.*, **72**, 6531.
81. S. R. Leone and F. J. Wodarczyk (1974), *J. Chem. Phys.*, **60**, 314.
82. A. B. Petersen and I. W. M. Smith (1978), *Chem. Phys.*, **30**, 407.
83. T. G. Slanger and G. Black (1978), *J. Chem. Phys.*, **68**, 1844.
84. G. H. Atkinson, M. E. McIlwain and C. G. Venkatesh (1978), *J. Chem. Phys.*, **68**, 726.

85. G. H. Atkinson, M. E. McIlwain, C. G. Venkatesh and D. M. Chapman (1978), *J. Photochem.*, **8**, 307.
86. J. Maya (1979), *IEEE J. Quantum Electron.*, **QE-15**, 579.
87. T. G. Slanger, R. L. Sharpless and G. Black (1977), *J. Chem. Phys.*, **67**, 5317.
88. C. J. Sam and J. T. Yardley (1978), *J. Chem. Phys.*, **69**, 4621.
89. L. C. Lee (1980), *J. Chem. Phys.*, **72**, 6414.
90. L. C. Lee, L. Oren, E. Phillips and D. L. Judge (1978), *J. Phys. B*, **11**, 47.
91. K. Y. Tang, P. W. Fairchild and E. K. C. Lee (1979), *J. Phys. Chem.*, **83**, 569.
92. J. S. Chang, J. R. Barker, J. E. Davenport and D. M. Golden (1979), *Chem. Phys. Lett.*, **60**, 385.
93. T. F. Hunter and K. S. Kristjansson (1978), *Chem. Phys. Lett.*, **58**, 291.
94. W. B. Lewis and A. H. Zeltmann (1980), *J. Photochem.*, **12**, 51.
95. M. P. Roellig and P. L. Houston (1978), *Chem. Phys. Lett.*, **57**, 75.
96. R. E. Rebbert and P. Ausloos (1975), *J. Photochem.*, **4**, 419.
97. G. Nathanson, B. Gitlin, A. Rosan and J. T. Yardley (1981), *J. Chem. Phys.*, **74**, 361.
98. J. T. Yardley, B. Gitlin, G. Nathanson and A. Rosan (1981), *J. Chem. Phys.*, **74**, 370.
99. J. W. Simons and R. Curry (1976), *Chem. Phys. Lett.*, **38**, 171.
100. T. R. Loree, J. H. Clark, K. B. Butterfield, J. L. Lyman and R. Engleman Jr (1979), *J. Photochem.*, **10**, 359.
101. T. F. Deutsch, D. J. Ehrlich and R. M. Osgood Jr (1979), *Appl. Phys. Lett.*, **35**, 175.
102. R. J. Cody, M. J. Sabety-Dzvonik and W. M. Jackson (1977), *J. Chem. Phys.*, **66**, 2145.
103. M. J. Sabety-Dzvonik, R. J. Cody and W. M. Jackson (1976), *Chem. Phys. Lett.*, **44**, 131.
104. R. J. Cody and M. J. Sabety-Dzvonik (1976), *J. Chem. Phys.*, **64**, 4794.
105. R. J. Cody and M. J. Sabety-Dzvonik (1977), *J. Chem. Phys.*, **66**, 125.
106. A. P. Baronavski and J. R. McDonald (1977), *Chem. Phys. Lett.*, **45**, 172.
107. J. R. McDonald, R. G. Miller and A. P. Baronavski (1977), *Chem. Phys. Lett.*, **51**, 57.
108. A. P. Baronavski, R. G. Miller and J. R. McDonald (1978), *Chem. Phys.*, **30**, 119.
109. W. S. Drozdoski, A. P. Baronavski and J. R. McDonald (1979), *Chem. Phys. Lett.*, **64**, 421.
110. M. J. Molina, T. Ishiwata and L. T. Molina (1980), *J. Phys. Chem.*, **84**, 821.
111. J. E. Butler, L. P. Goss, M. C. Lin and J. W. Hudgens (1979), *Chem. Phys. Lett.*, **63**, 104.
112. H. Zacharias, R. Schmiedl, R. Böttner, M. Geilhaupt, U. Meier and K. H. Welge (1979), *Springer Ser. Opt. Sci.*, **21**, 329.
113. K. G. Spears and S. P. Brugge (1978), *Chem. Phys. Lett.*, **54**, 373.
114. R. K. Lengel and R. N. Zare (1978), *J. Am. Chem. Soc.*, **100**, 7495.
115. J. Danon, S. V. Filseth, D. Feldmann, H. Zacharias, C. H. Dugan and K. H. Welge (1978), *Chem. Phys.*, **29**, 345.
116. D. Feldmann, K. Meier, H. Zacharias and K. H. Welge (1978), *Chem. Phys. Lett.*, **59**, 171.
117. N. Sanders, J. E. Butler, L. R. Pasternak and J. R. McDonald (1980), *Chem. Phys.*, **48**, 203.
118. M. Asscher, Y. Haas, M. P. Roellig and P. L. Houston (1980), *J. Chem. Phys.*, **72**, 768.
119. M. P. Roellig, P. L. Houston, M. Asscher and Y. Haas (1981), *J. Chem. Phys.*, **73**, 5081.

120. M. S. Jhon and J. S. Dahler (1978), *J. Chem. Phys.*, **69**, 819.
121. P. L. Houston and C. B. Moore (1976), *J. Chem. Phys.*, **65**, 757.
122. M. E. Umstead, R. G. Shortridge and M. C. Lin (1978), *J. Phys. Chem.*, **82**, 1455.
123. L. N. Krasnoperov, V. R. Braun and V. N. Panfilov (1978), *Kinet. Katal.*, **19**, 1610.
124. L. N. Krasnoperov and V. N. Panfilov (1979), *Kinet. Katal.*, **20**, 540.
125. M. T. Macpherson, J. P. Simons and R. N. Zare (1979), *Mol. Phys.*, **38**, 2049.
126. G. A. Chamberlain and J. P. Simons (1975), *J. Chem. Soc. Faraday Trans. 2*, **71**, 2043.
127. M. T. Macpherson and J. P. Simons (1978), *J. Chem. Soc. Faraday Trans. 2*, **74**, 1965.
128. M. T. Macpherson and J. P. Simons (1979), *J. Chem. Soc. Faraday Trans. 2*, **75**, 1572.
129. M. T. Macpherson and J. P. Simons (1977), *Chem. Phys. Lett.*, **51**, 261.
130. E. D. Poliakoff, S. H. Southworth, D. A. Shirley, K. H. Jackson and R. N. Zare (1979), *Chem. Phys. Lett.*, **65**, 407.
131. J. Husain and J. R. Wiesenfeld and R. N. Zare (1980), *J. Chem. Phys.*, **72**, 2479.
132. D. L. Feldman and R. N. Zare (1976), *Chem. Phys.*, **15**, 415.
133. E. W. Rothe, U. Krause and R. Düren (1980), *Chem. Phys. Lett.*, **72**, 100.
134. R. J. Van Brunt and R. N. Zare (1968), *J. Chem. Phys.*, **48**, 4304.
135. P. F. Zittel and D. D. Little (1979), *J. Chem. Phys.*, **71**, 713.
136 S. M. Papernov, G. V. Shlyapnikov and M. L. Yanson (1978), *Sov. Phys.–Dokl.*, **23**, 58.
137. P. F. Zittle and D. D. Little (1980), *J. Chem. Phys.*, **72**, 5900.
138. O. Kajimoto and R. J. Cvetanović (1976), *Chem. Phys. Lett.*, **37**, 533.
139. C. L. Creel and J. Ross (1976), *J. Chem. Phys.*, **64**, 3560.
140. S. L. Baughcum and S. R. Leone (1978), *Proc. Soc. Photo-Opt. Instrum. Eng.*, **158** (*Laser Spectrosc.–Appl. Tech.*) 29.
141. H. W. Hermann and S. R. Leone, to be published.
142. S. L. Baughcum and S. R. Leone, to be published .
143. A. B. Callear and H. E. Van den Bergh (1970), *Chem. Phys. Lett.*, **5**, 23.
144. K. Chen and E. S. Yeung (1980), *J. Chem. Phys.*, **72**, 4723.
145. W. M. Jackson, J. B. Halpern and C.-S. Lin (1978), *Chem. Phys. Lett.*, **55**, 254.
146. W. M. Jackson and J. B. Halpern (1979), *J. Chem. Phys.*, **70**, 2373.
147. C. Tai and F. W. Dalby (1978), *Can. J. Phys.*, **56**, 183.
148. T. A. Cool, J. A. McGarvey, Jr and A. C. Erlandson (1978), *Chem. Phys. Lett.*, **58**, 108.
149. J. R. McDonald, A. P. Baronavski and V. M. Donnelly (1978), *Chem. Phys.*, **33**, 161.
150. F. B. Wampler, J. J. Tiee, W. W. Rice and R. C. Oldenberg (1979), *J. Chem. Phys.*, **71**, 3926.
151. C. L. Sam and J. T. Yardley (1979), *Chem. Phys. Lett.*, **61**, 509.
152. H. Komine and R. L. Byer (1977), *J. Appl. Phys.*, **48**, 2505.
153. T. F. Deutsch, D. J. Ehrlich and R. M. Osgood, Jr (1979), *Opt. Lett.*, **4**, 378.
154. R. N. Zare and D. R. Herschbach (1965), *Appl. Opt. Suppl.*, **2**, 193.
155. J. V. V. Kasper and G. C. Pimentel (1964), *Appl. Phys. Lett.*, **5**, 231.
156. L. D. Mikheev (1978), *Sov. J. Quantum Electron.*, **8**, 677.
157. S. J. Davis (1978), *Appl. Phys. Lett.*, **32**, 656.
158. J. C. White (1978), *Appl. Phys. Lett.*, **33**, 325.
159. R. Burnham (1977), *Appl. Phys. Lett.*, **30**, 132.
160. D. J. Ehrlich, J. Maya and R. M. Osgood, Jr (1978), *Appl. Phys. Lett.*, **33**, 931.

161. H. T. Powell and J. J. Ewing (1978), *Appl. Phys. Lett.*, **33**, 165.
162. H. T. Powell, D. Prosnitz and B. R. Schleicher (1979), *Appl. Phys. Lett.*, **34**, 571.
163. D. J. Ehrlich and R. M. Osgood, Jr (1979), *Appl. Phys. Lett.*, **34**, 655.
164. H. Hemmati and G. J. Collins (1979), *Appl. Phys. Lett.*, **34**, 844.
165. M. J. Berry (1974), *J. Chem. Phys.*, **61**, 3114.
166. G. A. West and M. J. Berry (1978), *Chem. Phys. Lett.*, **56**, 423.
167. J. T. Knudtson and M. J. Berry (1978), *J. Chem. Phys.*, **68**, 4419.
168. E. B. Gordon, A. K. Kurnosov and S. A. Sotnichenko (1978), *Kvantovaya Elektron. (Moscow)*, **8**, 653.
169. E. J. Schimitschek, J. E. Celto and J. A. Trias (1977), *Appl. Phys. Lett.*, **31**, 608.
170. N. G. Basov, V. S. Zuev, L. D. Mikheev, D. B. Stavrovskii and V. I. Yalovoi (1977), *Sov. J. Quantum Electron.*, **4**, 1401.
171. N. G. Basov, V. S. Zuev, A. V. Kanaev, L. D. Mikheev and D. B. Stavrovskii (1979), *Sov. J. Quantum Electron.*, **6**, 629.
172. Y. B. Band, M. D. Morse and K. F. Freed (1979), *Chem. Phys. Lett.*, **67**, 294.
173. B. Ritchie (1976), *Phys. Rev.*, **A14**, 1396.
174. K. E. Holdy, L. C. Klotz and K. R. Wilson (1970), *J. Chem. Phys.*, **52**, 4588.
175. J. P. Simons and P. W. Tasker (1973), *Mol. Phys.*, **26**, 1267.
176. M. J. Berry (1974), *Chem. Phys. Lett.*, **27**, 73.
177. M. J. Berry (1974), *Chem. Phys. Lett.*, **29**, 329.
178. Y. B. Band and K. F. Freed (1975), *J. Chem. Phys.*, **63**, 3382.
179. Y. B. Band and K. F. Freed (1975), *J. Chem. Phys.*, **63**, 4479.
180. Y. B. Band and K. F. Freed (1976), *J. Chem. Phys.*, **64**, 4329.
181. M. D. Morse, K. F. Freed and Y. B. Band (1977), *Chem. Phys. Lett.*, **49**, 399.
182. Y. B. Band and K. F. Freed (1977), *J. Chem. Phys.*, **67**, 1462.
183. M. D. Morse, K. F. Freed and Y. B. Band (1976), *Chem. Phys. Lett.*, **44**, 125.
184. K. F. Freed, M. D. Morse and Y. B. Band (1979), *Faraday Discuss. Chem. Soc.*, **67**, 297.
185. M. D. Morse, K. F. Freed and Y. B. Band (1979), *J. Chem. Phys.*, **70**, 3604.
186. M. D. Morse, K. F. Freed and Y. B. Band (1979), *J. Chem. Phys.*, **70**, 3620.
187. S. Mukamel and J. Jortner (1974), *J. Chem. Phys.*, **60**, 4760.
188. S. Mukamel and J. Jortner (1976), *J. Chem. Phys.*, **65**, 3735.
189. O. Atabek, J. A. Beswick R. Lefebvre, S. Mukamel and J. Jortner (1976), *J. Chem. Phys.*, **65**, 4035.
190. O. Atabek, J. A. Beswick, R. Lefebvre, S. Mukamel and J. Jortner (1977), *Chem. Phys. Lett.*, **45**, 211.
191. O. Atabek and R. Lefebvre (1977), *J. Chem. Phys.*, **67**, 4983.
192. M. Shapiro (1972), *J. Chem. Phys.*, **56**, 2582.
193. M. Shapiro (1973), *Isr. J. Chem.*, **11**, 691.
194. M. Shapiro (1977), *Chem. Phys. Lett.*, **46**, 442.
195. J. A. Beswick, M. Shapiro and R. Sharon (1977), *J. Chem. Phys.*, **67**, 4045.
196. M. Shapiro and R. Bersohn (1980), *J. Chem. Phys.*, **73**, 3810.
197. M. Tamir, U. Halavee and R. D. Levine (1974), *Chem. Phys. Lett.*, **25**, 38.
198. D. F. Heller, M. L. Elert and W. M. Gelbart (1978), *J. Chem. Phys.*, **69**, 4061.
199. R. L. Jaffe and S. R. Langhoff (1978), *J. Chem. Phys.*, **68**, 1638.
200. S. R. Langhoff and R. L. Jaffe (1979), *J. Chem. Phys.*, **71**, 1475.
201. W. R. Wadt (1980), *J, Chem. Phys.*, **72**, 2469.
202. R. Runau, S. D. Peyerimhoff and R. J. Buenker (1977), *J. Mol. Spectrosc.*, **68**, 253.
203. J. D. Goddard and H. F. Schaefer III (1979), *J. Chem. Phys.*, **70**, 5117.
204. Y. Osamura, K. Nishimoto, S. Yamabe and T. Minato (1979), *Theor. Chim. Acta*, **52**, 257.

205. S. Yamabe, T. Minato and Y. Oamura (1979), *J. Am. Chem. Soc.*, **101**, 4525.
206. S. Yamabe and K. Morokuma (1978), *J. Am. Chem. Soc.*, **100**, 7551.
207. J. A. Beswick and J. Jortner (1977), *Chem. Phys.*, **24**, 1.
208. E. J. Heller (1978), *J. Chem. Phys.*, **68**, 2066.
209. E. J. Heller (1978), *J. Chem. Phys.*, **68**, 3891.
210. R. T. Pack (1976), *J. Chem. Phys.*, **65**, 4765.
211. J. E. Mentall and P. M. Guyon (1977), *J. Chem. Phys.*, **67**, 3845.
212. I. V. Komarov and V. N. Ostrovsky (1978), *JETP Lett.*, **28**, 413.
213. I. V. Komarov and V. N. Ostrovsky (1979), *J. Phys. B*, **12**, 2485.
214. M. S. Child and R. B. Bernstein (1973), *J. Chem. Phys.*, **59**, 5916.
215. R. J. LeRoy, Calculations contained in refs. 72 and 79.
216. C. P. Hemenway, T. G. Lindemann and J. R. Wiesenfeld (1979), *J. Chem. Phys.*, **70**, 3560.
217. M. S. DeVries, N. J. A. Van Veen and A. E. DeVries (1978), *Chem. Phys. Lett.*, **56**, 15.
218. M. S. Child (1976), *Mol. Phys.*, **32**, 1495.
219. S. L. Baughcum, H. Hofmann, S. R. Leone and D. J. Nesbitt (1979), *Faraday Discuss. Chem. Soc.*, **67**, 306.
220. M. S. Child, private communication.
221. M. A. A. Clyne and M. C. Heaven (1978), *J. Chem. Soc. Faraday Trans. 2*, **74**, 1992.
222. M. A. A. Clyne, and M. C. Heaven and E. Martinez (1980), *J. Chem. Soc. Faraday Trans. 2*, **76**, 177.
223. M. A. A. Clyne and M. C. Heaven (1980), *J. Chem. Soc. Faraday Trans. 2*, **76**, 49.
224. M. A. A. Clyne and I. S. McDermid (1978), *J. Chem. Soc. Faraday Trans. 2*, **74**, 1935.
225. M. A. A. Clyne and I. S. McDermid (1980), *J. Chem. Soc. Faraday Trans. 2*, **76**, 1677.
226. L. C. Lee and T. G. Slanger (1978), *J. Chem. Phys.*, **69**, 4053.
227. L. C. Lee, T. G. Slanger, G. Black and R. L. Sharpless (1977), *J. Chem. Phys.*, **67**, 5602.
228. E. W. Rothe, U. Krause and R. Düren (1980), *J. Chem. Phys.*, **72**, 5145.
229. E. K. Kraulinya and M. L. Yanson (1979), *Opt. Spectrosc.*, **46**, 629.
230. P. S. Julienne, D. D. Konowalow, M. Krauss, M. E. Rosenkrantz and W. J. Stevens (1980), *Appl. Phys. Lett.*, **36**, 132.
231. J. C. White and G. A. Zdasiuk (1978), *J. Chem. Phys.*, **69**, 2256.
232. B. L. Earl, R. R. Herm, S. -M. Lin and C. A. Mims (1972), *J. Chem. Phys.*, **56**, 867.
233. B. L. Earl and R. R. Herm (1974), *J. Chem. Phys.*, **60**, 4568.
234. R. Bersohn (1979), *Alkali Halide Vapors*, Academic Press, New York, p. 345.
235. R. N. Zare and D. R. Herschbach (1965), *J. Mol. Spectrosc.*, **15**, 462.
236. W. R. Anderson, B. M. Wilson and T. L. Rose (1977), *Chem. Phys. Lett.*, **48**, 284.
237. R. J. Donovan and J. Konstantatos (1972), *J. Photochem.*, **1**, 75.
238. S. T. Amimoto, J. R. Wiesenfeld and R. H. Young (1979), *Chem. Phys. Lett.*, **65**, 402.
239. U. Halavee and M. Shapiro (1977), *Chem. Phys.*, **21**, 105.
240. O. Atabek and R. Lefebvre (1977), *Chem. Phys. Lett.*, **52**, 29.
241. W. M. Jackson, G. Miller and J. B. Halpern (1978), *J. Photochem.*, **9**, 137.
242. G. E. Miller, W. M. Jackson and J. B. Halpern (1979), *J. Chem. Phys.*, **71**, 4625.
243. H. Okabe (1980), *J. Chem. Phys.*, **72**, 6642.
244. T. Carrington (1964), *J. Chem. Phys.*, **41**, 2012.
245. K. H. Becker, W. Groth and D. Kley (1965), *Z. Naturforsch.*, **A20**, 748.

246. A. Mele and H. Okabe (1969), *J. Chem. Phys.*, **51**, 4798.
247. I. Yamashita (1975), *J. Phys. Soc. Jpn.*, **39**, 205.
248. I. P. Vinogradov and F. I. Vilesov (1978), *Opt. Spectrosc.*, **44**, 653.
249. R. Akamatsu and K. O-ohata (1977), *J. Phys. Soc. Jpn.*, **43**, 264.
250. R. Akamatsu and K. O-ohata (1978), *J. Phys. Soc. Jpn.*, **44**, 589.
251. W. G. Hawkins and P. L. Houston (1980), *J. Chem. Phys.*, **73**, 297.
252. C.-L. Lin and W. B. DeMore (1973–74), *J. Photochem.*, **2**, 161.
253. G. K. Moortgat and P. Warneck (1975), *Z. Naturforsch.*, **30a**, 835.
254. O. Kajimoto and R. J. Cvetanović (1979), *Int. J. Chem. Kinet.*, **11**, 605.
255. T. G. Slanger, R. L. Sharpless, G. Black and S. V. Filseth (1974), *J. Chem. Phys.*, **61**, 5022.
256. E. Phillips, L. C. Lee and D. L. Judge (1977), *J. Chem. Phys.*, **66**, 3688.
257. C. Y. R. Wu, E. Phillips, L. C. Lee and D. L. Judge (1978), *J. Geophys. Res.*, **83**, 4869.
258. C. Y. R. Wu and D. L. Judge (1979), *Chem. Phys. Lett.*, **68**, 495.
259. H. Okabe (1972), *J. Chem. Phys.*, **56**, 4381.
260. L. C. Lee and D. L. Judge (1975), *J. Chem. Phys.*, **63**, 2782.
261. A. Freedman, S. -C. Yang and R. Bersohn (1979), *J. Chem. Phys.*, **70**, 5313.
262. R. Gilpin and K. H. Welge (1971), *J. Chem. Phys.*, **55**, 975.
263. A. B. Harker, W. Ho and J. J. Ratto (1977), *Chem. Phys. Lett.*, **50**, 394.
264. A. J. Grimley and P. L. Houston (1980), *J. Chem. Phys.*, **72**, 1471.
265. J. G. Eden and R. W. Waynant (1979), *Appl. Phys. Lett.*, **34**, 324.
266. J. G. Eden and R. W. Waynant (1978), *Opt. Lett.*, **2**, 13.
267. J. Maya (1977), *J. Chem. Phys.*, **67**, 4976.
268. M. Kawasaki and R. Bersohn (1978), *J. Chem. Phys.*, **68**, 2105.
269. T. L. Andreeva, G. N. Birich, V. N. Sorokin and I. I. Struk (1976), *Sov. J. Quantum Electron.*, **6**, 781.
270. S. De Silvestri, O. Svelto and F. Zaraga (1980), *Appl. Phys.*, **21**, 1.
271. A. Andreoni, R. Cubeddu, S. De Silvestri and F. Zaraga (1980), *Chem. Phys. Lett.*, **69**, 161.
272. O. De Witte, R. Dumanchin, M. Michon and J. Chatelet (1977), *Chem. Phys. Lett.*, **48**, 505.
273. W. B. Lewis, F. B. Wampler, E. J. Huber and G.C. Fitzgibbon (1979), *J. Photochem.*, **11**, 393.
274. D. W. Trainor and S. A. Mani (1978), *Appl. Phys. Lett.*, **33**, 31.
275. A. Freedman and R. Bersohn (1978), *J. Am. Chem. Soc.*, **100**, 4116.
276. M. R. Levy and J. P. Simons (1975), *J. Chem. Soc. Faraday Trans. 2*, **71**, 561.
277. L. S. Ershov, V. Yu. Zalesskii and V. N. Sokolov (1978), *Sov. J. Quantum Electron.*, **8**, 494.
278. A. M. Pravilov, A. S. Kozlov and F. I. Vilesov (1978), *Sov. J. Quantum Electron.*, **8**, 666.
279. A. M. Pravilov, F. I. Vilesov, V. A. Elokhin, V. S. Ivanov and A. S. Kozlov (1978), *Sov. J. Quantum Electron.*, **8**, 355.
280. H. Pummer, J. Eggelston, W. K. Bischel and C. K. Rhodes (1978), *Appl. Phys. Lett.*, **32**, 427.
281. L. G. Karpov, A. M. Pravilov and F. I. Vilesov (1977), *Sov. J. Quantum Electron.*, **7**, 457.
282. W. L. Ebenstein, J. R. Wiesenfeld and G. L. Wolk (1978), *Chem. Phys. Lett.*, **53**, 185.
283. W. H. Pence, S. L. Baughcum and S. R. Leone, to be published.
284. J. B. Koffend and S. R. Leone (1981), *Chem. Phys. Lett.*, **81**, 136.

285. M. Kawasaki, S. J. Lee and R. Bersohn (1975), *J. Chem. Phys.*, **63**, 809.
286. M. Kawasaki, S. J. Lee and R. Bersohn (1977), *J. Chem. Phys.*, **66**, 2647.
287. A. Freedman, S. C. Yang, M. Kawasaki and R. Bersohn (1980), *J. Chem. Phys.*, **72**, 1028.
288. G. Marconi (1979), *J. Photochem.*, **11**, 385.
289. M. N. S. Rayo, C. Fotakis and R. J. Donovan (1978), *J. Photochem.*, **9**, 433.
290. D. M. Shold and P. J. Ausloos (1979), *J. Photochem.*, **10**, 237.
291. L. J. Colcord and M. C. Lin (1978), *J. Photochem.*, **8**, 337.
292. C. Hubrich and F. Stuhl (1980), *J. Photochem.*, **12**, 93.
293. H. Okabe and M. Lenzi (1967), *J. Chem. Phys.*, **47**, 5241.
294. R. A. Back and S. Koda (1977), *Can. J. Chem.*, **55**, 1387.
295. I. P. Vinogradov and F. I. Vilesov (1977), *Zh. Fiz. Khim.*, **51**, 2017.
296. J. H. Glownia and S. J. Riley (1980), *Chem. Phys. Lett.*, **71**, 429.
297. D. Coulter, D. Dows, H. Reisler and C. Wittig (1978), *Chem. Phys.*, **32**, 429.
298. J. P. Reilly, J. H. Clark, C. B. Moore and G. C. Pimentel (1978), *J. Chem. Phys.*, **69**, 4381.
299. J. H. Clark, C. B. Moore and N. S. Nogar (1978), *J. Chem. Phys.*, **68**, 1264.
300. W. L. Hase and P. M. Kelley (1977), *J. Chem. Phys.*, **66**, 5093.
301. R. J. Gill and G. H. Atkinson (1979), *Chem. Phys. Lett.*, **64**, 426.
302. H. Okabe (1977), *J. Chem. Phys.*, **66**, 2058.
303. I. P. Vinogradov and F. I. Vilesov (1977), *Khim. Vys. Energ.*, **11**, 21.
304. G. Inoue, Y. Nakata, Y. Usui, H. Akimoto and M. Okuda (1979), *J. Chem. Phys.*, **70**, 3689.
305. W. Sharfin, K. E. Johnson, L. Wharton and D. H. Levy (1979), *J. Chem. Phys.*, **71**, 1292.

Dynamics of the Excited State
Edited by K. P. Lawley
© 1982 John Wiley & Sons Ltd.

COLLISIONAL QUENCHING OF ELECTRONICALLY EXCITED METAL ATOMS

W. H. BRECKENRIDGE AND H. UMEMOTO

*Department of Chemistry, University of Utah,
Salt Lake City, UT 84112, USA*

CONTENTS

I. INTRODUCTION

The collisional quenching of electronically excited metal atoms has long been of interest to chemical physicists and to chemists. The discovery of the first photosensitized process, the scission of the H_2 bond when a mixture of hydrogen gas and mercury vapour was irradiated with light from a mercury resonance lamp,[1] was followed by a series of basic studies in the 1920s and 1930s of the quenching of resonance emissions from metal vapours when molecular gases were added.[2,3] After a period of relative neglect, the research area became active again in the late 1950s and the 1960s. In the last decade, interest in excited metal atom quenching has increased greatly, for two main reasons. On the practical side, many current and potential laser systems involve electronically excited states of metal atoms, and it is important to know the rates and mechanisms by which these excited states are collisionally deactivated. Secondly, major breakthroughs in experimental and theoretical techniques have created the possibility that excited metal atom quenching studies can provide fundamental information for our understanding of dynamical collision processes in general. For example, molecular beam and laser techniques are providing increasingly detailed 'state-to-state' information about collisional quenching of atomic excited states,[4-12] and new theoretical methods are being developed which are capable of treating collisional processes at the required level of detail.[13-17]

This review is an attempt to assess our current knowledge of the quenching of excited states of metal atoms by simple molecules and the inert gases. As is always true in this sort of endeavour, it has been necessary to limit our scope, so that the review is of manageable size and has some depth of treatment of those experimental and theoretical developments which, in our opinion, offer the most insight into the dynamics of collisional quenching processes. All experimental and theoretical literature from 1960 to late 1980 on the quenching of the low-lying excited states of group IA, IIA, and IIB elements (i.e. Li, Na, K, Rb, Cs, Mg, Ca, Sr, Ba, Zn, Cd, and Hg) has been examined, although we make no claim that our search was exhaustive.[18] Work earlier than 1960 is mentioned only if there is no later experimental data on that particular subject or if, in our opinion, the results are sufficiently reliable for comparison with modern studies.

The emphasis of this review is on experimental results and their interpretation rather than on experimental techniques or detailed theoretical models. Therefore, experimental techniques are not discussed in any depth, and theoretical models are only discussed qualitatively.

Two main areas have been neglected in the review. First of all, we have chosen to concentrate on *net* quenching of a metal atom excited state. Intramultiplet quenching (i.e. the collisional quenching of one *J*-multiplet of an electronic state to another *J*-multiplet of the same electronic state) is only discussed when it is necessary to ascertain whether it is an exit channel competitive with net quenching. We believe, in fact, that enough information on collisional intramultiplet quenching has become available recently that a separate review of this subject will be warranted in the near future.

Secondly, we have chosen not to review the substantial amount of accumulating information on electronic energy transfer, excimer formation, and net quenching of excited metal atoms by ground-state metal atoms. There is, of course, a great deal of activity in this area because of projected high-power excimer laser systems involving group I and II metal atoms, but a separate review would be more appropriate in two or three years when a more reliable database will be available.

There are several valuable earlier reviews which are pertinent to the material discussed here. The best source of information on experiments before 1953 is the book by Steacie.[3] An early review of Cvetanovic[19] is an excellent summary of work on mercury-photosensitized reactions through to 1962. A general discussion of excited atom quenching results (mostly Hg* states) may also be found in the book by Calvert and Pitts.[20] Two valuable reviews by Callear appeared in 1965 and 1969 in which excited atom quenching and energy transfer processes were discussed.[21,22] Phillips has reviewed the experimental results on Hg-photosensitized emission processes through to 1973.[23] Cadmium-photosensitization work prior to 1977 has been reviewed by Tsunashima and Sato.[24] Several reviews now exist which treat various aspects of the quenching of electronically excited alkali-metal atoms.[13,14,25,26] A review of the reverse of electronic quenching, the excitation of metal atom excited states by collision between ground-state metal atoms and molecules or atoms, has been published by Kempter.[27] Good general treatments of excited atom quenching[28] and electronic-to-vibrational energy transfer[29] may be found in two recent articles in *Annual Reviews of Physical Chemistry*.

This review is organized into six sections. Sec. II gives pertinent data about the particular excited states treated. In Sec. III, the efficiency of collisional deactivation of the excited metal atom states of interest is discussed. Compilations of absolute quenching cross sections are given in several tables, with the idea of comparing values for one quencher versus another, and for different excited states with the same quenching species. Sec. IV is devoted to the information available on the identity and relative importance of the major exit channels for dissipation of the electronic energy (i.e. electronic-to-electronic, electronic-to-vibrational, electronic-to-rotational transfer, chemical reaction, sensitized bond scission, etc.). Sec. V has for its subject matter the limited number of experiments in which state-resolved measurements of

TABLE I. Properties of the electronically excited metal atom states of interest.

Atom	State	Energy (kcal mol^{-1})	Difference in energy from state immediately above (kcal mol^{-1})	Ionization potential (kcal mol^{-1})	Radiative lifetime	References (lifetimes)	Vacuum wavelength of radiative transition to ground state (Å)
Mg	$3s3p\ {}^1P_1$	100·2	—	76·1	$(2·01 \pm 0·15) \times 10^{-9}$	31–37	2853·4
	${}^3P_2$	62·6	37·6	113·7	metastable	—	—
	${}^3P_1$	62·5	0·12	113·8	$(4·5 \pm 0·5) \times 10^{-3}$	38	4572·4
	${}^3P_0$	62·5	0·06	113·8	metastable	—	—
Zn	$4s4p\ {}^1P_1$	133·6	—	83·1	$(1·40 \pm 0·10) \times 10^{-9}$	39–45	2139·3
	${}^3P_2$	94·0	39·6	122·7	metastable	—	—
	${}^3P_1$	92·9	1·11	123·8	$(2·0 \pm 0·4) \times 10^{-5}$	46	3076·8
	${}^3P_0$	92·4	0·54	124·3	metastable	—	—
Cd	$5s5p\ {}^1P_1$	125·0	—	82·3	$(1·7 \pm 0·2) \times 10^{-9}$	47	2288·7
	${}^3P_2$	91·0	34·0	116·3	metastable	—	—
	${}^3P_1$	87·6	3·35	119·7	$(2·4 \pm 0·1) \times 10^{-6}$	48–52	3262·0
	${}^3P_0$	86·1	1·55	121·2	metastable	—	—
Hg	$6s6p\ {}^1P_1$	154·6	—	86·1	$(1·37 \pm 0·15) \times 10^{-9}$	42, 53–58	1849·5
	${}^3P_2$	126·0	28·6	114·7	metastable	—	—
	${}^3P_1$	112·7	13·2	128·0	$(1·19 \pm 0·03) \times 10^{-7}$	59–62	2537·3
	${}^3P_0$	107·7	5·0	133·0	metastable	—	—

	State					References	Wavelength
Ca	$4s4p\ ^1P_1$	67·6	—	73·4	$(5·1 \pm 0·7) \times 10^{-9}$	32–36, 39, 40, 63–70	4227·9
	$4s3d\ ^1D_2$	62·5	5·1	78·5	$2·3 \times 10^{-3}$	203	—
	$4s3d\ ^3D_3$	58·2	4·3	82·8	$\sim 4 \times 10^{-7}$	106	—
	$^3D_2$	58·2	0·06	82·8	$\sim 4 \times 10^{-7}$	106	—
	$^3D_1$	58·1	0·04	82·9	$\sim 4 \times 10^{-7}$	106	—
	$4s4p\ ^3P_2$	43·8	14·3	97·2	metastable	—	—
	$^3P_1$	43·5	0·30	97·5	$(3·6 \pm 0·4) \times 10^{-4}$	71, 72	6574·6
	$^3P_0$	43·4	0·15	97·6	metastable	—	—
Sr	$5s5p\ ^1P_1$	62·0	—	69·4	$(5·4 \pm 0·7) \times 10^{-9}$	39, 40, 63, 65, 66, 68, 70, 73–75	4608·6
	$5s4d\ ^1D_2$	57·6	4·4	73·8	metastable	—	—
	$5s4d\ ^3D_3$	52·4	5·2	79·0	?	—	—
	$^3D_2$	52·1	0·29	79·3	?	—	—
	$^3D_1$	51·9	0·17	79·5	?	—	—
	$5s5p\ ^3P_2$	42·6	9·3	88·8	metastable	—	—
	$^3P_1$	41·5	1·13	89·9	$(2·1 \pm 0·3) \times 10^{-5}$	76, 77	6894·5
	$^3P_0$	41·0	0·53	90·4	metastable	—	—
Ba	$6s6p\ ^1P_1$	51·6	—	68·6	$(8·5 \pm 1·0) \times 10^{-9}$	31, 65, 68, 75, 78–80	5537·0
	$6s6p\ ^3P_2$	38·6	13·0	81·6	?	—	—
	$^3P_1$	36·1	2·51	84·1	$(2 \pm 1) \times 10^{-6}$	66, 81	7913·5
	$^3P_0$	35·1	1·06	85·1	?	—	—
	$6s5d\ ^1D_2$	32·6	2·5	87·6	metastable	—	—

TABLE I. (Contd.)

Atom	State	Energy (kcal mol⁻¹)	Difference in energy from state immediately above (kcal mol⁻¹)	Ionization potential (kcal mol⁻¹)	Radiative lifetime	References (lifetimes)	Vacuum wavelength of radiative transition to ground state (Å)
	$6s5d\ ^3D_3$	27·4	5·2	92·8	metastable	—	—
	$^3D_2$	26·3	1·09	93·9	metastable	—	—
	$^3D_1$	25·8	0·52	94·4	metastable	—	—
Li	$2p\ ^2P_{3/2}$	42·6	—	81·7	$(2\cdot7 \pm 0\cdot4) \times 10^{-8}$	82–87	6709·6
	$^2P_{1/2}$	42·6	0·001	81·7	$(2\cdot7 \pm 0\cdot4) \times 10^{-8}$	82–87	6709–8
Na	$3p\ ^2P_{3/2}$	48·5	—	70·0	$(1\cdot62 \pm 0\cdot10) \times 10^{-8}$	65, 66, 88–97	5891·6
	$^2P_{1/2}$	48·5	0·05	70·0	$(1\cdot62 \pm 0\cdot10) \times 10^{-8}$	65, 66, 88–97	5897·6
K	$4p\ ^2P_{3/2}$	37·3	—	62·8	$(2\cdot7 \pm 0\cdot15) \times 10^{-8}$	66, 92, 98–101	7667·0
	$^2P_{1/2}$	37·1	0·16	63·0	$(2\cdot78 \pm 0\cdot15) \times 10^{-8}$	66, 92, 98–101	7701·1
Rb	$5p\ ^2P_{3/2}$	36·7	—	59·7	$(2\cdot7 \pm 0\cdot2) \times 10^{-8}$	92, 102	7802·4
	$^2P_{1/2}$	36·0	0·68	60·4	$(2\cdot9 \pm 0\cdot2) \times 10^{-8}$	92, 102	7949·8
Cs	$6p\ ^2P_{3/2}$	33·6	—	56·3	$(3\cdot0 \pm 0\cdot3) \times 10^{-8}$	92, 103–105	8523·4
	$^2P_{1/2}$	32·0	1·58	57·9	$(3\cdot5 \pm 0\cdot4) \times 10^{-8}$	92, 102, 104, 105	8946·0

energy distributions within a given exit channel have been made. Sec. VI consists of brief concluding remarks.

II. DATA ON METAL ATOM EXCITED STATES

Shown in Table I is a compilation of relevant information concerning the excited states of interest for this review. The excited-state energies and ionization potentials were taken from Moore.[30a] The lifetimes are averages of experimental values published in the references listed. Exclusion of other published values in calculating the averages was based on our judgement of possible errors in the techniques used, and in some cases on the disparity between one value and several others. The uncertainties listed are conservative, and in most cases the final range of uncertainty includes all the averaged values.

Shown in Table II are a consistent set of atomic diameters of the ground-state metal atoms, taken from Slater's paper.[30b] The hard-sphere Lennard-Jones diameter for Hg is 2·9 Å,[30c] consistent with the set of values in the table. Also, Köhler, Feltgen, and Pauly[30d] have determined the Lennard-Jones hard-sphere collision diameter for $Hg(^1S_0)$ scattered off H_2 to be 2·91 Å. Since the collision diameter of H_2 is known to be $2·92 \pm 0·05$ Å,[30c] this is further corroborating evidence that the Slater values are quite reasonable. It is expected that the P excited states of these metal atoms will have collision diameters approximately 10–20% greater than the groundstates. Corroboration of this estimate is provided by scattering measurements of $Hg(^3P_2)$ off CO, N_2, and CO_2,[30e] where repulsive collision diameters of $3·6 \pm 0·3$, $3·5 \pm 0·3$, and $3·8 \pm 0·3$ Å, respectively, were determined, compared with hard-sphere diameters for CO, N_2, and CO_2 alone of $3·75 \pm 0·15$, $3·74 \pm 0·10$,

TABLE II. Atomic diameters of ground-state-metal atoms.[30b]

Atom	Atomic diameter (Å)
Li	2·9
Na	3·6
K	4·4
Rb	4·7
Cs	5·2
Mg	3·0
Zn	2·7
Cd	3·1
Hg	3·0
Ca	3·6
Sr	4·0
Ba	4·3

and $4.0 \pm 0.10 \text{Å}$.[30c] A consistent $Hg(^3P_2)$ collision diameter of $3.4 \pm 0.2 \text{Å}$ can be extracted from the measurements, which is in fact about 15% higher than the 2·9 Å diameter for $Hg(^1S_0)$.

III. ABSOLUTE QUENCHING CROSS SECTIONS

Tables III–VII are compilations of quenching cross sections for the excited states treated in this review. The quenching cross section is defined as

$$\sigma = k_Q/\bar{v}$$

where k_Q is the rate constant for the bimolecular quenching process and $\bar{v}$ is the mean relative velocity of the excited metal atom and quencher. When an experiment is performed under conditions of thermal equilibrium,

$$\bar{v} = (8kT/\pi\mu)^{1/2},$$

where μ is the reduced mass. It should be noted that there is also a convention in which cross sections are reported as the square of the effective collision diameter for the quenching process,[19] resulting in values lower by a factor of π. Most authors have now adopted the logical definition of cross section as given above, however.

Because of the wide variation in temperatures at which measurements of the quenching cross sections of group I states have been made, the temperature of each measurement has been listed specifically in Tables III and IV. For the most part, the temperature variation in the measurements of quenching cross sections of group IIA and IIB excited states is not great, so only the range of temperatures is given in Tables V–VII.

Cross sections provide a more convenient way of comparing quenching efficiencies than rate constants, since differences in gas kinetic collision frequencies due to reduced mass and temperature variation are automatically factored out by dividing by the mean velocity. For most of the excited metal atom states and quenching molecules treated in this review, 'hard-sphere' gas kinetic collision cross sections are in the 20–100Å^2 range. Thus a quenching cross section of this magnitude or greater indicates that quenching occurs at virtually every collision, while a cross section of, say, 5Å^2 or lower can be interpreted to mean that several collisions are required before quenching can take place.

In a collision, the excited metal atom M* and quenching molecule Q approach under the influence of some intermolecular potential. If penetration is achieved to one or more critical regions of the M*–Q potential surface(s) where there is coupling of the initial electronic state to other final states, then quenching may take place. The separating products can consist of different electronic states of M and Q, with the energy defect taken up by recoil translational energy and by internal vibrational and rotational energy of Q, or

TABLE III. Absolute quenching cross sections (Å²) for the ²P excited states of the group IA atoms. Cross sections are for literature citations in which either the $J = \frac{1}{2}$ or $J = \frac{3}{2}$ state is not specified, or under conditions in which an effectively equilibrated $^2P_{1/2}$, $^2P_{3/2}$ mixture is assumed to be quenched. Values which, in our opinion, are suspect are shown in parentheses. Values in square brackets have been obtained for non-thermal velocity conditions by using photolysis of alkali halides at different wavelengths and at the temperature listed to produce ^2P states with different initial velocities.

Quencher	Li(^2P_J) T(K)	σ_Q(Å²)	Na(^2P_J) T(K)	σ_Q(Å²)	K(^2P_J) T(K)	σ_Q(Å²)	Rb(^2P_J) T(K)	σ_Q(Å²)	Cs(^2P_J) T(K)	σ_Q(Å²)
He	890	<0·05[130]	403–483 470 573 873–973 1400–1800	<0·01[107] (0·34)[108] 0·1[109] [<0·07][110] <0·3[111,112]	873–973 1400	[<0·04][110] <0·25[131b]	1400	<0·35[131b]	873–973 1400	[<0·04][110] <1·3[131b]
Ne			403–483 873–973	<0·01[107] [<0·52][110]						
Ar	1400	<0·9[131a]	403–483 873–973 1400–1800 2070–2210 2300–3600	<0·01[107] [<0·14][110] <0·3[111,112] (2·3)[113] <0·004[114,115]	873–973 1400 2070–2210	[<0·07][110] <0·6[131b] (0·94)[113]	873–973 1400	[<0·14][110] <0·9[131b]	1400	<2·8[131b]
Kr			403–483 873–973	<0·01[107] [<0·15][110]					873–973	[<0·16][110]
Xe			403–483 873–973	<0·01[107] [<0·16][110]			873–973	[<0·21][110]		
N₂			403 400 470 (~500?)	45·6[129] (b)40·3[116] 43[108] 36·6[117]	353	34[135]				

TABLE III. (Contd.)

Quencher	Li(²P_J) T(K)	σ_Q(Å²)	Na(²P_J) T(K)	σ_Q(Å²)	K(²P_J) T(K)	σ_Q(Å²)	Rb(²P_J) T(K)	σ_Q(Å²)	Cs(²P_J) T(K)	σ_Q(Å²)
	890	39.1[130]	538–638	30[109]						
			913	22.6[118]						
			911–914	[12–21][118]						
			?	22[119]						
	1400	21.2[131a]	911–914	[11–21][120]	—	[10–21][136]				
			1400–1800	21.8[111,112]	1400	17.6[131b]	1400	19[131b]	1400	79[131b]
			1500–2500	22[121]	1720	19[137]	1720	18.5[132]	1500	40[134]
			1750	17[122]	1760	19[124]	1900	25[133]	1720	35[134]
			1900	21[123]	1900	19[123]				
					1940	18[124]			2000	37[134]
			2100	14[122]						
			2700–2900	(48)[114]						
			2600–2900	(49)[115]						
CO	890	41.6[130]	403	88[129]						
			573	41[109]						
	1400	39.6[131a]	901	[24–35][118]	1400	39.0[131b]	1400	37[131b]		
			1400–1800	37.4[111,112]						
			1700	(85)[124]	1700	47[124]				
			1900	41[123]	1900	44[123]				
			2700–2900	(70)[114]						
H₂	890	[24.3][130]	403	23.2[129]	353	9.4[135]				
			400	16.2[116]						
			573	12.2[109]						
			1873	[10][125]						
			919	[7.7–8.0][118]	893	[3][138]				

			910	[6·5][120]						
	1400	16·3[131a]	1500	9·3[126]	1400	3·2[131b]	1400	1·9[131b]	1400	5·3[131b]
			1600–1800	9·0[111,112]						
			1750	6·5[122]	1720	3·3[137]	1720	2·1[132]	1720	5·0[134]
			1850	(15·5)[124]	1850	3·7[124]				
			1900	8[123]	1900	3·4[123]	1900	3·6[133]		
			2095	(14·0)[124]	2095	3·4[124]				
			2500	6·8[126]						
HD			400	11·6[116]	353	11·9[135]				
			573	11·5[109]						
D_2			400	9·8[116]	353	8·0[135]				
			573	10·2[109]						
	890	[21][130]	873	[8][125]	893	[3][138]				
			898–902	[6·2–8·7][118]						
O_2					1400	48·7[131b]	1400	79[131b]		
			1720	39[126]	1720	63[132]	1720	90[132]	1720	134[134]
			1750	24[122]						
			1800	38·6[111,112]						
			1885	(66)[124]	1885	31·5[124]				
			1900	34[123]	1900	49[123]	1900	83[133]		
			1990	(62)[124]						
			2100	20[122]						
			2500	31[126]						
I_2	890	[129][130]	888	[140–190][127]						
			900	[140–230][128]						
CO_2	890	53[130]	∼900	[40–65][128]	—	[27–65][136]				
			909–914	[27–52][118]						
	1400	29[131a]	1400–1800	53[111,112]	1400	67[131b]	1400	75[131b]		
			1670	(113)[124]	1670	80[124]	1720	74[204]	1720	86[204]
			1790	(106)[124]	1900	66[123]				
			1900	50[123]						
			2015	(94)[124]						

TABLE III. (Contd.)

Quencher	Li(^2P_J) T(K)	σ_Q(Å²)	Na(^2P_J) T(K)	σ_Q(Å²)	K(^2P_J) T(K)	σ_Q(Å²)	Rb(^2P_J) T(K)	σ_Q(Å²)	Cs(^2P_J) T(K)	σ_Q(Å²)
H_2O	1400	6·0[131a]	820	[3][125]	890	[1·5][139]				
			1500–2500	2·2[131b]	893	[2][138]	1400	4·0[131b]	1400	17[131b]
			1600–1800	1·6[111,112]	1400–1800	2·8[131b]			1500	9·0[134]
			2100	2·2[123]	1720	3·6[137]	1720	3·9[132]	1720	9·0[134]
			2070–2210	1·0[124]	2100	2·6[123]	2100	4·0[133]	2000	9·0[134]
					2070–2210	(0·42)[124]				
SO_2			900	[80–140][128]	893	[140][138]				
CH_4			403	0·35[129]	890	[<1][139]				
			573	0·7[109]	893	[<1][138]				
			873	[<1][125]						
			900	0·5[118]						
C_2H_6	890	[1·8][130]	403	0·53[129]						
			873	[<1][125]						
C_3H_8			403	0·6[129]						
			(~500?)	0·58[117]						
			573	0·9[109]						
$n\text{-}C_4H_{10}$			403	0·9[129]						
C_2H_4	890	56[130]	403	138[129]	873	[51][138]				
			920–961	[40–65][128]						
			903	[53][118]						
C_3H_6			403	163[129]						

1-C$_4$H$_8$		403	182[129]
2-C$_4$H$_8$		403	182[129]
C$_6$H$_6$		403	236[129]
		573	95[109]
		~900	[70–100][128]
C$_6$H$_5$CH$_3$		(~500?)	103[117]
CF$_4$	890–893 [<1][138,139]	873	[<1][125]
C$_2$F$_6$	890 [11:2][130]		
CH$_3$OH	893 [24][138]	873	[25][125]
CH$_3$CN		861–920	[55–100][128]
CF$_3$Cl		920	[60–90][128]

TABLE IV. Absolute quenching cross sections for the $^2P_{1/2}$ and $^2P_{3/2}$ excited states of the group IA atoms (Å^2). Cross sections represent all processes causing destruction of one spin–orbit multiplet state without the production of the other multiplet state. Values which, in our opinion, are suspect are shown in parentheses.

Quencher	Li(^2P_J)			K(^2P_J)			Rb(^2P_J)			Cs(^2P_J)		
	T(K)	$\sigma_{1/2}(\text{Å}^2)$	$\sigma_{3/2}(\text{Å}^2)$	T(K)	$\sigma_{1/2}(\text{Å}^2)$	$\sigma_{3/2}(\text{Å}^2)$	T(K)	$\sigma_{1/2}(\text{Å}^2)$	$\sigma_{3/2}(\text{Å}^2)$	T(K)	$\sigma_{1/2}(\text{Å}^2)$	$\sigma_{3/2}(\text{Å}^2)$
H_2	564	40·7[141]	29·0[141]	363	7[140]	4[140]	340	6[142] (10)[143]	3[142] —	313 315	11·7[145] 7[146]	6·5[145] 5[146]
HD				363	11[140]	14[140]	340	6[142]	5[142]	315	4[146]	3[146]
D_2	564	48·7[141]	47·7[141]	363	2±1[140]	1±0·5[140]	340	3[142]	5[142]	315	8[146]	7[146]
N_2				363	35[140]	39[140]	304 340	37[144] (75)[143] 58[142]	36[144] — 43[142]	313 315	55[145] 77[146]	64[145] 69[146]
CH_4							340	<1[142]	3±2[142]			
CD_4							340	2±2[142]	3±3[142]			
C_2H_6							340	2±2[142]	6±3[142]			
C_2H_4							340	139[142]	95[142]			

TABLE V. Absolute quenching cross sections for the $^1P_1$ excited states of Hg, Cd, Zn, and Mg. Values in square brackets are relative cross sections which have been converted to absolute values by assuming that the cross sections for quenching by CO were 40 Å² and 30 Å² for Hg($^1P_1$) and Zn($^1P_1$), respectively. Values in parentheses we regard as suspect.

Quencher	Hg($^1P_1$) (296 K)	Cd($^1P_1$) (368–568 K)	Zn($^1P_1$) (578–625 K)	Mg($^1P_1$) (375–475 K)
He	[~0][153]	<0·03[147], (0·30)[148]		<1[156], <1·5[157]
Ar	[(44)][153]	(1·1)[148]		
Xe				<1[156]
N_2	[40][153]	35[149], 50[148], 52[147]	19[154], [25][155]	80[156], 42[157]
CO	[40][153]	47[149], 56[147], (138)[148]	30[154]	70[156], 55[157]
NO		86[147]		
H_2	[8][153]	(10·7)[148], 24[147]	12[154], [23][155]	
HD		24[150], 26[149]	13[154]	40[156], 31[157]
D_2		27[147]	19[154], [21][155]	29[157]
CO_2	[36][153]	30[147], 31[149], 47[150]	[40][155]	90[156], 68[157]
N_2O	[52][153]	64[149], 86[147], (150)[148]		
NH_3	[~48][153]	(149)[148]	[48][155]	
SF_6		48[147], 58[148]		
CH_4	[12][153]	147[147]	34[154], [34][155]	58[157]
C_2H_6		38[149]	68[154]	88[157]
C_3H_8		81[151]	99[154]	33[157]
$i\text{-}C_4H_{10}$		110[149]	106[154]	61[157]
$n\text{-}C_4H_{10}$		131[149], 155[147]		73[157]
$neo\text{-}C_5H_{12}$		113[152], (267)[148]		89[157]
$i\text{-}C_5H_{12}$		102[151]		97[157]
$n\text{-}C_5H_{12}$		150[151]		108[157]
C_2H_4	[~60][153]			60[157]
C_3H_6		(234)[148]		76[157]
C_2H_2				90[156], 56[157]
CF_4				<9[157]
C_2F_6				<9[157]

TABLE VI · Absolute quenching cross sections for the ^3P_J excited states of Hg, Cd, Zn, and Mg (Å²). Values for Hg($^3P_1$) and Cd($^3P_1$) correspond to all processes which do not yield Hg($^3P_0$) or Cd($^3P_0$), respectively. The cross sections for Hg($^3P_1$) were estimated from available experimental determinations of Hg($^3P_1$) to Hg($^3P_0$) intramultiplet quenching cross sections.[158-164]

It was also assumed that the $^3P_1$ to $^3P_0$ quenching cross section for both Cd and Hg was negligible compared with the cross section for overall quenching of the ($^3P_1$) states for all unsaturated organic compounds,[162,163,181] CCl₄, and CF₃Br.

The cross sections listed as Cd($^3P_{0,1}$), Zn(^3P_J), and Mg(^3P_J) are actually population-weighted averages (not sums) of the ^3P_J cross sections at the temperature of the experiment (e.g. a cross section listed under Cd($^3P_{0,1}$) would correspond to $0.58\sigma_{3P_0} + 0.42\sigma_{3P_1}$ if the experiment had been performed at 553 K. Values in square brackets were obtained in molecular beam experiments. Values in parentheses we regard as suspect.

Quencher	Hg($^3P_1$) (292–298 K)	Hg($^3P_0$) (292–298 K)	Cd($^3P_1$) (458–486 K)	Cd($^3P_0$) (458 K)	Cd($^3P_{0,1}$) (458–557 K)	Zn(^3P_J) (578–625 K)	(a) Mg(^3P_J) (~450 K)	(800 K)
He			<0·001[182]		<0·001[186]			
Ar			<0·002[182]		<0·003[186]			
Xe	<0·002[165]	8·4 × 10⁻⁵[177]	<0·003[182]					
N₂	<0·1[165]	<2 × 10⁻⁶[160]			0·010[186] 0·012[184] 0·015[187]	0·09[154] 0·09[189] 0·54[155]	0·083[192]	0·036[191]
CO	<7[159,165,167]	1·2[177], 1·8[163] 2·1[161,180] (7·7)[178]	0·9[183]	2·3[183]	2·2,[186] 2·4[187] 1·9[183]	1·1,[154] 1·5[155]	0·23[192]	0·088[191]
NO	~80[167]	37,[163] 40[177] 41,[178] 51[161,180] 55,[177] 57[178]	44[183]	47[183]	45,[183] 24[186]			
O₂	61,[165] 75[167]	35,[163] 38[161,180] 55,[177] 57[178]	6·6[184]	4·9[184]	5·2,[186] 5·5[184]	19·8[155,190]	(a)(3·6)[192]	
H₂	24·6,[165] 30·8[168] 31·4,[169] 33·9[166] 34·9,[167] 36·9[170]	2·9,[177] 3·0[161,180] 5·2,[178] 6·0[163]					0·0046[192]	0·024[193] 0·030[91]

HD	31·4,[168] 34·2[169]			5·1[186]	
D₂	22·7,[165] 27·0[168] 8·5[163]	2·5[184] 2·4[184]	2·9,[186] 2·4[184] 16·4[155,190]		
CO₂	10·2[165] 10·8[158,159]	0·09,[163] 0·10[177] 0·11,[161,180]	27·7[183] 9·7[183]	15·8,[183] 13·6[187] 64[155] 15·9,[186]	(1·5)[192]
N₂O	67[167]	(14),[178] 27[161,180] 28,[177] 35[163]			(3·2),[192] [71][12]
NH₃	0·038[163] 0·049[161,180] 0·049[177]		0·058[185] 0·059[186] 0·021[148] 3·0[186]	4·7[155]	
SF₆				[92][12]	
TeF₆				[128][12]	
WF₆				[194][12]	
Cl₂	28[179]			[85],[194] [102][12]	
C₂H₄	124,[167] 116[19] 151[171]	68,[177] 83[161,180] 129[163]	78,[185] 83[181] 68[181]	58,[186] 73[181]	
C₃H₆	118,[168] 145[19] 152[167]	135[163]	91,[185] 103[181] 92[181]	78,[186] 96[181]	
i-C₄H₈	167[19]				
cis-2-C₄H₈	176,[19] 219[167]		96,[185] 103[181] 89[181]	94[181]	
l-C₄H₈	174[168]		111[185]		
C₂H₂	104,[19] 124[168]	91[163]	69,[185] 76[181] 70[181]	72[181]	3·4[192]

3·0[191]

TABLE VI. (Contd.)

Quencher	$Hg(^3P_1)$ (292–298 K)	$Hg(^3P_0)$ (292–298 K)	$Cd(^3P_1)$ (458–486 K)	$Cd(^3P_0)$ (458 K)	$Cd(^3P_{0,1})$ (458–557 K)	$Zn(^3P_J)$ (578–625 K)	(~450 K)	$Mg(^3P_J)$ (800 K)
C_6H_6	124,[172] 151[173] 188[19]		89[185]		90[186]			11·9[191]
CH_4	0·40,[170] 0·27[19]	3×10^{-4} [177] 4×10^{-4} [163] 9×10^{-4} [161,180]			0·0017[188]	0·23[155,190]		$2·1 \times 10^{-3}$ [191]
C_2H_6	<0·8[19,159,170,174]	0·0053[163] 0·018[161,180]			<0·007[188]			
C_3H_8	3–11[19,159,170,174]	0·11[161,180] 0·15[163]			0·0035[188]			
$n\text{-}C_4H_{10}$	9–20[19,159,167,170]	2·0[163]						
$i\text{-}C_4H_{10}$	3–21[19,159]	4·2[163]			0·0046[188]			
neo-C_5H_{12}	0–6[19,170,174]	0·30[163]						
CH_2CHF			87[181]	63[181]	71[181]			
CH_2CF_2	~90[175]		57[181]	37[181]	44[181]			
CF_2CHF			37[181]	28[181]	31[181]			
C_2F_4	35[19]		21[181]	13[181]	15[181]			
CCl_4	207[176]							[122][12]
CF_3Br	112[176]							[32][12]

TABLE VII. Absolute quenching cross sections ($Å^2$) for low-lying excited states of Ca, Sr, and Ba. Values in brackets were determined in molecular beam experiments.

	Ca(^3P) (970 K)	Sr(^3P) (913–962 K)	Ca($^1P_1$) (900 K)	Sr($^1P_1$) (inert gases, 800 K; others, 1765–2740 K)	Ca($^1D_2$)	Ba($^1D_2$) (385 K)
He	9×10^{-5} [195]	$1·1 \times 10^{-4}$ [195]	$0·025$ [199]	$0·38$ [199]		
Ne	$1·8 \times 10^{-4}$ [195]	$3·1 \times 10^{-4}$ [195]	$0·028$ [199]	$0·61$ [199]		
Ar	$4·8 \times 10^{-4}$ [195]	$4·5 \times 10^{-4}$ [195]	$0·046$ [199]	$1·6$ [199]		
Kr	$7·0 \times 10^{-4}$ [195]	$5·2 \times 10^{-4}$ [195]	$0·064$ [199]	$1·4$ [199]		
Xe	$8·1 \times 10^{-4}$ [195]	$7·8 \times 10^{-4}$ [195]	$1·15$ [199]	$0·25$ [199]		
N₂				16 [200]		$2·0$ [202]
CO				49 [200]		$15·2$ [202]
H₂				22 [201]		$5·4$ [202]
D₂						$5·3$ [202]
N₂O		[68] [198]		30 [200]		
Cl₂	[65],[12] [64][196] [124],[197] [100][194] [128][12]	[180] [194]		152 [200]	[112][12]	
CO₂	[37][7]				[85][12]	
O₂	[60][12]				[83][7]	
HCl	[29][12]				[90],[197] [98][12]	
HF	[78][12]				[36][12]	
SF₆					[90][12]	
TeF₆	[170][12]				[190][12]	
WF₆	[204][12]				[178][12]	
CCl₄	[117][12]				[86][12]	
CF₃Br	[12][12]				[34][12]	
CF₄	[20][12]				[34][12]	

can be the result of nuclear rearrangement (i.e. chemical reaction). For very low values of quenching cross sections, then, it might be postulated that no critical region for coupling initial to final states is accessible for the available range of translational energies. The initial and final surfaces may, for example, be essentially parallel and not approach each other except on high-energy, low-internuclear-distance regions of the surface(s). Even in cases where an exothermic chemical exit channel is available, there may simply be an activation barrier (i.e. locally repulsive portions of the potential surface beyond which most trajectories cannot penetrate). It is convenient, then, to separate quenching processes into two types, *entrance-channel controlled*, where one or more efficient coupling mechanisms to final states result in high quenching cross sections, and *exit-channel controlled*, where inaccessibility of a coupling region (and/or inefficient coupling) result in cross sections at least a factor of 5 lower than a typical gas kinetic collision cross section.

We first discuss processes which are entrance-channel controlled and attempt to rationalize the magnitudes of the cross sections and their dependence on the properties of the quencher and on the relative collision energy. Coupling mechanisms which differentiate entrance- from exit-channel control cases will then be discussed specifically, with a view towards explaining some of the low quenching cross sections which have been observed.

A. Entrance-Channel Control

For entrance-channel controlled quenching, net coupling to final states must occur at greater than 'hard-sphere' collision distances, where the potential is attractive due either to long-range forces (usually the dispersion interactions) or to avoided crossings with very attractive surfaces which correlate to higher-energy products (e.g. ion pairs). Successful quenching trajectories are those which surmount the barrier in the effective potential obtained by adding the centrifugal repulsion to the true long-range attractive potential, and thereby reach the critical distance R_c necessary for quenching. This 'orbiting' model[138,205-206] leads to the following expression for the cross section for an entrance-channel controlled process:

$$\sigma = \left(\frac{3\pi}{2}\right)\left(\frac{2C_6}{E}\right)^{1/3} \tag{1}$$

where E is the relative collision energy and C_6 is the proportionality constant for the attractive part of the Lennard-Jones 6–12 potential, $V(R) = -C_6/R^6$. For a generalized long-range potential of the form $-C_n/R^n$ the cross section would be proportional to $(C_n/E)^{2/n}$. The validity of the assumption of a strictly $1/R^6$ potential for excited states has been discussed by Herm[138] as well as by Setser.[205] Eq. (1) is convenient, however, because C_6 coefficients can easily be calculated using the Slater–Kirkwood approximation.[205] Thus, the dependence of the cross sections on the 1/3 power of the C_6 coefficients divided by the relative collision energy allows a test of this postulated form of long-range control. But, as pointed out by Herm,[138] for energies greater than some critical energy E^*, where the orbiting barrier moves in to the critical separation for quenching, the above equation is no longer valid. All that is required for $E > E^*$ is that E also be greater than the effective potential at some critical distance R_c, leading to the following equation for the quenching cross section (the so-called 'absorbing-sphere' model):[118,130,138,207]

$$\sigma = \pi R_c^2 \left(1 - \frac{V(R_c)}{E} \right), \tag{2}$$

which has an energy dependence different from Eq. (1) and of course has no parametrized dependence on long-range forces since $V(R_c)$ is molecule-specific (although for purely dispersive forces the cross section would be proportional to C_6, since $V(R_c) = -C_6/R_c^6$).[205] Examples of specific 'absorbing-sphere' models are modifications by Barker[208] and by Lijnse[209] of the classic Bauer–Fisher–Gilmore charge-transfer surface-crossing model[210] for quenching of excited alkali atoms. The attractive potential at R_c is considered to result from the outermost avoided crossing, with quenching taking place either on the diabatic[209] or lower adiabatic[208] surface.

Examination of the quenching cross sections in Tables III–VII shows that for virtually all quenching species other than the inert gases the quenching of $^1P_1$ excited states of Hg, Cd, Zn, and Mg is entrance-channel controlled. The same is probably true for the $^1P_1$ excited states of Ca, Sr, and Ba, but there is insufficient data to make a generalization. For the ^2P_J levels of group I metals, entrance-channel control can be inferred for unsaturated quenchers (i.e. those with at least one π-bond) and quenchers containing I, Br, or Cl atoms. We examine first any evidence regarding a possible relationship between the magnitude for the quenching cross sections and the long-range dispersion forces as represented by the C_6 coefficient for M*–quencher interactions, as implied for example by Eq. (1). We then look at the dependence of the value of the quenching cross section on initial collision energy for a number of individual cases, which provides a more critical test of Eqs. (1) and (2).

Only recently has there been sufficient cross-section information to attempt any reliable correlation with C_6 parameter for the excited states treated in this review. It should be noted at the outset that Setser[205] has shown that the quenching cross sections above 25Å^2 for the $^3P_2$ metastable states of Ar, Kr, and Xe do correlate well with the C_6 coefficients, but to a power of approximately 1·0 instead of 0·33, thus showing Eq. (1) to be totally invalid for inert-gas metastable states. For a limited number of quenching species, plots of $\log \sigma_Q$ versus $\log C_6$ for quenching of Hg($^1P_1$),[205] Zn($^1P_1$),[154] and Cd($^1P_1$)[152] give slopes of $\sim 1·0, 0·9 \pm 0·1$, and $0·7 \pm 0·1$, respectively, also inconsistent with Eq. (1). Recent data for Mg($^1P_1$) quenching[157] is shown plotted in the same

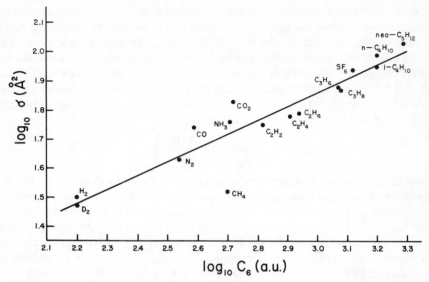

Fig. 1. A log–log plot of the C_6 coefficient versus the cross section for quenching of $Mg(^1P_1)$ by several gases.[241] Induced-dipole, permanent-dipole contributions to the C_6 coefficient were included for molecules with dipole moments.[157,241]

manner in Fig. 1. The slope of the best-fit straight line in Fig. 1 is 0.48 ± 0.05, which is still too high for an orbiting model with entirely dispersive long-range forces falling off as $1/R^6$, but could be consistent with an effective potential with a radial dependence between $1/R^6$ and $1/R^4$. Interestingly enough, the limited amount of data available for large cross-section quenching of ^2P states of group I metals shows a dependence on the C_6 coefficient to the power 0.4 ± 0.1.[147] It is tempting to make something out of the fact that the C_6 correlation power seems to drop off from 1 to nearly 1/3 as the excited-state electronic energy drops in the following order: inert-gas metastable, $Hg(^1P_1)$, $Zn(^1P_1)$, $Cd(^1P_1)$, $Mg(^1P_1)$, $Na(^2P)$, and $Li(^2P)$; but the metal atom data are not sufficiently extensive for any definitive conclusions to be drawn.

A potentially more sensitive test of entrance-channel control mechanisms is the dependence of the cross section on initial relative collisional translational energy for a series of specific metal atom state and quencher combinations. Fortunately, a convenient method exists for producing electronically excited alkali atoms with varying amounts of initial translational energy. Alkali halides can be photolysed at increasing photon energies within an absorption continuum which produces a dissociative state correlating with a halogen atom and an excited alkali atom. In this manner, excited alkali atoms with known and variable initial translational energy can be prepared, and quenching cross sections can be measured using Stern–Volmer[128,138] or direct-lifetime, single-photon counting[118,130] methods. Measurements of this kind have been carried out for $Li(^2P)$, $Na(^2P)$, and $K(^2P)$. Unfortunately, over the range of translational energies available, the precision of the data is not sufficient to distinguish clearly between an E^{-1} and an $E^{-1/3}$ dependence, and the data have been interpreted using Eq. (1)[138] as well as Eq. (2).[118]

The quenching pair for which the most information regarding temperature dependence is available is $Na(^2P)$–N_2, where several cross-section values have been obtained at temperatures corresponding to translational energies lower than that accessible in the NaCl photolysis experiments (see Table III). Shown in Fig. 2 is a plot of all the available cross-section data as a function of the inverse of the mean relative translational energy for the quenching of $Na(^2P_J)$ by N_2. The data are more consistent with Eq. (2) than Eq. (1), indicating that a 'breathing-sphere' model may be a reasonable representation of the quenching in this case. Further, from the slope-to-intercept ratio in Fig. 2, $V(R_c)$ can be calculated from Eq. (2) to be approximately $-2 \cdot 5 \, kcal \, mol^{-1}$, in good agreement with the theoretical value of $-3 \cdot 2 \, kcal \, mol^{-1}$ obtained by Barker[208] using the Hasted–Chong empirical correlation. Therefore, for the classic case of quenching of $Na(^2P)$ by N_2, the simple modified charge-transfer curve-crossing model of Barker,[208] in which the quenching is assumed to result from trajectories on the *adiabatic* potential resulting from the outermost

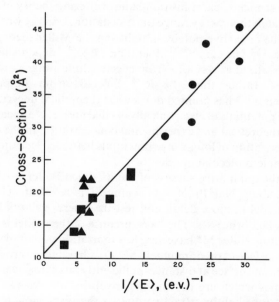

Fig. 2. A plot of the cross section for the quenching of $Na(^2P)$ by N_2 versus the reciprocal of the mean relative collision energy. Values obtained from Table III.

curve crossing, is approximately correct despite acknowledged shortcomings.[208] Further consistent evidence regarding product-state vibrational distributions will be discussed in Sec. V. While there are insufficient data concerning the translational energy dependence of cross sections for any other excited metal atom–quencher pair to generalize the success of the Barker 'breathing-sphere' treatment, the available information is certainly consistent with the predictions of the model.

However, if the $Na(^2P)$–N_2 data are plotted in the form $\ln \sigma$ versus $\ln(1/\langle E \rangle)$, a least-squares slope of 0·6 is obtained. This result might be expected, since neither Eq. (1) nor (2) are strictly correct over the range of $\langle E \rangle$ studied, and the precision of the data does not justify a more sophisticated treatment. The molecular beam cross-section data of Krause, Datz, and Johnson[211] on quenching of $Hg(^3P_2)$ to $Hg(^3P_1)$ by N_2, NO, CH_4, H_2, and D_2 over a similar range of $\langle E \rangle$, 0·02–0·20 eV, were fitted best by an $\langle E \rangle^{-1/2}$ potential, but the data of Liu and Parson[207] for the same gases over an $\langle E \rangle$ range greater by a factor of 2 could readily be fitted to Eq. (2), consistent with an $\langle E \rangle^{-1}$ dependence. A variety of molecular beam experiments by Martin and coworkers[206,212,213] and others[214] on the energy dependence of ionization or dissociative exit-channel cross sections for the deactivation of inert-gas metastable excited states have also been fitted to an $\langle E \rangle^{-n}$ dependence in the 0·02–0·20 eV region, with $n \approx 0·3$–0·7. Many of these data were interpreted in terms of the semiclassical transition-state scattering theory of Miller.[215] It was claimed that in this energy range the translational energy was less than the well depth of the attractive potential, reducing the Miller theory to a simple 'orbiting' model.[212] The $\langle E \rangle^{-1/2}$ rather than $\langle E \rangle^{-1/3}$ dependence for several cases was rationalized as resulting from effective long-range potentials which fell off as R^{-4} rather than the R^{-6} expected from simple dispersive interactions. Setser[205] has pointed out several theoretical uncertainties about the long-range potentials of electronically excited atoms. It appears that a very useful area of theoretical and experimental concentration in the future would be the characterization of long-range potentials between electronically excited atoms and simple molecular species.

Because of the qualitative success of the modified Bauer–Fisher–Gilmore model in describing $Na(^2P)$–N_2 quenching, it is relevant at this stage to describe the model in more detail and test its general validity for entrance-channel controlled processes. The basic premise of the model is that because excited metal atom states M* have very low ionization potentials, it is possible for a potential surface of predominantly M^+Q^- character to cross the diabatic M*–Q interaction surface at distances sufficiently large (for quenchers Q with electron affinities which are not too negative) that the M–Q surface is still essentially flat (or slightly attractive due to the dispersion potential). The simplest way of constructing the charge-transfer potential surface is to assume a strictly Coulombic M^+–Q^- interaction with asymptotic energy

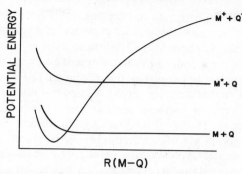

Fig. 3. A schematic representation of the potential curves in the Bauer–Fisher–Gilmore charge-transfer model for the quenching of an excited metal M* by a molecule Q.

$\Delta = IP(M^*) - EA(Q)$ relative to the energy of neutral $M^* + Q$, although it has been found necessary to add a polarizability term to the potential to obtain agreement with experimental cross-section values.[210] Simple representations of the neutral and ionic surfaces in a typical case are shown in Fig. 3. Note that for the purposes of this section, it does not matter if in fact quenching occurs to the ground state as shown or via some other exit channel, such as chemical reaction. The important point is that the charge-transfer surface avoided crossing *facilitates* net quenching and determines the crucial distance R_c in 'breathing-sphere' terminology. Net reformation of $M^* + Q$ can be accounted for by a constant probability factor.

Actually, the potentials for the electronic-to-vibrational (E–V) case consist of families of $M^+-Q^-(v)$ and $M^*-Q(v)$ curves, allowing a grid of possible vibrational surface crossings. The curve-crossing probability at each crossing is independently evaluated using the Landau–Zener method, with transition matrix elements obtained as a product of an electronic component (derived from the Hasted–Chong correlation) and the vibrational overlap at the crossing point (obtained from the appropriate Franck–Condon factors).[208,210] The model therefore not only predicts the total cross section for net quenching but also the final distribution of $Q(v)$ states for the E–V case. The details of the computational procedures may be found in ref. 208.

There is no question that the model is a drastic oversimplification of the true quenching process. For example, most quenching molecules do not have positive electron affinities, so that extrapolation from $M^+ + Q^-$ asymptotic limits using a simple Coulombic potential to construct the charge-transfer surface is incorrect and at best a qualitative estimate. In these cases, it is naively assumed that the very short-lived electron-resonance states[216] of molecules like N_2, CO, CO_2 are stabilized by simple Coulombic forces generated by the presence of M^+. Possible contributions of covalent exchange interactions to the M^+Q^- potential, of orientational dependence of both the charge-transfer and neutral surfaces, or the inevitable presence of minima in the M^+Q^- potentials at sufficiently short M^+-Q^- distances are all completely ignored. The application of the Landau–Zener treatment to atom–polyatomic transitions is also questionable. Despite all this, the model provides a picture which is bound to be at least qualitatively correct,[141] and is of course much simpler to apply than more exacting methods such as diatomics-in-molecules surface-hoping calculations[17,217] or close-coupling calculations,[15] which require accurate *ab initio* potential surfaces. In this review, we use the modified Bauer–Fisher–Gilmore model as a starting point for a discussion of quenching mechanisms, acknowledging its inadequacies but tentatively accepting the rudiments of the model as being essentially correct for many cases of excited metal atom quenching.

The first question to ask, of course, is whether the quantity Δ is sufficiently

TABLE VIII. Electron affinities of quenching molecules.

Molecule	Adiabatic electron affinity (kcal mol^{-1})	Vertical electron Affinity (kcal mol^{-1})
N_2		(-44 ± 10)[216]
H_2		(-75 ± 25)[216]
CO		(-35 ± 10)[216]
NO		$+1$[218,219]
O_2	$+10$[220,221]	
Cl_2	$+56$[222]	$+23$[222]
I_2	$+58$[222]	$+40$[222]
CO_2	-14[223]	
N_2O	-3[224]	
H_2O		$<(-70)?$[225]
CH_4		$<(-70)?$[225]
CF_4		(negative?)[226]
SF_6	$+12$[227]	
TeF_6	$+77$[227]	
WF_6	$+104$[228]	
C_2H_4	-36[229]	-41[230]
C_3H_6		-46[230]
cis-C_4H_8		-51[230]
trans-C_4H_8		-48[230]
i-C_4H_8		-50[230]
C_2H_2		-60[230]
C_6H_6		-27[230]
CH_2CHF		-44[231]
trans-CHFCHF		-42[231]
cis-CHFCHF		-50[231]
CF_2CH_2		-55[231]
CF_2CHF		-56[231]
CF_2CF_2		-69[231]

small in all the entrance-channel controlled cases for the charge-transfer curve to cross the M*–Q interaction region at distances larger than the repulsive wall, as shown in Fig. 3. Shown in Table VIII are the measured electron affinities of several of the quenching molecules considered here. For the group I excited states the M* ionization potentials are very low, ranging from $82\,kcal\,mol^{-1}$ for the smallest atom, $Li(^2P)$, to $60\,kcal\,mol^{-1}$ for the largest atom, $Cs(^2P)$. Among the quenchers for which the group I cross sections appear to be entrance-channel controlled (i.e. N_2, CO, CO_2, NO, O_2, I_2, and the alkene hydrocarbons), N_2 has the most negative electron affinity, $-44\,kcal\,mol^{-1}$ [216] Since the Bauer–Fisher–Gilmore (BFG) model has been shown to be qualitatively adequate in predicting the large cross sections for quenching of all the alkali ^2P states by N_2,[232] it is obvious for all the quenching molecules which are comparable in size to N_2(CO, CO_2, NO, O_2) that the model should also be valid. In the I_2 case, the highly positive electron affinity more than makes up for the increase in size of the quencher.[128] The validity of the BFG model for the alkenes is somewhat questionable, however, since the electron affinities are negative and the molecules are much larger, but it may be that the repulsive potential wall for the *double-bond-specific* interaction occurs at a relatively constant distance for all the alkenes, increasing only slightly as the overall size of the molecule increases. Another factor which could be favourable to the quenching of all the unsaturated molecules is that because of possible bonding interactions between the M* p-orbital and the π-density on the alkenes, at least one of the incoming neutral potential surfaces may have substantial attractive character at distances greater than calculated for the simple Coulombic curve crossing. Thus, it appears that the modified BFG model is at least qualitatively successful in explaining all the large cross-section processes for quenching of excited group I metals.

However, the modified BFG model does not, we believe, account for all the large cross-section cases for the $^1P_1$ states of Mg, Zn, Cd, and Hg. Since the ionization potentials and radii of these states are comparable with those of $Li(^2P)$ (see Tables I and II), the modified BFG model can certainly be invoked for N_2, CO, NO, CO_2, and the alkenes, but the large cross sections for all *alkane* hydrocarbons for the $^1P_1$ states are in stark contrast to the small cross sections for quenching of $Na(^2P)$ by all the alkanes (see Table III). Since the electron affinities of alkane hydrocarbons are probably more negative than $-3\,eV$,[225] it is not surprising that the charge-transfer surface crossing could be on the repulsive covalent M*–RH wall, even for $Li(^2P)$–C_2H_6 or $Na(^2P)$–CH_4. Thus some explanation other than charge-transfer surface crossings must be invoked to account for the large $M(^1P_1)$–alkane cross sections. A strong chemical interaction of $M(^1P_1)$ levels with CH bonds (insertion of the high-energy $^1P_1$ states into a C–H bond corresponds to formation of excited singlet states of alkyl metal hydrides) has been postulated previously to provide the necessary attractive potential.[147] The $Cd(^1P_1)$ case has been investigated extensively by Breckenridge and coworkers,[9,10,147,149] and will be discussed in detail in Sec. V below. The analogous chemical interaction of group I excited states with C–H bonds is expected to be much weaker, since insertion of a low-energy monovalent atom such as $Na(^2P)$ into a C–H bond is not expected to be a favourable bonding situation, and end-on attack is likely to be repulsive if the charge-transfer surface is too high in energy.

B. Exit-Channel Control

Turning now to cross sections in Table III–VII which are obviously substantially *less* than gas kinetic, it is perhaps a more important test of the modified BFG model to see if there are cases in which the cross sections *should* be entrance-channel controlled but apparently are not. This sort of examination of cross sections has been presented before by Breckenridge and Renlund[147,233] for some of the ^3P and ^1P quenching cases, and the details need not be repeated here. It was shown in those papers, for example, for the P excited states of Li, Cd, and Hg (all of which have estimated hard-sphere diameters in the 3·35 to 3·55 Å range), that an empirical transition from moderate to large cross sections appears to occur for values of $\Delta \approx 135$ kcal mol^{-1} for quenchers like N_2, NO, and CO_2 with hard-sphere diameters in the 3.6–4.0 Å range. With an assumed polarizability of 40 Å^3 for the M^+Q^- intermediate in these cases, the BFG model predicts the ionic curve crossing to occur at $\sim 3·6$ Å for $\Delta = 135$. This is to be compared with the estimated hard-sphere M^*–Q repulsion distances of 3·5–3·8 Å for these excited states and quenchers. Thus the modified BFG model is successful in predicting the transition from exit-channel to entrance-channel control for several representative examples, although the choice of 40 Å^3 as the polarizability of the M^+Q^- complex is admittedly rather arbitrary.

It is also possible to rationalize the more extensive data of Tables III, V, and VI. For the quenching of the ^3P_J states of Hg, Cd, Zn, and Mg by N_2 and CO, $\Delta > 149$ and the cross sections are all quite small. For the $^1P_1$ states of Hg, Cd, Zn, and Mg, as well as $Li(^2P)$, $\Delta < 130$ and the cross sections are all in the gas kinetic range. For NO, $\Delta < 132$ for *all* the ^3P states and all the cross sections are large. For CO_2, Δ is in the intermediate range of 134–147 for the ^3P states of Hg, Cd, and Zn, and the cross sections do vary from 0·1 Å^2 for $Hg(^3P_0)$, where $\Delta = 147$, to 28 Å^2 for $Cd(^3P_1)$, where $\Delta = 134$.

From Table VI, the other molecular quenching gases for which there are small cross sections in certain ^3P cases are H_2, NH_3, SF_6, and alkane hydrocarbons. The alkanes are known to have very negative electron affinities,[225] and the electron affinity of NH_3 is also expected to be quite negative, analogous to that of H_2O.[225] Rough estimates for SF_6^{147} and H_2^{233} are consistent with them being 'intermediate' BFG cases analogous to

CO_2. Quenching of the ^3P states by H_2 and the alkane hydrocarbons is discussed in more detail in the next section.

Small cross sections for group I excited states are observed for H_2O, NH_3, alkane hydrocarbons, and CF_4, (see Table III), all of which are nicely rationalized by the very negative electron affinities either measured or expected for these species. Cross sections for H_2 for the larger alkali-atom states also drop off substantially, but again rough Bauer–Fisher–Gilmore estimates put these cases into the borderline or even unfavourable category. The H_2^- resonance state is very short-lived and ill defined,[216] however, injecting considerable uncertainty into simplistic BFG estimates.

From Table VII, the small cross sections for quenching of $Ba(^1D_2)$ by N_2 and H_2 are consistent with BFG estimates because of the large size of $Ba(^1D_2)$.

We have seen, then, that for molecular quenchers there are no serious exceptions to the rough BFG-type correlation attempted here in all the data in Tables III–VII, in the sense that cross sections are never small when the BFG model predicts them to be large. Larger cross sections than predicted by BFG arguments are sometimes observed, but can usually be attributed to chemically attractive potentials which are *not* charge-transfer-like in nature.

Quenching by the inert gases deserves special mention as the last part of this section. Early claims of moderate cross sections for the quenching of group I and II excited states by the inert gases are now known to be due to experimental difficulties with impurities and with pressure broadening of absorption profiles. As can be seen from Tables III, VI, and VII, cross sections for net quenching of the ^2P and ^3P states by all the inert gases are extremely small. In the ^2P cases, Speller, Standenmayer and Kempter[234] have shown, by studying the reverse excitation processes in molecular beams and invoking microscopic reversibility, that for the ^2P states of Li, Na, K, and Rb the quenching cross sections for *all* the inert gases, including Xe, were less than 0.01Å^2. This observation is consistent with theoretical estimates of the potential curves which indicate that the $M(^2P)$–rare gas excited-state potentials parallel the $M(^2S)$–rare gas ground-state potentials up to very high energies.[235,236] This must be the case for the ^3P states of the group II metals as well. Data for the $^1P_1$ excited states of Mg, Zn, Cd, and Hg is too sparse as yet for any conclusions to be made, but the latest time-resolved data for $Cd(^1P_1)$ and $Mg(^1P_1)$ show small cross sections for He and Ar.[9,157]

In the $Sr(^1P_1)$ and $Ca(^1P_1)$ cases, Wright and Balling[199] have recently shown in careful experiments that the quenching cross sections range from 0.03Å^2 up to 1.6Å^2 (see Table VII). These moderate cross sections can be rationalized by the presence of a plethora of molecular Sr–rare gas and Ca–rare gas states correlating to the energetically close-lying ^3D and ^1D states. In the $Mg(^1P_1)$, $Zn(^1P_1)$, $Cd(^1P_1)$, and $Hg(^1P_1)$ cases, the ^3D and ^1D states lie at high energies and the closest lower-lying atomic states are the ^3P_J states at much lower energy.

IV. EXIT CHANNELS FOR ELECTRONIC ENERGY DISSIPATION

There are several ways in which the electronic energy of an excited metal atom can be dissipated in a quenching collision. Various exit channels for the quenching of excited state M* by polyatomic molecule AB are shown below. Note that M** is an electronically excited state of M which is lower in energy than M*, AB(V,R) is ground-state AB with a distribution of vibrational and rotational quantum states, AB*(V,R) is an electronically excited state of AB with a distribution of vibrational and rotational quantum states, A and B are either atoms or polyatomic fragments, MA and MB are bound ground-state molecules, BA is a geometric isomer of AB, and (MAB)* is an exciplex of M* and AB which can decay by emission of light quanta with less energy than the M* electronic energy.

$$
\left.
\begin{aligned}
M^* + AB &\to M + AB(V,R) \\
&\to M^{**} + AB(V,R) \\
&\to M + AB^*(V,R) \\
&\to M^{**} + AB^*(V,R)
\end{aligned}
\right\} \text{Physical}
$$

$$
\left.
\begin{aligned}
M^* + AB &\to M + A + B \\
&\to MA + B \\
&\to MB + A \\
&\to M + BA(V,R)
\end{aligned}
\right\} \text{Chemical}
$$

$$
M^* + AB \to (MAB)^* \to M + AB(V,R) + h\nu\} \quad \text{Exciplex emission}
$$

The exit channels accessible depend on the properties of the quenching molecule as well as the energy of the excited state of the metal atom. In this

TABLE IX. Bond strengths.

Species	D_0^0 (kcal mol^{-1})	Species	D_0^0 (kcal mol^{-1})
LiH	56·01[250]	LiO	81[250]
NaH	47[250]	NaO	60[312]
KH	43[250]	KO	56[313]
RbH	39[250]	RbO	60[313]
CsH	41,[250] 41·7[311]	CsO	66[313]
MgH	29·3[314]	MgO	94,[250] 82[315]
CaH	< 39[250]	CaO	95[7]
SrH	38·0[250]	SrO	97,[250] 95[315]
BaH	41[250]	BaO	133,[250] 125[315]
ZnH	19·5[250]	ZnO	65[250]
CdH	15·6[250]	CdO	< 88[250]
HgH	8·6[250]		

section, for a representative set of quenchers, we discuss the possible exit channels, summarize experimental evidence for the major exit channels in individual cases, and list the few quantitative experimental determinations of branching ratios into competitive exit channels.

In Table IX are listed the bond strengths of possible metal hydride and metal oxide products in chemical exit channels. The energetics of metal hydride formation in the reactions of the excited states with molecular hydrogen are shown in Table X.

A. N_2 and CO

Because the lowest-lying excited states of N_2 and CO are at 142·1 and 138·5 kcal mol^{-1}, respectively,[237] for all the metal atom states except Hg($^1P_1$) considered in this review, quenching must necessarily result in vibrational, rotational, and/or translational excitation of the ground-state molecules. The efficiency of this sort of energy dissipation appears to depend on the accessibility of facilitating charge-transfer surfaces, as discussed above in

TABLE X. Energetics of metal hydride formation in the quenching of excited metal atom states by H_2.

State	$\Delta E(M^* + H_2 \rightarrow MH + H)$ (kcal mol^{-1})
Li(^2P)	+4·6
Na(^2P)	+7·8
K($^2P_{1/2}$)	+23
Rb($^2P_{1/2}$)	+28
Cs($^2P_{1/2}$)	+30
Mg(^3P_J)	+11·6
Ca($^3P_1$)	> +20·8
Sr($^3P_1$)	+24
Ba($^3P_1$)	+26
Zn($^3P_1$)	−9·2
Cd($^3P_1$)	0·0
Hg($^3P_1$)	−18·1
Mg($^1P_1$)	−26·1
Ca($^1P_1$)	> −3·3
Sr($^1P_1$)	+3·3
Ba($^1P_1$)	+10·7
Zn($^1P_1$)	−49·9
Cd($^1P_1$)	−37·4
Hg($^1P_1$)	−60·0

Sec. III. Recent experiments in which initial distributions of vibrational energy have been determined are discussed in Sec. V.

In the quenching of $Cd(^1P_1)$ by N_2 and CO, the predominant exit channel is the spin-forbidden production of $Cd(^3P_J)$ rather than ground-state $Cd(^1S_0)$:[152,238]

$$Cd(^1P_1) + M \rightarrow Cd(^3P_J) + M. \tag{3}$$

The lower limits for the branching ratios for $Cd(^3P_J)$ production have been measured to be 0·83 and 0·90, respectively.[152,238] Process (3) is also known to be a major exit channel in the quenching of $Hg(^1P_1)$ by N_2 and CO.[239,240] In marked contrast to the $Hg(^1P_1)$ and $Cd(^1P_1)$ states, $Zn(^1P_1)$ is quenched directly to ground-state $Zn(^1S_0)$ by CO and N_2, with little or no $Zn(^3P_J)$ production.[154] It was necessary, in the Breckenridge and Renlund study[154] in which this conclusion was made, to assume that CO and N_2 quenched $Zn(^3P_J)$ inefficiently, but independent measurements have now confirmed this assumption.[155,189] Very recent experiments have shown that production of $Mg(^3P)$ is also a minor exit channel (branching ratio much less than 0·01) in the efficient quenching of $Mg(^1P_1)$ by N_2.[157] The decrease in spin–orbit coupling for Hg, Cd, Zn, and Mg (the $^3P_0$–$^3P_1$ energy gaps are 5·0, 1·55, 0·54, and 0·06 kcal mol^{-1}, respectively) apparently results in a decreasing probability of singlet-to-triplet surface crossings, although the sudden switch-over in exit channels from Cd to Zn seems rather abrupt for a change in spin–orbit coupling of only a factor of 3.

B. NO

The lowest $(a^4\Pi)$ excited state of NO lies at 109·4 kcal mol^{-1},[242] so that for all the states in Table I except the $^1P_1$ levels of Zn and Cd, and the $^1P_1$, $^3P_2$, and $^3P_1$ states of Hg, only electronic-to-vibrational, or electronic-to-rotational (E–V, R) transfer can occur, analogous to N_2 and CO (see Sec. V).

Electronic energy transfer producing the $NO(A^2\Sigma^+)$ state at 125·9 kcal mol^{-1} has been identified as an exit channel in the quenching of $Cd(^1P_1)$,[243] $Zn(^1P_1)$,[244,245] and $Hg(^3P_2)$[207] by NO. For $Cd(^1P_1)$, $NO(A^2\Sigma^+)$ production is slightly endothermic, but nevertheless occurs[243] with a branching ratio of 0·04 and a cross section of $\sim 3\,\text{Å}^2$. No branching ratio or absolute cross section for $NO(A^2\Sigma^+)$ production in the quenching of $Zn(^1P_1)$ by NO has been measured, but arguments can be made which bracket the specific electronic energy transfer cross section in the 1–10 Å^2 range.[245] For $Hg(^3P_2)$, an upper limit for the cross section for production of $NO(A^2\Sigma^+)$ was measured in a molecular beam experiment to be 1·4 Å^2 at $E = 1·9$ kcal mol^{-1}.[207] It was shown in the same series of experiments that intramultiplet quenching of $Hg(^3P_2)$ to $Hg(^3P_1)$ was about 10 times more rapid than formation of $NO(A^2\Sigma^+)$, so that the branching ratio for $NO(A^2\Sigma^+)$ production is 0·10 or less for $Hg(^3P_2)$ quenching. The higher-lying $NO(B^2\Pi)$ state (129·7 kcal) was also detected when the relative translation energy of the crossed $Hg(^3P_2)$ and NO beams was increased.[207]

Production of the $NO(a^4\Pi)$ state was postulated in the quenching of $Hg(^3P_1)$ by NO,[246] but has never been observed directly in the quenching of any of the states of interest here. Because of its metastable character, detection by emission is virtually impossible, so that absorption or laser-induced fluorescence measurements will be required to confirm $NO(a^4\Pi)$ as a product.

In the $Cd(^1P_1)$ case, recent experiments have shown that production of the $Cd(^3P_J)$ levels occurs with a branching ratio of nearly 1·0.[152] It is entirely possible that $Zn(^3P_J)$ production is the major exit channel in the quenching by NO of $Zn(^1P_1)$ as well, since the process is spin-allowed.

C. O_2

The oxygen molecule is an unusual diatomic quencher in that there are two low-lying electronic states,[237] $O_2(^1\Delta)$ at 22·5 kcal mol^{-1} and $O_2(^1\Sigma^+)$ at 37·5 kcal mol^{-1}. Electronic energy transfer is therefore a possible exit channel for every excited state considered here. The only evidence for $O_2(^1\Delta)$ production is indirect. In crossed-beam, crossed-laser experiments, Hertel, Hofmann and Rost[247] measured the velocity distribution of product Na atoms resulting from the quenching of $Na(^2P)$ by several molecules. For O_2, there appeared to be structure in the velocity distribution which occurred at just the right thresholds for $O_2(^1\Delta_g)$ and $O_2(^1\Sigma_g^+)$ production. Silver, Blais, and Kwei,[4] who performed similar experiments but detected the molecular products, found no evidence for $O_2(^1\Delta)$ or $O_2(^1\Sigma_g)$ production in their velocity distributions. They point out, however, that fragmentation of $O_2(^1\Delta)$ and/or $O_2(^1\Sigma^+)$ into atomic ions may have occurred in their electron bombardment analyser.[4]

Chemical reaction is known to be a major exit channel in several cases:

$$M^* + O_2 \rightarrow MO + O. \tag{4}$$

The high exothermicity of process (4) for many states also allows the possibility of electronic excitation in the metal oxide products. Dagdigian, Alexander, and coworkers have conducted a series of experimental and theoretical studies of the reactions of the alkaline-earth metastable states with oxidants such as O_2.[7,196,248,249] Chemiluminescence of CaO excited states from the interaction of a molecular beam of $Ca(^3P)$ and $Ca(^1D_2)$ metastables with O_2 has recently been shown to be due entirely to the $Ca(^1D_2)$ state,[7] while reaction of $Ca(^3P_J)$ with O_2 produces ground-state CaO, probably for reasons of energy conservation (the dissociation energy of CaO is still a matter of controversy). Absolute cross sections for the production of $CaO(A'^1\Pi)$ and $CaO(A^1\Sigma^+)$ excited states in the reaction of $Ca(^1D_2)$ with O_2 were determined to be 3·2 and

$1.1 Å^2$, respectively.[7] The dynamics of these reactions are dominated by the intermediate formation of the stable $Ca^+O_2^-$ charge-transfer complex.[248]

The energy of $Hg(^3P_1)$ is sufficient for electronic-to-electronic (E–E) transfer to several higher-lying electronic states of O_2. Early workers postulated the formation of various states based entirely on indirect evidence.[251–254] Hippler, Wendt and Hunziker[255] have recently obtained direct evidence for a long-lived ($\geqslant 5 \mu s$) intermediate in the quenching of $Hg(^3P_1)$ by O_2 which they believe is likely to be one of the A, C, or c metastable excited states of O_2. Highly vibrationally excited O_2, also detected as a product, was efficiently quenched by CO_2 addition, but the long-lived intermediate was not.

D. CO_2

The lowest excited electronic state of CO_2 is the bent $^3B_2$ state at 80 ± 10 kcal mol^{-1}.[256] For the alkali ^2P states, where the chemical channels to form alkali oxides are probably all endothermic (see Table IX), E–V, R transfer is the only accessible exit channel at low relative velocities. For all the excited states of the alkaline earths, metal oxide production is exothermic. In molecular beam experiments, ground-state CaO was observed as a product in the quenching of $Ca(^3P)$ by CO_2,[249] and $Ca(^1D_2)$ was shown to react with CO_2 to produce the excited $A'^1\Pi$ and $a^3\Pi$ states of CaO.[250] Although reaction of $Hg(^3P_1)$ with CO_2 to form HgO and CO is probably exothermic, direct $Hg(^3P_1)$-sensitized decomposition of CO_2 does not occur,[19] so that quenching must proceed via excited triplet CO_2 or E–V, R transfer.

In the quenching of $Cd(^1P_1)$, CO_2 was one of the very few quenchers for which $Cd(^3P_J)$ production was not observed.[152] It was suggested that a competing chemical channel might be responsible, but no direct evidence was obtained. Direct scission of the double bond in CO_2 by $Cd(^1P_1)$ is just slightly endothermic.

E. N_2O

Dagdigian and coworkers[196,198] have shown that $Mg(^3P)$, $Ca(^3P)$, and $Sr(^3P)$ all react with N_2O to produce the corresponding metal oxides. Surprisingly large branching ratios were found for the production of fluorescent electronically excited metal oxide states in these studies:[196,198] 0·10 for $Ca(^3P)$–N_2O, 0·40 for $Sr(^3P)$–N_2O, and ~ 0.8 for $Mg(^3P)$–N_2O (the latter calculated using the cross section for total quenching from ref. 12). However, it now appears (Dagdigian, private communication) that the $Mg(^3P) + N_2O$ measurements were in error and that the chemiluminescence yield is essentially zero.

The only other process which has been studied is the well known sensitized production of $O(^3P)$ atoms in the quenching of $Hg(^3P_{0,1})$ by N_2O:[19]

$$Hg(^3P_{0,1}) + N_2O \rightarrow Hg(^1S_0) + N_2 + O(^3P). \tag{5}$$

Early monoisotopic sensitization[257] and chemical scavenging[258] experiments proved conclusively that little HgO was produced in the quenching, and that the branching ratio for $O(^3P)$ production was approximately 1·0.

F. H₂ and D₂

There are no low-lying, non-dissociative excited states of the H_2 molecule. The chemical channel

$$M^* + H_2 \rightarrow MH + H$$

is endothermic for all the alkali ^2P states, leaving only E–V, R transfer as an exit channel at low thermal energies for these species (see Table X). For $Li(^2P)$, the reaction to from LiH is only 4·6 kcal mol^{-1} endothermic and may be important at flame temperatures. Muller, Schofield, and Steinberg[259] have recently demonstrated that metal hydride formation is an exit channel in the quenching of $Na(^2P)$ and $Li(^2P)$ by H_2 in fuel-rich flames. For $Na(^2P)$, the branching ratio for NaH formation was only ~ 0.005 at 1925 K, compared with the ~ 0.10 expected if the only limitation to reaction is the 7·8 kcal mol^{-1} endothermicity. Even this inefficient channel provides a sodium chemical reservoir which can cause serious errors in laser saturation measurements of Na in flames, however.[259] For $Li(^2P)$ an upper limit for the LiH formation branching ratio was determined[259] which was roughly the same as expected from the thermal rate-constant upper limit at 1925 K.

The quenching of the $Mg(^3P_J)$, $Zn(^3P_J)$, $Cd(^3P_J)$, and $Hg(^3P_{0,1})$ states by H_2 has now been well studied, and it appears that the exit channels are entirely *chemical* in nature when the quenching is efficient. In the $Hg(^3P_1)$ and $Hg(^3P_0)$ cases, there are two possible exit channels:

$$Hg(^3P) + H_2 \rightarrow HgH + H, \tag{6a}$$

$$Hg(^3P) + H_2 \rightarrow Hg + H + H. \tag{6b}$$

Callear and Wood[260] have shown that the sum of the branching ratios for processes (6a) + (6b) is 0.93 ± 0.10, indicating that E–V, R transfer to H_2 is a minor channel. The branching ratio for process (6a) was measured as 0·62, and that for process (6b) as 0·31, indicating that HgH formation predominates over direct bond scission. In contrast to the Callear work, Lee *et al.*[261] have claimed that production of vibrationally excited H_2 is a major exit channel in the quenching of $Hg(^3P_1)$ by H_2. However, there was no apparent attempt made in those experiments to filter out 1849 Å resonance emission from a microwave Hg lamp, and vibrationally excited H_2 could have been formed by the following process of deactivation of the resultant $Hg(^1P_1)$:

$$Hg(^1P_1) + H_2 \rightarrow Hg(^3P_J) + H_2^{\ddagger}. \tag{7}$$

This E–V, R process occurs readily for the analogous $Cd(^1P_1)$ state (see below).

In the $Cd(^3P_J)$ and $Zn(^3P_J)$ cases, metal hydride formation is the only chemical channel open, and resonance radiation flash photolysis experiments[154,233] indicate that CdH and ZnH formation are the major exit channels, though it was not possible to measure branching ratios. In the $Cd(^3P_J)$ case, the CdH yield was shown to be nearly the same as the CdH yield for several alkane hydrocarbons, and it was argued that the only sensible interpretation of the data was that the yield was unity.[188,233] The production of CdH is just thermoneutral in the $Cd(^3P_1) + H_2$ case, and yet quenching occurs at one in three collisions,[233] showing that there cannot be a substantial activation barrier for the reaction ($\leqslant 1.0$ kcal mol^{-1}). (An earlier claim by Strausz and coworkers[262,263] of having measured an activation energy of ~ 5 kcal mol^{-1} appears to have been based on sparse and incorrect data.)

The isotope effect on the cross section for quenching of $Cd(^3P_1)$ by D_2 versus H_2 is also consistent with CdH(CdD) formation as the dominant exit channel.[233] The production of CdD from D_2 is 1.2 kcal mol^{-1} *endothermic* because of the differences in zero-point energies of reactant and product diatomic molecules. The ratio of the $Cd(^3P_1)$ cross section for D_2 versus H_2 quenching has just been measured by Umemoto et al.[184] to be 0.38 ± 0.10, which is consistent with the ratio of ~ 0.3 expected if the activation barrier to CdD formation is 1.2 kcal mol^{-1} higher than that for CdH formation. It should be noted that for the quenching of $Hg(^3P_0)$, $Hg(^3P_1)$, and $Zn(^3P_J)$ by H_2, where the chemical channel is exothermic for all isotopes, the H_2/D_2 isotope effect on the quenching cross section is ~ 1.0, within experimental error (see Table VI).

Recent experimental[193] and theoretical[264] studies of the quenching of $Mg(^3P)$ by H_2 provide striking confirmation of the chemical nature of the interaction of the ^3P levels with H_2, and also give hard evidence for a case of specific exit-channel control of quenching rates. In contrast to the highly efficient quenching of the $Hg(^3P_J)$, $Cd(^3P_J)$, and $Zn(^3P_J)$ levels by H_2, $Mg(^3P)$ is known to be quenched by H_2 only once in 10^3 to 10^4 collisions (see Table VI). This is easily rationalized by the fact that MgH production is 11.5 kcal endothermic.[193] Thus the chemical channel must have an activation barrier of at least 11.5 kcal mol^{-1} and will be inefficient even at 800 K. E–V, R transfer is also predicted to occur with low cross section in this case by the Bauer–Fisher–Gilmore charge-transfer considerations outlined in Sec. III. In addition, the spin-change restriction in the E–V, R process could play some role for an atom so light as magnesium.

Shown in Fig. 4 is an Arrhenius plot of the variation of the rate constant for

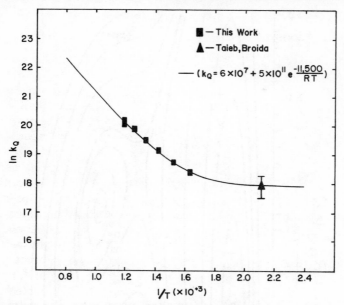

Fig. 4. An Arrhenius plot of the variation of the rate of quenching of $Mg(^3P_J)$ with temperature. From Breckenridge and Nikolai.[193] Reproduced by permission of the American Institute of Physics.

the quenching of $Mg(^3P)$ by H_2 with temperature.[193] The full curve is the result of a very simple model based on chemical exit-channel control at high temperatures, E–V, R control at low temperatures:

$$k(\text{l mol}^{-1}\text{ s}^{-1}) = 6 \times 10^7 + 5 \times 10^{11} \exp(-11\,500/RT). \qquad (8)$$
$$\text{(E–V, R)} \quad \text{(chemical)}$$

The chemical exit-channel expression was constructed merely by using the gas kinetic A factor (equivalent to a cross section of $30\,\text{Å}^2$) and an activation energy equal to the endothermicity of MgH formation. The data are fitted remarkably well by the model. The presence of MgH was also detected[193] by laser-induced fluorescence in the quenching of $Mg(^3P)$ by H_2 at ~ 800 K.

Ab initio theoretical calculations of the lowest $Mg(^3P_J) + H_2$ potential surface[264] have been performed which are entirely consistent with the postulated chemical pathway. Shown in Fig. 5 is a contour-line representation of the relevant C_{2v} (side-on approach) portions of the potential surface of the lowest ($^3B_2$) triplet state of MgH_2, on which $Mg(^3P) + H_2$ can be adiabatically converted to MgH + H. The SCF level calculations were adequate for three main conclusions to be made:

(1) The side-on approach provides a pathway by which $Mg(^3P)$ can react to

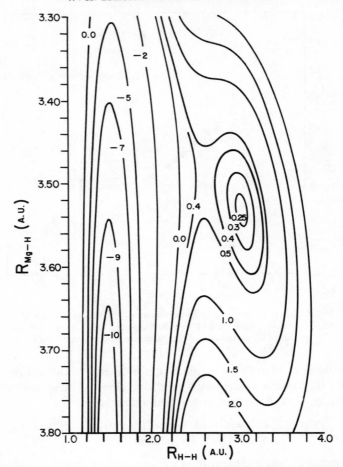

Fig. 5. A contour plot of the C_{2v} portion of the potential surface of the lowest triplet ($^3B_2$) state of MgH_2 on which the reaction $Mg(^3P_J)$ + $H_2 \rightarrow MgH + H$ takes place. Plot constructed from a fit of SCF-level calculations by Adams, Breckenridge, and Simons.[264]

yield $MgH + H$ with no substantial barrier ($\leqslant 1.5$ kcal) beyond the 11.5 kcal mol^{-1} endothermicity. A few points done on the surface at the (MSCF–CI) level are consistent with this conclusion. On the other hand end-on approach of $Mg(^3P) + H_2$ generates a substantial barrier (~ 15 kcal mol^{-1} at the (SCF) level) beyond the endothermicity. Thus the theoretical results provide justification for the assumption in the model above that $Mg(^3P)$ can react with H_2 with no activation barrier beyond the endothermicity.

(2) From orbital population analysis, the 'seam' of the C_{2v} potential surface

at $R_{H_2} \approx 1 \cdot 3 \text{Å}$ appears to correspond to an avoided crossing between a diabatic surface of $Mg(^3P) + H_2$ character and a surface which has substantial $Mg^+ H_2^-$ character. This is undoubtedly the realistic counterpart to the simple Coulombic charge-transfer surface of the BFG model. The fact that the crossing occurs at high energies (10 kcal mol^{-1} or more above the asymptotic $Mg(^3P) + H_2$ energy) is consistent with the rough predictions of the BFG theory that E–V, R transfer will be slow in this system.

(3) There is apparently no bond $MgH_2(^3B_2)$ surface, minimizing any possibility of overlap with the ground-state MgH_2 surface which might lead to E–V, R transfer.

It should be noted that because of the shape of the $Mg(^3P) - H_2$ potential surface, H_2 vibrational energy might be expected to be particularly effective in causing reaction. In fact, $H_2(v = 1)$ has $11 \cdot 6 \text{ kcal mol}^{-1}$ energy and could be the sole reacting species in the Boltzmann equilibrium. Experiments with D_2 and with $H_2(v = 1)$ are planned or in progress which should yield some information about the effectiveness of vibrational energy in this interesting case.[265]

Production of the metal hydride is also known to be an exit channel in the quenching of the $^1P_1$ levels of Cd,[11,233,266] Zn,[154] and Mg[157,241] by H_2. In the $Cd(^1P_1)$ case, however, direct competitive production of $Cd(^3P_J)$ also occurs with branching ratios[238,266] of $0 \cdot 28 \pm 0 \cdot 08$ for H_2 and $0 \cdot 33 \pm 0 \cdot 07$ for D_2. Combining earlier resonance-radiation data[233] with these results, it is possible to estimate that the branching ratio for direct CdH production in the quenching of $Cd(^1P_1)$ by H_2 is only $0 \cdot 18 \pm 0 \cdot 04$ (assuming the branching ratio for CdH production in the quenching of $Cd(^3P_J)$ by H_2 to be unity). A recent direct determination[266] of the branching ratio for $CdH(v = 0)$ initial production in the quenching of $Cd(^1P_1)$ by H_2, performed using the 'pump-and-probe' technique by comparing the integrated intensity of the LIF signals from $Cd(^3P_J)$ and $CdH(v = 0)$, yields a consistent value of $0 \cdot 12$. Thus it appears that the quenching of $Cd(^1P_1)$ by H_2 results in several competitive modes of electronic energy dissipation: $\sim 20\%$ into direct production of CdH, $\sim 30\%$ via an E–V channel to produce $Cd(^3P_J) + H_2^\ddagger$, and presumably $\sim 50\%$ into direct sensitized scission of H_2 to form $Cd(^1S_0) + H + H$.

In contrast, because of the lower energy of $Mg(^1P_1)$, direct scission of the H–H bond is endothermic and thus must occur with a branching ratio of $0 \cdot 05$ or less at moderate temperatures. Since $Mg(^3P)$ production has also been shown to be inefficient in recent experiments,[147] it is quite likely that the major exit channel in the quenching of $Mg(^1P_1)$ by H_2 is MgH + H (although production of $Mg(^1S_0)$ and vibrationally excited H_2 cannot be ruled out). Very recent 'pump-and-probe' experiments by Umemoto and Breckenridge do reveal a substantial yield of MgH in the quenching of $Mg(^1P_1)$ by H_2.[241]

G. Alkane Hydrocarbons

The alkane hydrocarbons also possess no low-lying non-dissociative electronically excited states. Except for reactions of $Li(^2P)$ with secondary and tertiary C–H bonds, and $Na(^2P)$ with tertiary C–H bonds, the metal hydride formation exit channels,

$$M^* + RH \rightarrow MH + R \qquad (9)$$

are closed for the alkali ^2P states, so that only inefficient E–V, R transfer is possible (see Table III).

The quenching of the ^3P levels of Mg, Zn, Cd, and Hg by alkanes offers an interesting variety of chemical exit channel possibilities. For the $Hg(^3P_{0,1})$ states, both direct scission and HgH formation are exothermic for all alkanes. For $Zn(^3P_J)$, direct C–H bond scission is exothermic only for tertiary C–H bonds, ZnH formation being exothermic for all hydrocarbons. In the $Cd(^3P_J)$ case, direct scission of C–H bonds is endothermic, CdH formation being exothermic for all alkanes. For $Mg(^3P_J)$, direct scission of all C–H bonds and MgH formation for all but tertiary C–H bonds are endothermic. For the higher hydrocarbons, C–C bond scission exit channels are exothermic for $Cd(^3P)$, $Zn(^3P)$, and $Hg(^3P)$.

There is a large body of evidence[19,170,171,174,268] which indicates that both $Hg(^3P_0)$ and $Hg(^3P_1)$ interact chemically with specific C–H bonds, and that the rate of reaction increases in the order, primary to secondary to tertiary C–H bonds. Holroyd and Klein[267] pointed out early that the $Hg(^3P_1)$ quenching cross sections for normal alkanes could be reproduced fairly well by assigning 'effective' cross sections for primary, secondary, and tertiary C–H bond quenching, then summing the cross sections for each C–H bond in the alkane. The cross sections needed are ~ 0.05, ~ 3.0, and $\sim 16 \text{Å}^2$ for primary, secondary, and tertiary C–H bonds, respectively, corresponding to ratios of 1 : 60 : 320. Several workers have proposed mechanisms in which the initial step is the formation of a weak complex between $Hg(^3P_1)$ and the alkane quencher. The resulting complex can then either radiate or decompose (if there is sufficient energy) by surmounting a potential barrier, the height of which is dependent on the type of C–H bond. Such a general model was originally proposed by Yang,[174] and estimates of the rate of passage over the barriers have been made by the use of Kassel,[174] RRKM,[268] and, BEBO[170] treatments.

Callear and McGurk have shown conclusively via the resonance radiation flash photolysis technique that, in contrast to H_2, alkanes quench $Hg(^3P_{0,1})$ by direct C–H bond scission and not HgH formation.[269] Phaseshift measurements confirm the lack of HgH production.[270] The HgH product seen by Vikis[271] is undoubtedly due to $Hg(^3P_J)$ quenching by product H_2 over the longer effective times characteristic of his flow-tube experiments.

In contrast to $Hg(^3P_{0,1})$, $Cd(^3P_J)$ is quenched by all the normal alkane hydrocarbons very inefficiently (see Table VI), and CdH is formed in yields comparable to the H_2 case.[188] Breckenridge and Renlund[188] propose that the mechanism for C–H bond scission in the $Hg(^3P_{0,1})$ case is merely energetically inaccessible for $Cd(^3P_J)$, and that by default an inefficient surface-crossing mechanism leading to CdH production is responsible for the quenching.

Product analysis studies of the quenching of both $Hg(^3P_{0,1})^{19}$ and $Cd(^3P_{0,1})^{24}$ by alkanes have provided no evidence for major primary exit channels involving sensitized C–C cleavage. This implies that the specific interaction of these excited atoms with C–C bonds is not chemically favourable, possibly due to steric hindrance. The electronic energy apparently does not delocalize effectively over the entire molecule either, since C–H bonds are always broken instead of the weaker C–C bonds.

Consistent with chemical exit channel control, quenching of both $Zn(^3P)$ and $Mg(^3P)$ by the only alkane studied, CH_4, occurs with small cross sections.

Just as in the cases of N_2 and CO as quenchers, the major exit channel for quenching of $Cd(^1P_1)$ and $Hg(^1P_1)$ by the alkane hydrocarbons appears to be spin-forbidden production of $Cd(^3P_J)$ and $Hg(^3P_J)$ states, respectively.[147,149,152,153] This is remarkable since both direct C–H bond scission as well as metal hydride formation are highly exothermic in both cases. Recent measurements for a variety of alkanes indicate that the branching ratios for $Cd(^3P_J)$ formation are all near unity.[152] This implies a very strong interaction between the singlet and triplet potential surfaces, and the probable explanation, as discussed in Sec. III above, is the bound character of the singlet surface which corresponds to the insertion of a $^1P_1$ atom into a C–H bond, resulting in crossings with repulsive $Cd(^3P)$–RH surfaces at moderate approach distances. Very recent measurements[272] have revealed that direct CdH production also occurs as a minor exit channel for $Cd(^1P_1)$, with branching ratios of the order of only 0·05.

For $Zn(^1P_1)$, again analogous to N_2 and CO, $Zn(^3P_J)$ formation is definitely negligible for CH_4 and probably for the higher alkanes.[154] Definite confirmation must await measurements of quenching cross sections for $Zn(^3P)$ by the higher homologues. Large yields of ZnH were observed in the quenching of $Zn(^1P_1)$ by CH_4.[154] The relative importance of direct C–H bond scission and ZnH formation is not known, however.

Recent experiments show that the quenching of $Mg(^1P_1)$ by CH_4 and C_3H_8 also produce MgH in good yield, but that the formation of $Mg(^3P_J)$ is extremely inefficient.[241] Thus it appears that the metal hydride formation channel becomes important by default when the singlet–triplet exit channel closes for the lighter atoms Zn and Mg.

H. Unsaturated Hydrocarbons

The energies of the lowest (triplet) excited states of simple alkene and alkyne hydrocarbons are not known precisely, partly because the geometries of the molecules are usually drastically different from the ground state.[273-275] It is unlikely, except for conjugated polyenes, that any of the triplet states lie below $50 \, \text{kcal} \, \text{mol}^{-1}$, however,[273] so that quenching of all the ^2P alkali levels by simple alkenes and alkynes must result in E–V, R transfer.

For the ^3P levels of Hg, Cd, and Zn, triplet–triplet energy transfer is probably exothermic for all the simple alkenes and alkynes, as well as for aromatic species such as benzene and toluene.[186] There is considerable evidence that formation of the triplet state is indeed the major exit channel in the quenching of the ^3P levels of Hg, Cd, and Zn by simple alkenes. Efficient isomerization of ethylene-d_2 and cis-butene have been observed, for example, in the quenching of $Hg(^3P_1)$ and $Cd(^3P_J)$.[276-282] Isomerization of propylene-1,3,3,3-d_4 has also been observed in the quenching of $Cd(^3P_J)$.[283] Burton and Hunziker[284] detected, via biacetyl fluorescence, an intermediate in the quenching of C_2H_2 by $Hg(^3P_1)$ which they identified as triplet acetylene. The phaseshift technique developed by Hunziker has since been utilized for direct spectroscopic detection of the first triplet state ($^3B_2$) of C_2H_2 as a product of the quenching of $Hg(^3P_0)$ by acetylene.[285] A study by Hunziker on the effect of benzene on the $Cd(^3P_1)$-photosensitized isomerization of butene-2 provided indirect evidence for production of $C_6H_6(\tilde{a}^3B_{1u})$ in the quenching process.[286] Subsequent studies using the phaseshift technique confirmed spectroscopically that $C_6H_6(\tilde{a}^3B_{1u})$ is formed in the quenching of $Hg(^3P_0)$ by benzene.[287] Triplet toluene and triplet anthracene were also detected in the quenching of $Hg(^3P_0)$ by the respective ground-state molecules.[287,288]

A novel study has been reported[289] of the quenching of $Cd(^3P_1)$ by the organic scintillator vapour POPOP, which is a candidate for a vapour-phase dye laser material. Since the absorption spectrum of POPOP vapour corresponding to a singlet–singlet transition has maximum intensity just at the $Cd(^1S_0 \rightarrow {}^3P_1)$ transition wavelength, the question arises as to the relative efficiency of near-resonant production of the singlet state versus non-resonant production of the triplet state of POPOP in the quenching process. Long-range dipole–dipole theory predicts a rate of transfer from $Cd(^3P_1)$ donor to POPOP acceptor (forming the singlet state) of about $3 \times 10^{10} \, \text{l} \, \text{mol}^{-1} \, \text{s}^{-1}$, in excellent agreement with the rate constant determined by measuring POPOP fluorescence quantitatively, of $2\cdot5 \times 10^{10} \, \text{l} \, \text{mol}^{-1} \, \text{s}^{-1}$.[289] The total rate of quenching, however, was determined to be $3\cdot5 \times 10^{11} \, \text{l} \, \text{mol}^{-1} \, \text{s}^{-1}$, thus indicating that triplet–triplet transfer still dominates (although E–V, R transfer to ground-state POPOP may occur to some extent).

The only $^1P_1$ level for which detailed exit-channel information is available for unsaturated hydrocarbons is $Cd(^1P_1)$. Recent evidence[152] shows that production of $Cd(^3P_J)$ occurs with a branching ratio near unity for a variety of alkenes, analogous to the alkane case. Quantum-state-resolved measurements of the $Cd(^3P_J)$ distributions have been interpreted using a mechanism involving competitive attack of $Cd(^1P_1)$ at both C–H bond and π-bond sites[152] (see Sec. V). A very recent experiment[241] shows that MgH is formed in the quenching of $Mg(^1P_1)$ by C_3H_6. Again, the metal hydride exit channel appears to open when the singlet-triplet pathway is closed.

I. Halogens and Halogen-Containing Species

Because the bond strengths of the ionic metal monohalides are fairly high compared with halogen–halogen or carbon–halogen bonds (except perhaps for C–F bonds), a chemical exit channel is usually available for the states of interest here in the quenching of halogen-containing compounds:

$$M^* + X_2 \rightarrow MX + X, \tag{10}$$

$$M^* + RX \rightarrow MX + R. \tag{11}$$

A wide variety of results indicate that chemical channels predominate, leading either to ground-state or electronically excited metal halide products.

For quenching of the ^2P group I levels by halogen-containing molecules, metal halide formation has not been observed directly (partly because the alkali halides are not easily detected spectroscopically), but the quenching has been interpreted[125,127,128,139,290] within the charge-transfer surface 'harpoon' mechanism framework developed for molecular beam results on ground-state alkali plus halogen reactions, where metal halides are detected as products.[291]

The only direct evidence, apparently, for ground-state metal halide production in the quenching of the ^3P levels of Zn, Cd, and Hg is the detection of CdF as a major product of the quenching of $Cd(^3P_J)$ by SF_6 and PF_3.[186] Because of the recent interest in mercury halide laser systems, however, several workers have reported the production of excited $HgX(B^2\Sigma^+)$ in the quenching of $Hg(^3P_J)$ states by halogen-containing species. The systems include: $Hg(^3P_0)$ + Cl_2;[179] $Hg(^3P_0)$ + Br_2;[292] $Hg(^3P_1)$ + Cl_2, I_2, Br_2, ICl;[290] $Hg(^3P_2)$ + Cl_2, halogenated methanes.[293] Collins and coworkers[176] have observed that $HgBr(B^2\Sigma^+)$ is *not* produced in the quenching of $Hg(^3P_1)$ by CBr_4 and $CHBr_3$, however, even though the reaction is definitely exothermic.

TABLE XI. Cross sections for production of $HgCl(B^2\Sigma^+)$ and $Hg(^3P_1)$ in the quenching of $Hg(^3P_2)$ by several molecules. From Krause et al.[293] Reproduced by permission of Amer. Inst. Physics

Quencher	Cross section for $Hg(^3P_1)$ production ($Å^2$)	Cross section for $HgCl(B^2\Sigma^+)$ production ($Å^2$)
Cl_2	0·13	90 (± 25)
CCl_4	0·22	34 (± 10)
$CHCl_3$	1·0	7·9 ($\pm 1·8$)
CH_2Cl_2	3·8	~ 0
CH_3Cl	6·3	~ 0

Shown in Table XI are the cross sections for $HgCl(B^2\Sigma^+)$ production compared with the cross sections for $Hg(^3P_1)$ production via intramultiplet relaxation in the quenching of $Hg(^3P_2)$ by several halogen-containing molecules.[293] (Note that only the cross sections for exit channels with *emitting* species were measured; the total quenching cross sections could be much higher than the sum of the two specific exit-channel cross sections.) It can be seen that production of $HgCl(B^2\Sigma^+)$ and $Hg(^3P_1)$ seem to be negatively correlated, as if $HgCl(B^2\Sigma^+)$ is produced only at the expense of $Hg(^3P_1)$ as C–Cl bonds sequentially replace C–H bonds. It should also be noted that with such a large cross section for $HgCl(B^2\Sigma^+)$ production in the Cl_2 case, the branching ratio must necessarily be near unity. In contrast, Wodarczyk and Harker[179] have shown that the branching ratio for $HgCl(B^2\Sigma^+)$ production in the quenching of $Hg(^3P_0)$ by Cl_2 is only ~ 0.01.

Several studies have provided evidence for ground-state and excited metal halide production in the quenching of excited alkaline-earth atoms. In molecular beam experiments, Solarz and coworkers have detected,[294a,294b] by laser-induced fluorescence, ground-state barium halides and strontium halides as products of the quenching of $Ba(^3P_J)$ and $Sr(^3P_1)$ by a variety of halogen-containing species. Electronically excited metal halide products have been observed, by chemiluminescence in molecular beam systems, for the following: $Ca(^3P)$, $Ca(^1D_2) + HCl$, Cl_2;[197] $Mg(^3P)$, $Sr(^3P) + Cl_2$;[295] $Ba(^3D) + X_2$;[296] $Mg(^3P), Sr(^3P) + X_2$;[12] $Ca(^3P)$, $Ca(^1D_2) + X_2$;[297] $Ca(^3P) + HCl$;[298] $Sr(^3P) + Cl_2$.[298]

J. Quenching Molecules for which Exciplex Emission is a Major Exit Channel

For several molecules with unshared pairs of electrons, such as NH_3, H_2O, amines, and alcohols, exciplex emission has been observed to be an important exit channel in the quenching of $Hg(^3P_0)$ and $Cd(^3P_J)$. Work on $Cd(^3P_J)$ exciplexes has recently been reviewed,[24] and will not be discussed here. The experiments before 1973 on $Hg(^3P_J)$ exciplex emission have been reviewed by Phillips.[23]

Recent published papers in this area have concentrated on the determination of the detailed kinetics of the $Hg(^3P_0)$–NH_3 system, where bimolecular and termolecular formation of $Hg(NH_3)^*$ complexes is known to occur. The mechanism originally proposed by Callear and coworkers[299,300] is now generally acknowledged to be correct:[301,302,304]

$$Hg(^3P_1) + NH_3 \rightarrow Hg(^3P_0) + NH_3, \tag{12}$$

$$Hg(^3P_0) + NH_3 \rightarrow Hg(^1S_0) + NH_3 + h\nu(\text{short}), \tag{13}$$

$$Hg(^3P_0) + NH_3 + M \rightarrow HgNH_3^* + M, \tag{14}$$

$$HgNH_3^* \rightarrow Hg(^1S_0) + NH_3 + h\nu(3450\,\text{Å band}). \tag{15}$$

The proposal of Mori and coworkers[303] and Umemoto *et al.*[305] that the short-wavelength emission was due to a complex of NH_3 with $Hg(^3P_1)$ has now been shown conclusively to be incorrect by Callear and Kendall[306] using flash photolysis, by Umemoto, Tsunashima, and Sato[301] using phaseshift techniques, and by Hikida, Santoku, and Mori[302] using a high-repetition-rate light pulser. The radiative lifetime of the stabilized $HgNH_3^*$ complex is $(1.7 \pm 0.2) \times 10^{-6}$ s.

Recent studies of exciplex emission with H_2O,[307-309] CH_3OH,[308] and *t*-butylamine[310] as quenchers of $Hg(^3P_0)$ have also been reported and lifetimes of the exciplexes have been determined to be 3×10^{-8} s or less,[309] $(1.4 \pm 0.7) \times 10^{-8}$ s,[308] and 2.1×10^{-6} s,[310] respectively. For these quenchers, formation of the exciplexes is predominately bimolecular rather than termolecular, even at 1 atm total pressure. In the H_2O case, direct evidence for the products $HgH + OH$ has also been obtained.[309]

V. QUANTUM-STATE DISTRIBUTIONS OF ENERGY IN SPECIFIC EXIT CHANNELS

Advances in spectroscopic and dynamic techniques in the last few years have facilitated experiments in which the detailed distribution of initial internal quantum states of a given exit channel can be determined for an excited metal atom quenching process. Such information is obviously extremely valuable in understanding and modelling the dynamics of energy-transfer and chemical reaction processes by which excited metal atoms lose their energy collisionally.

To accomplish the goal of individual quantum-state determinations, a detection technique is required which is capable of separately measuring the states of interest. A kinetic system must also be designed to assure that the quantum states detected are indeed the initial states, i.e. that there are no secondary collisions which have skewed the distribution of states towards Boltzmann equilibrium conditions.

There are two main approaches to the determination of initial product state distributions in excited metal atom quenching. The first is to utilize low particle density to avoid secondary collisions, i.e. the molecular beam method. This approach is particularly useful for the metastable excited states which can be created by discharge techniques and do not decay appreciably before colliding with a quencher molecule. For short-lived excited states, molecular beam experiments are considerably more difficult, and require crossed-beam, crossed-laser configurations. On the other hand, detailed dynamical scattering information can in some cases be obtained which provides evidence for direct versus complex formation mechanisms.

The second main approach, developed more recently (but essentially an extension of the basic notion of flash photolysis developed originally by

Norrish and Porter), is to combine short-pulse excitation of excited states with rapid detection of products. This method is complementary to the molecular beam method since it is most readily applied to easily pumped (and thus short-lived) excited states. Common to both methods is the need for extremely sensitive detection techniques, since the necessity of preventing *secondary* collisions dictates that very few *primary* collisions can be allowed to occur. Laser techniques (particularly laser-induced fluorescence, as developed by Zare and coworkers) have been especially useful in this regard, although mass spectrometric and hot-wire detectors have been employed to good advantage in beam experiments.

In this section, we describe recent experiments involving state-resolved detection of products, and attempt to relate this information to models of the excited metal atom quenching process.

A. Electronic-to-Vibrational Energy Transfer

Several investigators have recently reported determinations of distributions of internal quantum states in molecular quenchers resulting from simple E–V transfer. Because E–V is discussed in some detail in another review in this book,[316] only a brief overview will be given here.

1. $Na\,(^2P) + CO$

Quenching of $Na(^2P)$ by CO has been studied by Hsu and Lin,[6] who determined the initial vibrational quantum-state distribution of product CO by absorption of CW carbon monoxide laser radiation following short-pulse laser excitation of $Na(^2P)$. Extrapolation to zero time eliminated the effects of rapid V–V relaxation processes in the product CO. The quantum-state distribution, very nearly a Poisson distribution with a maximum at $v = 2$, was shown to be consistent with the predictions of the simple BFG charge-transfer model. However, the distribution was also consistent, within the experimental error, with a model of simple impulsive energy release. Total conversion of electronic to vibrational energy was about 35%. The rotational quantum-state distribution could not be determined because of the rapid rotational relaxation at the high pressures of CO necessary for the experiments.

The internal energy of the product CO has also been studied in crossed-beam, crossed-laser experiments by two different groups. Kwei and coworkers[4] determined the translational energy distribution of scattered products using time-of-flight techniques with electron-impact detection of CO in a quadrupole mass spectrometer. The distribution of internal energy in vibrational and/or rotational states was similar to that determined by Hsu and Lin for rotationally *relaxed* CO products ($v_{max} \approx 3$ to 4) and the authors concluded that little energy goes into rotational excitation of CO. The Kwei experiments, performed at several scattering angles, also showed marked forward scattering, which the authors interpreted as indicating a direct mechanism in which the collision complex exists for a time considerably less than one rotational period.

The beam experiments of Hertel and coworkers[247] are similar, except that the velocity distribution of the ground-state sodium atom product is detected via a mechanical velocity selector and hot-wire detector. However, the distribution was measured at only one scattering angle. The Hertel results also follow a Poisson distribution, but with an apparent maximum at internal excitation energies which would correspond to vibrational excitation of $v' = 4,5$. The discrepancy between this distribution and the Hsu and Lin results was rationalized by these workers as indicating *appreciable* population of higher rotational quantum states of CO. It was also claimed that a Monte Carlo simulation of the kinematics of the experimental system indicated that, if only low rotational quantum states were being populated, vibrational structure would have been observed in the Na velocity distributions. Considering the possible experimental error in these difficult experiments, it is unfortunate that opposite conclusions have been drawn concerning rotational excitation from results which are really not that different. Experimental refinements and further work will be necessary to resolve these and other more substantial discrepancies between the Kwei and Hertel results (see below).

2. $Na(^2P) + N_2$

Results from the Kwei group[4] indicate little difference between N_2 and CO as quenchers with regard to the distribution of internal energy ($v_{max} \approx 3$ when zero rotational excitation is assumed). The Hertel results,[5,247] similar to CO, are peaked at slightly higher internal energies ($v_{max} \approx 4$), but in addition the apparent distribution of internal energy is narrower than a Poisson function. Both the Kwei data and recent angle-resolved data from the Hertel laboratories[316] indicate that the scattered products are strongly forward-peaked, just like CO, again implying that the lifetime of any intermediate complex must be several times less than a rotational period.

It should be noted also that for an analogous reverse process:

$$K(^2S) + N_2 \rightarrow K(^2P) + N_2, \tag{16}$$

studied by Fluendy and coworkers[317] at high translational energies, the experimental results were in excellent agreement with a primitive model using a predominantly $K^+ N_2^-$ intermediate state analogous to the BFG state, and diabatic neutral potentials determined from scattering data.

3. $Na(^2P) + NO$

In the NO case, the scattering data of the Kwei group[4] show definite forward–backward symmetry, providing direct evidence for an intermediate complex lifetime of the order of one rotational period. This is perhaps not surprising, since NO^- is stable and the Na^+NO^- complex is known to be bound even with respect to ground-state $Na + NO$.[4] With regard to the distribution of internal energy in the product NO, the Kwei and Hertel results again disagree. The Kwei data peak at $NO(v \approx 7)$, again neglecting rotational energy. The Hertel data,[316,318] peaking at $NO(v \approx 3)$ have higher signal-to-noise ratio but have been extracted at only one scattering angle, while because of forward–backward peaking in the angle-resolved work of Kwei, essentially two independent measurements were made of the internal energy distribution. It is interesting to note (see figure 19 of ref. 316) that the 'prior' distribution which would result from statistical partitioning of energy in a long-lived complex is intermediate between that of the Kwei and the Hertel c.m. recoil energy distributions, so that it is possible, given the uncertainty in the data, that equipartitioning of internal energy is actually being approached in the $Na(^2P)$–NO case.

4. $Na(^2P) + O_2$

Because the $Na^+O_2^-$ complex is also stable with respect even to ground-state $Na + O_2$, it might have been expected that symmetric forward–backward angular distributions would have been observed analogous to the NO case, but Kwei and coworkers[4] observe strong forward scattering as for N_2 and CO. However, for O_2 the low-lying $^1\Delta_g$ and $^1\Sigma_g^+$ electronically excited states are accessible and this may affect the dynamics. The Kwei and Hertel[247] internal energy distributions are in serious disagreement, and there is an added uncertainty in the Kwei experiments as to whether the electron-impact efficiencies for the excited states of O_2 are the same as that for vibrationally excited ground-state O_2, or whether fragmentation of the excited states may occur in the detector.

5. $Na(^2P) + H_2, D_2$

The Hertel results[247,316] at one scattering angle show vibrational distributions which peak at $v = 2$ for H_2 and $v = 3$ for D_2. However, the internal *energy* distributions peak at the same value and, since the potential energy surfaces are exactly the same in both cases, Hertel argues that the isotopic internal energy partitioning is determined entirely by surface crossings, not by dynamical mass effects. *Ab initio* calculations[316] indicate that there is an attractive $^2B_2$ surface in C_{2v} symmetry with substantial $Na^+ - H_2^-$ charge-transfer character which can cross the ground-state surface when the H–H distance lengthens upon the approach of $Na(^2P)$. This is, of course, the realistic counterpart to the crude charge-transfer state postulated in the BFG model. Close-coupling calculations by McGuire and Bellum[15] indicate production of predominantly $H_2(v = 2)$, in qualitative agreement with the sharp-peaked internal energy distribution found by Hertel. A 'bond-stretch attraction' model, discussed by Hertel,[316] which is based on the latest *ab initio* calculations, also predicts internal energy distributions in qualitative agreement with experiment.

6. $Na(^2P) + polyatomic\ molecules$

Hertel has also reported single-scattering-angle measurements[316,318] for several triatomic and polyatomic quenchers with double or triple bonds. Total energy transferred into the polyatomics appears higher than for diatomics, and the energy distributions are consistent with a statistical model in which the participation of two to three atoms in the vicinity of the interaction of $Na(^2P)$ with the π-bond are involved (see figure 20 of ref. 316). Full angular scattering measurements for one or more of the polyatomic cases would be very informative.

7. $Hg(^3P_0) + NO$

Horiguchi and Tsuchiya[319] have measured the initial distribution of vibrational states of NO in the quenching of $Hg(^3P_0)$ by using a modulated mercury arc, and phaseshift detection of infrared fluorescence of $NO(v)$ to allow for V–V relaxation effects. The distribution shows a maximum at $v = 6$ and was shown to be consistent with impulsive release of energy, as modelled by the 'impulsive half-collision' model of Simons and Tasker,[320] except that minor population of vibrational levels above the impulsive limit was observed. Conversion efficiency of electronic to vibrational energy was approximately equal to 30%.

8. $Hg(^3P_0) + CO$

The Tsuchiya method[319] gave a distribution of $CO(v)$ which peaked at $v = 5$, but which could not be made to fit the Simons–Tasker impulsive model. It should be pointed out that in comparison to quenching of $Hg(^3P_0)$ by NO, or to $Na(^2P)$ by CO, where the cross sections are large (as predicted by BFG theory), the cross section for quenching of $Hg(^3P_0)$ by CO is only $1-2\,\text{Å}^2$. In the $Hg(^3P_0)$–CO case, the charge-transfer-like intermediate is too high in energy for efficient quenching (see Sec. III), and the mechanism for E–V transfer may be entirely different. A 'golden rule' model by Berry,[321] whereby the CO oscillator is 'dressed' by the presence of $Hg(^3P_0)$ and the distribution of final states is determined by the Franck–Condon overlap integrals between the vibrational wavefunctions of the 'dressed' CO and ground-state CO, can reproduce the distribution, but only if the CO internuclear distance is

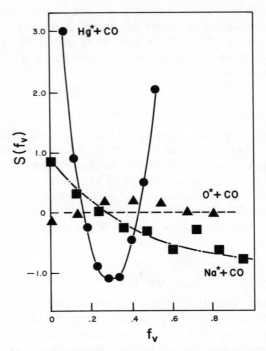

Fig. 6. Surprisal plots of the distribution of initial vibrational states in the quenching by CO of $Hg(^3P_0)$,[319] $Na(^2P_J)$,[6] and $O(^1D_2)$.[322]

increased substantially from $1 \cdot 13$ to $1 \cdot 32$ Å.[244] The Franck–Condon factors are of the order of $\sim 0 \cdot 1$, however, consistent with the value of the total cross section.[244] To illustrate the differences in quenching by CO of $Hg(^3P_0)$ as compared with two other cases, surprisal plots of the vibrational energy distribution in the quenching of $Na(^2P)$, $Hg(^3P_0)$, and $O(^1D_2)$ by CO are shown in Fig. 6. The quenching of $O(^1D)$ by CO is known to proceed through long-lived ground-state CO_2 in which energy is completely randomized to all degrees of freedom as the products form.[322]

9. $Hg(^3P_1) + CO, N_2, NO$

Quenching of $Hg(^3P_1)$ by CO[319] and N_2[323] is now known to occur by exclusive production of $Hg(^3P_0)$ with concomitant excitation of the first vibrational level of N_2 or CO, even though the processes are endoergic by $564\,cm^{-1}$ and $376\,cm^{-1}$, respectively, for zero rotational energy. In contrast, quenching of $Hg(^3P_1)$ by NO produces predominantly ground-state $Hg(^1S_0)$ and vibrationally excited NO, despite the fact that the production of $Hg(^3P_0)$ +

$NO(v = 1)$ is only $109\,\mathrm{cm}^{-1}$ endoergic.[319] This can readily be understood within the BFG model framework, since the $Hg^{+}NO^{-}$ surface provides efficient coupling directly to the ground-state surface, while the surfaces of $Hg^{+}N_{2}^{-}$ and $Hg^{+}CO^{-}$ character lie too high in energy for efficient transfer directly to the ground-state surface.

B. Product Atomic Electronic States in Electronic-to-Vibrational Energy Transfer

Breckenridge and Malmin[152] have recently reported measurements of the initial J state distributions of $Cd(^{3}P_{J})$ formed in the process:

$$Cd(^{1}P_{1}) + Q \rightarrow Cd(^{3}P_{J}) + Q^{\ddagger}, \tag{17}$$

where Q represents a variety of quenching molecules. A schematic representation of the energy levels involved in the measurements is shown in Fig. 7. The 'pump-and-probe' method developed in the Breckenridge laboratories utilizes short-pulse (6 ns) creation of excited metal atoms with detection of initial quenching products after a very short delay ($\sim 20\,\mathrm{ns}$) by laser-induced fluorescence, so that initial product distributions can be determined in the

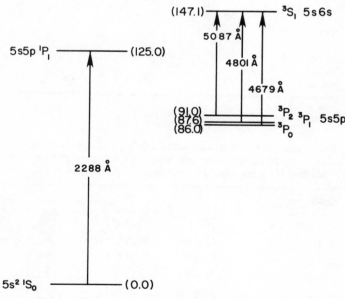

Fig. 7. The energy levels of the cadmium relevant to the determination of the initial $Cd(^{3}P_{0,1,2})$ quantum states in the process $Cd(^{1}P_{1}) + Q \rightarrow Cd(^{3}P_{0,1,2}) + Q^{\#}$ using the laser pump-and-probe technique.[152] Shown by arrows are the transitions for pumping $Cd(^{1}P_{1})$ and for probing the $Cd(^{3}P_{0,1,2})$ concentrations by laser-induced fluorescence.

complete absence of secondary relaxation.[9,11,151,152] Experimental details of the new method and proof of the absence of secondary collisions even when effective relaxation cross sections are over $100\,\text{Å}^2$ may be found in refs. 151 and 152.

The initial distributions of $Cd(^3P_{0,1,2})$ electronic states in the E–V, R process (17) exhibit surprisingly large and quite interesting variations with quencher molecule, as shown in Tables XII–XIV. The variations were interpreted by a qualitative bond-specific model of the quenching mechanism.[152] The results in Table XII for atomic and diatomic quenchers illustrate the major points of the model. The $Cd(^1P_1)$ state is three-fold

TABLE XII. Initial distributions of $Cd(^3P_{2,1,0})$ in process (17) for atomic and diatomic quenchers, in per cent.

Quencher	$^3P_2$	$^3P_1$	$^3P_0$
Ar[152]	80 ± 3	16 ± 2	4 ± 1
H_2[266]	61 ± 2	36 ± 2	4 ± 1
D_2[266]	62 ± 1	35 ± 1	3 ± 1
N_2[152]	55 ± 2	34 ± 2	11 ± 1
CO[152]	56 ± 2	32 ± 2	12 ± 1
NO[152]	49 ± 2	37 ± 2	14 ± 1
$2J + 1$	56	33	11
'Prior' (see text)	49	37	14

TABLE XIII. Initial distributions of $Cd(^3P_{2,1,0})$ in process (17) for alkane hydrocarbons in per cent. From Breckenridge and Malmin.[152] Reproduced by permission of *Amer. Inst. Physics.*

Quencher	$^3P_2$	$^3P_1$	$^3P_0$
CH_4	76 ± 2	19 ± 2	5 ± 1
C_2H_6	67 ± 1	25 ± 1	8 ± 1
C_3H_8	65 ± 1	27 ± 1	8 ± 1
$i\text{-}C_4H_{10}$	56 ± 1	33 ± 1	11 ± 1
$n\text{-}C_rH_{10}$	55 ± 1	34 ± 1	11 ± 1
$neo\text{-}C_4H_{12}$	62 ± 2	30 ± 2	8 ± 1
$i\text{-}C_5H_{12}$	58 ± 3	31 ± 3	11 ± 1
$2J + 1$	56	33	11

TABLE XIV. Initial distributions of $Cd(^3P_{2,1,0})$ in process (17) for unsaturated hydrocarbons, in per cent. From Breckenridge and Malmin.[152] Reproduced by permission of Amer. Inst. Physics

Quencher	Structure	$^3P_2$	$^3P_1$	$^3P_0$
perfluoropropene	(F_6)	~25	~60	~15
butadiene		17 ± 2	65 ± 1	18 ± 1
acetylene		36 ± 2	56 ± 2	14 ± 1
ethylene		37 ± 2	47 ± 2	16 ± 1
propylene		44 ± 3	42 ± 3	14 ± 1
trans-2-butene		55 ± 2	34 ± 2	11 ± 1
cis-2-butene		54 ± 2	35 ± 2	11 ± 1
isobutylene		57 ± 2	33 ± 2	10 ± 1
$2J + 1$		56	33	11

degenerate and will correlate with three Cd–quencher potential surfaces. The product states $Cd(^3P_2)$, $Cd(^3P_1)$, and $Cd(^3P_0)$ have degeneracies of 5, 3, and 1, respectively, and each will correlate with this number of Cd–quencher surfaces. Net crossings between the three $Cd(^1P_1)$–quencher entrance surfaces and nine $Cd(^3P_J)$–quencher exit surfaces will determine the final quantum-state distributions. One of the simplest cases, of course, is totally non-preferential production of the $Cd(^3P_J)$ states, which would result in an electronically statistical distribution of $^3P_2:^3P_1:^3P_0$ states in the ratios 5:3:1, or 56:33:11 in per cent. As can be seen from Table XII, N_2 and CO behave in this manner, and the results are consistent with the BFG model predictions that states of $Cd^+ N_2^-$ and $Cd^+ CO^-$ character strongly couple the entrance and exit channels in these cases.

Another mechanistic possibility is that a long-lived intermediate complex is formed, leading to statistical distribution of the initial energy to *all* degrees of freedom. For the rigid-rotor, harmonic oscillator approximation for a diatomic product, the 'prior' statistical distribution in such a case is proportional to the $^1P_1-^3P_J$ energy difference to the 5/2 power,[324] leading to a $^3P_2:^3P_1:^3P_0$ distribution of 0·49:0·37:0·14. Obviously, quenching by NO is such a case, and the stability of the Cd^+NO^- intermediate state must be responsible, analogous to the long-lived Na^+NO^- complex identified in the quenching of $Na(^2P)$ by NO by Kwei and coworkers.[4]

In the case of quenching of $Cd(^1P_1)$ by Ar, there is no strong coupling between entrance and exit channels, and repulsive interactions must play a large role. The interaction of the p-orbital on the excited Cd–atom with the Ar filled-shell electronic distribution will lead to Σ and Π Cd–Ar states.[325] It is likely that the distributions of ^3P_J states observed are produced by crossings of entrance CdAr($^1\Pi$) curves with exit CdAr($^3\Sigma$) curves, since the Σ states will be more highly repulsive due to the p-density along the internuclear axis. That the distribution is skewed away from statistical towards the $^3P_2$ state can be understood, then, since the most repulsive $^3\Sigma$ states (which will cross the $^1\Pi$ curves at the largest internuclear separation) correlate adiabatically with Cd($^3P_2$) product.

All that remains to be explained in Table XII are the recently determined distributions for H_2 and D_2. In these cases, the $^3P_2{:}^3P_1$ distributions are in almost exactly the 5:3 statistical ratio expected for strong coupling, say via an intermediate surface of $Cd^+H_2^-$ character, but the $^3P_0$ population is extremely low, indicating that the $^3P_0$ exit channel is nearly closed. One way of rationalizing the results is to assume that the molecular CdH_2 state correlating with $Cd(^3P_0) + H_2$ is more strongly bound than the other triplet surfaces, thereby not intersecting the $Cd(^1P_1)$–H_2 surfaces or the charge-transfer surface in the energetically accessible regions. However, it should be pointed out that H_2 is a rather special case in the quenching of $Cd(^1P_1)$, in that all the other quenchers in Tables XII–XIV produce predominantly $Cd(^3P_J)$, while the branching ratio for $Cd(^3P_J)$ production for H_2 is $\sim 0{\cdot}30$, with chemical channels (CdH + H, Cd + H + H) dominating.[266] It is also possible, then, that the lack of $Cd(^3P_0)$ is due to a particularly favourable surface for chemical decomposition of the initially formed $Cd(^3P_0)$–H_2 complex as compared with $Cd(^3P_2)$–H_2 or $Cd(^3P_1)$–H_2 complexes.

Shown in Table XIII are the initial ^3P_J distributions for the alkane hydrocarbons. These are readily explained[152] by assuming that the interactions with stronger C–H bonds are similar to that with Ar (note that the wavefunctions for Cd–CH_4 excited surfaces, for example, will resemble those for Cd–Ne surfaces at long range). The $Cd(^1P_1)$–RH potential is assumed to become more and more attractive as the C–H bond strength decreases (note that insertion of $Cd(^1P_1)$ into a C–H bond would form a singlet excited state of a cadmium alkyl hydride, which may have bound character at some geometry), and as the polarizability of the hydrocarbon increases. This allows stronger and stronger coupling with the $Cd(^3P_J)$–RH repulsive exit surfaces. Thus the distributions for the butanes and iso-C_5H_{12} are truly statistical, and statisticality is approached in the sequence CH_4, C_2H_6, and C_3H_8. The neopentane distribution is also consistent with these considerations, since it possesses only strong *primary* C–H bonds.

In Table XIV, it is seen that strikingly different $^3P_2{:}^3P_1{:}^3P_0$ distributions are observed for some of the alkenes, but that the butenes exhibit the statistical distribution. This has been qualitatively rationalized in the following way.[152]

For the quenchers perfluoropropene, butadiene, acetylene, and ethylene, the main interaction of $Cd(^1P_1)$ is with the π-electron density (note that C–F bonds are inert to $Cd(^1P_1)$ attack). It is quite likely in these cases that an intermediate π-complex is formed with a lifetime sufficiently long for some equilibration of the initial energy over the molecular framework before $Cd(^3P_J)$ product formation. While the prior distribution of $Cd(^3P_J)$ cannot be calculated with an equivalently simple approximation for a polyatomic as for a diatomic molecule, the effective functional dependence on the $^1P_1-^3P_J$ energy differences can be shown to be of powers higher than $5/2$.[152] Thus the fact that the distributions are severely skewed from statistical towards preferential production of $^3P_1$ and $^3P_0$ can be understood as a specific interaction with the unsaturated sites in which energy is at least partially randomized into the other degrees of freedom of the molecule before dissociation to $Cd(^3P_J)$. The recent beam experiments of Hertel[318] on the quenching of $Na(^2P)$ by alkenes provide support for this kind of model. As the alkenes get larger, however, there are carbon centres at which efficient C–H bond attack by $Cd(^1P_1)$ can occur simultaneously. Thus for propylene, C–H bond and π-bond attack is probably roughly competitive, and, in the butene cases, C–H bond interaction completely dominates, with six weak C–H bonds on non-double-bond carbon atoms. Thus a simple, competitive two-site mechanism can account qualitatively for all the data in Table XIV.

C. Electronic Energy Transfer

There have been a few studies which utilize short-lived fluorescence of electronically excited diatomic molecular products to determine initial vibrational and/or rotational quantum-state distributions in electronic energy transfer exit channels.

Breckenridge and coworkers[244,245] have described the production of $NO(A^2\Sigma^+)$ by energy transfer from $Zn(^1P_1)$. The relevant energy levels are shown in Fig. 8. A low-pressure stream of Zn vapour and NO gas in argon was irradiated with an electrodeless discharge lamp to excite $Zn(^1P_1)$. The well known NO gamma-band emission, recorded photographically, was utilized to monitor $NO(A^2\Sigma^+)$ production in the following cases:

$$Zn(^1P_1) + NO(X^2\Pi_J) \rightarrow Zn(^1S_0) + NO(A^2\Sigma^+,\ v'=1), \qquad (18a)$$

$$\rightarrow Zn(^1S_0) + NO(A^2\Sigma^+,\ v'=0). \qquad (18b)$$

It was shown that the $NO(A^2\Sigma^+)$ radiative lifetime was sufficiently short under the conditions of the experiment that no secondary relaxation of vibrational *or* rotational quantum states occurred. A long-exposure experiment allowed sufficient resolution to determine the complete vibrational and rotational initial quantum-state distribution for process (18), via a spectral simulation procedure.

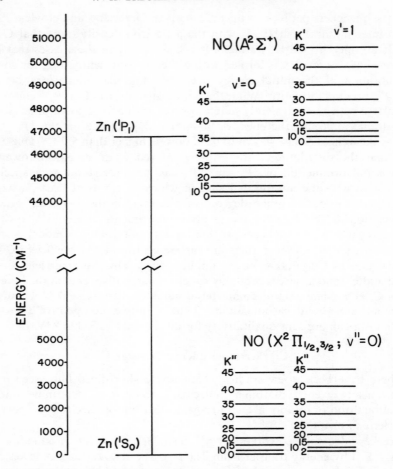

Fig. 8. Energy levels of NO($A^2\Sigma^+$), showing the possibility of dipole-allowed near-resonant electronic energy transfer transitions between Zn($^1P_1$) and NO($^2\Pi_J$) to form NO($A^2\Sigma^+$, $v' = 1$). From Breckenridge et al.[245] Reproduced by permission of the American Institute of Physics.

The initial population distribution of NO($A^2\Sigma^+$, $v' = 1$) rotational levels was remarkably consistent with that predicted by the Cross–Gordon version[326] of near-resonant dipole–dipole theory, except that a minor ($\sim 7\%$) Boltzmann-like population had to be included for satisfactory simulation of the sensitized NO $\gamma(1, X)$ band spectra in the frequency regions dominated by transitions originating from lower rotational levels. A plot of the unusual initial distribution of rotational quantum states is shown in Fig. 9. The initial rotational level population distribution of NO($A^2\Sigma^+$, $v' = 0$) was very similar to a Boltzmann population distribution at ~ 1400 K. The rate of production

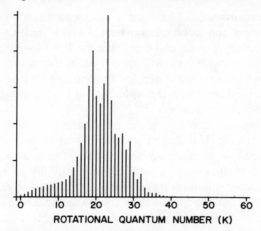

Fig. 9. The initial quantum-state distribution of rotational levels of $NO(A^2\Sigma^+, v = 1)$ produced by the process $Zn(^1P_1) + NO(^2\Pi_J) \rightarrow Zn(^1S_0) + NO(A^2\Sigma^+, v' = 1)$. From Breckenridge et al.[245] Reproduced by permission of the American Institute of Physics.

of $NO(A^2\Sigma^+, v' = 1)$ was about 10 times that of $NO(A^2\Sigma^+, v' = 0)$. Efficient 'near-resonant' dipole–dipole interactions appear, therefore, to constitute the dominant mechanism for electronic energy transfer in this unusual case. Production of $NO(A^2\Sigma^+, v' = 0)$, and possibly the lower rotational levels of $NO(A^2\Sigma^+, v' = 1)$, are thought to result from a minor competing mechanism. A Zn^+NO^- charge-transfer surface crossing was postulated as a likely possibility.

There have been several cases reported of atom–diatomic or diatomic–diatomic electronic energy transfer in which near-resonance has been shown to play an important role (initial and final vibrational and/or rotational levels are not shown, for simplicity):

$$Ar(^1P_1) + H_2(X^1\Sigma_g^+) \rightarrow Ar(^1S_0) + H_2(B^1\Sigma_u^+),^{327} \tag{19}$$

$$Ar(^3P_1) + H_2(X^1\Sigma_g^+) \rightarrow Ar(^1S_0) + H_2(B^1\Sigma_u^+),^{327} \tag{20}$$

$$Kr(^3P_1) + CO(X^1\Sigma^+) \rightarrow Kr(^1S_0) + CO(A^1\Pi),^{328} \tag{21}$$

$$LiH(A^1\Sigma^+) + Li(^2S_{1/2}) \rightarrow LiH(X^1\Sigma^+) + Li(^2P_J),^{329} \tag{22}$$

$$NO(A^2\Sigma^+) + NO^*(X^2\Pi) \rightarrow NO(X^2\Pi) + NO^*(A^2\Sigma^+).^{330} \tag{23}$$

(where * indicates isotopically labeled NO)

In all these cases: (i) the donor and acceptor transitions are dipole-allowed; (ii) there are state-to-state energy-transfer processes for which the off-resonance energy is 100 cm^{-1} or less; and (iii) the initial molecular states for such resonant processes have at least a moderate Boltzmann population. These are precisely the conditions which should lead to high effective cross sections, according to dipole–dipole theory. It is obvious, of course, that such energy-level coincidences will rarely occur for atoms and/or small molecules, so that near-resonant dipole–dipole energy transfer will certainly be the exception rather than the rule, but an interesting and possibly important exception, nevertheless.

Similar experiments involving sensitized emission of several free-radical diatomic molecules by $Hg(^3P_0)$ in a flow system have been reported by Vikis.[331–333] Because of the uncertainty about secondary collisional relaxation in most of the cases, as well as the possibility of non-Boltzmann distributions of initial quantum states of the radicals due to their production by $Hg(^3P)$ sensitization, the quantitative conclusions reached in these studies should be viewed with some caution. However, energy transfer of $Hg(^3P_0)$ to CN and OH must certainly be greatly affected by the very stable $Hg^+ CN^-$ and $Hg^+ OH^-$ charge-transfer states. In the most careful study, that of the process

$$Hg(^3P_0) + OH(X^2\Pi_J) \rightarrow Hg(^1S_0) + OH(A^2\Sigma^+), \tag{24}$$

the measured vibrational and rotational population of $OH(A^2\Sigma^+)$ was consistent with that expected from partial relaxation by secondary collisions of a statistical distribution from a long-lived intermediate complex.

D. Chemical Reaction

Initial vibrational, and in some cases, rotational quantum-state distributions have been determined for several systems involving the chemical reactions of excited metal atoms to form metal monohydrides, monohalides, or monoxides.

Breckenridge and Oba,[11] using the 'pump-and-probe' laser technique have determined the initial distribution of rotational quantum states in the following chemical exit channel:

$$Cd(^1P_1) + H_2 \rightarrow CdH(v=0) + H. \tag{25}$$

Spectral simulation was required to deconvolute the complicated $\Sigma-\Pi$ LIF spectra of $CdH(v=0)$, and the 'best-fit' rotational distribution resembles closely a Boltzmann distribution at an effective temperature of approximately 5000 K. Shown in Fig. 10 are two experimental 'best-fit' distributions for a particular experiment. Also shown are two versions of the statistically expected distributions for this very exothermic process, the 'prior' distribution and the distribution calculated according to phase-space theory. It is obvious that there is *less* rotational energy than expected statistically even though the $CdH(v=0)$ is produced quite 'hot', rotationally. Production of $CdH(v = 1, 2, 3, 4)$ was also observed experimentally, with a distribution which

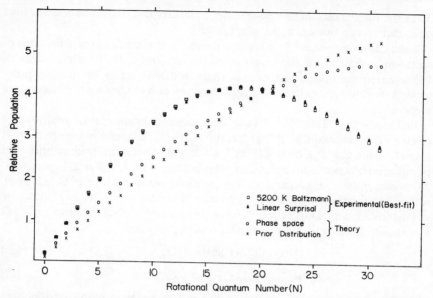

Fig.10. The initial quantum-state distribution of rotational levels of CdH($v = 0$) produced in the process $Cd(^1P_1) + H_2 \rightarrow CdH(v = 0) + H$. Shown are two different single-parameter 'best-fit' distributions from computer simulation of CdH($v = 0$) LIF spectra. Also shown are the 'prior' and phase-space statistical distributions for comparison. From Breckenridge and Oba.[11] Reproduced by permission of North Holland Publishing Co.

appears qualitatively to drop off continuously from a maximum at CdH($v = 0$), as expected statistically. Quantitative measurements were impossible because of severe spectral overlap problems. Spectral overlap has also prevented quantitative determinations of rotational distributions in any of the other vibrational levels. It should be noted that $CdH(X^2\Sigma^+) + H(^2S)$ does *not* correlate adiabatically with $Cd(^1P_1) + H_2(^1\Sigma_g^+)$, so that a surface crossing must occur.[233]

Solarz and coworkers[294b] find that a high percentage ($\sim 70\%$) of the available energy is channelled into vibrational energy of SrCl in the reaction of $Sr(^3P)$ with HCl, and that the vibrational quantum-state distribution is unusually narrow. They explain this result in terms of early release of energy as the triplet interaction surface rapidly acquires Sr^+-Cl^--H character. In the reaction of $Sr(^3P)$ with HF, 40% of the available energy is partitioned into vibration. The reactions of Sr ground-state atoms with HCl and HF are endothermic, but the exothermic reactions of ground-state barium atoms with HCl and HF deposit only 33% and 12% of the available energy into vibration, respectively. Solarz *et al.* postulate that the singlet surface, at lower energies with respect to the asymptotic $Sr^+ + HX^-$ states, acquires ionic character only late in the reaction path, resulting in less release of energy into vibration.

Thus the importance of charge-transfer character in excited metal atom plus quencher interactions is again emphasized.

Solarz and coworkers[294a] have also determined vibrational distributions of metal monohalides in the reaction of $Ba(^3D)$ and $Sr(^3P)$ with a series of halogenated methanes, and discuss their results in terms of electron-jump models used to rationalize analogous experimental studies with ground-state metal atoms.

Brinkman and Telle[298,334] have determined vibrational distributions of electronically excited $CaCl(A^2\Pi_J)$ and $CaCl(B^2\Sigma^+)$ formed in the reaction of $Ca(^1D_2)$ and $Ca(^3P_J)$ with Cl_2 and HCl. The results indicate vibrational population inversions in the $CaCl(A^2\Pi_J)$ but not in the $CaCl(B^2\Sigma^+)$ products. An argument similar to the early-release charge-transfer idea of Solarz et al. was used to rationalize the high vibrational excitation in the $A^2\Pi$ state.

Pasternack and Dagdigian[249] have determined the distribution of CaO vibrational quantum states in the following processes:

$$Ca(^3P) + O_2 \rightarrow CaO(v) + O, \tag{26}$$

$$Ca(^3P) + CO_2 \rightarrow CaO(v) + CO. \tag{27}$$

In both cases, most of the population was in the $v = 0$ level, with sharp declines from $v = 0$ to $v = 3$. A lower limit for the rotational 'temperature' of the $v = 0$ state was also determined to be about 800 K. Only a very small fraction of the available energy was found in CaO internal excitation. The authors suggest that the reactions proceed through $Ca^+O_2^-$ and $Ca^+CO_2^-$ charge-transfer intermediates which resemble Ca^+O^- *excited* states in electronic structure more than the ground-state $CaO(X^1\Sigma^+)$ product, thus leading to avoided crossings and exit-channel barriers which could account for the cold internal energy distribution.

VI. CONCLUDING REMARKS

A great deal of information has been and is continuing to be accumulated on the quenching of electronically excited metal atoms. There is no question that in many cases excited-state potential surfaces with substantial metal-to-quencher charge-transfer character play significant roles in the quenching process. In this review, we have shown that the modified Bauer–Fisher–Gilmore treatment, simplistic as it is, provides a remarkably good zero-order model for such cases, one which must of course be replaced with higher-order, more accurate theories as *ab initio* calculations become available for excited-state surfaces, and quantum scattering dynamical treatments are developed. Caution must be exercised, however, in attempting to generalize from quite special cases such as Na^*-H_2 to the whole range of excited metal atom plus quenching molecule combinations.

The developing experimental techniques which are allowing quantum-state resolution of quenching products (or at least total internal energy distributions) undoubtedly hold the greatest promise of providing new insights into excited metal atom quenching, but it will be important in the future not to ignore other experimentally accessible measurements such as the values of quenching cross sections and branching ratios for different exit channels, and their dependence on initial translational (or vibrational) energy. As in any developing field, experiments should be designed with a view towards their utility in understanding the phenomenon of interest, not because techniques happen to exist which are elegant, expensive, and/or fashionable. 'There are many pathways to the truth, but the simplest often require the most wisdom.'[335]

Acknowledgements

The authors gratefully acknowledge the continuing support of the National Science Foundation for research on the dynamics of excited metal atom collisional processes being conducted in our laboratories. Acknowledgement is also made to the Department of Energy and to the donors of the Petroleum Research Fund, administered by the American Chemical Society, for partial support of our research.

One of us (W. H. B.) would like to thank all the present and former graduate students and postdoctoral fellows who have made such valuable contributions to the research in this area.

References

1. G. Cario and J. Franck, (1923) *Z. Phys.*, **17**, 202.
2. A. C. G. Mitchell and M. W. Zemansky (1934, reprinted 1961), *Resonance Radiation and Excited Atoms*, Cambridge University Press, London, and references therein.
3. E. W. R. Steacie (1954), *Atomic and Free Radical Reactions*, 2nd edn., vols. I and II, Reinhold, New York.
4. J. A. Silver, N. C. Blais and G. H. Kwei (1979), *J. Chem. Phys.*, **71**, 3412.
5. I. V. Hertel, H. Hofmann and K. A. Rost (1979), *J. Chem. Phys.*, **71**, 674.
6. D. S. Y. Hsu and M. C. Lin (1980), *J. Chem. Phys.*, **73**, 2188.
7. J. A. Irvin and P. J. Dagdigian (1980), *J. Chem. Phys.*, **73**, 176.
8. J. Elward-Berry and M. J. Berry (1980), *J. Chem. Phys.*, **72**, 4500.
9. W. H. Breckenridge and O. Kim Malmin (1981), *J. Chem. Phys.*, **74**, 3307.
10. W. H. Breckenridge and O. Kim Malmin (1979), *Chem. Phys. Lett.*, **68**, 341.
11. W. H. Breckenridge and D. Oba (1980), *Chem. Phys. Lett.*, **72**, 455.
12. A. Kowalski and M. Menzinger, submitted for publication.
13. E. E. Nikitin (1975), *Adv. Chem. Phys.*, **28**, 317.
14. I. V. Hertel (1981), *Adv. Chem. Phys.*, **45**, 341.
15. P. McGuire and J. C. Bellum (1979), *J. Chem. Phys.*, **71**, 1975.
16. H. S. Taylor (1979), *Chem. Phys. Lett.*, **64**, 17.
17. D. Truhlar, private communication.

18. A bibliography of the 1960–1980 literature related to quenching of group IA, IIA, and IIB excited states is available from the authors.
19. R. J. Cvetanović (1964), *Prog. React. Kinet.*, **2**, 39.
20. J. G. Calvert and J. N. Pitts, Jr (1966), *Photochemistry*, Wiley, New york.
21. A. B. Callear (1965), *Appl. Opt. Suppl., Chemical Lasers*, 145.
22. A. B. Callear and J. D. Lambert (1969), *Comprehensive Chemical Kinetics*, vol. 3, Elsevier, Amsterdam.
23. L. F. Phillips (1974), *Acc. Chem. Res.*, **7**, 135.
24. S. Tsunashima and S. Sato (1979), *Rev. Chem. Intermed.*, **2**, 201.
25. P. L. Lijnse (1972), *Review of Literature on Quenching, Excitation, and Mixing Cross-Sections for the First Resonance Doublets of the Alkalis*, Report 1398, Fysisch Laboratorium, Fijksuniversiteit, Utrecht, Netherlands.
26. L. Krause (1975), *Adv. Chem. Phys.*, **28**, 267.
27. V. Kempter (1975), *Adv. Chem. Phys.*, **30**, 417.
28. D. L. King and D. W. Setser (1976), *Annu. Rev. Phys. Chem.*, **27**, 407.
29. S. Lemont and G. W. Flynn (1977), *Annu. Rev. Phys. Chem.*, **28**, 261.
30a. C. E. Moore (1971), *Atomic Energy Levels*, vols. I, II, and III, *Natl. Bur. Stand. Ref. Data Ser.*, **35**, US Government Printing Office, Washington, DC.
30b. J. C. Slater (1964), *J. Chem. Phys.*, **41**, 3199.
30c. J. O. Hirschfelder, C. F. Curtiss and R. B. Bird (1964), *Molecular Theory of Gases and Liquids*, Wiley, New York.
30d. K. A. Köhler, R. Feltgen and H. Pauly (1977), *Phys. Rev. A*, **15**, 1407.
30e. J. Costello, M. Fluendy and K. Lawley (1977), *Faraday Discuss. Chem. Soc.*, **62**, 291.
31. A. Lurio (1964), *Phys. Rev. A*, **136**, 376.
32. W. W. Smith and A. Gallagher (1966), *Phys. Rev.*, **145**, 26.
33. W. L. Wiese, M. W. Smith and B. M. Miles (1969), *Natl. Bur. Stand. Ref. Data Ser.*, **22**.
34. W. H. Smith (1970), *Nucl. Instrum. Meth.*, **90**, 115.
35. T. Andersen, J. Desesquelles, K. A. Jessen and G. Sorensen (1970), *J. Quant. Spectrosc. Radiat. Transfer*, **10**, 1143.
36. W. H. Smith and H. S. Liszt (1971), *J. Opt. Soc. Am.*, **61**, 938.
37. F. M. Kelly and M. S. Mathur (1978), *Can. J. Phys.*, **56**, 1422.
38. P. S. Furcinitti, J. J. Wright and L. C. Balling (1975), *Phys. Rev. A*, **12**, 1123.
39. R. de Zafra, R. J. Goshen, A. Landman and A. Lurio (1962), *Bull. Am. Phys. Soc.* **7**, 433.
40. A. Lurio, R. L. de Zafra and R. J. Goshen (1964), *Phys. Rev. A*, **134**, 1198.
41. A. Landman and R. Novick (1964), *Phys. Rev. A*, **134**, 56.
42. T. Andersen and G. Sorensen (1973), *J. Quant. Spectrosc. Radiat. Transfer*, **13**, 369.
43. R. Abjean and A. Johannin-Gilles (1975), *J. Quant. Spectrosc. Radiat. Transfer*, **15**, 25.
44. J. Kowalski and F. Träger (1976), *Z. Phys. A*, **278**, 1.
45. I. Martinson, L. J. Curtis, S. Huldt, U. Litzēn, L. Liljeby, S. Mannervik and B. Jelenkovic (1979), *Phys. Scr*, **19**, 17.
46. F. W. Byron, Jr, M. N. McDermott, R. Novick, B. W. Perry and E. B. Saloman (1964), *Phys. Rev. A*, **134**, 47.
47. A. Lurio and R. Novick (1964), *Phys. Rev. A*, **134**, 608.
48. J. P. Barrat and J. Butaux (1961), *C. R. Acad. Sci.*, **253**, 2668.
49. F. W. Byron, Jr, M. N. McDermott and R. Novick (1964), *Phys. Rev. A*, **134**, 615.
50. N. L. Moise (1966), *Astrophys. J.*, **144**, 763.

51. T. M. Bieniewski, T. K. Krueger and S. J. Czyzak (1968), *Adv. Quantum Chem.*, **4**, 141.
52. A. R. Schaefer (1971), *J. Quant. Spectrosc. Radiat. Transfer*, **11**, 197.
53. A. Lurio (1965), *Phys. Rev. A*, **140**, 1505.
54. P. D. Lecler (1968), *J. Physique*, **29**, 611.
55. A. Skerbele and E. N. Lassettre, (1970), *J. Chem. Phys.*, **52**, 2708.
56. D. Gebhard and W. Behmenburg (1975), *Z. Naturforsch. A*, **30**, 445.
57. R. Abjean and A. Johannin-Gilles (1976), *J. Quant. Spectrosc. Radiat. Transfer*, **16**, 369.
58. C. Bousquet and N. Bras (1980), *J. Physique*, **41**, 19.
59. J. S. Deech and W. E. Baylis (1970), *Can. J. Phys.*, **49**, 90.
60. J. N. Dodd, W. J. Sandle and O.M. Williams (1970), *J. Phys. B*, **3**, 256.
61. M. Popp, G. Schafer and E. Bodenstedt (1970), *Z. Phys.*, **240**, 71.
62. G. C. King and A. Adams (1974), *J. Phys. B*, **7**, 1712.
63. Yu. I. Ostrovskii and N. P. Penkin (1961), *Opt. Spectrosc.*, **11**, 307.
64. A. I. Odintsov (1963), *Opt. Spectrosc.*, **14**, 172.
65. E. Hulpke, E. Paul and W. Paul (1964), *Z. Phys.*, **177**, 257.
66. N. P. Penkin (1964), *J. Quant. Spectrosc. Radiat. Transfer*, **4**, 41.
67. M. Chenevier, J. Dufayard and J. C. Pebay-Peyroula (1967), *Phys. Lett. A*, **25**, 283.
68. N. P. Penkin and L. N. Shabanova (1969), *Opt. Spectrosc.*, **26**, 191.
69. M. D. Havey, L. C. Balling and J. J. Wright (1977), *J. Opt. Soc. Am.*, **67**, 488.
70. D. W. Fahey, W. F. Parks and L. D. Schearer (1979), *Phys. Lett. A*, **74**, 405.
71. P. S. Furcinitti, L. C. Balling and J. J. Wright (1975), *Phys. Lett. A*, **53**, 75.
72. P. G. Whitkop and J. R. Wiesenfeld (1980), *Chem. Phys. Lett.*, **69**, 457.
73. L. O. Dickie, F. M. Kelly, T. K. Koh, M. S. Mathur and F. C. Suk (1973), *Can. J. Phys.*, **51**, 1088.
74. F. M. Kelly, T. K. Koh and M. S. Mathur (1974), *Can. J. Phys.*, **52**, 795.
75. N. M. Erdevdi and L. L. Shimon (1976), *Opt. Spectrosc*, **40**, 443.
76. I.-J. Ma, G. Zu Putlitz and G. Schutte (1968), *Z. Phys.*, **208**, 276.
77. M. D. Havey, L. C. Balling and J. J. Wright (1976), *Phys. Rev. A*, **13**, 1269.
78. L. O. Dickie and F. M. Kelly (1970), *Can. J. Phys.*, **48**, 879.
79. P. Schenck, R. C. Hilborn and H. Metcalf (1973), *Phys. Rev. Lett.*, **31**, 189.
80. F. M. Kelly and M. S. Mathur (1977), *Can. J. Phys.*, **55**, 83.
81. H. Bucka and H. H. Nagel (1961), *Ann. Phys.*, **8**, 329.
82. K. C. Brog, T. Eck and H. Wieder (1967), *Phys. Rev.*, **153**, 91.
83. J.-P. Buchet, A. Denis, J. Desesquelles and M. Dufay (1967), *C. R. Acad, Sci. B*, **265**, 471.
84. W. S. Bickel, I. Martinson, L. Lundin, R. Buchta, J. Bromander and I. Bergstrom (1969), *J. Opt. Soc. Am.*, **59**, 830.
85. S. -M. Lin and R. E. Weston, Jr (1976), *J. Chem. Phys.*, **65**, 1443.
86. N. V. Karlov, B. B. Krynetskii and O. M. Stel'makh (1977), *Pis'ma Zh. Tekh. Fiz.*, **3**, 716.
87. W. Nagourney, W. Happer and A. Lurio (1978), *Phys. Rev. A*, **17**, 1394.
88. Yu. I. Ostrovskii and N. P. Penkin (1961), *Opt. Spectrosc.*, **11**, 1.
89. W. Demtroder (1962), *Z. Phys.*, **166**, 42.
90. G. V. Markova and M. P. Chaika (1964), *Opt. Spectrosc.*, **17**, 170.
91. H. Ackermann (1966), *Z. Phys.*, **194**, 253.
92. J. K. Link (1966), *J. Opt. Soc. Am.*, **56**, 1195.
93. B. P. Kibble, G. Copley and L. Krause (1967), *Phys. Rev.*, **153**, 9.
94. C. Bastlein, G. Baumgartner and B. Brosa (1969), *Z. Phys.*, **218**, 319.

95. T. Andersen, J. Desesquelles, K. A. Jessen and G. Sorensen (1970), *J. Opt. Soc. Am.*, **60**, 1199.
96. L. E. Brus (1970), *J. Chem. Phys.*, **52**, 1716.
97. T. Andersen, O. H. Madsen and G. Sorensen (1972), *Phys. Scr.*, **6**, 125.
98. Yu. I. Ostrovskii and N. P. Penkin (1962), *Opt. Spectrosc.*, **12**, 379.
99. R. W. Schmieder, A. Lurio and W. Happer (1968), *Phys. Rev.*, **173**, 76.
100. G. Copley and L. Krause (1969), *Can. J. Phys.*, **47**, 533.
101. D. Zimmerman (1975), *Z. Phys. A*, **275**, 5.
102. A. Gallagher (1967), *Phys. Rev.*, **157**, 68.
103. G. Markova, G. Khvostenko and M. Chaika (1967), *Opt. Spectrosc.*, **23**, 456.
104. J. N. Dodd, E. Enemark and A. Gallagher (1969), *J. Chem. Phys.*, **50**, 4838.
105. R. J. Exton (1976), *J. Quant. Spectrosc. Radiat. Transfer*, **16**, 309.
106. E. M. Anderson, V. A. Zilitis and E. S. Sorokina (1967), *Opt. Spectrosc.*, **23**, 279.
107. G. Copley, B. P. Kibble and L. Krause (1967), *Phys. Rev.*, **163**, 34.
108. W. Demtroder (1962), *Z. Phys.*, **166**, 42.
109. C. Bastlein, G. Baumgartner and B. Brosa (1969), *Z. Phys.*, **218**, 319.
110. M. G. Edwards (1969), *J. Phys. B*, **2**, 719.
111. D. R. Jenkins (1966), *Chem. Commun.*, 171.
112. D. R. Jenkins (1966), *Proc. R. Soc. A*, **293**, 493.
113. H. P. Hooymayers and C. T. J. Alkemade (1966), *J. Quant. Spectrosc. Radiat. Transfer*, **6**, 847.
114. S. Tsuchiya (1964), *Bull. Chem. Soc. Jpn.*, **37**, 828.
115. S. Tsuchiya and K. Kuratani (1964), *Combust. Flame*, **8**, 299.
116. B. P. Kibble, G. Copley and L. Krause (1967), *Phys. Rev.*, **159**, 11.
117. E. Hulpke, E. Paul and W. Paul (1964), *Z. Phys.*, **177**, 257.
118. J. R. Barker and R. E. Weston, Jr (1976), *J. Chem. Phys.*, **65**, 1427.
119. K. J. Mintz and D. J. LeRoy (1971), *Chem. Phys. Lett.*, **10**, 280.
120. J. R. Barker and R. E. Weston, Jr (1973), *Chem. Phys. Lett.*, **19**, 235.
121. P. L. Lijnse and R. J. Elsenaar (1972), *J. Quant. Spectrosc. Radiat. Transfer*, **12**, 1115.
122. N. S. Ham and P. Hannaford (1979), *J. Phys. B*, **12**, L199.
123. H. P. Hooymayers and P. L. Lijnse (1969), *J. Quant. Spectrosc. Radiat. Transfer*, **9**, 995.
124. H. P. Hooymayers and C. T. J. Alkemade (1966), *J. Quant. Spectrosc. Radiat. Transfer*, **6**, 847.
125. B. L. Earl and R. R. Herm (1974), *J. Chem. Phys.*, **60**, 4568.
126. P. L. Lijnse and C. van der Maas (1973), *J. Quant. Spectrosc. Radiat. Transfer*, **13**, 741.
127. L. E. Brus (1970), *J. Chem. Phys.*, **52**, 1716.
128. B. L. Earl, R. R. Herm, S.-M. Lin and C. A. Mims (1972), *J. Chem. Phys.*, **56**, 867.
129. R. G. W. Norrish and W. M. Smith (1940), *Proc. R. Soc. A*, **176**, 295.
130. S.-M. Lin and R. E. Weston, Jr (1976), *J. Chem. Phys.*, **65**, 1443.
131a. D. R. Jenkins (1968), *Proc. R. Soc. A*, **306**, 413.
131b. D. R. Jenkins (1968), *Proc. R. Soc. A*, **303**, 453.
132. P. L. Lijnse, P. J. T. Zeegers and C. T. J. Alkemade (1973), *J. Quant. Spectrosc. Radiat Transfer*, **13**, 1033.
133. H. P. Hooymayers and G. Nienhuis (1968), *J. Quant. Spectrosc. Radiat. Transfer*, **8**, 955.
134. P. L. Lijnse, P. J. T. Zeegers and C. T. J. Alkemade (1973), *J. Quant. Spectrosc. Radiat. Transfer*, **13**, 1301.
135. G. Copley and L. Krause (1969), *Can. J. Phys.*, **47**, 533.

136. J. Gatzke (1963), *Z. Phys. Chem.*, **223**, 321.
137. P. L. Lijnse and J. C. Hornman (1974), *J. Quant. Spectrosc. Radiat. Transfer*, **14**, 1079.
138. B. L. Earl and R. R. Herm (1974), *J. Chem. Phys.*, **60**, 4568.
139. B. L. Earl and R. R. Herm (1973), *Chem. Phys. Lett.*, **22**, 95.
140. D. A. McGillis and L. Krause (1968), *Can. J. Phys.*, **46**, 25.
141. J. Elward-Berry and M. J. Berry (1980), *J. Chem. Phys.*, **72**, 4510.
142. E. S. Hrycyshyn and L. Krause (1970), *Can. J. Phys.*, **48**, 2761.
143. B. R. Bulos (1969), *Bull. Am. Phys. Soc.*, **14**, 618.
144. J. A. Bellisio, P. Davidovits and P. J. Kindlmann (1968), *J. Chem. Phys.*, **48**, 2376.
145. D. A. McGillis and L. Krause (1967), *Phys. Rev.*, **153**, 44.
146. D. A. McGillis and L. Krause (1968), *Can. J. Phys.*, **46**, 1051.
147. W. H. Breckenridge and A. M. Renlund (1978), *J. Phys. Chem.*, **82**, 1474.
148. P. D. Morten, C. G. Freeman, R. F. C. Claridge and L. F. Phillips (1974/75), *J. Photochem.*, **3**, 285.
149. W. H. Breckenridge, R. J. Donovan and O. K. Malmin (1979), *Chem. Phys. Lett.*, **62**, 608.
150. R. Pepperl (1970), *Z. Naturforsch. A*, **25**, 927.
151. W. H. Breckenridge and O. K. Malmin (1979), *Chem. Phys. Lett.*, **68**, 341.
152. W. H. Breckenridge and O. K. Malmin (1981), *J. Chem. Phys.*, **74**, 3307.
153. A. Granzow, M. Z. Hoffman and N. N. Lichtin (1969), *J. Phys. Chem.*, **73**, 4289.
154. W. H. Breckenridge and A. M. Renlund (1979), *J. Phys. Chem.*, **83**, 1145.
155. S. Yamamoto, T. Takei, N. Nishimura and S. Hasegawa (1980), *Bull. Chem. Soc. Jpn.*, **53**, 860.
156. R. Bleekrode and W. van Benthem (1973), *Philips Res. Rep.*, **28**, 130.
157. W. H. Breckenridge and H. Umemoto (1981), *J. Chem. Phys.*, **75**, 698.
158. A. C. Vikis, G. Torrie and D. J. LeRoy (1970), *Can. J. Chem.*, **48**, 3771.
159. A. C. Vikis, G. Torrie and D. J. LeRoy (1972), *Can. J. Chem.*, **50**, 176.
160. J. Pitre, K. Hammond and L. Krause (1972), *Phys. Rev. A*, **6**, 2101.
161. A. B. Callear and J. C. McGurk (1973), *J. Chem. Soc. Faraday Trans. 2*, **69**, 97.
162. H. Horiguchi and S. Tsuchiya (1977), *Bull. Chem. Soc. Jpn.*, **50**, 1657.
163. H. Horiguchi and S. Tsuchiya (1977), *Bull. Chem. Soc. Jpn.*, **50**, 1661.
164. H. Horiguchi and S. Tsuchiya (1975), *J. Chem. Soc. Faraday Trans. 2*, **71**, 1164.
165. J. S. Deech, J. Pitre and L. Krause (1971), *Can. J. Phys.*, **49**, 1976.
166. J. V. Michael and C. Yeh (1970), *J. Chem. Phys.*, **53**, 59.
167. J. V. Michael and G. N. Suess (1974), *J. Phys. Chem.*, **78**, 482.
168. S. D. Gleditsch and J. V. Michael (1975), *J. Phys. Chem.*, **79**, 409.
169. J. H. Hong and G. J. Mains (1973), *J. Photochem.*, **1**, 463.
170. H. E. Gunning, J. M. Campbell, H. S. Sandhu and O. P. Strausz (1973), *J. Am. Chem. Soc.*, **95**, 746.
171. K. Yang (1966), *J. Am. Chem. Soc.*, **88**, 4575.
172. G. J. Mains and M. Trachtman (1970), *J. Phys. Chem.*, **74**, 1647.
173. G. J. Mains and M. Trachtman (1972), *J. Phys. Chem.*, **76**, 2665.
174. K. Yang (1967), *J. Am. Chem. Soc.*, **89**, 5344.
175. A. J. Yarwood, O. P. Strausz and H. E. Gunning (1964), *J. Chem. Phys.*, **41**, 1705.
176. T. Shay, H. Hemmati, T. Stermitz and G. J. Collins (1980), *J. Chem. Phys.*, **72**, 1635.
177. L. F. Phillips (1977), *J. Chem. Soc. Faraday Trans. 2*, **73**, 97.
178. C. G. Freeman, M. J. McEwan, R. F. C. Claridge and L. F. Phillips (1971), *Trans. Faraday Soc.*, **67**, 2004.

179. F. J. Wodarczyk and A. B. Harker (1979), *Chem. Phys. Lett.*, **62**, 529.
180. A. B. Callear and J. McGurk (1970), *Chem. Phys. Lett.*, **7**, 491.
181. H. Umemoto, T. Kyogoku, S. Tsunashima and S. Sato (1980), *Chem. Phys.*, **52**, 481.
182. H. Umemoto, S. Tsunashima and S. Sato (1980), *Chem. Phys.*, **47**, 263.
183. H. Umemoto, S. Tsunashima and S. Sato (1980), *Chem. Phys.*, **47**, 257.
184. H. Umemoto, S. Tsunashima and S. Sato (1979), *Chem. Phys.*, **43**, 93.
185. E. W. R. Steacie and D. J. LeRoy (1943), *J. Chem. Phys.*, **11**, 164.
186. W. H. Breckenridge, T. W. Broadbent and D. S. Moore (1975), *J. Phys. Chem.*, **79**, 1233.
187. S. Yamamoto, M. Takaoka, S. Tsunashima and S. Sato (1975), *Bull. Chem. Soc. Jpn.*, **48**, 130.
188. W. H. Breckenridge and A. M. Renlund (1979), *J. Phys. Chem.*, **83**, 303.
189. M. Czajkowski and L. Krause (1976), *Can. J. Phys.*, **54**, 603.
190. S. Yamamoto, T. Takei, N. Nishimura and S. Hasegawa (1976), *Chem. Lett.*, 1413.
191. R. P. Blickensderfer, W. H. Breckenridge and D. S. Moore (1975), *J. Chem. Phys.*, **63**, 3681.
192. G. Taieb and H. P. Broida (1976), *J. Chem. Phys.*, **65**, 2914.
193. W. H. Breckenridge and W. L. Nikolai (1980), *J. Chem. Phys.*, **73**, 2763.
194. A. Kowalski (1979), *Z. Naturforsch. A*, **34**, 459.
195. R. J. Malims and D. J. Benard (1981), *Chem. Phys. Lett.*, **74**, 321:
     R. J. Malims, A. D. Logan and D. J. Benard (1981) *Chem. Phys. Lett.*, **83**, 605
196. P. J. Dagdigian (1978), *Chem. Phys. Lett.*, **55**, 239.
197. H. Telle and U. Brinkmann (1980), *Mol. Phys.*, **39**, 361.
198. B. E. Wilcomb and P. J. Dagdigian (1978), *J. Chem. Phys.*, **69**, 1779.
199. J. J. Wright and L. C. Balling (1980), *J. Chem. Phys.*, **73**, 1617.
200. B. J. Jansen, T. Hollander and H. Van Helvoort (1977), *J. Quant. Spectrosc. Radiat. Transfer*, **17**, 193.
201. T. Hollander, P. L. Lijnse, L. P. L. Franken, B. J. Jansen and P. J. T. Zeegers (1972), *J. Quant. Spectrosc. Radiat. Transfer*, **12**, 1067.
202. J. D. Eversole and N. Djeu (1979), *J. Chem. Phys.*, **71**, 148.
203. L. Pasternack, D. Silver, D. Yarkony and P. Dagdigian (1980), *J. Phys. B*, **13**, 2231.
204. C. T. J. Alkemade, private communication.
205. J. E. Velazco, J. H. Kolts and D. W. Setser (1978), *J. Chem. Phys.*, **69**, 4357.
206. T. P. Parr and R. M. Martin (1978), *J. Phys. Chem.*, **82**, 2226.
207. K. Liu and J. M. Parson (1976), *J. Chem. Phys.*, **65**, 815.
208. J. R. Barker (1976), *Chem. Phys.*, **18**, 175.
209. P. J. Lijnse (1973), *Chem. Phys. Lett.*, **18**, 73.
210. E. Bauer, E. R. Fisher and F. R. Gilmore (1969), *J. Chem. Phys.*, **51**, 4173.
211. H. F. Krause, S. Datz and S. G. Johnson (1973), *J. Chem. Phys.*, **58**, 367.
212. R. M. Martin and T. P. Parr (1979), *J. Chem. Phys.*, **70**, 2220.
213. P. B. Foreman, T. P. Parr and R. M. Martin (1977), *J. Chem. Phys.*, **67**, 5591.
214. E. Illenberger and A. Niehaus (1975), *Z. Phys. B*, **20**, 33.
215. W. H. Miller (1970), *J. Chem. Phys.*, **52**, 3563.
216. G. J. Schulz (1973), *Rev. Mod. Phys.*, **45**, 423.
217. J. C. Tully (1973), *J. Chem. Phys.*, **59**, 5122.
218. P. D. Burrow (1974), *Chem. Phys. Lett.*, **26**, 265.
219. M. W. Siegel, R. J. Celotta, J. L. Hall, J. Levine and R. A. Bennett (1972), *Phys. Rev. A*, **6**, 607.
220. M. J. W. Boness and G. J. Schulz (1970), *Phys. Rev. A*, **2**, 2182.

221. R. J. Celotta, R. A. Bennett, J. L. Hall, M. W. Siegel and J. Levine (1972), *Phys. Rev. A*, **6**, 631.
222. J. A. Aten and J. Los (1977), *Chem. Phys.*, **25**, 47.
223. R. N. Compton, P. W. Reinhardt and C. D. Cooper (1975), *J. Chem. Phys.*, **63**, 3821.
224. S. J. Nalley, R. N. Compton, H. C. Schweinler and V. E. Anderson (1973), *J. Chem. Phys.*, **59**, 4125.
225. L. Sanche and G. J. Schulz (1973), *J. Chem. Phys.*, **58**, 479.
226. R. W. Fessenden and K. M. Bansal (1970), *J. Chem. Phys.*, **53**, 3468.
227. R. N. Compton and C. D. Cooper (1973), *J. Chem. Phys.*, **59**, 4140.
228. J. D. Webb and E. R. Bernstein (1978), *J. Am. Chem. Soc.*, **100**, 483.
229. P. D. Burrow and K. D. Jordan (1975), *Chem. Phys. Lett.*, **36**, 594.
230. K. D. Jordan and P. D. Burrow (1978), *Acc. Chem. Res.*, **11**, 341.
231. N. S. Chiu, P. D. Burrow and K. D. Jordan (1979), *Chem. Phys. Lett.*, **68**, 121.
232. E. R. Fisher and G. K. Smith (1971), *Appl. Opt.*, **10**, 1803.
233. W. H. Breckenridge and A. M. Renlund (1978), *J. Phys. Chem.*, **82**, 1484.
234. E. Speller, B. Standenmayer and V. Kempter (1979), *Z. Phys. A*, **291**, 311.
235. W. E. Baylis (1969), *J. Chem. Phys.*, **51**, 2665.
236. J. Pascale and J. Vandeplanque (1974), *J. Chem. Phys.*, **60**, 2278.
237. G. Herzberg (1950), *Spectra of Diatomic Molecules*, 2nd edn., Van Nostrand Reinhold, New York.
238. W. Nikolai (1980), *PhD Thesis*, University of Utah.
239. V. Madhavan, N. N. Lichtin and M. Z. Hoffman (1973), *J. Phys. Chem.*, **77**, 875.
240. A. L. Rockwood and E. A. McCullough, Jr (1977), *J. Phys. Chem.*, **81**, 2050.
241. W. H. Breckenridge and H. Umemoto, *J. Chem. Phys.*, **75**, 4153.
242. R. P. Frueholz, R. Rianda and A. Kupperman (1978), *J. Chem. Phys.*, **68**, 775, and references cited therein.
243. W. H. Breckenridge and J. FitzPatrick (1976), *J. Phys. Chem.*, **80**, 1955.
244. W. H. Breckenridge, R. P. Blickensderfer and J. FitzPatrick (1976), *J. Phys. Chem.*, **80**, 1963.
245. W. H. Breckenridge, R. P. Blickensderfer, J. FitzPatrick and D. Oba (1979), *J. Chem. Phys.*, **70**, 4751.
246. O. P. Strausz and H. E. Gunning (1961), *Can. J. Chem.*, **39**, 2549.
247. I. V. Hertel, H. Hofmann and K. A. Rost (1977), *Chem. Phys. Lett.*, **47**, 163.
248. M. H. Alexander and P. J. Dagdigian (1978), *Chem. Phys.*, **33**, 13.
249. L. Pasternack and P. J. Dagdigian (1978), *Chem. Phys.*, **33**, 1.
250. A. G. Gaydon (1968), *Dissociation Energies and Spectra of Diatomic Molecules*, 3rd edn., Chapman and Hall, London.
251. D. H. Volman (1954), *J. Am. Chem. Soc.*, **76**, 6034.
252. E. K. Gill and K. Laidler (1958), *Can. J. Chem.*, **36**, 79.
253. R. J. Fallon, J. T. Vanderslice and E. A. Mason (1960), *J. Phys. Chem.*, **64**, 505.
254. A. B. Callear, C. R. Patrick and J. C. Robb (1959), *Trans. Faraday Soc.*, **55**, 280.
255. H. Hippler, H. R. Wendt and H. E. Hunziker (1978), *J. Chem. Phys.*, **68**, 5103.
256. H. H. Mohammed, J. Fournier, J. Deson and C. Vermeil (1980), *Chem. Phys. Lett.*, **73**, 315, and references therein.
257. H. E. Gunning (1958), *Can. J. Chem.*, **36**, 89.
258. R. J. Cvetanovic (1955), *J. Chem. Phys.*, **23**, 1203.
259. C. H. Muller III, K. Schofield and M. Steinberg (1980), *J. Chem. Phys.*, **72**, 6620.
260. A. B. Callear and P. M. Wood (1972), *J. Chem. Soc. Faraday Trans. II*, **68**, 302.
261. P. H. Lee, H. P. Broida, W. Braun and J. T. Herron (1973/74), *J. Photochem.*, **2**, 165.

262. P. J. Young, G. Greig and O. P. Strausz (1970), *J. Am. Chem. Soc.*, **92**, 413.
263. P. Young, E. Hardwidge, S. Tsunashima, G. Greig and O. P. Strausz (1974), *J. Am. Chem. Soc.*, **96**, 1946.
264. N. Adams, W. H. Breckenridge, and J. Simons (1981), *Chem. Phys.*, **56**, 327.
265. W. H. Breckenridge, J. Stewart and H. Umemoto, experiments in progress.
266. W. H. Breckenridge, J. Beckerle W. Nikolai, D. Oba and H. Umemoto (1982), submitted for publication.
267. R. A. Holroyd and G. W. Klein (1963), *J. Phys. Chem.*, **67**, 2273.
268. A. C. Vikis and H. C. Moser (1970), *J. Chem. Phys.*, **53**, 2333.
269. A. B. Callear and J. C. McGurk (1972), *J. Chem. Soc. Faraday Trans.* 2, **68**, 289.
270. H. E. Hunziker, private communication.
271. A. Vikis and D. J. LeRoy (1973), *Can. J. Chem.*, **51**, 1207.
272. W. H. Breckenridge, J. Beckerle and D. Oba (1982), submitted for publication.
273. A. J. Merer and R. S. Mulliken (1969), *Chem. Rev.*, **69**, 639.
274. W. M. Flicker, O. A. Mosher and A. Kuppermann (1975), *Chem. Phys. Lett.*, **36**, 56.
275. R. W. Wetmore and H. F. Schaefer III (1978), *J. Chem. Phys.*, **69**, 1648.
276. H. E. Hunziker (1969), *J. Chem. Phys.*, **50**, 1288.
277. S. Tsunashima, S. Hirokami and S. Sato (1968), *Can. J. Chem.*, **46**, 995.
278. T. Terao, S. Hirokami, S. Sato and R. J. Cvetanović (1966), *Can. J. Chem.*, **44**, 2173.
279. S. Hirokami and S. Sato (1967), *Can. J. Chem.*, **45**, 3181.
280. A. B. Callear and R. J. Cvetanović (1956), *J. Chem. Phys.*, **24**, 873.
281. D. W. Setser, B. S. Rabinovitch and D. W. Placzek (1963), *J. Am. Chem. Soc.*, **85**, 862.
282. R. J. Cvetanović and L. C. Doyle (1962), *J. Chem. Phys.*, **37**, 543.
283. S. Hirokami and S. Sato (1970), *Bull. Chem. Soc. Jpn.*, **43**, 2389.
284. C. S. Burton and H. E. Hunziker (1972), *J. Chem. Phys.*, **57**, 339.
285. H. R. Wendt, H. Hippler and H. E. Hunziker (1979), *J. Chem. Phys.*, **70**, 4044.
286. H. E. Hunziker (1969), *J. Chem. Phys.*, **50**, 1294.
287. C. S. Burton and H. E. Hunziker (1970), *Chem. Phys. Lett.*, **6**, 352.
288. H. E. Hunziker (1969), *Chem. Phys. Lett.*, **3**, 504.
289. D. J. Ehrlich and J. Wilson (1977), *Opt. Commun.*, **20**, 314.
290. L. A. Gundel (1975), *PhD Thesis*, University of California, Berkeley.
291. D. R. Hershbach (1966), *Adv. Chem. Phys.*, **10**, 319.
292. S. Hayashi, T. M. Mayer and R. B. Bernstein (1978), *Chem. Phys. Lett.*, **53**, 419.
293. H. F. Krause, S. G. Johnson, S. Datz and F. K. Schmidt-Bleek (1975), *Chem. Phys. Lett.*, **31**, 577.
294a. R. W. Solarz and S. A. Johnson (1979), *J. Chem. Phys.*, **70**, 3592.
294b. R. W. Solarz, S. A. Johnson and R. K. Preston (1978), *Chem. Phys. Lett.*, **57**, 514.
295. A. Kowalski and J. Heldt (1978), *Chem. Phys. Lett.*, **54**, 240.
296. R. C. Estler and R. N. Zare (1978), *Chem. Phys.*, **28**, 253.
297. A. Kowalski and M. Menzinger (1981), submitted for publication.
298. U. Brinkmann, V. H. Schmidt and H. Telle (1980), *Chem. Phys. Lett.*, **73**, 530.
299. A. B. Callear and J. McGurk (1970), *Chem. Phys. Lett.*, **7**, 491.
300. A. B. Callear and J. H. Connor (1974), *J. Chem. Soc. Faraday Trans.* 2, **70**, 1667.
301. H. Umemoto, S. Tsunashima and S. Sato (1978), *Chem. Phys.*, **35**, 103.
302. T. Hikida, M. Santoku and Y. Mori (1978), *Chem. Phys. Lett.*, **55**, 280.
303. T. Hikida, T. Ichimura and Y. Mori (1974), *Chem. Phys. Lett.*, **27**, 548.
304. T. Hikida, T. Ishihara and Y. Mori (1977), *Chem. Phys. Lett.*, **52**, 43.
305. H. Umemoto, S. Tsunashima and S. Sato (1978), *Chem. Phys. Lett.*, **53**, 521.

306. A. B. Callear and D. R. Kendall (1978), *Chem. Phys. Lett.*, **58**, 31.
307. A. B. Callear and J. H. Connor (1974), *J. Chem. Soc. Faraday Trans.* 2, **70**, 1767.
308. K. Luther, H. R. Wendt and H. E. Hunziker (1975), *Chem. Phys. Lett.*, **33**, 146.
309. A. B. Harker and C. S. Burton (1975), *J. Chem. Phys.*, **63**, 885.
310. A. B. Callear, D. R. Kendall and L. Krause (1978), *Chem. Phys.*, **29**, 415.
311. U. Ringström (1970), *J. Mol. Spectrosc.*, **36**, 232.
312. D. L. Hildenbrand and E. Murad (1970), *J. Chem. Phys.*, **53**, 3403.
313. L. Brewer and G. M. Rosenblatt (1969), *Adv. High Temp. Sci.*, **2**, 1.
314. W. J. Balfour and B. Lindgren (1978), *Can. J. Phys.*, **56**, 767.
315. K. Schofield (1967), *Chem. Rev.*, **67**, 707.
316. I. V. Hertel (1982), *Progress in electronic-to-vibrational energy transfer*, in *Dynamics of the Excited State*, ed. K. P. Lawley, this volume.
317. G. W. Black, M. A. D. Fluendy and D. Sutton (1980), *Chem. Phys. Lett.*, **69**, 260.
318. I. V. Hertel and W. Reiland (1981), *J. Chem. Phys.*, **74**, 6757.
319. H. Horiguchi and S. Tsuchiya (1979), *J. Chem. Phys.*, **70**, 762.
320. J. P. Simons and T. W. Tasker (1973), *Mol. Phys.*, **26**, 1267.
321. M. J. Berry (1974), *Chem. Phys. Lett.*, **29**, 329.
322. R. G. Shortridge and M. C. Lin (1976), *J. Chem. Phys.*, **64**, 4076.
323. J. Degani, E. Rosenfeld and S. Yatsiv (1978), *J. Chem. Phys.*, **68**, 4041.
324. R. D. Levine and A. Ben-Shaul (1977), in *Chemical and Biochemical Applications of Lasers*, vol. 2, ed. C. B. Moore, Academic Press, New York.
325. A. R. Malvern (1978), *J. Phys. B*, **11**, 831.
326. R. J. Cross, Jr and R. G. Gordon (1966), *J. Chem. Phys.*, **45**, 3571.
327. E. H. Fink, D. Wallach and C. B. Moore (1972), *J. Chem. Phys.*, **56**, 3608.
328. A. C. Vikis (1978), *Chem. Phys. Lett.*, **57**, 522.
329. K. G. Ibbs, P. H. Wine, K. J. Chung and L. A. Melton (1981), *J. Chem. Phys.*, **74**, 6212.
330. L. A. Melton and W. Klemperer (1973), *J. Chem. Phys.*, **59**, 1099.
331. A. C. Vikis and D. J. LeRoy (1973), *Chem. Phys. Lett.*, **21**, 103.
332. A. C. Vikis and D. J. LeRoy (1973), *Chem. Phys. Lett.*, **22**, 587.
333. A. C. Vikis (1975), *Chem. Phys. Lett.*, **33**, 506.
334. U. Brinkmann and H. Telle (1977), *J. Phys. B*, **10**, 133.
335. F. Capra (1975), *The Tao of Physics*, Bantam, New York.

Dynamics of the Excited State
Edited by K. P. Lawley
© 1982 John Wiley & Sons Ltd.

REACTION DYNAMICS AND STATISTICAL MECHANICS OF THE PREPARATION OF HIGHLY EXCITED STATES BY INTENSE INFRARED RADIATION

MARTIN QUACK

Institut für Physikalische Chemie der Universität, Tammannstrasse 6, D-3400 Göttingen, West Germany

CONTENTS

'Schon jetzt führt jede Untersuchung über die Struktur einer physikalischen Theorie unvermeidlich auf die Frage nach der Natur der Wahrscheinlichkeitshypothesen.'

P. Ehrenfest and T. Ehrenfest (1911)

I. INTRODUCTION

Unimolecular reactions induced by monochromatic infrared radiation (URIMIR) from pulsed lasers have received much attention since their discovery in the early 1970s (Isenor *et al.*, 1973; see also Bordé *et al.*, 1966). Briefly, these reactions can be summarized (1) by

$$M \xrightarrow{nh\nu} \text{Products} \tag{1}$$

with $h\nu \simeq 12$ kJ mol^{-1} and n between 10 and 40 for the CO_2 laser as an example. Several excellent reviews of experimental progress are available, which also include discussions of various theoretical approaches (Letokhov and Moore, 1976; Ambartzumian and Letokhov, 1977; Bloembergen and Yablonovitch, 1978; Cantrell *et al.*, 1979; Schulz *et al.*, 1979; Lyman *et al.*, 1980; Golden *et al.*, 1980; Ashfold and Hancock, 1981; see also the articles of King and of Lau in the present volume). With such a new phenomenon, however, many questions, both experimental and theoretical, remain to be answered. In particular, there is as yet no completely agreed theoretical scheme of the mechanism of multiphoton excitation and the various theories differ in more than just formal aspects of the mathematical approach. Owing to the limitations of space, only a very brief discussion of the various possible approaches will be given here (in Sec. III E).

In the present article we shall mainly discuss the foundations of the statistical mechanical treatment of URIMIR (Quack, 1978) and some of its applications. One ramification of this treatment is the theoretical justification of the use of rate equations for URIMIR, which were first applied phenomenologically (Lyman, 1977; Grant et al., 1978), and are widely accepted now because of their simplicity (Baldwin et al., 1979; Duperrex and van den Bergh, 1979a, b; Schek and Jortner, 1979; Mukamel 1979; Fuss, 1979; Carmeli et al., 1980; Carmeli and Nitzan, 1980; Barker, 1980; Stone et al., 1980; Troe, 1980; Lawrence et al., 1981; Yahav et al., 1980). It appears, however, that the use of the simple rate equations is becoming more widespread than the understanding of the underlying physical principles and of the inherent limitations. A review dealing with the latter aspects and including more general treatments might therefore be adequate at the present time.

Experimentally, the breakdown of the simple rate equation picture has been highlighted by measuring increased product yields with short, intense pulses instead of long pulses of the same fluence ($F = \int I \, dt$) in the URIMIR of $CH_3 NH_2$ (Ashfold et al., 1979) Naaman and Zare 1979 and, using related ideas, for a few other, small molecules (Rossi et al., 1979; King and Stephenson, 1979). This is an unambiguous proof of the nonlinear intensity dependence required under certain conditions by the general statistical-mechanical theory (Quack, 1978, 1979a). Another, even more striking consequence of the nonlinearity predicted for very high intensity is the intensity fall-off (weak within the ordinary rate equation picture and stronger for 'case D', see below). This would imply that under certain, specified conditions excitation with higher intensity gives less yield than at the same fluence and lower intensity (Quack, 1978). This effect has not yet been verified experimentally. Furthermore, the effects of static magnetic and electric fields on product yields recently discovered by Duperrex and van den Bergh (1980) and Gozel and van den Bergh (1980) very probably require an interpretation by means of the general theory beyond the simple rate equation picture (Quack and van den Bergh, 1981).

The outline of the review is as follows. In Sec. II we shall establish the quantum-mechanical equations of motion for the excitation of polyatomic molecules with coherent infrared radiation and also the rate equations for excitation with incoherent, quasimonochromatic and with thermal radiation. The section concludes with a discussion of measurable quantities in URIMIR both from a fundamental and a practical point of view. Sec. III deals with the physical foundations of the statistical-mechanical treatment of URIMIR, covering both the linear regime (Pauli equation and simple rate equations) and the nonlinear regime. The following short sections deal with unimolecular decay, solutions of the master equations, and a brief, critical discussion of the comparison of experiment and theory.

It is clear that a complete coverage of the already large literature in this field is not possible and cannot be the aim of an article in the present series of *Advances in Chemical Physics*. We have tried, however, to make at least brief reference to both alternative and related treatments wherever a comparison is possible. These references and the reviews already quoted should enable the reader to obtain a fairly complete picture of the present status of the theory of URIMIR.

II. THE EXCITATION OF POLYATOMIC MOLECULES BY INTENSE INFRARED RADIATION

We shall summarize in this section the quantum-mechanical equations of motion for a polyatomic molecule interacting with strong radiation fields. Although other approaches are possible, we think that the problem is best attacked by splitting it into two parts to be solved separately:

(i) The solution of the time-independent Schrödinger equation (Eq. (2)) for the isolated molecule in the absence of radiation (and neglecting spontaneous emission):

$$\hat{H}_{is}\phi_k = E_k\phi_k = \hbar\omega_k\phi_k, \tag{2}$$

where the existence of decay into continua may be accounted for by allowing for complex energies with decay widths γ_k:

$$\omega_k = \mathrm{Re}(\omega_k) - i\gamma_k/2. \tag{3}$$

(ii) the coupling of the states ϕ_k by the radiation field.

It will be assumed that the molecular problem in the absence of radiation is already solved, say, by the means of ultrahigh-resolution spectroscopy or *ab initio* calculations or both. Then the molecular (spectroscopic) energies E_k and widths γ_k and the corresponding time-dependent wavefunctions are known. Obviously, the molecular problem is a vast field of research by itself and it is best discussed separately. We restrict our major attention here to the truly dynamical interaction problem. The idealized dynamical problem of two states coupled by a monochromatic radiation field is a standard textbook example, although some important questions remain unsolved even for this simple problem (Dion and Hirschfelder, 1976). The extension to polyatomic systems with many states is 'in principle' straightforward by matrix techniques. The new physical structures arising from the very large numbers of states involved will be considered in some detail in Sec. 3.

A. Excitation with monochromatic, coherent radiation

1. *Exact treatment*

For an idealized situation we start from the following assumptions:

(i) The radiation field can be treated by a z-polarized, monochromatic, classical wave of circular frequency $\omega = 2\pi\nu$ propagating in the y direction. We then have for the electric field:

$$E_z(y, t) = \text{Re}\{E_0 \exp[\text{i}(\omega t - k_\omega y + \eta)]\}. \tag{4}$$

Doppler shifts are neglected (zero velocity component v_y of the molecules, for instance in a molecular beam, otherwise add $k_\omega v_y$ in the exponent with k_ω being the wavenumber of the field). The phaseshifts $k_\omega y + \eta$ will be omitted hereafter for brevity of notation (otherwise one makes a phase average at the end of the calculation, which can be shown to be unimportant for the final equations of this section). The amplitude E_0 of the electric field is assumed to be constant or slowly varying in the considered time range. The classical approximation to the field is excellent for the case of laser excitation (number of photons in one mode very much greater than unity) and the short timescales of interest here (Messiah, 1964). The creation of photons, which are not in the original wave, is not treated here.

(ii) Only the electric dipole interaction is retained, assuming the wavelengths to be much larger than the molecular size, which is perfectly satisfied in the IR region. Other interactions can be treated similarly. Thus with the molecular dipole operator

one has:
$$\boldsymbol{\mu} = \sum e_i \mathbf{r}_i,$$

$$\hat{H} = \hat{H}_{\text{is}} + \hat{H}_1, \tag{5a}$$

$$\hat{H}_1 = -\boldsymbol{\mu}.\mathbf{E}(y, t). \tag{5b}$$

For the time-dependent molecular wavefunction $\psi(t)$ one has the Schrödinger equation

$$\text{i}\hbar\dot{\psi} = \hat{H}\psi. \tag{6}$$

The dot denotes derivation with respect to time. In the Schrödinger picture one substitutes for $\psi(t)$ the expansion in the basis ϕ_k from Eq. (2):

$$\psi(t) = \sum b_k \phi_k \tag{7}$$

and one obtains the matrix Schrödinger equation for the amplitude vector $\mathbf{b}(t)$:

$$\text{i}\dot{\mathbf{b}} = [\mathbf{W} + \mathbf{V}\cos(\omega t)]\mathbf{b} \tag{8}$$

with the diagonal matrix $\mathbf{W}$:

$$W_{kk} = \text{Re}(\omega_k) - \text{i}\gamma_k/2 \tag{9}$$

and the off-diagnol matrix $\mathbf{V}$:

$$V_{kj} = - M_{kj}^z |E_0|/\hbar, \tag{10}$$

$$M_{kj}^z = \langle \phi_k | \boldsymbol{\mu} \cdot \mathbf{e}_z | \phi_j \rangle. \tag{11}$$

M_{kj}^z is the z-component of the electric dipole transition moment ($\mathbf{e}_z$ is the unit vector in the z-direction of the laboratory frame). We note that the diagonal elements M_{kk} are zero for symmetry reasons if the ϕ_k have a well defined parity. Disregarding weak interactions this is in principle true for the eigenstates of H_{is}. For the practical 'spectroscopic' eigenstates it is true only for planar and easily inverting molecules. The V_{kj} can be made to be real in an appropriate basis ϕ_k.

The solution of Eq. (8) is formally given by

$$\mathbf{b}(t) = \mathbf{U}(t, t_0)\mathbf{b}(t_0). \tag{12}$$

$\mathbf{U}$ satisfies $\mathbf{U}(t_0, t_0) = \mathbf{1}$ and

$$i\dot{\mathbf{U}} = [\mathbf{W} + \mathbf{V}\cos(\omega t)]\mathbf{U}. \tag{13}$$

$\mathbf{U}$ is of the general form:

$$\mathbf{U}(t, 0) = \mathbf{L}\exp(\mathbf{A}t) \tag{14}$$

with a time-independent $\mathbf{A}$ and a Liapounoff matrix $\mathbf{L}$, satisfying

$$\mathbf{L}(t + 2\pi/\omega) = \mathbf{L}(t). \tag{15}$$

There is not, so far, any general direct method of finding $\mathbf{L}$ and $\mathbf{A}$ and there appears to be little work in this direction (see Quack, 1978). Approaches for an exact solution of the two-state problem ($\mathbf{U}$ and $\mathbf{b}$ of order 2) have been discussed by Fontana and Thomann (1976) and reviewed by Dion and Hirschfelder (1976). Exact, stepwise numerical solutions are straightforward for any low order of $\mathbf{U}$. They can, for instance, be formulated in terms of the Magnus expansion, which solves the general set of first-order linear differential equations with the time-dependent coefficient matrix here being

$$\mathbf{C}(t) = \mathbf{W} + \mathbf{V}\cos(\omega t)$$

(Magnus, 1953: Pechukas and Light, 1966):

$$\mathbf{U}(t, t_0) = \exp(\mathbf{B}), \tag{16}$$

$$\mathbf{B} = \sum_{n=0}^{\infty} \mathbf{B}_n. \tag{17}$$

The first two terms are

$$i\mathbf{B}_0 = \int_{t_0}^{t} dt' \, \mathbf{C}(t') \tag{18}$$

$$i\mathbf{B}_1 = -\frac{1}{2}\int_{t_0}^{t} dt'' (\int_{t_0}^{t''} dt' [\mathbf{C}(t'), \mathbf{C}(t'')]). \tag{19}$$

For Eq. (8) this becomes

$$i(\mathbf{B}_0 + \mathbf{B}_1) = \mathbf{W}t + \mathbf{V}[\sin(\omega t)/\omega]$$

$$+ \tfrac{1}{2}[\mathbf{W}, \mathbf{V}]\{(4/\omega^2)[\sin(\omega t/2)]^2 - t\sin(\omega t)/\omega\}. \tag{20}$$

The exponential function in Eq. (16) can now be evaluated for a given time step t, for example by diagonalizing $\mathbf{B} = \mathbf{B}_0 + \mathbf{B}_1$ (see also appendix). A simpler approximation for sufficiently short t, corresponding to the lowest order, is obtained by expanding $\sin x \simeq x$, giving

$$\mathbf{U}(t, 0) = \exp[-it(\mathbf{W} + \mathbf{V})]. \tag{21}$$

This is also equivalent to setting $\cos(\omega t) \simeq 1$ for $(\omega t) \ll 1$ in Eq. (13). In addition a phase average is needed (cf. Eq. (4)). Using the simple Eq. (21) the excitation of an anharmonic model oscillator has been discussed in connection with classical trajectory calculations by Walker and Preston (1977). The second-order approximation, Eq. (20), has been given by Quack (1978) and model calculations with a third-order Magnus approximation for an anharmonic oscillator with up to 27 states have been presented by Schek et al. (1979) (no explicit expressions for the particular expansion used were given there). No systematic investigation upon an optimum truncation of the Magnus expansion is available. Often, Eq. (21) with a short time step will be adequate. We note that Eqs. (14) and (15) should be used in connection with the numerical results in order to simplify the calculations (see also Schek et al., 1981, and Leasure et al., 1981).

Eqs. (8)–(17) constitute in essence an exact quantum-mechanical solution of the idealized problem, subject to the restrictions mentioned, but including all kinds of 'true multiphoton processes' of high and low order. The numerical calculations are not difficult but in practice a much simpler approximation with an analytical one-step solution is quite adequate for the excitation conditions in present-day experiments.

2. Quasiresonant or Rotating-Wave Approximation

Eqs. (8) and (13) are matrix differential equations with coefficient matrices $\mathbf{C} = \mathbf{W} + \mathbf{V}\cos(\omega t)$ that depend upon time in a non-commutative way ($[\mathbf{C}(t''), \mathbf{C}(t')] \neq 0$). There is no general analytical (one-step) solution known for this case. It is the aim of the present section to remove the time dependence of $\mathbf{C}$ by a suitable approximate transformation. We start by substituting the interaction picture amplitudes into Eq. (8) (see Messiah, 1964):

$$c_k = b_k \exp(i\omega_k t) \tag{22}$$

and define the matrix

$$D_{kj} = \omega_k - \omega_j - \omega\,\text{sgn}(\omega_k - \omega_j) \tag{23}$$

$(\text{sgn}\,(x) = +1\,(-1)$ for $x > 0\,(<0))$. This gives

$$i\dot{c}_k = \frac{1}{2}\sum_j V_{kj}c_j \exp{(itD_{kj})}\{1 + \exp{[2i\omega t\,\text{sgn}\,(\omega_k - \omega_j)]}\}. \tag{24}$$

If D_{kj} and V_{kj} are much smaller than ω, we can estimate the relative contribution from the two terms in $\{\Box\}$ by integrating over a short time interval Δt, which satisfies $\Delta t \ll |V_{kj}|^{-1}$ and $\Delta t \ll |D_{kj}|^{-1}$ (therefore $c_j \simeq$ constant), but with $\Delta t \gg \omega^{-1}$. The average relative contribution of the oscillatory term in $\{\Box\}$ is given by:

$$|f(\Delta t)| \leqslant (\Delta t\omega)^{-1} \ll 1. \tag{25}$$

Neglecting this small contribution one obtains

$$i\dot{c}_k = \frac{1}{2}\sum_j V_{kj}c_j exp\,(itD_{kj}). \tag{26}$$

The conditions for the validity of Eq. (26) can be summarized by

$$V_{kj} \ll \omega, \tag{27}$$

$$D_{kj} \ll \omega. \tag{28}$$

For typical excitation conditions with the CO_2 laser and intensities in the MW cm^{-2} and GW cm^{-2} range, the inequality (27) is easily satisfied. The inequality (28) is satisfied for transitions that are reasonably close to resonance, but not for off-resonance transitions with $D_{kj} \simeq \omega$. The latter do not lead to appreciable excitation compared with near-resonance transitions (see below) and may therefore be neglected altogether. This procedure will not work when the near-resonance transitions have a very small V_{kj} whereas the transitions that are far off resonance have a very large V_{kj}. In this case one must use Eq. (8). Discarding this situation, we can introduce the level scheme shown in Fig. 1, in which each state is characterized by an integer level index $n_k \geqslant 0$,

$$\omega_k - \omega_0 = n_k\omega + x_k, \tag{29}$$

$$-\omega/2 < x_k < +\omega/2, \tag{30}$$

$$D_{kj} = [n_k - n_j - \text{sgn}\,(\omega_k - \omega_j)]\omega + x_k - x_j. \tag{31}$$

Now, Eq. (26) becomes with the appropriate substitutions:

$$i\dot{c}_k = \frac{1}{2}\sum_{j,\,|n_k - n_j| = 1} V_{kj}c_j \exp{[i(x_k - x_j)t]}$$

$$+ \frac{1}{2}\sum_{j,\,|n_k - n_j| = 0,\,2} V_{kj}c_j \exp{[i(x_k - x_j)t]}\exp{(\pm i\omega t)} \tag{32}$$

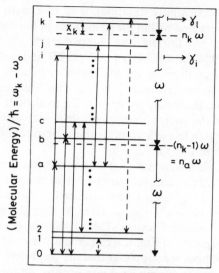

Fig. 1. Level scheme for radiative excitation indicating true molecular frequencies and near-resonant ($\leftrightarrow$) and off-resonant ($\leftarrow\!\dashv\!\dashv\!\rightarrow$) couplings.

Consistent with the neglect of far off-resonance transitions, we have $|x_k - x_j| \ll \omega$ and we therefore can neglect all terms with $\exp(\pm mi\omega t)$, $m = 1, 2, \ldots$, using the reasoning leading to Eq. (26). Making a further transformation to the 'resonance picture amplitudes'

$$a_k = \exp(-ix_k t)c_k, \tag{33}$$

one gets the final result

$$i\dot{\mathbf{a}} = \{\mathbf{X} + \tfrac{1}{2}\mathbf{V}\}\mathbf{a} \tag{34}$$

($X_{kj} = \delta_{kj}x_k$, with the Kronecker delta $\delta_{kj} = 1$ for $j = k$ and zero otherwise). The exact solution of Eq. (34) is given by

$$\mathbf{a}(t) = \mathbf{U}(t, t_0)\mathbf{a}(t_0), \tag{35}$$

$$\mathbf{U}(t, t_0) = \exp[-i(t - t_0)(\mathbf{X} + \tfrac{1}{2}\mathbf{V})]. \tag{36}$$

Since the time evolution is negligible during one period $2\pi/\omega$ of the field, one can omit the phase average (cf. Eq. (4)) in this case, and Eqs. (35) and (36) are the final results. These equations are good approximations for weak fields (intensity $\propto |V_{kj}|^2$) and for transitions that are reasonably close to resonance. We have therefore called this approximation the weak field quasiresonant approximation (WFQRA) (see Quack, 1978, where one can also find a detailed discussion). For the two-state problem, it is derived in many textbooks (Rabi problem, rotating-wave approximation). The case with several excitation stages has also been treated within the framework of the classical treatment for the radiation field by Larsen and Bloembergen (1976), Larsen (1976) Goodman et al. (1976), and Eberly et al. (1977) and with an approximate quantum treatment of the field by Mukamel and Jortner (1976), using the dressed atom approach (Haroche, 1971). The final expressions for the molecular state amplitudes are essentially equivalent in the various treatments. In particular, the approximate quantum field treatment does not yield a more exact result (it is in fact less accurate than our exact Eq. (8) but equivalent to our approximate Eq. (34)).

Eq. (34) has a simple interpretation if we consider the coefficient matrix $\mathbf{C} = \mathbf{X} + \mathbf{V}/2$ as an effective, time-independent Hamiltonian matrix. The true molecular frequencies W_{kk} of Eq. (8) are replaced by the resonance defects X_{kk}, and the couplings $V_{kj} \cos(\omega t)$ are replaced by $V_{kj}/2$ (states in adjacent levels) or zero (states in non-adjacent levels). This is shown schematically in Fig. 2, which exactly corresponds to Fig. 1. Again, the full lines are couplings that are retained, the dashed lines are couplings that are neglected. This kind of picture (without states 2, 3, etc.) has been used notably also by Jortner and coworkers (see, for example, Carmeli et al., 1980). In this picture transitions between adjacent, effectively isoenergetic levels occur by a small coupling $V_{kj}/2$. There is no direct coupling between states in non-adjacent levels (including states in the same level). However, such 'forbidden' transitions are mediated, of course, by states in adjacent levels, see for example the routes $0 \leftrightarrow c \leftrightarrow 2$ or $0 \leftrightarrow a \leftrightarrow i$. The picture becomes even more suggestive in connection with the approximate quantum field treatment (Mukamel and Jortner, 1976). There, the level index is identified with the number of photons removed from the field and

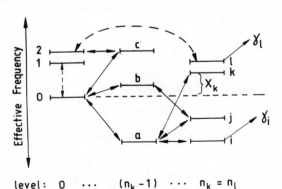

Fig. 2. Level scheme for radiative excitation indicating effective molecular frequencies x_k, Eqs. (29)–(31), and 'allowed' resonant couplings ($\leftrightarrow$) and neglected off-resonant couplings ($\leftarrow$---$\rightarrow$). There is a one-to-one correspondence to Fig.1.

absorbed by the molecule. The sum of molecular energy and field energy corresponds to the effective frequency in Fig. 2, that is, transitions occur indeed 'on the energy shell'. One must remember, however, that this picture is only approximately true since neither can the monochromatic field be described by a well defined number of photons nor are 'energy non-conserving transitions' such as $2 \leftrightarrow 1$ strictly forbidden. Our detailed mathematical argument leading from the exact Eq. (8) to the approximate Eq. (34) is more useful for quantitative purposes and it is quite adequate to consider Fig. 2 as a *representation* of the approximate Eq. (34) without ever referring to the state of the electromagnetic field.

3. *The Time Dependence of the Statistical*
Matrix and of State Populations

The amplitude vectors **b** and **a** are connected by the transformation with a unitary, diagonal matrix $S_{kk} = \exp(in_k \omega t)$:

$$\mathbf{a} = \mathbf{Sb}. \tag{37}$$

It is immediately seen that the populations of molecular states p_k can be computed from either amplitude vector:

$$p_k = |a_k|^2 = |b_k|^2. \tag{38}$$

More generally we can introduce the statistical (density) matrix **P** in either basis:

$$P_{kj}^{(a)} = \langle a_k a_j^* \rangle, \tag{39}$$

$$P_{kj}^{(b)} = \langle b_k b_j^* \rangle, \tag{40}$$

$$\mathbf{P}^{(a)} = \mathbf{S} \mathbf{P}^{(b)} \mathbf{S}^+. \tag{41}$$

The unitary transformation in Eq. (41) has particularly useful properties, since for states in the same *level* ($n_j = n_k$) one has $P_{kj}^{(a)} = P_{kj}^{(b)}$ and since a diagonal $\mathbf{P}^{(b)}$ always becomes diagonal $\mathbf{P}^{(a)}$. For the thermal molecular ensemble $\mathbf{P}^{(b)}$ is diagonal with

$$P_{jj}^{(b)} = \frac{\exp(-E_j/kT)}{\Sigma \exp(-E_j/kT)}.$$

This corresponds to the usual experimental (mixed) initial state in the basis of molecular eigenstates, which have all been counted individually in these expressions. The time evolution of **P** in either basis is given by the integrated form of the Liouville–von Neumann equation:

$$\mathbf{P}(t) = \mathbf{U}(t,0)\mathbf{P}(0)\mathbf{U}^+(t,0). \tag{42}$$

With a diagonal $P(0)$ one obtains for the populations of molecular states

$$p_k(t) = P_{kk}(t) = \sum_1 |U_{k1}|^2 P_{11}(0).$$ (43)

One can also discuss the time evolution in the Heisenberg picture by considering the representative matrix Q of a dynamical variable Q in the basis ϕ_k. The Heisenberg equation of motion for Q is, again in integrated form:

$$Q(t) = U^+(t, 0)Q(0)U(t, 0).$$ (44)

Eqs. (42) and (44) constitute the most general solution of the dynamical problem of coherent optical excitation in terms of the time evolution matrix U, which is either computed for the exact Eq. (8) with Eqs. (14)–(16) or for the approximate Eq. (34) with Eq. (36) (for the calculation of the exponential function, see appendix).

We conclude with the solution of the two-state problem, originally derived in the context of the radiative excitation of a spin system (Rabi, 1937). If we have the initial condition $p_1(0) = 1$, the time-dependent population of the second state becomes (from Eq. (43)):

$$p_2(t) = [V^2/(D^2 + V^2)]\{\sin[(t/2)(V^2 + D^2)^{1/2}]\}^2.$$ (45)

For the two-state problem there is only one resonance defect $D \equiv X_{22}(X_{11} = 0)$ and one coupling matrix element (real $V = V_{12} = V_{21}$). Eq. (45) is the well known Rabi oscillation of populations. The amplitude and the time-averaged populations correspond to Lorentzians of width $2V$:

$$\langle p_2(t) \rangle_{\Delta t} = \frac{1}{2} \frac{V^2}{(\omega - \omega_{12})^2 + V^2}.$$ (46a)

This width is often called the 'power-broadening width', which must not be confused with true widths which are due to coupling to continua. We have written $D^2 = (\omega - \omega_{12})^2$ with the explicit notation for the excitation frequency and the resonance frequency ω_{12}. Eq. (46a) shows why in general the excitation of states with $D^2 \gg V^2$ can be neglected. The frequency of motion in Eq. (45) is often called the 'Rabi frequency'

$$\omega_R = (V^2 + D^2)^{1/2}.$$ (46b)

Sometimes a Rabi frequency is defined by $\omega_R' = V$ (resonance case), but the actual frequency of motion is of course given by Eq. (46b). Note that $D \ll V$ is *not* required for the validity of the WFQRA. For systems with many states, one still gets an oscillatory (complicated) time dependence of the state populations as a general consequence of the fact that U is the exponential function of a complex matrix. If the complex frequencies W_{kk} contain a large imaginary decay term $i\gamma_k$, the effective motion can become non-oscillatory as happens in damped vibrations. A few problems of low order or very special structure can be solved 'on the back of an envelope' as can the Rabi problem, but in general one will evaluate Eq. (36) by appropriate algorithms on a computer (see appendix). Still, the solution is an analytical one-step solution which is easily obtained as long as the matrices are not too large (of the order of a few hundred states at most).

B. Excitation with Incoherent Radiation

Excitation with incoherent radiation leads to an equation of motion for the population vector $\mathbf{p}$ of molecular states, which is of the form

$$\dot{\mathbf{p}} = \mathbf{K}\mathbf{p}, \tag{47}$$

$$\mathbf{p}(t) = \mathbf{Y}(t, t_0)\mathbf{p}(t_0), \tag{48}$$

$$\mathbf{Y}(t, t_0) = \exp\left[\mathbf{K}(t - t_0)\right]. \tag{49}$$

This is a rate equation or generalized first-order kinetics (Jost, 1947, 1950) with rate coefficients K_{fi} connecting the molecular states. Eq. (49) is valid if $\mathbf{K}$ does not depend upon time. We shall come back to this equation in Sec. III and state here only briefly the result for K_{fi} for two important conditions of incoherent irradiation, to be contrasted with coherent monochromatic excitation.

1. Quasimonochromatic, Incoherent Radiation

For excitation with z-polarized, incoherent, quasimonochromatic light with a bandwidth that is large compared with the widths for spontaneous and stimulated transitions, and with intensity I_ν^z (say from a thermal light source and a monochromator or filter), one has (see Quack, 1979b, for a discussion):

$$K_{fi} = \frac{8\pi^3}{h^2}|M_{fi}^z|^2\left(\frac{I_\nu^z}{c}\right). \tag{50}$$

$|M_{fi}^z|$ is as defined in Eq. (11) and c is the speed of light. One has obviously

$$K_{fi} = K_{if}. \tag{51}$$

In these equations we have assumed that, on the timescale of interest, spontaneous emission can be neglected. Otherwise one must take care of the asymmetric $(E_f < E_i)$ addition of the Einstein coefficient for spontaneous emission, which is in the electric dipole approximation:

$$A_{fi} = \frac{64\pi^4|E_f - E_i|^3}{3h^4c^3}|M_{fi}|^2. \tag{52}$$

Furthermore one has for the diagonal elements

$$-K_{ii} = k_i + \sum_{j \neq i} K_{ji}. \tag{53}$$

The k_i correspond to rate coefficients for irreversible decay from the state i, for example by predissociation or chemical reaction in general ($\Gamma_i = \hbar\gamma_i$)

$$k_i = \Gamma_i/\hbar. \tag{54}$$

2. Thermal Radiation

The rate coefficients for radiative transitions between molecular states i and f under the influence of thermal (black-body) radiation of temperature T can be given in one equation, including both stimulated and spontaneous processes (Quack, 1979b):

$$K_{fi} = A_{fi}\frac{\text{sgn}(E_f - E_i)}{\exp[(E_f - E_i)/kT] - 1}. \tag{55}$$

Eq. (55) is exact, if the exact Einstein coefficient A_{fi} is used. Fig. 3 shows representatives results for the excitation ($E_f > E_i$) of 'typical' rotational, vibrational, and electronic transitions (indicated by typical energies in cm^{-1}) with an assumed transition moment of 1 D. It is seen that the radiative transitions are slow and cannot compete, at ordinary pressures, with collisional (de)activation except for electronic transitions at high radiation temperatures, since the collisional relaxation of electronic states is often inefficient. Therefore, for thermal reactions, ordinarily the effect of vibrational heating by thermal radiation is neglected with good justification, in contrast to the early hypothesis concerning the role of thermal radiation in chemical reactions (Perrin, 1922). Nevertheless it has been suggested more recently that in a molecular beam guided through a heated tube radiative (vibrational)

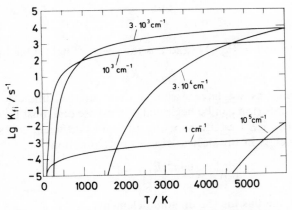

Fig. 3. Rate coefficients for the excitation by thermal (black-body) radiation of temperature T. From Quack (1979b). Reproduced by permission of Verlag Chemie

heating and subsequent reaction may be observable for polyatomic molecules Quack, 1979b).

It is of interest to consider the qualitative differences of the dynamics of radiative transitions with various kinds of radiation for the simple two-state system without decay. Neglecting spontaneous emission, coherent radiation gives an oscillation of the populations according to Eq. (45). There is no 'relaxed' limiting population but the time average of the population p_2 with the initial condition $p_1(0) = 1$ is given by the Lorentzian in Eq. (46).

Intense, incoherent, quasimonochromatic radiation gives on a short timescale (neglect of spontaneous emission) an exponential relaxation to a state of equipartition, independent of the initial state:

$$(p_2/p_1)_{\text{relaxed}} = 1. \tag{56}$$

Thermal radiation gives an exponential relaxation on a long timescale with a Boltzmann distribution as the final state, independent of the initial condition:

$$(p_2/p_1)_{\text{relaxed}} = \exp\left[(E_1 - E_2)/kT\right]. \tag{57}$$

Obviously, these three physical situations are quite different and one should not transfer even qualitative results from one to the other, as is sometimes done in the literature. Additional considerations arise, however, if one does not consider the ideal two-state problem but polyatomic molecules with finite energy resolution.

C. Measurable Quantities and Experimental Coarse Graining

The starting point in any physical theory should be a discussion of the quantities that are considered to be measurable. We shall consider an ideal experiment, namely a thin molecular beam, propagating in x-direction, excited by a monochromatic, coherent, z-polarized laser beam, which propagates in the y-direction and has a large geometrical cross section in the x–z plane. The duration of the irradiation is assumed to be governed by the length of the laser pulse of constant intensity I during $t \gg v_{\text{laser}}^{-1}$. The initial state of the molecules prior to irradiation should be characterized by a Boltzmann distribution with a well defined internal temperature T (including possibly $0\,\text{K}$). Most of the following considerations can easily be transferred to less ideal situations including bulk experiments.

In experiments with or without reaction, the first experimental quantity to be considered is the energy of the molecules after excitation. The *molecular energy* can be defined and measured after the crossing of the laser beam *in the field-free region* (during irradiation the molecular energy is not well defined separately from the field energy). To within the linewidths for spontaneous emission (say less than $1\,\text{kHz}$ in the IR corresponding to longer than $1\,\text{ms}$ flight time), this corresponds to the energy of true molecular eigenstates or spectroscopic states. In principle, E could be measured by any fine-grained technique such as high-resolution molecular spectroscopy, deflection techniques, etc. (a fundamental discussion of the energy measurement is given by Heisenberg, 1958). Furthermore the full set of commuting variables could be measured (angular momentum, parity, etc.). The result of such an experiment would be the populations p_k of all molecular eigenstates labelled by their energy and other good quantum numbers.

In this article, we shall be satisfied with a less fine-grained experiment, in which the energy of the molecules arriving at a detector just behind the interaction region is measured to within an energy uncertainty ΔE satisfying

$$\Delta E \ll h\nu^{\text{Laser}} , \tag{58}$$

$$\Delta E \gg \rho(E, J)^{-1}, \tag{59}$$

with the density of molecular states $\rho(E, J)$ for a given set of good quantum numbers of the discretely quantized variables that may be measured in addition (collectively denoted by J for brevity). This coarse-grained experiment could be performed with a bolometer, which is able to detect individual molecules and their energy to within ΔE (a 'molecular energy counter', not available at present). The result of the experiment would be the population vector $\mathbf{p}$ (components p_N) of *molecular levels*. The levels are defined in practice by the initial energy (to within ΔE) and the 'number of photons absorbed', if the conditions for the WFQRA are satisfied or more generally by the energy within ΔE and a set of discrete, good quantum numbers. Note that the *level* population vector $\mathbf{p}$ is of reasonably low order, whereas the population vector of molecular eigenstates would be of huge order for polyatomic molecules (often $> 10^{20}$). We shall discuss below the equations of motion for the level populations P_N.

The population vector of the coarse-grained levels of molecules (and products) cannot be measured at present, but other more easily measurable quantities can be derived from it. If our detector can measure the average energy of many arriving molecules we have with the duration of the irradiation $t(\gg \nu_{\text{laser}}^{-1})$

$$\langle E \rangle = \sum_N E_N p_N(t). \tag{60}$$

Such measurements have been made in a different context in beams with bolometers (Gough et al., 1977) and have been reviewed for more straightforward absorption experiments in bulk situations in connection with URIMIR by Lyman et al. (1980). Often, the equivalent quantity $\langle n \rangle = (\langle E \rangle - E_{\text{initial}})/h\nu$ the 'average number of photons absorbed', is reported.

If we have a means of measuring the fraction of molecules that disappear from the initial beam (say by dissociation), we obtain direct information on the reaction yields (an example of this approach is the experiment of Brunner and

Proch, 1978). Several quantities can be defined, depending upon the experimental conditions:

(i) If the flight time between leaving the irradiated volume and arriving at the detector is made very small, all reactant molecules remaining immediately after irradiation will arrive at the detector, even if their energy exceeds the reaction threshold energy. The fraction of these remaining reactant molecules is thus

$$F_R = \sum_N p_N(t). \tag{61}$$

F_R can, in principle, always be measured by placing the detector arbitrarily close to the interaction volume. The corresponding product yield is $F_P = 1 - F_R$.

(ii) If the flight time between irradiation and detection is very long, all molecules that were above the reaction threshold will be dissociated (fraction F_R^{**}, the double asterisk standing for highly excited, above threshold). We then measure the *stable* reactant molecules at the detector (fraction F_R^*, the asterisk standing for partly excited, but below threshold):

$$F_R^* = \sum_{N=1}^{N_m} p_N(t) = F_R - F_R^{**}. \tag{62}$$

The sum in Eq. (62) is restricted to levels with $E(N_M) < E_T$ (threshold energy). The corresponding product yield is

$$F_P^* = 1 - F_R^* = F_P + F_R^{**}.$$

A typical situation for URIMIR is sketched in Fig. 4. F_R^* can only be measured

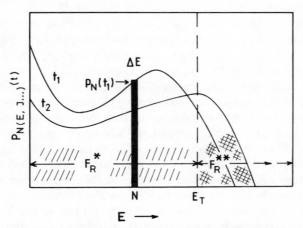

Fig. 4. Sketch of typical, time-dependent, coarse-grained level populations with fractions F_R^* (below threshold E_T) and F_R^{**} (above threshold).

unambiguously if the lifetime of the molecules above threshold is sufficiently short (at least shorter than the lifetime for spontaneous IR emission). Otherwise some quantity intermediate between F_R and F_R^* would be measured, which could be calculated and defined through $p_N(t)$.

By a variation of the irradiation time t one can furthermore obtain the differential quantities:

$$k(t) = -\frac{d \ln F_R}{dt}, \tag{63}$$

$$k^*(t) = -\frac{d \ln F_R^*}{dt}. \tag{64}$$

The 'rate coefficients' $k(t)$ can be measured even if sometimes the absolute $F_R (F_R^*)$ are not known, because we need only to know the *ratio* dF_R/F_R. On the other hand, it is sometimes possible to obtain F_R at a few times without obtaining $k(t)$. Therefore it is useful to consider both quantities, although theoretically one implies the other if the complete time dependence of either $F_R(t)$ or $k(t)$ is known.

Since the rearrangement during chemical reaction is profound, F_R, etc., can be measured and defined even during irradiation, in contrast to the fine-grained molecular energy. The coarse-grained quantities discussed so far can be defined and measured also in bulk experiments, giving proper consideration to some small changes in the definitions that are not important for practical purposes. As in ordinary kinetics, the most common observable will be the time-dependent reactant concentration $c_R(t)$ with $F_R(t) = c_R(t)/c_R(0)$ and related quantities as defined above.

In beam experiments it is natural also to consider differential quantities as the flux $f(\Theta)$ of product molecules arriving at an angle Θ with respect to the beam (laboratory system) or the corresponding quantity in the centre-of-mass system, which is not so easily obtained unambiguously for polyatomic systems (Fluendy and Lawley, 1973; Faubel and Toennies, 1977; Schulz et al., 1979). One can ideally derive the centre-of-mass product translational energy distribution of the dissociation fragments $P(E_t)$, and by combination with laser-induced fluorescence one might measure to some extent internal state distributions of the reaction products (Schultz et al., 1972; King and Stephenson, 1979). Theoretically these can all be derived from the $p_N(t)$ if the dissociation dynamics of the unstable molecular levels are also known. Since even the proper modelling of these dynamics by the current detailed statistical theories of unimolecular reactions is quite involved (Quack and Troe, 1977; Quack, 1980c), we shall not discuss these quantities in the limited space available for this review. We stress, however, that the $p_N(t)$ are the fundamental quantities for all the 'coarse-grained' observables.

The independent, externally controlled experimental variables in URIMIR are the irradiation intensity I, which we assume for theoretical purposes to be constant in time, and the duration of the irradiation t. A third variable is the fluence

$$F(t) = \int_0^t I(t')\,dt'. \tag{65}$$

For pulses with $I(t) = $ constant, these three variables correspond to two degrees of freedom. More generally, the infinitely many possible functions $I(t)$ may be considered to create an infinite number of degrees of freedom. Therefore it is a primordial requirement to create irradiation experimentally with constant and well defined intensity. There are, however, special situations in which it is sufficient to specify one degree of freedom (F). In another theoretically important situation (steady state), the only remaining effective degree of freedom is the intensity. These questions have been discussed by Quack *et al.* (1980) (see also Sec. V). At present we shall consider the general idealized situation with a uniform intensity and the time as independent variables.

III. STATISTICAL MECHANICS OF THE COHERENT EXCITATION OF POLYATOMIC MOLECULES

Eqs. (34) to (44) provide a complete, and for many purposes practically exact, solution of the problem of the coherent excitation of polyatomic molecules. The exponential matrix in Eq. (36) is in principle easily calculated

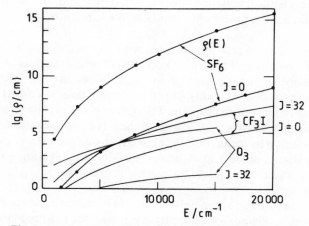

Fig. 5. Ro-vibrational densities of states $\rho(E, J)$ for O_3, CF_3I and SF_6. The upper curves for O_3 and SF_6 are $\rho(E)$, otherwise J is indicated. The 'spectroscopic energy unit' $1\,cm^{-1} \triangleq 11.962\,J\,mol^{-1}$.

(see appendix). The difficulty arises from the order of the matrices being much larger than one could ever hope to treat. This is illustrated in Fig. 5, where we show some relevant densities of ro-vibrational states $\rho(E, J)$, etc., that is the number of states in a 1 cm^{-1} interval. (We use $\tilde{v}$ or the 'spectroscopic energy unit', with $1\,\text{cm}^{-1} \hat{=} 11\cdot962\,\text{J mol}^{-1}$, corresponding to $E = hc\tilde{v}$.) At high excitation energies $\rho(E, J)$ gives about the order of magnitude of the number of states coupled by the strong radiation field at each level of excitation. An upper bound for the total number of coupled states would be $\int_0^{E>E_T} \rho(E)\,dE$. Only for O_3 is this is less than 10^{10}. Although the selection rules will greatly reduce the number of non-zero matrix elements, only exceptional cases such as O_3 (at best) may become amenable to a reasonably complete treatment using Eqs. (34)–(44). It would be quite erroneous to assume that the selection rules are so restrictive that the coupling problem is reduced to a small size in general (see below).

On the other hand, the experiments provide only much less detailed information, say at best the coarse-grained *level* population vector p_N, which is of low order. For this vector we shall discuss in this section simplified equations of motion which are in the simplest case of the general form

$$\dot{\mathbf{p}} = \mathbf{K}\mathbf{p}. \tag{66}$$

This looks similar to Eq. (47) but with a different definition for $\mathbf{p}, \mathbf{K}$ and a different physical origin of the equation. The fundamental idea in obtaining Eq. (66) is the experimental coarse graining discussed in Sec. II C together with a probability reasoning (Quack, 1978). In reviewing the derivation of the statistical-mechanical equation (Eq. 66)) we shall avoid four current misconceptions that can be found abundantly in the literature.

Misconception 1: 'The incoherent equations, (47) and (50), can be used directly for coherent radiation if somehow very many states are coupled, and coarse graining is either unnecessary or is just an additional averaging and simplification.' We have already shown by formulating the correct Eqs. (34) to (44) that this is not true and that the description of the in principle measurable fine-grained populations p_k discussed in Sec. II C would indeed require the complete solution by means of Eq. (36). We shall discuss furthermore that even for the description of the coarse-grained variables Eqs. (47) and (50) do not, in general, give the correct result for coherent excitation.

Misconception 2: 'Eq. (66) can only be derived by assuming stochastic external perturbations (collisions for example) that render the density matrix diagonal at all times.' In fact the statistical equation (66) can be derived as a coarse-grained approximation to the dynamical equation (34) without invoking any external perturbations.

Misconception 3: 'Coherent excitation leads in general to oscillatory level populations. Eq. (66) can be derived from Eq. (34) only by making some very special assumptions about the effective coupling Hamiltonian $(\mathbf{X} + \frac{1}{2}\mathbf{V})$, each assumption giving quite different results for the $\mathbf{K}$ matrix.' We shall see, however, that the simplifications leading to Eq. (66), with the same $\mathbf{K}$ for many conditions, are expected to be the rule (with overwhelming probability) rather than the exception.

Misconception 4: 'The validity of Eq. (66) can be rejected by providing specific counter-examples (from theoretical models, not from experiment).' We shall start our discussion with a simple mathematical model that already shows the essential features of the statistical reasoning and also shows the fallacy of the fourth statement.

A. Statistics, Dynamics, Mathematics

Let us consider the following adaption of the Ehrenfest and Ehrenfest (1907) urn model for the transitions between two *levels*. A set of M_1 balls is contained in urn no. 1, another set of M_2 balls in urn no. 2. M_1 and M_2 represent 'level (or urn) populations'. $M = M_1 + M_2$ remains constant in the following (say $M = 100$). In consecutive drawings we now generate numbers $0 \leqslant k \leqslant M - 1$ according to a prescription to be discussed below. If K is less than M_1 we shall put one ball from urn 1 into urn 2 and inversely otherwise. We shall consider the populations of the urns as a function of the number N of drawings. The observed evolution of the urn populations will depend upon the initial population and upon the prescription used for generating the K.

In the ordinary discussions of the Ehrenfest model the drawing prescription is taken to be 'stochastic', say the drawing of cards numbered 0 to 99 from a well shuffled deck or a similar procedure (see e.g. Kohlrausch and Schrödinger, 1926, or the interesting application by Eigen, 1971). These procedures may be replaced on a computer by drawing random numbers between 0 and 99 from a 'random number generator' (see below). The open circles (curve 'r') in Fig. 6 show the corresponding evolution of $p_1(N) = M_1(N)/M$ with $M_1(0) = M = 100$. As expected the result is a 'relaxation' to an equilibrium state near $p_1 = 0.5$ with some fluctuations ($p_2(N) = 1 - p_1(N)$ is *not* shown for this example but for a different prescription to be discussed now). In a seemingly minor change of the conventional discussion we can replace the 'stochastic' prescription by a 'dynamical' prescription, say by drawing consecutively pairs of digits from the decimal representation of a real number, for example:

$$\mathrm{e} = \sum_{n=0}^{\infty} (n!)^{-1} = 2{\cdot}718\,281\,828\,459\,045\,23\ldots \qquad (67)$$

The first drawings would be 27, 18, 28, 18, 28, 45, etc., which is now well defined

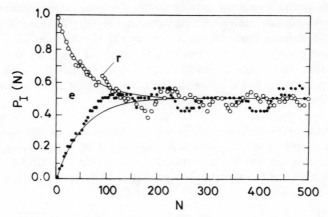

Fig. 6. Urn populations in the Ehrenfest urn model with drawings
from a random number generator ('r', urn 1) and from e ('e', urn 2)
and the same initial condition. The full curves correspond to
Eq. (68).

and certainly not 'random'. The corresponding result for $p_2(N) = M_2/M$ is
shown in Fig. 6 by the curve labelled 'e' (full points). The behaviour is hardly
less statistical than the result from the random number generator. Clearly, this
observation depends upon certain 'statistical' properties of the number e
(Quack, 1981a; Hardy and Wright, 1958). We may, however, conjecture that
almost all real numbers will provide a similar, though not identical, result (A
simple proof is immediate upon inspection of possible sequences of numbers
and a rigorous number theoretical discussion is not attempted here.) In that
sense we may say that drawing subsequent digits from an arbitrary real
number leads to a coarse behaviour of $p_1(N)$, which with $p_1(0) = 1$ is described
by (lines in Fig. 6):

$$p_1(N) = 1 - p_2(N) \simeq 0.5[1 + \exp(-kN)]. \tag{68}$$

This coarse behaviour and some properties of the fluctuations could also be
simulated by stochastic procedures.

The fine behaviour cannot be simulated since it does depend to some extent
upon the particular sequence of digits, but the less the greater M. For the
coarse behaviour the deterministic Eq. (68) can be considered as adequate as a
stochastic simulation. It is clear that our statistical statement is not valid for all
individual numbers, because the counter-examples giving a different behaviour are abundant ($1.\overline{00}, 9/10$, etc.). But these infinitely many counter-
examples are the exception rather than the rule, and if we avoid the selection of
exceptions, considering a great sample of 'naturally provided numbers' such as
$e, \pi, \int_0^\infty x^3(e^x - 1)^{-1}dx$, or the mass of an iodine atom in units of 10^{-24}g, there

will be an overwhelming probability to observe a coarse behaviour similar to the one in Fig. 6, although it may be exceedingly difficult to provide a rigorous mathematical proof in any particular instance.

We should also point to the fact that a typical (pseudo)random number generator is not all that random. Many of them, including the one used for Fig. 6, calculate a sequence of numbers by recursion (Knuth, 1969):

$$I_{n+1} = (AI_n + B) \bmod C, \tag{69}$$

where A, B, and C are appropriate constants. Subsequent manipulations of the integers I_{n+1} generate real numbers in the interval 0 to 1 or integers between 0 and 99 as in our example. In that sense Eq. (69) is similar to Eq. (67). Eq. (69) is inferior, though, because of a finite cycle length. Even if one reconsiders more closely the random numbers obtained from card shuffling, roulette wheels, etc., one will find that these are determined by classical mechanics and dynamics (possibly also cheating) and therefore are again similar (although possibly inferior) to the sequence created by the well defined solution of the mathematical problem in Eq. (67), which is difficult to solve if the number of digits required is very large. The use of the simulation with Eq. (69) or the deterministic Eq. (68) has the advantage of providing the right coarse behaviour at low cost.

We now summarize the discussion of this section:

(1) The dynamical problem at hand is completely and uniquely specified by the solution of a difficult mathematical problem, Eq. (67), which is one example of a general class (for example, sequences of digits in real numbers).

(2) The coarse behaviour can be described approximately by the simple deterministic equation, Eq. (68), or by a simple stochastic simulation using Eq. (69), which is much easier to evaluate than Eq. (67)).

(3) Although the mathematical proof that the approximation is valid is difficult in any particular instance, and although there are (infinitely many) exceptions, inspection shows that almost all of the infinitely many problems of the same general kind are well approximated. In that sense there is an overwhelming probability for each particular problem to show approximately this same coarse behaviour provided that care is taken specifically to avoid selecting the exceptions.

It is not our intention to go into any details of the mathematical theory of probability as applied to physics (see, for example, von Mises, 1931, Kolmogoroff, 1933, Popper, 1980, Rice, 1980, and Primas, 1980, for various points of view). But we shall find the above ideas useful in clarifying the derivation of the statistical-mechanical equation of motion for URIMIR.

B. The Pauli Equation

1. *Derivation*

A large class of problems, including the one formulated in Sec. II for coherent excitation (Eq. (34)), can be described by the Schrödinger equation (70) with a time-independent **H** (in frequency units):

$$i\dot{\mathbf{a}} = \mathbf{Ha}. \tag{70}$$

The reader may keep in mind in particular a two-level system similar to the two-urn model in Sec. III A, with N_K and N_J states in each level K and J ('states' are the basis set states for the amplitude vector **a**), which are coupled by the radiation field (off-diagonal matrix $\frac{1}{2}\mathbf{V}$ in Eq. (34)). This defines again a well defined class of mathematical problems, which we shall now discuss in statistical terms similar to Sec. III A. We shall discuss *level* populations (see coarse graining in Sec. II C):

$$p_K = \sum_k{}' p_{k(K)} = \sum_{k=x+1}^{k=x+N_k} a_{k(K)} a_{k(K)}^*. \tag{71}$$

N_K is assumed to be large (cf. Fig. 5), otherwise Eq. (70) could be solved exactly. Hereafter we shall write Σ' for sums over indices of states in one level and we use capital indices for levels. From Eq. (35) we obtain:

$$p_K(t) = \sum_k{}' \left(\sum_J \sum_j{}' U_{kj} a_j(0) \right)\left(\sum_J \sum_j{}' U_{kj}^* a_j^*(0) \right), \tag{72a}$$

$$= \sum_k{}' \sum_J \sum_j{}' |U_{kj}|^2 |a_j(0)|^2 + \sum_k{}' \sum_1 \sum_{j \neq 1} U_{kj} U_{k1}^* a_j(0) a_1^*(0). \tag{72b}$$

The terms in the first sum in Eq. (72b) are necessarily all real and non-negative. We assume that the state at $t = 0$ is characterized by many non-zero $a_j(0)$ with uncorrelated 'irregular' phases and also many non-zero U_{kj} (sufficiently strong coupling). Then with high probability (in the sense of Sec. III A) the second sum in Eq. (72b) contributes negligibly due to cancellation. Therefore we have approximately:

$$p_K(t) = \sum_J \sum_j{}' \sum_k{}' |U_{kj}|^2 \, p_j(0). \tag{73}$$

Only the total level population $p_J = \Sigma' p_J$ is measurable in a coarse-grained manner. On the other hand, we may assume that at any time $t = 0$ one has with high probability (in the sense of Sec. III A):

$$p_j(0) \simeq N_J^{-1} p_J(0). \tag{74}$$

Strictly, in statistical models there would be a maximum probability close to

Eq. (74), but in reality an obvious requirement is that the off-diagonal interaction H_{ij} effectively couples states sufficiently strongly within some energy range, and outside this range Eq. (74) will not hold. This causes no difficulties if the initial energy spread is large and if the p_j in Eq. (73) are not actually required but only formally replaced by Eq. (74) to yield

$$p_K(t) = \sum_J p_J(0)\left(N_J^{-1} \sum_j{}' \sum_k{}' |U_{kj}|^2 \right) \tag{75a}$$

$$= \sum_J p_J(0) Y_{kJ}(t). \tag{75b}$$

More generally, the term in large parentheses in Eq. (75a), identical to $Y_{KJ}(t)$, has to be replaced by other, appropriately weighted expressions without changing the essence of our derivation. The matrix $\mathbf{Y}(t)$ depends only upon time differences, as does $\mathbf{U}(t)$ (cf. Eq. (36)). Therefore we have:

$$\mathbf{p}(t) = \mathbf{Y}(t)\mathbf{p}(0), \tag{76}$$
$$p(t_1 + t_2) = \mathbf{Y}(t_1 + t_2)\mathbf{p}(0) \tag{77a}$$
$$= \mathbf{Y}(t_2)\mathbf{p}(t_1) \tag{77b}$$
$$= \mathbf{Y}(t_2)\mathbf{Y}(t_1)\mathbf{p}(0).$$

Hence

$$\mathbf{Y}(t_1 + t_2) = \mathbf{Y}(t_2)\mathbf{Y}(t_1) = \mathbf{Y}(t_1)\mathbf{Y}(t_2). \tag{78}$$

Subject to wide conditions, the only function satisfying Eq. (78) is the exponential function with a constant $\mathbf{K}$:

$$\mathbf{Y}(t) = \exp(\mathbf{K}t). \tag{79}$$

Differentiating Eq. (76) or Eq. (79) with respect to time, one obtains the customary form of the Pauli equation

$$\dot{\mathbf{p}}(t) = \mathbf{K}\mathbf{p}(t). \tag{80}$$

We proceed now by expanding the exponential function for short times:

$$\mathbf{Y} = \mathbf{1} + \mathbf{K}t + \dots. \tag{81}$$

For the off-diagonal terms of $\mathbf{K}$ we therefore have

$$K_{KJ} = t^{-1} N_J^{-1} \sum_j{}' \sum_k{}' |U_{kj}|^2. \tag{82}$$

Making use of first-order perturbation theory and writing $\omega_{kj} = X_{kk} - X_{jj}$ (cf. Eq. (34)), we obtain (see Messiah, 1964; Löwdin, 1967):

$$|U_{kj}|^2 = 4|H_{kj}|^2 \omega_{kj}^{-2}[\sin(\omega_{kj}t/2)]^2. \tag{83}$$

Allowing for a reasonably distributed frequency spacing of states in each level with an average δ_K (a density of states replaces δ^{-1} in the case of a true continuum), we can replace the sums in a standard way by integration over a frequency range Δ satisfying

$$\Delta > 2\pi/t$$

and obtain

$$K_{KJ} = 2\pi < |H_{KJ}|^2 > /\delta_K. \tag{84a}$$

$\langle |H_{KJ}|^2 \rangle$ stands for the average square coupling matrix element between the two levels K and J, which can be taken out of the sums and integrals assuming the H_{kj} to be reasonably distributed. The long-time or 'relaxed' equilibrium distribution between the two levels is given by the symmetry relationship $|H_{KJ}|^2 = |H_{JK}|^2$ and thus:

$$\frac{p_K}{p_J} = \frac{K_{KJ}}{K_{JK}} = \frac{\delta_J}{\delta_K} = \frac{\rho_K}{\rho_J}. \tag{84b}$$

Note that it is the densities ρ of states or spacings δ that appear in these equations and *not* level degeneracies. Combining the various dynamical conditions used in addition to the probability assumptions, we obtain Eq. (85a) for the effective level width Δ over which the coupling must be strong and without systematic trends:

$$\Delta \gg 4\pi^2 \langle |H|^2 \rangle /\delta, \tag{85a}$$

$$\Delta/\delta \gg 1, \tag{85b}$$

$$\langle |H| \rangle /\delta > 1. \tag{85c}$$

The use of first-order perturbation theory is expected to be appropriate if all transitions are strongly allowed in first order. If they are forbidden, higher-order expressions have to be used (Dirac, 1967).

2. Discussion

We now summarize the derivation of the previous section:

(1) The dynamics are completely specified by the equation of motion for the state vector **a**, Eq. (70), depending on **H** and **a**(0), which is difficult to solve if the number of states involved is large.

(2) The coarse behaviour of *level* populations (many *states*) can be described approximately by the simpler Eq. (76), neglecting terms that are almost always small (second sum in Eq. (72b) and deviations from Eq. (74)).

(3) Although the mathematical proof that the approximation is valid is difficult in any particular instance (**H** and **a**(0) given), and although it is a trivial matter to find exceptions, inspection of the possible values of the neglected terms in Eq. (72b) shows that almost all (out of infinitely many) problems of the same general kind satisfying certain general dynamical conditions, Eq. (85), are well approximated by Eq. (76). In that sense there is an overwhelming probability for each particular

problem to show approximately this same behaviour, provided that care is taken specifically to avoid selecting the exceptions.

(4) Making the same probability assumption at several times $0, t_1, t_2$, etc., one obtains Eqs. (79) and (80) (essential), and finally by perturbation theory Eq. (84) (not essential).

Our derivation of the Pauli equation is an adaption of Pauli's (1928) original method (see Quack, 1981a). The important assumption concerns the *irregular phases* for the elements of the state vector in Eq. (72b) leading to Eq. (73). Eq. (74) is also related to this, since the phases should remain irregular under a unitary transformation (change of basis set) between states in the same level. The irregular phase assumption is valid with the same probability at $t = 0, t_1, t_2$, etc. It is not necessary to invoke the concept of an ensemble. A slightly different route has been taken in the first discussion of the Pauli equation for URIMIR (Quack, 1978). There, Eq. (73) was obtained directly by coarse-graining Eq. (43), which results from a diagonal $\mathbf{P}(0)$. We have called this *random phases* for the initial state, applicable, for instance, if before irradiation one has a thermal ensemble. However, one must invoke Eq. (73) at several times and it is mathematically obvious that $\mathbf{P}(t_1)$, etc., cannot be diagonal (see Eq. (42)). Therefore we have previously introduced the concept of *irregular phases* for the elements of the non-diagonal density matrices $\mathbf{P}(t)$ on the same probability grounds as discussed above. Indeed, irregular phase state vectors correspond to density matrices with irregular phases (never random phases) and in this sense the two derivations are equivalent. The equation of motion for a diagonal $\mathbf{P}(0)$ is just the ensemble average over many quantum-mechanical trajectories calculated with Eq. (70). Therefore the Pauli equation results from Eq. (42) in the same way as it results from Eq. (70). The important distinction (Quack, 1978) between random and irregular phases has been very often missed in the literature and has led to misunderstandings and unjustified criticism of 'Pauli's repeated random-phase assumption', which are not to the point.

Other efforts to derive the Pauli equation for URIMIR have used the concept of 'random coupling', i.e. assumptions concerning the matrix $\mathbf{H}$. Indeed a 'random' distribution of the H_{kl} has been found to give the Pauli equation (Schek and Jortner, 1979; Carmeli and Nitzan, 1980; Carmeli *et al.*, 1980). Since the irregular phase assumption for $\mathbf{a}$ specifies the properties of $\mathbf{U}$ via the mapping in Eq. (35), and $\mathbf{U}$ specifies $\mathbf{H}$ via Eq. (36), it is clear that the irregular phase assumption implicitly contains assumptions on $\mathbf{H}$. We can in fact consider the following relationship:

$$
\begin{array}{ccc}
\text{irregular phases of} & & \text{irregular ('random') couplings} \\
\mathbf{a}(t) = \mathbf{U}(t)\, a(0) & & H_{kj} \\
\downarrow & & \downarrow \\
\text{most probable state and} & \leftrightarrow & \text{most probable } \mathbf{H} \text{ and subset} \\
\text{subset of } \mathbf{U} = \exp(-i\mathbf{H}t) & & \text{of } \mathbf{U} = \exp(-i\mathbf{H}t)
\end{array}
$$

Although there is no one-to-one correlation at the level of **a** and **H**, for both quantities the most probable subset shows the required irregular properties. Therefore, *as a rule* the two assumptions on **a** and **H** are correlated. The repeated irregular phase assumption may imply irregular couplings (this remains to be proven rigorously), whereas an irregular ('random') **H** obviously does *not* imply irregular phases for the $\mathbf{a}(t)$. In that sense the derivation of the Pauli equation using the irregular phase assumption is more general than the random coupling models. In the work on random coupling models an implicit, very special assumption was made concerning the initial state being characterized by $a_1(0) = 1$. This limit of 0 K initial temperature is too restrictive and hardly ever corresponds to experimental situations. We would suggest that it could be replaced by a random or irregular phase assumption for the initial state, but certainly some assumption is necessary since even an irregular **H** does not always give a time evolution consistent with the Pauli equation. We should not conclude this section without making reference to the considerable literature which is available on the Pauli equation in the more general context of macroscopic equations of motion (Tolman, 1938; Jancel, 1963; Prigogine and Résibois, 1961: Résibois, 1963; van Hove, 1957, 1962; van Kampen, 1962; Zwanzig, 1964; van Vliet, 1978). There are also some discussions of the Pauli equation for intramolecular processes (Rice, 1975; Kay, 1974, 1976; Fleming *et al.*, 1974; Ramaswamy *et al.*, 1978), and collisions (Augustin and Rabitz, 1976). Still a different route has been taken in Mukamel's (1979) derivation of the Pauli equation for URIMIR

3. *Quantum-mechanical Trajectories*

We shall illustrate the general results with some quantum-mechanical model calculations. Since we shall quote such calculations repeatedly, it is useful first to discuss the basic ideas. The method of 'quantum-mechanical trajectories' (Quack, 1978) consists of solving the Schrödinger equation (Eq. (70)) for a given **H**, which may be either derived from a particular molecular example or chosen according to some general rules. We shall do the latter here. **H** is then to be characterized by the dynamical boundary conditions specifying the *structure* of the matrix (see Eq. (34) and the coupling schemes in Figs. 1 and 2) and by the distribution functions $F(H_{kj})$ or $F_X(X_{kk})$ and $F_V(V_{kj})$ for the matrix elements. This specifies an ensemble of **H**, from which examples are 'drawn', in practice using a random number generator such as in Eq. (69). This does not mean that randomness is somehow introduced from the beginning. Indeed, this procedure guarantees rather that a typical member of the ensemble of **H** is drawn (i.e. a 'typical molecule'), the dynamics being fully contained in the boundary conditions for the structure of **H**, which can be specified in as detailed a manner as suitable for the problem. The method is to be contrasted with another very often used procedure, namely selecting a member **H** of the ensemble, which has a special structure such that closed ('analytical') solutions for the eigenvalues and eigenvectors of **H** or directly for the time evolution are obtained. However, the cases for which closed solutions can be obtained are rather special atypical members of the ensemble of possible **H** and the conclusions that are derived are often not useful for generalizations because they are too special. Many examples of such fallacies could be given but sometimes the conclusions may be insensitive to the particular method used. The two methods are complementary, and particularly the method of quantum-mechanical trajectories should be used in connection with a general theory which we shall always do here (for example, the theory of the Pauli equation and cases C and D below). (It may be noted that neither obtaining closed solutions for special examples nor calculating quantum-mechanical trajectories constitute a 'theory' by themselves.)

If the solutions of Eqs. (34) and (70) are evaluated with respect to the coarse-grained level populations in Eq. (71), we shall speak of 'coarse-grained quantum-mechanical trajectories'. If we consider solutions of the Liouville–von Neumann equation (42) with a diagonal **P**(0), i.e. Eq. (43), we shall speak of (coarse-grained) quantum-statistical trajectories. One (coarse-grained) quantum-statistical trajectory is just an ensemble average over (coarse-grained) quantum-mechanical trajectories. In molecular experiments (chemistry and low-energy physics) one is in general only able to measure at best such an ensemble average, unless one has a well defined initial state of the molecules (or an ensemble at 0 K initially). Nevertheless it is sometimes useful to picture the motion of one molecule in the ensemble as corresponding to a quantum-mechanical trajectory, the motion of the ensemble as corresponding to the quantum-statistical trajectory.

Fig. 7(a) shows a coarse-grained quantum-mechanical trajectory for level populations in a two-level system with 59 states in each level, coupled by the radiation field and satisfying approximately the dynamical boundary conditions for the Pauli equation. The distribution functions for the diagonal elements are uniform (i.e. constant average density of states, but the individual spacings are selected in an irregular fashion). For the off-diagonal elements they are uniform with $-H_m \leqslant H_{kj} \leqslant +H_m$. The points in Fig. 7(a) correspond to the exact solution of Eq. (70) and the full lines to the solution of the Pauli equation. There is clearly good agreement even for such a small system (cf. Fig. 5!). The short-time behaviour is spoiled somewhat because the levels cannot be made wide enough for computational reasons (width Δ in Eq. (85a)). One should note in particular the great contrast to the periodic solution of the analogous two-state system in Eq. (45). Coarse graining leads to solutions that behave *almost always* as shown in Fig. 7(a) (exceptions see below) and practically never as in Eq. (45).

The H_{kj} are in general complex,

$$H_{kj} = |H_{kj}| \exp(i\alpha_{kj}) = H_{jk}^*.$$

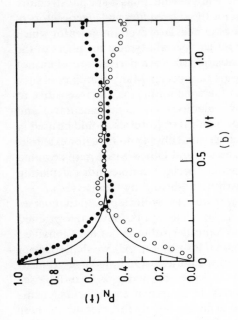

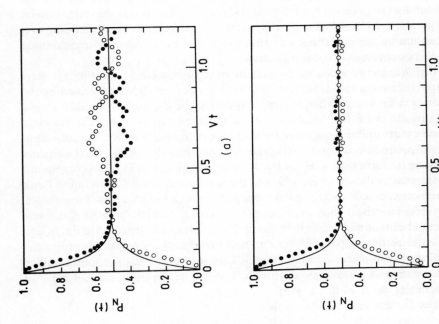

Fig. 7. Coarse-grained trajectories for radiative excitation in a two-level system with 59 states in each level. Points are exact quantum-mechanical solutions, full curves are from the Pauli equation. (a) Quantum-mechanical trajectory (Eq. (70)), with uniform distributions of off-diagonal matrix elements $-H_m \leq H_{kj} \leq +H_m$. (b) As (a) but $|H_{kj}| = $ constant and irregularly distributed signs of H_{kj}. (c) As (a), but quantum-statistical trajectory (Eq. (43)).

For the problem at hand, one can, however, without loss of generality, transform to a real form for the H_{kj}. The absolute magnitudes of the H_{kj} may be made subject to stringent conditions, say $|H_{kj}| = $ constant (not probable in reality). Still, the arbitrary phase factors α_{kj} are reflected in the real form by arbitrary signs of the H_{kj}. Fig. 7(b) shows a corresponding trajectory for such a distribution function of H_{kj}. Pauli behaviour is obtained to within a good approximation. The assumption $|H_{kj}| = $ constant is sufficient for the derivation of the perturbation result in Eq. (83) but is *not necessary*. Actual coupling matrices will not in general correspond to $|H_{kj}| = $ constant.

In the previous examples the initial state vector $a(t_0)$ was chosen such that $\langle |a_k(t_0)|^2 \rangle = N_1^{-1}$ for the states in level 1 with an arbitrary choice of phases of $a(t_0)$. The ensemble average over infinitely many trajectories of this kind is the coarse-grained quantum-statistical trajectory (solution in Eq. (43)), shown in Fig. 7(c) for the Hamiltonian of Fig. 7(a). The time evolution is now smoother but still shows some fluctuations that are characteristic of the Hamiltonian. At this point it is useful to establish the relationship to the Ehrenfest urn model in Fig. 6, which treats in essence the same dynamical problem by 'stochastic simulation'. One might think that the stochastic simulation is somewhat more correct than the use of the deterministic and smooth solution of the Pauli equation, because it can also provide for some 'fluctuations'. One must remember, however, that the 'fluctuations' in Fig. 7 are not the consequence of a random perturbation but are the exact consequence of the *deterministic* equations (34), (42) or (70). They are characteristic of the particular dynamical situation, only the coarse behaviour being described by the Pauli equation. The stochastic simulation will not properly describe the characteristic deviations from the Pauli equation, since the stochastic fluctuations are of a different origin (they are in fact characteristic of the random number generator used). Although all three results are equivalent in their coarse behaviour, no two of them are equivalent in fine behaviour and it would be misleading to consider the fluctuations to be of stochastic origin (or due to quantum-mechanical indeterminacy). The best philosophy appears to be to take the *Pauli equation as a straightforward deterministic approximation to the deterministic quantum-mechanical equations of motion.* Nevertheless, in certain circumstances it may be permissible to simulate the size of the deviations from the average coarse behaviour by stochastic methods, but we shall not make use of this procedure. We note that, of course, the stochastic simulation is also one method of solving the Pauli equation, but an inefficient one (see appendix for efficient methods).

The examples shown in Fig. 7 illustrate how, in general, Pauli behaviour emerges from the Schrödinger equation if certain conditions are satisfied. The interesting feature of this result is the new, simple *structure* in the time evolution that emerges as a rule out of the many dynamical possibilities that are, in principle, consistent with the Schrödinger equation and the dynamical boundary conditions imposed. However, in appropriate special cases non-Pauli behaviour may occur. We present in Fig. 8(a) one example, again for a two-level structure with 59 states in each level, coupled by the radiation field. We have assumed now a uniform distribution with $-a \leqslant H_{kj} \leqslant +b$ (real), with $a < b$ and a uniform diagonal structure as before. This structure arises as a subset in the distribution used for Fig. 7(a), $-H_m \leqslant H_{kj} \leqslant +H_m$, with a low probability for the subset if a and b are different. Furthermore in Fig. 8(a) it is assumed that the elements of the initial state vector are $a_k(0) = N_1^{-1/2}$ for the states in level 1, i.e. all amplitudes are in phase. The resulting time evolution is

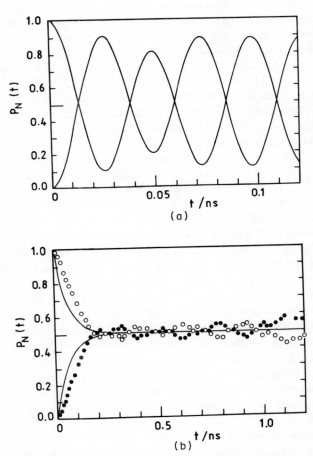

Fig. 8. Coarse-grained quantum-mechanical trajectories for a two-level system with 59 states in each level and uniform distributions $-a \leqslant H_{kj} \leqslant +b$. (a) Initial state vector in phase. (b) Irregularly distributed phases of initial state vector.

oscillatory in spite of the coarse graining, similar to the fine-grained result in Eq. (45). The period of motion is determined by the average $\langle H_{kj} \rangle$ and is essentially independent of $\langle |H_{kj}|^2 \rangle$. Therefore no oscillations are observed for a 'probable' distribution of matrix elements with $\langle H_{kj} \rangle = 0$.

If we assume now the same Hamiltonian as in Fig. 8(a) but an initial state vector with $a_k(0) = N_k^{-1/2} \exp(i\alpha_k)$ with irregularly distributed phases, this is already sufficient to remove the oscillations, as is shown in Fig. 8(b). Destructive interference leads to Pauli behaviour with a rate coefficient

$$K_{IJ} \simeq 2\pi(\langle |H_{kj}|^2 \rangle - \langle H_{kj} \rangle^2)/\delta_I.$$

This is equal to the ordinary rate coefficient if $\langle H_{kj} \rangle \simeq 0$, which corresponds to the probable distribution function of matrix elements. Non-Pauli behaviour and oscillatory solutions have been discussed by Quack (1978, 1981a). The necessity of vanishing $\langle H_{kj} \rangle$ for ordinary Pauli behaviour and the role of the variance ($\langle |H_{kj}|^2 \rangle - \langle H_{kj} \rangle^2$) has been noted in particular in the context of radiationless intramolecular processes and also in URIMIR (Delory and Tric, 1974; Kay, 1974, 1976; Schek and Jortner, 1979). Using averaging techniques and also quantum-mechanical trajectory methods similar to the ones of Quack (1978), Carmeli et al. (1980) and Carmeli and Nitzan (1980) have obtained a most interesting equation of the form of Eq. (80) but with rate coefficients that are different from the Pauli result and which include direct transitions between levels that are not directly coupled for some special coupling models ('constant coupling', 'separable random coupling', 'asymmetric random coupling'). We do not discuss these in detail, since they constitute a highly improbable subset of general coupling Hamiltonians. They are therefore not expected to be of great practical importance, excluding special situations, that might, of course, also be produced experimentally.

The result of the quantum-mechanical trajectory calculations can be summarized as follows: (i) Eq. (80) provides a good description of the coarse-grained quantum-mechanical and quantum-statistical behaviour for all typical situations satisfying Eq. (85). (ii) The exact solutions are consistent with the Pauli rate coefficient, Eq. (84), obtained from perturbation theory. (iii) Exceptions with oscillatory solutions and relaxation with different rate coefficients can be constructed but are 'improbable' in the same sense as the irregular phase assumption leading to the Pauli equation is 'probable'.

4. The Golden Rule for the Decay of a Single Initial State

If in the coupling scheme of Fig. 1 the initial (lowest) level is characterized by a single state, the next level being characterized by many closely spaced states, one has the well known example of an irreversible decay according to the golden rule (Messiah, 1964; Robinson, 1974). Fig. 9 shows the more general scheme for optical excitation in URIMIR. Effectively, condition (85c) is now

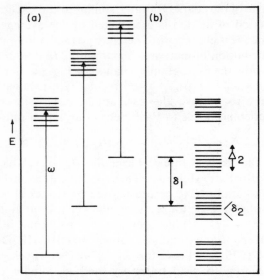

Fig. 9. Level scheme for radiative excitation and golden rule conditions: (a) corresponds to Fig. 1 and (b) to Fig. 2. From Quack (1978). Reproduced by permission of Amer. Inst. Phys.

replaced by

$$|H_{12}| \ll \delta_1, \tag{87a}$$

$$|H_{12}| \gg \delta_2. \tag{87b}$$

Eq. (87a) leads to effectively isolated, single initial states, whereas Eq. (87b) provides the 'continuum' in the upper state. Owing to the sharp rise of the densities of states with energy (cf. Fig. 5), such a situation may sometimes arise in URIMIR. The exponential decay of the initial state is described by the rate coefficient

$$K_{21} = 2\pi \langle |H_{12}|^2 \rangle / \delta_2. \tag{88}$$

The form of this result is well known from the theory of radiationless transitions (Jortner *et al.*, 1969; Róbinson, 1974; Rice, 1975) and is similar to the result for the coupling to true continua, radioactive decay, predissociation, autoionization, and related situations (Messiah, 1964). A particularity of the coupling to a quasicontinuum of discrete states is illustrated in Fig. 10 by the trajectory for the decay of one state coupled to 61 states. The initial decay corresponds well to an exponential law with a rate coefficient from Eq. (87). However, at longer times the initial state 'recurs' to some extent. Such recurrences can be regular and periodic if special assumptions are made of the Hamiltonian as is known

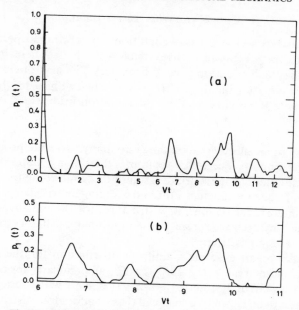

Fig. 10. Exponential decay and recurrence of one state coupled to 61 states in level 2 (see Fig. 9). From Quack (1978). Reproduced by permission of Amer. Inst. Phys.

from the Bixon–Jortner model of radiationless transitions (Bixon and Jortner, 1968, 1969). In practice they are not expected to be important for URIMIR because of irregular couplings and further up-pumping on a short timescale, which depletes the upper level and prevents recurrences of the initial state.

C. The Statistical Mechanics of Case C and Case D

The Pauli equation is applicable with coherent excitation if, in addition to the (statistical) irregular phase assumption, the dynamical boundary conditions specified in Eq. (85) are also satisfied. If this is not the case one has to return to Eq. (70). We have to look for statistical simplifications, if the number of coupled states is large. Depending upon the spectral properties of **H** many different cases can arise. We shall give the results for two important limiting cases, which have been established much more recently than the Pauli equation and which are of some relevance for the coherent optical excitation of polyatomic molecules. For a detailed discussion, which again uses as a starting point coarse graining and probability assumptions, we refer to Quack (1978). Here we shall illustrate the results with some new quantum-mechanical trajectories and present useful practical approximations for the transition between the limiting cases.

1. Case C (Low-intensity Irradiation)

We have already seen two sets of conditions for spectral properties, namely Eq. (87) giving the golden rule for irreversible decay (case A hereafter), and Eq. (85) giving the Pauli equation (case B hereafter). The next obvious situation applies at low radiation intensities ($|H_{kj}|^2 \propto I$), when we have for the frequency spacings in both levels connected by the radiation field ($J = K$ or L):

$$\langle|H_{KL}|\rangle \ll \delta_J. \tag{89}$$

This will lead to an effective decoupling into many two-state problems which behave according to Eq. (45). Assuming that many states are initially populated and making appropriate assumptions concerning the distribution functions for frequency spacings $F(\delta)$, and coupling matrix elements $G(H)$, one can derive interestingly simple, new structures for the time-dependent level populations, which emerge from the complicated oscillatory behaviour of individual state populations.

Let us take as an example a uniform distribution for the off-diagonal elements in the interval $-H_m \leqslant H_{kj} \leqslant +H_m$ and similarly for off-resonance energies D (Eq. (45)) in the interval $-\delta/2 < D < +\delta/2$ with average spacings δ between states in one level. We shall call these 'flat random distributions'. The initial condition is assumed to be $p_M(0) = 1$ and we consider transitions between adjacent levels. Then one finds an effectively irreversible relaxation with an approximate rate coefficient (see Quack, 1978, 1981b)

$$K^{(C)}_{M+1, M} = 2\pi \langle|H_{M+1, M}|^2\rangle/\delta_{M+1}; \tag{90}$$

a 'relaxed' distribution at long times:

$$\left(\frac{p_{M+1}}{p_M}\right)_{\text{relaxed, case C}} = \frac{\pi(3\langle|H_{M+1, M}|^2\rangle)^{1/2}}{2}\left(\frac{\delta_M}{\delta_{M+1}}\right); \tag{91}$$

and correspondingly

$$K^{(C)}_{M, M+1} = K_{M+1, M}\left(\frac{\delta_{M+1}}{\delta_M}\right)\left(\frac{\pi(3\langle|H_{M+1, M}|^2\rangle)^{1/2}}{2 \quad \delta_{M+1}}\right)^{-1}. \tag{92}$$

Two deviations from a simple rate law are, however, important. First, there are some small oscillations that are damped to zero amplitude as $t \to \infty$. This is illustrated in Fig. 11, which shows also that the initial rates in case B and case C are equal on a reduce timescale $t' = t\langle|H|^2\rangle/\delta$ as required by Eqs. (90) and (84). Secondly, Eqs. (91) and (92) are only valid for the specific initial condition. They are, for instance, to be reversed if $p_{M+1} = 1$. Therefore the rate coefficient matrix **K** itself depends upon the previous history of the level populations and the time evolution in case C is in this sense non-Markoffian. This is in complete contrast to the Pauli equation, where **K** depends only upon the Hamiltonian. Still, the differential equation (66) can be used in connection with Eqs. (90)–

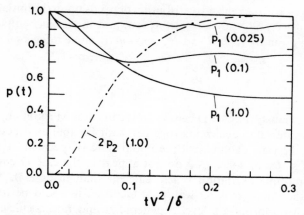

Fig. 11. Time evolution for the initially populated level in a two-level structure similar to Fig. 7(a) but with radiative coupling matrix elements changing with intensity in the transition range between case B and case C ($\langle|H|^2\rangle = 1.384\,V^2$. Parameter V/δ indicated in parentheses. From Quack (1979b). Reproduced by permission of Verlag Chemie

(92) as an approximation to the true time evolution, *if the initial state is specified* $(p_M(0) = 1)$.

2. Case D (High-intensity Irradiation)

Even if the radiation intensity is high and Eq. (85c) is amply satisfied, the Pauli equation may not be applicable if Eq. (85a) does not apply. In practice this corresponds to the situation where the density of effectively coupled states is concentrated in a small frequency range Δ. Even if the coupled states are very numerous (Eq. (85b)), the coarse vibrational absorption spectrum will show a relatively sharp absorption 'line' or band of width Δ, which may satisfy the condition for case D:

$$|H_{KJ}| \gg \Delta. \tag{93}$$

Assuming a flat random distribution for the H_{KJ} one finds an approximate rate law, Eq. (66), with rate coefficients valid for $p_I(0) = 1$ and $p_J(0) = 0$ (Quack, 1978)

$$K_{JI} = \alpha\pi(\langle|H_{JI}|^2\rangle)^{1/2}, \tag{94}$$

with the factor $1 \lesssim \alpha \lesssim 2$. For the relaxed distribution one finds in the case that $\delta_J < \delta_I$ (usual case for excitation in URIMIR):

$$\left(\frac{p_I}{p_J}\right)_{\text{relaxed}} = 1. \tag{95}$$

Similarly to case C the relaxed distribution depends upon the initial state. If we assume $p_I(0) = 0$ and $p_J(0) = 1$ with $\delta_J < \delta_I$ and the *numbers* of states in the levels being given by $N_I = n$ and $N_J = n + m$ (positive integers n and m), one has:

$$\left(\frac{p_I}{p_J}\right)_{\text{relaxed}} = \frac{n/2}{m + n/2}. \tag{96}$$

Although obviously the case D behaviour is truly non-Markoffian, one can use the appropriate rate coefficients together with Eq. (66) as a good approximation for the time evolution if the initial state is specified. It is seen that densities of states or spacings δ do not appear in the case D equation. The effective long-time excitation is less efficient than in case B. The case D equations show in a striking way how incorrect it would be to use a rate equation with appropriate level degeneracies and rate coefficients as they would result from Eq. (50). Such misleading formulations of rate equations can, however, be found abundantly in the literature of the problem.

3. Coarse-grained Quantum-mechanical Trajectories for Coherent Optical Excitation as a Function of Radiation Intensity

For coherent radiative excitation, the matrix elements $|H_{KL}|$ are proportional to the square root of the radiation intensity. Therefore, for a given level structure and transition moments in a polyatomic molecule, the experimentalist can control to some extent whether case B, C or D applies according to Eqs. (85), (86), and (93). The time evolution changes in a characteristic manner as the radiation intensity is changed.

We illustrate this intensity dependence for a three-level structure (cf. the coupling schemes of Figs. 1 and 2) with a total of 119 states and frequency spacings $\delta_1 = 10^9 \, \text{s}^{-1}$, $\delta_2 = 0.5 \times 10^9 \, \text{s}^{-1}$, $\delta_3 = 0.25 \times 10^9 \, \text{s}^{-1}$ in the three levels, which correspond to a 'typical' situation in which the density of states increases with the level of excitation. Fig. 12(a) shows the time evolution for a situation approaching case C at low intensities. The points are the result of a coarse-grained quantum-statistical trajectory calculation and the full lines are the solution of Eq. (66) with the rate coefficients being calculated from the case C equations. The differences are mainly due to the fact that the case C conditions are not completely satisfied. One may note in particular that the long-time 'relaxed' populations are by no means proportional to the densities of states in the levels. Fig. 12(b) shows the time evolution for a radiation intensity increased by a factor of 100. The timescale is shortened correspondingly and the full lines are now calculated with the Pauli (case B) rate coefficients, which gives good agreement with the quantum-mechanical results (points). Now the long-time relaxed populations are, indeed, proportional to

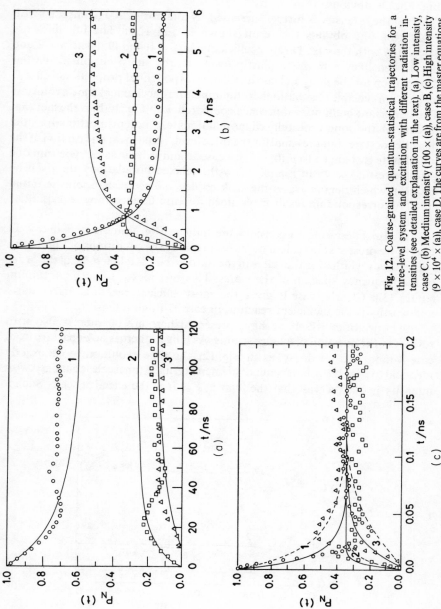

Fig. 12. Coarse-grained quantum-statistical trajectories for a three-level system and excitation with different radiation intensities (see detailed explanation in the text). (a) Low intensity, case C. (b) Medium intensity ($100 \times$ (a)), case B, (c) High intensity ($9 \times 10^4 \times$ (a)), case D. The curves are from the master equations.

the densities of states and the dramatic qualitative change compared with Fig. 12(a) is obvious.

When the intensity is further increased by a factor of 900 (compared with Fig. 12(b)), one obtains the result shown in Fig. 12(c). The full lines are calculated with the case D rate coefficients, Eqs. (94) and (95) with $\alpha = 2$ and the dashed lines with $\alpha = 1$. Partly owing to the sharp cut-off in the flat distribution for the matrix elements the non-exponential properties of case D lead to appreciable fluctuations in the quantum-mechanical time evolution. The important qualitative difference compared with Fig. 12(b) is the fact that in case D the long-time relaxed populations are not proportional to the densities of states but are *equal* for the three levels (Eq. (95) and $p_1(0) = 1$). If the initial state is changed to $p_3(0) = 1$, for case D and still the same spectrum one gets the result shown in Fig. 13. Now the relaxed populations are given by Eq. (96), which demonstrates the striking dependence of **K** upon the initial state. A corresponding result is obtained for case C, according to Eqs. (90)–(92).

It is also interesting to consider the influence of unimolecular decay. In Fig. 14 we present some calculations for precisely the conditions of Fig. 12 but with a decay width γ for the states in the third level; γ is taken to be of the order of the frequency spacing of the states. This provides a smooth continuum (cf. Eq. (3)). Clearly, case B gives the most efficient reaction. Particularly noteworthy is the inefficient reaction in case D (high intensity).

The trajectories which we have presented here are of interest also with respect to the question of fluence scaling. As long as there is no departure from case B (and also no decay, as in Fig. 12), the time evolution looks exactly identical to Fig. 12(b), independent of intensity, if the timescale is changed with intensity in such a way that the fluence $F = It$ is the effective scale. Such a

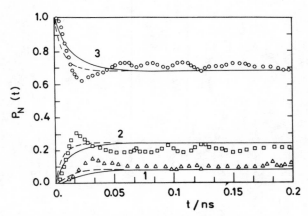

Fig. 13. Coarse-grained quantum-statistical trajectory for case D as in Fig. 12(c), but with $p_3(0) = 1$ (see text).

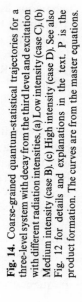

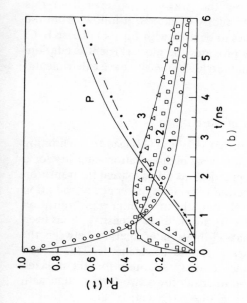

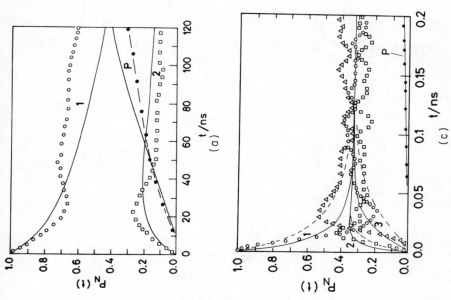

Fig. 14. Coarse-grained quantum-statistical trajectories for a three-level system with decay from the third level and excitation with different radiation intensities. (a) Low intensity (case C). (b) Medium intensity (case B). (c) High intensity (case D). See also Fig. 12 for details and explanations in the text. P is the product formation. The curves are from the master equations.

behaviour is implicitly assumed also in the phenomenological rate equation treatments. Figs. 12(a) and 12(c) show the drastic departures from this behaviour as one leaves the range of validity of case B. It follows from the structure of the rate coefficient matrices to be used with Eq. (66) in case B, C, and D that *at a given* fluence case B provides the most efficient excitation. Therefore in this sense both too low and too high intensity have a detrimental effect on URIMIR.

4. *Unified Treatment of Case B and Case C*

The conditions specified in Eqs. (85), (86), (89), and (93) lead to the limiting behaviour at very low, intermediate, and very high radiation intensity for a given molecular 'spectrum'. There is a practical need also to treat the transition between these limiting cases as a function of coupling strength or radiation intensity. We discuss here quantitatively only the more important transition between case C (low intensity) and case B (intermediate intensity). It has been suggested (Quack, 1979a, 1981b) that a simple interpolation between the analytical results for case B and case C should be adequate. This is best understood by considering the exact case C solution for the relaxed populations as a function of radiation intensity for a two-level system with frequency spacing δ in each level with $p_I(0) = 0$ (Quack, 1978):

$$p_I(\text{relaxed}) = \frac{\pi V_m}{8 D_m} + \frac{1}{4}\left[1 - \frac{(V_m^2 + D_m^2)}{V_m D_m}\arctan\left(\frac{V_m}{D_m}\right)\right]. \tag{97}$$

Here we have used the parameters $V_m = 12\langle|H_{12}|^2\rangle$ and $D_m = \delta/2$. Eq. (97)

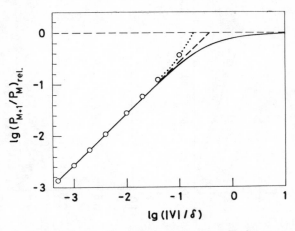

Fig. 15. Relative level populations in a two-level system according to Eq. (97) (full line), Eq. (91) (dashed line) and Eq. (98) ($\bigcirc$ and dotted line). See discussion in the text.

reduces to Eq. (91) if $V_m \ll D_m$, which is the physical condition for case C. Nevertheless, one can formally calculate the case C solution also for $V_m \gg D_m$, which corresponds to the physical condition for case B. Eq. (97) then gives equipartition as in case B. The complete function for the relative populations of adjacent levels according to Eq. (97) is shown in Fig. 15 (full line). The sloping dashed line corresponds to Eq. (91) and the points correspond to an alternative approximation to Eq. (97):

$$\left(\frac{p_{M+1}}{p_M}\right)_{\text{relaxed}} = \left(\frac{2}{\pi} \frac{\delta_{M+1}}{(3\langle |H_{M+1,M}|^2 \rangle)^{1/2}} - 1\right)^{-1}. \tag{98}$$

All three equations give the same result in case C ($V_m \ll D_m$). None of the equations is physically valid if $|V|/\delta > 0.1$, but we know that the constant dashed line $\lg(p_{M+1}/p_M) = 0$ must be approached as $\lg(|V|/\delta) > 0$ (case B). One would therefore think of using the case C result in Eq. (97) together with Eq. (90). This provides a smooth connection with case B. One can also use the simpler expressions Eqs. (90) and (91) until one has.

$$\frac{\pi(3\langle |H_{M+1,M}|^2 \rangle)^{1/2}}{2 \quad \delta_{M+1}} = 1. \tag{99}$$

This corresponds to making a sharp transition between case C and case B at the crossing point of the two dashed lines in Fig. 15. This can be expressed by the unified rate coefficient

$$K_{M,M+1}^{(B,C)} = \text{Max}\left\{K_{M,M+1}^{(B)}, K_{M,M+1}^{(C)}\right\}, \tag{100}$$

where Max $\{x, y\}$ is the larger quantity of x, y and $K^{(B)}$ and $K^{(C)}$ are calculated with Eqs. (84) and (92) respectively.

Fig. 16 shows how well this simple approximation works, indeed, for the time evolution in a realistic five-level structure with unimolecular decay from the highest level, only the lowest level being populated initially. The frequency spacings of states in the levels for increasing level index are $1:0.56:0.42:0.36:0.3$ (in units of 10^9 s^{-1}), with a total number of 119 states. In this example the intensity is changed by a factor of 9 just in the transition domain between case B and case C with the full lines computed using Eq. (66) with the approximation in Eq. (100) for the rate coefficients. One may note in particular also the change in the *relative* level populations when changing the radiation intensity, going from Fig. 16(a) to Figs. 16(b) and (c).

5. Coupling the Schrödinger Equation to Master Equations

For small molecules the situation arises that for several excitation steps at low energies the number of effectively coupled states is too small to justify any of the statistical approximations by coarse graining. For these low levels one can then solve the Schrödinger equation, but it is necessary to make the

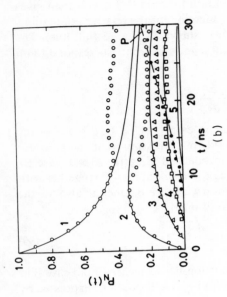

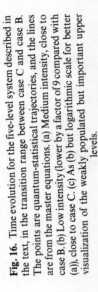

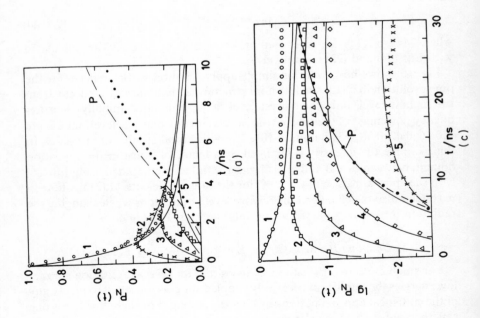

Fig. 16. Time evolution for the five-level system described in the text, in the transition range between case C and case B. The points are quantum-statistical trajectories, and the lines are from the master equations. (a) Medium intensity, close to case B. (b) Low intensity (lower by a factor of 9 compared with (a)), close to case C. (c) As (b) but logarithmic scale for better visualization of the weakly populated but important upper levels.

connection with the master equations at higher levels of excitation, where the number of coupled states becomes too large to be included in the Schrödinger equation. This can be done by first including levels that approximately satisfy the statistical conditions and adding calculated effective rates above the truncation point. Then one can reduce the size of the problem by simulating the previous result with a master equation which now includes all levels at high excitations. This procedure has been discussed (Quack, 1978), but only simple model applications are available, so far.

D. Spectral Structures and Practical Rate Coefficients

The theory presented in Sec. II and its statistical-mechanical generalization in Sec. III is based upon observable spectroscopic quantities, namely the line positions and transition moments from high-resolution spectroscopy. We have also formulated in terms of these spectroscopic structures the conditions determining which equations from cases A, B, C, and D are to be used for calculating the effective rate coefficients. It is important to note here that the densities of states shown, for instance, in Fig. 5 are not in general equal to the densities of effectively coupled states or proportional to δ^{-1} in the equations of this section. This can lead to difficulties since the spectroscopic resolution is often insufficient to resolve individual lines. Furthermore, even the coarse structures of the vibrational spectra are not known for highly excited polyatomic molecules. In brief, it is necessary to obtain more information on the following quantities:

(i) The average frequency spacing δ of effectively coupled ro-vibrational states at each level of excitation and the distribution functions for the spacings, in particular whether there are accumulations leading to case D. For reasonably uniform distribution functions and in the range of case B only the products $\langle |H_{KJ}|^2 \rangle / \delta_K$ are needed. These are more easily obtained from simple models (Quack, 1978, 1979b).

(ii) The integrated band strength

$$G = \int \sigma(v) \mathrm{d} \ln v \qquad (101)$$

to the extent that sufficiently well isolated vibrational bands are still occurring. As discussed elsewhere in detail (Quack, 1979b), G is approximately independent of the level of excitation and is proportional to the vibrational transition moment of the fundamental vibration that is pumped by the laser.

(iii) The coarse vibrational bandshape function (transition moments), see Eq. (11): (see also Sec V A.)

$$\langle |M_v(\omega', \omega'')|^2 \rangle_{\Delta\omega} \simeq |M(1_0^1)|^2 g_v(\omega', \omega'') \qquad (102)$$

and rotational contributions $g_r(\omega', \omega'')$. The latter are easy to compute as long as there is approximate separation of rotation and vibration. This separation fails at high excitations for large vibrational amplitudes. The approximate equality in Eq. (102) with a normalized bandshape function $g(\omega', \omega'')$ is meaningful for isolated bands only and thus is often inadequate for polyatomic molecules. We shall concentrate here on the general appearance of $\langle |M_v(\omega', \omega'')|^2 \rangle_{\Delta\omega}$, the transition moment which is averaged over a small frequency range $\Delta\omega$ containing many absorption lines. For a microcanonical ensemble of molecules at an initial energy ω'', disregarding rotation, this can be visualized through the 'absorption cross section' that would be observed with enough pressure broadening to make the spectrum independent of the resolution of the spectrometer but without changing the coarse structure. Strictly, at high excitation such an absorption cross section is not easily measurable because absorption and stimulated emission occur simultaneously. We therefore prefer the theoretical definition as an averaged transition moment, which can be derived

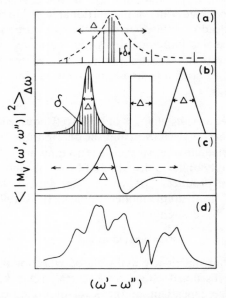

Fig. 17. Coarse vibrational spectrum of polyatomic molecules. (a) Multiple resonance interactions in a small polyatomic molecule. (b) Idealized dense structures (Lorentzian, rectangular, triangular) in larger polyatomic molecules or at high excitations. (c) Simple interference structure. (d) Complicated interference structure

from the true ultrahigh-resolution cross section, which is in principle always measurable.

The dependence of $|M_v|^2$ on ω' and ω'' is not well known for highly excited polyatomic molecules. Fig. 17 summarizes a small selection of such coarse bandshapes that are consistent with theoretical intuition and, to some extent, experimental observations. The figure is largely self-explanatory with the relevant parameters being indicated approximately. The level widths Δ of the previous paragraphs roughly correspond to the widths of the bands. The multiple resonance profile in Fig. 17(a) corresponds to the situation in small polyatomic molecules at low energies and such a structure has for instance been observed for CF_3H (Dübal and Quack, 1980). The Lorentzian bandshape in Fig. 17(b) can be derived from simple, idealized models of a single optically active zero-order state coupled to a quasicontinuum (see e.g. Bixon and Jortner, 1968). Bandshapes consistent with such a simple profile for a fundamental transition have been observed for the C–H vibration in $(CF_3)_3CH$, which is well separated from other transitions in this molecule (Dübal and Quack, 1980). Rectangular and triangular bandshapes may sometimes be useful for model purposes, without a simple physical background for such a drastic idealization. More general bandshapes include simple and complicated interference patterns (Figs. 17(c) and (d)). There may furthermore be finer substructures in the coarse profile possibly including clusters for highly symmetric molecules (Harter, Patterson and Galbraith, 1978). This may lead to a complicated combination of several cases (Quack, 1978).

Table 1 contains a summary of the statistical-mechanical rate coefficients for the various simple limiting cases as a function of spectral parameters. The transition behaviour between the limiting cases can be very complicated for the more complicated bandshapes in Fig. 17. It should also be clear from the table and from our previous discussion that only for case B (or for incoherent polychromatic excitation is it possible to use rate coefficients which are proportional to radiation intensity with a constant 'cross section' as a factor of proportionality. Therefore the use of such a cross section may be quite misleading. We shall hereafter reserve the term 'cross section' for the true molecular absorption cross section σ which is measurable at high resolution with low intensity. This quantity is also used in Eq. (101). For full applicability of the simple rate coefficients in Table 1 it is necessary that the average coupling matrix elements do not vary systematically over the range Δ. In that sense the Δ given in Fig. 17 are high estimates.

We conclude this section by considering far off-resonant excitation, which is not included in Table 1. In many cases, even in the wings of the absorption, the effective absorption process will be dominated by the resonant part of the absorption. There is, however, also an off-resonant contribution which can

TABLE 1. Summary of practical rate coefficients as a function of spectral structures.

Case	Monochromatic coherent excitation[a]			
	A	B[f]	C	D
$K_{M+1,M}$	$2\pi|H|^2/\delta_{M+1}$	$2\pi|H|^2/\delta_{M+1,}$ [e]	$2\pi|H|^2/\delta_{M+1}$	$\alpha\pi\sqrt{(|H|^2)}$ [c,e]
$K_{M+1,M}/K_{M,M+1}$	∞	$\left(\dfrac{\delta_M}{\delta_{M+1}}\right)$	$\left(\dfrac{\delta_M}{\delta_{M+1}}\right)\left(\dfrac{\pi\sqrt{(3|H|^2)}}{\delta_{M+1}}\right)$	1 [d]
Does K depend upon $\mathbf{p}(0)$?	No	No	Yes[b]	Yes[b]
Range of validity for spectral parameters	$\delta_M \gg |H|$ $\delta_{M+1} \ll |H|$ $\Delta_{M+1} \gg 4\pi^2|H|^2/\delta_{M+1}$	$\delta_N \ll |H|$ $\Delta_N \gg 4\pi^2|H|^2/\delta_N$ $N \equiv M$ and $M+1$	$\delta_N \gg |H|$ $\delta_N \ll \omega$ $N = M$ and $M+1$	$\Delta_N \ll |H|$ $N = M$ and $M+1$
Nature of the statistical assumption	—	Irregular phases and piecewise equal populations	Piecewise equal populations and smoothly distributed irregular H_{kl}	Irregular phases and piecewise equal populations

[a] All transitions refer to coarse-grained level populations. $|H|$ stands for $\sqrt{(\langle|H_{M+1,M}|^2\rangle)}$, with the effective coupling matrix elements in frequency units. Thus for z-polarized light

$$|H|^2 = \langle|M^z_{M+1,M}|^2\rangle_{\Delta\omega}\,|E_0|^2/(4\hbar^2)$$

in terms of the transition moment, Eqs. (10) and (11), with $|E_0|^2 = fI$ (laser intensity I) and $f = 2/(c\varepsilon_0)$ in the SI system and $f = 8\pi/c$ in the Gaussian system.

[b] For the results shown it is assumed that $P_{M+1}(0) = 0$.

[c] Approximate result with $1 \leqslant \alpha \leqslant 2$, see Sec. III C.2.

[d] Valid if $\delta_{M+1} < \delta_M$, see Sec. III C.2.

[e] The validity can be extended to some extent and approximately only by using

$$(\langle|H_{M+1,M}|^2\rangle - \langle H_{M+1,M}\rangle^2) \quad \text{if} \quad \langle H_{M+1,M}\rangle \neq 0,$$

$$(\langle|H_{M+1,M}|^2\rangle - \langle H_{M+1,M}\rangle^2)$$

see Quack (1978) for case D and Carmeli and Nitzan (1980) and Carmeli et al. (1980) for case B, where one also finds rate coefficients for more special cases.

[f] The effective rate coefficients for excitation with z-polarized quasimonochromatic, incoherent light of bandwidth $\chi < \Delta$ but $\chi \gg \delta$ are the same if the intensity integrated over the bandwidth is the same as the monochromatic laser intensity (Quack, 1979b).

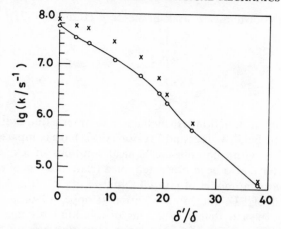

Fig. 18. Steady-state rate coefficient k for overall decay in a three-level system with 39 states in each level and decay from the highest level. δ is the spacing of states in each level and (δ'/δ) indicates the detuning of the radiation frequency from central resonance. $\times$, k at 10 ns; $\circ$, k at 60 ns; the relatively small difference indicates the non-exponential properties of case C. From Quack *et al.* (1980). Reproduced by permission of Amer. Inst. Phys.

easily be investigated in theory by using a bandshape with a sharp edge as for the rectangular or triangular functions in Fig. 17. Fig. 18 shows a result for the effective irreversible decay rate coefficient k in a three-level structure with rectangular bandshapes and an unstable last level. The steady-state decay rate coefficient k is a good measure for the effective rate of excitation to the unstable third level (ground level populated initially). The coefficient k is shown as a function of the detuning δ'/δ between the laser frequency and central resonance. At $\delta'/\delta \geqslant 20$ there is no resonant connection from the ground level to the reacting level. However, the effective k decreases quite smoothly through this point. This illustrates first that off-resonance excitation may provide an efficient route for the excitation of 'molecules that do not absorb the laser light'. Secondly, the off-resonance mechanism will compete with the resonance mechanism only if there is a difference of several orders of magnitude between the resonant wing absorption and the off-resonant peak absorption.

E. Alternative Treatments of the Infrared Excitation

1. *Treatments Based Upon the Solution of the Classical Equations of Motion*

One of the more straightforward approaches to the theoretical treatment of URIMIR is the solution of the classical mechanical equations of motion for the conjugate coordinates q_i and momenta p_i (Bunker, 1971):

$$\frac{\partial H}{\partial p_i} = \dot{q}_i, \tag{103a}$$

$$\frac{\partial H}{\partial q_i} = -\dot{p}_i. \tag{103b}$$

The Hamiltonian function H includes the interaction with a classical electromagnetic field. Walker and Preston (1977) have compared solutions of these equations with quantum-mechanical solutions of a diatomic model. Calculations have also been presented for a triatomic model of O_3 (Hänsel, 1979), a five-atomic model of CD_3Cl (Noid et al., 1977) and even a seven-atomic model of SF_6 including rotation (Poppe, 1980a, b, 1981). It can therefore be expected that calculations of this kind become feasible for a variety of realistic models of molecular systems that are not too large. The major restriction is that the equations of motion as a rule cannot be solved accurately for times exceeding picoseconds, whereas the experimental process usually occurs on the nanosecond timescale. The other obvious restriction is related to the limited validity of classical mechanics for molecular states. Roughly speaking one may estimate that classical mechanics becomes a good approximation if the total molecular energy is large compared with the *total* molecular zero-point energy. This is not often the case. The validity of classical mechanics could be greatly extended by using semiclassical techniques (Miller, 1974, 1975) which are, however, more complicated.

2. Treatments Based upon the Solution of Simplified Quantum Models

The quantum-mechanical equations of motion discussed in Sec. II allow for a straightforward solution if the numbers of coupled states are small. One way of handling the problem is thus to retain only a few coupled states in highly simplified molecular models. This has been done, for instance, by Larsen (1976) and Larsen and Bloembergen (1976) for low levels of excitation, disregarding higher excitations. The models advanced by Mukamel and Jortner (1976) and Goodman et al. (1976) include higher excitations as resonance transitions to single states or to degenerate states (representing the increased density of states). The solutions of the equations of motion for these simple models lead to highly oscillatory level populations which were popular prior to the advent of the statistical-mechanical treatment. Few calculations of this kind have appeared more recently. Among the simple quantum models one may also quote the perturbation calculations by Pert (1973) and Faisal (1976) and related treatments.

3. Treatments Based Upon the Optical Excitation of One IR-active Vibration Weakly Coupled to Other Degrees of Freedom

A number of treatments are based upon the idea that one molecular degree of freedom is strongly coupled to the radiation field and in first order separable from the other molecular degrees of freedom. The two-ladder model of Mukamel and Jortner (1976) may be classified in this category. Of particular interest in this class of models is the treatment of Hodgkinson and Briggs (1976, 1977), which is historically the first straightforward statistical-mechanical approach to the problem. Briefly, an equation of motion for the reduced density matrix of the separable, single mode is used with explicit coupling terms for the field interaction and with a phenomenological treatment of the coupling to other degrees of freedom, including also perturbations by collisions, etc. The equation of motion could formally be written in exact form (Cantrell et al., 1979) and would then be equivalent to the general equation of motion for the density matrix in the eigenstate basis (Eq. (42)) but transformed to a new basis, which is separable with respect to the excited vibration. Elaborations of this treatment have appeared in the work of Friedmann (1979) and of Stone and Goodman (1978) who in this work as in other work prior to 1978 discuss the great importance of oscillatory solutions in the absence of collisions. This is due to the model assumptions in the treatment and not generally adequate.

Although the exact theory is clearly basis independent, the following disadvantages of the separable basis as compared with the eigenstate basis must be noted:

(i) The initial density matrix in the eigenstate basis is of a well defined, diagonal form for typical (thermal) initial conditions. The initial density matrix in the separable basis is of a more complicated structure to be derived explicitly by carrying out the transformation (this has not been done so far).

(ii) The treatment of rotations is automatically included in the eigenstate basis. No consistent treatment has, to our knowledge, been advanced for the coupling of the separable vibrational degree of freedom to *both* the other vibrations and external rotation.

(iii) The treatment in the eigenstate basis involves molecular properties that are measurable (line positions and strengths) and experimental quantities that are actually measured or measurable (reaction yields and the energy content of the excited molecules; see Sec. II C). In the separable basis one invokes quantities that are not directly measurable, such as the energy content and level populations of the IR-active mode, the energy levels and the transition moments for the separable mode, and the phenomenological, model-dependent intramolecular coupling terms. We stress that the absorption features in the high-resolution IR spectrum, which are often loosely attributed to 'a vibrational mode', correspond to molecular eigenstates (or 'spectroscopic states', disregarding special effects due to hyperfine structure and isomers). The absorption lines *do not* correspond to separable modes. The off-diagonal coupling energies are often the order of cm^{-1} even at the zero-point level or in the region of the fundamental transition and cannot be considered to be small perturbations (coupling timescales of the order of picoseconds!).

This last point leads us to the important consideration of practical approximations that can be derived from the two approaches. Although the approach based upon the eigenstate basis is, in principle, general, the simplest useful approximations (case B and case C) arise if the absorption bandwidth Δ is large compared with the coupling bandwidth $|H_{KL}|$ with the radiation field. On the other hand, the separable basis might provide simple and adequate approximations if the absorption bandwidth Δ is small or if the intramolecular mode–mode coupling can be considered to be a small perturbation compared with the strong coupling $|H_{LK}|$ with the radiation field. Somewhat loosely, a large absorption bandwidth Δ can be identified with a fast intramolecular decay of a local excitation in the separable IR-active mode (for example, in a

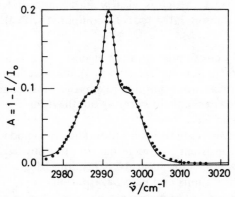

Fig. 19. Spectrum of the C–H fundamental transition in $(CF_3)_3CH$. Experiment: full line, fit assuming a homogeneous, Lorentzian *vibrational* structure with width $\Gamma = 3\,cm^{-1}$ represented by the points. Without vibrational structure the Q branch would be extremely narrow and well separated from the P and R branches. See discussion in the text. From Dübal and Quack (1980). Reproduced by permission of North Holland Publ. Co.

hypothetical very-short-pulse experiment). There are hardly any systematic investigations of the relevant spectroscopic properties. We may refer to the vibrational coupling bandwidth of about $3\,cm^{-1}$ for the fundamental C–H transition in $(CF_3)_3CH$ (Dübal and Quack, 1980). In this particularly simple case one has a well separated absorption and a simple rotational structure (sharp Q branch) that can be accounted for easily. The spectrum is reproduced in Fig. 19. Other examples lead to similar and often larger couplings, including small molecules such as CF_3H. These bandwidths correspond to decay times of the order of picoseconds. The coupling strength with the radiation field is much weaker, corresponding to nanosecond timescales. Therefore it would appear that the simple approximations which are based in first order upon vibrational separability are not adequate for the description of current experiments. Our approach, which is based upon eigenstates (or more strictly spectroscopic states) appears to be both more useful from a practical point of view and more fundamental from a theoretical point of view. Nevertheless it might still be of some theoretical interest to formulate a similarly general treatment in the separable basis, which, to our knowledge, has not been done so far.

4. *Rate Equation Treatments*

Phenomenological rate equation treatments have been formulated both for the states of a separable oscillator coupled to other degrees of freedom (Tamir and Levine, 1977) and for individual spectroscopic energy *states* (Black *et al.,* 1977) and also somewhat loosely for energy *levels* as defined here for the molecule as a whole (Lyman, 1977; Grant *et al.,* 1978; Fuss, 1979). The last of these treatments uses intensity-independent absorption cross sections and level degeneracies derived from vibrational densities of states. This finds its justification to some extent in the statistical-mechanical treatment of case B. Whereas prior to 1978 the quantum-mechanical models with oscillatory solutions enjoyed great popularity, there is in the current literature a certain convergence to the practical use of these rate equations, particularly after the advent of the more detailed theoretical justifications (Quack, 1978; Schek and Jortner, 1979; Mukamel, 1979; Carmeli and Nitzan, 1980). 'Simple derivations' of all kinds have appeared as well that we need not consider in detail here. We must stress, however, the limited range of validity of the rate equations with intensity-proportional rate coefficients (i.e. using 'cross sections') and we refer to Table 1 for more generally useful practical approximations and to the general theoretical discussion of the previous Sec. III A to III D. There have been attempts phenomenologically to include nonlinear intensity dependences by introducing rates for coherent multiphoton excitation (Schulz, 1979; Schulz *et al.,* 1980) or by allowing for intensity-dependent 'fractions of reacting molecules' (Bagratashvili *et al.,* 1979, and related work by several groups). These do not quite properly describe the mechanism of the nonlinear intensity dependences as discussed here (see Quack, 1979a), and none of the phenomenological treatments has been justified in a consistent manner.

IV. THE STATISTICAL THEORY OF SPECIFIC RATE COEFFICIENTS FOR UNIMOLECULAR DISSOCIATION

The decay of metastable molecules above the reaction threshold by 'vibrational predissociation' (Herzberg, 1966) has been characterized in Sec. II, Eq. (3), by complex frequencies with widths γ_k or energy widths Γ_k connected to monomolecular decay rate coefficients k by

$$k_j = \Gamma_j/\hbar. \tag{104}$$

Apart from difficulties arising with this equation in the case of overlapping resonances of non-Lorentzian shape, a more consistent treatment in connection with the statistical-mechanical theory of excitation would be based upon a statistical-mechanical theory of the monomolecular reaction. A well established theory in this connection is RRKM theory (Marcus and Rice, 1951; Rosenstock et al., 1952; Forst, 1973; Troe, 1975). Recent advances in statistical unimolecular rate theory have been reviewed (Quack and Troe, 1977). The statistical rate coefficient is given by

$$k(E, J) \leqslant \frac{W(E, J)}{h\rho(E, J)}. \tag{105}$$

$W(E, J)$ is the number of dynamically accessible reaction channels (or 'adiabatically open channels' in the framework of the adiabatic channel model of Quack and Troe (1974)). $\rho(E, J)$ is the average density of ro-vibrational states at a given energy, total angular momentum J and other good quantum numbers. A non-trivial dependence of k upon J has been noted by Quack and Troe (1974). All quantities are taken to be averaged over an energy range ΔE containing many resonances $(\Delta E \gg \rho(E, J)^{-1})$. One notes the inequality in Eq. (105). From the present point of view this inequality can be derived both from scattering theory (Quack and Troe, 1975) and from a master equation treatment of intramolecular processes (Kay, 1976; Quack, 1981a). We shall not elaborate here upon the theoretical background of Eq. (105) nor upon the practical evaluation of $W(E, J)$ and $\rho(E, J)$, for which we refer to the books and reviews quoted above. As an example, Fig. 20 shows, however, the typical behaviour of $k(E, J = 0)$ for a series of model molecules of various sizes and threshold energies. As is quite well known, long lifetimes may be observed substantially above threshold for the larger polyatomic systems. This has some consequences for the time-dependent dissociation of highly excited molecules above threshold after the end of optical excitation. Otherwise the

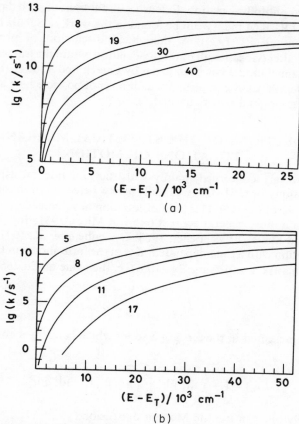

Fig. 20. Typical specific rate constants $k(E, J = 0)$ for a simple bond fission reaction calculated with the adiabatic channel model. (a) A five-atomic model with different reaction threshold energies indicated in units of $10\,000\,cm^{-1}$. (b) A series of model molecules with an increasing number of atoms as indicated. The threshold energy is about $19\,000\,cm^{-1}$. From Quack (1979e). Reproduced by permission of Verlag Chemie.

$k(E, J)$ are only of moderate importance for the overall rate processes in URIMIR.

When different physical and chemical product channels are labelled in a more detailed way by their quantum numbers, the correspondingly generalized Eq. (105) can also be used for the calculation of branching ratios. Some considerations for populations of different chemical channels in URIMIR can be found in Quack (1978). A very important application of the calculation of branching ratios for physical channels is the evaluation of product translational energy distributions $P(E_t)$ of dissociation fragments in molecular beam infrared photodissociation. Extensive discussions of such data using an adaptation of RRKM theory can be found in the work of Schulz et al. (1979) and of Grant et al. (1977). A theoretical discussion of product state distributions after (vibrational) photodissociation within the framework of the adiabatic channel model has been given by Quack and Troe (1975) (see also Quack, 1979c, 1980c). A complete discussion is not possible here and we refer to the various original papers quoted.

V. SOLUTIONS OF THE STATISTICAL-MECHANICAL EQUATIONS OF MOTION

Because we are concerned mainly with the fundamental aspects of the master equations for URIMIR, we briefly review here only the basic properties of the solutions of Eq. (66). This kind of equation has a considerable literature of its own in a more general context (see von Mises, 1931; Jost, 1947, 1950; Montroll and Shuler, 1958; Feller, 1968; Oppenheim et al., 1977). The most general solution with a time-dependent K is of the form shown in Eq. (48) with the square matrix Y satisfying Eq. (106) (i.e. the same as Eq. (66)):

$$\dot{Y} = K Y, \tag{106a}$$

$$Y(t_0, t_0) = 1. \tag{106b}$$

Y can be represented in the form of a series which converges absolutely and uniformly:

$$Y(t, t_0) = 1 + \int_{t_0}^{t} K(t')dt' + \int_{t_0}^{t} K(t')(\int_{t_0}^{t'} K(t'')dt'')dt' \dots . \tag{107}$$

Alternatively one can use the Magnus expansion

$$Y(t, t_0) = \exp(Z), \tag{108}$$

$$Z = \sum_{n=0}^{\infty} Z_n, \tag{109}$$

$$Z_0 = \int_{t_0}^{t} K(t')dt', \tag{110}$$

$$Z_1 = -\frac{1}{2} \int_{t}^{t} \int_{t_0}^{t''} [K(t'), K(t'')]dt'dt'', \tag{111}$$

and further terms with more complicated commutators.

If K depends on time in a general way, the only solutions available are such series expansions or else stepwise, numerical solutions. If K depends upon time in a commutative way, i.e.

$$[K(t'), K(t'')] = 0 \tag{112}$$

in an interval (t_0, t), then the Magnus expansion terminates after the first term (Magnus, 1953). The solution is then

$$\mathbf{Y}(t, t_0) = \exp\left(\int_{t_0}^{t} \mathbf{K}(t')dt'\right). \tag{113}$$

Such a situation can approximately be realized in URIMIR as follows. We may write

$$\mathbf{K} = \mathbf{K}^\circ + \mathrm{Diag}(-k_N). \tag{114a}$$

The k_N are the specific rate constants for the monomolecular decay from level N calculated from Eq. (105), for example. The $\mathbf{K}^\circ$ is only due to optical pumping if there are no collisions and no spontaneous emission. In cases A and B (but *not* C and D) one finds from Table 1 that

$$\mathbf{K}^\circ = \mathbf{K}_I^\circ \cdot I(t). \tag{114b}$$

$\mathbf{K}_I^\circ$ is independent of time and intensity. Therefore $\mathbf{K}^\circ$ satisfies Eq. (112). The k_N can effectively be neglected, if either of the following conditions applies:

(i) During (t_0, t) all the $p_{N>L}$ and $k_{N \leq L}$ are zero (no unstable states populated). This applies, for instance, if the molecules are not sufficiently excited to reach threshold or if the k_N immediately above threshold are so much larger than the optical pumping rates at threshold that all $p_{N>L} \simeq 0$ at all times.

(ii) Some $p_{N>L}$ are different from zero during (t_0, t) but the corresponding k_N are so small that

$$\sum p_N(t) = \sum p_N(t_0).$$

Under these conditions the solution of Eq. (66) becomes:

$$\mathbf{p}(t) = \exp\left(\mathbf{K}_I^\circ \int_{t_0}^{t} I(t')\,dt'\right)\mathbf{p}(t_0). \tag{115}$$

We note that the new variable $F = \int I\,dt$ effectively replaces time (see II C).

In all other cases (case C, case D, or case B with k_N that are not negligible in the above sense, or also if there are collisions) a simple solution is only obtained if $I \neq f(t)$, which we shall assume hereafter:

$$\mathbf{Y}(t, t_0) = \exp\left[\mathbf{K}(t - t_0)\right]. \tag{116}$$

A practical solution is obtained by noting the symmetry relationship ('detailed balance'):

$$K_{IJ}p_J^e = K_{JI}p_I^e. \tag{117}$$

The p_J^e are positive parameters that are different for the various cases. They are easily evaluated from Table 1 or from the corresponding general equations including incoherent excitation also. They can be interpreted as generalized 'equilibrium' populations, i.e. the long-time populations if there were no decay. For instance in case B we have

$$p_J^e = \delta_J^{-1}/Z \tag{118}$$

where $Z = \Sigma \delta_J^{-1}$ is a generalized partition function. Because of Eq. (117) one knows that $\mathbf{K}$ is similar to a real symmetric matrix $\mathbf{S}$:

$$\mathbf{F}^{-1}\mathbf{K}\mathbf{F} = \mathbf{S} \tag{119}$$

with the real diagonal matrix $\mathbf{F}$

$$\mathbf{F} = \text{Diag}(\sqrt{p_J^e}). \tag{120}$$

The transformation in Eq. (119) multiplies each element K_{IJ} by the factor $(p_J^e/p_I^e)^{1/2}$. The symmetric matrix $\mathbf{S}$ can be made diagonal by an orthogonal transformation

$$\mathbf{G}^\mathrm{T}\mathbf{S}\mathbf{G} = \mathbf{\Lambda}. \tag{121}$$

Thus we have finally for $\mathbf{Y}$:

$$\mathbf{Y}(t, t_0) = \mathbf{F}\mathbf{G} \exp[\mathbf{\Lambda}(t - t_0)]\mathbf{G}^\mathrm{T}\mathbf{F}^{-1}. \tag{122}$$

$\mathbf{\Lambda}$ is diagonal and real since $\mathbf{S}$ is symmetric. Furthermore, all the λ_k are negative (or zero) if all the k_N are positive (or zero) (see, for instance, Quack, 1979b). The matrix $\mathbf{G}$ of the eigenvectors of $\mathbf{S}$ is easily calculated and therefore Eqs. (122) and (116) constitute suitable solutions both for explicit calculations and for more general analytical investigations. With the population vector $\mathbf{p}(t)$ one can derive all the other experimental quantities of interest (see Sec. II C).

A. The Basic Mechanism of Multiphoton Excitation and Reaction in Polyatomic Molecules

URIMIR have been observed in polyatomic molecules that have strong vibrational resonance transitions close to the laser light. If the laser frequency is far from resonance, no reaction is observed, in general. A suitable description would therefore be based upon level schemes and stepwise transitions as shown in Figs. 1 and 2. If only a few spectroscopic states are involved in the absorption, the particularities of every molecule will determine the mechanism of the rate process and there is no general basic mechanism other than the time-dependent amplitudes of spectroscopic states as described by the Schrödinger equation. We shall not discuss this case hereafter. If the states involved in the process are numerous, a statistical approximation according to Table 1 will be suitable. In order to obtain a simple model for $\mathbf{K}$

one must have some idea about the coupling matrix elements as a function of molecular excitation. A good starting point may be a separable (possibly harmonic) model, for which the average square coupling $\langle |V_{KL}^0|^2 \rangle$ is easily calculated. One can then show that Eq. (123) holds approximately (Quack, 1978, 1979b)

$$\langle |V_{KL}|^2 \rangle_{\Delta\omega} \simeq \langle |V_{KL}^0|^2 \rangle_{\Delta\omega}. \tag{123}$$

This is a valid approximation for the true $\langle |V_{KL}|^2 \rangle$ if the interaction removing the separability is not too strong, of the order of $\Delta\omega \ll \omega$ (not unreasonable) and the identity of individual, possibly broad, bands is preserved. One can show then that the optical up-pumping rate coefficient for resonance pumping very roughly, and disregarding all spectroscopic details, is of the form (see Quack, 1979b):

$$K_{M+1,M} \simeq C\Delta E^{-1} GI(\rho_M^{0\prime}/\rho_M) \tag{124a}$$

$$\simeq (s-1)h|V(1_0^1)|^2 \frac{\varepsilon_1(E_M + E_Z')^{s-2}}{\Delta E(E_M + E_Z)^{s-1}}. \tag{124b}$$

C is a constant that may incorporate spectroscopic details as a function of M, but shall be taken to be independent of M here, ΔE is a bandwidth parameter, G from Eq. (101) is the integrated band strength of the pumped vibrational fundamental transition, ρ_M is the average vibrational density of states at the level M, and $\rho_M^{0\prime}$ is the density of states with the degree of freedom corresponding to the pumped vibration being removed from the count. Eq. (124b) follows by introducing the semiclassical approximations for ρ, s is the number of vibrational degrees of freedom, $\varepsilon_1 = \hbar\omega_1$ is the quantum in the pumped vibration, E_Z and E_Z' are the total zero-point energies with and without the pumped vibration, and $|V(1_0^1)|$ is the coupling matrix element for the vibrational fundamental transition $(|V(1_0^1)|^2 \propto |M_v(1_0^1)|^2 I$, Eq. (10)). Eq. (124) describes a weakly decreasing function of the level index M (increasing molecular excitation). We stress that spectroscopic details (cf. Fig. 17) have been neglected on purpose in order to obtain a general view of the pumping process. For specific model calculations molecular spectroscopic particularities must be introduced into the model.

Fig. 21 shows a level scheme with weakly decreasing up-pumping rates indicated by the widths of the arrows. These rates are *the same* for case C and case B (Table 1) and we disregard the practically less important case D here. The down-pumping rates are, however, different in case B and case C. At moderate excitation energies (small densities of states) and low radiation intensities, case C will be applicable. Then the effective down-pumping rates are *much faster* than the up-pumping rates as indicated by the broad arrows in Fig. 21. The reader may be puzzled by this, but it is a theoretically well established fact (see Table 1 and Quack, 1978, 1979a, 1980b). At higher levels of excitation the increased effective density of states will require the

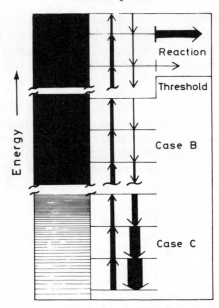

Fig. 21. Scheme for the basic mechanism of URIMIR. The widths of the arrows indicate the magnitude of the rate coefficients. (See detailed discussion in the text.)

application of the case B rate coefficients (Table 1) with down-pumping rates being *smaller than* the up-pumping rates. Above the reaction threshold there will be competition between optical pumping and the monomolecular reaction. As indicated, the monomolecular rate coefficients increase rapidly with excitation energy (see also Fig. 20). With increasing radiation intensity the number of steps obeying case C will decrease until finally only case B applies. This transition in the range of case C is easily understood by noting that the up-pumping increases in proportion to I, whereas the initially faster down-pumping increases only in proportion to the square root of I until it is equal to the slower case B pumping (see Sec. III C. 4). For large molecules at thermal initial energies, case B alone may often be adequate starting with the first step. The transition to case D at high I would have to complement the picture but is excluded from discussion here. Of course there can also arise more special situations without a steady decrease in $K_{M+1,M}$ but with minima and maxima in the up-pumping rates as a function of the level index M according to special spectroscopic properties that cannot be dealt with in a general way. Having in mind such finer details, Fig. 21 can be considered to be an adequate description of the basic mechanism of URIMIR in the range of the validity statistical-mechanical cases B and C, which covers many of the current experimental conditions.

The mechanism described here by Fig. 21 is different from a popular mechanism often invoked in the literature (see, for instance, the review by Ashfold and Hancock (1981) and the references quoted there). In this popular mechanism a coherent three- to five-photon transition in one separable mode is assumed and subsequent fast intramolecular energy transfer with incoherent one-step pumping at high levels of excitation. In our mechanism, coherent multiphoton transitions are absent, since they are expected to be inefficient indeed (Quack, 1979a). Furthermore, explicit assumptions on intramolecular vibrational relaxation on the nanosecond timescale appear to be misleading (see Sec. III E. 3) and are unnecessary as long as the absorption bandwidths are large compared with the pumping bandwidth, which is usually expected to be the case. The mechanism presented in Fig. 21 is shown as an approximation for visualization. It does invoke approximate concepts (bandwidths, matrix element models, etc.) that are less generally valid than the theory of Sec. III. However, we think that there is good reason to believe that the mechanism gives a qualitatively and semiquantitatively correct picture for the present experimental conditions and moderately large polyatomic molecules.

Two closely related concepts have to be added to the above, simple mechanism: symmetry and conservation laws, and the reducibility of the rate coefficient matrix. A more detailed, general theoretical discussion of these has been given by Quack (1978, 1979b). In practice the most important symmetry leads to the angular momentum dipole selection rules $\Delta J = 0, \pm 1$, which adds a second dimension to the structure of the rate coefficient matrix, which remains, however, irreducible with respect to J. Furthermore, molecular energy to within multiples of $\hbar \omega$ (X_{kk} in the effective Hamiltonian of Eq. (34)) is conserved in a band of width approximately equal to $4\pi^2 \langle |H_{KL}|^2 \rangle / \delta$. Usually, many such bands are populated at thermal energies. Therefore the rate coefficient matrix $\mathbf{K}$ becomes *reducible* with respect to these energy quantum numbers, i.e. it can be decomposed into disconnected blocks. This presents an additional complication, which is, in principle, easily dealt with by solving the master equation separately for the disconnected blocks and superimposing the solutions. This has been done for simple models by Quack (1979b) and some further consequences have been discussed by Quack et al. (1980). The distinction of two ensembles of reactive molecules' and 'non-reactive molecules' (Bagratashvili et al., 1979) is a somewhat crude, special case of the reducibility with one $\mathbf{K}$ with large rates and one $\mathbf{K}$ with negligible rates. We shall hereafter discuss solutions only for one irreducible $\mathbf{K}$ at a time, the general solution being obtained simply by superposition.

B. Steady State

For one irreducible $\mathbf{K}$ one has a non-degenerate maximum eigenvalue λ_1 close to zero (or zero in the absence of decay). Therefore at sufficiently long times the evolution can be characterized by this eigenvalue and eigenvector, which corresponds to the steady-state limit of Eq. (122) (all other exponentials contribute negligibly as $t \to \infty$):

$$p_N(t) = \exp\left[\lambda_1(t - t_0)\right] F_{NN} G_{N1} \sum_M G_{M1} F_{MM}^{-1} p_M(0). \tag{125}$$

At steady state the relative level populations $p_N/\Sigma p_N$ do not change any more, according to Eq. (125). When the concept of steady state was first used for URIMIR it was already pointed out that, in contrast to ordinary thermal reactions, steady state is reached only rather late after completion of a considerable fraction of the reaction (Quack, 1978). Nevertheless, steady state is of great importance because at steady state only one degree of freedom, intensity, is needed to describe the relative populations and rate coefficients independent of the previous history of the reaction. Therefore, if the role of intensity in URIMIR is to be investigated unambiguously, this is best done at steady state.

Complementary to this simplification of the behaviour at steady state, which occurs for *all* cases (A, B, C, D) described by master equations, one has the simplification already mentioned in Eq. (115). This simplification allows reduction to one degree of freedom, fluence F, for all times in the interval considered also prior to steady state. The validity of Eq. (115) is, however, restricted to an intensity-proportional $K^\circ = K_I^\circ I(t)$ as a necessary condition (only cases A and B or incoherent pumping).

C. Level Populations as a Function of Time and Intensity

Fig. 22 shows a solution of the master equation (case C and B and transition range) for a five-atomic model system irradiated with an intensity corresponding to typical experimental conditions. Fig. 22(a) indicates the steady-state distributions as a function of intensity. The shape with two maxima depends dramatically upon intensity at the lower intensities, where case C is important for the first excitation steps. Once case B is dominant at higher intensities, only one weakly intensity-dependent maximum occurs at high energies. (This latter, simpler behaviour is also typical of the phenomenological rate equation treatments.) Fig. 22(b) shows how the steady-state distribution at an intermediate, constant intensity is established as a function of time. We refer to Quack (1979b, 1981b) for more details concerning these distributions.

D. Time- and Intensity-dependent Rate Coefficients and Product Yields

Well defined level populations p_N have not, so far, been reported in experiments. The more straightforward experimental quantities are the

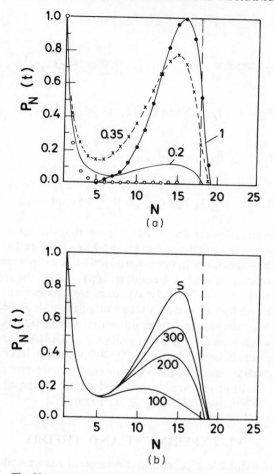

Fig. 22. (a) Relative steady-state level populations for a five-atomic model molecule in the transition range between case B and case C as a function of radiation intensity (proportional to the numbers indicated). (b) Time-dependent establishment of steady state for an intermediate intensity (0.35) in (a), times being indicated in nanoseconds.

reactant fractions F_R, F_R^* or the complementary product yields F_P and F_P^*, and the rate coefficients $k(t)$ and $k^*(t)$ (see Sec. II C). From the general solution of the master equations in Eq. (122) these can be written explicitly (Quack, 1979b):

$$F_R = 1 - F_P = \sum_K \Phi_K \exp(\lambda_K t), \tag{126}$$

$$\Phi_K = \sum_N F_{NN} G_{NK} \sum_M F_{MM}^{-1} G_{MK} \, p_M(0), \tag{127}$$

$$k(t) = - \left(\sum_K \lambda_K \Phi_K \exp(\lambda_K t) \right) \left(\sum_K \Phi_K \exp(\lambda_K t) \right)^{-1}, \tag{128}$$

$$F_R^* = 1 - F_P^* = \sum_K \Phi_K^* \exp(\lambda_K t), \tag{129}$$

$$\Phi_K^* = \sum_N{}' F_{NN} G_{NK} \sum_M F_{MM}^{-1} G_{MK} \, p_M(0), \tag{130}$$

$$k^*(t) = - \left(\sum_K \lambda_K \Phi_K^* \exp(\lambda_K t) \right) \left(\sum_K \Phi_K^* \exp(\lambda_K t) \right)^{-1}. \tag{131}$$

The Σ' in these equations is restricted to levels that do not react during the observation time. The weighting factors Φ_K^* have been called characteristic populations. They depend both upon K and $p(0)$. It is easily seen that the long-time limit of both $k(t)$ and $k^*(t)$ is equal to $k(\text{st}) = -\lambda_1$, the *steady-state rate coefficient* (the rate 'constant' in ordinary reactions). We present as an example only one calculation for $k(t)$ and F_P, for two intensities, for a five-atomic model molecule. It is seen that for the lower intensities steady state is reached quite early (case C), whereas at the higher intensities (case B) $k(t)$ or $k^*(t)$ approaches its steady value only after considerable (20–30%) reaction has occurred. Quite a few detailed discussions of the time- and intensity-dependent rate coefficients and product yields have appeared by now and we refer to these for more details (Quack, 1978, 1979b, 1981b; Quack *et al.*, 1980).

VI. EXPERIMENT AND THEORY

There is now available a general and consistent theoretical framework for the quantum-mechanical and statistical-mechanical description of the preparation of highly excited molecular states by intense, coherent infrared radiation and subsequent reactions. Apart from internal consistency, the criterion of validity of the theory is provided by experiment. We cannot review here the available experiments in detail and we refer to the excellent recent reviews of Ashfold and Hancock (1981) and Lyman *et al.* (1981) (see also King's article in the present volume). However, with respect to theory, several questions are pertinent: (i) Which qualitative experimental effects are (not) properly explained or predicted? (ii) Which new experiments or evaluations are suggested by theory? (iii) Is there quantitative agreement between experiment and theory? The last question is the most ambitious one, but can be answered at present only very roughly, since only few quantitative experiments are available. We shall deal here with collisionless URIMIR. The

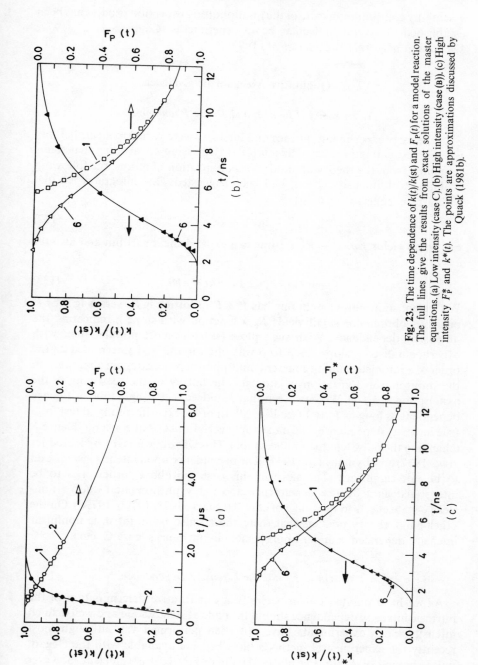

Fig. 23. The time dependence of $k(t)/k(\text{st})$ and $F_P(t)$ for a model reaction. The full lines give the results from exact solutions of the master equations. (a) Low intensity (case C). (b) High intensity (case B). (c) High intensity F_P^* and $k^*(t)$. The points are approximations discussed by Quack (1981b).

essentially collisionless nature of the multiphoton absorption process has been established in several molecular beam experiments (Coggiola *et al.*, 1977; Diebold *et al.*, 1977; Brunner *et al.*, 1977).

A. Qualitative Mechanistic Evidence

1. *Intensity, Fluence, and Delay Phenomena*

In the early experiments, fluence and intensity were varied in a parallel way when investigating product yields in URIMIR. Using a reasoning that stems originally from the theory of multiphoton ionization of atoms, the following evaluation of 'intensity effects' had been suggested. One integrates Eq. (63) with a rate 'constant' $k \neq f(t)$:

$$-\ln F_R = kt. \tag{132}$$

For small yields $F_P = 1 - F_R \ll 1$ one can expand the logarithm and also set $k = aI^n$:

$$-\ln(1 - F_P) \simeq F_P = kt = aI^n t. \tag{133}$$

With a constant pulse length one has $I^n \propto F^n$. Plotting $\lg F_P$ versus $\lg F$ (or I) one thus obtains for small yields F_P a function with slope n indicating the 'intensity' dependence. With such plots (still the most popular ones with experimentalists) a slope $n \simeq 3$ to 6 was often found and interpreted as the order of a rate-determining coherent multiphoton process (again analogous to the multiphoton ionization of atoms). The fallacy in the reasoning is the assumption $k \neq f(t)$. We have seen that k, indeed, is strongly time or fluence dependent as long as $F_P \ll 1$ (see Fig. 23). In order experimentally to test for a true intensity dependence, one must apply variations of I at constant fluence F (changing the pulse length or pulse shape). This was done for SF_6 by Kolodner *et al.* (1977) and Lyman *et al.* (1977). The dependence upon intensity was found to be not significant. Therefore in this case the above 'order' n is to be interpreted as a fluence dependence and delay phenomenon due to non-steady-state effects in the range of case B (Quack, 1978, 1979b, 1979d). On the other hand, theory predicts that there should also be an intrinsic nonlinear intensity dependence under appropriate circumstances (case C, etc.).

2. *Intrinsic Nonlinear Intensity Dependence*

As we have pointed out in Sec. V B, a quantitative determination of the intrinsic intensity dependence should be made at steady state and well defined intensity. Such experiments have not been performed yet, although most recently the experimental methods have become available with the elegant pulse shaping of Ashfold *et al.* (1981). Qualitatively it is sufficient, however to prove a dependence of yield upon average intensity after irradiation with the same fluence. The first experiment showing unambiguously higher yields at the higher average intensities and the same fluence was carried out for CH_3NH_2 (Ashfold *et al.*, 1979). Further, related experiments have appeared for CF_3I (Rossi *et al.*, 1979) and CF_2HCl (King and Stephenson, 1979), and $(CF_3)_2CO$ (Naaman and Zare, 1979). We note that neither the phenomenological rate equations nor case B would be able to predict such an effect. Case C, however, predicts precisely this behaviour, and the molecules for which intensity effects have been found are expected to fall into the range of case C at low excitation energies (Table 1!). A coherent n-photon process might also explain the intensity dependence, but we do not think that this is the correct explanation, for quantitative reasons (Quack, 1979a). In simple terms one can write for the steady-state part of the reaction and pulses of constant I:

$$-\ln(F_R/F_R^{(s)}) = k(st)(t - t_s)$$
$$= aI^n(t - t_s)$$
$$= aI^{n-1}(F - F_s). \qquad (134)$$

Therefore *at a given fluence* there will be no intensity dependence of the yields if $n = 1$ (large k_N and simplification, Eq. (115)). There will be larger yields for the higher intensities for $n > 1$ (case C) and *lower* yields for the higher intensities if $n < 1$ (case D and case B with 'fall-off'). The latter, somewhat striking effect (Quack, 1978) has not so far been seen experimentally. The above reasoning remains true during the period preceding steady state for all practically important conditions, as can be seen easily by inspection of the $k(t)$.

3. Intramolecular Relaxation: Product Translational Energy Distributions

Although not essential for the general theory, one practically important mechanistic ingredient is the assumption of broad absorption structures corresponding to relaxations of local excitations that are fast on the timescale of excitation. This fast relaxation has been tested indirectly and extensively by comparing product translational (or internal) energy distributions measured in molecular beam experiments with statistical calculations assuming fast relaxation. The evidence has been reviewed in detail (Schulz *et al.*, 1979). No example contradicting fast relaxation has been found.

4. Intramolecular Relaxation: Bond-specific Chemistry

There have been reports on bond-specific chemistry with I R lasers. Intuitively this is suggested if the relaxation of a 'local' vibrational mode excited by the laser light were slow. A critical discussion of the fallacies in interpreting bond-specific I R laser chemistry has been given by Ashfold and Hancock (1981). We are not aware of any bond-specific chemistry proving slow relaxation of local vibrational excitations in molecules. There is some evidence that the intramolecular vibrational relaxation time is often of the order of picoseconds (Rynbrandt and Rabinovitch, 1971; Wolters et al., 1980; Quack, 1980; Dübal and Quack, 1980). Therefore, in principle, with fast pumping the competition between excitation and relaxation may be observable. The *experimentum crucis* remains to be done. We note that the observation of slow intramolecular processes *per se* is no proof of slow *vibrational relaxation*. Indeed, theory requires the existence of slow intramolecular processes on characteristic timescales.

$$\tau \gtrsim h\rho(E, J).$$

With high average densities $\rho(E,J)$, τ becomes large, in the case of true dengeneracies, infinite. However, these slow time-dependent processes do not correspond to vibrational relaxation.

B. Quantitative Evidence

1. *Rate Coefficients for URIMIR*

One of the more important quantitative questions in URIMIR concerns the parameters which best characterize the *rates* of the laser-induced reactions. The complete functional dependence of product yields upon fluence might be taken for such a description. However, this information is complicated and not even complete, because different intensities at the same fluence will often give different yields, particularly at low intensities. It has been discovered (Quack, 1979d), that from the limiting slope of a graph of $-\ln F_R^*$ versus the fluence F (logarithmic reactant fluence plot), one can obtain the steady-state rate coefficient $k(\text{st}) = -\lambda_1$, the long-time limit of Eqs. (128) and (131). This result can be understood formally by using the approximation in Eq. (115) together with Eqs. (128) and (131) for pulses of arbitrary shape. Somewhat less formally one has for pulses of constant I at steady state:

$$-\frac{d\ln F_R}{I\,dt} = -\frac{d\ln F_R^*}{I\,dt} = -\frac{d\ln F_R^*}{dF} = \frac{k(\text{st})}{I}. \tag{135}$$

The first equation follows, because at steady state the relative level populations do not change. Therefore one can write

$$F_R^* = F_R(1 - F_R^{**}/F_R) \tag{136}$$

with a *time-independent* F_R^{**}/F_R. This situation is illustrated in Fig. 24 with the solution of the master equation for a realistic model. The upper function gives $-\ln F_R^*$, corresponding to much higher yields than the lower function $-\ln F_R$,

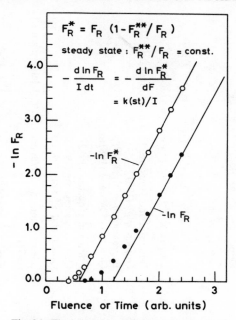

Fig. 24. The experimental determination of steady-state rate coefficients from logarithmic reactant fluence plots (see detailed discussion in the text).

which does not include the dissociation of metastable molecules after the pulse (see Sec. II C). The limiting slopes are the same as required by Eq. (135). It can also be shown that the limiting slope of the upper curve can be obtained either by changing the fluence through intensity at constant pulse duration or through pulse duration at constant intensity, as long as case B applies approximately and $k(st)$ is *approximately* proportional to intensity over small intensity ranges. It can be shown furthermore that the parameters $k(st)/I$ thus obtained should be reasonably independent of uncontrolled laboratory variables, such as the laser pulse shape. A significant result in this evaluation is also that an essentially *kinetic* quantity, $k(st)$, which can also be measured in direct time-resolved experiments, can be obtained in an indirect manner without any time-resolved measurements. For a detailed discussion we refer to Quack (1979b) and Quack *et al.* (1980).

Table 2 summarizes some rate coefficients that have been determined by evaluating the fluence dependence of product yields with this method. Logarithmic reactant fluence plots have recently also been used in the evaluation of the laser-induced desorption of CH_3F (Heidberg *et al.*, 1980, 1981) and for the URIMIR of negative ions (CH_3OHF^-) (Rosenfeld *et al.*, 1980).

TABLE 2. Experimental rate coefficients for URIMIR obtained from logarithmic reactant fluence plots.

Reaction	$k(st)/s^{-1}$ $I/MW\,cm^{-2}$	Conditions	References
$SF_6 \rightarrow SF_5 + F$	1.0×10^5	bulk near 300 K	Campbell et al. (1976)
$SF_6 \rightarrow SF_5 + F$	0.84×10^5	bulk 223 K	Duperrex and van Bergh (1979a, b)
	0.68×10^5	bulk 293 K	
	0.49×10^5	bulk 343 K	
$SF_6 \rightarrow SF_5 + F$	$(2 \pm 1) \times 10^5$	beam 150 K	Brunner and Proch (1978) (see Quack, 1979b)
$(CF_3)_2CO \rightarrow ?$	3.3×10^5	bulk	Fuss et al. (1979) see Luther and Quack, (1979)
$CF_3I \rightarrow CF_3 + I$	1.1×10^5	bulk	Bittenson and Houston, 1977 (see Quack, 1979d)
	4.0×10^5	VLPϕ	Golden et al. (1981)
	7.8×10^5	bulk	Bagratashvili et al. (1979)
	1.6×10^6	bulk	Quack and Seyfang (1981a)
$C_2F_4S_2 \rightarrow 2CF_2S$	2.3×10^6	bulk ($1076\,cm^{-1}$)	Plum and Houston (1982)
	3.2×10^6	bulk ($1076\,cm^{-1}$)	Quack and Seyfang (1981)
	7.9×10^5	bulk ($955\,cm^{-1}$)	Quack and Seyfang (1981)
$N_2F_4 \rightarrow 2NF_2$	1.0×10^6	bulk	Kleinermanns and Wagner (1979), Kleinermanns (1978)
$UO_2(HFACAC)_2THF$ $\rightarrow UO_2(HFACAC)_2 + THF$	1.5×10^7	bulk	Kaldor et al. (1979) (see Quack et al., 1980)
$SF_6 \rightarrow SF_5 + F$	$(4.5 \pm 3) \times 10^5$	classical trajectory calculations	Poppe (1981)

[a] HFACAC stands for hexafluoroacetylacetonate and THF for tetrahydrofurane.

The question of the dependence of rate coefficients upon molecular parameters has been investigated theoretically (Quack, 1979e). More empirical correlations, in which somewhat less well defined kinetic parameters were used, should be mentioned as well (Lyman *et al.*, 1981). Also a linear representation of yields as a function of lg F on 'probability graph paper' has been suggested on the basis of numerical experiments (Barker, 1980), however without a general theoretical justification.

2. *The Road Ahead*

We think that the accurate experimental evaluation of rate coefficients and comparison with theory will play an important role in the future. Indeed, the rate coefficients presented in Table 2 are of the right order of magnitude expected from the statistical-mechanical theory in connection with simple models of **K** (Quack, 1979e). To some extent this is a success of the theory. Assuming, say, a rate-determining coherent three- or five-photon process to an overtone of the separable vibration, one would obtain absolute rates that are orders of magnitude different from the ones in Table 2. Also the constant coupling models (Carmeli and Nitzan, 1980) would completely fail the quantitative test. Unfortunately, there are few quantitative data available and even the examples in Table 2 probably do not correspond to definite experimental results in all cases. Two otherwise very powerful methods such as molecular beam photodissociation (Brunner *et al.*, 1977; Coggiola *et al.*, 1977; Diebold *et al.*, 1977) or the laser-induced fluorescence detection of products (Filseth *et al.*, 1979; Ashfold *et al.*, 1979; King and Stephenson, 1979) apparently have great difficulties in providing accurate absolute product yields. More work should be directed towards this goal.

To conclude, we list a few urgent problems for the future: (i) Fluence must be controlled better than is commonly done now (usually one has no uniform fluence, often ill defined spatial fluence profiles). (ii) Intensity must be controlled. Most experiments are done with multimode lasers of irreproducible intensity–time profiles. (iii) The best quantitative experiment should have uniform spatial fluence and constant temporal intensity. The latter is of particular importance in the quantitative investigation of the intrinsic nonlinear intensity dependence, which is of great mechanistic significance. It will be possible and meaningful, then, to compare experiment and theory at a detailed level. (iv) The infrared absorption profiles of highly excited molecules should be studied in detail both at low and at high resolution in order to obtain information on the applicability of the statistical–mechanical cases and their natural extension to more complex band structures.

Acknowledgements

It is a pleasure to acknowledge the generous support from J. Troe, from the Fonds der chemischen Industrie, and from the Deutsche Forschungsgemeinschaft (SFB 93 Photochemie mit Lasern). Help from S. Hoff and A. Nadler, and discussions with H. R. Dübal and G. Seyfang, are also gratefully acknowledged. Computing facilities were provided by the GWD-Göttingen and the RRZN Hannover. R. G. Gilbert and E. Sutcliffe kindly read and criticised the manuscript.

Appendix: The Calculation of the Exponential Function of Square Matrices

Both the time-dependent Schrödinger equation and the master equations require the solution of a set of first-order, linear, ordinary differential equations (in matrix notation)

$$\dot{\mathbf{c}}(t) = \mathbf{D}\mathbf{c}(t). \tag{A.1}$$

If $\mathbf{D}$ does not depend upon time, the solution is given by

$$\mathbf{c}(t) = \exp{(\mathbf{D}t)}\mathbf{c}(0) \tag{A.2}$$

$$= \exp{(\mathbf{A})}\mathbf{c}(0). \tag{A.3}$$

This requires the calculation of the exponential function of the square matrix $\mathbf{A}$. Numerous methods for doing this, or equivalently for solving (A.1), are available. We shall discuss here two, one very general but not very efficient, the second efficient but not general, although usually sufficient for cases treated in this article.

The general method is based upon the series expansion, which converges for all $\mathbf{A}$

$$\exp{(\mathbf{A})} = \sum_{n=0}^{\infty} \mathbf{A}^n/n! \tag{A.4}$$

It may be difficult to obtain sufficiently accurate results from Eq. (A.4). Sometimes, it is useful to start from the 'small' matrix

$$\mathbf{M} = \mathbf{A}/2^n, \tag{A.5}$$

$$\exp{(\mathbf{A})} = [\exp{(\mathbf{M})}]^{2^n}. \tag{A.6}$$

$\exp{(\mathbf{M})}$ converges quickly and afterwards one obtains its power (2^n) by n matrix multiplications.

If a similarity transformation exists, which makes $\mathbf{A}$ diagonal, one has

$$\mathbf{X}^{-1}\mathbf{A}\mathbf{X} = \mathbf{\Lambda}, \tag{A.7}$$

$$\exp{(\mathbf{A})} = \mathbf{X}\exp{(\mathbf{\Lambda})}\mathbf{X}^{-1}, \tag{A.8}$$

$$\exp{(\mathbf{\Lambda})} = \text{Diag}(\exp{(\lambda_1)}, \exp{(\lambda_2)}, \ldots). \tag{A.9}$$

Here it is necessary to find the matrix $\mathbf{X}$ of linearly independent eigenvectors and the eigenvalues λ_k. This is always possible for matrices that are real, symmetric or similar to real symmetric matrices. This is the case for the master equation and has been discussed in Sec. V. It is also the case for Hermitian **H** in the time-dependent Schrödinger equation (Secs. II, III). Efficient algorithms exist, and therefore these problems are best solved using Eqs. (A.7)–(A.9). Other methods, such as direct, numerical, stepwise integration of Eq. (A.1) or stochastic simulation of the master equations, etc., are in general less efficient.

A difficulty arises with the Schrödinger equation, when the matrix **H** is not Hermitian due to complex diagonal elements, Eq. (9). Then it is not always possible to find a similarity transformation as in Eq. (A.7). A simple example for this is the resonant two-level problem with decay, characterized by the Hamiltonian (in arbitrary units):

$$\mathbf{X} + \tfrac{1}{2}\mathbf{V} = \begin{pmatrix} 0 & 1 \\ 1 & -2i \end{pmatrix}. \tag{A.10}$$

There is only one linearly independent eigenvector $(i, 1)^T$ for the degenerate eigenvalue $(-i)$. The exponential function, Eq. (A.4), still exists, and therefore the time evolution matrix still can be computed, but not from the eigenvectors and eigenvalues of the effective Hamiltonian (as is sometimes believed).

However, Eq. (A.7) can be applied for general matrices if all eigenvalues or all elementary divisors are different, since then a complete set of linearly independent eigenvectors exists. A few useful properties of the **X** for special cases may be mentioned here. For Hermitian **A**: $\mathbf{X}^+ = \mathbf{X}^{-1}$; for real, symmetric $\mathbf{A} : \mathbf{X}^T = \mathbf{X}^{-1}$; for complex symmetric **A** (applicable to the problem in Sec. II) one has (*if* Eq. (A.7) holds!):

$$\mathbf{B}^T \mathbf{A} \mathbf{B} = \mathbf{\Lambda}. \tag{A.11}$$

X can be calculated with any of the standard algorithms for finding the eigenvectors of general matrices. **B** can then be obtained in the following way with a diagonal matrix **C**:

$$\mathbf{B} = \mathbf{X}\mathbf{C}, \tag{A.12}$$

$$\mathbf{X}^T\mathbf{X} = (\mathbf{C}^2)^{-1} = \mathrm{Diag}(1/C_{jj}^2), \tag{A.13}$$

as can easily be verified. This avoids the calculations of the inverse $\mathbf{X}^{-1}$, which in general is *different* from X^T. It must be stressed that the procedure works only for cases in which no catastrophe such as in Eq. (A.10) occurs. One must also avoid matrices close to such a catastrophical situation, which would lead to inaccuracies. Therefore for non-Hermitian **A**, Eq. (A.8) must be appropriately checked after the calculation.

References

R. V. Ambartzumian, and V. S. Letokhov (1977), Selective dissociation of polyatomic molecules by intense infrared laser fields, *Acc. Chem. Res.*, **10**, 61.

M. N. R. Ashfold, and G. Hancock (1981), infrared multiple photon excitation and dissociation: reaction kinetics and radical formation, in *Gas Kinetics and Energy Transfer*, vol. 4, eds. R. Donovan and P. G. Ashmore, The Chemical Society, London.

M. N. R. Ashfold, G. Hancock, and G. Ketley (1979), Infrared multiple photon excitation and dissociation of simple molecules, *Faraday Discuss. Chem. Soc.*, **67**, 204.

M. N. R. Ashfold, C. G. Atkins, and G. Hancock (1981), *Chem. Phys. Lett.*, **80**, 1.

S. D. Augustin, and H. Rabitz, (1976), *J. Chem. Phys.*, **64**, 1223.

V. N. Bagratashvili, V. S. Doljikov, V. S. Letokhov and E. A. Ryabov (1979), Study of primary characteristics of multiple IR-photon excitation and dissociation of CF_3I, in *Laser Induced Processes in Molecules*, eds. K. L. Kompa and S. D. Smith, Springer, Berlin, p. 179.

A. C. Baldwin, J. R. Barker, D. M. Golden, R. Duperrex and H. van den Bergh (1979), infrared multiphoton chemistry, comparison of theory and experiment, *Chem. Phys. Lett.*, **62**, 178.

J. R. Barker (1980), Infrared multiphoton decomposition, *J. Chem. Phys.*, **72**, 3686.

S. Bittenson and P. L. Houston (1977), Carbon isotope separation by multiphoton dissociation of CF_3I, *J. Chem. Phys.*, **67**, 4819.

M. Bixon and J. Jortner (1968), Intramolecular radiationless transitions, *J. Chem. Phys.*, **48**, 715.

M. Bixon and J. Jortner (1979), Long radiative lifetimes of small molecules, *J. Chem. Phys.*, **50**, 3284.

J. G. Black, E. Yablonovitch, N. Bloembergen and S. Mukamel (1977), Collisionless multiphoton dissociation of SF_6, *Phys. Rev. Lett.*, **38**, 1131.

N. Bloembergen and E. Yablonovitch (1978), Infrared-laser-induced unimolecular reactions, *Phys. Today*, **31**, 23.

M. C. Borde, Anne Henry and M. L. Henry (1966), *C. R. Acad. Sci: Paris*, **263B**, 619.

F. Brunner, T. P. Cotter, K. L. Kompa and D. Proch (1977), *J. Chem. Phys.*, **67**, 1547.

F. Brunner and D. Proch (1978), The selective dissociation of SF_6, *J. Chem. Phys.*, **68**, 4936.

D. L. Bunker (1971), in *Methods in Computational Physics*, vol. 10, ed. B. Alder, Academic Press, New York.

J. D. Campbell, G. Hancock and K. H. Welge (1976), Energy dependence of SF_6 dissociation, *Chem. Phys. Lett.*, **43**, 581.

C. D. Cantrell, S. M. Freund and J. L. Lyman (1979), Laser induced chemical reactions and isotope separation, in *Laser Handbook*, vol. IIIb, ed. E. L. Stitch, North-Holland, Amsterdam.

B. Carmeli, and A. Nitzan (1980), Random coupling models for intramolecular dynamics, *J. Chem. Phys.*, **72**, 2054, 2070.

B. Carmeli, I. Schek, A. Nitzan and J. Jortner (1980), Numerical simulations of molecular multiphoton excitation models, *J. Chem. Phys.*, **72**, 1928.

M. J. Coggiola, P. A. Schulz, Y. T. Lee and Y. R. Shen (1977), Molecular beam study of multiphoton dissociation of SF_6, *Phys. Rev. Lett.*, **38**, 17.

J. M. Delory and C. Tric (1974), Time interferences and nonexponential decay of quasi-isolated molecules, *Chem. Phys.*, **3**, 54

G. J. Diebold, F. Engelke, D. M. Lubman, J. C. Whitehead and R. N. Zare (1977), Infrared multiphoton dissociation of SF_6 in a molecular beam, *J. Chem. Phys.*, **67**, 5407.

D. R. Dion and J. O. Hirschfelder (1976), Time dependent perturbation of a two state quantum system by a sinusoidal field, *Adv. Chem. Phys.*, **35**, 265.

P. A. M. Dirac (1967), *The Principles of Quantum Mechanics*, Oxford University Press, London.

H. R. Dúbal and M. Quack (1980), Spectral bandshape and intensity of the C–H chromophore in the infrared spectra of CF_3H and C_4F_9H, Chem. Phys. Lett., **72**, 342.

R. Duperrex and H. van den Bergh (1979a), Competition between optical pumping and collisions in unimolecular reactions induced by monochromatic infrared radiation, Chem. Phys., **40**, 275.

R. Duperrex and H. van den Bergh (1979b), Temperature dependence in the multiphoton dissociation of SF_6, J. Chem. Phys., **70**, 5672.

R. Duperrex and H. van den Bergh (1980), Magnetic field effects in the infrared multiphoton dissociation of CF_2HCl, J. Chem. Phys., **73**, 585.

J. H. Eberly, B. W. Shore and Z. Bialynicka-Birula and I. Bialynicki-Birula (1977), Coherent dynamics of N-level atoms and molecules, Phys. Rev., **A16**, 2038, 2048.

P. Ehrenfest and T. Ehrenfest (1907), Über zwei bekannte Einwände gegen das Boltzmannsche H-Theorem, Phys., Z., **8**, 311.

P. Ehrenfest and T. Ehrenfest (1911), Begriffliche Grundlagen der statistischen Auffassung in der Mechanik, Encykl, Math. Wiss., **4**, No. 32.

M. Eigen (1971), Selforganization of matter and the evolution of Macromolecules, Naturwissenschaften, **33a**, 465.

F. H. M. Faisal (1976), A model for dissociation of polyatomic molecules by multiple photon absorption, Opt. Commun., **19**, 404.

M. Faubel and J. P. Toennies (1977), Adv. Atom. Mol. Phys., **13**, 229.

W. Feller (1968), An Introduction to Probability Theory and its Applications, John Wiley & Sons, New York.

S. V. Filseth, J. Danon, D. Feldman, J. D. Campbell and K. H. Welge (1979), Infrared multiple photon dissociation of N_2H_4 and CH_3NH_2, Chem. Phys. Lett., **63**, 615.

G. R. Fleming, O. L. J. Gijzeman and S. H. Lin (1974), Theory of intramolecular vibrational relaxation in large systems, J. Chem. Soc. Faraday Trans. II, **70**, 37.

M. A. D. Fluendy and K. P. Lawley (1973), Chemical Applications of Molecular Beam Scattering, Chapman and Hall, London.

P. R. Fontana and P. Thomann (1976), Stimulated absorption and emission of strong monochromatic radiation by a two-level atom, Phys. Rev., **A13**, 1512.

W. Forst (1973), Theory of Unimolecular Reactions, Academic Press, New York.

H. Friedmann (1979), Nonthermal theory of threshold behaviour of collisionless multiphoton dissociation, in Laser Induced Processes in Molecules, eds. K. L. Kompa and S. D. Smith, Springer-Verlag, Berlin, p. 149.

W. Fuss (1979), Rate equations approach to the infrared collisionless multiphoton excitation, Chem. Phys., **36**, 135.

W. Fuss, K. L. Kompa and F. M. G. Tablas (1979), Wavelength dependence of multiphoton absorption and dissociation of hexafluoroacetone, Faraday Discuss. Chem. Soc., **67**, 180

D. M. Golden, M. J. Rossi, A. C. Baldwin and J. R. Barker (1981), Infrared photochemistry and photophysics, Acc. Chem. Res., **14**, 56.

M. F. Goodman, J. Stone and D. A. Dows (1976), Laser induced rate processes in gases, J. Chem. Phys., **65**, 5052, 5062.

T. E. Gough, R. E. Miller and G. Scoles (1977), Infrared laser spectroscopy of molecular beams, Appl. Phys. Lett., **30**, 338.

P. Gozel and H. van den Bergh (1981), Electric field effects in the infrared multiphoton dissociation of CF_2HCl, J. Chem. Phys., **74**, 1724.

E. R. Grant, M. J. Coggiola, Y. T. Lee, P. A. Schulz, A. S. Sudbo and Y. R. Shen (1977), The extent of energy randomization in the infrared multiphoton dissociation of SF_6, Chem. Phys. Lett., **52**, 595.

E. R. Grant, P. A. Schulz, A. S. Sudbo, Y. R. Shen and Y. T. Lee (1978), Is multiphoton dissociation of molecules a statistical thermal process?, *Phys. Rev. Lett.*, **40**, 115.

K. D. Hänsel (1979), On the dynamics of multiphoton dissociation of polyatomic molecules II. Application to O_3, in *Laser Induced Processes in Molecules*, eds. K. L. Kompa and S. D. Smith, Springer, Berlin, p. 145.

G. H. Hardy and E. M. Wright (1958), *An Introduction to the Theory of Numbers*, Oxford University Press., London.

S. Haroche, (1971), L'atome habillé, *Ann. Phys. Paris*, **6**, 189.

W. G. Harter, C. W. Patterson and H. W. Galbraith (1978), *J. Chem. Phys.*, **69**, 4896.

J. Heidberg, H. Stein and E. Riehland (1981), Desorption by resonant multiphoton excitation of internal adsorbate vibration, in *Proc. EPS Conf. on Vibrations at Surfaces, Namur* 1980, Plenum Press, New York (in press).

J. Heidberg, H. Stein, E. Riehland and A. Nestmann (1980), Evaporation and desorption by resonant excitation of molecular normal vibrations with laser infrared, *Z. Phys. Chem., N. F.*, **121**, 145.

W. Heisenberg (1958), Die physikalischen Prinzipien der Quantentheorie, B. I. Wissenschaftsverlag, Mannheim.

G. Herzberg (1966), *Molecular Spectra and Molecular Structure*, Vol. III, Van Nostrand, Toronto.

D. P. Hodgkinson and J. S. Briggs (1976), Dissociation of polyatomic molecules by intense infrared laser radiation, *Chem. Phys. Lett.*, **43**, 451.

D. P. Hodgkinson and J. S. Briggs (1977), Theory of excitation of polyatomic molecules by intense infrared laser radiation, *J. Phys. B*, **10**, 2583.

N. R. Isenor, V. Merchant, R. S. Hallsworth and M. C. Richardson (1973), CO_2 laser induced dissociation of SiF_4, *Can. J. Phys.*, **51**, 1281.

R. Jancel (1963), 'Les fondements de la méchanique statistique classique et quantique, Gauthier Villars, Paris.

J. Jortner, S. A. Rice and R. M. Hochstrasser (1969), Radiationless transitions in photochemistry, *Adv. Photochem.*, **7**, 149.

W. Jost (1947), Über den Ablauf zusammengesetzter chemischer Reaktionen. Systeme von Reaktionen I. Ordnung, *Z. Naturforsch.*, **2a**, 159.

W. Jost (1950), Über den Ablauf zusammengesetzter chemischer Reaktione II. Offene Systeme von Reaktionen I. Ordnung, *Z. Phys. Chem.*, **195**, 21.

A. Kaldor, R. B. Hall, D. M. Cox, J. A. Horsley, P. Rabinowitz and G. M. Kramer (1979), Isotope selective IR photodissociation of a volatile uranyl compound, *J. Am. Chem. Soc.*, **101**, 4465.

K. G. Kay (1974), *J. Chem. Phys.*, **61**, 5205.

K. G. Kay (1976), *J. Chem. Phys.*, **65**, 3813.

D. S. King and J. C. Stephenson (1979), Laser intensity effects in the IR-multiphoton decomposition of CF_2HCl, *Chem. Phys. Lett.*, **66**, 33.

C. Kleinermanns (1978), *IR-laser eingeleitete Dissoziation von N_2F_4 und Isomerisierung von CH_3NC*, Bericht 12, Max Planck Institut für Strömungsforschung.

C. Kleinermanns and H. G. Wagner (1979), Geschwindigkeitskonstante der Photodissoziation von N_2F_4 mit CO_2-Laserstrahlung, *Z. Phys. Chem., N.F.*, **118**, 1.

D. E. Knuth (1969), *The Art of Computer Programming*, vol. 2, Addison-Wesley, London.

K. W. F. Kohlrausch and E. Schrödinger (1926), Das Ehrenfestsche Modell der H-Kurve, *Phys. Z.*, **27**, 306.

A. Kolmogoroff (1933), Grundbegriffe der Wahrscheinlichkeitsrechung, in *Ergebnisse der Mathematik und ihrer Grenzgebiete*, vol. 2, Springer Verlag, Berlin), p. 195.

P. Kolodner, C. Winterfeld and E. Yablonovitch (1977), Molecular dissociation of SF_6 by ultrashort CO_2-laser pulses, *Opt. Commun.*, **20**, 119.

D. M. Larsen (1976), Frequency dependence of the dissociation of polyatomic molecules by radiation, *Opt. Commun.*, **19**, 404.

D. M. Larsen and N. Bloembergen (1976), Excitation of polyatomic molecules by radiation, *Opt. Commun.*, **17**, 254.

W. D. Lawrence, A. E. W. Knight, R. G. Gilbert and K. D. King (1981), *Chem. Physics*, **56**, 343.

S. C. Leasure, K. F. Milfeld and R. E. Wyatt (1981), *J. Chem. Phys.*, **74**, 6197.

V. S. Letokhov and B. C. Moore (1976), *Sov. J. Quantum Electron.*, **6**, 129, 259.

P. O. Löwdin (1967), Quantum theory of time dependent phenomena treated by the evolution operator technique, *Adv. Quantum Chem.*, **8**, 323.

K. Luther and M. Quack (1979), *Faraday Disc. Chem. Soc.*, **67**, 229.

J. L. Lyman (1977), A model for unimolecular reaction of SF_6, *J. Chem. Phys.*, **67**, 1868.

J. L. Lyman, G. P. Quigley and O. P. Judd (1981), Single-infrared-frequency studies of multiple-photon excitation and dissociation of polyatomic molecules, in *Multiple Photon Excitation and Dissociation of Polyatomic Molecules*, ed. C. Cantrell, Springer, Berlin and Heidelberg.

J. L. Lyman, S. D. Rockwood and S. M. Freund (1977), Multiple photon isotope separation in SF_6, *J. Chem. Phys.*, **67**, 4545.

W. Magnus (1953), On the exponential solution of differential equations for a linear operator, *Commun. Pure Appl. Math.*, **7**, 649.

R. A. Marcus and O. K. Rice (1951), *J. Phys. Colloid. Chem.*, **55**, 894.

A. Messiah (1964), *Quantum Mechanics*, North-Holland, Amsterdam.

W. H. Miller (1974), Classical-limit Quantum mechanics and the theory of molecular collisions, *Adv. Chem. Phys.*, **25**, 69.

W. H. Miller (1975), The classical S-matrix in molecular collisions, *Adv. Chem. Phys.*, **30**, 77.

E. W. Montroll and K. E. Shuler (1958), The application of the theory of stochastic processes to chemical kinetics, *Adv. Chem. Phys.*, **1**, 361.

S. Mukamel (1979), On the derivation of rate equations for collisionless molecular multiphoton processes, *J. Chem. Phys.*, **71**, 2012.

S. Mukamel and J. Jortner (1976), Multiphoton molecular dissociation in intense laser fields, *J. Chem. Phys.*, **65**, 5204.

R. Naaman and R. N. Zare (1979), *Far. Disc. Chem. Soc.*, **67**, 242.

D. W. Noid, M. L. Koszykowski, R. A. Marcus and J. D. McDonald (1977), Classical trajectory study of infrared multiphoton photodissociation, *Chem. Phys. Lett.*, **51**, 540.

I. Oppenheim, K. E. Shuler and G. H. Weiss (1977), *Stochastic Process in Chemical Physics*, MIT Press, Cambridge, MA.

W. Pauli (1928), in *Probleme der Modernen Physik, Festschrift zum 60, Geburtstage A. Sommerfelds*, Hirzel, Leipzig.

P. Pechukas and J. C. Light (1966), On the exponential form of time displacement operators in quentum mechanics, *J. Chem. Phys.*, **44**, 3897.

J. Perrin (1922), On the radiation theory of chemical action, *Trans. Faraday Soc.*, **17**, 546.

G. J. Pert (1973), The dissociation of molecules by intense infrared radiation, *IEEE J. Quantum Electron.*, **QE-9**, 435.

C. N. Plum and P. L. Houston (1980), Infrared photolysis of $C_2F_4S_2$, *Chem. Phys.*, **45**, 159.

D. Poppe (1980a), Multiphoton absorption of SF_6: a classical trajectory study, *Chem. Phys.*, **45**, 371.

D. Poppe (1980b), On the role of angular momentum in the multiphoton absorption of polyatomic molecules, *Chem. Phys. Lett.*, **75**, 264.

D. Poppe (1981), *J. Chem. Phys.*, **74**, 5326.

K. R. Popper (1980) *The Logic of Scientific Discovery* (Appendix), Hutchinson, London.

H. Primas (1980), in *Quantum Dynamics of Molecules* (R. G. Wooley, ed.), Plenum Press, London.

L. Prigogine and P. Résibois (1961), *Physica* (*Utrecht*), **27**, 629.

M. Quack (1978), 'Theory of unimolecular reactions induced by monochromatic infrared radiation, *J. Chem. Phys.*, **69**, 1282.

M. Quack (1979a), Nonlinear intensity dependence of the rate coefficient in unimolecular reactions induced by monochromatic infrared radiation, *Chem. Phys. Lett.*, **65**, 140.

M. Quack (1979b), Master equations for photochemistry with intense infrared light, *Ber. Bunsenges. Phys. Chem.*, **83**, 757.

M. Quack (1979c), Quantitative comparison between detailed (state selected) relative rate data and averaged (thermal) absolute rate data for complex forming reactions, *J. Phys. Chem.*, **83**, 150.

M. Quack (1979d), On the determination of rate constants from the dependence of product yields upon laser energy fluence, *J. Chem. Phys.*, **70**, 1069.

M. Quack (1979e), Master equations for photochemistry with intense infrared light III, *Ber. Bunsenges. Phys. Chem.*, **83**, 1287 (1979).

M. Quack, Statistical models for product energy distributions, *Chem. Phys.*, **51**, 353.

M. Quack (1981a), Statistical mechanics and dynamics of molecular fragmentation, *Il Nuovo Cimento*. **38**, 358.

M. Quack (1981b), Master equations for photochemistry with intense infrared light IV, *Ber. Bunsenges. Phys. Chem.*, **85**, 318.

M. Quack, P. Humbert and H. van den Bergh (1980), The dependence of rate coefficients and product yields upon fluence, intensity, and time in unimolecular reactions induced by monochromatic infrared radiation, *J. Chem. Phys.*, **73**, 247.

M. Quack and G. Seyfang (1982), Absolute rate parameters for infrared photochemistry, *J. Chem. Phys.*, in press.

M. Quack and G. Seyfang (1981b), for forrate coefficient for IR-band intensities and IR-photochemistry $C_2 F_4 S_2$, *Chem. Phys. Lett.*, **84**, 541.

M. Quack and J. Troe (1974), Specific rate constants of unimolecular processes, *Ber. Bunsenges. Phys. Chem.*, **78**, 240.

M. Quack and J. Troe (1975), Product state distributions after dissociation, *Ber. Bunsenges. Phys. Chem.*, **79**, 469.

M. Quack and J. Troe (1977), in *Gas Kinetics and Energy Transfer*, vol. 2, eds. P. G. Ashmore and R. J. Donovan, The Chemical Society, London.

M. Quack and H. van den Bergh (1981), Theory of the influence of static electric and magnetic fields in unimolecular reactions induced by monochromatic infrared radiation, to be published.

I. I. Rabi (1937), Space quantization in a gyrating magnetic field, *Phys. Rev.*, **51**, 652.

R. Ramaswamy, S. Augustin and H. Rabitz (1978), Stochastic theory of intramolecular energy transfer, *J. Chem. Phys.*, **69**, 5509.

P. Résibois (1963), *Physica* (*Utrecht*), **29**, 721.

S. A. Rice (1975), in *Excited States*, ed. E. C. Lim, Academic Press, New York.

S. A. Rice (1980), in *Quantum Dynamics of molecules* (R. G. Wooley, ed.), Plenum Press, London.

G. W. Robinson (1974), Molecular electronic radiationless transitions, in *Excited States*, ed. E. C. Lim, Academic Press, New York.

R. N. Rosenfeld, J. M. Jasinski and J. I. Brauman (1980), Saturation effects on the fluence dependence of the infrared photodissociation of CH_3OHF^{-1}, *Chem. Phys. Lett.*, **71**, 400.

H. M. Rosenstock, M. B. Wallenstein, A. L. Wahrhaftig and H. Eyring (1952), *Proc. Natl. Acad. Sci., USA*, **38**, 667.

M. Rossi, J. R. Barker and D. M. Golden (1979), Infrared multiphoton dissociation yields via a versatile new technique: intensity, fluence, and wavelength dependence for CF_3I, *Chem. Phys. Lett.*, **65**, 523.

J. D. Rynbrandt and B. S. Rabinovitch (1971), Intramolecular energy relaxation, nonrandom decomposition of hexafluorobicyclopropyl, *J. Phys. Chem.*, **75**, 2164.

I. Schek and J. Jortner (1979), Random coupling model for multiphoton photofragmentation of large molecules, *J. Chem. Phys.*, **70**, 3016.

I. Schek, M. L. Sage and J. Jortner (1979), Validity of the rotating wave approximation, *Chem. Phys. Lett*, **63**, 230.

I. Schek, J. Jortner and M. Sage (1981), Application of the Magnus Expansion for high order multiphoton excitation, *Chem. Phys.*, **59**, 11.

P. A. Schulz (1979), Dissertation, University of California, Berkeley.

P. A. Schulz, A. S. Sudbo, D. J. Krajnovitch, H. S. Kwok, Y. R. Shen and Y. T. Lee (1979), Multiphoton dissociation of polyatomic molecules, *Annu. Rev. Phys. Chem.*, **30**, 379.

P. A. Schulz, A. S. Sudbo, E. R. Grant, Y. R. Shen and Y. T. Lee (1980), Multiphoton dissociation of SF_6 by a molecular beam method, *J. Chem. Phys.* **72**, 4985.

A. Schultz, H. W. Cruse and R. N. Zare (1972), *J. Chem. Phys.*, **57**, 1354.

J. Stone and M. F. Goodman (1978), Laser induced rate processes in gases, *Phys. Rev.*, **A18**, 2618, 2642.

J. Stone and M. F. Goodman (1979), A re-examination of the use of rate equation for laser driven polyatomic molecules, *J. Chem. Phys.*, **71**, 408.

J. Stone, E. Thiele, M. F. Goodman, J. C. Stephenson and D. S. King (1980), Collisional effects in the multiphoton dissociation of CF_2CFCl, *J. Chem. Phys.*, **73**, 2259.

M. Tamir and R. D. Levine (1977), *Chem. Phys. Lett.*, **46**, 208.

R. C. Tolman (1938), *The Principles of Statistical Mechanics*, Oxford University Press, London.

J. Troe (1975), in *Physical Chemistry: an Advanced Treatise*, vol. VI B, ed. W. Jost, Academic Press, New York.

J. Troe (1980), Solutions of the master equation for multiphoton dissociation, *J. Chem. Phys.*, **73**, 3205.

L. Van Hove (1957), *Physica (Utrecht)*, **23**, 441.

L. Van Hove (1962), in *Fundamental Problems in Statistical Mechanics*, ed. E. G. D. Cohen, North-Holland, Amsterdam.

N. G. Van Kampen (1962), in *Fundamental Problems in Statistical Mechanics*, ed. E. G. D. Cohen, North-Holland, Amsterdam.

K. M. Van Vliet (1978), *Can. J. Phys.*, **56**, 1204.

R. Von Mises (1931), *Wahrscheinlichkeitsrechung und ihre Anwendung in der Statistik und Theoretischen Physik*, Leipzig.

R. B. Walker and R. K. Preston (1977), Quantum versus classical dynamics in the treatment of multiple photon excitation, *J. Chem. Phys.*, **67**, 2017.

F. C. Wolters, B. S. Rabinovitch and A. N. Ko (1980), Initial state selection and intramolecular vibrational relaxation in reacting polyatomic molecules, *Chem. Phys.*, **49**, 65.

G. Yahav, Y. Haas, B. Carmeli and A. Nitzan (1980), Incubation times in the multiphoton dissociation of polyatomic molecules, *J. Chem. Phys.*, **72**, 3410.

R. Zwanzig (1964), On the identity of three generalized master equations, *Physica (Utrecht)*, **30**, 1109.

Dynamics of the Excited State
Edited by K. P. Lawley
© 1982 John Wiley & Sons Ltd.

PROGRESS IN ELECTRONIC-TO-VIBRATIONAL ENERGY TRANSFER

I. V. HERTEL

Institut für Molekülphysik, Fachbereich Physik der Freien Universität Berlin, Arnimallee 14, 1000 Berlin 33, West Germany

CONTENTS

I. INTRODUCTION

Electronic-to-vibrational, -rotational, -translational (E–V, R, T) energy transfer involving an excited atom and a molecule can be considered as one of the most elementary non-adiabatic processes in molecular dynamics. It has been the subject of numerous studies since 1927 under the title of quenching of resonance radiation, and our knowledge of these processes now belongs in standard textbooks (e.g. Massey and Burhop, 1971; Nikitin, 1974; Davidovits and McFadden, 1979; Yardley, 1980).

Alkali atoms, in particular, have been the object of intensive investigation, and the process

$$\text{Na}(3^2P) + \text{AB}(n,j) \rightarrow \text{Na}(3^2S) + \text{AB}(n',j') \qquad (1)$$

can be considered to be a model case. Both the vibrational (n) and the rotational (j) quantum numbers of the quenching molecule AB may change. Thereby the electronic energy $E_{ex} \sim 2 \cdot 1$ eV (20·3 kJ) of the excited sodium can be transferred partially into relative translational energy $E'_{c.m.}$ of the receding particles (all energies are given with respect to the centre-of-mass (c.m.) frame) and partially into internal energy of the molecule, which is changed by ΔE_{vibrot} during the process. The energy balance of reaction (1) may thus be written as

$$E_{ex} + E_{in} = \Delta E_{vibrot} + E'_{c.m.} \qquad (2)$$

where E_{in} refers to the initial relative kinetic energy of the reactants. For the present discussion, we restrict ourselves to small (essentially thermal) initial energies E_{in}, since those are of central interest to molecular reaction dynamics. Processes at higher translational energies have been reviewed, e.g. by Kempter (1975). The older material on the quenching process (1) at thermal energies has been thoroughly reviewed on several occasions, e.g. by Alkemade and Zeegers (1971), Lijnse (1972, 1973), Nikitin (1975), Krause (1975), and Barker and Weston (1976). Also, the more recent findings including a variety of other species in (1) have been compiled and discussed, e.g. by Lemont and Flynn (1977), Bersohn (1979), and Donovan (1979). An up-to-date survey of the whole field of quenching of electronically excited metal atoms is given by Breckenridge and Umemoto (1981) elsewhere in this volume. In spite of intensive studies of the process (1) since 1926, it has only been during the last few years that one has actually been able to measure the energy ΔE_{vibrot} transferred to the molecule according to (2). Two approaches have been successful: laser absorption studies of the molecule after collision (Hsu and Lin, 1976, 1980) and molecular beam studies of the kinetic energy $E'_{c.m.}$ after collision (Hertel et al., 1976, 1977a,b; Silver et al., 1977, 1979; Hertel and Reiland, 1981). Some of the recent results are summarized and critically compared with earlier findings and theoretical predictions in Hertel (1981). The rapid progress in the field, in particular in view of the theoretical understanding gained within the last two years, makes it necessary to modify or completely revise some of the earlier ideas taken as the basis of discussion in the last-mentioned review, which was originally written in 1977. The impact of the experimental data, the molecular beam work in particular, has stimulated strong theoretical efforts towards *ab initio* calculations of the potential energy surfaces involved in the E–V,R,T process (1). Results have now emerged for N_2 (Habitz, 1980) and H_2 (Botschwina et al., 1981) as quenching molecules. They are quite surprising when compared with the so-called 'ionic intermediate' complex which has long been believed to be responsible for process (1). At the same time, the new potential surfaces allow a very simple and direct qualitative interpretation of the experimentally observed energy transfer ΔE_{vibrot} by a mechanism which we shall name 'bond-stretch attraction', and this leads us to designate all dynamical calculations to date as semistatistical. Based on the new potential surfaces, a full dynamical attack on the problem seems to be within reach either quantum mechanically or semiclassically. However, although the adiabatic dynamics of three atoms on one potential surface, on the one hand (see e.g. Loesch, 1980), and non-adiabatic transitions for diatomic collision processes, on the other (see e.g. Barat, 1979, and further references therein), are now fairly well understood, the non-adiabatic problem under consideration here, involving at least three atoms, is of considerably higher complexity. So it will certainly need a few more years, guided by adequate experimental progress, before a full *ab initio* calculation of the E–V, R, T process may be possible even in the simplest cases like $Li^* + H_2$ or $Na^* + H_2$, N_2.

The above arguments make it obvious that at present no complete review of the subject is required or possible. However, the rapid development of the field necessitates a progress report, which we shall try to give here. We start by briefly recalling earlier data and theoretical models. We then compare different methods to obtain information on the energy distribution after a quenching reaction and critically compare results of the last few years. We then discuss as a model case the quenching of $Na(3^2P)$ by H_2 and N_2 in terms of the new 'bond-stretch attraction' mechanism as it becomes apparent from *ab initio* potential surface calculations. We also analyse older measurements on temperature dependences of the quenching cross section in terms of the new model. After that, we discuss some most recent beam studies on angular distributions for the quenching cross sections. In connection with investigations of larger quenching molecules, we introduce a statistical classification of the process. Finally, a comparison is made between the new model and the old ionic intermediate picture. At the end, some open questions are pointed out, especially in connection with quenching studies on aligned sodium atoms.

II. QUENCHING CROSS SECTIONS AND THE IONIC INTERMEDIATE

A. Experimental Findings

Most investigations before 1976 on the quenching of resonance radiation of excited alkali atoms were concerned exclusively with the determination of total rate constants. By studying either the fluorescense intensity as a function of quenching gas pressure (Stern–Volmer relation) at steady-state conditions or by measuring the effective lifetime of the excited atoms in time-dependent

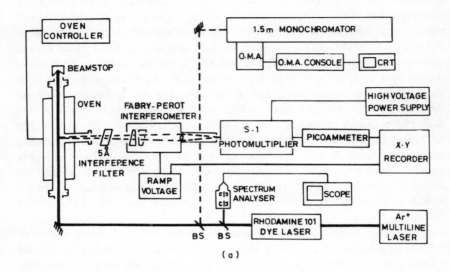

(a)

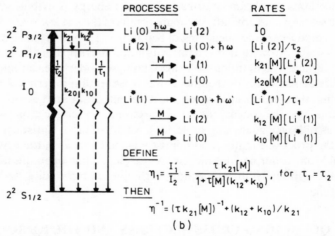

(b)

Fig. 1. (a) Apparatus for dye laser excitation of atomic lithium and for interferometric detection of resonance and sensitized (i.e. after fine-structure transition) fluorescences. Thick solid line segments: Ar^+ and dye laser outputs. BS, beam splitters for dye laser output monitoring; OMA, optical multichannel analyser; CRT, cathode ray tube display. From Elward-Berry and Berry (1980a). Reproduced by permission of the American Institute of Physics. (b) Kinetic model for the determination of quenching and fine-structure transition rate coefficients (k_{ij}) from the ratio of resonance (I_2) to sensitized (I_1') fluorescence intensities. The figure is drawn for the case of $Li^*(2^2P_{3/2})$ excitation. Subscripts 0, 1, 2, denote $2^2S_{1/2}$, $2^2P_{1/2}$, and $2^2P_{3/2}$ states, respectively. From Elward-Berry and Berry (1980b). Reproduced by permission of the American Institute of Physics.

studies, a wealth of data has been obtained in bulk gas experiments. Both gas cells and flames have been used for investigation of temperature dependences. The excitation of the alkali atom is achieved either by optical excitation (dye lasers are used in more recent experiments) or by photodissociation of alkali halides. For details, the reader is referred to the reviews mentioned in the introduction. As a typical steady-state experiment, using advanced modern techniques, we show schematically in Fig. 1(a) a recent experiment by Elward-Berry and Berry (1980a, b). There both fine-structure transition and quenching rates can be measured. The kinetic scheme adopted for the Li $(2^2P_{1/2,3/2} + H_2, D_2)$ interaction is illustrated in Fig. 1(b).

The total cross sections found experimentally for the quenching of the first excited ^2P state in alkali atoms by molecules are large, ranging from 8 Å^2 for Na* + D$_2$ up to several hundred Å^2 for larger organic molecules. All diatomic molecules are good quenchers, as are a variety of polyatomic molecules with double or triple bonds. We shall discuss larger molecules in Sec. V B. Here we note that, in contrast to molecules, rare-gas atoms are very poor quenchers with cross sections far below 1 Å^2.

The experimental data show significant scatter, and when comparing data one has to be aware of two problems. First, it is not always possible to evaluate the rate equations for all possible processes uniquely without additional assumptions. Secondly, one usually measures velocity-averaged quenching cross sections $\sigma_Q = k_Q/\langle v \rangle$ derived from an average rate constant k_Q and an average velocity $\langle v \rangle$. Usually, a significant temperature dependence of the quenching cross sections is found (see e.g. Barker and Weston, 1976; Lin and Weston, 1976) and the experimental conditions thus determine critically the result of the averaging. Discrepancies between results from optically excited atoms in gas cells and flames, on the one hand, and from atoms produced in photodissociation, on the other, indicate that the initial rotational population of the quenching molecules also plays an important role. Nevertheless, important information is given by studies of energy dependences of σ_Q which will be discussed in Sec. IV B in the light of the new bond-stretch attraction model.

B. Earlier Theoretical Interpretations

The negligible cross section for alkali quenching by rare-gas atoms is easily understood in terms of the Massey adiabatic criterion (see e.g. Nikitin, 1974). The effective collision time $\tau_{col} = a/v$ is here typically of the order of 10^{-12} to 10^{-13} s and is thus much longer than the atomic timescale for the electronic transition, $\tau_{tr} = h/E_{ex} \sim 3 \times 10^{-16}$s.

Thus the process is expected to be adiabatic with a Massey parameter of $\xi = E_{ex}a/hv \sim 10^3$ to 10^4, where a is a typical interaction range, E_{ex} is the electronic excitation energy of the alkali atom, and v is the relative velocity of the reactants.

In contrast, the origin of the large cross section for quenching by molecular species is not immediately obvious. It is apparent that the possibility of depositing energy in internal degrees of freedom of the molecule (eqs. (1) and (2)) may be responsible. Early speculations on a resonant energy transfer mechanism, where the electronic excitation energy would have to be nearly equal to some vibrational energy level of the molecule, had to be discarded for quenching of excited alkali atoms on the grounds of isotope studies which showed no significant effect (Bästlein et al., 1969). It should be mentioned at this point that more recent observations of resonant E–V, R energy transfer in the quenching of atoms like $Te(5^3P_{1,0})$, $I(5^2P_{1/2})$, $Te(6^2P_{3/2})$ to their respective ground states 5^3P_2, $5^2P_{3/2}$, $6^2P_{1/2}$ (see e.g. Donovan, 1979; Donovan et al., 1979) seems an entirely different matter. The quenching cross sections observed in this case are orders of magnitude less than for quenching of the first excited alkali $2^2P_{1/2,3/2}$ levels or of mercury $Hg(6^3P_{1,0})$ (Horiguchi and Tsuchiya, 1979) and of other atoms where a change in atomic orbital angular momentum is involved. The quenching in these resonant cases is explained as an effect of long-range multipolar interactions (French and Lawley, 1977) and the 'electronic' transitions observed are between different levels of the same multiplet, split by a very large spin–orbit interaction. Therefore we shall not discuss these processes any further here.

The first semiquantitative attempts to understand the quenching of alkali resonance radiation were made by Bjerre and Nikitin (1967), who recognized that a curve-crossing mechanism must be responsible for the large quenching cross sections. They adopted the harpoon model originally put forward by Polanyi (1932, 1949), Magee (1940), Magee and Ri (1941), and Laidler (1942), and fully developed for alkali halide reactions by Herschbach (1966). One assumes that a so-called 'ionic intermediate' complex provides the attractive potential necessary to mediate between excited- and ground-state potentials, both of which are assumed to be essentially repulsive. Typical potential energy curves for reaction (1) in this model are shown in Fig. 2. A quenching process in this picture would thus be seen as a two-step process

$$Na^* + AB \overset{a}{\rightarrow} Na^+AB^- \overset{b}{\rightarrow} Na + AB^\#,$$

the transitions to and from the ionic intermediate occurring at the crossings a and b, respectively. The transition probabilities at the crossings may be obtained by a Landau–Zener formula, but when performing a full dynamical calculation one has to take account of all internuclear coordinates. In the case of a diatomic quencher, e.g. in the $Na^* + N_2$ problem, three coordinates are important: the distance R between Na and N_2, the vibrational coordinate r between the two N atoms, and the angle γ between the N_2 internuclear axis and the N_2–Na axis. The latter, connected with molecular rotation, has been completely ignored in all calculations hitherto reported. The remaining dynamical problem has been treated by Bjerre and Nikitin (1967) in a classical

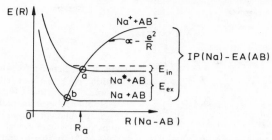

Fig. 2. Potential energy scheme for old quenching model. The 'ionic intermediate' potential is essentially taken as a Coulomb attraction with an asymptotic energy given by the alkali ionization potential IP and the electron affinity EA of the molecule.

trajectory calculation on the two-dimensional surface, and by Bauer, Fisher, and Gilmore (1969) quantum mechanically in a one-dimensional manner also by assuming vibrational adiabaticity. The vibrational wavefunctions of the molecule are factored and one separate potential curve belongs to each vibrational state of the ionic intermediate as well as to the electronic ground state. Fig. 2 now becomes a whole grid of curves and crossings along which the internuclear motion evolves similar to the fall of a ball on a Galton's board. The model of Bauer, Fisher, and Gilmore (1969) (the BFG model) has been expanded and modified (see e.g. Fisher and Smith, 1970, 1971, 1972; Barker, 1976; and references therein). A somewhat radical consequent step along these lines has been taken by Bottcher and Sukumar (1977) who treated the whole problem essentially as scattering of the sodium electron at the quenching molecule, making use of electron scattering data (Schulz, 1973) where low-lying electron + molecule resonances are found (i.e. the electron affinity is less than zero, EA < 0, in Fig. 2).

All these approaches were reasonable steps towards an understanding of the E–V transfer. Their possibilities are discussed together with methods for computation of excited state potentials by Bottcher (1980). Apart from the total neglect of molecular rotation, the major problem with the BFG model was the completely speculative nature of the potential surfaces and coupling elements. Molecular ions such as N_2^-, H_2^-, CO^- are known only as short-lived resonances, and one was forced to argue about the stabilizing influence of the alkali ion nearby. The outcome of these early calculations may be summarized as follows:

(a) The order of magnitude of the quenching cross sections was reproduced correctly, its large value being given essentially by πR_a^2, where R_a is the location of the crossing a in Fig. 2.
(b) Without substantial modifications (Barker, 1976) the observed tempera-

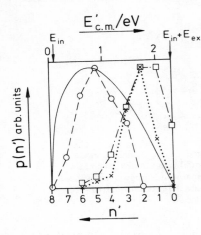

Fig. 3. Predictions for the distribution $p(n')$ of final vibrational quantum numbers n' for $Na(3^2P)$ + N_2 quenching at thermal energies E_{in} from different adoptions of the ionic intermediate model: BFG, —$\bigcirc$——$\bigcirc$—; Bjerre and Nikitin, ...X......X....; Bottcher and Sukumar, —$\square$—·—$\square$—. For comparison, a statistical prior translational distribution $p^0(E'_{c.m.})$ is shown (——) as a function of final translational energy $E'_{c.m.}$.

ture dependence of the quenching cross section could not be explained.

(c) Non-resonant distributions for the final vibrational state population of the diatomic quenching molecules were predicted. Their shape, location, and width were strongly dependent on the assumptions for the potential curves used.

As an illustration we show in Fig. 3 a comparison of different theoretical predictions for the final vibrational distribution in $Na^* + N_2$ quenching. For clarity we show only some of the numerous results which, by adopting slightly different potentials, span the whole field of Fig. 3. Thus one might as well compare with predictions from purely statistical theory (e.g. Bernstein and Levine, 1975; and Sec. V B). Since rotation is neglected in the computations, we have to assume that all remaining energy is translational $E'_{c.m.}$ as given by Eq. (2). We have thus plotted a prior translational distribution in Fig. 3 which, lacking any more detailed information, is as convincing as the other predictions.

So much for the older concepts to explain E–V, R, T transfer. We know now that these models have to be revised substantially as discussed in Sec. IV.

III. MEASUREMENTS OF THE ENERGY DISTRIBUTION AFTER QUENCHING

From the above discussion, it is apparent that the crucial experimental test of all calculations is the measurement of the partition of the initially available energy $E_{ex} + E_{in}$ between the final vibrational, rotational, and translational degrees of freedom after the quenching process. A complete analysis has so far not been performed. There are, however, three complementary experimental approaches which have led to detailed information on the quenching dynamics during the last few years. One may analyse the final state distribution of the molecule by spectroscopic techniques. Mainly for intensity reasons such studies have only been carried out in the gas phase and are up to now restricted to IR-active quenching molecules. One may, according to Eq. (2), determine the final kinetic energy $E'_{c.m.}$ contained in the relative motion between the centre of gravity of the molecule and the atom (centre-of-mass (c.m.) system). $E'_{c.m.}$ can be measured in crossed-beam experiments. Finally, one may choose a third way by inverting reaction (1) and observe the fluorescence from atoms excited in collisions with vibrationally hot molecules. We shall now briefly discuss these three schemes.

A. Spectroscopic Studies

The first spectroscopic studies of infrared emission from the $AB^{\#}$ molecule vibrationally excited in quenching collisions were reported by Karl and Polanyi (1963) on the system Hg* + CO. These studies were later refined (Karl *et al.*, 1967a) and extended (Karl *et al.*, 1967b; Heydtmann *et al.*, 1971) but were difficult to interpret quantitatively in terms of vibrational state populations. A complex set of rate equations has to be solved in all these studies and intermolecular vibration–vibration (V–V) transfer influences the vibrational state distribution observed. Fushiki and Tsuchiya (1973) applied a modulation technique in such IR emission studies which allows a time-dependent study of the product state distribution and an extrapolation to the initial time after the collision process. Using this method and studying overtone IR emission, Horiguchi and Tsuchiya (1979) were able to derive vibrational distributions of $CO^{\#}$ and $NO^{\#}$ after quenching reactions with $Hg(6^{3}P_{1})$ and $Hg(6^{3}P_{0})$. The reported vibrational populations are widely distributed over the accessible range of energies, are clearly non-resonant, and show a maximum at intermediate levels. A typical example is shown in Fig. 4 which resembles the type of distributions predicted theoretically for the Na* +

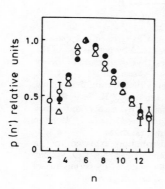

Fig. 4. The determined sets of vibrational distribution of NO in Hg*–NO collisions: ○, determined from the data in the mixture of 1·0 Torr Ar + 0·02 Torr NO with 1 kHz modulation; ●, from the data in the mixture of 1·0 Torr He + 0·02 Torr NO with 1 kHz modulation; Δ, from the data in the mixture of 1·0 Torr Ar + 0·02 Torr NO with 500 Hz modulation. From Horiguchi and Tsuchiya (1979). Reproduced by permission of the American Institute of Physics.

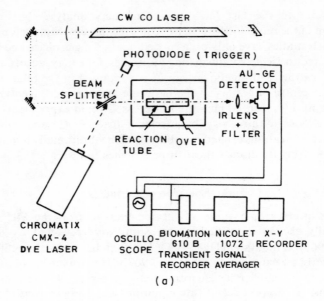

CW CO LASER

PHOTODIODE (TRIGGER)

BEAM
SPLITTER

AU-GE
DETECTOR

IR LENS
+
FILTER

REACTION OVEN
TUBE

CHROMATIX
CMX-4
DYE LASER

OSCILLO- BIOMATION NICOLET X-Y
SCOPE   610 B   1072 RECORDER
           TRANSIENT SIGNAL
           RECORDER AVERAGER

(a)

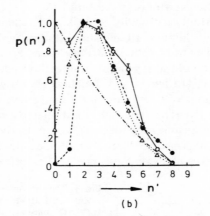

(b)

Fig. 5. (a) Schematic diagram of experimental apparatus of Hsu and Lin (1980). Reproduced by permission of the American Institute of Physics. (b) Comparison of theory and experiment for the Na* + CO reaction: —○—○—, average CO vibrational energy distribution determined experimentally by Hsu and Lin at 528 K; — — ● — — — ● —, distribution calculated by Fisher and Smith according to the BFG model; ...Δ......Δ...., Poisson distribution fitted to experimental data; —·—·— a statistical prior vibrational distribution expected for an energetically completely randomized Na–CO system. From Hsu and Lin (1980). Reproduced by permission of the American Institute of physics.

N_2 case (Fig. 3). The results show that populations reported earlier by Polanyi and coworkers from steady-state observations were strongly determined by V–V transfer after the genuine quenching process.

An elegant alternative approach (Fig. 5(a)) has been followed by Hsu and Lin (1976, 1980). They investigate the vibrational distribution in $CO^\#$ after quenching of $Na(3^2P_{1/2,3/2})$ by infrared laser absorption which allows a high sensitivity. The sodium is excited by a pulsed dye laser, and thus a detailed study of the time dependence of the $CO^\#$ vibrational population after quenching is possible for times greater than 1 μs. Again the evaluation of the rate equations necessitates a backward extrapolation of the data to time zero. The results from Hsu and Lin, shown in Fig. 5(b), display again the clearly non-resonant character of the quenching process, a broad vibrational distribution with a maximum probability for population of the $n' = 2$ vibrational level. For later comparison with the crossed-beam results, we note at this point that all the IR experiments discussed here were performed in the presence of a relatively high (5 Torr) Ar buffer gas pressure. In view of large translational–rotational cross sections, even for the shortest time of observation ($\sim 1\,\mu$s) one may assume a completely thermalized rotational energy distribution for each vibrational level in the quenching molecule. Thus the IR *absorption* experiment monitors a signal proportional to the *whole population* of each vibrational level. It is thus complementary to the *crossed-beam* experiments where the final translational energy of the reactants is measured which corresponds to a defined *total internal energy transfer* ΔE_{vibrot} to the molecule.

The comparison with a statistical prior vibrational distribution $p^0(n')$ in Fig. 5(b) (Bernstein and Levine, 1975) gives no agreement with the experiment. Again we point out that $p^0(n')$ has to be distinguished from $p^0(E'_{\text{c.m.}})$ as displayed in Fig. 3. In contrast, a Poisson distribution as predicted by the linearly forced harmonic oscillator model (see e.g. Nikitin, 1974; Wilson and Levine, 1974) may be used to fit the experimental data by one free parameter, the mean energy transferred. This indicates a sudden energy transfer when after a surface hop the molecule is no longer at its equilibrium distance and is thus 'forced' to oscillate.

The good agreement between Hsu and Lin's data and the results of the BFG model (Fisher and Smith, 1971) must be considered purely accidental since, as discussed in Sec. II, both the potential surfaces and the coupling constants used are completely speculative. Nevertheless, one may take the non-resonant nature of the process together with the overall agreement between measurement, the linearly forced oscillator model, and the BFG model as a strong indication for an underlying curve-crossing mechanism.

Finally, as a practical implication of the above results, we mention the observed population inversion between vibrational levels of the molecule immediately after the quenching process which in principle might be used for an E–V IR laser. As we shall see in Sec. VB, other quenching molecules may be even more favourable in this respect.

B. Crossed-Beam Experiments

The obvious advantages of crossed-beam experiments are single-collision conditions, well defined initial kinetic energies E_{in}, and the possibility of studying the angular ($\Omega_{c.m.}$) dependence as well as the dependence on final kinetic energy $E'_{c.m.}$ of the quenching cross section $d^2\sigma/dE'_{c.m.}\, d\Omega_{c.m.}$. Initial investigations for the prototype system $Na(3^2P_{3/2}) + N_2$ have been reported by Hertel et al. (1976) and were subsequently improved and extended to quenching by H_2, D_2, CO (Hertel et al., 1979) and applied to a number of small molecules (Hertel et al., 1977a). These first studies were restricted to one fixed laboratory angle corresponding to approximately zero-angle scattering in the c.m. system ($\vartheta_{c.m.} \sim 0°$). We do not intend to describe these experiments in detail here (see Hertel et al., 1979; Hertel, 1981). The main features are a supersonic sodium beam crossed at right-angles with a molecular beam effusing from a capillary array, excitation of the Na atoms in the scattering region into the $3^2P_{3/2}\ F = 3$ state by a single-frequency CW dye laser, velocity analysis of the scattered sodium atoms by a mechanical selector, and high-efficiency detection by a hot-wire detector and particle multiplier at one fixed laboratory angle which corresponds to approximately zero c.m. scattering angle $\vartheta_{c.m.}$. The quenching signal (difference 'light on' − 'light off') measured as a function of the sodium velocity after collision is converted by simple kinematic relations into relative c.m. quenching cross sections $d^2\sigma/dE'_{c.m.}\, d\Omega_{c.m.}$ as a function of the translational energy $E'_{c.m.}$ after the quenching process. Some typical results are shown for the model process $Na^* + N_2$, CO and H_2, D_2 in Figs. 6 and 7 respectively.

Kwei and collaborators (Silver et al., 1977, 1979) use an effusive Na oven, multimode laser excitation, supersonic molecular beams seeded with H_2, and a time-of-flight (TOF) measurement of the scattered molecules detected by electron-impact ionization. (For typical data, see Sec. V A, Fig. 19.) In total, the results unfortunately have rather poor statistics. This, together with some uncertainty in the detection efficiency for different vibrational states by electron-impact ionization, necessitates some caution in the interpretation of the deconvoluted data of Silver et al. (1979). In addition, great care has to be taken when analysing beam results for small $E'_{c.m.}$ where quenching cannot be distinguished from totally elastic scattering (the cross section for elastic and rotationally inelastic scattering from excited $Na(^3P)$ may typically be a factor of 10 larger than the quenching cross section (see e.g. Bottcher, 1975). The beam results are therefore only unique for non-resonant energy transfer. This confidence limit is indicated by shaded areas in Fig. 6. Thus we find it difficult to compare the results from our laboratory and Kwei's laboratory. We shall

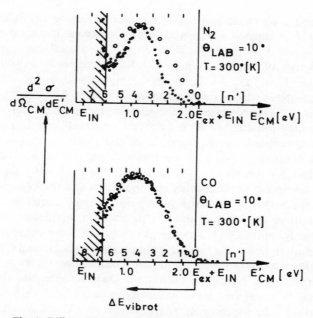

Fig. 6. Differential cross sections for the quenching of Na* by CO and N_2. Comparison of a Poisson distribution (open circles) with experimental results (small dots). From Hertel *et al.* (1979).[1]

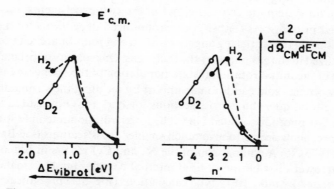

Fig. 7. Relative differential quenching cross section for Na* + H_2 and Na* + D_2 normalized to equal maximum plotted as a function of $\Delta E_{vib\,rot}$ (the amount of energy transferred to the molecule) and as a function of final vibrational quantum number n'. From Hertel *et al.* (1979).

however discuss their pioneering angular distribution measurements in Sec. V A.

Figs. 6 and 7 illustrate some of the essential findings: (1) Clearly non-resonant energy transfer is observed in all four cases, the maximum cross section being found where approximately half of the available energy goes into internal degrees of freedom, and half into translational motion. (2) For quenching by H_2 and D_2, the significance of vibrational modes may be tested since the potential surfaces are identical in both cases. Apparently the energy transfer ΔE_{vibrot} characterizes the dynamics more adequately than the vibrational quantum number. As will become apparent from the next section, ΔE_{vibrot} is determined by the location of the surface crossing. The final vibrational–rotational state population just follows accordingly. Unfortunately, for theoretically the most tractable system H_2, $D_2 + Na^*$ the kinematics are most unfavourable (light molecule) and the systematic errors may be large in Fig. 7, experimental points having been obtained by a deconvolution of the original data.[2] In contrast, no deconvolution was necessary for N_2 and CO. (3) The energy resolution of the experiment would allow structure to be seen in the quenching results from N_2 and CO (Fig. 6) if pure vibrational excitation were dominant. The absence of such structure has to be interpreted as rotational excitation. One can estimate that around $\sim 0 \cdot 1$ eV may be transferred into rotational degrees of freedom. Since this corresponds to $\Delta J \sim 20$, we consider it large in the sense that such changes must strongly influence the collision dynamics. (4) Comparison with Hsu and Lin's (1980) results in the gas phase for CO (Fig. 5(b)) shows a pronounced deviation: while they observe a maximum for $n' = 2$, the beam results for c.m. zero scattering angle in Fig. 6 shown maximum cross section for $n' \sim 4$. Assuming no systematic error in either of the two experiments, this may be interpreted in two ways: either by a pronounced dependence of the energy distributions on the scattering angle or by a strong rotational excitation which is thermalized in the gas phase by the buffer gas. One may also speculate on the influence of the initial rotational state population of the quenching molecule. This view obtains some additional support by recalling the strong discrepancies in the total quenching cross sections for $Na^* + CO$ measured in flames and with fast photodissociated Na^*. The lower rotational temperature in the latter experiments seems to give much smaller cross sections (see Barker and Weston, 1976). (5) A comparison of the N_2 and CO results in Fig. 6 shows a much narrower distribution of final internal energies for N_2, smaller in fact than predicted by most BFG-type calculations (Fig. 3). This indicates a more distinct dynamical process as, for example, assumed by the linearly forced oscillator model (Poisson distribution). In contrast, the CO data are well fitted with such a model distribution

$$p(n') \propto \varepsilon^{n'} \exp(-\varepsilon)/(n'!),$$

where ε gives the average energy transferred to the molecule measured in units of the vibrational energy quantum. The beam data are best fitted by $\varepsilon = 4\cdot8$, while $\varepsilon = 2\cdot8$ reproduces the gas cell data best (the total energy available corresponds to $\varepsilon = 9$).

C. Electronic Excitation by Inverse Quenching (V, R, T–E Transfer)

Excitation of resonance radiation in inverse processes to Eq. (1), i.e. of the type

$$Na(3^2S) + AB(j', n') \rightarrow Na(3^2P) + AB(j, n), \tag{3}$$

has been observed in many experiments. When fast atoms are used (several eV to some keV), the processes are of a rather different nature to those just discussed in that several surface crossings are involved, the translational motion is faster than vibration, etc. The interested reader is referred to Kempter's (1975) review. For recent sophisticated work, see e.g. Blattman et al. (1980), Black et al. (1981) and Loesch and Brieger (1977). In direct connection to quenching are those experiments where hot seeded beams of vibrationally excited molecules are used with kinetic energies not much above the excitation threshold for (3). Such experiments were first performed by Krause, Fricke, and Fite (1972) with the $Na + N_2$ system, were later extended by Larsen et al. (1973) and Loesch (1977) to $K + N_2$ (CO), and were finally complemented by Buck et al. (1980). The problem in all these studies is the averaging over different initial vibrational states and a range of kinetic energies. Thus some

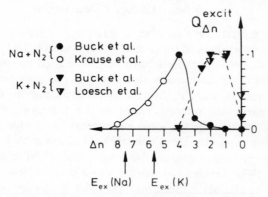

Fig. 8. Electronic excitation (inverse quenching) of Na*(3^2P) and K*(4^2P) in the reaction Na, K + $N_2(n')$ → Na*, K* + $N_2(n)$. Normalized data adopted from Buck et al. (1980). The excitation cross section $Q_{\Delta n}^{excit}$ is dominated by processes $n' = \Delta n \rightarrow n = 0$. The corresponding electronic excitation energies of Na (2·1 eV) and K (1·6 eV) are indicated.

reasonable threshold laws for different $n' \to n$ processes are assumed, while rotational transitions are ignored in the evaluation procedure. The latter consists of a least-squares fit to fluorescence intensities from $Na^*(K^*)$ measured as a function of molecular beam temperature and velocity. The different experiments have been compared by Buck *et al.* (1980) and the results for $K + N_2$ and $Na + N_2$ are shown in Fig. 8. The experiments only allow one to derive averaged cross sections for different combinations $n' \to n$ (with $\Delta n = n' - n$). One estimates however that the $n = 0$ vibrational channel is dominating the experiment. In that sense Fig. 8 can be compared with the data given in Sec. II and III A, B (Figs. 3–7) for the inverse process, Δn corresponding to n' in the quenching experiments. Similarly as in the quenching experiments, one recognizes here a clearly non-resonant energy transfer process. Care should be taken, nevertheless, with a direct quantitative comparison, since experimental averaging over various processes may obscure the final result $Q_{\Delta n}^{\text{excit}}$. We just note that similar to Fig. 6 for $Na + N_2$ a maximum is found at $\Delta n = 4$ $(n' = 4)$ and point out the distinctly smaller energy transfer in the $K + N_2$ system where no quenching crossed-beam experiment has yet been performed. For the $K + CO$ system (not shown in Fig. 8), a behaviour similar to Fig. 7, adequately scaled in energy, seems possible but not conclusive from Buck's results.

IV. POTENTIAL ENERGY SURFACES AND THE BOND-STRETCH ATTRACTION

A. *Ab Initio* Surface Computations

Although potentials for a number of triatomic systems are known today with good accuracy, material on excited states is scarce. The possibilities and difficulties in such computations have been discussed recently by Bottcher (1980). The experimental results described in the previous section have prompted interest in realistic potential surface calculations for some model systems. Of particular interest seems to be the question whether the speculative assumptions on the 'ionic intermediate' gain support or have to be dismissed. Some early Hartree–Fock (HF) calculations by Krauss (1968) for $Li + H_2$ have remained practically unnoticed, while DIM calculations for this system (Tully, 1973; see also Elward-Berry and Berry, 1980b) have still only a semiquantitative character. (See also DIM calculations for $Na + H_2$ (Ivanov, 1979).) Progress in quantum chemistry and a substantial increase in the availability of computational facilities have allowed recent *ab initio* calculations of the system $Na + H_2(D_2)$ (Botschwina, 1979, 1980; Botschwina *et al.*, 1981) and of $Na + N_2$ (Habitz, 1980). While in the former case restricted Hartree–Fock (RHF) calculations and, more rigorously, the correlated electron pair approximation (CEPA) (Meyer, 1973) have been applied, the much bigger system Na + N_2 has been treated by closed-shell HF calculations and Koopman's theorem together with some semiempirical inclusion of the correlations. Without describing the details of these calculations, we shall now discuss the resulting potential surfaces qualitatively as far as is relevant to E–V, R, T energy transfer at thermal initial energies. Both systems are very similar in their essential features. Thus Figs. 9 and 10 are only qualitative (the reader is referred to the original work for numbers) and Figs. 11 and 12 show only the most important parts of the surfaces quantitatively. The potential surfaces for our triatomic Na + M_2 systems depend on three coordinates: the distance R between the sodium atom and the centre of gravity of the M_2 molecule, the distance r between the atoms M–M in the molecule, and the angle γ between the two coordinates. The ground-state Na(3^2S) atom is connected with the $\tilde{X}^2A'$ state, and the excited Na(3^2P) leads to three possible states $\tilde{A}^2A'$, $\tilde{A}'^2A''$, and $\tilde{B}^2A'$ (we have added $\tilde{X}$, $\tilde{A}$, $\tilde{A}'$, and $\tilde{B}$ to the designations $^2A'$, $^2A''$ for the elements of the C_s symmetry group in order to avoid confusion, in agreement e.g. with Elward-Berry and Berry (1980b)). In the limiting case of $\gamma = 0°$ (collinear approach of Na and M_2), the corresponding states in $C_{\infty v}$ symmetry are $\tilde{X}^2\Sigma$, $\tilde{A}^2\Pi$ (the $\tilde{A}$ and $\tilde{A}'$ state are degenerate), and $\tilde{B}^2\Sigma$. For $\gamma = 90°$ in C_{2v} symmetry, these states are $\tilde{X}^2A_1$, $\tilde{A}^2B_2$ (asymptotically the $3p\pi_x$ orbital is parallel to the M_2 molecular axis), $\tilde{A}'^2B_1$ ($3p\pi_y$ perpendicular to both R and r), and $\tilde{B}^2A_1$ (leading to $3p\sigma$ asymptotically).

The calculations show that the C_{2v} symmetry gives the lowest-lying crossing between excited and ground states, which is the relevant one for our process.

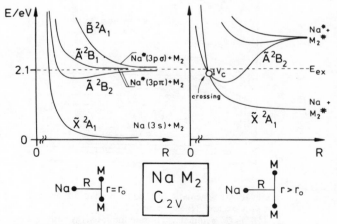

Fig. 9. Qualitative behaviour of NaM$_2$ potentials ($M_2 = H_2$, D_2, N_2) in C_{2v} symmetry as a function of the Na + M_2 distance. Left: Equilibrium distance ($r = r_0$) for the molecule; no crossing is accessible. Right: Stretched molecular bond ($r > r_0$); the $\tilde{X}^2A_1$–$\tilde{A}^2B_2$ crossing is below the asymptotic excitation energy (2·1 eV) for Na(3^2P).

However, as illustrated in the left part of Fig. 9, no crossing can be reached at
thermal initial energies for the equilibrium distance of the molecule. Also, there
is definitely no ionic intermediate to help us in this situation. Nevertheless, a
crossing between the lowest excited state $\tilde{A}^2B_2$ and the ground state becomes
accessible when the molecular bond is stretched (Fig. 9 (right), $r > r_0$).
Although this stretching of course brings us higher up on the potential surface
for large internuclear distances R, for small R the net result of the *bond*

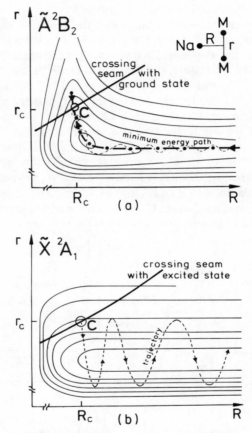

Fig. 10. (a) Schematic of a contour map for the
lowest excited NaM_2 state ($M_2 = H_2$, D_2, N_2)
$\tilde{A}^2B_2$ in C_{2v} symmetry. The minimum-energy path
($-\cdot-$) and the crossing seam ($-$) with the $\tilde{X}^2A_1$
ground state are indicated. The actual particle
trajectory may look as indicated by the dotted line.
(b) Schematic of a contour map for the ground
state of NaM_2 system. The trajectories (.........)
start at the crossing C with the excited state.

stretching is an *attraction* leading to a crossing *below* the asymptotic $Na(3^2P)$ excitation energy. (It should be pointed out that this effect is a consequence of correlation terms properly accounted for by CEPA while the pure RHF calculations show a crossing lying somewhat too high.) Again, there is no ionic intermediate for both the NaH_2 and NaN_2 systems.

It is this crossing which has to be held responsible for the effective transfer of electronic-to-vibrational energy at thermal initial energies. For NaH_2 all other crossings are at substantially higher energies. The mechanism, in a semiclassical picture, is much simpler than the one postulated in terms of the old ionic intermediate model. Since initial kinetic energies are low, the approaching excited three-particle system will essentially follow the minimum-energy path on the $\tilde{A}^2B_2$ surface. There it reaches the crossing with the ground state along an essentially attractive path when the M–M bond is stretched. It will jump with a high probability to the ground state $\tilde{X}^2A_2$, in which the molecular potential however is practically unchanged as the calculations show. Thus, after the surface hop, the molecule M_2 is in a non-equilibrium position and will start to vibrate with an amplitude corresponding to the bondstretching at the crossing point. The nuclear motion is illustrated in more detail by Fig. 10(a), where one can see the minimum-energy path on a contour map for the excited $\tilde{A}^2B_2$ state. The electronic transition may occur near the crossing seam, in particular in the region marked C where the crossing has its minimal energy. For this point, the calculations are summarized in Table I.

In Fig. 10(b) the motion of the particles on the ground-state surface $\tilde{X}^2A_1$ is illustrated schematically. The trajectory starts at the lowest point of the crossing seam C where it has left the $\tilde{A}^2B_2$ surface. One imagines from the shape of the $\tilde{X}^2A_1$ contour plot that trajectories may look as indicated by the dotted line, thus maintaining the bond-stretching distance at C as an amplitude for oscillation in the molecular r coordinate. Thus, the final internal energy of the molecule is determined strongly by the location of the crossing.

A cut through the $\tilde{X}^2A_1$ and $\tilde{A}^2B_2$ surfaces along the minimum-energy path is illustrated *quantitatively* for the NaH_2 CEPA calculations in Fig. 11. This figure has now to replace the old idea of an ionic intermediate (Fig. 2). An alternative way of displaying this is quantitatively shown in Fig. 12 for NaN_2,

TABLE 1 Position R_c, r_c and energy $E_{ex} - V_c$ of the lowest crossing point between $\tilde{X}^2A_1$ and $\tilde{A}^2B_2$ states from *ab initio* calculations. Also indicated are the approximate vibrational M_2 quantum numbers $n(r_c)$ to which r_c at the crossing seam corresponds in the ground state.

System	Reference	R_c(au)	r_c(au)	V_c(eV)	$n(r_c)$
NaH_2 }	{ Botschwina	3·5	2·17	0·1	2
NaD_2 }	{ *et al.* (1981)	3·5	2·17	0·1	2–3
NaN_2	Habitz (1980)	4	2·2	0·35	1–5

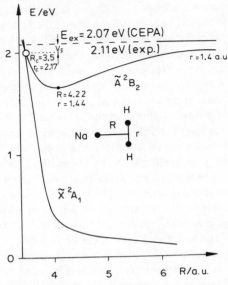

Fig. 11. Potential energy along excited state $\tilde{A}^2B_2$ minimum-energy path for the NaH$_2$ system in C$_{2v}$ symmetry from CEPA calculations. Note that the asymptotic Na(3^2P) excitation energy is 2·11 eV while CEPA gives the slightly different value 2·07 eV. From Botschwina *et al.* (1981).

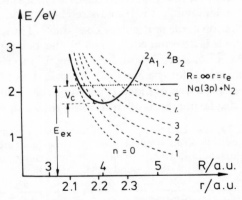

Fig. 12. Potential energy of $\tilde{A}^2B_2$ and $\tilde{X}^2A_1$ states along the crossing seam (———) for the NaN$_2$ system. Also shown is the potential energy of the ground state $\tilde{X}^2A_1$ for $r = r_0$, shifted up by an energy corresponding to the vibrational quantum number n. From Habitz (1980). Reproduced by permission of North-Holland Publ. Comp.

where a cut along the crossing seam (see Fig. 10) is shown. Also indicated are the vibrational states of the molecule, corresponding to the bond stretching of N_2 on the ground-state surface. One sees from Fig. 12 that the transition region is well localized for initially thermal energies to $R = 4 \pm 0.5$ au and $r = 2.2 \pm 0.1$ au. This immediately suggests vibrational excitation in the N_2 due to this transition in the range of $n = 1$ to 5 with a maximum probability for $n \sim 1$ to 3, whereas the experiment gives 3 to 4 (see Fig. 6). This is still a surprisingly good qualitative agreement without a dynamical calculation. Similarly, the corresponding plots for H_2 and D_2 suggest maximum probabilities for $n \sim 2$ and 3, respectively, again in agreement with experiment (Fig. 7).

So far we have only discussed C_{2v} symmetry, which is energetically the most favourable for quenching. In the C_s configuration, the $\tilde{A}^2 A'$, $\tilde{A}'^2 A''$, and $\tilde{B}^2 A'$ states become strongly repulsive near the crossing region as soon as the angular coordinate γ differs significantly from $90°$ (C_{2v}) and the crossing becomes inaccessible at thermal initial kinetic energies. Since most collisions do not occur in or near C_{2v} for approaching collision partners, one might be led to argue that only a very small quenching cross section is thus expected, contrary to observation. However, for small initial kinetic energies, one has to accept in a classical trajectory model that the three-particle system moving near the minimum-energy path will tend to the C_{2v} configuration with the lowest energy. The rotational motion through the angle γ initiated in order to reach this preferable geometry could easily explain the rotational excitation deduced from experiment (Sec. III B). Thus we expect no reduction of the quenching cross section due to initial geometries other than C_{2v}. At larger internuclear distances, finer details of the quenching dynamics may be determined (see Sec. V C).

In $C_{\infty v}$ geometry, the NaH_2 surfaces are most repulsive since a σ_u $1s$ molecular orbital ($2p\sigma_u$ in the united-atom limit) is responsible for the bond-stretch attraction. In the NaN_2 case, however, it is a $\pi_g 2p$ molecular orbital ($3d\pi_g$) instead, and the situation becomes somewhat more complicated: It appears that in C_s geometry an $A' - A''$ crossing occurs which also may contribute significantly to quenching (Archiel and Habitz (1982)). For NaCO on the other hand the collinear approach of Na* to either side of the CO molecule is responsible for the process as shown by recent CI calculations (Bonačić-Koutecký et al. (1982)).

We summarize the *bond-stretch attraction model* for quenching outlined here as follows. E–V, R, T transfer in Na* + M_2 collisions arises from crossings between the excited $\tilde{A}^2 A'$ and the ground state $\tilde{X}^2 A'$ near C_{2v} (and C_s for NaN_2) geometry. These crossings are reached by an essentially attractive minimum-energy path on the excited-state surfaces due to bond stretching of the molecule. After a surface hop, this stretching leads to vibrational excitation, while rotational excitation is induced by a strong anisotropy of the potentials near this crossing.

B. Simple Model for the Magnitude of the Quenching Cross Section

The 'ionic intermediate' model explained the quenching cross sections σ_Q observed experimentally by the large crossing distance R_a (Fig. 2). The crossing of true potential surfaces described above is found at significantly smaller distances R_c. However, since the excited-state potential along the minimum-energy path is attractive even for large distances, it is not R_c exclusively that determines σ_Q. Rather, for small kinetic energies, the attraction will lead most of the trajectories to the crossing where, with a certain probability w, a quenching transition occurs unless the centrifugal potential $E_{in}(b/R)^2$ prevents such close approach for impact parameters $b > b_{max}$. Such simple models have been discussed, e.g. by Lijnse (1973), Levine and Bernstein (1974), Barker and Weston (1976), and Bersohn (1979). Although strictly they apply only to diatomic systems, one may obtain a crude estimate of the quenching cross sections and even of their energy dependence which was not predictable from the original BFG model. We have to distinguish two cases. The centrifugal barrier at large internuclear distances R may determine the maximal impact parameter. In this case

$$\sigma_Q = w\pi \left(\frac{C_s(s-2)}{2E_{in}} \right)^{2/s} \left(\frac{s}{s-2} \right) \tag{4}$$

(see e.g. Bersohn, 1979) if the asymptotic potential is given by C_s/r^s. This will yield an upper limit to the cross section for very low kinetic energies E_{in}.

At higher E_{in}, the centrifugal potential will move the crossing at R_c (lying an amount V_c below the asymptotic Na(3^2P) energy) up, so that it becomes inaccessible for impact parameters $b > b_{max}$ as given by $-V_c + E_{in}(b_{max}/R_c)^2 > E_{in}$ which leads to an estimated quenching cross section in this so-called absorbing-sphere model of

$$\sigma_Q = w\pi R_c^2 \left(1 + \frac{V_c}{E_{in}} \right). \tag{5}$$

Botschwina et al. (1981) are able to analyse[3] the data compiled by Barker and Weston (1976) in terms of their values for R_c and V_c. Fig. 13(a) shows the results for Na* + H$_2$, D$_2$ quenching neglecting E$_{no}$. By closer inspection of the potential curves (Fig. 11), one is led to use Eq. (5) in the energy range of Fig. 13 for this estimate. It is very satisfactory that the experimental data are reproduced well by reasonable transition probabilities $w = 0.45$ for H$_2$ and 0.33 for D$_2$ as quenching molecules. For Na* quenching by N$_2$, we approximate the potentials of Habitz (1980) for large R by either C_6/R^6 or C_5/R^5 potentials. This can certainly give only a very crude estimate of the true potentials, in particular because the ab initio calculations are not considered accurate for large R. Nevertheless, Fig. 13(b) again illustrates the good qualitative agreement between measured cross sections and model

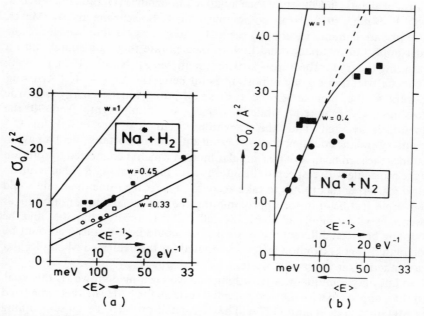

Fig. 13 (a) Energy dependence of quenching across sections. Experimental data adopted from Barker and Weston (1976): $H_2(D_2) + Na^*$ photodissociation, ●(O); cells and flames, ■(□). The full lines indicate the absorbing-sphere model $\sigma_Q = w\pi R_c^2(1 + V_c)/E_{in}$ with $R_c = 3.5$ au and $V_c = 0.1$ eV see Botschwina *et al.* (1981), and w suitably chosen to fit the experimental points. **(b)** Quenching cross section for the $N_2 + Na^*$ system: experimental points, as in Fig. 13(a); straight full line, absorbing-sphere model with $R_c = 4.1$ au and $V_c = 0.35$ eV from Habitz (1980). The bent full curve for small energies is derived from the height of the centrifugal well of larger R by fitting C/R^5 or C/R^6 potentials to the $\tilde{A}^2B_2$ potentials of Habitz.

estimates at different initial kinetic energies. We have used Eqs. (4) or (5) for lower energies depending on which gave the lowest b_{max}. The good agreement with the experimental results using only one fitting parameter $w = 0.4$ is again surprising.

C. Scattering Dynamical Calculations

Certainly, the semiqualitative models discussed above are too simple, in particular when energy transfer measurements or even angular scattering distributions are to be predicted. A full dynamical quantum-mechanical treatment including all coordinates R, r and the rotational coordinate γ has not yet been attempted. A prerequisite for any quantitative computation is the knowledge of not only the potential surfaces but also the nature and magnitude of coupling potentials near the surface crossings, especially for the $\tilde{X}^2A'$ and $\tilde{A}^2A'$ states. Work is still in progress to determine these coupling matrix elements and to study their relative importance (Habitz and Vovota, 1980; P. Habitz, 1980 private communication; P. Botschwina and W. Meyer, 1980 private communication). It seems by no means obvious which are the most important couplings and how to incorporate them adequately into a scattering theory for this three-particle system (see e.g. Nikitin, 1974, p. 148ff). We focus attention on one important point here. The $\bar{X}^2 A' - \bar{A}^2 A'$ crossing relevant to the quenching process at small initial kinetic energy is a true crossing in C_{2v} symmetry, while for angles γ deviating slightly from $90°$ the crossing is avoided. Thus the transitions occur near, not exactly in, C_{2v} geometry and the coupling due to $\delta/\delta\gamma$ must play an important role. This again outlines the significance of this angular motion which is connected to rotation of the molecule with respect to the internuclear motion along R. Thus a full *ab initio* scattering theory has to take account of γ. At the same time, one could find some justification for approximating the molecular motion along r as essentially vibrationally adiabatic (as well as electronically adiabatic), and the coupling between different vibrational states could be taken into account by adequate Franck–Condon factors computed for the deformed M_2 molecular wavefunctions in the NaM_2 system.

So far, only semiquantitative attempts have been made towards this goal. The first approach with the new potential surfaces for NaH_2 has been reported by McGuire and Bellum (1979). They treat the problem by close-coupling calculations in a one-dimensional manner using potential curve matrix elements

$$V_{nn}(R) = \langle n|E(R,r)|n \rangle$$

for each vibrational level n, thus assuming vibrational adiabaticity. The resulting 'grid' of curve crossings is reproduced in Fig. 14. This picture is equivalent to the old BFG multiple crossings, however now using quantum mechanics and realistic potentials. There are, however, some serious shortcomings in this treatment, even if one accepts the vibrational parametrization. The main objection is the complete neglect of molecular rotation (sudden approximation), which according to all that has been said above is not justified, especially in view of both the strong asymmetry of the potentials $E(\gamma = 0°) - E(\gamma = 90°) \gg E_{in}(!)$ near the crossing and the timescales involved ($t_{rot} \sim 10^{-13}-10^{-14}$ and $t_{col} \sim 10^{-12}-10^{-13}$ s). Other limitations of these calculations are the unknown coupling constants and potential matrix elements (which had to be guessed from reasonable assumptions) and the limited number of vibrational states incorporated in the close-coupling system. Furthermore, at the time of these computations, only the RHF surfaces were available and these give a crossing position too high, thus too small $V_{nn'}$, and consequently too low quenching cross sections. In fact, the results give only $\sigma_Q \sim 0.2 \text{Å}^2$ but lead to a pronounced maximum probability for $n' = 2$ excitation, in fair agreement with the experimental observations.

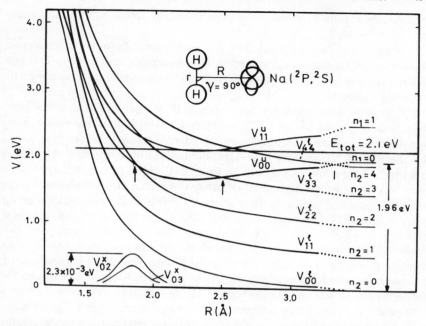

Fig. 14. Diagonal vibrational matrix elements of the $\tilde{X}^2 A_1$ groundstate and $A^2 B_2$ excited-state surfaces for the C_{2v} geometry. The matrix elements of the lower, $V^l_{n'n'}(R)$, and upper, $V^u_{nn}(R)$, surfaces go asymptotically to the internal energy in the respective channels, so that the local kinetic energy at any point R for each channel can be read down from the E_{tot} = 2.1 eV line. The Na($S-P$) excitation energy is 1.96 eV in the Hartree–Fock calculation, and the collision energy in the initial excited channel is 0.1 eV. The arrows mark points of kinetic energy resonance, and two examples of the non-adiabatic electronic plus vibrational coupling $V^x_{nn'}(R)$ are given. From McGuire and Bellum (1979) (note $n_1 = n, n_2 = n'$ in our notation). Reproduced by permission of the American Institute of Physics.

One sees this immediately from Fig. 14, since only for the $n = 0, n' = 2$ crossing is the coupling $V_{nn'}$ non-vanishing.

In spite of its preliminary character, it is fair to say that McGuire and Bellum's (1979) work is the first major step towards a realistic scattering calculation for this problem since Bauer *et al.* (1969), based of course on the pioneering *ab initio* calculations for the potential surfaces (Botschwina, 1980). The scattering calculations are currently being improved and extended (P. Botschwina and W. Meyer, 1980 private communication) and lead to a number of interesting results. For example, closed channels in the excited state as high as $n = 4$ play a dominant role by changing the close-coupling results by up to four orders of magnitude! Another finding is that not only does the location of the surface crossing determine the final vibrational state distribution but so also does the magnitude of the coupling. Another approach is currently being followed by Habitz (1980 private communication), who uses classical trajectory calculations with surface hopping for the system NaN_2.

Preliminary results reproduce the experimentally found vibrational populations of the N_2 fairly well.

It is apparent that only the fruitful collaboration between experimental physics, quantum chemistry, and scattering theory can lead to a gradual and steady improvement of our understanding of this important non-adiabatic three-body process.

Yet another new approach for treating E–V energy transfer has been proposed by Gislason et al. (1979, 1981). Again, vibrational decoupling from translational motion is assumed and the latter is treated classically whereas the vibration is followed by time-dependent quantum mechanics. At surface crossings where a hop can occur, the Franck–Condon principle is applied to construct the new time-dependent vibrational wavefunction which leads to the final state populations. Gislason et al. (1979) were able to interpret interesting energy-dependent effects in E–V transfer. Quantitative use of the new potential surfaces should allow great insight into the quenching processes discussed here, especially when compared with the close-coupling calculations mentioned above. The time-dependent approach resembles somewhat the boomerang model successfully applied to electron–molecule resonance scattering (Birtwistle and Herzenberg, 1971), and thus reminds us not to forget completely the connection between $e + M_2$ resonance scattering and the E–V, R, T energy transfer process, which the old ionic intermediate model relates simply via the M_2^- ion.

D. The Bond-Stretch Attraction and the 'Ionic Intermediate'

In spite of the great success of the ab initio calculations discussed above, in the near future they will be limited to a few not too complicated systems. For heavier or larger quenching molecules, semiquantitative viewpoints will have to guide our understanding for the time being. Thus it seems useful to discuss the possibilities and shortcomings of the earlier ionic intermediate model in the light of the new ab initio potential surfaces. It is clear that the ionic model was and still is the appropriate description of processes involving molecules for which a stable molecular ion exists. This is probably most pronounced for collisions between an alkali atom and a halogen molecule, which have been described very successfully in this way (see e.g. Los and Kleyn (1979) for an excellent review on ion-pair formation in such systems). What remains of the ionic intermediate character in our model systems NaH_2, NaN_2? Clearly, there is no $1/r$ attraction found as illustrated by Fig. 11. Nevertheless, the calculations show that the bond-stretch attraction is caused by significant charge transfer from Na* to M_2. At the lowest point of the $\tilde{X}^2A_1 - \tilde{A}^2B_2$ crossing seam, one finds that 50% of the $3p_x$ orbital is localized near the molecule in the Na + H_2 case. This is illustrated by Fig. 15, a contour plot of equal charge densities. While for the minimum-energy configuration of the

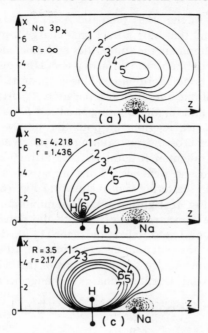

Fig. 15. Contour maps of the charge density distribution for the NaH$_2$ molecular orbital of the outer electron in the $\tilde{A}^2B_2$ state at different internuclear separations R and intramolecular spacings r. At infinite distance between Na and H$_2$, the orbital converges to sodium $3p_x$ (for convenience we have just plotted the upper half part). Near the minimum of the $\tilde{A}^2B_2$ state, a charge transfer to the H$_2$ becomes significant, while at the lowest crossing point $(R = 3\cdot5, r = 2\cdot17)$ the main part of the p_x shape seems to be localized near the stretched H$_2$. Note however that only $\sim 50\%$ of the electron charge is transferred. From Botschwina *et al.* (1981).

$\tilde{A}^2A'$ surface (middle, Fig. 15) the charge transfer is already significant, at the lowest crossing (bottom, Fig. 15) the p_x character of the charge distribution around the molecule becomes clearly visible. We also recall that the well known $\sigma_u 1s\,H_2^-$ resonance ($2p\sigma_u$ in the united-atom limit) seen in $e + H_2$ scattering between 2 and 5 eV (see the fundamental review by Schulz, 1973) clearly correlates with the $3p_x$ orbital in the $\tilde{A}^2B_2$ state (x being parallel to the H$_2$ axis). The situation is similar for the Na* + N$_2$ system: the $\pi_g 2p\,N_2^-$ resonance ($3d\pi_g$ in the united-atom limit) has a significant overlap with the $3p_x$

orbital connected with the $\tilde{A}^2B_2$ (and the $\tilde{A}^2\Pi$) excited state. We should, however, bear in mind that only a fractional charge transfer occurs.

Another way to illuminate the similarities and differences between the old ionic intermediate and the *ab initio* surfaces is to cut through the CEPA surface along the molecular bond coordinate r. This is illustrated in Fig. 16 (full and dashed curves) for infinite internuclear separation R (left) reproducing the H_2 molecular potential and for the crossing situation $R = R_c$ (right). While the ground state $\tilde{X}^2A_1$ leaves the H_2 potential nearly unchanged in the neighbourhood of the crossing, just shifting the energy upwards, the excited state deforms the H_2 potential. In addition to the attraction (shift downwards), the bond is stretched and weakened. For comparison, the H_2^- potential as derived from electron scattering is also shown in Fig. 16 (dotted curve) and shifted downwards (Fig. 16, right) to give equal minimum energy to the $\tilde{A}^2B_2$ cut. We observe the same tendencies in both the potential curves but no quantitative agreement. The 'stabilization' of the H_2^- ion due to the presence of the Na^+ core moves the potential energy minimum to smaller $r = r_{min}$ and increases the force constant, to speak in terms of the old ionic intermediate model, in addition to only fractional charge exchange between Na^* and H_2. A

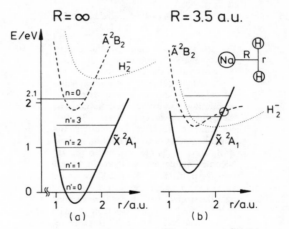

Fig. 16. Cuts through the surface $\tilde{X}^2A_1$ (——) and $\tilde{A}^2B_2$ (— — —) relevant for Na* quenching by H_2 along the molecular bond distance r. While the ground state H_2 potential is nearly undeformed in the presence of the Na even near the crossing ($R = 3.5$ au) and is only shifted upwards, the excited Na* deforms the H_2 potential substantially by bond-stretch attraction, so that a low-lying crossing (circled) becomes accessible for quenching (adapted from CEPA calculation by Botschwina *et al.* (1981). For comparison the H_2 potential is shown (.....) as derived by Schulz (1973) from experimental e + H_2 scattering (Ehrhardt *et al.*, 1968) at original position (left) and aligned with the minimum of the $\tilde{A}^2B_2$ curve on the crossing geometry (right).

similar behaviour is observed for the $Na^* + N_2$ system with a significantly better agreement, however, between N_2^- potential and the cut through $\tilde{A}^2 B_2$. Whereas for N_2^- the bond stretching with respect to the N_2 equilibrium r_e is $r_{min} - r_e = 0.17$ au (Huber and Herzberg, 1979), Habitz (1980) finds ~ 0.15 au in the excited-state surface for $R = R_c$.

In conclusion, one has to realise that the ionic intermediate model in connection with electron–molecule resonance scattering (M_2^-) gives only a first, not necessarily reliable, estimate for the shape of the excited-state molecular potential. It is probably more adequate for lower-lying resonances or even slightly bound negative molecular ions. In any case, one has to be careful with a too literal application of such models based on estimates of the 'ionic intermediate' and direct application of Franck–Condon factors derived from vertical transitions in the $e + M_2$ scattering (Rost, 1977; Taylor, 1979, 1980; Bottcher and Sukumar, 1977).[4] Apart from the deformation of the M_2^- in the presence of the Na^+ core, one cannot assume the full 2.1 eV electronic excitation energy to be available at the crossing due to the upward shift of the ground state and the downward shift of the excited state. The main difficulty with such models is, however, that the bond stretching on the excited-state surface occurs essentially adiabatically and the transitions are localized to the crossing region (at $r \sim 2.17$ au for the H_2 case, Fig. 16) rather than to the ground-state equilibrium distance ($r_e \sim 1.4$ au for H_2) assumed by application of vertical Franck–Condon transitions as in $e–M_2$ scattering. We do not fail to recognize the connection between electron–molecule scattering and E–V, R, T transfer, but quantitative predictions do not seem possible in a direct way.

V. ANGULAR DISTRIBUTIONS AND OTHER NEW DATA

In this final section, we briefly discuss some more recent experimental observations in terms of the bond-stretch attraction model for E–V, R, T energy transfer described above.

A. Angular Dependence of the Differential Quenching Cross Section

Experience from molecular beam work for a variety of scattering problems leads us to expect the most detailed information from a study of the differential quenching cross section $d\sigma/d\Omega_{c.m.}$ as a function of scattering angle $\vartheta_{c.m.}$ or even of $d^2\sigma/dE'_{c.m.} \, d\Omega_{c.m.}$ as a function of both c.m. scattering angle $\vartheta_{c.m.}$ and final translational c.m. energy $E'_{c.m.}$. Silver et al. (1979) were the first to report studies of $d\sigma/d\Omega_{c.m.} = I(\vartheta)$. They evaluated their data by assuming no dependence of $I(\vartheta)$ on $E'_{c.m.}$. As mentioned in Sec. III, the statistics of the data points were not too convincing, but a number of different quenching gases were investigated. Recent experiments in our laboratory on $Na^* + N_2$ (Reiland et al., 1982c), using the technique described by Hertel et al. (1979) but

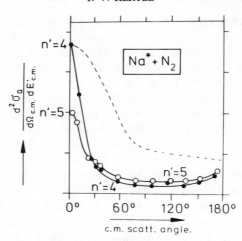

Fig. 17. Measured double differential quenching cross section $d^2\sigma/d\Omega_{c.m.}dE'_{c.m.}$ for the $Na(3^2P) + N_2$ system as a function of c.m. scattering angle $\vartheta_{c.m.}$ at two different values of energy transfer $\Delta E_{vibrot} = n'\hbar\omega_0$. Points ● and circles ○ indicate the values obtained from direct laboratory to c.m. conversion of the raw data. See Reiland *et al.* (1982c). For comparison, the best fit of the average angular distribution from Silver *et al.* (1979) is also indicated (dashed line).

with a supersonic molecular target beam as well as a Na jet beam, confirm the forward peaking of $d^2\sigma/dE_{c.m.}d\Omega_{c.m.}$ reported by Silver *et al.* Our beam geometry and the good statistics of the raw data allow the direct conversion from laboratory frame to c.m. frame without significant error. Fig. 17 shows some of our results for the $Na + N_2$ system. We observe a strongly forward-peaked differential quenching cross section for all values of the energy transfer ΔE_{vibrot}, much stronger than that reported by Silver *et al.* Also, we see that the angular distribution is not independent of ΔE_{vibrot}. While at $\vartheta_{c.m.} = 0$ the maximum differential cross section is found for $\Delta E_{vibrot} \sim 3\hbar\omega_0$, for backward scattering $\Delta E_{vibrot} \sim 4\hbar\omega_0$ dominates.

At present, it does not seem immediately obvious how to explain these angular distributions by the bond-stretch attraction in a direct way. Since the centrifugal potential allows classically only impact parameters up to $b_{max} \sim (2-3)R_c$ (see Sec. IVB) and since after the surface hop the potential is repulsive (see Figs. 9–12), one cannot immediately explain the strong forward peak. In fact approximate preliminary trajectory studies (Habitz, 1980 private communication) predict a maximum quenching cross section for $\vartheta_{c.m.} > 90°$. Even McGuire and Bellum's (1979) close-coupling results show no strong forward scattering for NaH_2. On the other hand, Silver *et al.* (1979) observe forward peaking not only for the Na* N_2 system but also for quenching by O_2, CO, and NO.[5]

Since both the experimental work and theoretical interpretation are still in progress, we do not want to speculate too much about possible explanations. However, we can see no 'direct' mechanism by which this process is

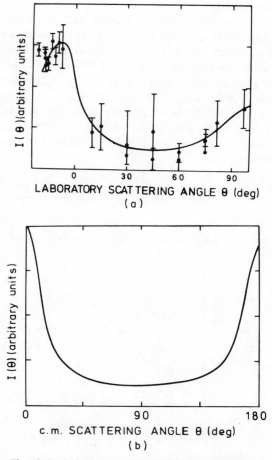

Fig. 18. (a) Laboratory angular distribution of NO scattered inelastically from Na*. The full curve is an analytic fit. Error bars indicate one standard deviation. From Silver *et al.* (1979). Reproduced by permission of the American Institute of Physics. (b) Best-fit c.m. angular distribution of NO scattered inelastically from Na*. From Silver *et al.* (1979). Reproduced by permission of the American Institute of Physics.

responsible for the forward peak as presumed by Silver *et al.* (1979). Rather, one might discuss the angular distributions as caused by a short-lived complex (rotational time a few times longer than complex lifetime). In this model the different behaviour of angular distribution for $\Delta E_{\text{vibrot}} = 4\hbar\omega_0$ and $5\hbar\omega_0$ would indicate a longer complex lifetime in the latter case, which seems reasonable in view of the smaller total probabilities to find $\Delta E_{\text{vibrot}} = 5\hbar\omega_0$. One may even suspect an indication of a rise in the backward direction for the latter distribution in Fig. 17.

NO is a special case. The angular distributions determined by Silver *et al.* (1979) clearly illustrate (Figs. 18(a) and (b)) by the strong backward peak that here a complex is formed with a lifetime similar to its rotational time.[6] Silver *et al.* discuss this complex formation in detail. We simply note that NO seems to be a case where the ionic character at the crossing region is dominant (stable NO^-) and thus a deep well is formed. This also explains why NO is predominantly populated in low vibrational states after the quenching process (Fig. 19). The strong attraction leads to a very low crossing of excited and ground states even without strong bond stretching, and consequently low vibrational states are excited.

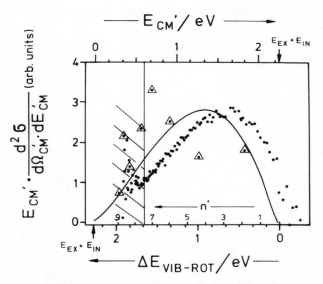

Fig. 19. Energy-transfer spectrum for the quenching process

$$\text{Na}^*(3^2P) + \text{NO}(n = 0, j) \rightarrow \text{Na}(3^2S) + \text{NO}(n', j')$$

at forward scattering in the c.m. system. Experimental data (full points) are shown together with those of Silver *et al.* (1979) (triangles) suitably scaled for comparison. The full curve gives the prior distribution for a diatomic molecule. Adapted from Hertel and Reiland (1981).

B. Triatomic and Polyatomic Quenching Molecules

Energy transfer to a number of other quenching molecules has been studied (Hertel *et al.*, 1977a). Recently a systematic study of several organic molecules with double bonds has been published (Hertel and Reiland, 1981). Organic molecules with double or triple bonds are known to be extremely good quenchers of alkali resonance radiation (Lijnse, 1973). The bond-stretch attraction force operating on the double (triple) bond can again be held responsible. A comparison with statistical prior distributions for diatomic or

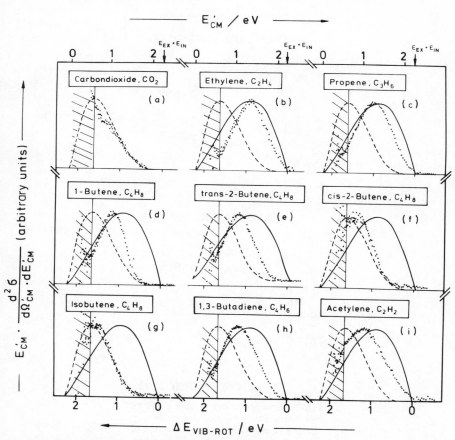

Fig. 20. Relative differential quenching cross sections $Na(3P) + M$ multiplied by $E'_{c.m.}$ as a function of $E'_{c.m.}$ (relative kinetic energy in the c.m. system after collision) and of ΔE_{vibrot} (internal energy transferred to the molecule during the collision), respectively, for different molecules. E_{ex} indicates the excitation energy of the atom and E_{in} the relative kinetic energy before collision; the c.m. scattering angle is approximately $0°$. Experimental data (points) may be compared with prior distributions for diatomic (solid curve) and triatomic (broken curve) quenching molecules. From Hertel and Reiland (1981).

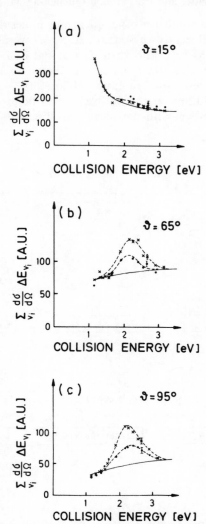

Fig. 21. Energy transfer of kinetic energy by electron scattering into internal energy of *cis-* and *trans*-2-butene as a function of the primary energy for electron scattering angles (a) $\vartheta = 15°$, (b) $\vartheta = 65°$, (c) $\vartheta = 95°$. The crosses represent the results for *cis*-2-butene, the dots are for *trans*-2-butene. From Jung and Kadisch (1981). Reproduced by permission of the Institute of Physics.

triatomic quenchers illustrates that mainly the double (triple) bond and only a fraction of the neighbouring bonds participate in this process (Fig. 20).

At this point, it is interesting to compare the electron scattering data of Jung and Kadisch (1981) for cis- and trans-butene. There, a low-lying resonance was observed and a study of individual vibrational excitations in electron + butene scattering was performed. By summing the energy-weighted differential cross sections, one obtains a measure for the total energy transfer to the molecule. Fig. 21 shows that trans-butene accepts less internal energy than cis-butene by resonant e + M scattering. This appears plausible if one assumes the resonance to be connected with the double bond, since cis-butene has a lower symmetry than trans-butene with respect to the $C=C$ axis. For Na* + cis-(trans-)butene quenching (Figs. 20(f) and 20(e)), one observes a similar behaviour. Whereas the maximum in $E'_{c.m.} d^2\sigma/d\Omega_{c.m.} dE'_{c.m.}$ is observed at $\Delta E_{vibrot} \sim 1.2\,eV$ for trans-butene, the corresponding value $\sim 1.6\,eV$ for cis-butene is significantly larger. Thus we have here a nice case for the qualitative correspondence of e + M resonant scattering and Na* + Mol E–V, R, T transfer.

C. Polarization Studies

The excitation of the Na*(3P) by laser optical pumping allows yet another interesting type of study. By choosing linearly polarized laser light, one prepares sodium predominantly in its $3p_z$ state, z being parallel to the electric vector of the light. By rotating the electric vector with respect to the c.m. scattering system, one can prepare the excited sodium predominantly in either the $3p\sigma_{c.m.}$, the $3p\pi_x$ or the $3p\pi_y$ state. In electron–atom scattering (Hertel and Stoll, 1978), a rotation of the electric vector of the laser light through 90° could ideally give a 100% modulation of the scattering intensity. The $\sigma_{c.m.}$ preparation is however not pure since the electron orbital angular momentum is coupled to electron and nuclear spin. Thus an anisotropy of up to 2·5:1 for $\sigma_{c.m.}$ and $\pi_{c.m.}$ preparation, respectively, can be (and have been) observed in e + Na* scattering. (See also Fisher and Hertel (1982)).

For quenching experiments, such studies have also been reported (Hertel et al., 1977b). The anisotropy is much less than in electron scattering but is significant. From our previous discussion which holds the $\tilde{A}^2 B_2$ state responsible for quenching, one should expect that the Na($3p\pi_x$) state, which correlates with this surface when prepared by laser polarization perpendicular to the c.m. direction in the scattering plane, should give the maximum cross section. The opposite is observed. Both π_x and π_y preparation in the c.m. system give smaller quenching cross sections than the $\sigma_{c.m.}$ state correlating to the $\tilde{B}^2 A'$ state which should not be involved in the quenching (see Fig. 9). A qualitative interpretation of this apparent contradiction given by Hermann and Hertel (1980) is difficult to maintain on quantitative grounds.

Long-range interactions probably have to be held responsible, which induce a strong $\Sigma-\Pi$ coupling. It should be pointed out that this is important for the absolute magnitude of the quenching cross section. A small anisotropy for different alignments of the laser polarization is equivalent to saying that the populations of all three surfaces $\tilde{A}^2 A'$, $\tilde{A}^2 A''$, and $\tilde{B}^2 A'$ contribute to the quenching, and the cross section is not reduced by a factor $1/3$ which would be applicable if only the initial $\tilde{A}^2 A'$ population was effective for quenching. Work is still in progress (Reiland et al., 1982b) and is expected to yield further insight into details of the quenching mechanism.[7]

D. Energy Oscillations in the $Ar^* + N_2$ E–V Transfer Reaction

Before concluding this progress report, we want to draw attention to some related important work which has recently been reported by Cutshall and Muschlitz (1979). They have studied quenching of metastable argon in the reaction

$$Ar^*(^3P_{2,0}) + N_2(X, n = 0) \rightarrow Ar(^1S_0) + N_2^*(O^3\Pi_u, n') \qquad (6)$$

in an angularly integrated crossed-beam study by detecting the fluorescence in the $N_2^* \, C \rightarrow B$ and $C \rightarrow A$ emission band. This reaction had been studied before by using a flowing afterglow (Setser and Stedman, 1970; Kolts et al., 1977; Tonzeau et al., 1977), by pulse radiolysis (Yokohama et al., 1977; Sadeghi and Nguyen, 1977), as well as in molecular beam studies (Lee and Martin, 1975; Parr and Martin, 1978; Sanders et al., 1976; Schweid et al., 1976; Krenos and BelBruno, 1976, 1977). It is known to lead to a product state distribution with a maximum at $n' = 0$ and a high degree of rotational excitation. The recent work reported by Cutshall and Muschlitz (1979) is the first E–V study which not only resolves the product states but also has a well defined initial kinetic energy E_{in} which has been varied between 0·06 eV and 0·41 eV. Their important finding is that the measured ratio between $n' = 0$ and $n' = 1$ populations shows oscillatory behaviour as a function of E_{in}. This process has usually been interpreted in terms of time-dependent perturbation theory ('golden rule') which, however, does not predict such oscillations. Thus one might expect that the process proceeds in a manner similar to our bond-stretch attraction model used for $Na^* + M_2$ quenching. In fact, Gislason et al. (1979) were able to analyse this process by a curve-crossing scheme shown in Fig. 22(a), which is completely equivalent to our Fig. 11 as far as the excited and lower state crossing is concerned. Gislason et al. (1979) propose a sudden change of the molecular bond distance near R_2, and a time-dependent treatment of the vibronic wavefunction then allows a nearly perfect fit to the data of Cutshall and Muschlitz, as seen in Fig. 22(b). These results shed new light on the E–V process and illustrate the need for energy-dependent studies, if possible even angularly resolved studies. The sudden change in the

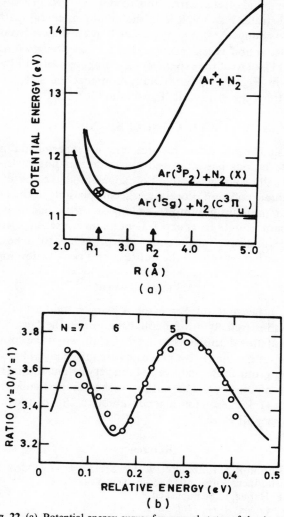

Fig. 22. (a) Potential energy curves for several states of the $Ar + N_2$ system plotted against the Ar–N_2 separation. The zero of energy corresponds to ground state Ar and N_2 at infinite separation. At $R = R_2$ the equilibrium bond length of $N_2(X)$ changes as indicated in Table 1. The avoided crossing between the $Ar(^3P_2) + N_2(X)$ and $Ar(^1S_g) + N_2(C^3\Pi_u)$ curves occurs at $R = R_1$. From Gislason *et al.* (1979). Reproduced by permission of North-Holland Publ. Comp. (b) Population ratio $(n' = 0)/(n' = 1)$ for the $N_2(C^3\Pi_u)$ product as a function of relative collision energy. The circles are the experimental values, the full curve is the theoretical result. From Gislason *et al.* (1979). Reproduced by permission of North-Holland Publ. Comp.

molecular bond coordinate postulated by Gislason *et al.* is easily understood in terms of our bond-stretch attraction model. A view on the excited-state surface plot illustrates how for $R \gtrsim R_c$ the minimum-energy path bends and leads to a rapid change Δr in the molecular bond distance. Even though the bond-stretch attraction is less pronounced for reaction (6) (Gislason *et al.* estimate $\Delta r = 0.17$ au to fit the experiment as compared with 0.77 au for Na* + H_2 indicated in Fig. 10(a)), we readily convince ourselves of the close similarities in this type of E–V, R, T reactions.

VI. CONCLUSION

It has been shown that recent progress in both experimental methods and theoretical calculations has improved our understanding of E–V, R energy transfer drastically and generally. For some model cases of these non-adiabatic triatomic collision processes, it seems possible in the near future to reach a truly *ab initio* insight into these long-studied quenching processes in terms of comparable data from experimental and scattering theory based on *ab initio* surfaces and coupling potentials. Surprises are still to be expected and one may look with interest into the future of this rapidly developing subject.

Acknowledgement

The author wishes to thank Professor W. Meyer and Dr P. Botschwina for an intensive and extremely fruitful collaboration on the Na* + H_2 problem. Similarly, stimulating discussions with Dr P. Habitz and the communications of calculations on the Na* + N_2 processes are greatfully acknowledged. The members of our group in Berlin have all been very helpful during the process of preparing this manuscript. The author wishes in particular to thank W. Reiland and Dr G. Jamieson for their continuous efforts on the progress of the difficult experimental work.

References

C. T. J. Alkemade and P. J. T. Zeegers (1971), in *Spectrochemical Methods of Analysis*, ed. I. D. Winefordner, Wiley-Interscience, New York.
P. Archiel and P. Habitz, to be published
M. Barat (1979), *Comm. Atom. Mol. Phys.*, **8**, 73.
I. R. Barker (1976), *Chem. Phys.*, **18**, 175.
I. R. Barker and R. E. Weston (1976), *J. Chem. Phys.*, **65**, 1443.
C. Bästlein, G. Baumgartner and B. Brosa (1969), *Z. Phys.*, **218**, 315.
E. Bauer, E. R. Fisher and F. R. Gilmore (1969), *J. Chem. Phys.*, **51**, 4173.
R. B. Bernstein and R. D. Levine (1975), *Adv. Atom. Mol. Phys.*, **11**, 215.
R. Bersohn (1979), in *Alkali Halide Vapors*, eds. P. Davidovits and D. L. McFadden, Academic Press, New York, p. 345.
D. T. Birtwistle and A. Herzenberg (1971), *J. Phys. B*, **4**, 53.
A. Bjerre and E. E. Nikitin (1967), *Chem. Phys. Lett.*, **6**, 438.
V. Bonacić-Koutecký, M. Persiko, I. V. Hertel, W. Reiland and U. Tittes (1982), to be published

G. W. Black, M. A. D. Fluendy and D. Sutton (1980), *Chem. Phys. Lett.*, **69**, 260.
K. H. Blattmann, B. Menner, W. Schäuble, B. Staudenmeyer, L. Zehnle and V. Kempter (1980), *J. Phys. B*, **13**, 3635.
P. Botschwina (1979), *Verh. Dtsch. Phys. Ges.*, **2**, 529.
P. Botschwina (1980), *PhD Thesis*, University of Kaiserslautern.
P. Botschwina, W. Meyer, I. V. Hertel and W. Reiland (1981), *J. Chem. Phys.*, November.
C. Bottcher (1975), *Chem. Phys. Lett.*, **35**, 367.
C. Bottcher (1980), *Adv. Chem. Phys.*, **42**, 169.
C. Bottcher and C. V. Sukumar (1977), *J. Phys. B*, **10**, 2853.
W. H. Breckenridge and H. Umemoto (1981), this volume.
U. Buck, E. Lessner and D. Pust (1980), *J. Phys. B*, **13**, L125.
E. R. Cutshall and E. E. Muschlitz (1979), *J. Chem. Phys.*, **70**, 3171.
P. Davidovits and D. L. McFadden (eds.) (1979), *Alkali Halide Vapors: Spectra, Structure and Reactions*, Academic Press, New York.
R. J. Donovan (1979), *Prog. React. Kinet.*, **10**, 253.
R. J. Donovan, H. M. Gillespie, H. Breckenridge and C. Fotakis (1979), *J. Chem. Soc. Faraday Trans. II*, **75**, 1557.
H. Ehrhardt, L. Langhans, F. Linder and H. S. Taylor (1968), *Phys. Rev.*, **173**, 222.
J. Elward-Berry and M. J. Berry (1980a), *J. Chem. Phys.*, **72**, 4500.
J. Elward-Berry and M. J. Berry (1980b), *J. Chem. Phys.*, **72**, 4510.
A. Fischer and I. V. Hertel (1982), *Z. Phys.*, A304, in press.
E. R. Fisher and G. K. Smith (1970), *Chem. Phys. Lett.*, **6**, 438.
E. R. Fisher and G. K. Smith (1971), *Appl. Opt.*, **10**, 1083.
E. R. Fisher and G. K. Smith (1972), *Phys. Lett.*, **13**, 448.
N. French and K. P. Lawley (1977), *Chem. Phys.*, **22**, 105.
Y. Fushiki and S. Tsuchiya (1973), *Chem. Phys. Lett.*, **22**, 47.
E. A. Gislason, A. W. Kleyn and J. Los (1979), *Chem. Phys. Lett.*, **67**, 252.
E. A. Gislason, A. W. Kleyn and J. Los (1981), *Chem. Phys.*, **59**, 91.
P. Habitz (1980), *Chem. Phys.*, **54**, 131.
P. Habitz and C. Vovota (1980), *J. Chem. Phys.*, **72**, 5532.
H. W. Hermann and I. V. Hertel (1980), in *Coherence and Correlation in Atomic Collisions*, eds. H. Kleinpoppen and J. F. Williams, Plenum; New York, p. 625.
D. R. Herschbach (1966), *Adv. Chem. Phys.*, **10**, 319.
I. V. Hertel and W. Stoll (1978), *Adv. Atom. Mol. Phys.*, **13**, 113.
I. V. Hertel and W. Reiland (1981), *J. Chem. Phys.*, **74**, 6757.
I. V. Hertel (1981), in *The Excited State in Chemical Physics* (J. W. McGowan (ed.), Wiley, Chichester.
I. V. Hertel, H. Hofmann and K. A. Rost (1976), *Phys. Rev. Lett.*, **36**, 861.
I. V. Hertel, H. Hofmann and K. A. Rost (1977a), *Chem. Phys. Lett.*, **47**, 163.
I. V. Hertel, H. Hofmann and K. A. Rost (1977b), *Phys. Rev. Lett.*, **38**, 343.
I. V. Hertel, H. Hofmann and K. A. Rost (1979), *J. Chem. Phys.*, **71**, 674.
H. Heydtmann, J. C. Polyani and R. T. Taguchi (1971), *Appl. Opt.*, **10**, 244.
H. Horiguchi and S. Tsuchiya (1979), *J. Chem. Phys.*, **70**, 762.
D. S. Y. Hsu and M. C. Lin (1976), *Chem. Phys. Lett.*, **42**, 78.
D. S. Y. Hsu and M. C. Lin (1980), *J. Chem. Phys.*, **73**, 2188.
K. P. Huber and G. Herzberg (1979), *Constants of Diatomic Molecules*, Van Nostrand-Reinhold, New York.
G. K. Ivanov (1979), *Teor. Eksp. Khim. (USSR)*, **15**, 115.
K. K. Jung and A. M. Kadisch (1981), *J. Phys. B*, submitted.
G. Karl, P. Kruus and J. C. Polanyi (1967a), *J. Chem. Phys.*, **46**, 224.
G. Karl, P. Kruus and J. C. Polanyi (1967b), *J. Chem. Phys.*, **46**, 244.

G. Karl and J. C. Polanyi (1963), *J. Chem. Phys.*, **38**, 271.

V. Kempter (1975), *Adv. Chem. Phys.*, **30**, 417.

J. H. Kolts, H. C. Brashears and D. W. Setser (1977), *J. Chem. Phys.*, **67**, 2931.

L. Krause (1975), *Adv. Chem. Phys.*, **28**, 267.

H. F. Krause, J. Fricke and W. L. Fite (1972), *J. Chem. Phys.*, **56**, 4593.

M. Krauss (1968), *J. Res. Natl. Bur. Stand.*, **72A**, 553.

J. R. Krenos and J. BelBruno (1976), *J. Chem. Phys.*, **65**, 5017.

J. R. Krenos and J. BelBruno (1977), *Chem. Phys. Lett.*, **49**, 447.

K. J. Laidler (1942), *J. Chem. Phys.*, **10**, 34.

R. A. Larsen, H. J. Loesch, J. R. Krenos and D. R. Herschbach (1973), *Faraday Disc. Chem. Soc.*, **55**, 229.

W. Lee and R. M. Martin (1975), *J. Chem. Phys.*, **63**, 962.

S. Lemont and G. W. Flynn (1977), *Annu. Rev. Phys. Chem.*, **28**, 261.

R. D. Levine and R. B. Bernstein (1974), *Molecular Reaction Dynamics*, Clarendon Press, Oxford.

P. L. Lijnse (1972), *Review of Literature on Quenching, Excitation and Mixing Cross Sections for the First Resonance Doubletts of the Alkalies*, Report 398, Fysisch Laboratorium, Rijksuniversiteit Utrecht.

P. L. Lijnse (1973), *PhD Thesis*, Utrecht.

S.-M. Lin and R. E. Weston (1976), *J. Chem. Phys.*, **65**, 1443.

H. Loesch (1977), *Proc. 6th Int. Symp. on Molecular Beams. Noordwijkerhout*, Abstracts II, p. 302.

H. Loesch, (1980), *Adv. Chem. Phys.*, **42**, 421.

H. Loesch and M. Brieger (1977), *Proc. 10th Int. Conf. on the Physics of Electronic and Atomic Collisions, Paris, Commisariat à l'Energy Atomique*, Abstracts, p. 220.

J. Los and A. W. Kleyn (1979), in *Alkali Halide Vapors*, eds. P. Davidovits and D. L. McFadden, Academic Press, New York.

J. L. Magee (1940), *J. Chem. Phys.*, **8**, 687.

J. L. Magee and T. Ri (1941), *J. Chem. Phys.*, **9**, 638.

H. S. W. Massey and E. H. S. Burhop (1971), *Electronic and Ionic Impact Phenomena*, vol. 3, Clarendon Press, Oxford.

P. McGuire and J. C. Bellum (1979), *J. Chem. Phys.*, **71**, 1975.

W. Meyer (1973), *J. Chem. Phys.*, **58**, 1017.

E. E. Nikitin (1974), *Theory of Elementary Atomic and Molecular Processes in Gases*, Oxford University Press, London.

E. E. Nikitin (1975), *Adv. Chem. Phys.*, **28**, 317.

T. P. Parr and R. M. Martin (1978), *J. Chem. Phys.*, **69**, 11613.

M. Polanyi (1932), *Atomic Reactions*, Williams and Norgate, London.

M. Polanyi (1949), *Endeauvour*, **8**, 3.

W. Reiland, U. Tittes and I. V. Hertel (1981a), *Phys. Rev. Lett.*, submitted.

W. Reiland, G. Jamieson, U. Tittes and I. V. Hertel (1981b), *J. Chem. Phys.*, in press.

W. Reiland, C. P. Shultz, U. Tittes and I. V. Hertel (1982c), *Chem. Phys. Lett.*, to be submitted.

K. A. Rost (1977), *PhD Thesis*, Kaiserslautern.

N. Sadeghi and T. D. Nguyen (1977), *J. Phys. B.*, **8**, L283.

R. A. Sanders, A. N. Schweid and E. E. Muschlitz (1976), *J. Chem. Phys.*, **65**, 270.

G. Schulz, (1973), *Rev. Theor. Phys.*, **45**, 47.

A. N Schweid, M. A. D. Fluendy and E. E. Muschlitz (1976), *Chem. Phys. Lett.*, **42**, 103.

D. W. Setser and D. H. Stedman (1970), *J. Chem. Phys.*, **53**, 1004.

J. A. Silver, N. C. Blais and G. H. Kwei (1977), *J. Chem. Phys.*, **67**, 839.

J. A. Silver, N. C. Blais and G. H. Kwei (1979), *J. Chem. Phys.*, **71**, 3412.

M. Tonzeau, D. Pagnon and A. Ricard (1977), *J. Phys.*, **71**, 3412.

H. S. Taylor (1979), *Chem. Phys. Lett.*, **64**, 17.

H. S. Taylor (1980), in *Electronic and Atomic Collisions*, eds. N. Oda and K. Takayanagi, North-Holland, Amsterdam.

J. C. Tully (1973), *J. Chem. Phys.*, **59**, 5122.

A. D. Wilson and R. D. Levine (1974), *Mol. Phys.*, **27**, 1197.

J. T. Yardley (1980), *Introduction to Molecular Energy Transfer*, Academic Press, New York.

A. Yokohama, T. Ueno and Y. Hatano (1977), *Chem. Phys.* **22**, 459.

Notes

1 (P487)

Recent experiments with improved energy and angular resolution for Na* + N_2 show an even sharper structure at c.m. scattering angle $0°$, peaking for $n' = 3$. In the Na* + CO case the forward scattering cross section shows a marked structure. In both cases backward scattering differs significantly. (Reiland *et al.*, 1982c and Bonacić-Koutecký *et al.*, 1982)

2 (P488)

Note added in proof: Recent experiments with supersonic molecular beams of H_2, D_2 and HD underline the points made here. Due to improved energy and angular resolution even the vibrational structure is resolved (Reiland *et al.*, 1982a)

3 (P496)

Note added in proof: In order to take account of the zero point vibrational energy E_{no} of the M_2 molecule one replaces V_c by $V_c + E_{no}$ which is not done in Fig. 13.

4 (P503)

Note added in proof: It therefore is not surprising that even more advanced calculations based on this model (Gislason *et al.* (1981)) fail to explain the experimental observations e.g. for the system Na* + N_2: While a broad distribution of the energy in the molecule is predicted peaking at $\Delta E_{vibrot} \sim 2\hbar\omega_0$ our measurements show a sharply peaked structure at around $3\hbar\omega_0$ in forward and $4\hbar\omega_0$ in backward scattering whereas the integrated cross section is nearly statistical with a maximum at $5\hbar\omega_0$ (Reiland *et al.* (1982c)).

5 (P505)

Note added in proof: In the mean time, energy and angular resolved measurements with good resolution and statistics have been obtained for Na* + H_2, D_2 (Reiland *et al.*, 1982a), Na* + N_2 (Reiland *et al.*, 1982c) Na* + CO (Bonacić-Koutecky *et al.*, 1982) and Na* + O_2, NO in our laboratory. The experiments are qualitatively well understood but show only vague agreement with Silver *et al.* (1979).

6 (P506)

Note added in proof: Our current results show a much smaller backward peak but a drastic shift of the maximum in the energy transfer at $\Theta_{c.m.} = 180°$ towards higher n'.

7 (P510)

Note added in proof: These studies are now considered as completed for NaN_2 and have lead to a very detailed insight into the understanding of the formation of the molecular system in the course or a collision. The critical internuclear distances are $R = 10$ to 20 au.

Dynamics of the Excited State
Edited by K. P. Lawley
© 1982 John Wiley & Sons Ltd.

THE CALCULATION OF POTENTIAL ENERGY SURFACES FOR EXCITED STATES

DAVID M. HIRST

Department of Chemistry and Molecular Sciences, University of Warwick, Coventry CV4 7AL, UK

CONTENTS

I. INTRODUCTION

In order to understand fully the spectroscopic and photochemical proper-ties of a molecule, one needs to know details of the potential surfaces of the states in question. Theoretical treatments of reaction dynamics, whether by classical, semiclassical or quantum-mechanical methods, take as their starting points the relevant potential energy surfaces. At a more qualitative level, potential surfaces are extremely valuable in the interpretation of experimental data for molecular scattering and chemical reactions. Thus, potential surfaces play a fundamental role in our understanding of chemical phenomena. Advances in computer technology and in computational techniques in recent years have enabled research workers to perform increasingly ambitious calculations of potential surfaces. In many cases, theory can give reliable information, which may be very difficult to obtain experimentally, about the structure and spectroscopy of a species. Potential surface calculations, particularly when accompanied by the calculation of vibrational energy levels and transition moments, are of great value in the assignment of spectral data.

It is the purpose of this chapter to review recent progress in the calculation of potential surfaces for excited states. There are a number of reasons for focusing attention on excited states. The availability of a wide range of lasers has led to increasing interest in the reaction dynamics of atoms in excited states, in the spectroscopy of free radicals and ions, and in the detailed photochemistry of small molecules. The use of spectroscopic methods in molecular beam scattering work is yielding valuable information about the quantum states of reaction products. The recent discovery of rare-gas excimer lasers has led investigators to be interested in excited-state potentials for a wide variety of rare-gas 'molecules'. In ion–molecule reactions, there are often several potential surfaces which are quite close in energy, and non-adiabatic processes are of considerable importance. Apart from the desire of the theoretician to provide an understanding of many fascinating phenomena, the calculation of excited-state potential surfaces poses many challenges. In various ways, such calculations are more difficult than those for ground-state potentials.

Comprehensive reviews of potential surface calculations have been given by Balint-Kurti (1975) and by Bader and Gangi (1975). Thomson (1975) has given a detailed review of calculations on small molecules (containing up to four atoms), many of which refer to potential surfaces. Two articles by Kaufman (1975a, 1979) are concerned with potential surfaces for ion–molecule reactions, and a third article (Kaufman, 1975b) is concerned with potential surfaces for excited states. The Chemical Society devoted a *Faraday Discussion* (1977) to the subject of potential energy surfaces, and the published proceedings contain several papers concerned with excited-state potentials. A book by Mulliken and Ermler (1977) discusses the application of *ab initio* methods to diatomic molecules. The proceedings of a NATO Advanced Study Institute on excited states in quantum chemistry have been published (Nicolaides and Beck, 1978) and, in particular, contain useful articles by Buenker and Peyerimhoff (1978), by Buenker *et al.* (1978), and by Peyerimhoff and Buenker (1978). More recently, Schaefer (1979) has discussed the calculation of atom–molecule potentials by *ab initio* methods, and Kuntz (1979) has given an account of semiempirical methods. In a recent volume in this series (Lawley, 1980), Tully (1980) has reviewed the use of the diatomics-in-molecules (DIM) method for the calculation of potential surfaces for reactive systems. In the same volume, Bottcher (1980) has outlined methods for the calculation of excited-state potentials. However, the emphasis of his article is on methods other than those based on the usual procedures of quantum chemistry. The proceedings of a recent symposium on computational methods in chemistry (Bargon, 1980) includes papers on the *ab initio* calculation of properties for small molecules (Meyer *et al.*, 1980) and on the calculation of electronically excited states (Peyerimhoff and Buenker, 1980). An article by Bagus *et al.* (1980) discusses the ability of *ab initio* methods to make useful predictions. An article by Hay *et al.* (1979), devoted to theoretical aspects of molecular electronic transition lasers, discusses the relevant potential surfaces. Murrell (1978) has reviewed methods developed by him and his coworkers for the representation by analytical functions of potential surfaces for small polyatomic molecules.

This chapter will present, in Sec. II, an outline of the quantum-chemistry techniques currently being used to calculate excited-state potential surfaces. It will also include recent theoretical developments that will almost certainly be applied to the problem in the near future. Recent calculations of excited-state potentials will be reviewed in Sec. III (for diatomic molecules) and IV (for triatomic species). One cannot, in the space and time available, cover these aspects exhaustively. In general, the discussion of calculations for specific molecules is restricted to papers which appeared between January 1977 and July 1980. The invaluable bibliography of Richards *et al.* (1978) documents *ab initio* molecular wavefunction calculations published during the period 1974–1977. It is very much hoped that further volumes in the series will be forthcoming.

II. METHODS FOR THE CALCULATION OF POTENTIAL ENERGY SURFACES

In order to obtain potential energy surfaces for molecular systems, we seek solutions $\psi(\mathbf{R}, \mathbf{r})$ to the Schrödinger equation

$$(H_{\text{nuc}}(\mathbf{R}) + H_{\text{el}}(\mathbf{R}, \mathbf{r}))\psi(\mathbf{R}, \mathbf{r}) = E\psi(\mathbf{R}, \mathbf{r}), \tag{1}$$

where H_{nuc} represents the contribution to the Hamiltonian from the nuclei (at positions given by $\mathbf{R}$) and H_{el} describes the motion of the electrons (at positions $\mathbf{r}$) relative to the nuclei, and E is the energy of the system. Because nuclei are much heavier than the electrons, we can invoke the Born–Oppenheimer approximation in which the wavefunction $\psi(\mathbf{R}, \mathbf{r})$ is written as a product of an electronic wavefunction $\Phi(\mathbf{R}, \mathbf{r})$ for a given nuclear configuration $\mathbf{R}$ and a nuclear wavefunction $\chi(\mathbf{R})$:

$$\psi(\mathbf{R}, \mathbf{r}) = \Phi(\mathbf{R}, \mathbf{r})\chi(\mathbf{R}). \tag{2}$$

$\Phi(\mathbf{R}, \mathbf{r})$ is the solution of the electronic Schrödinger equation for a fixed nuclear configuration

$$H_{el}\Phi(\mathbf{R}) = W(\mathbf{R})\Phi(\mathbf{R}, \mathbf{r}) \tag{3}$$

and $\chi(\mathbf{R})$ is an approximate solution to the equation

$$(H_{nuc}(\mathbf{R}) + W(\mathbf{R}))\chi(\mathbf{R}) = E\chi(\mathbf{R}). \tag{4}$$

The potential energy is given by the sum of $W(\mathbf{R})$ and the nuclear repulsion energy. To map out the potential energy surface for a molecular system, we require sufficiently reliable or useful solutions and energies $W(\mathbf{R})$ from Eq. (3) for sets of nuclear configurations $\mathbf{R}$ spanning the regions of interest. It is only possible to solve Eq. (3) exactly for one-electron systems, and for all other systems it is necessary to adopt a method which will approximate to the exact solution. This involves two processes—namely the choice of a model followed by the actual calculation of the wavefunction and the energy. Balint-Kurti (1975) has usefully classified methods for the calculation of potential surfaces into four categories. *Ab initio* methods are those in which, having chosen a model for the wavefunction, the subsequent calculation is carried out exactly, with no further approximations. Semitheoretical methods either aim to reproduce the results of *ab initio* methods with the introduction of some simplifying features or to correct errors inherent in the *ab initio* method by the use of some experimental data. Semiempirical methods are derived from *ab initio* methods by making a series of, usually, quite drastic approximations and usually including the use of some experimental data. Entirely empirical methods are usually not derived from rigorous quantum-mechanical methods but aim to represent the surface by a suitable analytical function or physical model. We will follow this classification in what follows.

A. *Ab Initio* Methods

1. *The Self-Consistent Field Molecular Orbital (SCF MO) Method*

Although, for reasons to be discussed below, the molecular orbital method is generally not suitable for the calculation of molecular potential surfaces for a wide range of nuclear configurations, it is very frequently the starting point for a more rigorous calculation. We therefore discuss it in some detail here.

If it were possible to ignore the electronic repulsion terms in the electronic Hamiltonian $H_{el}(\mathbf{R}, \mathbf{r})$, it would then be possible to separate the electronic

Schrödinger equation (3) into a set of one-electron Schrödinger equations, and the electronic wavefunction could be written as a product of one-electron functions or spin orbitals $\varphi_i(\mathbf{r}_i)$:

$$\Phi(\mathbf{R},\mathbf{r}) = \varphi_1(\mathbf{r}_1)\varphi_2(\mathbf{r}_2)\ldots\varphi_n(\mathbf{r}_n). \tag{5}$$

Writing each spin orbital as a product of a spatial function $\psi_i(\mathbf{r}_i)$ and a spin function, α or β, we obtain, for an even number of electrons,

$$\Phi(\mathbf{R},\mathbf{r}) = \psi_1(\mathbf{r}_1)\alpha\psi_1(\mathbf{r}_2)\beta\ldots\psi_{n/2}(\mathbf{r}_{n-1})\alpha\psi_{n/2}(\mathbf{r}_n)\beta. \tag{6}$$

However, our electronic wavefunction must be antisymmetric with respect to the interchange of any two electrons in order to satisfy the Pauli principle. This can be achieved by taking the appropriate linear combination of functions derived from (6) by permuting the electrons among the orbitals. The result can be most conveniently written as a Slater determinant

$$\Phi(\mathbf{R},\mathbf{r}) = \frac{1}{\sqrt{(n!)}} \begin{vmatrix} \psi_1(\mathbf{r}_1)\alpha & \psi_1(\mathbf{r}_1)\beta & \psi_2(\mathbf{r}_1)\alpha & \ldots & \psi_{n/2}(\mathbf{r}_1)\beta \\ \psi_1(\mathbf{r}_2)\alpha & \psi_1(\mathbf{r}_2)\beta & \psi_2(\mathbf{r}_2)\beta & & \vdots \\ \vdots & & & \vdots & \\ \psi_1(\mathbf{r}_n)\alpha & \ldots & \ldots & & \psi_{n/2}(\mathbf{r}_n)\beta \end{vmatrix}. \tag{7}$$

The *orbitals* ψ_i can be determined by applying the variation theorem to minimize the expectation value of $\Phi(\mathbf{R},\mathbf{r})$ with respect to the electronic Hamiltonian $H_{el}(\mathbf{R},\mathbf{r})$. This gives an upper bound W_{approx} to the electronic energy $W(\mathbf{R})$:

$$W_{approx} = \frac{\int\Phi(\mathbf{R},\mathbf{r})^*H_{el}(\mathbf{R},\mathbf{r})\Phi(\mathbf{R},\mathbf{r})d\tau}{\int\Phi(\mathbf{R},\mathbf{r})^*\Phi(\mathbf{R},\mathbf{r})d\tau} \tag{8}$$

where $d\tau$ represents integration with respect to all the electronic coordinates. Applying this condition leads to the Hartree–Fock equations, a set ot pseudo-eigenvalue equations for the orbitals ψ_i. Although these coupled integro-differential equations can be solved numerically for atoms, this is not possible for molecules because of the absence of spherical symmetry.

For molecules the *molecular orbitals* ψ_i are expanded in terms of a suitable *basis set* of functions $\{\chi_r\}$ (Roothaan, 1951):

$$\psi_i = \sum_r c_{ir}\chi_r, \tag{9}$$

and the resulting equations are solved iteratively. If the basis set $\{\chi_r\}$ is sufficiently well chosen, the wavefunction will approximate to the Hartree–Fock wavefunction. In addition to yielding the $n/2$ occupied molecular orbitals, a calculation using a basis set of dimension m will give $m - n/2$ *virtual* or unoccupied orbitals which are orthogonal to the occupied orbitals $\psi_1,\ldots,\psi_{n/2}$.

Roothaan (1960) extended his formalism to the calculation of open-shell wavefunctions. Methods of implementation have been discussed by Rose and McKoy (1971), by Guest and Saunders (1974), and by Bobrowicz and Goddard (1977).

The choice of an appropriate basis set $\{\chi_r\}$ is crucial for the calculation of molecular orbitals which approximate reasonably well to the Hartree–Fock orbitals. Indeed, this is true of all *ab initio* methods. One can only expect to obtain reliable results if the basis set is capable of giving an accurate representation of the wavefunction. Basis sets have been discussed in some detail by Schaefer (1972). In molecular calculations, two types of functions—Slater-type functions (STF) and Gaussian-type functions (GTF)—are used. Slater-type functions are nodeless exponential functions of the type

$$\chi_{nlm} = N_{nlm} r^n e^{-\xi r} Y_{lm}(\theta, \varphi), \tag{10}$$

where N_{nlm} is a normalization factor and Y_{lm} is a spherical harmonic. Although these functions form a suitable basis for molecular calculations, three- and four-centre electron repulsion integrals are difficult to evaluate for nonlinear molecules. This difficulty has led to the widespread use of Gaussian functions (Boys, 1950) of the type

$$\chi_{nlm} = N_{lmn} e^{-\alpha r^2} Y_{lm}(\theta, \varphi) \tag{11}$$

in molecular calculations because integral evaluation is much faster for GTF than for STF. However, these functions are less suitable than STF because the Gaussian function does not have a cusp at the nucleus. Consequently, many more GTF are needed in the basis set in order to yield wavefunctions of comparable quality to those obtained using STF. Huzinaga (1965, 1971, 1973) and Huzinaga and Saki (1969) have published extensive sets of GTF obtained by energy optimization in atoms.

In order to avoid excessive computation times during the iterative calculation of the orbitals, contracted Gaussian basis sets are used in which each term in the basis set (Eq. (9)) is written as a linear combination of Gaussian functions:

$$\chi_r = \left(\sum_{i=1}^{n} d_{ir} e^{-\alpha_i r^2} \right) Y_{lm}(\theta, \varphi). \tag{12}$$

Thus, one can use a large number of GTF without increasing the actual number of basis functions used in the SCF calculation. If $l\,s, m\,p$, and $n\,d$ functions are contracted to x combinations of s type, y of p type, and z of d type, the basis is written $(ls\,mp\,nd)[xs\,yp\,zd]$. Dunning and Hay (1977a) have reviewed the use of contracted Gaussian basis sets for molecular calculations and have given recommended contraction schemes. More extensive basis sets for second-row atoms have been published by McLean and Chandler (1980). Raffenetti (1973) has published a more general contraction scheme. For other contraction schemes, see Ditchfield *et al.* (1971).

Orbital exponents α_i chosen on the basis of atomic calculations may not be optimum for molecular calculations. However, the optimization of many exponents is a very time-consuming and expensive process. Ruedenberg *et al.* (1973) have proposed the use of 'even-tempered' basis sets in which exponents are written as a geometric sequence in terms of two parameters a, b:

$$\alpha_k = ab^k; \qquad a, b > 0, \quad b \neq 1, \quad k = 1, \ldots, n,$$

so that optimization can be carried out in terms of just two parameters (a, b) per atom. Subsequent developments of this idea are discussed by Feller and Ruedenberg (1979) and by Schmidt and Ruedenberg (1979).

In addition to including in the basis set functions capable of representing the wavefunction of an isolated atom, it is also necessary to include additional functions to describe effects which occur on molecule formation. This can be achieved by the addition of polarization functions (p functions for hydrogen atoms, d functions for heavy atoms). There is abundant evidence that the inclusion of such functions is essential for an accurate description of geometries and potential surfaces for molecules. An effective alternative is to include a set of s and p functions at the midpoints of bonds, though perhaps this should be done with caution (Bauschlicher, 1980b). For many problems a double zeta plus polarization (DZP) basis is reasonably adequate. A DZP basis consists of two basis functions per core and valence atomic orbital, plus polarization functions. In studies of excited states, it is also essential that the basis should be adequate to describe the excited state. If the state is of Rydberg character, then diffuse Rydberg-type functions must be included. For states which yield a negative ion on dissociation, appropriate diffuse functions are essential for a correct asymptotic description. Dunning and Hay (1977a) give some guidance on the choice of orbital exponents for diffuse functions.

As mentioned above, the molecular orbital method is, in general, inadequate for the calculation of potential energy surfaces. A realistic potential surface must dissociate correctly and should give a comparable description of the system for all geometries. The molecular orbital function does not usually meet either of these criteria. It is well known, for example, that the MO wavefunction for H_2 dissociates to two ions rather than to two ground-state atoms. The molecular orbital wavefunction is of the form appropriate to an independent-particle model. Correlation between electrons of the same spin is accounted for by requiring the wavefunction to be antisymmetric with respect to the interchange of two particles, but correlation between electrons of opposite spins is not accounted for. Thus, the energy of a Hartree–Fock wavefunction will differ from the exact non-relativistic energy by an amount known as the correlation energy. This, in general, will be dependent on geometry, so even if the wavefunction dissociates correctly the calculated surface will not be parallel to the exact surface. In subsequent sections, we shall discuss methods which yield wavefunctions that dissociate correctly and which make allowance for the correlation energy. A recent book by Hurley (1976) gives a detailed account of some current methods for treating electron correlation.

There are, however, a number of cases where the molecular orbital function does show the correct asymptotic behaviour and for which MO potential surfaces are a useful first approximation. One such case is when the dissociation products have closed-shell electronic structures. In such systems the number of electron pairs remains constant and large changes in correlation energy are not expected. The SCF MO method is also a reasonably good description for diatomic systems which dissociate into an atom with a closed shell and an atom with a half-filled shell. It is not, however, suitable for a system in which an unshared electron and a bonded pair are exchanged between reactants and products.

2. The Generalized Valence Bond Method

The inability of the molecular orbital function to describe dissociation properly stems from the constraint that the wavefunction is written in terms of doubly occupied orbitals. Thus if dissociation should lead to the two electrons in orbital ψ_i, in the Slater determinant represented by

$$\Phi = |\psi_1\alpha\psi_1\beta\psi_2\alpha\psi_2\beta\ldots\psi_i\alpha\psi_i\beta\ldots| \tag{13}$$

ending up on different fragments A and B, then clearly a function in the form of (13) will be unable to describe this. This difficulty is removed in the generalized valence bond (GVB) method in which the wavefunction for the pair of electrons $\psi_i\alpha\psi_i\beta$ in (13) is rewritten in terms of two non-orthogonal orbitals ψ_{ia}, ψ_{ib} whose spins are singlet coupled:

$$\Phi = |\psi_1\alpha\psi_1\beta\psi_2\alpha\ldots\psi_{ia}\psi_{ib}(\alpha\beta - \beta\alpha)\ldots|. \tag{14}$$

The GVB method has been reviewed by Bobrowicz and Goddard (1977), so we shall only give a brief outline here. In a wavefunction of the form of (14), ψ_{ia}, ψ_{ib} can adjust to the correct asymptotic form as the molecule is pulled apart. Also, the form of the wavefunction allows for some correlation between the electrons in the pair described by ψ_{ia} and ψ_{ib} and will thus give a lower energy than the SCF MO wavefunction. It is, of course, possible to introduce further correlation effects by replacing other orbital products $\psi_j\alpha\psi_j\beta$ by $\psi_{ja}\psi_{jb}(\alpha\beta - \beta\alpha)$. From the computational point of view, it is useful to introduce the condition of *strong orthogonality* between the functions $\psi_{ia}\psi_{ib}$ and the other molecular orbitals in function (13).

Additional flexibility can be introduced into the GVB scheme by recognizing that there may be other linearly independent singlet functions than that arising from the *perfect pairing* (PP) coupling $\psi_{ia}\psi_{ib}(\alpha\beta - \beta\alpha)\psi_{ja}\psi_{jb}$

$(\alpha\beta - \beta\alpha)$. An improved wavefunction can be obtained by optimizing a linear combination of the independent spin functions.

The relationship between the GVB PP function and the SCF MO plus configuration interaction wavefunction can be seen by transforming the orbitals $\psi_{ia}\psi_{ib}$ to natural orbitals $\varphi_{ia}, \varphi_{ib}$ using the relations

$$\psi_{ia} = \frac{c_{1i}^{1/2}\varphi_{ia} + c_{2i}^{1/2}\varphi_{ib}}{(c_{1i} + c_{2i})^{1/2}}; \qquad \psi_{ib} = \frac{c_{1i}^{1/2}\varphi_{ia} - c_{2i}^{1/2}\varphi_{ib}}{(c_{1i} + c_{2i})^{1/2}} \tag{15}$$

where $\varphi_{ia}, \varphi_{ib}$ are orthogonal, $c_{1i}, c_{2i} > 0$, and $(c_{1i}^2 + c_{2i}^2) = 1$. This transformations leads to a function of the form

$$\Phi = |\psi_1\alpha\psi_1\beta\psi_2\alpha\psi_2\beta \ldots (c_{1i}\varphi_{1a}\varphi_{1a} - c_{2i}\varphi_{1b}\varphi_{1b})\alpha\beta \ldots|. \tag{16}$$

Thus the wavefunction is equivalent to a linear combination of two Hartree–Fock-type Slater determinants and is, in fact, a two-term configuration interaction wavefunction. This form is computationally more convenient than (14).

The GVB method thus has the advantage of being able to describe dissociation correctly and can therefore lead to potential surfaces which are qualitatively correct. The GVB function is often a better starting point for a more refined calculation than the Hartree–Fock wavefunction.

3. *Configuration Interaction*

The method of *configuration interaction* (CI) or superposition of configurations is perhaps the most straightforward way of constructing a wavefunction which does not have the deficiencies of the Hartree–Fock single determinant. The CI method usually takes as its starting point the SCF MO wavefunction Φ_0. One then constructs further Slater determinants by excitation of electrons from the occupied orbitals in Φ_0 to the virtual orbitals. From these determinants, one forms, where necessary, linear combinations Φ_i which have the same spin and symmetry as the starting function Φ_0. Such combinations are known as configuration state functions (CSF) or configuration functions (CF). The CI wavefunction ψ is then obtained by solving the linear variational problem

$$\psi = \sum_{i=0}^{n} c_i\Phi_i. \tag{17}$$

Thus the concept is very simple. However, the implementation, especially for potential surfaces for excited states, can be very demanding. A useful survey of the CI method has been given by Schaefer (1972), and the method has been comprehensively reviewed by Shavitt (1977a). We will confine ourselves here, first, to the consideration of the criteria to be fulfilled in the calculation of excited-state potential surfaces, secondly, to discussion of the methods most appropriate to such calculations, and finally, to a brief discussion of recent developments.

At the outset, we reiterate the point made earlier that it is important to choose a basis set which is sufficiently flexible to describe the system being considered. CI cannot remedy deficiencies in the basis set. Some consideration should be given to the choice of starting function Φ_0. If all possible functions are included in the expansion of Eq. (17), the result will be independent of the initial choice for Φ_0. However, as we shall see below, it is necessary to work with a truncated set of functions $\{\Phi_i\}$. In such circumstances, an appropriate starting point is the SCF MO function for the state in question. Another problem is that the virtual orbitals from a SCF MO calculation are not particularly well suited for the construction of a rapidly convergent CI expansion. An improved set of orbitals can be obtained by performing a multiconfiguration SCF (MCSCF) calculation (see Sec. II A.4), by the use of natural orbitals (see Schaefer, 1972; Hurley, 1976; Jafri and Whitten, 1977), by the use of improved virtual orbitals (IVO) (e.g. Hunt and Goddard, 1969) or modified virtual orbitals (MVO) (Bauschlicher, 1980a; Cooper and Pounder, 1979).

The conventional approach to CI, as implemented, for example, in the SPLICE library on the Science Research Council computers in the UK (Guest and Rodwell, 1977), is to construct a set of formulae for the matrix elements

$$H_{ij} = \int \Phi_i^* H \Phi_j \, d\tau, \qquad S_{ij} = \int \Phi_i^* \Phi_j \, d\tau$$

and diagonalize the resulting secular determinant. If we include all possible excitations from Φ_0 in Eq. (17), we get a *full* or *complete* CI function. Except for calculations on very small molecules or for calculations using small basis sets, such a procedure will yield a very large number of configurations. If the number of basis functions n is considerably larger than the number of electrons N, the total number of linearly independent CFs goes up roughly as n^N. Thus, in general, it is necessary to limit the excitations included. An analysis by perturbation theory (Shavitt, 1977a) indicates that double excitations (i.e. where two electrons are excited to virtual orbitals) play a dominant role. For the accurate calculation of one-electron properties, single excitations are important. They are also very important in open-shell molecules. Thus a very useful truncation of the configuration space is achieved by limiting the expansion in Eq. (17) to single and double excitations with respect to Φ_0 (SDCI). However, there is increasing evidence from comparative studies using many-body perturbation theory (see Sec. II A.6) that the effects of triple and quadruple excitations are not negligible (Bartlett and Shavitt, 1977a, b, 1978; Siegbahn, 1978; Wilson and Silver, 1979; Wilson and Saunders, 1979; Wilson, 1979a; Guest and Wilson, 1980b) if one is interested in 'chemical' accuracy. There is also the problem of size extensiveness. An

SDCI calculation for an assembly of N isolated molecules will give an energy which is different from N times the energy of a single molecule. This can be attributed to the neglect of quadruple excitations. Langhoff and Davidson (1974) have proposed the formula

$$\Delta E_Q = (1 - c_0^2)E_{SD} \qquad (18)$$

to estimate the effect of quadruple excitations ΔE_Q in terms of the SDCI correlation energy E_{SD} and the coefficient c_0 of the dominant term in the CI expansion.

However, for potential surface calculations there is the problem that the Hartree–Fock function will not, in general, give a good description of the molecule at all points on the surface. Indeed, the Hartree–Fock function may be grossly in error at the dissociation limits. Also, excited states of the same symmetry as the ground state will not be calculated to the same accuracy as the ground state.

In order to obtain a balanced treatment of potential surfaces of a given symmetry, it is necessary to generalize the SDCI to include single and double excitations from all functions $\Phi_0 \dots \Phi_k$ which have significant coefficients at any point on the surfaces of interest. The configurations $\Phi_0 \dots \Phi_k$ are referred to as *reference* or *root* configurations. For accurate work, the set of root configurations should include all configurations having a coefficient ≥ 0.1 in the final CI vector. The use of a multireference set of configurations does mean that some of the more important higher excitations are included.

Even with the restriction to single and double excitations, the configuration list may be prohibitively long for processing by conventional techniques. There are four possible ways of rendering the calculation tractable:

(i) reduce still further the excitations considered;
(ii) retain the full number of configurations but neglect the less important Hamiltonian matrix elements H_{ij};
(iii) select the most important configurations;
(iv) develop different approaches to CI which will permit the handling of larger numbers of configurations.

The orbital space can be partitioned into three subspaces—(1) the core orbitals, (2) the valence orbitals, which include both the occupied orbitals and the corresponding antibonding orbitals constructed from the 'valence' atomic orbitals in the basis set, and (3) the external orbitals. A very common restriction is the frozen core approximation in which the core orbitals are kept doubly occupied and at the same time the corresponding high-energy virtual orbitals (inner-shell complement) are eliminated from the orbital space. This restriction assumes that core correlation effects are independent of geometry. The *first-order wavefunction* (Schaefer *et al.*, 1969) is more drastic.

It includes all single excitations with respect to the Hartree–Fock function Φ_0 but restricts double excitations to those in which not more than *one* electron occupies an orbital in the external space. A more general version of this idea is the POLCI wavefunction of Hay *et al.* (1975) which takes a GVB wavefunction as the starting point rather than the Hartree–Fock wavefunction.

The second approach recognizes that only a relatively small number of configurations will play a dominant role in the final CI wavefunction. Thus, the configuration space can be partitioned into a core subspace $\{\Phi_0 \dots \Phi_k\}$ which contains the dominant configurations and a remainder subspace $\{\Phi_{k+1} \dots\}$. If a limited CI using only the core subspace yields a zeroth-order wavefunction Φ_c, then first-order perturbation theory can be used to estimate the second-order contribution to the energy of a configuration Φ_r from the remainder subspace by

$$\Delta E_r = |H_{cr}|^2/(H_{cc} - H_{rr}). \tag{19}$$

Clearly, any CF making a substantial second-order contribution to the energy should be included in the core subspace. In the A_k procedure of Gershgorn and Shavitt (1968) and in other similar procedures (see Shavitt, 1977a), the configurations in the remainder subspace are screened successively for possible inclusion in the core subspace. In the B_k method (Gershgorn and Shavitt, 1968), the CI Hamiltonian matrix is approximated by including only: (i) all matrix elements between core CFs, (ii) diagonal elements with respect to remainder configurations, and (iii) all matrix elements between core CFs and remainder CFs. Thus, off-diagonal elements between CFs in the remainder subspace are set to zero. Provided one revises the choice of functions for the core subspace to include any configurations found to interact significantly with the original core subspace, the method can give a reasonable approximation to the CI function. Segal *et al.* (1978) have proposed and tested an iterative scheme in which a nucleus of dominant configurations (up to 5) is chosen. This is then augmented by other functions to form a core of up to 50 configurations. The Hamiltonian matrix includes all core–core and core–remainder (or tail) elements and diagonal elements for remainder configurations. The neglected interactions are taken into account through an effective potential derived from the solution of this restricted matrix.

Perhaps the most widely used approach, in cases in which the length of the configuration list is prohibitive, is the use of a selection procedure whereby the configuration space is truncated by the elimination of the less important configurations. There are two ways in which the energy contribution of a particular configuration can be estimated. One way is to use perturbation theory, as in the A_k approach (Gershgorn and Shavitt, 1968), or, alternatively, to solve the secular problem for the core configurations plus the configuration being tested (Buenker and Peyerimhoff, 1974). All configurations in the remainder subspace are screened in this way. For calculations in which one is interested in the m lowest roots of a particular symmetry, it is essential that the energy lowerings be computed with respect to at least the m lowest roots of the core CI matrix. One then rejects all configurations which contribute less than a certain threshold energy T where T typically lies between $10^{-4}E_h$ and $10^{-5}E_h$. Having selected a set of configurations $\{\Phi_s\}$, one then proceeds to calculate all the necessary matrix elements and to solve the secular problem. From the energy lowerings of individual configurations, it is possible to estimate the contribution to the energy of the rejected configurations. By performing the calculation at several thresholds, a reasonably reliable extrapolation to zero threshold can be made (Buenker and Peyerimhoff, 1974, 1975; Buenker et al., 1978). Buenker et al. call their approach 'multi-reference double-excitation CI' (MRD CI). For potential surface calculations, one may raise the objection that for each point on the surface a different configuration space is being used and that not all points will be treated with comparable accuracy. Inclusion of estimates of the contributions from rejected terms and extrapolation to zero threshold should reduce any errors to negligible proportions. A cumulative selection scheme has been used (Raffenetti et al., 1977) in which the separate energy contributions are sorted into order and summed, starting at the smallest. When a predetermined threshold energy is reached, the remaining configurations are retained for a CI calculation. This method aims to produce potential curves parallel to and equally displaced from those that would be obtained in an unselected calculation.

Huron et al. (1973) have proposed a CIPSI procedure whereby a zeroth-order multiconfigurational wavefunction is obtained by an iterative perturbation selection process. The second-order contribution to the energy arising from the remaining configurations is then estimated.

Although selection and extrapolation procedures are very effective, it is rather unsatisfying not to be able to solve the full SDCI problem. There has been considerable effort during the past five years in the development of techniques for performing large CI calculations without the necessity of producing and manipulating large formula files. Initially, direct CI or CIMI (CI from molecular integrals) methods (Roos, 1972; Roos and Siegbahn, 1977a) were restricted to SDCI with respect to a single closed-shell determinant, but more recently methods have been devised for triplets (Pakiari and Handy, 1975; Lucchese and Schaefer, 1978; Ferguson and Handy, 1979), doublets, quartets, and open-shell singlets (Handy et al., 1979), and to a multiconfiguration reference state containing several closed-shell determinants (Roos and Siegbahn, 1980). In the latter extension, an application involved 76 471 configurations. Ferguson and Handy (1980) have extended the method to include a specific set of triple and quadruple excitations. A general algorithm for the evaluation of the coupling constants for CIMI has been devised by Duch and Karwowski (1979). Thus, we can look forward to the generalization to a multiconfiguration space containing open-shell configurations as is necessary for potential surface calculations. Direct CI methods have been implemented on a minicomputer by Brooks and Schaefer (1978). Handy (1980) has shown that by storing integrals and CI coefficients as integers, large direct CI calculations can be done in core. Using this method, Saxe et al. (1981a) have performed a complete CI calculation, using a double zeta basis set, for H_2O, in which 256473 configurations were included.

The development of more efficient methods for CI calculations continues to be a very active field of research. The proceedings of a recent conference on electron correlation (Guest and Wilson, 1980a) contain several papers surveying current trends. Of particular promise is the unitary group approach to electronic structure problems of Paldus (1974, 1975, 1976a, b) and Shavitt (1977b, 1978, 1979). Applications to CI calculations have been discussed by Paldus et al. (1977), Adams et al. (1977), Hegarty and Robb (1979), Ruttink (1978), Wormer and Paldus (1979), Paldus and Wormer (1979), Paldus and Boyle (1980). Calculations using the graphical unitary group method have been reported for the vertical electronic spectrum of ketene (Brooks and Schaefer, 1979), the singlet–triplet splitting in dioxymethane (Siegbahn, 1979), the triplet methylene abstraction reaction (Siegbahn, 1980), and for BH_3, CH_2, C_2H_4, and SO_2 (Brooks et al., 1980).

4. The Multiconfiguration SCF Method (MCSCF)

We saw above that in the CI method it is usually necessary to include a large number of configurations in order to obtain satisfactory results. One reason for this slow convergence is that the configurations are constructed by considering excitations from the occupied molecular orbitals, obtained in an MO calculation, to the virtual orbitals which have not been subjected to any optimization. The aim of the MCSCF method is to calculate a compact multiconfiguration wavefunction of the form

$$\psi = \sum_k a_k \psi_k, \tag{20}$$

where, in addition to optimizing the linear variation parameters a_k, the expansion coefficients in the one-electron functions used in the construction of the configurations ψ_k are simultaneously optimized. This is clearly a much more difficult computational task than the separate stages of the SCF MO plus CI procedure, and it is therefore necessary to restrict the number of configurations included.

Wahl and Das (1970) proposed the optimized valence configuration (OVC) scheme for the choice of configurations. The philosophy is that, in order to calculate reliable potential curves, it is only necessary to include those configurations required to describe changes in electronic structure which occur on molecule formation. These are of two types, namely those which have to be added to the Hartree–Fock function to ensure correct dissociation (together these form the base function), and those which are important in the molecule but formally vanish at the dissociation limit. These latter configurations allow for the molecular extra correlation energy (MECE). Billingsley and Krauss (1974) and Stevens et al. (1974) have discussed the required configurations in some detail. The MECE terms are of three types. (i) Bond correlation pair excitations are double excitations out of doubly occupied bonding orbitals into non-valence orbitals. These allow for angular and radial correlation in the region of R_e and for the interatomic dispersion interaction at large R. (ii) Double excitations to valence orbitals are known as valence charge transfer terms, and, if they have not already been included in the base function, contribute a substantial fraction of the correlation energy. (iii) A third class of excitation in which one electron is excited to a valence orbital and the other to a non-valence orbital (split shell charge transfer with excitation) accounts for R-dependent components of the valence-shell correlation energy.

Computational procedures for the MCSCF method have recently been reviewed by Wahl and Das (1977). Alternative schemes based on the generalized Brillouin theorem (Levy and Berthier, 1968) have been discussed by Ruttink and van Lenthe (1977), Chang and Schwarz (1977), and Rueden-berg et al. (1979). A method for excited states has been developed by Grein and Banerjee (1975) and applied to excited states of NH (Banerjee and Grein, 1977). The selection of configurations for an excited state for which there are lower roots of the same symmetry is not as straightforward as the OVC procedure for the lowest state of a given symmetry. The base wavefunction contains those configurations necessary to describe the molecule near R_e, at the asymptotic limits and in the region of any curve crossings that may be present. The MECE configurations are determined by testing the contribution of all the configurations obtained from single and double excitations from the base set, by a procedure analogous to that of Buenker and Peyerimhoff (1974). Those resulting in an energy lowering larger than a given value are included in an MCSCF calculation. The procedure is repeated using all the configurations in the previous MCSCF step until additional configurations make a negligible contribution or until the MCSCF function has reached the maximum size that one can handle. Bauschlicher and Yarkony (1980) have proposed a scheme for the calculation of excited states of polar molecules which is also based on the generalized Brillouin theorem and an iterative procedure for the selection of configurations to be included in the MCSCF calculation.

Alternative optimization schemes have recently been proposed by Dalgaard and Jorgensen (1978), Yeager and Jorgensen (1979), and Roothaan *et al.* (1979). Unitary transformations of the molecular orbitals and the CI coefficients are generated by the use of exponential operators. Expansion of the energy expression to second order in the parameters defining the unitary transformation is then followed by minimization of the energy by a Newton–Raphson procedure. Lengsfield (1980) has developed a density matrix method for the efficient construction of the required Hessian which leads to rapid convergence for both ground and excited states. The ability of the method to cope with single excitations is demonstrated by calculations on BeO. Roos, Taylor and Siegbahn (1980) have devised a density matrix method for the efficient calculation of MCSCF wavefunctions which include the complete set of configurations from a given set of active orbitals.

5. *Valence Bond Methods*

Valence bond theory plays an important role in our understanding of molecular electronic structure but has not, because of computational difficulties, been widely used for *ab initio* calculations of potential surfaces, despite recent advances which make the method more tractable (Gerratt, 1974). Two features of the valence bond method make it attractive for potential surface calculations. The wavefunction is made up of terms which are products of atomic orbitals, thus making it easy to ensure correct dissociation. Secondly, in a multiconfiguration calculation it is much easier than in the SCF MO CI scheme to use chemical intuition to decide which configurations are of importance.

One approach (Balint-Kurti and Karplus, 1974), which has been used in a variety of potential surface calculations, starts with the construction of approximate atomic eigenfunctions for all the atomic states to be considered in the calculation. Let these be denoted Φ_i^P (for atom P). Antisymmetrized products Φ_i (composite functions, CF) of these approximate atomic eigenfunctions are then formed,

$$\Phi = \mathcal{A}[\Phi_i^P \Phi_j^Q \ldots], \tag{21}$$

where $\mathcal{A}$ is a partial antisymmetrizer. These are combined in such a way as to give functions Φ_i^{SM} which are eigenfunctions of the S^2 operator. The functions Φ_i^{SM} form the basis for a variational calculation

$$\psi = \sum_i c_i \Phi_i^{SM}. \tag{22}$$

It is possible to obtain good descriptions of ground and excited potential surfaces by what is, by CI standards, a relatively modest number of functions Φ_i^{SM}. The non-orthogonality problem, which has inhibited the widespread use of valence bond methods, is overcome by employing a transformation to orthogonal orbitals. Errors introduced by the use of approximate atomic eigenfunctions can be compensated for by use of the orthogonalized Moffitt method.

Van Lenthe and Balint-Kurti (1980) have proposed a VB SCF method in which MCSCF orbital optimization techniques, using the generalized Brillouin theorem, are applied to multistructure valence bond functions. Preliminary calculations for the ground-state potential function for OH are presented, and the method appears to have considerable promise.

Another variant of the valence bond method which has been applied to potential surface calculations is the spin-coupled valence bond method formulated by Pyper and Gerratt (1977) and by Gerratt and Raimondi (1980). In its simplest form, the wavefunction for N electrons is written

$$\psi_{SM} = \sum_k c_{Sk} \sqrt{N!} \mathscr{A}(\varphi_1 \varphi_2 \ldots \varphi_N \Theta_{SMk}^N), \qquad (23)$$

where $\varphi_i (i = 1, \ldots, N)$ represent normalized orbitals, Θ_{SMk}^N is an N-electron spin function in which the spins are coupled to give a total resultant S and z-component M. The summation with respect to k is over all such couplings. Optimization is performed with respect to both the orbitals φ_i and the coefficients c_{Sk}. The function can be made more flexible by taking a linear combination of the function (23) with a small number of double replacements of orbitals φ_i, φ_j.

In addition to the *ab initio* formulation of valence bond theory, it forms the basis of two semiempirical schemes—the diatomics-in-molecules (DIM) method and the London–Eyring–Polanyi–Sato (LEPS) method—which are widely used for calculating potential surfaces.

6. Many-Body Perturbation Theory

Although well designed variational calculations using CI, MCSCF or valence bond techniques are capable of giving very accurate results for potential surfaces, the drawbacks of long configurations lists and lack of size consistency have led to the consideration of perturbation theory as an alternative route to accurate wavefunctions and energies. During the past few years, there has been intense activity by some research groups in the application of diagrammatic many-body perturbation theory (MBPT) to molecular problems. This approach has been reviewed recently by Bartlett and Purvis (1980) and by Wilson (1980, 1981). These methods do yield wavefunctions which are size consistent.

It has been shown that it is possible to devise efficient non-iterative algorithms to evaluate third- and fourth-order contributions to the energy, but the method has not been applied, as far as the reviewer is aware, to the calculation of excited-state potential curves. There have been calculations for ground-state curves for CO (Wilson, 1977), HF (S. Wilson, 1978), CH^+ (Wilson, 1979b), CO, BF (Wilson and Silver, 1980), BH, F_2, N_2 (Urban and Kellö, 1979), Be_2 (Bartlett and Purvis, 1978), Mg_2 (Purvis and Bartlett, 1978), HCO (Adams et al., 1979), H_2O (Bartlett et al., 1979), and He + LiH (Silver, 1980). These calculations, which take single configuration Hartree–Fock wavefunctions as their starting points, have, with the exception of the HCO and He + LiH calculations, been restricted to the portion of the surface close to equilibrium. The HCO calculations used a UHF determinant as the zeroth-order wavefunction. In order to calculate potential surfaces which are accurate over a wide range of geometries, it will be necessary to generalize the method to permit efficient calculations using a multi-reference set of configurations. Hose and Kaldor (1979) have formulated an approach applicable to general model spaces.

Although MBPT methods for potential surface calculations are still in their infancy compared with CI methods, the calculations performed to date have shown that the method is capable of giving results for which the accuracy is comparable to or better than that for CI. These calculations have led to a deeper understanding of the size consistency defect of CI and have made it possible to estimate the effects of neglecting triple and quadruple excitations in CI (Wilson, 1979a; Wilson and Saunders, 1979; Wilson and Silver, 1979; Guest and Wilson, 1980b; Krishnan et al., 1980).

Detailed examination of the various contributions to the third- and fourth-order energies indicates that it is not possible to restrict consideration to just one type of excitation and to ignore the others. For meaningful results, it is clear that all contributions to a given order must be included. In a recent calculation on water, Wilson and Silver (1979) suggested that it was probably better to improve the basis set and to go to third order in perturbation theory than to calculate fourth-order and higher terms. At best, such terms will only recover a few millihartree in fourth order.

7. Pair Correlation Methods

Some of the most accurate calculations of ground-state potential curves for first- and second-row hydrides and their positive and negative ions have been made by Meyer and Rosmus (1975) and Rosmus and Meyer (1977, 1978) using pseudo-natural orbital (PNO) + CI and coupled electron pair methods. Hurley (1976) and Kutzelnigg (1977) discuss in detail the formulation and application of pair theories to the calculation of molecular wavefunctions. These methods assume that the dominant contribution to the correlation energy arises from interactions between pairs of electrons. In the independent electron pair approximation (IEPA), the total correlation energy is approximated as the sum of independently calculated pair contributions ε^{IEPA}. The correlation energy for pair ij is obtained by considering all double excitations from the pair ij to orbitals ab:

$$\psi_{ij} = c_0\Phi + \sum_{a<b} c_{ij}^{ab}\Phi_{ij}^{ab}, \tag{24}$$

where Φ is the starting wavefunction, Φ_{ij}^{ab} represents the excitation, and c_{ij}^{ab} is a variational parameter. For n electrons this would lead to $n(n-1)/2$ variational calculations. Reformulation of the problem in terms of excitation spin-adapted pairs ψ_{RR} (excitation from $\varphi_R^\alpha \varphi_R^\beta$), $^s\psi_{RS}$ (excitation from singlet coupled orbitals $\varphi_R \varphi_S$), and $^t\psi_{RS}$ (excitation from triplet coupled $\varphi_R \varphi_S$) leads to the following expression for the total energy for a closed-shell state:

$$E = \langle \Phi|H|\Phi \rangle + \sum_R \varepsilon_{RR} + \sum_{R<S} (^s\varepsilon_{RS} + {}^t\varepsilon_{RS}), \tag{25}$$

where ε_{RR}, etc., are the pair correlation energies given by expressions of the form

$$\varepsilon_{RR} = E_{RR} - \langle \Phi|H|\Phi \rangle, \tag{26}$$

with E_{RR} being the energy corresponding to ψ_{RR}. The calculation involves optimizing pair excitation functions corresponding to (24). One of the approximations made by IEPA is the neglect of coupling terms between excitations of different pairs. Relaxation of this approximation leads to the coupled electron pair approximation (CEPA) (Meyer, 1973; Ahlrichs et al., 1975; Kutzelnigg, 1977). These approaches are very powerful if the pair correlation functions are expanded in terms of pseudo-natural orbitals (PNO) (Kutzelnigg, 1977; Meyer, 1977). The use of PNOs allows the use of large basis sets without the disadvantage of the generation of an excessive number of configurations.

The disadvantage of the CEPA method is that it does not result in a variational upper bound to the energy. It does, however, give size-consistent results. The use of PNOs in CI calculations results in a very effective CI method with much shorter configuration lists than in the conventional CI method. However, the wavefunction in not size consistent.

The CEPA method is related to the coupled many-electron method (CPMET) of Čížek (1966, 1969) by the use of a simplified expression for the unlinked cluster contribution in CPMET. The CPMET has recently been reviewed briefly by Čížek and Paldus (1980). A version of CPMET, the coupled cluster method with double excitations (CCD), has been used for the calculation of potential surfaces for the ground state of Be_2 (Bartlett and Purvis, 1978), Mg_2 (Purvis and Bartlett, 1979), and H_2O (Bartlett et al., 1979). In a comparison of CI and coupled pair calculations for the potential curve of CN^-, Taylor et al. (1979) found that the inclusion of unlinked cluster excitations leads to significant changes in the values of the calculated spectroscopic constants. Pople et al. (1978) have used CCD for the calculation of potential surfaces for 1,2 hydrogen shifts in C_2H_2, HCN, H_2CO, and N_2H_2, and give a comparison with perturbation theory calculations. Adams *et al.* (1979) compare CCD and MBPT calculations for the ground-state potential surface of HCO. Bartlett and Purvis (1980) review CCD and MBPT methods and specifically discuss potential surface calculations. Recently, Lindgren (1978) has presented a coupled cluster approach applicable to open-shell systems.

A method which has considerable promise for the efficient calculation of correlated wavefunctions for potential surfaces is the self-consistent electron pair (SCEP) method proposed by Meyer (1976) in which the wavefunction ψ is written in terms of a closed-shell determinant ψ_0 and a sum of doubly substituted functions ψ_P

$$\psi = \psi_0 + \sum_P \psi_P. \tag{27}$$

By an iterative procedure, different sets of optimal external orbitals are calculated for each electron pair in ψ_0. First-order perturbation theory is used in each iteration to improve the wavefunction. Single excitations were included by Dykstra *et al.* (1976), and Dykstra (1978) has considered the effects of higher-order excitations. Parsons and Dykstra (1979) have applied the method to rearrangements of the N_2H_2 system with calculations involving up to 170 000 configurations. A full CI calculation using SCEP has been performed for the ground-state potential surface of H_3^+ by Dykstra and Swope (1979). Computational methods have been discussed by Saunders (1978), and the theory has been further developed by Dykstra (1977, 1980).

B. Semitheoretical Methods

1. *Pseudopotentials and Effective Core Potentials*

The computation time and computer storage requirements for *ab initio* calculations go up very rapidly as the number of basis functions increases. Although such calculations are feasible for simple molecules containing heavy atoms, it is not possible to treat more complex systems at the same level without introducing some simplifying features. Traditionally, the chemist divides electrons into 'core' electrons (in completely filled inner shells) and 'valence' electrons, which are responsible for the bonding and chemical properties of the molecule. If it is possible to devise a suitable potential to represent the effect of the essentially non-bonding core electrons, then 'valence only' calculations can be performed with only a fraction of the effort required for an all-electron calculation. This can be achieved by the use of pseudopotentials or effective core potentials. The use of pseudopotentials in molecular calculations has been reviewed by Dixon and Robertson (1978), and Kahn *et al.* (1976) have given a detailed account of the theory of effective core potentials Writing the total electronic wavefunction as

$$\psi = \mathscr{A}(\psi_{\text{core}}\psi_{\text{valence}}), \tag{28}$$

where $\mathscr{A}$ is the antisymmetrizer, it is possible, provided that the core and valence orbitals are strongly orthogonal, to write the total energy as

$$E_{\text{tot}} = E_{\text{core}} + \langle \psi_{\text{valence}}|H_{\text{V}}|\psi_{\text{valence}} \rangle. \tag{29}$$

H_{V} is the valence Hamiltonian

$$H_{\text{V}} = \sum_v \left(h + \sum_c (2J_c - K_c) \right) + \sum_{v > v'} \frac{1}{r_{vv'}} \tag{30}$$

where v and c represent summation over valence and core orbitals respectively, h is the one-electron operator, J_c and K_c are Coulomb and exchange operators for the core electrons, and $1/r_{vv'}$ is the interelectronic distance. However, since the two-electron integrals with respect to both core and valence electrons are required for the evaluation of these expressions, there is no saving at the SCF stage. It is possible, though, to reduce the computational effort for a CI calculation in which the core orbitals are frozen.

The explicit orthogonalization constraint can be removed by projecting core states from the valence Hamiltonian. It is necessary to introduce an approximation at this stage and limit the projection to the one-electron operators to yield an approximate valence Hamiltonian of the form

$$H_{\text{PV}} \approx \sum_v (h + U^{\text{core}}(v)) + \sum\sum_{v > v'} \frac{1}{r_{vv'}}, \tag{31}$$

in which the one-electron operators are modified by the *effective core potential* $U^{\text{core}}(v)$. In the simplest approach the effective core potential is represented by a suitable functional form or *model potential* which may be just a radial function or may be a semi-local potential which is angular dependent. Details of commonly used model potentials are given by Dixon and Robertson (1978) and by Kahn *et al.* (1976). The adjustable parameters in the model potentials are chosen so that the calculations agree with all-electron calculations or reproduce experimental data. Simple model potentials have been used in potential curve calculations by Watson *et al.* (1977), Valance (1978a, b), Janoschek and Lee (1978), and Nemukhin and Stepanov (1979).

A more flexible approach is to determine the effective core potential directly from atomic all-electron calculations without constraining the functional form of the effective core potential. One approach (Kahn *et al.*, 1976) is to write the effective core potential in terms of a local potential $U_{\text{L}}^{\text{core}}(r)$, which includes the effect of Coulomb and exchange potentials common to all valence electrons, and terms $U_l^{core}(r)$ appropriate to each symmetry l:

$$U^{core} = U_L^{core}(r) + \sum_{l=0}^{L-1} \sum_{m=l}^{-l} |lm\rangle [U_l^{core}(r) - U_L^{core}(r)]\langle lm|, \tag{32}$$

where L is one greater than the largest angular momentum quantum number in the core orbitals. This choice of L has been shown to be adequate by Topiol et al. (1977) in a calculation of the ground-state potential curve for HF. The terms $U_i^{core}(r)$ are obtained numerically from the eigenvalue equation for the valence pseudo-orbital χ_{nl},

$$(h + U_i^{core} + W^{val})\chi_{nl} = \varepsilon_{nl}\chi_{nl}, \tag{33}$$

where W^{val} represents the two-electron interactions in the valence space. The valence pseudo-orbital is a mixture of the valence orbital φ_{nl} with core orbitals φ_c

$$\chi_{nl} = \varphi_{nl} + \sum_c a_c\varphi_c, \tag{34}$$

and is chosen to satisfy the conditions that it should have no radial nodes, that it should be as close as possible to the original all-electron Hartree–Fock orbital, and that it should have the minimum number of undulations.

Kahn et al. (1976) expanded the valence pseudo-orbital in terms of a linear combination of Gaussian orbitals. The numerical effective core potential is fitted to a set of Gaussian functions using the equation

$$r^2\left(U_i^{core}(r) - \frac{N_c}{r}\right) = \sum_k d_{kl}[r^{n_{kl}}\exp(-\xi_{kl}r^2)] \tag{35}$$

where N_c is the number of core electrons. Valence-electron calculations were made for ground-state potential curves for HF, HCl, HBr, HI, F_2, Cl_2, Br_2, and I_2. Good agreement with all-electron calculations was obtained for HF, HBr, and F_2. However, in subsequent work (Hay et al., 1978a) significant discrepancies were noted for F_2 and Cl_2 between the valence-electron calculations and all-electron calculations with identical basis sets. These discrepancies were attributed to a long-range attractive tail in the potential. Corrections were applied which resulted in improved agreement. Christiansen et al. (1979) suggest that the problem is perhaps more fundamental. They propose a method in which the expansion of the valence pseudo-orbital χ_v, according to Eq. (34), is replaced by

$$\chi_v = \varphi_v + f_v \tag{36}$$

where f_v is zero outside the core region and cancels the oscillations of the valence orbital φ_v in the core region. Thus χ_v matches φ_v exactly outside the core region. The use of this procedure results in much better agreement with all-electron calculations for F_2, Cl_2, and LiCl.

Topiol *et al.* (1978) also expand the effective core potential in terms of a set of functions $r^{n_i}\exp(-\alpha_i r^2)$. The pseudo-orbital is a linear combination of Hartree–Fock valence orbitals chosen to give the best fit to a Slater orbital with $n = 3$. The parameters of the effective potential are adjusted to give the best fit between the matrix elements for the coreless Hartree–Fock Hamiltonian and those of the Hartree–Fock Hamiltonian.

Barthelat *et al.* (1977) use what is essentially a model potential given by

$$U^{\text{core}} = \sum_l \sum_m W_l(r)|lm\rangle\langle lm|, \tag{37}$$

where the functions $W_l(r)$ are either a three-parameter function

$$\left(\frac{c_1}{r^2} + c_2 r^2\right)\exp(-\alpha r^2)$$

or a two-parameter function

$$\frac{c_2}{r}\exp(-\alpha r^2).$$

Systematic comparisons for F_2 and Cl_2 are given by Teichteil *et al.* (1977), and further studies are presented by Pelissier and Durand (1980).

An alternative expression to Eq. (33) is obtained by writing U^{core} as the sum of two terms

$$U^{\text{core}} = \sum_c (2J_c - K_c) + V^{\text{GPK}} \tag{38}$$

where V^{GPK} is known as the generalized Phillips–Kleinmann potential which can be expressed in terms of core projection operators. The core Coulomb operators give rise to a spherically symmetrical local potential. In order to retain the non-local character of the exchange and core projection operators, Dixon *et al.* (1977) model U^{core} in terms of a Coulomb potential $J(\rho)$ of the spherical core density $\rho(r)$ and an expansion of the exchange and core projection operators over three or four Gaussian orbitals with expansion coefficients A_{ij}:

$$U^{\text{core}} = J(\rho) + \sum_l \sum_m \sum_i \sum_j A_{ij}|r^l\exp(-\alpha_i r^2)Y_{lm}\rangle\langle Y_{lm}r^l\exp(-\alpha_j r^2)|. \tag{39}$$

The expansion coefficients are determined by fitting to the kernels of the exchange and projection operators. The spherical core density $\rho(r)$ is expanded in terms of $1s$ Gaussian functions. The pseudopotentials derived are used in a multistructure valence bond calculation of excited-state potential curves for SO. An improved parametrization atom scheme has been given by Dixon and Robertson (1979). They compare the results of calculations using the potential of Eq. (39) with those obtained using Eq. (32) and show that the two methods give good agreement for Fe and FeH^+.

The eigenvalue equation for the pseudo-orbital χ_v can also be written in the form

$$(F + V^{PP} + W_{nl})\chi_{nl} = \varepsilon_{nl}\chi_{nl}, \tag{40}$$

where F is the Fock operator for the valence interactions, and W_{nl} represents the core–valence interactions. V^{PP} is obtained from the generalized Phillips–Kleinmann potential by assuming that the core functions φ_{ck}^i in the projection operators are close to the true core solutions so that we can write

$$V^{PP} = \overset{\text{nuclei}}{\underset{k}{\sum}}\ \overset{\text{orbitals}}{\underset{l}{\sum}}\ |\varphi_{ck}^i\rangle(\varepsilon_v - \varepsilon_{ck}^i)\langle \varphi_{ck}^i|. \tag{41}$$

The term W_{nl} will be different for each symmetry type of valence function. Ewig et al. (1977) calculate W_{nl} from the atomic Coulomb and exchange integrals. For molecular calculations they use a weighted average $\bar{W}$, for s and p functions, expanded in terms of Gaussian functions

$$\bar{W}(r) \approx \frac{N}{r}\left(1 - \sum_i c_i\exp(-\alpha_i r^2)\right), \tag{42}$$

where N is the number of core electrons. A similar formalism is used by Wahlgren (1978a, b), but he determines the potential W by fitting matrix elements of W to the matrix elements obtained from atomic calculations. In the calculations of Gropen et al. (1980) on Cl_2, Br_2, and I_2, the parameters of the pseudopotential (42) are determined by requiring the model calculations for atoms to yield similar orbital energies and orbital shapes to those given by all-electron calculations.

The use of a functional form for the core potential W is avoided in the iterative pseudopotential MCSCF of Das and Wahl (1978) which has been used in the calculation of excited-state potential curves for I_2. They conclude that the method is capable of making chemical predictions of the same quality as other methods. However, Das (1980) has pointed out that errors may arise in the description of excited states which have a strong admixture of charge transfer states.

Relativistic effects are important for heavy atoms (see Clugston and Pyper, 1978; Pyykkö and Desclaux, 1979), and valence-electron calculations for molecules containing heavy atoms should employ pseudopotentials or effective core potentials which incorporate these effects. The method of Kahn et al. (1976) has been extended by Kahn et al. (1978) to include relativistic effects. The effective core potential is calculated from approximate relativistic atomic Hartree–Fock wavefunctions obtained from a non-relativistic Hamiltonian to which the mass velocity and Darwin terms of the Pauli equation have been added. Calculations of excited-state potentials using these relativistic effective core potentials have been reported by Wadt *et al.* (1978) and by Hay *et al.* (1978). Spin–orbit effects are taken into account subsequently by perturbation theory (Cohen and Schneider, 1974).

In the work of Das and Wahl (1978), relativistic effects are incorporated by using numerical relativistic Hartree–Fock orbitals in the construction of the core potential, and spin–orbit coupling is again taken into account by perturbation theory.

Lee *et al.* (1978) have extended the methods of Kahn *et al.* (1976) to the calculation of relativistic effective core potential from Dirac–Hartree–Fock atomic wavefunctions. The method yields effective potentials which can be used in the *jj*-coupling scheme for valence-electron calculations on heavy atoms and molecules containing heavy atoms. Alternatively, for cases where *LS* coupling is dominant, averaged relativistic effective core potentials can be constructed and perturbation theory used subsequently to allow for spin–orbit coupling. Applications have been made to potential curve calculations for Xe_2, Xe_2^+ (Ermler *et al.*, 1978) and Au_2 (Lee *et al.*, 1979; Ermler *et al.*, 1979). More recently, Lee *et al.* (1980) have presented calculations in which averaging is avoided and molecular calculations are made in ω–ω coupling.

Basch and Topiol (1979) also derive relativistic effective core potentials from Dirac–Hartee–Fock wavefunctions. Purely *l*-dependent pseudo-valence orbitals are obtained by averaging the two *j*-valued pseudo-valence orbitals belonging to a given *l* value. Relativistic analogues to the pseudo-potentials of Ewig *et al.* (1977) have been calculated from Dirac–Hartee–Fock wavefunctions by Dutta *et al.* (1978).

Relativistic pseudopotential theories have been appraised critically by Pyper (1980) who indicates that pseudopotentials should be derived from the Fock–Dirac equations themselves.

2. *Local Density Methods*

A method of somewhat limited applicability to excited states is the electron-gas model of Gordon and Kim (1972) which has been found to be successful for the calculation of the interactions between closed-shell species. The basic assumption of the model is that the total electron density of the species can be represented by the sum of the Hartree–Fock electron densities of the separated species. The Coulomb contribution to the interaction energy is calculated directly from the electron densities, and the electronic kinetic energy, electron exchange effects, and electron correlation are obtained by using a uniform electron-gas model. The method has been extended by Clugston and Gordon (1977a, b) to include interactions between one open-shell atom and a closed-shell species. In calculations on rare-gas halides

(Clugston and Gordon, 1977a), the open shell is a p orbital for both Rg^+ and X. The electron density is taken as one-half of that for the closed shell. Densities for Rg and X are taken for the covalent states correlating with ground-state atoms, and for the ionic excited states, densities for Rg^+ and X^- are used. The lack of spherical symmetry in the open-shell atom is taken into account by treating it as though it were a diatomic molecule. Σ molecular states are obtained by orienting the half-filled p orbital along the internuclear axis. For the Π states the half-filled orbital is of π symmetry. Comparison with *ab initio* calculations for KrF show that the model works well for the Π states, but agreement for the Σ states is rather poor. The assumption of additive electron density is less good for the Σ states in which there is some bonding between the two species. A similar situation applies to the alkali oxides (Clugston and Gordon, 1977b) in which the model fails to predict Σ ground states for the heavier alkali metals. The model is unable to account for the covalent bonding which occurs in the Σ state.

Nielson *et al.* (1977) have used the method to calculate A' and A'' potential surfaces for the Ar–NO interaction. A comparative study of variants of the electron-gas model has been made by Clugston and Pyper (1979).

A number of calculations for ground- and excited-state potential curves for diatomic molecules have been made using a density functional approach (Gunnarsson *et al.*, 1977). The total electronic energy $E[n]$ is written as a functional of the electron density $n(r)$:

$$E[n] = T_0[n] + E^{\text{ext}}[n] + E^{\text{es}}[n] + E^{\text{xc}}[n]. \tag{43}$$

T_0 represents the kinetic energy of N non-interacting electrons, E^{ext} arises from the interaction of the electron density with the external potential due to the nuclei, and E^{es} and E^{xc} are the electrostatic and exchange correlation parts of the interelectronic interaction E^{int}. The exchange correlation part E^{xc} is approximated by the local density expression

$$E^{\text{xc}}[n] = \int d\mathbf{r}\, n(\mathbf{r})\varepsilon^{\text{xc}}(n(\mathbf{r})), \tag{44}$$

where $\varepsilon^{\text{xc}}(n)$ is the exchange correlation energy per particle of a homogeneous electron gas of density n. The total electronic energy E and the ground-state density $n_0(r)$ correspond to the minimum of the functional (43). It can be shown that the same density minimizes the expression

$$E[n] = T_0(n) + \int d\mathbf{r}\, n(\mathbf{r})V^{\text{eff}}(\mathbf{r}) \tag{45}$$

with

$$V^{\text{eff}}(\mathbf{r}) = V^{\text{ext}}(\mathbf{r}) + \varphi(\mathbf{r}) + \varepsilon^{\text{xc}}(n(\mathbf{r})) + n(\mathbf{r})\frac{\partial}{\partial n}\varepsilon^{\text{xc}}(n(\mathbf{r})), \tag{46}$$

where $\varphi(\mathbf{r})$ is the Coulomb potential of the electrons. The density $n(\mathbf{r})$ is given

by

$$n(\mathbf{r}) = \sum_i f_i |\psi_i(\mathbf{r})|^2, \tag{47}$$

where f_i is an occupation number and ψ_i is a solution of the independent-particle wave equation

$$[-\tfrac{1}{2}\nabla^2 + V^{\mathrm{eff}}(\mathbf{r}) - \varepsilon_i]\psi_i(\mathbf{r}) = 0. \tag{48}$$

The scheme can be modified to yield the lowest energy for a wavefunction specified by a given set of quantum numbers. Splittings between multiplets can be incorporated into the theory by use of a spin-unrestricted formalism which regards the energy as a functional of the spin density as well as of the electron density. The method has been used to calculate spectroscopic parameters for C_2 multiplets (Gunnarsson et al., 1977), potential curves for the $^1\Sigma_g^+$ and $^3\Sigma_u^+$ states of alkali dimers (Harris and Jones, 1978), potential curves for transition-metal dimers (Harris and Jones, 1979) and for the $^2\Sigma^+$ and $^2\Pi$ states of LiBe, LiMg, and LiCa (Jones, 1980). An improved formalism has been presented by Gunnarsson and Jones (1980). March (1981) has reviewed the application of local density methods to molecular calculations.

C. Semiempirical Methods, Diatomics-in-Molecules

Impressive advances have been made in recent years in the application of *ab initio* methods to potential surface calculations. However, it is still extremely difficult and expensive to generate *ab initio* potential surfaces suitable for scattering studies for triatomic systems. It is even more difficult to do this for systems containing more than three atoms because of the additional degrees of freedom. Thus it is natural that much effort has been put into the development of semiempirical methods which can yield chemically useful information with modest computational effort. Balint-Kurti (1975) and Kuntz (1979) have discussed many of the semiempirical methods that have been used in potential surface calculations. For many of the methods, there has been little further development in recent years. Consequently, in this chapter, discussion is restricted to just one of the methods, namely diatomics-in-molecules (DIM). This method does have many features which make it well suited to the calculation of potential surfaces for reactive systems. Detailed accounts of the method have been given by Tully (1973, 1977) and Kuntz (1979), and Tully (1980) has reviewed the application of the method to the calculation of potential energy surfaces.

The fundamental principle of the method is that the n-electron Hamiltonian operator for a polyatomic molecule containing N atoms can be partitioned exactly into terms which are Hamiltonian operators for diatomic and atomic subsystems (Ellison, 1963):

$$H = \sum_{K}^{N} \sum_{L > K}^{N} H^{(KL)} - (N - 2) \sum_{K}^{N} H^{(K)}, \qquad (49)$$

where $H^{(KL)}$ is the Hamiltonian operator for molecule KL, and $H^{(K)}$ is the Hamiltonian for atom K. The wavefunction is constructed from a set of polyatomic basis functions Φ_m which are antisymmetrized products of N atomic functions:

$$\Phi_m(1 \ldots n) = \mathscr{A}_n \varphi_m(1 \ldots n) \qquad (50)$$

where $\mathscr{A}$ is the antisymmetrizer and

$$\varphi_m(1 \ldots n) = \xi_m^{(A)}(1 \ldots n_A)\xi^{(B)}(n_A + 1 \ldots n_A + n_B) \ldots \xi_m^{(N)}(n - n_N + 1 \ldots n). \qquad (51)$$

The function $\xi_m^{(I)}$ represents a many-electron basis function for atom I. The total wavefunction ψ_1 is written as a linear combination of the polyatomic basis functions

$$\psi_1(1 \ldots n) = \sum_m C_{1m} \Phi_m(1 \ldots \ ). \qquad (52)$$

The total Hamiltonian matrix required for the solution of the variational problem for Eq. (52) can be written in terms of atomic and diatomic fragment Hamiltonian matrices,

$$H_{ij}^{(K)} = \langle \Phi_i | \mathscr{A} H^{(K)} | \varphi_j \rangle \qquad (53)$$

$$H_{ij}^{(KL)} = \langle \Phi_i | \mathscr{A} H^{(KL)} | \varphi_j \rangle \qquad (54)$$

thus

$$\mathbf{H} = \sum_{K}^{N} \sum_{L > K}^{N} \mathbf{H}^{(KL)} - (N - 2) \sum_{K}^{N} \mathbf{H}^{(K)}. \qquad (55)$$

The atomic and diatomic fragment Hamiltonian matrix elements can be expressed in terms of atomic and diatomic energies. If the expansion in Eq. (52) is in terms of a complete set of polyatomic basis functions, then the treatment is exact. However, as in all theoretical methods, it is necessary to use a truncated basis set. The basis assumption of the DIM approach is that, even with a truncated set of polyatomic basis functions, the Hamiltonian matrix elements can still be constructed entirely from atomic and diatomic data.

The use of the DIM method for potential surface calculations is appealing for a number of reasons. Since the Hamiltonian matrix is calculated from atomic and diatomic data, the method ensures correct dissociation. One can use the best available experimental or *ab initio* data as input. In cases where the required diatomic potentials are not available, it may be possible to generate *ab initio* curves without excessive effort. Excited potential surfaces are readily generated provided a sufficiently large basis set of atomic fragments and polyatomic basis functions is used. Spin–orbit effects are readily incorporated into the formalism. The computational effort required to obtain the energy for a particular configuration is sufficiently modest for it to be feasible to do this in the course of a dynamical calculation. It is also possible to obtain the gradient of the potential directly.

The DIM method has been reformulated in terms of projections onto diatomic and atomic subspaces by Faist and Muckerman (1979a). They discuss the approximations inherent in the DIM formalism. One problem with the DIM method is the inclusion of overlap between polyatomic basis functions. Overlap is commonly neglected. Proper inclusion of overlap results in a non-Hermitian DIM matrix (Tully and Truesdale, 1976). Eaker (1978) has shown that the problem can be avoided by adopting a Hermitian formulation and fully optimizing the elements of the orthogonal diatomic mixing matrix. An alternative DIM method based on a generalized Hamiltonian has been presented by Wu (1979, 1980). A new parametrization scheme is suggested and the H_3 surface is fitted, using just one parameter.

Recent calculations of potential surfaces for excited states using the DIM method include applications to FH_2^+ (Kendrick et al., 1978), $NeHe_2^+$ (Kendrick and Kuntz, 1979), and triatomic rare-gas halides (Huestis and Schlotter, 1978).

D. Empirical Methods

Dynamical calculations for molecular scattering require as input analytical potential functions. The methods described in the previous sections yield potential surfaces in the form of a table of energies evaluated for particular geometries. Thus, before these can be used in a scattering calculation, it is necessary to fit the points to some suitable analytical function. Such functions are usually empirical, although in some cases they may be based on an approximate valence bond model. Some of the functions used for fitting potential surfaces have been documented by McLaughlin and Thompson (1979), and Murrell (1978) has reviewed the use of analytic potential functions for small polyatomic molecules.

However, the use of empirical analytical functions is not just limited to fitting a grid of *ab initio* points. Quite a lot of work has been done on methods for the construction of analytical functions which can be parametrized with information from spectroscopy and from *ab initio* calculations. A series of papers by Murrell and his coworkers (see Murrell, 1978) has demonstrated how a relatively simple analytical function can give a good description of potential surfaces for a variety of different types of triatomic molecule. The potential function for the molecule ABC is written as the sum of two- and three-body terms (Sorbie and Murrell, 1975)

$$V(R_1, R_2, R_3) = V_{AB}(R_1) + V_{BC}(R_2) + V_{AC}(R_3) + V_1(R_1, R_2, R_3) \qquad (56)$$

where R_1, R_2, and R_3 are the AB, BC, and AC internuclear distances respectively. The three-body term V_1 is chosen so that it becomes zero at all dissociation limits. The form of the potential thus ensures correct dissociation. The three-body potential is written, in terms of displacements s_i from a suitably chosen configuration, as

$$V_1(R_1, R_2, R_3) = A P(s_1, s_2, s_3) \prod_{i=1}^{3} (1 - \tanh(\gamma_i s_i | 2), \qquad (57)$$

where A is a constant, P is a polynomial containing up to quartic terms, and γ_i are adjustable parameters. The coefficients in the polynomial P are also varied to give the best fit to the surface. The number of terms retained in the polynomial P depends on the amount of data available.

The function has been shown to give a realistic description of the potential surface for H_2O (Sorbie and Murrell, 1975), to satisfy the permutation symmetry of O_3 (Murrell et al., 1976), to predict a second minimum in the SO_2 surface (Farantos et al., 1977), and to be applicable to linear molecules (Murrell et al., 1978). A modified three-body term was used by Murrell and Farantos (1977) to describe correctly a metastable symmetric minimum for O_3. The choice of internal coordinates for surfaces which have more than one mimimum was discussed by Murrell et al. (1979a). They also discussed the choice of terms in the polynomial P in cases where the data are insufficient to determine all the coefficients up to a given order. Comparison with *ab initio* calculations for HOF showed that the potential function correctly predicts metastable minima (Murrell et al., 1979b).

The method has been extended by Murrell et al. (1981) to the calculation of a two-valued surface for the $\tilde{X}^1 A$ and $\tilde{B}^1 A$ states of H_2O. The adiabatic energies are obtained by calculating the eigenvalues of a 2×2 matrix. The diagonal elements are given by functions of the form of Eq. (56) and the off-diagonal term is of the form

$$V_{12} = C \sin(\text{HÔH}) \qquad (58)$$

where C is also a three-body term. This approach was shown to give an acceptable fit to a set of *ab initio* points and to describe correctly the $\Sigma-\Pi$ conical intersection for linear configurations. By including four-body terms, Carter et al. (1980) were able to obtain an analytical potential function for the ground state of formaldehyde which reproduced exactly the experimental geometry, the energy, and the quadratic force constants. The geometries, energies, and potential surfaces of all the dissociation products were also given correctly.

A function of the form of (56) is also used by Vance and Gallup (1978) for the fitting of *ab initio* surfaces. Tricubic polynomials are used to represent the three-body term. Potential surfaces which reproduce satisfactorily the presently available data for LiO_2 and NaO_2 have been obtained by Alexander

(1978) who also expressed the potential as the sum of two- and three-body terms. Clary (1979) has proposed an iterative direct semiclassical method for the calculation of potential functions for polyatomic molecules. The potential function is expanded in terms of normal coordinates with linear coefficients. Excitation energies are forced to have given values at one point in the angle variable space. The method is shown to work well for NO_2, SO_2, and ClO_2.

A number of empirical potential surfaces have been published in which a particular model is chosen to suit the system being considered. For rare-gas halide excimers, Krauss (1977) has used a truncated Rittner potential for the short-range interaction and has represented the long-range interaction by terms derived from perturbation theory. Siska (1979) has used a one-electron model potential to calculate potential curves for the interaction of $He(2^1S, 2^3S)$ with Ne, Ar, Kr, and Xe. Interactions between excited and ground-state rare-gas atoms have also been considered by Vallée et al. (1978). The Born formula is used to describe the short-range exchange potential and the long-range interaction potential is treated by perturbation theory.

III. CALCULATED POTENTIAL CURVES FOR DIATOMIC MOLECULES

A. H_2, HeH, He_2, and Their Ions

The Schrödinger equation can be solved exactly for the problem of one electron in the field of two nuclei. T. G. Winter et al. (1977) have calculated, in spheroidal coordinates, the exact eigenvalues (accurate to one part in 10^{13}), electronic wavefunctions, and their derivatives with respect to internuclear separation for the lowest 20 states of HeH^{2+}. States arising from $2p\sigma$, $4f\sigma$, and $4f\pi$ orbitals are predicted to be bound.

For molecules containing two electrons, the electron correlation problem can be tackled directly by incorporating the interelectronic separation directly into the wavefunction. Kolos and Wolniewicz have extended their accurate studies on H_2 by considering Π states (Kolos and Rychlewski, 1977), $^1\Sigma_g^+$ states (Wolniewicz and Dressler, 1977, 1979), and the $a^3\Sigma_g^+$ state (Kolos and Rychlewski, 1978). They expand the wavefunction in functions of elliptic coordinates $\xi_1, \xi_2, \eta_1, \eta_2$ of the form

$$\xi_1^p \eta_1^q \xi_2^r \eta_2^s r_{12}^n \exp(-\alpha_1\xi_1 - \alpha_2\xi_2 + \beta_1\eta_1 + \beta_2\eta_2), \qquad (59)$$

where $\alpha_1, \alpha_2, \beta_1, \beta_2$ are nonlinear parameters to be optimized. For the Π and $^1\Sigma_g^+$ states, $n = 0, 1, 2$. In the case of the $a^3\Sigma_g^+$ study, terms in $n = 3$ were included in the wavefunction. The Π states are Rydberg states arising from the configurations $1\sigma_g 2p\pi$ near R_e for the $C^1\Pi_u$ and $c^3\Pi_u$ states. The $I^1\Pi_g$ and $i^3\Pi_g$ states have the configuration $1s2p\pi$ at large R, but an avoided intersection with the attractive configuration $1\sigma_g 3d\pi$ leads to a minimum in the region of $R = 2a_0$. There is little experimental data for these states. Discrepancies between theory and experiment are probably due to adiabatic effects. The $^1\Sigma_g^+$ states are interesting because their potential curves have double minima. The 2, 3, and 4 $^1\Sigma_g^+$ states are designated EF, GK, and $H\bar{H}$ respectively, with the first letter referring to the inner minimum. These states cannot be described by conventional sets of rotation–vibration constants. The inner minimum of the EF curve corresponds to the configuration $1s\sigma_g 2s\sigma_g$ which changes to $2p\sigma_u^2$ for the outer minimum. For the GK state the configuration changes from $1s\sigma_g 3d\sigma_g$ near the inner minimum through $2p\sigma_u^2$ to $1s2s$ near the outer minimum, whereas the $H\bar{H}$ curve is described by the configurations $1s\sigma_g 3s\sigma_g$ and $1s\sigma_g 3d\sigma_g$. The inner and outer minima of the GK state can each support one vibrational level. Comparison with experiment indicates that this is a useful first-order description. Non-adiabatic coupling effects in the EF and GK states have been calculated by Dressler $et\ al.$ (1979) using the curves of Wolniewicz and Dressler (1977).

Glover and Weinhold (1977) have also calculated the potential curve for the GK state using a 20-term generalized James–Coolidge function in which the exponential function in (59) is replaced by

$$\exp(-\alpha_1\xi_1 - \alpha_2\xi_2)\cosh(\beta_1\eta_1 + \beta_2\eta_2).$$

For the $a^3\Sigma_g^+$ state, Kolos and Rychlewski (1978) found that the inclusion of r_{12}^3 terms improved the value of D_0 by only $0.147\,\mathrm{cm}^{-1}$.

A different approach has been adopted by Kehl $et\ al.$ (1977) who use a demi-H_2 model in which the wavefunction is written in terms of configurations

$$\psi = \frac{1}{\sqrt{2}}(|\chi_1\bar{\chi}_2| \pm |\bar{\chi}_1\chi_2|). \tag{60}$$

χ_1 and χ_2 are exact solutions for the ground and excited states of a one-electron, two-centre problem with nuclear charges 1 and 0·5 respectively. They include all the diatomic orbitals χ_1, χ_2 necessary for correct asymptotic behaviour and include all configurations from a given basis set. Potential curves are calculated for the states $2, 3\,^1\Sigma_g^+$, $1, 2\,^3\Sigma_g^+$, $1, 2\,^1\Sigma_u^+$, $1–3\,^3\Sigma_u^+$, $1\,^1\Pi_g$, $1\,^1\Pi_u$, $1\,^3\Pi_g$, $1\,^3\Pi_u$. A natural orbital analysis of the wavefunctions is given.

Green $et\ al.$ (1978) conclude their studies on HeH by calculating potential curves for the triplet Σ, Π, and Δ states. They use ellipsoidal orbitals for $R \leqq 3a_0$ and a mixture of Slater and ellipsoidal orbitals for large R. All states correlating with $n = 2$ and $n = 3$ are included. There is a qualitative similarity between the curves for singlet and triplet states.

For He_2^{2+}, using a wavefunction of 153 terms of the type

$$\Phi(1, 2) = \exp(-\alpha\xi_1 - \alpha\xi_2)[\exp(\beta\eta_1 + \beta\eta_2) \pm (-1)^{k_i + l_i}$$
$$\exp(-\beta\eta_1 - \beta\eta_2)\xi_1^{m_i}\xi_2^{n_i}\eta_1^{k_i}\eta_2^{l_i}\rho^{q_i} \tag{61}$$

where $\rho = 2r_{12}/R$, Bishop and Cheung (1979) have derived spectroscopic parameters for six bound levels of $^1\Sigma_g^+$ symmetry and eight of $^1\Sigma_u^+$ symmetry. They believe the calculated B_e values to be within a few wavenumbers of the true values.

Potential curves for H_2^- are of interest for the interpretation of electron scattering data, and there have been two recent studies of resonant states of H_2^-. Using a basis of Slater and elliptical orbitals in a nonlinear variational wavefunction, Bardsley and Cohen (1978) calculated potential curves for the $A^2\Sigma_g^+, B^2\Sigma_g^+, C^2\Pi_u, D^2\Delta_g$, and $E^2\Pi_u$ states. They believe their calculations to be accurate to 0·2–0·3 eV for narrow H_2^- resonances at $R \approx 2a_0$ and more accurate for larger R. Van der Hart and Mulder (1979), using a Slater basis and CI, calculated 18 states for both g and u symmetry. Both associate the $B^2\Sigma_g^+$ state with observed resonances.

Hickman and Morgner (1977) have obtained HeH potential curves dissociating to $H(1^2S)$ and $He(2^1S, 2^3S)$ for the calculation of associative and Penning ionization. They performed almost a full CI calculation in which they included all $^2\Sigma$ configurations constructed from three σ orbitals and all double excitations to π orbitals from the configuration $1\sigma^2 2\sigma$.

Saxon et al. (1977a) report an MCSCF calculation for $^1\Sigma_g^+$ states of He_2 in which they include all distributions of four electrons among the orbitals $1\sigma_g, 1\sigma_u, 2\sigma_g, 2\sigma_u$. An avoided crossing between the two highest states correlating with He $(2^3S) + He(2^3S)$ and $He^+(1^2S) + He^-(2^2S)$ results in the doubly excited state being bound by 0·4 eV relative to $He(2^3S) + He(2^3S)$.

B. Hydrides

In view of the wealth of spectral data, it is rather surprising that there have not been more calculations of potential curves for excited states of hydrides than those reported below.

Gerratt and Raimondi (1980) have applied the spin-coupled valence bond method to the $^2\Sigma^+$ states of BeH using a large Slater basis. An interesting feature is that a large change in the spin coupling coefficients occurs in the region of $R = 4a_0$. This can be attributed to an avoided crossing between two curves correlating with $Be(^3P) + H$ and $Be(^1S) + H$. An improved description is obtained by including replacements of pairs of valence orbitals by π orbitals, and it yields ground-state energies lower than those previously obtained in large-scale CI calculations.

The $^2\Sigma^+$ potential curves for BH^+ are qualitatively similar to those for BeH. On the basis of extensive CI calculations using a large Gaussian basis set, Guest and Hirst (1981) predict $R = 1·903$ Å and $\omega_e = 1257·9$ cm^{-1} for the $B'^2\Sigma^+$ state thought to be responsible for chemiluminescence in the $B^+ + H_2$ system.

Interest in radiative charge transfer processes between H and C^{2+}, C^{3+},

N^{2+}, and N^{3+} has led to calculations by Butler *et al.* (1977) and by Butler (1979) of the relevant potential curves. The relevant avoided crossings have been analysed. Calculations for CH^{2+}, CH^{3+} involved full CI, but quadruple excitations were omitted for NH^{2+}. The later calculations for NH^{2+}, NH^{3+} include both the testing of larger basis sets and CI with respect to the set of configurations required for correct dissociation.

Potential curves for the $a^3\Pi$ and $b^3\Sigma^-$ states of CH^+ have been calculated by Kusunoki *et al.* (1980) using SDCI with a large Gaussian basis. The calculated spectroscopic constants are in good agreement with experiment.

In a study on the potential curves for the valence states of NH^+ Guest and Hirst (1977) used a DZP basis and included SDCI with respect to a set of reference configurations appropriate both for R_e and the asymptotes. The calculated R_e values for the excited states were generally in good agreement with experiment. However, the calculations demonstrated that it is often very difficult to reproduce correctly small energy separations. The $a^4\Sigma^-$ state was calculated to be $0.0032E_h$ below the $X^2\Pi$ state, whereas experiment puts the $X^2\Pi$ state lower by $0.0026E_h$. This is probably because the calculation was not recovering a sufficiently high proportion of the valence-shell correlation energy due to deficiencies in the basis set. I. D. L. Wilson (1978), in a study on Λ-doubling in NH^+, reports spectroscopic constants for NH^+ derived from CI calculations using a Slater basis. In these calculations the $a^4\Sigma^-$ state is found to be above the $X^2\Pi$ state, but no details of the potential curves are given.

Banerjee and Grein (1977, 1978) have applied their MCSCF technique for excited states to the $b^1\Sigma^+$, $d^1\Sigma^+$, $A^3\Pi$, and $2^3\Pi$ states of NH. A Slater DZP basis was augmented with orbitals necessary for describing proper dissociation and orbital exponents were optimized. Excellent agreement with experiment is obtained for the known b, d, and A states, and the $2^3\Pi$ state is predicted to be bound with a dissociation energy of 0.51 eV.

Valance (1978a), using a pseudopotential method, has performed one-electron calculations in prolate spheroidal coordinates for the four lowest states of the alkali hydride ions NaH^+, KH^+, RbH^+, and CsH^+, yielding curves considered to be sufficiently accurate for a description of elastic and inelastic scattering. Olson *et al.* (1980) have considered several *ab initio* CI approaches to the five lowest $^2\Sigma^+$ potential curves for NaH^+ and KH^+. Spectroscopic constants are obtained for the $A^2\Sigma$ state. By comparison, valence pseudopotential calculations overestimate the equilibrium separation and underestimate the dissociation energy.

The potential curves for 10 states of MgH^+ have been obtained by Olson and Liu (1979b) using an extensive Slater orbital basis and SDCI. The potential well characteristics for the $X^1\Sigma^+$, $A^1\Sigma^+$, and $B^1\Pi$ states are in good agreement with experiment. Saxon *et al.* (1978), using an extensive basis, have calculated $X^2\Sigma^+$, $A^2\Pi$, and $B'^2\Sigma^+$ potential curves for MgH. They used

MCSCF orbitals as the basis for extensive CI calculations. In their most extensive calculations, including single, double, and triple excitations resulting from three electrons in the valence and virtual orbitals, they obtained values for R_e, D_e, and T_e within 0.03Å, 0.09eV, and 0.05eV of the spectroscopic values. SDCI gave larger discrepancies for the energies but yielded the same spectroscopic constants, indicating that triple excitations have a negligible effect, in this case, on the shape of the potential curve in the region of equilibrium. Comparison with the work of Sink and Bandrauk (1979) indicates that the more flexible MO basis of Saxon et al. (1978) is better able to account for changes in electronic structure occurring as R increases. The $^2\Sigma^+$ potential curves for MgH are qualitatively very similar to those for BeH and BH$^+$.

Using a pseudopotential obtained from a DZ atomic Hartree–Fock calculation for AlH, Pelissier and Malrieu (1977) have applied the CIPSI method to 12 states of AlH. For the ground state at R_e they obtain 77% of the PNO CI and CEPA correlation energies. Satisfactory results are obtained for the $X^1\Sigma^+, a^3\Pi, A^1\Pi$, and $b^3\Sigma^-$ states, which dissociate to give Al in a valence ^2P or ^4P state. States dissociating to excited $^2S, ^2D$, and ^2P states are not well described. AlH (and its positive ion AlH$^+$) have also been considered at the SCF level by Sabelli et al. (1978). A united-atom Si basis set was used for internuclear separations less than R_e. The prime motivation for these calculations seems to be comparison with the Thomas–Fermi model at short R (where correlation corrections are small) rather than the accurate calculation of spectroscopic parameters for the excited states. The ground-state curve does not dissociate correctly but repulsive Σ and Π states do show the correct long-range behaviour.

HCl is a molecule which has an extensive and complicated electronic spectrum. Hirst and Guest (1980) have calculated ab initio potential curves for the $X^1\Sigma^+, t^3\Sigma^+, a^3\Pi$, and $A^1\Pi$ valence states at the SDCI level using a DZP basis augmented with diffuse functions. They found that diffuse functions play a non-negligible role in the highly repulsive region of the $t^3\Sigma^+$ curve. They also made perturbation selection CI calculations for the $B^1\Sigma^+$ state for which the bond length is abnormally long (2.43Å compared with the ground state R_e value of 1.2745Å). The rotational constants and vibrational spacings for this state are very irregular. This state is predicted to have two very shallow minima. A particular difficulty in this calculation is the occurrence of four asymptotes (H$^+$ + Cl$^-$, H$(2s, 2p)$ + Cl(^2P), and H(^2S 1s) + Cl(^2D)) which are very close in energy. Thus, in the basis it was necessary to include hydrogen $2s$ and $2p$ functions as well as diffuse chlorine functions appropriate for a description of Cl$^-$. Goldstein et al. (1978) have used the CI partitioning method of Segal et al. (1978) to calculate resonance states of HCl$^-$. Five $^2\Sigma$ states and one $^2\Pi$ state are considered. As R increases, some of the resonant states merge into the continuum and it is impossible to follow the CI to larger R. Care is taken to check that the responances correspond to physical reality and are not artefacts. However, Krauss and Stevens (1981), in calculations using more flexible basis sets, question this conclusion.

C. Homonuclear Molecules

Konowalow and Rosenkrantz (1979) have used the restricted Hartree–Fock method to calculate potential curves for the $X^2\Sigma_g^+, {}^2\Sigma_u^+, A^2\Pi_u$, and ${}^2\Pi_g$ states of Li_2^+ with an expected accuracy of one or two per cent. A subsequent paper (Konowalow et al., 1979) presents similar calculations for the $2^2\Sigma_g^+$ and $2^2\Sigma_u^+$ states as well as a two configuration function for the $2^2\Sigma_g^+$ state. The long-range behaviour of the excited states is considered in terms of inverse power functions and an exponential exchange potential. Nemukhin and Stepanov (1979) have used a model potential for the ${}^2\Sigma$ and ${}^2\Pi$ states of Li_2^+, Na_2^+, K_2^+, Rb_2^+, and Cs_2^+ and obtain results in harmony with other model potentials. However, for an exact description of the curves, a more flexible multiparameter function is required.

Using an extensive Slater basis set and the OVC MCSCF technique, Olson and Konowalow (1977a, b) and Konowalow and Olson (1979) have calculated potential curves for the lowest $X^1\Sigma_g^+, A^1\Sigma_u^+, b^3\Sigma_g^+, {}^1\Pi_g, {}^3\Pi_g, {}^1\Pi_u, {}^3\Pi_u,$ ${}^3\Sigma_u^+$ states of Li_2. It was found that the inclusion of d functions in the basis has a crucial effect on the shallow potential well of the ${}^3\Sigma_u^+$ state, and the inclusion of f functions may possibly give a further lowering. For the strongly bound $X, A,$ and b states, the calculated spectroscopic constants are thought to be accurate to within 1–3%. The calculations for the Π states were less complete because of the inability to include configurations with one electron in a π orbital and one in a δ orbital. Olson and Konowalow (1977b) were only able to obtain 51% of the experimental D_e for the ${}^1\Pi_u$ state. On the other hand, Kahn et al. (1977), using a less extensive basis set but including π–δ terms in the MCSCF calculation, obtained 68% of the experimental value. Kahn et al. (1977) obtain 111% of the long-range barrier height compared with 125% obtained by Olson and Konowalow (1977b). Watson et al. (1977) have calculated Li_2 potential curves for 10 states using a model potential method. Comparison with the results of Olson and Konowalow indicates that the model potential chosen yields results of useful accuracy. Density functional calculations by Harris and Jones (1978) show weak but definite minima in the potential curves for the ${}^3\Sigma_u^+$ states of $Li_2, Na_2, K_2, Rb_2,$ and Cs_2.

Dupuis and Liu (1978) have calculated potential curves for the ${}^3\Sigma_g^-, {}^5\Sigma_u^-,$ and ${}^3\Sigma_u^-$ states of B_2 to clarify the identity of the ground state. Starting with a large basis set (including f functions), they performed an MCSCF calculation for the ${}^5\Sigma_u^-$ state. This is followed by a valence CI calculation (all configurations obtained by distributing six electrons in the six valence orbitals) and a first-order CI calculation (obtained by adding all single excitations to the configurations of the valence CI). The first-order calculation is believed to be of near-quantitative accuracy, whereas the valence CI does not account for all of the difference in correlation energy between the $^3\Sigma_g^-$ and $^5\Sigma_u^-$ states. The valence semi-internal correlation and polarization effects, which form the second largest contribution to the correlation energy, are neglected in the valence CI. On the basis of the calculations it is concluded that the ground state is $^3\Sigma_g^-$.

Two different approaches to the calculation of potential curves are exemplified by recent calculations on C_2. Kirby and Liu (1979) obtained curves for the 62 states in the valence manifold using MCSCF orbitals with limited CI. Their aim was to obtain a reasonable description of all the valence-state curves rather than to treat a particular spectroscopic problem with 'state of the art' accuracy. Indeed, these calculations do not make adequate allowance for the correlation energy in singlet states, and, as a consequence, result in the $a^3\Pi_u$ state being lower in energy than the $X^1\Sigma_g^+$ state. T_e values are too large for singlets and too small or about right for triplets. Also, there is likely to be substantial valence–Rydberg mixing in some of the upper states. Langhoff et al. (1977) have considered the perturbations between the $X^1\Sigma_g^+$ and $b^3\Sigma_g^-$ states using potential curves obtained from a DZP basis and including CI with single and selected (on the basis of perturbation theory) double excitations. Their calculation of spin–orbit matrix elements demonstrates the inadequacy of the SCF function. Zeitz et al. (1978), using an extensive spd basis augmented with s, p, and d bound functions, have made MRDCI calculations for the $X^1\Sigma_g^+$, $a^3\Pi_u$, and $d^3\Pi_g$ potential curves. R_e values and ω_e values for the X and a states are in excellent agreement with experiment. Zeitz et al. (1979) have made a similar study of the bound $X^2\Sigma_g^+$, $B^2\Sigma_u^+$, $^2\Pi_u$, $^4\Sigma_u^+$, and $^4\Delta_u$ states of C_2^-. All three doublets are bound with respect to C_2, but for the quartet states only the long-range portions of the potential curves are below the energy of C_2. The calculations give a realistic description of relative energy spacings and of the vibrational characteristics of the $^2\Sigma_g^+$ and $^2\Sigma_u^+$ states.

Gunnarsson et al. (1977) have applied the density functional method to the four lowest multiplets of C_2. Although the method reproduces trends in R_e values and in vibration frequencies correctly, it does not give the correct ordering of states.

SCF calculations, using large Gaussian basis sets, have been made by Cobb et al. (1980) for the five lowest states of N_2^{2+}. Comparison with near Hartee–Fock calculations for N_2 and N_2^+ indicates that calculated spectroscopic constants should be relatively accurate. However, correlation effects will be important in the determination of the nature of the ground state. In an MCSCF study, with limited CI, for the $^5\Pi_u - {}^5\Sigma_g^+$ transition in N_2, Krauss and Neumann (1979) observe a rapid variation of transition moment with R. However, the CI is inadequate to give potential curves of spectroscopic accuracy.

Honjou (1978) has calculated potential curves for eight states of O_2^+ for a limited range of internuclear distances, using a minimal Slater basis and full valence CI. The ordering of states is in agreement with experiment, but there are some rather larger deviations from the experimental excitation energies. In an investigation of the predissociation of the $c^4\Sigma_u^-$ state of O_2^+, Tanaka and Yoshimine (1979) have made CI calculations at several levels using MCSCF orbitals derived from an extensive Slater basis. The reliability of different levels of CI is discussed. A quasi-bound state arises from an avoided intersection between the attractive $2\sigma_u 3\sigma_g^2 1\pi_u^4 1\pi_g^2$ configuration and the repulsive $2\sigma_u^2 3\sigma_g 1\pi_u^3 1\pi_g^3$ configuration.

A comprehensive treatment of the 62 valence states of O_2 has been made by Saxon and Liu (1977). A Slater *spd* basis was employed, and Rydberg states and valence–Rydberg interactions ignored. The molecular orbitals were obtained from an MCSCF calculation for the ground state followed by a first-order CI calculation from which natural orbitals were derived. The potential curves were calculated in a first-order CI calculation in which the reference configurations included those necessary to describe dissociation to neutral atoms and to ion pairs. Near-quantitative results are obtained for low-lying states. Ten additional bound states are predicted but five of them are only weakly bound. The possibility of Rydberg–valence interaction makes their energies rather uncertain. More recently, Saxon and Liu (1980a, b) have extended their calculations to Rydberg states. In calculations for the nine lowest $^3\Pi_g$ states. (Saxon and Liu, 1980a), in which different levels of CI were considered, they concluded that 2σ and Rydberg semi-internal excitations have little effect on the Rydberg interaction potential. In the absence of significant Rydberg–valence mixing, a relatively small calculation should be sufficient for the diabatic Rydberg curve. The second paper (Saxon and Liu, 1980b) presents calculations for the $^3\Sigma_g^-$, $^3\Sigma_u^-$, $^3\Pi_g$, $^1\Pi_g$ and $^1\Sigma_g^+$ states. A larger basis set is used and internal and semi-internal excitations for valence and ion-pair states are included. There are quite large discrepancies between the calculated Rydberg asymptotes and the spectroscopic limits. Thus, excitation energies to Rydberg states relative to valence states will not be quantitative, but the energies of Rydberg states relative to each other should be more reliable.

In order to interpret the dissociative recombination of O_2^+, Guberman (1979) has made a more limited study in the region of intersection of excited O_2 curves with the $O_2^+ X^2\Pi_g$ curve. No spectroscopic constants are presented, but energies relative to the asymptotic limits are in agreement with Saxon and Liu (1977). Tatewaki *et al.* (1979) have used a basis which includes some diffuse functions in their CI study of the lower excited states of O_2 and O_2^+, and obtain good agreement with experimental excitation energies. They were particularly interested in Rydberg–valence mixing in the B and $E^3\Sigma_u^-$ states. In their CI calculation they found substantial mixing and concluded that it was essential to include both of the reference configurations $1\pi_u^4 1\pi_g 2\pi_u$ and $1\pi_u^3 1\pi_g^3$. Their calculations for other states only employed single reference configurations and were therefore inadequate for values of R outside the range $2 \cdot 2a_0$ to $2 \cdot 5a_0$.

Das *et al.* (1978) have used the MCSCF method with an extensive Slater basis, including f functions, to calculate potential curves for the $X^2\Pi_g$, $A^2\Pi_u$, $a^4\Sigma_u^-$, and $^2\Sigma_u^-$ states of O_2^-. Care was taken to ensure that the basis gave an adequate description of O^-. The MCSCF calculation included all configurations making a significant contribution to the valence-shell correlation energy. Furthermore, the calculations for the X and A states included additional terms to allow for the dispersion correlation. Excellent agreement with experiment was obtained for ground-state spectroscopic constants. The calculations for the excited Σ states were less accurate than those for the X and A states. The $A^2\Pi_u$ state was subsequently re-examined by Das *et al.* (1980) in calculations for the X, A and B states which were found to exhibit extensive valence–Rydberg mixing. For $R \geq 2 \cdot 75a_0$ the A state is of valence character, but for $R \leq 2 \cdot 5a_0$ the $B^2\Pi_u$ state is the lowest valence state.

Cartwright and Hay (1979, 1980) calculated potential curves for the 12 states of F_2 correlating with $F(^2P) + F(^2P)$, using the POLCI approach. With the exception of the $^3\Pi_u$ state, all excited states are repulsive. The calculated properties for the $^3\Pi_u$ state are in agreement with the results of near-UV photoabsorption studies.

Potential curves for the $X^1\Sigma_g$, $A^1\Sigma_u$, and $B^1\Pi_u$ states of Na_2 have been calculated by the MCSCF method by Stevens, Hessel, Bertoncini and Wahl (1977b). The ground-state curve is in good agreement with the experimental RKR curve. A larger number of states was included in the MCSCF OVC calculations of Konowalow *et al.* (1980) who used a Slater basis which included diffuse s, p, and d functions. The potential curves are generally very similar to those obtained for Li_2. The Σ curves are in satisfactory agreement with experiment but, as in the Li_2 calculations, the calculations are less accurate for the Π states. Valance (1978b) has used a model potential approach for calculations for Na_2^+, K_2^+, Rb_2^+, and Cs_2^+.

Dimers of group II elements have strongly bound excited states and weakly bound or repulsive ground states. Potential curves for states of Mg_2 correlating with $Mg(^1S) + Mg(^1S, ^3P, ^1P)$ and for the $^2\Sigma_u^+$ and $^2\Pi_u$ states of Mg_2^+ have been calculated with the MCSCF method by Stevens and Krauss (1977) using an extensive Slater basis. There is the possibility of valence–Rydberg mixing because molecular Rydberg states may be very low lying. At the Hartree–Fock level the $X^1\Sigma_g^+$ state is predicted to be repulsive, but the MCSCF calculation overestimates the well depth by 40%. The calculation of the ground-state curve is complicated by strong coupling between inter- and intra-atomic electron correlation. The experimental R_e value for the $^1\Sigma_u^+$ state is reproduced, but the calculated D_e value is too small by about

$1800 \, \text{cm}^{-1}$. The calculated curves were subsequently used by Mies *et al.* (1978) as the basis for model calculations for the electronic spectrum of Hg_2.

Potential curves for 13 low-lying states of S_2 have been calculated by Swope *et al.* (1979a) in a SDCI calculation based on single reference configurations. The basis set was chosen to give reasonable ground-state properties, and it was assumed that the excited states were valence in character. Neglect of configurations required for correct dissociation resulted in a rather poor ground-state dissociation energy. Rydberg mixing is probably important for states lying above $50\,000 \, \text{cm}^{-1}$. Inclusion of the effect of quadruple excitations reduced the discrepancy between theory and experiment for excitation energies from about 13% to 5%. Vibration frequencies were calculated to within 10% and R_e values to about 1%.

Liu and Olson (1978) have calculated potential curves for the $^2\Sigma_u^+$ and $^2\Sigma_g^+$ states of Ca_2 using a Slater basis and four different levels of CI. Internal and semi-internal correlation effects of the M-shell electrons and external correlations of the valence electrons were found to have little influence on the shapes of the potential curves.

Harris and Jones (1979) have applied the density functional method to the series of diatomic molecules K_2 to Cu_2. The method is shown to be good for simple σ bonds and satisfactory for s–p hybrid bonds. It is less reliable for s–d bonds because of the tendency of the functional to favour d configurations. The bonding between two nickel atoms has been considered by Shim *et al.* (1979) in *ab initio* CI calculations in which bands of states were obtained. Two sets of calculations were made with nickel atom occupancies $3d^8 4s^2$ and $3d^9 4s^1$. In the former case there is no chemical bonding because the partially occupied d orbitals are located deep in the atom. Interactions between the σ electrons for atoms with the configuration $3d^\circ 4s^1$ does give rise to bonding. There is hardly any d–s hybridization, and the d orbitals do not contribute to the bond.

The diatomic molecules Zn_2, Cd_2, and Hg_2 are of interest as possible laser systems. Bender *et al.* (1979) have made *ab initio* calculations, with full CI for the outer four valence electrons of Zn_2 and Cd_2, for states correlating with one atom in the ground ^1S state and the other in the ^1S, ^3P or ^1P state. The potential curves for the two systems are quite similar apart from the order of the $^1\Sigma_u^+$ and $^1\Pi_u$ states. For systems containing heavy atoms, it is important to make allowance for spin–orbit coupling. It is usually assumed that the spin–orbit coupling does not depend strongly on internuclear distance and that the spin–orbit matrix elements can be calculated from atomic data (Cohen and Schneider, 1974; Hay *et al.*, (1976). For Zn_2 and Cd_2, the spin–orbit curves are generally parallel to the unperturbed curves, but there is a large avoided crossing between the attractive $1_g(^1\Pi_g)$ and the repulsive $1_g(^3\Sigma_g^+)$ spin–orbit states. Potential curves for Mg_2 and Zn_2 can be used as a rough guide to the laser spectroscopy of Hg_2. In a study of the ground states of inert-gas and group IIB dimers, Clugston and Pyper (1978) have shown that, whereas relativistic effects are negligible for Xe_2 and Rn_2, they are increasingly significant for Zn_2, Cd_2, and Hg_2.

Das and Wahl (1978) have used a pseudopotential method in MCSCF calculations for the $X^1\Sigma_g^+$, $^3\Pi_u$, $^3\Pi_g$, and $D^1\Sigma_u^+$ states of I_2. In addition, they calculated curves for the states that will mix with the above states when spin–orbit coupling is included. The accuracy of their pseudopotential method is demonstrated by a comparison of ground-state properties obtained in relativistic calculations with experimental values. The $D^1\Sigma_u^+$ state, which is a strong mixture of $^1\Sigma_u^+$ and $^3\Pi$ states, is well described, but poor results are obtained for the $^3\Pi_{0^+u}$ and $^3\Pi_{0^+g}$ states. In the case of the $^3\Pi_{0^+u}$ state this is attributed primarily to basis set deficiencies and the exclusion of excitations from $5s$ orbitals. Near Hartree–Fock calculations of potential curves for the ground states of I_2, I_2^-, and the lowest $^2\Pi$ and $^2\Sigma^+$ states of I_2^+ are reported by McLean et al. (1980).

Potential curves for Cs_2^{2+} have been calculated by Das and Raffenetti (1980) by CI methods. The curves are repulsive and the ground-state curve is essentially given by the electrostatic repulsion of two Cs^+ ions. There is strong mixing in the excited states, and it is necessary to extend the calculations to $R = 20a_0$ in order to make a proper identification of the asymptotes.

Averaged relativistic effective core potentials were used in first-order CI calculations for potential curves for the ground state and excited states of Au_2 (Lee et al., 1979, Ermler et al., 1979). The calculated dissociation energies for the X, A, and B states were thought to be fortuitously good. The R_e and ω_e values were in reasonable agreement with experiment.

D. Heteronuclear Molecules

Potential curves for the $X^2\Sigma^+$ and $^2\Pi$ states of LiBe, LiMg, and LiCa have been obtained by the density functional method by Jones (1980). The SCF and MCSCF methods have been used by Stevens (1980) to calculate potential curves for the $X^2\Sigma^+$ and $A^2\Pi$ states of LiF^-. In the case of the ground state, there is no chemical bond, and the attractive potential is dominated by charge-induced effects. The SCF method should give adequate results for this state. The interaction between $Li(^2P)$ and F^- in the $A^2\Pi$ state is dominated at long range by charge-quadrupole and charge-induced dipole terms. The results of using this model are in agreement with the SCF calculation. Habitz et al. (1977) have used a model potential in MCSCF calculations for nine low-lying states of LiNa. The ground-state potential is in good agreement with other work. The $^3\Sigma^+$ state has a relatively deep van der Waals well which is due entirely to correlation effects. Correlation effects also give rise to a well in the $^1\Pi$ curve.

There is some doubt about the nature of the $A^2\Pi$ state of BeF. In order to calculate the Λ-doubling for this state, Prosser and Richards (1980) have made SCF MO calculations for the potential curves of nine states of BeF.

There has recently been a lot of interest in the ion CN^+ and, in particular, in the identity of the ground state. Shikamura et al. (1978) calculated potential curves for a number of excited states using a minimal Slater basis with full valence CI. They found that the lowest state was $^1\Sigma^+$, but the error limits on calculated T_e values were too large for a definitive assignment of the ground state. Wu (1978) concluded from the results of SCF calculations and the experimental $a^1\Sigma^+-b^1\Pi$ separation that the ground state was $^3\Pi$. A CI calculation by Ha (1979), using a single reference configuration and selection of configurations, found the $^3\Pi$ state to be lower by 1·02 eV. Extrapolation and inclusion of the effect of quadruple excitations reduced the separation to 0·41 eV. Hirst (1979a), in a CI study using a $(9s5p1d)[3s2p1d]$ basis and several reference configurations, showed that the $^1\Sigma^+$ and $^3\Pi$ states were very close in energy and that the ordering of the states depended on how the calculation was done. In a series of multi-reference CI calculations (with configuration selection and extrapolation to zero threshold), using a $(10s5p1d)[5s3p1d]$ basis. Murrell, Al-Dezi, Tennyson and Guest (1979c) found that the $^1\Sigma^+$ state was $0·0005E_h$ below the $^3\Pi$ state. Finally, in an exhaustive MRDCI calculation with a basis set which included an f function at the midpoint of the bond, Bruna et al. (1980a, c) showed that when the reference set accounts for more than 89% of the final wavefunction (on the basis of Σc_i^2), the energy of the $^1\Sigma^+$ state is consistently below that of the $^3\Pi$ state. They conclude that the $^1\Sigma^+$ state is more stable by 0·1–0·2 eV.

Thus, limited calculations for some problems may give the wrong answer or an unreliable 'correct' answer. In cases like this it is often necessary to perform very extensive CI in order to obtain a meaningful and consistent result.

Full valence CI calculations using a minimal Slater basis set have been made by Honjou and Sasaki (1979) in order to interpret the photoelectron spectrum of CO. These calculations give a useful picture of the excited states of CO^+ and predict 21 new bound states. In an attempt to interpret the satellite structure in the valence x-ray spectrum of CO, Bagus and Viinikka (1977) have calculated limited portions of the CO potential curves for the $A^2\Pi$ state and 19 states of $^2\Sigma^+$ symmetry. They conclude that it is improbable that any structure is due to the $^2\Pi$ state and that many-electron correlation effects are mainly responsible for the observed satellite structure.

There have been several calculations of dipole moment functions for CO. Limited portions of the potential curves of the $a^3\Pi$ and $A^1\Pi$ states have been obtained in a CI calculation by Norbeck et al. (1977). They noted that the SCF 3σ and 4σ MO, as well as the 5σ and 2π orbitals, are different for the a and A states, and that correlation effects are different in the two states.

The ground state has been studied by Kirby-Docken and Liu (1977) and, using the same basis, Kirby and Liu (1978) performed valence and first-order CI calculations for the $d^3\Delta$ and $a'^3\Sigma^+$ states, which were assumed to be of purely valence nature. The calculated R_e values are in excellent agreement with experiment, and the first-order vibrational energies are within 30–40 cm^{-1} of the experimental values, but neglect of L-shell correlation leads to D_e being overestimated. The first-order CI dipole derivatives are thought to be more accurate than the experimental data.

MCSCF calculations, using a limited number of configurations, have been made by Dunning et al. (1979) for low-lying valence states of CF. A second paper (White et al., 1979) considers the ground state of CF$^+$ and, at the Hartree–Fock level, Rydberg states of CF. Potential maxima are reported for the $^4\Sigma^-$, $B^2\Delta$, and $^2\Sigma^+$ curves. Calculated valence-state excitations are about 0.5 eV too high, whereas the CF$^+$ curve is too low. The $B^2\Delta$ curve crosses many of the Rydberg curves, giving rise to perturbations in the spectra.

Although not strictly within the terms of this review, mention should be made of an interesting paper by Albritton et al. (1979) in which evidence from many different sources is used in the construction of potential curves for states of NO lying below 24 eV.

Janoschek and Lee (1978) have compared fully ab initio calculations with the results of pseudopotential curves for NaK. Limited CI (single excitations only) was employed. The pseudopotentials were parametrized to reproduce the ionization potentials of Na and K within 0.1 eV. The ab initio calculations gave potential curves in better agreement with experiment than the pseudopotential calculations. There is a hump on the ab initio curve for the $c^3\Sigma^+$ state which is not reproduced by the pseudopotential calculation.

Calculations, using a new model potential, for the $X^2\Sigma$, $A^2\Pi$, and $B^2\Sigma$ states of NaNe yield potential curves for the X and A states which are in good agreement with spectroscopic data (Masnou-Seeuws et al., 1978). Valance (1978b) has calculated potential curves for four Σ states of the ions NaK$^+$, NaRb$^+$, NaCs$^+$, KRb$^+$, KCs$^+$, and RbCs$^+$, using a simple model potential. The electron-gas model has been applied to the $^2\Sigma$ and $^2\Pi$ states of alkali monoxides by Clugston and Gordon (1977b). The method is less successful for the Σ states than for Π states.

Pelissier and Malrieu (1979) have used a pseudopotential method in a CIPSI calculation of potential curves for states of AlN correlating with Al(^2P) and N(^4S, ^2D). Their calculations indicate that the ground state is $^3\Pi$. Use of $^1\Sigma^+$ molecular orbitals in the CIPSI calculation resulted in a T_e value for the $A^3\Pi$ state which was in poor agreement with experiment. However, the use of $^3\Pi$ orbitals resulted in much better agreement.

Bruna et al. (1980c) have studied electronic states arising from the configurations $\pi_u^4 3\sigma_g^0$, $\pi_u^3 3\sigma_g^1$, and $\pi_u^2 3\sigma_g^2$ for the isovalent series of molecules CN$^+$, Si$_2$, SiC, CP$^+$, and SiN$^+$. A DZP basis was used in most cases and the MRDCI procedure, with configuration selection and extrapolation to zero threshold, was employed. In some cases almost 150 000 configurations were considered. The case of CN^+ has been discussed above. In Si_2, the ground state is $^3\Sigma_g^-$, but the state is almost degenerate with the $^3\Pi_u$ state ($T_e = 0.02\,eV$). SiN^+ also has a $^3\Sigma^-$ ground state, whereas the ground states of CSi and CP^+ are $^3\Pi$. There are little experimental data for Si_2 and none for SiC, CP^+, and SiN^+. The results of these calculations provide good estimates for spectroscopic constants.

The CI method, with selection by perturbation theory, has been used by Langhoff and Arnold (1979), for the calculation of potential curves for the $X^1\Sigma^+, A^1\Pi, C^1\Sigma^-$, and $E^1\Sigma^+$ states of SiO. The CI curve for the ground state was in quantitative agreement with the RKR curve. In the case of the excited states, the calculated curves were several tenths of an electronvolt too high, R_e values were $0.02\,\text{Å}$ too large, and ω_e values $40–50\,cm^{-1}$ too small.

Gottscho et al. (1978) have calculated, at the Hartree–Fock level, portions of potential curves for PN arising from the configurations $2\pi^4 7\sigma^2, 2\pi^4 7\sigma 3\pi$, and $2\pi^3 7\sigma^2 3\pi$. Although the spectroscopic constants are in poor agreement with experiment, adequate estimates of spin–orbit parameters and rotation–electronic interaction energies are obtained.

Two sets of calculations for potential curves of SO have been published recently. Dixon et al. (1977) have used the multistructure valence bond approach with a double zeta Gaussian basis for the valence electrons and a pseudopotential for the core electrons. The lack of d functions in the basis is a serious deficiency which leads to substantial errors in the calculated dissociation energies, even when the orthogonalized Moffitt method is used to correct for atomic errors. They predict six strongly bound states correlating with ground-state atoms and a loosely bound $^3\Sigma^-$ state correlating with $O(^3P) + S(^1D)$. Swope et al. (1979b) have made SDCI calculations, with orbitals $1–6\sigma$ and 1π kept doubly occupied, for the seven lowest states. In most cases a single reference configuration was used, but for the $B^3\Sigma^-$ state it was necessary to use two configurations in order to get satisfactory results. For the $b^1\Sigma^+, a^1\Delta$, and $X^3\Sigma^-$ states, the differences from experiment for T_e, R_e, ω_e are within 15%, 1%, and 8% respectively. The ab initio CI results are closer to experiment than the pseudopotential calculations. On the basis of the discrepancies in excitation energies, a scaling procedure is suggested.

Potential curves for the $X^2\Pi$ and $A^2\Pi$ states of ClO have been calculated by Arnold et al. (1977) using a Slater basis. A 61-term MCSCF calculation was followed by a CI calculation in which configurations were selected by perturbation theory. The calculated curve for the $A^2\Pi$ state is not a good description for $R > 4a_0$ because of the omission of three configurations necessary to describe this state at large R.

CaO is another molecule for which there has been some uncertainty about the nature of the ground state. Experimentally it is thought to be $^1\Sigma^+$. In order to elucidate the electronic structure of the molecule, Bauschlicher and Yarkony (1978a) have started by making one- and two-configuration SCF calculations for 15 states, using a Gaussian basis. The lowest $^1\Sigma^+$ state is found to be highly correlated and, in fact, at this level of calculation, is predicted to lie above the lowest $^3\Pi$ state. It is not possible to give a proper description of the lowest $^1\Sigma^+$ states at this level. The Π states are better described and, because the orbitals are approximately the same for all four states considered, the SCF method is expected to give a reasonably good description. However, for the accurate prediction of R_e, a more extensive basis set is required. The $^3\Sigma^+$ and Σ^- states are also reasonably well described at this level.

Limited CI calculations have been made by Pouilly et al. (1978a, b) for several excited states of FeF. They concluded, on the basis of potential curve calculations for 15 states of $^6\Delta$, $^6\Pi$, $^6\Sigma^+$, $^4\Phi$, and $^4\Delta$ symmetry, that the ground state is $^6\Delta$ with the configuration $3d^6_{Fe}(\sigma\pi^2\delta^3)4s\sigma_{Fe}2p^6_F(\sigma^2\pi^4)$. The molecular orbitals preserve atomic-like character and the bonding is essentially ionic. Shim (1980) has considered nine states of the NiCu molecule arising from $Ni(3d^94s^1)$ and $Cu(3d^{10}4s^1)$. There is little participation of 4p and 3d orbitals in the bonding.

Relativistic effective core potentials have been used by Hay, Wadt, Kahn and Bobrowicz (1978b) in POLCI calculations of potential curves for AuH, AuCl, HgH, and $HgCl_2$. Atomic calculations gave excellent agreement between valence-electron calculations with relativistic core potentials and all-electron Hartree–Fock calculations provided that spin–orbit coupling is included. Quite different results were obtained in relativistic and non-relativistic GVB calculations for AuH and AuCl. The POLCI binding energies are 1 eV too low for AuH and AuCl, due to a combination of basis set deficiencies and the use of limited CI. Excited-state potentials are calculated in the case of HgH for states correlating with $H(^2S) + Hg(^1S, {}^3P, {}^1P)$. Bound states correlating with $Hg(^1P)$ are not given accurately because of avoided crossings with Rydberg states. Spin–orbit effects are found to have a critical effect on calculated R_e values. Calculated excitation energies are within 0·1–0·3 eV of experiment, but dissociation energies are underestimated by 0·6– 1 eV.

E. Diatomic Molecules Containing a Rare-Gas Atom

In reviewing the recent literature on potential surface calculations, one is struck by the large number of papers reporting work on molecules containing a rare-gas atom. One of the motivations for this interest has been the rapid development of electronic transition lasers in which a transition occurs between a bound ionic excited state and an unbound covalent ground state. The theoretical aspects of these lasers have been reviewed by Hay et al. (1979)

with particular emphasis on rare-gas halides and oxides. Excited states of rare-gas dimers and their positive ions Rg_2^+ are also of importance because of their possible participation in loss mechanisms in these lasers.

Stevens, Gardner, Karo and Julienne (1977a) have calculated potential curves for the $^2\Sigma_u^+, ^2\Pi_g, ^2\Pi_u$, and $^2\Sigma_g^+$ states of Ar_2^+, using single-configuration Hartree–Fock wavefunctions. Although these wavefunctions dissociate into the correct fragments, serious errors arise in the asymptotic energies because the same core orbitals are used for the atom and the ion. There may be some cancellation of this error with the molecular correlation energy. Near R_e, spin–orbit splitting is small compared with the Σ–Π separation.

Potential curves for the ground and excited states of Ar_2^+, Kr_2^+, and Xe_2^+ have been calculated by Wadt (1978), using Gaussian basis sets and the POLCI method. Spin–orbit coupling is taken into account by the usual atoms-in-molecules approach (Cohen and Schneider, 1974) in which the spin–orbit matrix elements are assumed to be independent of internuclear separation. Comparison of experimental and theoretical relative absorption cross sections lends some support to the validity of this approach. As one would expect, spin–orbit coupling is small for Ar_2^+ but increases for Kr_2^+ and Xe_2^+. Apart from spin–orbit coupling, relativistic effects are thought to be small. Dissociation energies are underestimated, possibly because of basis set limitations, but the theoretical R_e values should be more reliable than the experimental values. Michels et al. (1978) have compared ab initio CI calculations, using a minimal elliptic basis set, with the results of the $X\alpha$ method for potential curves for Ne_2^+, Ar_2^+, Kr_2^+, and Xe_2^+. The $X\alpha$ method would be expected to give a good description of these nearly closed-shell atom–ion systems in which the bonding is dominated by simple exchange forces. The $X\alpha$ curves are qualitatively similar to the ab initio curves and to those of Wadt (1978). However, the ab initio curves are more repulsive with the discrepancy increasing as R decreases towards R_e. The calculated curves have been used for the simulation of the absorption spectrum for the $A^2\Sigma_{\frac{1}{2}u} \to D^2\Sigma_{\frac{1}{2}g}$ transition (Michels et al., 1979).

Potential curves for the $X^2\Sigma$ and $A^2\Pi$ states of $ArHe^+$ have been obtained in extensive CI calculations by Olson and Liu (1978). Bender and Winter (1978) have calculated potential curves for $ArKr^+$ at the first-order CI level. Inclusion of spin–orbit coupling is more complicated in the case of a heteronuclear ion because the upper curves dissociate to $Ar^+ + Kr$, whereas the lower curves yield $Ar + Kr^+$. At shorter R the spin–orbit coupling may deviate significantly from that of the atomic ion. Inclusion of spin–orbit coupling worsens the agreement between calculated and experimental D_e values for the ground state. These calculations are not accurate enough to determine the binding energies of the excited states.

Two groups (Wadt et al., 1978; Ermler et al., 1978) have used effective potentials in calculations of potential curves for Xe_2^+ and Xe_2. Apart from the effects of spin–orbit coupling, there is little difference between the results using relativistic and non-relativistic potentials for these systems. The two sets of calculations, which used different methods for the calculation of the relativistic contributions, are in good agreement with each other. Comparison with all-electron calculations shows that the effective potentials reproduce dissociation energies satisfactorily but underestimate the repulsive contributions to the potentials.

Iwata (1979) has used the multireference CI method with configuration selection to calculate potential curves for excited states of Ne_2 in an attempt to assign the transient absorption spectrum of the excimer. The manifold of potential curves is complicated by many avoided crossings. A very different approach has been employed by Berman and Kaldor (1979) to the calculation of excited-state potentials for Ne_2 and Ar_2. A modified Fock operator is used to calculate a set of virtual orbitals which are then used in a random-phase approximation (RPA) calculation of excitation energies relative to the ground state. Finally, the curves are corrected by including the difference between the molecular extra-correlation energies for the ground and excited states. The Σ potentials agree well with CI calculations in the literature, but the Π potentials are less steep than the *ab initio* results. This may result from an inadequate $3p\pi$ basis in the *ab initio* calculations. The method is also applied to potential curves for the ions Ne_2^+ and Ar_2^+. Potential curves for excited states of Ar_2 have been calculated by Spiegelmann and Malrieu (1978) using an atomic pseudopotential and the CIPSI method. Well depths tend to be underestimated, and the positions of local maxima are overestimated.

Vallée *et al.* (1978) have represented the long-range interaction between excited and ground-state rare-gas atoms by a multipole expansion. The short-range potential is represented by the Born formula. Excitation of the electron to an s orbital gives a deeper well than for excitation to a p orbital. The interactions between $He(2^1S, 2^3S)$ and Ne, Ar, Kr, Xe have been represented by model potentials (Siska, 1979). The major limitation of the potential is the incomplete specification of the ion-atom contribution to the potential. The potentials were found to be very sensitive to the parameters in the ion–atom interaction term.

Potential curves for states of NeO, ArO, KrO, and XeO correlating with $Rg + O(^3P, {}^1D, {}^1S)$ have been calculated by Dunning and Hay (1977b) using Gaussian basis sets and the POLCI method. On the basis of qualitative considerations, none of the states are expected to be bound, and the degree of repulsiveness should be proportional to the number of $p\sigma$ electrons. However, the $1^1\Sigma^+$ curve will interact with the strongly bound potential which correlates with $Rg^+(^2P) + O^-(^2D)$. This interaction is sufficiently large in KrO and XeO to yield bound $1^1\Sigma^+$ states.

There has been a lot of interest in the rare-gas halides. For ArF, Michels *et al.* (1977) have shown that the density functional approach gives quantitative agreement with more conventional *ab initio* methods.

In these molecules the covalent Σ and Π states correlating with $Rg + F(^2P)$ are unbound, whereas the ionic states correlating with $Kr^+(^2P) + F^-$ are strongly bound. In addition to calculating potential curves, most workers also report calculated transition moments. N. W. Winter *et al.* (1977), using a Gaussian basis, have made first-order CI calculations for the four lowest states of NeF. However, their basis set is not adequate for a good description of the electron affinity of fluorine. Spin–orbit coupling was found to be small in the region of R_e but does have an effect on calculated transition moments. Potential curves for states of KrF correlating with $Kr + F(^2P)$, $Kr^+(^2P) + F^-$, and $Kr^*(^3P) + F(^2P)$ have been calculated by the POLCI method by Hay and Dunning (1977a). A subsequent paper (Dunning and Hay, 1978) gives a comprehensive discussion of the electronic structures of the four lowest states for the series of fluorides NeF, ArF, KrF, and XeF. Spin–orbit coupling is only important for the heavier atoms. A third paper (Hay and Dunning, 1978) considers the effect of varying the halogen atom from F to I in the xenon halides. Since the electron affinities vary over a limited range, the variation in spin–orbit coupling is the more important effect. The ionic states are quite similar because of the dominant effect of the spin–orbit coupling of Xe^+. In the covalent states, the increasing role of the spin–orbit coupling of the halogen atom is evident as one goes from fluorine to iodine. Using their *ab initio* curves, one is able to obtain a generally satisfactory description of the main spectral features. Clugston and Gordon (1977a) have applied the electron-gas model to the calculation of potential curves for rare-gas halides. The Π states appear to be described well by the model, but this method is less satisfactory for the Σ states. An alternative approach to the excited ionic states has been presented by Krauss (1977) who used perturbation theory to describe the long-range part of the potential and a truncated Rittner potential for the short-range interaction. The method is shown to fit the *ab initio* data reasonably well.

Potential curves for ArH have been obtained in SCF calculations by Matcha and Milleur (1978), but the calculations are of limited accuracy because of basis set deficiencies. Vance and Gallup (1980) have used a multiconfiguration valence bond method to calculate curves for the seven lowest excited states. Compared with unpublished CI calculations by Olson and Liu, the VB calculations underestimate dissociation energies and yield R_e values which are too high. This is attributed to basis set deficiencies and the use of a frozen $3s$ subshell for Ar. Gallup and Macek (1977) have calculated potential curves for the five lowest states of XeH^+ at the minimal basis set level using the valence bond method. Again, the results are of limited accuracy because of neglect of polarization functions.

The potential curves for the $X^2\Sigma$, $A^2\Pi$, and $B^2\Sigma$ states of NaAr have shallow van der Waals wells. Saxon *et al.* (1977b) have calculated potential curves for these states using a Slater basis chosen to describe both the long-range induction and dispersion forces and the short-range electrostatic repulsion. The CI calculations included only the interatomic valence correlation energy. A substantial well was obtained for the A state and shallow wells for the $^2\Sigma$ states. Comparison with pseudopotential calculations indicates that the use of a pseudopotential yields more repulsive curves.

POLCI calculations for low-lying electronic states of GaKr have been reported by Dunning et al. (1978). On qualitative grounds, none of the states are expected to be bound. For valence states, the degree of repulsiveness should be proportional to the number of $p\sigma$ electrons. Calculations show that the $1^2\Pi$ state is slightly bound ($D_e \approx 0.04\,\mathrm{eV}$) and that the $2^2\Sigma^+$ state is bound by $0.047\,\mathrm{eV}$. The calculations on GaKr are used to estimate potential curves for InKr and TlKr.

In SCF calculations for the $X^1\Sigma$ potential curve for Br^- and He and the $X^2\Sigma$ and $A^2\Pi$ curves for BrHe, Olson and Liu (1979a) obtain agreement with experiment for the threshold in the dominant electron detachment process.

Guest and Hillier, in collaboration with Ding, have calculated potential curves for the interactions between B^+ and Ne, Ar (Ding et al., 1978), C^+ and Ar (Hillier et al., 1979), and O^+ and Ar (Guest et al., 1979). SCF calculations for NeB$^+$ yield predominantly repulsive curves. For ArB$^+$ a shallow well is obtained for the $^1\Sigma^+$ state and a deeper well for the $^3\Pi$ state. Inclusion of electron correlation is important in the $^3\Pi$ state, and MCSCF calculations gave better agreement with experiment. Calculated binding energies were found to depend critically on the inclusion of polarization functions in the basis set. More extensive calculations were performed for states of ArC$^+$ correlating with $C^+(^2P) + Ar(^1S)$, $C(^3P) + Ar^+(^2P)$, and $C^+(^4P) + Ar(^1S)$. The POLCI method was used in conjunction with multiple root reference sets. Similar calculations were made for states of ArO$^+$ correlating with $O^+(^4S) + Ar(^1S)$ and $O(^3P) + Ar^+(^2P)$. The importance of electron correlation is illustrated in the case of the $X^4\Sigma^-$ state by the contrast between the essentially repulsive SCF curve and a well depth of $0.65\,\mathrm{eV}$ (in agreement with experiment) in the POLCI calculation.

IV. CALCULATED POTENTIAL SURFACES FOR TRIATOMIC MOLECULES

A. H$_3$

Mention should be made of the calculations of Liu (1973) and Siegbahn and Liu (1978) for the ground-state potential surface for H$_3$. These authors used large and carefully chosen basis sets in extensive CI calculations. The collinear surface (Liu, 1973) is estimated to lie between 0.8 and $3.3\,\mathrm{kJ\,mol^{-1}}$

of the exact clamped nuclei limit. The calculation of classical barrier heights to within $4 \, \text{kJ mol}^{-1}$ is a difficult and costly procedure, but accurate results for other properties of the saddle-point region, for the reaction path, and for the potential surface near the reaction path can be obtained with more modest CI.

Kulander and Guest (1979) have calculated slices of C_{2v} geometry of the excited electronic states of H_3 in order to interpret their role in the dissociative recombination of H_3^+. CI calculations yield potential curves for $^2A_1$ and $^2B_2$ states accurate to a few tenths of an electronvolt.

SCF calculations for the lowest singlet and triplet surfaces of H_3^- have been made by Garcia et al. (1979) for isosceles triangle configurations. It was necessary to use a projection operator technique since a completely variational calculation is not valid for non-stationary negative ion states.

B. AH$_2$ Molecules

Bell (1978) has made a useful study on the effects of basis set and of configuration interaction on the calculation of geometries for AH_2 molecules. The survey includes BH_2, CH_2, NH_2, H_2O, and H_2O^+. Similarities are noted for predictions for different species, making it possible to predict a geometry on the basis of a fairly modest calculation. For high accuracy, it is necessary to use a double-zeta basis with polarization functions and to include SDCI.

Potential surfaces for excited states of HeH_2^+ have been calculated, using CI methods with configuration selection, by Preston et al. (1978) and by McLaughlin and Thompson (1979). The latter paper gives careful consideration to the selection procedure and to the accuracy of calculations. The ground-state surface is thought to be accurate to a few kJ mol^{-1}, and the errors in the excited-state surfaces are about twice those for the ground state. Comparison of the SCF and CI ground-state surfaces indicates the correlation energy is fairly constant over the surface. Thus, the SCF method gives an adequate description of this surface. Approximate DIM surfaces for the second and third electronic states have been calculated by Scheider and Zülicke (1979). The surfaces are qualitatively correct up to about $4 \, \text{eV}$, but at higher energies the second excited surface is incorrect at the asymptotes because of neglect of hydrogen $2s$ and $2p$ functions from the basis. Extensive MRDCI calculations have been reported by Römelt et al. (1978a, 1979) for the six lowest potential surfaces of the $He + H_2$ system. Both collinear and perpendicular geometries are considered. On the basis of their surfaces, they interpret the quenching of $HD(^1\Sigma_u^+)$ fluorescence as arising from the perturbation of the $HD(B^1\Sigma_u^+)$ potential in the presence of He by the $HD(E^1\Sigma_g^+)$ state. Osherov and Polujanov (1978) have calculated potential curves for several states of BeH_2 and BH_2 using the DIM method as formulated by Tully (1973).

The importance of including polarization functions in the basis set was demonstrated quite dramatically in the calculations of Pearson and Roueff (1976) along the line of intersection of the $^2A_1$ and $^2B_2$ surfaces of CH_2^+. The inclusion of polarization functions lowers the energy of the intersection from $42 \, kJ \, mol^{-1}$ above that for isolated reactants $C^+ + H_2$ to $63 \, kJ \, mol^{-1}$ below.

Limited CI calculations, using a modest Gaussian basis set, have been made for the $\tilde{B}^2B_2$ and the first three $^4A_2$ surfaces of CH_2^+ by Galloy and Lorquet (1978). Although of limited accuracy, these calculations give useful information about the avoided crossing between the 2^4A_2 and 3^4A_2 surfaces and of predissociation processes for CH_2^+.

Over the past decade there has been a lot of interest in the methylene radical CH_2. This has been concerned with two problems: first, the nature of the ground state and, secondly, the splitting between the $^3B_1$ and $^1A_1$ surfaces. The identity of the ground state has now been established, but there is still controversy regarding the singlet–triplet splitting for which there is disagreement between the experimental value of $81 \cdot 6 \, kJ \, mol^{-1}$ and the results of several high-quality CI calculations. The results of early calculations for CH_2 have been reviewed by Schaefer (1972) and Harrison (1974). Four subsequent studies, using either equilibrium geometries or optimized geometries, report separations of about $46 \, kJ \, mol^{-1}$ (Lucchese and Schaefer, 1977a; Roos and Siegbahn, 1977b; Harding and Goddard, 1977a; Bauschlicher and Shavitt, 1978). Estimated error limits are such that these results are completely inconsistent with experiment. Harding and Goddard (1978) made a more extensive study of the regions of the potential surfaces close to equilibrium for the $^3B_1$, $^1A_1$, $^1B_1$, and $^3B_2$ states and reported rotational constants accurate to within 1%. Using the MRDCI method, Shih et al. (1978) have calculated potential surfaces for the $^3B_1$, $^1A_1$, and $^1B_1$ states of CH_2 and for the $^2B_1$ and $^2A_1$ states of CH_2^-. Vibrational wavefunctions and energy levels are calculated from the surfaces, and angular potential curves are presented for the five lowest states. Their value of $46 \, kJ \, mol^{-1}$ for the singlet–triplet splitting is consistent with the results of other workers. More recently, Saxe, Schaefer and Handy (1981b) have reported the results of CI calculations which include all single and double excitations with respect to all the configurations of a full valence CI. This resulted in configuration spaces of 57684 and 84536 for the singlet and triplet states respectively. This calculation yielded a singlet–triplet splitting of $43 \cdot 9 \, kJ \, mole^{-1}$. It seems unlikely that more extensive calculations will yield a separation which is significantly different, and the resolution of the discrepancy between theory and experiment awaits new experimental information.

MCSCF and CI calculations have been made by Bauschlicher and Yarkony (1978b) to establish a linear structure for the 2^1A_1 state of CH_2. Bauschlicher (1980c) reported an optimized geometry for the $^1B_1$ state and a value for the

$^1B_1-^1A_1$ splitting. Gervy and Verhaegen (1977) have used a semiempirical atoms-in-molecules approach to electron correlation in calculations of angular potential curves for the $^3B_1$, 1^1A_1, $^1B_1$, and 2^1A_1 states of CH_2.

The isoelectronic species NH_2^+ is of interest because of its relevance to the reaction of N^+ with H_2. Gittins *et al.* (1977), in a CI study using a double zeta Gaussian basis set, explored the surfaces correlating with $\tilde{X}N^+(^3P) + H_2$ for collinear and perpendicular geometries. As the geomtery is distorted slightly from C_{2v} symmetry, an avoided intersection between the $^3B_1$ and $^3A_2$ surfaces gives a low-energy route to the deep well of the $^3B_1$ ground state. The intersection between the $^3B_1$ and $^3A_2$ surfaces was considered in more detail by Bender *et al.* (1977) in a CI study in which polarization functions were included in the basis set. The inclusion of polarizatioń functions leads to a deepening of the well in the $^3A_2$ surface. A further study by Hirst (1978b) showed that the inclusion of polarization functions has little effect on the collinear and $^3B_1$ surfaces, and it confirmed the nature of the avoided crossing in the work of Gittins *et al.* (1977). SCF calculations (Hirst, 1978a) for quintet states of NH_2^+ correlating with $N(^4S) + H_2^+$ indicated that the $^5A_2-^5A''-^5\Sigma^-$ surface is repulsive for C_{2v} and near-C_{2v} geometries. However, for collinear and near-collinear geometries, there are no potential barriers to be surmounted for the formation of $NH^+(^4\Sigma^-)$. A more wide-ranging study of the excited states of NH_2^+ has been made by Peyerimhoff and Buenker (1979a), who made MRDCI calculations for the 14 lowest states. They present angular potential curves and potential curves for the symmetric stretch at two angles, and they consider asymmetric stretching in the region of potential minima. The electronic structures of the excited states are discussed, and excitation energies and intensities for electronic transitions calculated. The calculations indicate a need to revise correlation diagrams previously used for this system. Dunleavy *et al.* (1980) have made CI calculations for the vertical energies for the ionization of the $\tilde{X}^2B_1$ state of NH_2 to the $^3B_1$, $^1A_1$, $^1B_1$, and $^3A_2$ states of NH_2^+. Angular curves calculated at the SCF level are presented for a fixed value of R_{NH}.

Bell and Schaefer (1977) have made SDCI calculations, with various basis sets, for the potential surfaces of the $\tilde{X}^2B_1$, $\tilde{A}^2A_1$, and $^2B_2$ states of NH_2. The barrier height in shown to be sensitive to the correlation energy correction. The inclusion of polarization functions is essential for a correct description of the $^2B_2$ state which is predicted to have an equilibrium angle of $47.5°$. In addition to calculating angular potential curves for the $\tilde{X}$, $\tilde{A}$, and $^2B_2$ states, Peyerimhoff and Buenker (1979b) also considered Rydberg states of NH_2 arising from the configuration $\pi_u^2 3s$. Their results for the $\tilde{X}$ and $\tilde{A}$ states are in agreement with those of Bell and Schaefer (1977) and with experiment. They also predict a small bond angle $(50°)$ for the $^2B_2$ state, which is expected to have a relatively long lifetime. The potential curves for the Rydberg states are parallel to the corresponding curves for the ion.

A CIPSI calculation for the collinear potential surface for the reaction

$$O^+(^4S) + H_2(^1\Sigma_g^+) \rightarrow OH^+(^3\Sigma^-) + H \tag{62}$$

has been performed by Chambaud *et al.* (1978). No barrier to reaction was found. Surfaces for the H_2O^+ system correlating with $O^+(^4S) + H_2, O(^3P) + H_2^+, O^+(^2D) + H_2, OH^+(X^3\Sigma^-) + H, OH(X^2\Pi) + H^+, OH^+(^1\Delta, ^3\Pi) + H$ have been calculated by Balint-Kurti *et al.* (1980) with a double-zeta basis using the multistructure valence bond method. They find that there is no large barrier to reaction (62) even for highly bent geometries. Jackels (1980) has calculated potential surfaces for H_2O^+ in the region of equilibrium for the $\tilde{X}^2B_1, \tilde{A}^2A_1,$ and $\tilde{B}^2B_2$ states. Equilibrium geometries and vibrational frequencies are calculated. The crossing between the $^2A_1$ and $^2B_2$ surfaces is considered in detail. Several cuts through the 1 and $2\,^2A'$ surfaces are calculated and show that C_s distortions on the $1\,^2A'$ surface are favoured only in the immediate vicinity of the intersection. Dissociation pathways for the $\tilde{B}^2B_2$ state are discussed, and the reaction of $O(^3P)$ with H_2^+ to form $H_2O(\tilde{B}^2B_2, \tilde{A}^2A_2)$ is shown not to involve potential barriers.

Calculations of the potential surfaces for the $^1A', ^3A'', ^1A'',$ and $\tilde{B}^1A'$ states of H_2O have been made by Howard *et al.* (1979), using a basis set which includes both Rydberg and polarization functions. Correlation was taken into account by first-order CI with iteration over natural orbitals. The $^1A'$ surface shows no barrier for the insertion reaction $O(^3P) + H_2$. The $^3A''$ surface is totally repulsive, and there is a barrier for the insertion and collinear abstraction reactions. Details of the intersection of the $^1\Sigma^+$ and $^1\Pi$ surfaces (which becomes an avoided intersection in C_s symmetry) for hydrogen removal are given. Murrell *et al.* (1981) have calculated 150 points on the $\tilde{X}^1A'$ and $\tilde{B}^1A'$ surfaces using the MRDCI approach with configuration selection. The aim of these calculations was to generate surfaces of comparable accuracy for linear HOH, HHO, C_{2v}, and C_s geometries to provide data for the development of a two-valued analytical surface. Their basis set included Rydberg functions but omitted polarization functions. Consequently, dissociation energies were consistently underestimated. The use of empirical diatomic curves in the fitting procedure improved agreement with experiment. Angular potential curves for several excited states of $NH_2, NH_2^+, H_2O^+,$ and H_2O have been calculated by Wasilewski (1979) using a SCF method and a double zeta Gaussian basis.

CI calculations for surfaces for the entrance channels for the reactions $F^+ + H_2$ and $F + H_2^+$ have been made by Mahan *et al.* (1978). Triplet surfaces correlating with $F^+ + H_2$ in C_{2v} geometry are repulsive and unlikely to lead to reaction. The $^1A_1$ surface correlating with $F + H_2^+$ is strongly attractive and the $^3B_1$ and $^3B_2$ surfaces are also attractive. The FH_2^+ surface has also been considered by Kendrick *et al.* (1978) who performed both MCSCF CI and DIM calculations. As well as providing a comparison with the DIM results, the *ab initio* calculations were used to generate some of the diatomic potentials and the coupling parameters required for the DIM calculation. On the basis of the calculated surfaces, they concluded that the reaction

$$F^+(^3P) + H_2(X^1\Sigma_g^+) \rightarrow HF^+(X^2\Pi) + H \tag{63}$$

must occur by a non-adiabatic mechanism. The DIM surface reproduces the *ab initio* surface fairly well in the entrance channel and for the bottom of the exit channel. The shape of the collinear surface is reliably reproduced.

The theoretical treatment of the dynamics of the reaction $F + H_2$ is a multistate problem which requires potential surfaces for the excited states. In dynamical calculations by Komornicki *et al.* (1977) and Zimmerman *et al.* (1979), the ground state was represented by a LEPS surface, and empirical valence bond surfaces were used for the excited states. Faist and Muckerman (1979b) have applied their DIM method to the calculation of spin–orbit surfaces of the FH_2 system. The method yields ground-state surfaces which are qualitatively correct, and it is assumed that the excited-state surfaces are also qualitatively correct.

On the basis of angular potential curves for the $\tilde{X}^2B_1$ and $\tilde{A}^2A_1$ states of PH_2 calculated by the SCF method, with a minimal basis set which includes phosphorus $3d$ functions. So and Richards (1977) obtain structures consistent with the observed absorption and emission spectra. Perić *et al.* (1979) have made a thorough study of these two states in extensive MRDCI calculations. In addition to calculating angular potential curves, they also consider symmetric and asymmetric stretching. Their calculations of the vibrational intensity distribution reproduce the fairly detailed observed structure almost quantitatively.

The MRDCI method, using extensive Gaussian basis sets, has been used to investigate two aspects of the H_2S^+ system. A dissociation correlation diagram for the $\tilde{A}^2A_1$ and $\tilde{B}^2B_2$ states has been derived by Hirsch and Bruna (1980) on the basis of calculations for six different nuclear conformations. In their detailed discussion of dissociation pathways, they show that a correlation diagram constructed from a relatively small number of high-quality calculations can be a reliable guide to reaction dynamics. A study by Bruna, Hirsch, Perić Peyerimhoff and Buenker (1980b) is concerned with the calculation of the equilibrium geometries for the $\tilde{X}^2B_1$, $\tilde{A}^2A_1$, and $\tilde{B}^2B_2$ states, of angular potential curves, and of symmetric and asymmetric stretching curves in the region of the potential minima. Vibrational energy levels are also calculated. The paper includes a comparison with PH_2, NH_2, and H_2O^+

C. HAB Molecules

In the course of valence bond calculations for the potential surface for the reaction

$$Li + HF \rightarrow LiF + H, \tag{64}$$

which takes place on the ground-state surface, Balint-Kurti and Yardley (1977) also present cuts through the surfaces for a number of excited states. Collinear and perpendicular Li–F–H geometries were considered. The separation between the ground state $^2A'$ surface and the first excited state of that symmetry is smallest for linear geometries, and non-adiabatic transitions will take place preferentially in such configurations.

The 11 lowest states of the C_2H radical have been considered in an MRDCI calculation by Shih *et al.* (1979). Variation of the C–C bond length shows that nearly all the excited states prefer a longer bond length than that of the ground state. Angular potential curves are presented for $R_{CC} = 2.274a_0$ and $2.674a_0$, and there are significant differences between the absorption and emission spectra. Both are expected to have long progressions.

The vibrational structure of the third band in the photoelectron spectrum of HCN has been interpreted qualitatively in terms of a local maximum in the $\tilde{B}^2\Sigma^+$ surface of HCN^+. In limited CI calculations, using a double zeta basis set, for the $\tilde{B}^2\Sigma^+$ state, Hansoul *et al.* (1978) found no evidence for such a local maximum. However, Hirst (1979b), using a basis which included polarization functions, made extensive CI calculations and found evidence for a barrier of height 0.27 eV relative to $H + CN^+$. CI calculations for excited states of HCN have been made by Vazquez and Gouyet (1978, 1979), but their Gaussian basis set did not include polarization or Rydberg functions. The earlier paper contains angular potential curves for $R_{CN} = 2.185a_0$ and $2.451a_0$, and the second paper considers H–CN dissociation curves for two values of R_{CN} and bond angles of 180° and 125°. The spectroscopy and predissociation are discussed in terms of the calculated potential curves.

Potential surfaces for the four lowest doublet and the two lowest quartet states of HCO have been obtained in CI calculations, with a minimal basis set, by Tanaka and Davidson (1979). Optimized geometries for the $\tilde{X}^2A'$ and $\tilde{A}^2\Pi$ states are calculated with a better basis set. Both collinear and bent geometries are considered, and the conical intersection between the two lowest $^2A'$ surfaces is explored. The third $^2A'$ surface is more complicated and has three intersections. Hydrocarbon flame bands are interpreted in terms of the calculated surfaces.

The ground state and three excited states for HNO^+ and NOH^+ have been compared by Marian *et al.* (1977). SCF dissociation curves and angular curves are calculated for both species. Bruna and Marian (1979a) have made CI calculations for the angular potential curves for both species for a set of bond lengths near to the equilibrium values. The $1^2A'$ and $1^2A''$ states were considered by McLean *et al.* (1978) who compared different levels of CI. HNO has been studied by Nomura and Iwata (1979), who made CI calculations for the three lowest states of $^1A'$, $^3A'$, $^1A''$, and $^3A''$ symmetry. Use of a minimal basis set made full CI feasible. They present potential curves as a function of R_{NH} for values of R_{NO} and θ fixed at the ground-state values. The barrier height for the $\tilde{A}^1A''$ surfaces and $1^3A''$ surfaces are low enough for the hydrogen atom to pass over thermally. Many avoided crossings occur for the higher excited states. Equilibrium geometries obtained in MRDCI calculations are reported for NOH by Bruna and Marian (1979b).

Van Lenthe and Ruttink (1978) report RHF calculations for the $^3A''$ and $^1A''$ states of HO_2^+ and MCSCF calculations for the $^1A'$ state. The ion cannot be adequately described as a protonated oxygen molecule. Barriers for moving the hydrogen atom around the oxygen atoms are calculated. A multiconfiguration description is essential for the $^1A'$ state, and use of a single determinant would give rise to inconsistencies in the potential surface. An extensive MRDCI study of the potential surfaces of HO_2 has been made by Langhoff and Jaffe (1979). The two lowest states of $^2A''$ and $^2A'$ symmetry were considered, and the equilibrium geometries for the $1^2A''$ and $1^2A'$ states were in excellent agreement with experiment. Reaction of $O(^3P)$ with $OH(X^2\Pi)$ should lead to the $1^2A'$ and $1^2A''$ states being equally populated. The $2^2A'$ and $2^2A''$ states are ionic. Calculated photoabsorption cross sections and molar extinction coefficients are in good agreement with experiment at 300 K, but agreement is less good at 1100 K.

Ab initio CI calculations have been made for HOF by Murrell, Carter, Mills and Guest (1979c) in order to test the predictions of their analytical potential for this species. POLCI calculations, as a function of R_{OH}, for linear geometries of O...HF showed that the order of states is $^1\Delta < {}^1\Pi < {}^1\Sigma$ and that the linear configuration is preferred for the $^1\Delta$ state. Only the $^1\Delta$ state is bound relative to $O(^1D) + HF(X^1\Sigma^+)$. For the configuration HF...O, the order is $^1\Sigma < {}^1\Pi < {}^1\Delta$. However, in this case, the equilibrium geometry is a bent structure with a bond angle of about 95°. Studies with different basis sets indicated that the inclusion of a $3d$ function at the midpoint of the FO bond plays a significant role in the description of the FO bond breaking process.

Preuss *et al.* (1977) have considered the first three states of the HFF and FHF radicals. The bonding in these species is essentially described by the attraction of a fluorine atom to the HF dipole. The $^2A'$ ground state of HFF is bent, but the $^2A''$ state prefers a linear geometry. For FH...F, linear configurations are more stable. The angular potential curves for all states are very flat. Potential curves for 10 states of FHF^-, as a function of R_{FH}, are obtained in CI calculations by Loges and Sabin (1979). The lowest-lying excited states are not energetically accessible in biological processes.

In a comparative study of isomerization energies for HAB species, Buenker *et al.* (1980) present potential curves as a function of R_{CSi} for five states of $HCSi^+$ and $HSiC^+$ and for the $^2\Pi$ and $^2\Sigma$ states of HCSi and HSiC.

Potential curves for the HCS^+-CSH^+ system have been obtained in SCF calculations by Bruna *et al.* (1978). They present angular curves and stretching curves for the removal of the hydrogen atom, both for the linear configuration and for a bond angle of 130°. Whereas the ground state of HCS^+ is linear, excited states in which the $10a'$ orbital is occupied prefer bent geometries. Both the ground and excited states of CSH^+ have maxima for linear geometries, and there is no barrier to the rotation of the hydrogen atom from the sulphur atom to carbon. In contrast to HCO^+, excited states of HCS^+ dissociate to $CS^+(^2\Pi) + H$.

MRDCI calculations for 12 states of HNSi have been made by Preuss *et al.* (1979) for three bond angles. Most of the states lie within a range of 0·5 eV. The angular potential curves for $\pi\pi^*$ configurations are flat, and bent configurations are preferred in most cases. Variation of R_{SiN} for a bond angle of 150° showed that R_e values for $\pi\pi^*$ states are longer than for the ground state. States arising from $\sigma\pi^*$ configurations prefer intermediate bond lengths.

Some potential surface data for the excited $^2A'$ state of HSO are given by Sannigrahi *et al.* (1977c), and Sannigrahi *et al.* (1977a) compare the electronic structures of HSO and SOH. The HS_2 radical has been treated in a similar manner (Sannigrahi *et al.*, 1977b). The principal aim of these studies was the determination of equilibrium geometries and the calculation of transition probabilities.

In a theoretical study of the photodissociation of HOCl, Jaffe and Langhoff (1978) have used CI (with selection of configurations) to calculate angular potential curves for several R_{ClO} values for the $2^1A'$, $1^1A''$, and $1^3A'$ states. Transition moments for the $1^1A' \rightarrow 2^1A'$ and $1^1A' \rightarrow 1^1A''$ transitions are calculated from ClO stretching curves, obtained using a better basis set. Bruna *et al.* (1979) have made more extensive SCF calculations for potential curves for HOCl and HClO. Angular potential curves for eight states and potential curves for the removal of Cl from HOCl, for the removal of O from HClO, and for hydrogen abstraction are calculated. Excitation of HOCl to a valence state will result in dissociation to $Cl(^2P) + OH(^2\Pi)$. Similarly, the excited valence states of HClO are repulsive. Triplet states correlate with $O(^3P) + HCl(^1\Sigma^+)$, whereas the singlet states yield $O(^1D) + HCl$. Abstraction of a hydrogen atom is unlikely because of the presence of substantial barriers in the excited states.

Potential curves for the $B^2\Pi_u$ and $C^2\Sigma_u^+$ states of Ar_2H, calculated by Matcha and Milleur (1978), are similar in appearance to the curve for Ar_2H^+, suggesting that the cores of the excited neutral species are very similar to the electronic structure of the ion.

D. ABC Molecules

In extensive CI calculations for potential surfaces for Li_3, Kendrick and Hillier (1977) found that the minima for the $^2B_2$ and $^2A_2$ surfaces were very flat and close in energy. They were unable to decide which was the ground state. The equilibrium geometries correspond to small deviations from the equilateral geometry. The minima in the triangular configurations are of lower energy than the minimum for the linear $^2\Sigma_u^+$ state, and symmetric linear configurations are more stable than asymmetric structures. A DIM study by Companior (1978) gives the order of bonding energies as $^2B_2 > {}^2\Sigma_u^+ > {}^2A_1$ and also favours symmetric linear configurations. Alexander (1978) has proposed an empirical potential function for the lowest A' and A'' surfaces of LiO_2 and NaO_2. The function, which uses just a few adjustable parameters, is capable of reproducing satisfactorily the presently available experimental and theoretical data for these species. A semiempirical valence bond method developed by Zeiri and Shapiro (1978) gives good agreement with the orthogonalized Moffitt method for the $^2\Sigma^+$ excited state of LiF_2.

Potential surfaces for excited states of C_3 have been calculated by Perić-Radić *et al.* (1977) and by Römelt *et al.* (1978b). The earlier paper presents curves for the $^1\Pi_u$ and $^3\Pi_u$ states. Detailed consideration of the $^1\Sigma_u^+$ state and angular potential curves for states of $^1B_2$ symmetry, for C_{2v} geometries, are given in the second paper. The bending curves are complex because of several avoided crossings. Spectroscopic constants for five states of CCO have been obtained by Walch (1980) from POLCI calculations for limited portions of the potential surfaces.

The SCF method has been used by Rossi and Bartram (1979) for the calculation of potential curves for linear states of N_3^- obtained by excitation of an electron from the $2\pi_u$ orbital to the $1\pi_g$, $3\sigma_u$, and $4\sigma_g$ orbitals. Bent configurations were considered for triplet states arising from the configuration $2\pi_u 1\pi_g^3$. The calculations support previous assignments of the spectrum, but agreement is only qualitative.

The structure of the NCO^+ ion has been investigated by Wu (1977) and by Wu and Schlier (1978). The first paper presents potential curves for the first three $^3\Sigma^-$ states as a function of R_{NC} for a fixed value of R_{CO}. The second paper considers the three isomers NCO^+, CNO^+, and NOC^+ at the SCF level and concludes that NCO^+ is the more stable species. CI calculations were made for the $1^3\Sigma^-$, $2^3\Sigma^-$, $3^3\Sigma^-$, $1^3\Pi$, $2^3\Pi$, and $1^3\Delta$ states of NCO^+ as a function of R_{CN} and of angle. Avoided crossings occur for the $1^3\Pi$–$2^3\Pi$ and $2^3\Sigma^-$–$3^3\Sigma^-$ pairs of states.

A comprehensive discussion of nine states of O_3 has been given by Hay and Dunning (1977b). Using a $(9s5p1d)$ $[3s2p1d]$ basis, they made CI calculations using several variants of the POLCI method. Equilibrium geometries, force constants, fundamental frequencies, and excitation energies were calculated for all of the states. Angular potential curves were included, and more complete potential surface data were given for the 1^1A_1 and 1^3B_1 states. The 1^1B_2 state was shown to be unstable with respect to asymmetric distortions. Calculations indicate that the only bound excited state is the 1^3B_2 state. MCSCF/CI calculations by Wilson and Hopper (1981) support the existence of a bound $^3B_2$ state. One of the intriguing features of the excited states of ozone is the possibility of the existence of a cyclic D_{3h} structure of $1'A_1$ symmetry. A number of papers report an energy for this state, relative to the ground state. Hay and Dunning (1977b) predict that the cyclic state should be at $117·9 \, kJ \, mol^{-1}$. An SDCI calculation by Lucchese and Schaefer (1977b), which includes an estimate of the effects of quadruple excitations, gives a separation of $74·5 \, kJ \, mol^{-1}$. Harding and Goddard (1977b) made extensive GVB CI calculations and report a separation of $114·2 \, kJ$ mol^{-1}. On the basis of MCSCF calculations with CI, Karlström et al. (1978) obtain a separation of $120–128 \, kJ \, mol^{-1}$. More recently, Burton (1979) has made CEPA calculations with several different basis sets, and he put the energy separation in the range $40–88 \, kJ \, mol^{-1}$. He believes that the D_{3h} structure corresponds to a local minimum on the 1^1A_1 ground-state surface. However, Wilson and Hopper (1981) show that the ring state cannot be involved in the formation of O_3 from ground state asymptotic species. Carlsen and Schaefer (1977) have made SCF calculations for open (C_{2v}) and closed (D_{3h}) forms of S_3 and indicate that the open form is lower in energy.

Potential surfaces for four A' states and one A'' state of $NeHe_2^+$ have been obtained in an extended DIM calculation by Kendrick and Kuntz (1979). Comparison is made with calculations using a more limited basis. The ground-state surface indicates that there are low-energy pathways from $Ne + He_2^+$ to $NeHe^+ + He$ and $Ne^+ + He + He$ which do not have energy barriers. Schmidt et al. (1978) obtained approximate potential surfaces for the same five states of $NeHe_2^+$ by subtracting the appropriate orbital energies from the energies obtained in single-configuration SCF calculations. The surfaces are qualitatively similar to those of Kendrick and Kuntz (1979), but the DIM calculations give a better description of the asymptotic regions. An adequate description of the asymptotic regions is essential for a reliable treatment of avoided crossings. The mechanism of the $Ne + He_2^+$ reaction is discussed in terms of the calculated surfaces, and it is suggested that the reaction should be considered in terms of non-adiabatic transitions between the $1A'$ and $3A'$ surfaces.

Potential curves for the symmetric stretching of the linear $X^1\Sigma_g$ and $^1\Sigma_u$ states of Cl_3^-, Br_3^-, and I_3^- have been calculated by Tasker (1977). A valence bond method with pseudopotentials was used, and empirical ortho-gonalized Moffitt corrections were included.

Angular potential curves for 11 states of $HgCl_2$ have been calculated by Wadt (1980). The work used a non-relativistic effective core potential for Cl and a relativistic potential for Hg. States arising from the excitation of an electron from $2\pi_g$, $1\pi_u$, and $2\sigma_u$ orbitals to the $4\sigma_g$ orbital prefer nonlinear geometries, whereas linear geometries are favoured for $2\pi_g \to 2\pi_u$ excitations.

A number of papers have considered potential curves for the rare-gas halides Rg_2X. Bender and Schaefer (1978) have made first-order CI calculations for nine states of Ne_2F. The 2^2B_2 state is chemically bound and can radiate to lower $^2A_1$ and $^2B_2$ states. Preliminary calculations for Ar_2F have been made by Michels *et al.* (1977) who suggest that the 2^2B_2 state is the radiating state. A more thorough study of Ar_2F and Kr_2F has been made by Wadt and Hay (1978) who used the POLCI method. The covalent $^2A_1$, $^2B_2$, and $^2B_1$ states of Ar_2F are repulsive, whereas the ionic states are all bound relative to $Rg_2^+ + F^-$. Only the 2^2B_2 state is bound with respect to $Ar^+F^- + Ar$. The calculations included spin–orbit coupling which has a relatively small effect. As in the case of Ne_2F, there are dipole-allowed transitions to the 1^2B_2 and 1^2A_2 states. For Kr_2F only the 2^2B_2 state is stable with respect to $Kr^+F^- + Kr$. The spin–orbit coupling is more substantial than for Ar_2F. The DIM method has been used by Huestis and Schlotter (1978) for the calculation of the nine lowest states of Ne_2F, Ar_2F, Kr_2F, and Kr_2Cl. For the ionic states of Ar_2F, the DIM calculations agree with the POLCI calculations of Wadt and Hay (1978) to better than $0.15\,eV$. Good agreement with the *ab initio* bond lengths is also obtained.

V. CONCLUDING REMARKS

Several basic requirements have to be met for the accurate calculation, by *ab initio* methods, of potential energy surfaces for excited states. First, the basis set should be sufficiently flexible to describe the excited state and its dissociation products. Secondly, dissociation processes should be described correctly. This usually means that a method other than the self-consistent field method must be used. Thirdly, the calculations should be capable of describing changes in correlation energy, on going from the equilibrium configuration to the dissociation limits.

For many diatomic molecules, it is possible to make extensive *ab initio* calculations which yield results in good agreement with experiment. Calculations at this level have a very valuable predictive role for molecules for which there is no experimental data. Less complete calculations can give useful information about the electronic structures of excited states and make predictions for bound states which have not yet been observed. However, in some cases, great care must be taken in order to ensure that predictions are reliable.

In the prediction of spectroscopic parameters for triatomic molecules, it is necessary to consider only a limited portion of the potential surface. Such calculations can often be made with high accuracy and furnish much valuable information for species for which it is difficult to make measurements. It is much less easy to calculate *ab initio* surfaces which are suitable for dynamical studies. Many more points on the surface are required, and it is much more difficult to ensure that all regions of the surface are treated with comparable accuracy. There have been relatively few *ab initio* calculations for excited-state surfaces for reactive processes.

Developments in semitheoretical and semiempirical methods continue to provide computational schemes which are capable of giving useful information for systems for which high-quality *ab initio* calculations are not possible. However, it is always necessary to check that such methods yield results which are consistent with the best *ab initio* calculations for smaller systems or with experimental data.

References

B. G. Adams, J. Paldus and J. Čížek (1977), *Int. J. Quantum Chem.*, **11**, 849

G. F. Adams, G. D. Bart, G. D. Purvis and R. J. Bartlett (1979), *J. Chem. Phys.*, **71**, 3697

R. Ahlrichs, H. Lischka, V. Staemmler and W. Kutzelnigg (1975), *J. Chem. Phys.*, **62**, 1225

D. L. Albritton, A. L. Schmeltekopf and R. N. Zare (1979), *J. Chem. Phys.*, **71**, 3271.

M. H. Alexander (1978), *J. Chem. Phys.*, **69**, 3502

J. O. Arnold, E. E. Whiting and S. R. Langhoff (1977), *J. Chem. Phys.*, **66**, 4459.

R. F. W. Bader and R. A. Gangi (1975), in *Theoretical Chemistry*, vol. 2, eds. R. N. Dixon and C. Thomson, The Chemical Society, London, p. 1.

P. S. Bagus, B. Liu, A. D. McLean and M. Yoshimine (1980), in *Computational Methods in Chemistry*, ed. J. Bargon, Plenum, New York, p. 203.

P. S. Bagus and E. K. Viinikka (1977), *Phys. Rev.*, **A15**, 1486.

G. G. Balint-Kurti (1975), in *Molecular Scattering: Physical and Chemical Applications*, ed. K. P. Lawley, Wiley, London, p. 137.

G. G. Balint-Kurti and M. Karplus (1974), in *Orbital Theories of Molecules and Solids*, ed. N. H. March, Clarendon, Oxford; p. 250.

G. G. Balint-Kurti, O. J. Martinussen-Runde and A. C. S. Ramos (1980), in *Electron Correlation*, eds. M. F. Guest and S. Wilson, SRC Daresbury Laboratory, Warrington, p. 109.

G. G. Balint-Kurti and R. N. Yardley (1977), *Faraday Discuss. Chem. Soc.*, **62**, 77.

A. Banerjee and F. Grein (1977), *J. Chem. Phys.*, **66**, 1054.

A. Banerjee and F. Grein (1978), *Chem. Phys.*, **35**, 119.

J. N. Bardsley and J. S. Cohen (1978), *J. Phys. B*, **11**, 3645.

J. Bargon (ed.) (1980), *Computational Methods in Chemistry*, Plenum, New York.

J. C. Barthelat, P. Durand and A. Serafini (1977), *Mol. Phys.*, **33**, 159.

R. J. Bartlett and G. D. Purvis (1978), *Int. J. Quantum Chem.*, **14**, 561.

R. J. Bartlett and G. D. Purvis (1980), *Phys. Scr.*, **21**, 255.

R. J. Bartlett and I. Shavitt (1977a), *Int. J. Quantum Chem. Symp.*, **11**, 165.

R. J. Bartlett and I. Shavitt (1977b), *Chem. Phys. Lett.*, **50**, 190

R. J. Bartlett and I. Shavitt (1978), *Chem. Phys. Lett.*, **57**, 157

R. J. Bartlett, I. Shavitt and G. D. Purvis (1979), *J. Chem. Phys.*, **71**, 281.

H. Basch and S. Topiol (1979), *J. Chem. Phys.*, **71**, 802.

C. W. Bauschlicher (1980a), *J. Chem. Phys.*, **72**, 880.

C. W. Bauschlicher (1980b), *Chem. Phys. Lett.*, **74**, 277.

C. W. Bauschlicher (1980c), *Chem. Phys. Lett.*, **74**, 273.

C. W. Bauschlicher and I. Shavitt (1978), *J. Am. Chem. Soc.*, **100**, 739.

C. W. Bauschlicher and D. R. Yarkony (1978a), *J. Chem. Phys.*, **68**, 3990.

C. W. Bauschlicher and D. R. Yarkony (1978b), *J. Chem. Phys.*, **69**, 3875.

C. W. Bauschlicher and D. R. Yarkony (1980), *J. Chem. Phys.*, **72**, 1138.

S. Bell (1978), *J. Chem. Phys.*, **68**, 3014.

S. Bell and H. F. Schaefer (1977), *J. Chem. Phys.*, **67**, 5173.

C. F. Bender, J. H. Meadows and H. F. Schaefer (1977), *Faraday Discuss. Chem. Soc.*, **62**, 59.

C. F. Bender, T. N. Rescigno, H. F. Schaefer and A. E. Orel (1979), *J. Chem. Phys.*, **71**, 1122.

C. F. Bender and H. F. Schaefer (1978), *Chem. Phys. Lett.*, **53**, 27.

C. F. Bender and N. W. Winter (1978), *Appl. Phys. Lett.*, **33**, 29.

M. Berman and U. Kaldor (1979), *Chem. Phys.*, **43**, 375.

F. P. Billingsley and M. Krauss (1974), *J. Chem. Phys.*, **60**, 4130.

D. M. Bishop and L. M. Cheung (1979), *Mol. Phys.*, **38**, 1475.

F. W. Bobrowicz and W. A. Goddard (1977), in *Modern Theoretical Chemistry*, vol. 3, ed. H. F. Schaefer, Plenum, New York, p. 79.

C. Bottcher (1980), in *Potential Energy Surfaces*, ed. K. P. Lawley, Wiley, New York, p. 63.

S. F. Boys (1950), *Proc. R. Soc.* (*London*), **A200**, 542.

B. R. Brooks, W. D. Laidig, P. Saxe, N. C. Handy and H. F. Schaefer (1980), *Phys. Scr.*, **21**, 312.

B. R. Brooks and H. F. Schaefer (1978), *Int. J. Quantum Chem.*, **14**, 603.

B. R. Brooks and H. F. Schaefer (1979), *J. Chem. Phys.*, **70**, 5092.

P. J. Bruna, R. J. Buenker and S. D. Peyerimhoff (1980a), *Chem. Phys. Lett.*, **72**, 278.

P. J. Bruna, G. Hirsch, M. Perić, S. D. Peyerimhoff and R. J. Buenker (1980b), *Mol. Phys.*, **40**, 521.

P. J. Bruna, G. Hirsch, S. D. Peyerimhoff and R. J. Buenker (1979), *Can. J. Chem.*, **57**, 1839.

P. J. Bruna and C. M. Marian (1979a), *Chem. Phys.*, **37**, 425.

P. J. Bruna and C. M. Marian (1979b), *Chem. Phys. Lett.*, **67**, 109.

P. J. Bruna, S. D. Peyerimhoff and R. J. Buenker (1978), *Chem. Phys.*, **27**, 33

P. J. Bruna, S. D. Peyerimhoff and R. J. Buenker (1980c), *J. Chem. Phys.*, **72**, 5437.

R. J. Buenker, P. J. Bruna and S. D. Peyerimhoff (1980), *Isr. J. Chem.*, **19**, 309.

R. J. Buenker and S. D. Peyerimhoff (1974), *Theor. Chim. Acta* (*Berl.*), **35**, 33.

R. J. Buenker and S. D. Peyerimhoff (1975), *Theor. Chim. Acta* (*Berl.*), **39**, 217.

R. J. Buenker and S. D. Peyerimhoff (1978), in *Excited States in Quantum Chemistry*, eds. C. A. Nicolaides and D. R. Beck, Reidel, Dordrecht, p. 45.

R. J. Buenker, S. D. Peyerimhoff and W. Butscher (1978), *Mol. Phys.*, **35**, 771.

R. J. Buenker, S. D. Peyerimhoff and M. Perić (1978), in *Excited States in Quantum Chemistry*, eds. C. A. Nicolaides and D. R. Beck, Reidel, Dordrecht, p. 63.

P. G. Burton (1979), *J. Chem. Phys.*, **71**, 961.

S. E. Butler (1979), *Phys. Rev.*, **A20**, 2317.

S. E. Butler, S. Guberman and A. Dalgarno (1977), *Phys. Rev.*, **A16**, 500.

N. R. Carlsen and H. F. Schaefer (1977), *Chem. Phys. Lett.*, **48**, 390.

S. Carter, I. M. Mills and J. N. Murrell (1980), *Mol. Phys.*, **39**, 455.

D. C. Cartwright and P. J. Hay (1979), *J. Chem. Phys.*, **70**, 3191.

D. C. Cartwright and P. J. Hay (1980), *J. Chem. Phys.*, **72**, 1421.

G. Chambaud, P. Millie and B. Levy (1978). *J. Phys. B*, **11**, L211.

T. C. Chang and W. H. E. Schwarz (1977), *Theor. Chim. Acta* (*Berl.*), **44**, 45.

P. A. Christiansen, Y. S. Lee and K. S. Pitzer (1979), *J. Chem. Phys.*, **71**, 4445.

J. Čižek (1966), *J. Chem. Phys.*, **45**, 4256

J. Čížek (1969), *Adv. Chem. Phys.*, **14**, 35.

J. Čížek and J. Paldus (1980), *Phys. Scr.*, **21**, 251.

D. C. Clary (1979), *Chem. Phys.*, **41**, 387.

M. J. Clugston and R. G. Gordon (1977a), *J. Chem. Phys.*, **66**, 239.

M. J. Clugston and R. G. Gordon (1977b), *J. Chem. Phys.*, **66**, 244.

M. J. Clugston and N. C. Pyper (1978), *Chem. Phys. Lett.*, **58**, 457.

M. J. Clugston and N. C. Pyper (1979), *Chem. Phys. Lett.*, **63**, 549,

M. Cobb, T. F. Moran, R. F. Borkman and R. Childs (1980), *J. Chem. Phys.*, **72**, 4463.

J. S. Cohen and B. Schneider (1974), *J. Chem. Phys.*, **61**, 3236.

A. L. Companion (1978), *Chem. Phys. Lett.*, **56**, 500.

I. L. Cooper and C. N. M. Pounder (1979), *J. Chem. Phys.*, **71**, 957.

E. Dalgaard and P. Jorgensen (1978), *J. Chem. Phys.*, **69**, 3833.

G. Das (1980), *Chem. Phys. Lett.*, **71**, 202.

G. Das and R. C. Raffenetti (1980), *Chem. Phys. Lett.*, **71**, 198.

G. Das and A. C. Wahl (1978), *J. Chem. Phys.*, **69**, 53.

G. Das, A. C. Wahl, W. T. Zemke and W. C. Stwalley (1978), *J. Chem. Phys.*, **68**, 4252.

G. Das, W. T. Zemke and W. C. Stwalley (1980), *J. Chem. Phys.*, **72**, 2327.

A. Ding, J. Karlau, J. Weise, J. Kendrick, P. J. Kuntz, I. H. Hillier and M. F. Guest (1978), *J. Chem. Phys.*, **68**, 2206.

R. Ditchfield, W. J. Hehre and J. A. Pople (1971), *J. Chem. Phys.*, **54**, 724.

R. N. Dixon and I. L. Robertson (1978), in *Theoretical Chemistry*, vol. 3, eds. R. N. Dixon and C. Thomson, The Chemical Society, London, p. 100.

R. N. Dixon and I. L. Robertson (1979), *Mol. Phys.*, **37**, 1979.

R. N. Dixon, P. W. Tasker and G. G. Balint-Kurti (1977), *Mol. Phys.*, **34**, 1455.

K. Dressler, R. Gallusser, P. Quadrelli and L. Wolniewicz (1979), *J. Mol. Spectrosc.*, **75**, 205.

W. Duch and J. Karwowski (1979), *Theor. Chim. Acta (Berl.)*, **51**, 175.

S. J. Dunleavy, J. M. Dyke, N. Jonathan and A. Morris (1980), *Mol. Phys.*, **39**, 1121.

T. H. Dunning and P. J. Hay (1977a), in *Modern Theoretical Chemistry*, vol. 3, ed. H. F. Schaefer, Plenum, New York, p. 1

T. H. Dunning and P. J. Hay (1977b), *J. Chem. Phys.*, **66**, 3767.

T. H. Dunning and P. J. Hay (1978), *J. Chem. Phys.*, **69**, 134.

T. H. Dunning, M. Valley and H. S. Taylor (1978), *J. Chem. Phys.*, **69**, 2672.

T. H. Dunning, W. P. White, R. M. Pitzer and C. W. Matthews (1979), *J. Mol. Spectrosc.*, **75**, 297.

M. Dupuis and B. Liu (1978), *J. Chem. Phys.*, **68**, 2902.

S. N. Dutta, C. S. Ewig and J. R. Van Wazer (1978), *Chem. Phys. Lett.*, **57**, 83.

C. E. Dykstra (1977), *J. Chem. Phys.*, **67**, 4716.

C. E. Dykstra (1978), *J. Chem. Phys.*, **68**, 1829.

C. E. Dykstra (1980), *J. Chem. Phys.*, **72**, 2928.

C. E. Dykstra, H. F. Schaefer and W. Meyer (1976), *J. Chem. Phys.*, **65**, 2740.

C. E. Dykstra and W. C. Swope (1979), *J. Chem. Phys.*, **70**, 1.

C. W. Eaker (1978), *J. Chem. Phys.*, **69**, 1453.

F. O. Ellison (1963), *J. Am. Chem. Soc.*, **85**, 3540.

W. C. Ermler Y. S. Lee and R. S. Pitzer (1979), *J. Chem. Phys.*, **70**, 293.

W. C. Ermler, Y. S. Lee, K. S. Pitzer and W. W. Winter (1978), *J. Chem. Phys.*, **69**, 976.

C. S. Ewig, R. Osman and J. P. van Wazer, (1977), *J. Chem. Phys.*, **66**, 3557.

M. B. Faist and J. T. Muckerman (1979a), *J. Chem. Phys.*, **71**, 225.

M. B. Faist and J. T. Muckerman (1979b), *J. Chem. Phys.*, **71**, 233.

S. Farantos, E. C. Leisegang, J. N. Murrell, K. S. Sorbie, J. J. Texeira-Dias and A. J. C. Varandas (1977), *Mol. Phys.*, **34**, 947.

D. F. Feller and K. Ruedenberg (1979), *Theor. Chim. Acta (Berl.)*, **52**, 231.

I. Ferguson, and N. Handy (1979), *Theor. Chim. Acta (Berl.)*, **53**, 345.

W. I. Ferguson and N. C. Handy (1980), *Chem. Phys. Lett.*, **71**, 95.

C. Galloy and J. C. Lorquet (1978), *Chem. Phys.*, **30**, 169.

G. A. Gallup and J. Macek (1977), *J. Phys. B.*, **10**, 1601.

R. Garcia, A. R. Rossi and A. Russek (1979), *J. Chem. Phys.*, **70**, 5463.

J. Gerratt (1974), in *Theoretical Chemistry*, vol. 1, ed. R. N. Dixon, The Chemical Society, London, p. 60.

J. Gerratt and M. Raimondi (1980), *Proc. R. Soc.*, **A371**, 525.

Z. Gershgorn and I. Shavitt (1968), *Int. J. Quantum Chem.*, **2**, 751.

D. Gervy and G. Verhaegen (1977), *Int. J. Quantum Chem.*, **12**, 115.

M. A. Gittins, D. M. Hirst and M. F. Guest (1977), *Faraday Discuss. Chem. Soc.*, **62**, 67.

R. M. Glover and F. Weinhold (1977), *J. Chem. Phys.*, **66**, 303.

E. Goldstein, G. A. Segal and R. W. Wetmore (1978), *J. Chem. Phys.*, **68**, 271.

R. G. Gordon and Y. S. Kim (1972), *J. Chem. Phys.*, **56**, 3122.

R. A. Gottscho, R. W. Field and H. Lefebvre-Brion (1978), *J. Mol. Spectrosc.*, **70**, 420

T. A. Green, H. H. Michels and J. C. Browne (1978), *J. Chem. Phys.*, **69**, 101.

F. Grein and A. Banerjee (1975), *Int. J. Quantum Chem. Symp.*, **9**, 147.

O. Gropen, S. Huzinaga and A. D. McLean (1980), *J. Chem. Phys.*, **73**, 402.

S. L. Guberman (1979), *Int. J. Quantum Chem. Symp.*, **13**, 531.

M. F. Guest, A. Ding, J. Karlau, J. Weise and I. H. Hillier (1979), *Mol. Phys.*, **38**, 1427.

M. F. Guest and D. M. Hirst (1977), *Mol. Phys.*, **34**, 1611.

M. F. Guest and D. M. Hirst (1981), *Chem. Phys. Lett.*, **80**, 131.

M. F. Guest and W. R. Rodwell (1977), *SPLICE Reference Manual*, Science Research Council, Rutherford Laboratory, Chilton, Didcot.

M. F. Guest and V. R. Saunders (1974), *Mol. Phys.*, **28**, 819.

M. F. Guest and S. Wilson (eds.) (1980a), *Electron Correlation*, Science Research Council, Daresbury Laboratory, Warrington.

M. F. Guest and S. Wilson (1980b), *Chem. Phys. Lett.*, **72**, 49.

O. Gunnarsson, J. Harris and R. O. Jones (1977), *J. Chem. Phys.*, **67**, 3970.

O. Gunnarsson and R. O. Jones (1980), *J. Chem. Phys.*, **72**, 5357.

T.-K. Ha (1979), *Chem. Phys. Lett.*, **66**, 317.

P. Habitz, W. H. E. Schwarz and R. Ahlrichs (1977), *J. Chem. Phys.*, **66**, 5117.

N. C. Handy (1980), *Chem. Phys. Lett.*, **74**, 280.

N. C. Handy, J. D. Goddard and H. F. Schaefer (1979), *J. Chem. Phys.*, **71**, 426.

J. P. Hansoul, C. Galloy and J. C. Lorquet (1978), *J. Chem. Phys.*, **68**, 4105.

L. B. Harding and W. A. Goddard (1977a), *J. Chem. Phys.*, **67**, 1777.

L. B. Harding and W. A. Goddard (1977b), *J. Chem. Phys.*, **67**, 2377

L. B. Harding and W. A. Goddard (1978), *Chem. Phys. Lett.*, **55**, 217.

J. Harris and R. O. Jones (1978), *J. Chem. Phys.*, **68**, 1190.

J. Harris and R. O. Jones (1979), *J. Chem. Phys.*, **70**, 830.

J. F. Harrison (1974), *Acc. Chem. Res.*, **7**, 378.

P. J. Hay and T. H. Dunning (1977a), *J. Chem. Phys.*, **66**, 1306.

P. J. Hay and T. H. Dunning (1977b), *J. Chem. Phys.*, **67**, 2290.

P. J. Hay and T. H. Dunning (1978), *J. Chem. Phys.*, **69**, 2209.

P. J. Hay, T. H. Dunning and W. A. Goddard (1975), *J. Chem. Phys.*, **62**, 3912.

P. J. Hay, T. H. Dunning and R. C. Raffenetti (1976), *J. Chem. Phys.*, **65**, 2679.

P. J. Hay, W. R. Wadt and T. H. Dunning (1979), *Annu. Rev. Phys. Chem.*, **30**, 311.

P. J. Hay, W. R. Wadt and L. P. Kahn (1978a), *J. Chem. Phys.*, **68**, 3059.

P. J. Hay, W. R. Wadt, L. P. Kahn and F. W. Bobrowicz (1978b), *J. Chem. Phys.*, **69**, 984.

D. Hegarty and M. A. Robb (1979), *Mol. Phys.*, **38**, 1795.

A. P. Hickman and H. Morgner (1977), *J. Chem. Phys.*, **67**, 5484.

I. H. Hillier, M. F. Guest, A. Ding, J. Karlau and J. Weise (1979), *J. Chem. Phys.*, **70**, 864.

G. Hirsch and P. J. Bruna (1980), *Int. J. Mass Spectrom. Ion Phys.*, **36**, 37.

D. M. Hirst (1978a), *Chem. Phys. Lett.*, **53**, 125.

D. M. Hirst (1978b), *Mol. Phys.*, **35**, 1559.

D. M. Hirst (1979a), *Chem. Phys. Lett.*, **65**, 181.

D. M. Hirst (1979b), *Mol. Phys.*, **38**, 2017.

D. M. Hirst and M. F. Guest (1980), *Mol. Phys.*, **41**, 1483.

N. Honjou (1978), *Mol. Phys.*, **35**, 1569.

N. Honjou and F. Sasaki (1979), *Mol. Phys.*, **37**, 1593.

G. Hose and U. Kaldor (1979), *J. Phys. B*, **12**, 3827.

R. E. Howard, A. D. McLean and W. A. Lester (1979), *J. Chem. Phys.*, **71**, 2412.

D. L. Huestis and N. E. Schlotter (1978), *J. Chem. Phys.*, **69**, 3100.

W. J. Hunt and W. A. Goddard (1969), *Chem. Phys. Lett.*, **3**, 414.

A. C. Hurley (1976), *Electron Correlation in Small Molecules*, Academic, London.

B. Huron, J. P. Malrieu and P. Rancurel (1973), *J. Chem. Phys.*, **58**, 5745.

S. Huzinaga (1965), *J. Chem. Phys..*, **42**, 1293.

S. Huzinaga (1971), *Approximate Atomic Functions I, II*, Dept of Chemistry, The University of Alberta.

S. Huzinaga (1973), *Approximate Atomic Functions III*, Dept of Chemistry, The University of Alberta.

S. Huzinaga and Y. Saki (1969), *J. Chem. Phys.*, **50**, 1371.

S. Iwata (1979), *Chem. Phys.*, **37**, 251.

C. F. Jackels (1980), *J. Chem. Phys.*, **72**, 4873.

R. L. Jaffe and S. R. Langhoff (1978), *J. Chem. Phys.*, **68**, 1638.

J. A. Jafri and J. L. Whitten (1977), *Theor. Chim. Acta (Berl)*, **44**, 305.

R. Janoschek and H. U. Lee (1978), *Chem. Phys. Lett.*, **58**, 47.

R. O. Jones (1980), *J. Chem. Phys.*, **72**, 3197.

L. Kahn, P. Baybutt and D. G. Truhlar (1976), *J. Chem. Phys.*, **65**, 3826.

L. P. Kahn, T. H. Dunning, N. W. Winter and W. A. Goddard (1977), *J. Chem. Phys.*, **66**, 1135.

L. P. Kahn, P. J. Hay and R. D. Cowan (1978), *J. Chem. Phys.*, **68**, 2386.

G. Karlström, S. Engström and B. Jönsson (1978), *Chem. Phys. Lett.*, **57**, 390.

J. J. Kaufmann (1975a), in *Interactions Between Ions and Molecules*, ed. P. Ausloos, Plenum, New York, p. 185.

J. J. Kaufmann (1975b), in *The Excited State in Chemical Physics*, ed. J. W. McGowan, Wiley, New York, p. 113.

J. J. Kaufmann (1979), in *Kinetics of Ion–Molecule Reactions*, ed. P. Ausloos, Plenum, New York, p. 1.

S. Kehl, K. Helfrich and H. Hartmann (1977), *Theor. Chim. Acta (Berl.)*, **44**, 351.

J. Kendrick and I. H. Hillier (1977), *Mol. Phys.*, **33**, 635.

J. Kendrick and P. J. Kuntz (1979), *J. Chem. Phys.*, **70**, 736.

J. Kendrick, P. J. Kuntz and I. H. Hillier (1978), *J. Chem. Phys.*, **68**, 2373.

K. Kirby and B. Liu (1978), *J. Chem. Phys.*, **69**, 200.

K. Kirby and B. Liu (1979), *J. Chem. Phys.*, **70**, 893.

K. Kirby-Docken and B. Liu (1977), *J. Chem. Phys.*, **66**, 4309.

W. Kolos and J. Rychlewski (1977), *J. Mol. Spectrosc.*, **66**, 428.

W. Kolos and J. Rychlewski (1978), *Chem. Phys. Lett.*, **59**, 183.

A. Komornicki, K. Morokuma and T. F. George (1977), *J. Chem. Phys.*, **67**, 5012.

D. Konowalow and M. L. Olson (1979), *J. Chem. Phys.*, **71**, 450.

D. Konowalow and M. E. Rosenkrantz (1979), *Chem. Phys. Lett.*, **61**, 489.

D. D. Konowalow, M. E. Rosenkrantz and M. L. Olson (1980), *J. Chem. Phys.*, **72**, 2612.

D. D. Konowalow, W. J. Stevens and M. E. Rosenkrantz (1979), *Chem. Phys. Lett.*, **66**, 24.

M. Krauss (1977), *J. Chem. Phys.*, **67**, 1712.

M. Krauss and D. B. Neumann (1979), *Mol. Phys.*, **37**, 1661.

M. Krauss and W. J. Stevens (1981), *J. Chem. Phys.*, **74**, 570.

R. Krishnan, M. J. Frisch and J. A. Pople (1980), *J. Chem. Phys.*, **72**, 4244.

K. C. Kulander and M. F. Guest (1979), *J. Phys. B*, **12**, L501.

P. J. Kuntz (1979), in *Atom–Molecule Collision Theory*, ed. R. B. Bernstein, Plenum, New York, p. 79.

I. Kusunoki, S. Sakai, S. Kato and K. Morokuma (1980), *J. Chem. Phys.*, **72**, 6813.

W. Kutzelnigg (1977), in *Modern Theoretical Chemistry*, vol. 3, ed. H. F. Schaefer, Plenum, New York, p. 129.

S. R. Langhoff and J. O. Arnold (1979), *J. Chem. Phys.*, **70**, 852.

S. R. Langhoff and E. R. Davidson (1974), *Int. J. Quantum Chem.*, **8**, 61.

S. R. Langhoff and R. L. Jaffe (1979), *J. Chem. Phys.*, **71**, 1475.

S. R. Langhoff, M. L. Sink, R. H. Pritchard, C. W. Kern, S. J. Strickler and M. J. Boyd (1977), *J. Chem. Phys.*, **67**, 1051.

K. P. Lawley (ed.) (1980), *Potential Energy Surfaces*, Wiley, New York.

Y. S. Lee, W. C. Ermler and K. S. Pitzer (1978), *J. Chem. Phys.*, **67**, 5861.

Y. S. Lee, W. C. Ermler and K. S. Pitzer (1980), *J. Chem. Phys.*, **73**, 360.

Y. S. Lee, W. C. Ermler, K. S. Pitzer and A. D. McLean (1979), *J. Chem. Phys.*, **70**, 288.

B. H. Lengsfield (1980), *J. Chem. Phys.*, **73**, 382.

B. Levy and G. Berthier (1968), *Int. J. Quantum Chem.*, **2**, 307.

I. Lindgren (1978), *Int. J. Quantum Chem. Symp.*, **12**, 33.

B. Liu (1973), *J. Chem. Phys.*, **58**, 1925.

B. Liu and R. E. Olson (1978), *Phys. Rev.*, **A18**, 2498.

R. Loges and J. R. Sabin (1979), *Int. J. Quantum Chem.*, **16**, 273.

R. R. Lucchese and H. F. Schaefer (1977a), *J. Am. Chem. Soc.*, **99**, 6765.

R. R. Lucchese and H. F. Schaefer (1977b), *J. Chem. Phys.*, **67**, 848.

R. R. Lucchese and H. F. Schaefer (1978), *J. Chem. Phys.*, **68**, 769.

D. R. McLaughlin and D. L. Thompson (1979), *J. Chem. Phys.*, **70**, 2748.

A. D. McLean and G. S. Chandler (1980), *J. Chem. Phys.*, **72**, 5639.

A. D. McLean, O. Gropen and S. Huzinaga (1980), *J. Chem. Phys.*, **73**, 396.

A. D. McLean, G. H. Loew and D. S. Berkowitz (1978), *Mol. Phys.*, **36**, 1359.

B. H. Mahan, H. F. Schaefer and S. R. Ungemach (1978), *J. Chem. Phys*, **68**, 781.

N. H. March (1981), in *Theoretical Chemistry* (C. Thomson, ed.), The Royal Society of Chemistry, London, vol. 4, p. 92.

C. Marian, P. J. Bruna, R. J. Buenker and S. D. Peyerimhoff (1977), *Mol. Phys.*, **33**, 63.

F. Masnou-Seeuws, M. Philippe and P. Valiron (1978), *Phys. Rev. Lett.*, **41**, 395.

R. L. Matcha and M. B. Milleur (1978), *J. Chem. Phys.*, **69**, 3016.

W. Meyer (1973), *J. Chem. Phys.*, **58**, 1017.

W. Meyer (1976), *J. Chem. Phys.*, **64**, 2901.

W. Meyer (1977), in *Modern Theoretical Chemistry*, vol. 3, ed. H. F. Schaefer, Plenum, New York, p. 413.

W. Meyer, P. Botschwina, P. Rosmus and H.-J. Werner (1980), in *Computational Methods in Chemistry*, ed. J. Bargon, Plenum, New York, p. 157.

W. Meyer and P. Rosmus (1975), *J. Chem. Phys.*, **63**, 2356.

H. H. Michels, R. H. Hobbs and L. A. Wright (1977), *Chem. Phys. Lett.*, **48**, 158.

H. H. Michels, R. H. Hobbs and L. A. Wright (1978), *J. Chem. Phys.*, **69**, 5151.

H. H. Michels, R. H. Hobbs and L. A. Wright (1979), *J. Chem. Phys.*, **71**, 5053.
F. Mies, R. S. Stevens and M. Krauss (1978), *J. Mol. Spectrosc.*, **72**, 303.
R. S. Mulliken and W. C. Ermler (1977), *Diatomic Molecules: Results of ab initio Calculations*, Academic, New York.
J. N. Murrell (1978), in *Gas Kinetics and Energy Transfer*, vol. 3, eds. P. G. Ashmore and R. J. Donovan, The Chemical Society, London, p. 200.
J. N. Murrell, S. Carter and I. M. Mills (1979a), *Mol. Phys.*, **37**, 1885.
J. N. Murrell, S. Carter, I. M. Mills and M. F. Guest (1979b), *Mol. Phys.*, **37**, 1199.
J. N. Murrell, S. Carter, I. M. Mills and M. F. Guest (1981), *Mol. Phys.*, **42**, 605.
J. N. Murrell, S. Carter and A. J. C. Varandas (1978), *Mol. Phys.*, **35**, 1325.
J. N. Murrell, A. Al-Derzi, J. Tennyson and M. F. Guest (1979c), *Mol. Phys.*, **38**, 1755.
J. N. Murrell and S. Farantos (1977), *Mol. Phys.*, **34**, 1185.
J. N. Murrell, K. S. Sorbie and A. J. C. Varandas (1976), *Mol. Phys.*, **32**, 1359.
A. V. Nemukhin and N. F. Stepanov (1979), *Chem. Phys. Lett.*, **60**, 421.
C. A. Nicolaides and D. R. Beck (eds.) (1978), *Excited States in Quantum Chemistry*, Reidel, Dordrecht.
G. C. Nielson, G. A. Parker and R. T. Pack (1977), *J. Chem. Phys.*, **66**, 1396.
O. Nomura and S. Iwata (1979), *Chem. Phys. Lett.*, **66**, 523.
J. M. Norbeck, R. R. Merkel and P. R. Certain (1977), *Mol. Phys.*, **34**, 589.
M. L. Olson and D. D. Konowalow (1977a), *Chem. Phys.*, **21**, 393.
M. L. Olson and D. D. Konowalow (1977b), *Chem. Phys.*, **22**, 29.
R. E. Olson and B. Liu (1978), *Chem. Phys. Lett.*, **56**, 537.
R. E. Olson and B. Liu (1979a), *Phys. Rev.*, **A20**, 1344.
R. E. Olson and B. Liu (1979b), *Phys. Rev.*, **A20**, 1366.
R. E. Olson, R. P. Saxon and B. Liu (1980), *J. Phys. B*, **13**, 297.
V. I. Osherov and L. V. Polujanov (1978), *Theor. Chim. Acta (Berl.)*, **49**, 123.
A. H. Pakiari and N. C. Handy (1975), *Theor. Chim. Acta (Berl.)*, **40**, 17.
J. Paldus (1974), *J. Chem. Phys.*, **61**, 5321.
J. Paldus (1975), *Int. J. Quantum Chem. Symp.*, **9**, 165.
J. Paldus (1976a), *Phys. Rev.*, **A14**, 1620.
J. Paldus (1976b), in *Theoretical Chemistry: Advances and Perspectives*, vol. 2, eds. H. Eyring and D. J. Henderson, Academic, New York, p. 131.
J. Paldus, B. G. Adams and J. Čížek (1977), *Int. J. Quantum Chem.*, **11**, 813.
J. Paldus and M. J. Boyle (1980), *Phys. Scr.*, **21**, 295.
J. Paldus and P. E. S. Wormer (1979), *Int. J. Quantum Chem.*, **16**, 1321.
C. A. Parsons and C. E. Dykstra (1979), *J. Chem. Phys.*, **71**, 3025.
P. K. Pearson and E. Roueff (1976), *J. Chem. Phys.*, **64**, 1240.
M. Pelissier and P. Durand (1980), *Theor. Chim. Acta (Berl.)*, **55**, 43.
M. Pelissier and J. P. Malrieu (1977), *J. Chem. Phys.*, **67**, 5963.
M. Pelissier and J. P. Malrieu (1979), *J. Mol. Spectrosc.*, **77**, 322.
M. Perić, R. J. Buenker and S. D. Peyerimhoff (1979), *Can. J. Chem.*, **57**, 2491.
J. Perić-Radić, J. J. Römelt, S. D. Peyerimhoff and R. J. Buenker (1977), *Chem. Phys. Lett.*, **50**, 344.
S. D. Peyerimhoff and R. J. Buenker (1978), in *Excited States in Quantum Chemistry*, eds. C. A. Nicolaides and D. R. Beck, Reidel, Dordrecht, p. 403.
S. D. Peyerimhoff and R. J. Buenker (1979a), *Chem. Phys.*, **42**, 167.
S. D. Peyerimhoff and R. J. Buenker (1979b), *Can. J. Chem.*, **57**, 3182.
S. D. Peyerimhoff and R. J. Buenker (1980), in *Computational Methods in Chemistry*, ed. P. Bargon, Plenum, New York, p. 175.
J. A. Pople, R. Krishnan, H. B. Schlegel and J. S. Binkley (1978), *Int. J. Quantum Chem.*, **14**, 545.
B. Pouilly, J. Schemps, D. J. W. Lumley and R. F. Barrow (1978a), *J. Phys. B*, **11**, 2281.

B. Pouilly, J. Schemps, D. J. W. Lumley and R. F. Barrow (1978b), *J. Phys. B*, **11**, 2289.

R. K. Preston, D. L. Thompson and D. R. McLaughlin (1978), *J. Chem. Phys.*, **68**, 13.

R. Preuss, R. J. Buenker and S. D. Peyerimhoff (1979), *Chem. Phys. Lett.*, **62**, 21.

R. Preuss, S. D. Peyerimhoff and R. J. Buenker (1977), *J. Mol. Struct.*, **40**, 117.

S. J. Prosser and W. G. Richards (1980), *J. Phys. B*, **13**, 2767.

G. D. Purvis and R. J. Bartlett (1978), *J. Chem. Phys.*, **68**, 2114.

G. D. Purvis and R. J. Bartlett (1979), *J. Chem. Phys.*, **71**, 548.

N. C. Pyper (1980), *Mol. Phys.*, **39**, 1327.

N. C. Pyper and J. Gerratt (1977), *Proc. R. Soc.*, **A355**, 407.

P. Pyykkö and J.-P. Desclaux (1979), *Acc. Chem. Res.*, **12**, 276.

R. C. Raffenetti (1973), *J. Chem. Phys.*, **58**, 4452.

R. C. Raffenetti, K. Hsu and I. Shavitt (1977), *Theor. Chim. Acta (Berl.)*, **45**, 33.

W. G. Richards, P. R. Scott, E. A. Colbourn and A. F. Marchington (1978), *Bibliography of ab initio Molecular Wavefunctions*, Clarendon, Oxford.

J. Römelt, S. D. Peyerimhoff and R. J. Buenker (1978a), *Chem. Phys.*, **34**, 403.

J. Römelt, S. D. Peyerimhoff and R. J. Buenker (1978b), *Chem. Phys. Lett.*, **58**, 1.

J. Römelt, S. D. Peyerimhoff and R. J. Buenker (1979), *Chem. Phys.*, **41**, 133.

B. O. Roos (1972), *Chem. Phys. Lett.*, **15**, 153.

B. O. Roos and P. E. M. Siegbahn (1977a), in *Modern Theoretical Chemistry*, vol. 3, ed. H. F. Schaefer, Plenum, New York, p. 277.

B. O. Roos and P. M. Siegbahn (1977b), *J. Am. Chem. Soc.*, **99**, 7716.

B. Roos and P. E. M. Siegbahn (1980), *Int. J. Quantum Chem.*, **17**, 485.

B. O. Roos, P. R. Taylor and P. E. M. Siegbahn (1980), *Chem. Phys.*, **48**, 157.

C. C. J. Roothaan (1951), *Rev. Mod. Phys.*, **23**, 69.

C. C. J. Roothaan (1960), *Rev. Mod. Phys.*, **32**, 179.

C. C. J. Roothaan, J. Detrich and D. G. Hopper (1979), *Int. J. Quantum Chem. Symp.*, **13**, 93.

J. B. Rose and V. McKoy (1971), *J. Chem. Phys*, **55**, 5435.

P. Rosmus and W. Meyer (1977), *J. Chem. Phys.*, **66**, 13.

P. Rosmus and W. Meyer (1978), *J. Chem. Phys.*, **69**, 2745.

A. R. Rossi and R. H. Bartram (1979), *J. Chem. Phys.*, **70**, 532.

K. Ruedenberg, L. M. Cheung and S. T. Elbert (1979), *Int. J. Quantum Chem.*, **16**, 1069.

K. Ruedenberg, R. C. Raffenetti and R. C. Bardo (1973), in *Energy, Structure and Reactivity*, eds. D. S. Smith and W. B. McRae, Wiley, New York, p. 164.

P. J. A. Ruttink (1978), *Theor. Chim. Acta (Berl.)*, **49**, 223.

P. J. A. Ruttink and J. H. van Lenthe (1977), *Theor. Chim. Acta (Berl.)*, **44**, 97.

N. H. Sabelli, M. Kantor, R. Benedek and T. L. Gilbert (1978), *J. Chem. Phys.*, **68**, 2767.

A. B. Sannigrahi, S. D. Peyerimhoff and R. J. Buenker (1977a), *Chem. Phys.*, **20**, 381.

A. B. Sannigrahi, S. D. Peyerimhoff and R. J. Buenker (1977b), *Chem. Phys. Lett.*, **46**, 415.

A. B. Sannigrahi, K. H. Thunemann, S. D. Peyerimhoff and R. J. Buenker (1977c), *Chem. Phys.*, **20**, 25.

V. R. Saunders (1978), in *Correlated Wavefunctions*, (ed. V. R. Saunders), Science Research Council, Daresbury Laboratory, Warrington, p. 59.

P. Saxe, H. F. Schaefer and N. C. Handy (1981a), *Chem. Phys. Lett.*, **79**, 202.

P. Saxe, H. F. Schaefer and N. C. Handy (1981b), *J. Phys. Chem.*, **85**, 745.

R. P. Saxon, K. T. Gillen and B. Liu (1977a), *Phys. Rev.* **A15**, 543.

R. P. Saxon, K. Kirby and B. Liu (1978), *J. Chem. Phys.*, **69**, 5301.

R. P. Saxon and B. Liu (1977), *J. Chem. Phys.*, **67**, 5432.

R. P. Saxon and B. Liu (1980a), *J. Chem. Phys.*, **73**, 870.

R. P. Saxon and B. Liu (1980b), *J. Chem. Phys.*, **73**, 876.

R. P. Saxon, R. E. Olson and B. Liu (1977b), *J. Chem. Phys.*, **67**, 2692.

H. F. Schaefer (1972), *The Electronic Structure of Atoms and Molecules: A Survey of Rigorous Quantum Mechanical Results*, Addison-Wesley, Reading, MA.

H. F. Schaefer (1979), in *Atom–Molecule Collision Theory*, ed. R. B. Bernstein, Plenum, New York, p. 45.

H. F. Schaefer, R. A. Klemm and F. E. Harris (1969), *Phys. Rev.*, **181**, 137.

F. Scheider and L. Zülicke (1979), *Chem. Phys. Lett.*, **67**, 491.

H. M. Schmidt, H. von Hirschhausen and K. Helfrich (1978), *Chem. Phys.*, **29**, 219.

M. W. Schmidt and K. Ruedenberg (1979), *J. Chem. Phys.*, **71**, 3951.

G. A. Segal, R. W. Wetmore and K. Wolf (1978), *Chem. Phys.*, **30**, 269.

I. Shavitt (1977a), in *Theoretical Chemistry*, vol. 3, ed. H. F. Schaefer, Plenum, New York, p. 189.

I. Shavitt (1977b), *Int. J. Quantum Chem. Symp.*, **11**, 131.

I. Shavitt (1978), *Int. J. Quantum Chem. Symp.*, **12**, 5.

I. Shavitt (1979), *Chem. Phys. Lett.*, **63**, 421.

S.-K. Shih, S. D. Peyerimhoff and R. J. Buenker (1979), *J. Mol. Spectrosc.*, **74**, 124.

S.-K. Shih, S. D. Peyerimhoff, R. J. Buenker and M. Perić (1978), *Chem. Phys. Lett.*, **55**, 206.

N. Shikamura, H. Inouye, N. Honjou, M. Sagara and K. Ohno (1978), *Chem. Phys. Lett.*, **55**, 221.

I. Shim (1980), *Theor. Chim. Acta (Berl.)*, **54**, 113.

I. Shim, J. P. Dahl and H. Johansen (1979), *Int. J. Quantum Chem.*, **15**, 311.

P. E. M. Siegbahn (1978), *Chem. Phys. Lett.*, **55**, 386.

P. E. M. Siegbahn (1979), *J. Chem. Phys.*, **70**, 5391.

P. E. M. Siegbahn (1980), *J. Chem. Phys.*, **72**, 1647.

P. Siegbahn and B. Liu (1978), *J. Chem. Phys.*, **68**, 2457.

D. M. Silver (1980), *J. Chem. Phys.*, **72**, 6445.

M. L. Sink and A. D. Bandrauk (1979), *Can. J. Phys.*, **57**, 1178.

P. E. Siska (1979), *J. Chem. Phys.*, **71**, 3942.

S. P. So and W. G. Richards (1977), *Int. J. Quantum. Chem.*, **11**, 73.

K. S. Sorbie and J. N. Murrell (1975), *Mol. Phys.*, **29**, 1387.

F. Spiegelmann and J. P. Malrieu (1978), *Chem. Phys. Lett.*, **57**, 214.

W. J. Stevens (1980), *J. Chem. Phys.*, **72**, 1536.

W. J. Stevens, G. Das, A. C. Wahl, M. Krauss and D. Neumann (1974), *J. Chem. Phys.*, **61**, 3686.

W. J. Stevens, M. Gardner, A. Karo and P. Julienne (1977a), *J. Chem. Phys.*, **67**, 2860.

W. J. Stevens, M. M. Hessel, P. J. Bertoncini and A. C. Wahl (1977b), *J. Chem. Phys.*, **66**, 1477.

W. J. Stevens and M. Krauss (1977), *J. Chem. Phys.*, **67**, 1977.

W. C. Swope, Y.-P. Lee and H. F. Schaefer (1979a), *J. Chem. Phys.*, **70**, 947.

W. C. Swope, Y.-P. Lee and H. F. Schaefer (1979b), *J. Chem. Phys.*, **71**, 3761.

K. Tanaka and E. R. Davidson (1979), *J. Chem. Phys.*, **70**, 2904.

K. Tanaka and M. Yoshimine (1979), *J. Chem. Phys.*, **70**, 1626.

P. W. Tasker (1977), *Mol. Phys.*, **33**, 511.

H. Tatewaki, K. Tanaka, F. Sasaki, S. Obara, K. Ohno and M. Yoshimine (1979), *Int. J. Quantum Chem.*, **15**, 533.

P. R. Taylor, G. B. Bacskay, N. S. Hush and A. C. Hurley (1979), *J. Chem. Phys.*, **70**, 4481.

C. Teichteil, J. P. Malrieu and J. C. Barthelat (1977), *Mol. Phys.*, **33**, 181.

C. Thomson (1975), in *Theoretical Chemistry*, vol. 2, eds. R. N. Dixon and C. Thomson, The Chemical Society, London, p. 83.

S. Topiol, J. W. Moskowitz and C. F. Melius (1978), *J. Chem. Phys.*, **68**, 2364.
S. Topiol, M. A. Ratner and J. W. Moskowitz (1977), *Chem. Phys. Lett.*, **46**, 256.
J. C. Tully (1973), *J. Chem. Phys.*, **58**, 1396.
J. C. Tully (1977), in *Semi-Empirical Methods of Electronic Structure Calculation*, part A, ed. G. A. Segal, Plenum, New York, p. 173.
J. C. Tully (1980), in *Potential Energy Surfaces*, ed. K. P. Lawley, Wiley, New York, p. 63.
J. C. Tully and C. M. Truesdale (1976), *J. Chem. Phys.*, **65**, 1002.
M. Urban and V. Kellö (1979), *Mol. Phys.*, **38**, 1621.
A. Valence (1978a), *Chem. Phys. Lett.*, **56**, 289.
A. Valence (1978b), *J. Chem. Phys.*, **69**, 355.
O. Vallée, J. Glasser, P. Ranson and J. Chapelle (1978), *J. Chem. Phys.*, **69**, 5091.
J. A. Van der Hart and J. J. C. Mulder (1979), *Chem. Phys. Lett.*, **61**, 111.
J. H. van Lenthe and G. G. Balint-Kurti (1980), *Chem. Phys. Lett.*, **76**, 138.
J. H. van Lenthe and P. J. A. Ruttink (1978), *Chem. Phys. Lett.*, **56**, 20.
R. L. Vance and G. A. Gallup (1978), *J. Chem. Phys.*, **69**, 736.
R. L. Vance and G. A. Gallup (1980), *J. Chem. Phys.*, **73**, 894.
G. J. Vazquez and J. F. Gouyet (1978), *Chem. Phys. Lett.*, **57**, 385.
G. J. Vazquez and J. F. Gouyet (1979), *Chem. Phys. Lett.*, **65**, 515.
W. R. Wadt (1978), *J. Chem. Phys.*, **68**, 402.
W. R. Wadt (1980), *J. Chem. Phys.*, **72**, 2469.
W. R. Wadt and P. J. Hay (1978), *J. Chem. Phys.*, **68**, 3850.
W. R. Wadt, P. J. Hay and L. R. Kahn (1978), *J. Chem. Phys.*, **68**, 1752.
A. C. Wahl and G. Das (1970), *Adv. Quantum Chem.*, **5**, 261.
A. C. Wahl and G. Das (1977), in *Modern Theoretical Chemistry*, vol. 3, ed. H. F. Schaefer, Plenum, New York, p. 51.
U. Wahlgren (1978a), *Chem. Phys.*, **29**, 231.
U. Wahlgren (1978b), *Chem. Phys.*, **32**, 215.
S. P. Walch (1980), *J. Chem. Phys.*, **72**, 5679.
J. Wasilewski (1979), *J. Mol. Struct.*, **52**, 281.
D. K. Watson, C. J. Cerjan, S. Guberman and A. Dalgarno (1977), *Chem. Phys. Lett.*, **50**, 181.
W. P. White, R. M. Pitzer, C. W. Matthews and T. H. Dunning (1979), *J. Mol. Spectrosc.*, **75**, 318.
C. W. Wilson and D. G. Hopper (1981), *J. Chem. Phys.*, **74**, 595.
I. D. L. Wilson (1978), *Mol. Phys.*, **36**, 597.
S. Wilson (1977), *Int. J. Quantum Chem.*, **12**, 609.
S. Wilson (1978), *Mol. Phys.*, **35**, 1.
S. Wilson (1979a), *J. Phys. B*, **12**, L657.
S. Wilson (1979b), *J. Phys. B*, **12**, 1623.
S. Wilson (1980), in *Electron Correlation*, eds. M. F. Guest and S. Wilson, Science Research Council, Daresbury Laboratory, Warrington p. 112.
S. Wilson (1981), in *Theoretical Chemistry*, vol. 4, p. 1, ed. C. Thomson, The Royal Society of Chemistry, London.
S. Wilson and V. R. Saunders (1979), *J. Phys. B*, **12**, L403.
S. Wilson and D. M. Silver (1979), *Theor. Chim. Acta (Berl.)*, **54**, 83.
S. Wilson and D. M. Silver (1980), *J. Chem. Phys.*, **72**, 2159.
N. W. Winter, C. F. Bender and T. N. Rescigno (1977), *J. Chem. Phys.*, **67**, 3122.
T. G. Winter, M. D. Duncan and N. F. Lane (1977), *J. Phys. B*, **10**, 285.
L. Wolniewicz and K. Dressler (1977), *J. Mol. Spectrosc.*, **67**, 416.
L. Wolniewicz and K. Dressler (1979), *J. Mol. Spectrosc.*, **77**, 286.

P. E. S. Wormer and J. Paldus (1979), *Int. J. Quantum Chem.*, **16**, 1307.
A. A. Wu (1977), *Chem. Phys.*, **21**, 173.
A. A. Wu (1978), *Chem. Phys. Lett.*, **59**, 457.
A.-J. A. Wu (1979), *Mol. Phys.*, **38**, 843.
A.-J. A. Wu (1980), *Mol. Phys.*, **39**, 1287.
A. A. Wu and C. Schlier (1978), *Chem. Phys.*, **28**, 73.
D. Yeager and P. Jorgensen (1979), *J. Chem. Phys.*, **71**, 755.
Y. Zeiri and M. Shapiro (1978), *Chem. Phys.*, **31**, 217.
M. Zeitz, S. D. Peyerimhoff and R. J. Buenker (1978), *Chem. Phys. Lett.*, **58**, 487.
M. Zeitz, S. D. Peyerimhoff and R. J. Buenker (1979), *Chem. Phys. Lett.*, **64**, 243.
I. H. Zimmerman, M. Baer and T. F. George (1979), *J. Chem. Phys.*, **71**, 4132.

Dynamics of the Excited State
Edited by K. P. Lawley
© 1982 John Wiley & Sons Ltd.

FITTING LAWS FOR ROTATIONALLY INELASTIC COLLISIONS

TIMOTHY A. BRUNNER AND DAVID PRITCHARD

*Research Laboratory of Electronics and Department
of Physics, Massachusetts Institute of Technology,
Cambridge, Ma. 02139, USA*

CONTENTS

TABLE OF SYMBOLS

α	exponential parameter of the SPG and SEPG fitting laws
a	prefactor parameter of rates in all fitting laws
B	rotational constant
χ^2/ν	reduced chi-squared
D	weighted RMS fractional deviation between fit and data (Eq. (3.1))
ΔE	amount of energy transferred from rotation to translation (Eq. (1.1))
E	weighted RMS fractional error of data (Eq. (3.1))
ECS	energy-corrected sudden
EGL	exponential gap law
E^r	rotational energy of a molecule
E^t	initial translational kinetic energy of collisional system
ζ	Massey system adiabatic parameter (Eq. (2.10))
IC	impulsive classical
IOS	infinite-order sudden
j, j_i, j_f	rotational quantum number; i and f refer to initial and final
$k_{j_i \to j_f}$	rate constant for j_i going to j_f; optional superscript denotes scaling or fitting law used in prediction
$k_{l \to 0}$	basis rate constant for scaling laws
l^*	angular momentum beyond which power law turns into exponential

l_c	characteristic interaction length over which collision takes place (Eq. (2.2))
γ	exponential paramter of SPG and SEPG fitting laws
m_j, m_i, m_f	magnetic quantum numbers; i and f refer to initial and final
μ_m	reduced mass of diatomic molecule
μ_s	reduced mass of collisional system
$N, N_\Delta, N_0, N_\lambda$	angular momentum degenerating factor (Eq. (2.17)); subscripts Δ, 0, and λ denote assumptions concerning how magnetic sublevels are changed by collision (Eqs. (2.20), (2.21), (2.22))
ω_j	angular velocity of molecule's rotation (Eq. (2.3))
$\tilde{\omega}$	dephasing frequency (Eq. (2.12))
$Q_{j_i \rightarrow j_f}$	cross section for j_i going to j_f
r	interatomic separation of a diatom
$\hat{r}$	diatomic internuclear axis
R	molecule–atom coordinate
ρ_t	density of final translational states (Eq. (2.12))
RI	rotationally inelastic
$R(\Delta E)$	dimensionless translational phase space factor (Eq. (2.18))
S	surprisal (Eq. (2.23))
SEPG	statistical exponential–power gap
SPG	statistical power gap
T	absolute temperature; less frequently, time (duration) of collision
$T_{i \rightarrow j}$	T-matrix element
τ_d	reduced duration of collision (Eqs. (2.4) and (2.1))
T_d	duration of collision (Eq. (2.2))
θ	exponential parameter of SEPG and EGL fitting law
v	vibrational quantum number, or velocity of target atom relative to molecule
$-\Delta$	denotes use of N_Δ spin degeneracy factor in statistically based scaling laws
$-\lambda$	denotes use of N_λ spin degeneracy factor
-0	denotes use of N_0 degeneracy factor
$-P$	denotes use of power law to generate basis rate constants
$-EP$	denotes use of hybrid exponential–power law to generate basis rate constants

I. INTRODUCTION

In the last half decade it has become possible to make accurate measurements and calculations—but not often in the same system—of cross sections and thermally averaged rate constants for *rotationally inelastic* (RI) collisions on a level-to-level basis. In this subfield of molecular dynamics, we now have an exponentially growing (with time) body of level-to-level rates which numbers around 10^3 for both theory and experiment.

This review concentrates on mathematical relationships—we call them *fitting laws*—which can represent the many known rates in a particular atom–molecule system in terms of a few parameters. Although the work on fitting laws is predominantly very recent, we shall show that it is already successful in that it is capable of fitting the RI data sets nearly within error in the majority of systems considered using only three or four parameter.

This success leads naturally to the question of whether it is possible to understand the form of the successful fitting laws and especially to predict the fitting parameters directly from the potential. As we shall indicate, it is not yet possible to answer this question affirmatively: there are hopeful signs but also areas which need work. Before commenting further on the content and organization of this review, we first address the problems of the information explosion alluded to above, an explosion whose spread to other types of molecular collisions we must regard as desired progress.

Suppose, for the moment, that the dream of workers in state-to-state chemistry is realized: a complete set of state-to-state rate constants has been measured or calculated in a number of chemically similar systems. We can now set about isolating and studying the systematic variation of these rate constants with the changes of the systems under study; we can begin to educate our intuition on a state-to-state basis. We can, that is, if we first learn to deal with incredibly large masses of information (a problem not unique to this field). Even if we restrict our attention to thermally averaged *level*-to-*level* rate constants for inelastic processes in an atom–diatom system at a single temperature, and only to those levels which might be populated at room temperature—say 10 vibrational levels and 100 rotational levels—there will be roughly a million rate constants for each system (demanding state-to-state differential cross sections at 10 relative velocities in the same system would raise this number to approximately 10^{13}). Comparing a complete set of such thermally averaged rate constants for two systems is roughly equivalent to comparing the numbers in the Boston and San Francisco telephone directories! It should be clear that our restriction in this review to rotationally inelastic level-to-level cross sections still leaves us with potentially unmanageable databases.

It is instructive at this point to draw an analogy with a familiar field which has the same problem—molecular spectroscopy. This analogy is drawn out in Table I, where we contrast and compare 'photon spectroscopy' with 'collisional spectroscopy', the subject of this reivew. Both spectroscopies are characterized by massive amounts of data, most of it redundant and all of it difficult to relate directly (i.e. through the Schrödinger equation) to the intermolecular potentials which govern the systems under study.

TABLE I

Photon spectroscopy	Collisional spectroscopy

LEVEL-TO-LEVEL TRANSITIONS
CAUSED BY

Photons
$Na_2(v_i, j_i) + h\omega$
$\to Na_2^*(v_f, j_f)$

PROCESS *Perturbers*

$Na_2(v_i, j_i) + X$
$\to Na_2(v_f, j_f) + X$

PHYSICAL VARIABLE

Photon wavenumber
$v^*(v_i, j_i, v_f, j_f)$

Rate constant
$k(v_i, j_i, v_f, j_f, v_{rel})$

FITTING LAWS

Dunham coefficients
$B_e = t/(2\mu_m c_v r_e^2)$
$W = \rho^2 v \qquad 1/2$
$\quad _\rho r^2 \qquad \mu$
Long-range analysis

Energy transfer
Exponential gap law
Power law

Dynamical
Infinite-order sudden
Energy-corrected sudden
Impulsive classical

BOTH PROCESSES
GOVERNED BY
SCHRODINGER EQUATION
and
POTENTIAL

$V(R)$ and $V^*(R)$ $V(r, R, \theta)$

For photon spectroscopy there are mathematical relationships inter-mediate between databases and potentials which are known as spectroscopic energy level expansions (i.e. Dunham expansion (Dunham, 1932), long-range expansion (LeRoy, 1973: Stwalley, 1975). These are algebraic expressions which, given the initial and final quantum numbers and the spectroscopic constants, yield the photon energy for each transition.

We term mathematical expressions of this type *fitting laws* because the set of spectroscopic constants together with the fitting law completely characterize the data: the pattern of redundancy in the data is reflected in the mathematical form of the laws, and the spectroscopic constants represent the peculiarities of the particular system under study. A fitting law and the associated spectro-scopic constants represent the data in a far redundant and much more comprehensible form than the data themselves. They permit identification of incorrect data points, interpolation among the data points, and generally permit accurate extrapolation to regions of parameter space beyond those studied.

Besides representing the data, fitting laws may give direct information about the potential. For example, simple expressions relate the rotational and vibrational constants, B and ω, of the usual energy level expansion to the interatomic separation r of the molecule and the second derivative of the potential at that separation. The situation is not so tidy with respect to fitting laws for RI collision rates. We shall discuss the few beginnings of general theoretical treatments which pretend to the able to predict the fitting parameters directly from the potential; we shall also indicate places in which these important connections need to be developed.

We will be considering RI processes of the form

$$M(v,j_i) + X \rightarrow M(v, j_f = j_i + \Delta j) + X + \Delta E, \tag{1.1}$$

in which M is a diatomic molecule, with vibrational quantum number v and initial rotational quantum number j_i, which collides with target atom X such that the final rotational quantum number is j_f, and ΔE, is the amount of energy transferred from rotation to translation. Among the many other terms for this process (i.e. rotational energy transfer, rotation-changing collisions, etc.) perhaps the most accurate is 'R–T transfer' which emphasizes that energy is shifted between rotational (R) and translational (T) degrees of freedom.

The study of RI collisions is currently very active for a number of reasons. The recent development of highly monochromatic tunable lasers has made possible 'clean' measurements of RI rate constants with unprecedented accuracy. The laser has motivated the study of RI collisions in another way because the collisional degradation (or creation) of the population inversion in a molecular gas laser is often strongly dependent on RI processes; some of the most accurate RI work has centered on molecular systems which are potentially useful in lasers, e.g. N_2–Ar (Pack, 1975) or Na_2–rare gas (Henesian et al., 1976; Wellegehausen et al., 1977). Theoretical interest in RI collisions has intensified due to the development of several new ways of approximating the numerically tedious 'close-coupled' (i.e. exact) equations (Bernstein, 1979). Finally, of all the fundamental inelastic molecular processes, RI collisions are the most likely, with cross sections typically one to two orders of magnitude higher than either vibrational energy transfer or electronic energy transfer. It is therefore virtually impossible to understand these processes on a level-to-level basis without understanding the accompanying rotational inelasticity.

We have written this review with the primary objective of clarifying and advancing the present state of understanding of fitting laws; it is not intended to serve as a review of RI processes in general. The recent book *Atom–Molecule Collision Theory* (Bernstein, 1979) contains several reviews which specifically address the theory of RI collisions including treatments of the full quantal treatment (Secrest, 1979), approximations to this treatment, and classical trajectory methods (Pattengill, 1979); computational methods are discussed in Balint-Kurti, 1975. The situation with regard to reviews of experimental results is much poorer: several reviews of energy transfer in general include discussion of RI processes (Faubel and Toennies, 1977; Oka, 1973; Toennies, 1976), but we do not know of recent experimentally oriented RI reviews. Thus we must warn readers that the data sets to which we have applied RI fitting laws in this review do not include all recent data, but were selected for range of parameters, types of system, for the quality of data and error assessment and, where these criteria were insufficient to justify our selections, by mere whimsy.

In ending this introduction, we note that much of the material presented in this review is new and has not been filtered by the usual processes associated with referred journal articles. Included in this class is the impulsive classical calculation of $k_{l \rightarrow 0}$, the entire section on exponential–power hybrid laws (including the $I_2{}^*$–He data), data on alternations in $^6Li^7Li$, and the presentation of the systematic trend from power-law to exponential behaviour with decreasing anisotropy of the atom–molecule potential. Readers are warned to exercise their critical faculties in these sections.

II. THEORY

In this section we present the theory which underlies the interrelationships among the RI rate constants for a given system, as well as what theory there is which might enable one to predict the parameters of these interrelationships from some *a priori* basis (e.g. the intermolecular potential). We are dealing, in a typical atom–diatom system, with rate constants (either thermally averaged or at fixed energy) from typically 100 initial j values to the same 100 final values—a matrix with roughly 10^4 entries. We consider two distinct classes of simplified representation of this matrix: *scaling* laws and *fitting* laws. Scaling laws express any rate constant in terms of a subset of the matrix, typically one column or row (e.g. the subset $k_{l \rightarrow 0}$), which we term the basis rate constants. Fitting laws, on the other hand, express the entire matrix in terms of a set of parameters which generally bear no relationship to any particular rate constant. We shall discuss scaling laws first: those based on dynamical considerations in part II A, and those based on statistical considerations in part II B. In part II C we shall discuss fitting laws which incorporate both of these types of scaling laws, and in part II D we shall discuss ways in which the parameters of these fitting laws may be predicted or synthesized.

A. Dynamical Scaling Laws

We now consider the infinite-order sudden (IOS) and the energy-corrected sudden (ECS) scaling laws. Both laws may be derived by making certain dynamical approximations to the Schrödinger equation; the resulting expressions for the rate constants $k_{j_i \rightarrow j_f}$ involve a sum of products of angular momentum coupling coefficients and the basis rate constants $k_{l \to 0}$, as we shall see in the following discussion.

1. Reduced Duration

From a dynamical point of view, an important physical parameter in rotationally inelastic collisions is a dimensionless parameter called the *reduced duration* of the collision—the number of radians of rotation of the diatom during the collision:

$$\tau_j \equiv \omega_j \times T_d \tag{2.1}$$

where ω_j is the angular velocity of the molecule's rotation when it has angular momentum quantum number j and T_d is the duration of the collision,

$$T_d = l_c/v \tag{2.2}$$

where v is the velocity of the target atom relative to the molecule, and l_c is a characteristic interaction length over which the collision takes place. The angular velocity is defined (footnote 13 from Brunner et al., 1981) as

$$\omega_j = (E_{j+1}^r - E_{j-1}^r)\frac{1}{2h} = \frac{h}{\mu_m \bar{r}^2}(j + \tfrac{1}{2}) \tag{2.3}$$

where $E_j^r = BJ(j + 1)$ is the rotational energy of the molecule with rotational quantum number j, $B = h^2/(2\mu_m\bar{r}^2)$ is the rotational constant, μ_m is the reduced mass of the molecule, and $\bar{r}$ is the average internuclear separation of the diatom. We define the reduced duration of a collision in which the rotational quantum number changes from j_i to j_f as

$$\tau \equiv 0.5(\tau_{j_i} + \tau_{j_f}) = \frac{l_c}{r}\left[\frac{\mu_s}{\mu_m}\frac{B(\bar{j} + \tfrac{1}{2})^2}{E^t}\right]^{1/2} = \frac{l_c}{r}\left[\frac{\mu_s E_j^r}{\mu_m E^t}\right]^{1/2} \tag{2.4}$$

where μ_s is the reduced mass of the collisional system, $\bar{j} = (j_i + j_f)/2$ is the average rotational quantum number, and $E^+ = \mu_s v^2/2$ is the initial relative translational kinetic energy. The final expression in this equation emphasizes that τ^2 is related to the ratio of rotational to translational energy, and thus for a typical collision in a gas near thermal equilibrium τ^2 will be of order unity. This definition of τ estimates the total rotation during the collision by using the average of the initial and final angular velocity, and it is appropriate if the angular velocity changes at a uniform rate from ω_{j_i} to ω_{j_f}. An additional advantage of this definition is that the process $j_i \to j_f$ has the same τ as the process $j_f \to j_i$.

2. The Sudden Scaling Law

In the sudden limit, $\tau \to 0$, the diatom does not rotate during the collision, and a dramatically simple scaling law relates any general rate constant $k_{j_i \to j_1}$

to the basis rate constants $k_{l \to 0}$. The earliest such expression.

$$k_{j_i \to j_f} = (2j_f + 1) \sum_l \begin{bmatrix} j_i & j_f & l \\ 0 & 0 & 0 \end{bmatrix}^2 k_{0 \to l}, \tag{2.5}$$

where the large parentheses represent a 3-j symbol, was given by Mittleman *et al.* (1968, Eq. 11a) for collisions of electrons with polar molecules. Chu and Dalgarno (1975) later derived a similar formula for atom–molecule RI collisions, which depended only on the sudden limit. More recently, a similar expression has been obtained (Goldflam *et al.*, 1977a, b) from the IOS approximation to the fully quantal equations for atom–diatom scattering; we term this the IOS scaling law. The 'similar' expressions derived by these groups differ at most by factors which go to unity in the limit that the absolute value of energy transferred from translation to rotation is much less than the initial kinetic energy. The recent work of Khare (1978) has shown that the IOS scaling law depends only on the energy sudden (also called the l_z-conserving) approximation, and that the coupled states (or j_z-conserving) approximation is not required. For further details and references on quantum-mechanical derivations of the IOS scaling law, see the excellent review by Kouri (1979).

The 'simple physics' inherent in the IOS scaling law in masked by the complications of its quantum-mechanical derivation. We shall now demonstrate that Eq. (2.5) can be simply understood via a classical calculation where we assume that the diatom does not rotate during the collision (i.e. $\tau \ll 1$), a conclusion first demonstrated by Bhattacharrya and Dickinson (1979).

We define $\hat{z}$ to be the *fixed* direction of the internuclear axis of the diatom at the time of collision, $\mathbf{j}$ to be the classical angular momentum vector of the diatom (measured in units of h), and $Q_{j_i \to j_f} \, \mathrm{d}j_f$ to be the cross section that, given initial angular momentum $j_i = |\mathbf{j}_i|$, the final angular momentum lies between j_f and $j_f + \mathrm{d}j_f$. Under the assumption of a sudden collision, $|\boldsymbol{\delta}| \equiv |\mathbf{j}_f - \mathbf{j}_i|$ is independent of how fast the diatom is initially rotating, depending only on the orientation of the diatom with respect to the target atom trajectory. Therefore $Q_{0 \to \delta} \mathrm{d}\delta$ is the cross section that $\mathbf{j}_i$ will change by a vector whose length is between δ and $\delta + \mathrm{d}\delta$, *for any j_i*. Then $Q_{j_i \to j_f} \mathrm{d}j_f$ is the sum over all possible δ of the cross section for momentum transfer δ weighted by the probability $f(j_f; j_i, \delta)\mathrm{d}j_f$ that, for a given δ and j_i, the final angular momentum is between j_f and $j_f + \mathrm{d}j_f$, i.e.

$$Q_{j_i \to j_f} \mathrm{d}j_f = \int \mathrm{d}\delta \, Q_{0 \to \delta} f(j_f; j_i, \delta) \mathrm{d}j_f. \tag{2.6}$$

The vector change in the angular momentum $\boldsymbol{\delta}$ must be perpendicular to $\hat{z}$ because both j_i and j_f are perpendicular to the diatomic internuclear axis. (This is exactly true for molecules in Σ electronic states, and otherwise an excellent approximation.) The law of cosines relates j_f to j_i, δ, and the angle between $\mathbf{j}_i$

and δ. Assuming that collisions randomly orient δ with respect to j_i, it can be shown that

$$f(j_f; j_i, \delta) = \frac{2j_f}{\pi} [4j_i^2 \delta^2 - [j_i^2 + \delta^2 - j_f^2]^2]^{-1/2}, \qquad (2.7)$$

for $|j_i - \delta| \le j_f \le j_i + \delta$ and zero otherwise. Further, it can be demonstrated (Edmonds, 1974; Bhattacharyya and Dickinson, 1979) that $f(j_f; j_i, \delta)/(2j_f + 1)$ is the classical limit of the square of the 3-j coeffieients appearing in the IOS sum, and thus Eq. (2.6) is the classical analogue (Bhattacharyya and Dickinson, 1979) of the discrete sum in Eq. (2.5). The IOS scaling law may be completely understood as arising from the addition of angular momenta under the assumption that the collision is sudden. The fact that in a sudden collision all angular momenta are in a plane perpendicular to $\hat{z}$ explains why all the m's in the 3-j symbols of Eq. (2.5) are zero.

When we apply the IOS scaling law to thermally averaged rate constants we use the expression

$$k_{j_i \to j_f}^{IOS} = (2j_f + 1) \exp\left(\frac{E_{j_i}^r - E_{j_>}^r}{kT}\right) \sum_l \begin{bmatrix} j_i & j_f & l \\ 0 & 0 & 0 \end{bmatrix}^2 (2l + 1)k_{l \to 0} \qquad (2.8)$$

where kT is the average relative translational thermal energy and $j_>$ is the greater of j_i and j_f. Detailed balance is guaranteed by the two factors in front of the summation over l.

Eq. (2.8) represents the entire RI rate constant matrix $k_{j_i \to j_f}$ in terms of the single row $k_{l \to 0}$ (or by detailed balance the column $k_{0 \to l}$). It is expected to be valid for collisions in which the collisional time is short compared with the rotational period of the molecule. The IOS scaling law has been experimentally observed (Wainger et al., 1979; Brunner et al., 1981) in several collisional systems (e.g. Na_2^*–H_2, I_2^*–He) where the target atom is much lighter (and therefore much faster) than the diatom.

3. Corrections to the Sudden Scaling Law

The restriction to sudden collisions ($\tau \ll 1$) is very limiting, and it is important to extend and generalize the scaling relationships of the previous section beyond this restriction. The importance of such a generalization becomes evident by examining Eq. (2.4). If both the translational energy E^t and rotational energy E_j^r are assumed to be of order kT, and if l_c is several times larger than $\bar{r}$, as is typically observed (Brunner et al., 1981), then $\tau \gtrsim \sqrt{(\mu_s/\mu_m)}$. In most cases this reduced mass ratio will be comparable to or greater than unity, and therefore in most systems which are close to thermal equilibrium the sudden approximation is inadequate. In this section we discuss ways in which a generalization of the sudden scaling law can be achieved.

Assume for the moment that the intermolecular potential is known, and therefore the perturbing potential along the (classical) trajectory is known as a function of time. Then the first-order matrix element for a transition from state a to state b will have the form

$$M_{a \to b} \propto \int_{-\infty}^{\infty} dt \exp(i\omega_a t) V_{a \to b}(t) \exp(-\omega_b t) = \int_{-\infty}^{\infty} dt \exp(-i\Delta E t/h) V_{a \to b}(t),$$

(2.9)

where $V_{a \to b}(t)$ is the perturbing potential and $\Delta E = E_b^r - E_a^r = h(\omega_b - \omega_a)$. In general the intermolecular potential depends smoothly on the spatial coordinates with a maximum of width $\sim T_d^{-1}$ (see Eq. (2.2)) centred on the point of closest approach. One anticipates, therefore, that when $\Delta E/h \gg T_d^{-1}$ the exponential in Eq. (2.9) will oscillate rapidly relative to any variation in $V_{a \to b}(t)$, and $M_{a \to b}$ will be small in consequence.

This situation is termed 'adiabatic' (as opposed to 'sudden') and underlies the utility of the Massey adiabatic parameter,

$$\xi \equiv |E_{j_f}^r - E_{j_i}^r| T_d/h,$$

(2.10)

used in inelastic collision theory (e.g. Baylis, 1978). The reduced duration τ is closely related to the Massey adiabatic parameter; the combination of Eqs. (2.1) to (2.4) yields

$$\xi = |j_f - j_i|\tau = |\Delta j|\tau.$$

(2.11)

Note that, if $|\Delta j|$ is large, the collision can be sudden in a dynamical sense (i.e. $\tau \ll 1$) although adiabatic in the Massey sense (i.e. $\tau < 1$ but $\xi > 1$).

The relevance of the preceding discussion of adiabaticity to RI collisions is questionable because it is based on first-order perturbation theory. There are numerous reasons to believe that higher-order terms will make substantial contributions to RI rate constants: the cross sections are large, the cross sections do not decrease exponentially with increasing $|\Delta E|$ (more on this later), and large $|\Delta j|$ collisions are observed although the potential may not contain Legendre polynomials of correspondingly large order (in first-order perturbation theory $|\Delta j|$ can at most equal the order of the largest Legendre polynomial required to expand the dependence of the potential on the angle between the diatomic axis and the atom–molecule axis). It appears, therefore, that substantial contributions to RI processes are made by higher order terms—indeed it seems likely that large $|\Delta j|$ collisions result from repeated interactions by potential terms of low multipole order. Schinke (1981, private communication) has recently used a semiclassical approach to show graphically how the amplitudes for the intermediate j states vary continuously throughout the course of a large $|\Delta j|$ RI collision.

In view of the importance of the lowest-order terms in the potential, even in collisions with large $|\Delta j|$, most attempts to correct the sudden approximation for adiabatic effects have involved only the lowest-order non-zero transition. The dephasing frequency for this transition is

$$\tilde{\omega} = \Delta E_m / h \qquad (2.12)$$

where ΔE_m, the minimum energy defect, is the energy spacing between a given rotational level and the closest level which can be coupled by collision (i.e. those separated by $\Delta j = 2$ two for a homonuclear diatom, and $\Delta j = 1$ for a heteronuclear diatom). For a transition from j_i to j_f, ΔE_m has been chosen as the level spacing near the larger of j_i and j_f. (DePristo et al., 1979). A choice based on the spacing near the initial (or final) level could lead to corrections which violate microreversibility.

Once the value of $\tilde{\omega}$ is fixed, there are several approaches to calculating the adiabatic correction. One approach is simply to evaluate Eq. (2.9) with $\Delta E/h$ replaced by $\tilde{\omega}$ (Alexander and DePristo, 1979; Alexander, 1979a). This requires knowledge of $V(t)$, which may be replaced by $V(\mathbf{R}(t), \hat{r}(t))$ where $\mathbf{R}(t)$ and $\hat{r}(t)$ are the molecule–atom coordinate and diatomic axis respectively. $\mathbf{R}(t)$ and $\hat{r}(t)$ may be approximated by assuming a straight-line collision trajectory with negligible rotation during the collision. This approach will not yield adiabatic corrections which are independent of the specific potential and is therefore not useful in generating an adiabatically corrected scaling theory. One way around this difficulty is to expand the potential analytically in simpler functions (e.g. in a multipole expansion). If one of these dominates the potential, the adiabatic correction may be equivalent to a multiplicative correction of matrix element or impact parameter, but if the potential contains two such functions this simple scaling behaviour may no longer exist (Alexander, 1979b).

An alternative approach is to expand the exponential term in Eq. (2.9) rather than the potential (DePristo et al., 1979). Assuming that $V_{a \to b}(t)$ is symmetric about the time of closest approach*, the first non-zero correction term in this expansion is second order in $\tilde{\omega}$:

$$M_{a \to b} \propto \int_{-\infty}^{\infty} dt\, V_{a \to b}(t) + \int_{-\infty}^{\infty} dt (-\tilde{\omega}^2 t^2) V_{a \to b}(t) + \dots$$

$$= \int_{-\infty}^{\infty} dt\, V_{a \to b}(t)(1 - \tilde{\omega}^2 T^2 + \dots) \qquad (2.13)$$

where the time T is defined by the second line. This expression has an unphysical sign change when $\tilde{\omega}^2 T^2 \geq 1$, so DePristo et al. replaced the last

---

* In general $V_{a \to b}$ for an RI collision will not have this property. Instead, the time $t = 0$ may be chosen so that there is no first-order correction term for a particular ab pair. However, there is no guarantee that all $a'b'$ pairs will have no first-order term for this choice of $t = 0$. In addition, third-order terms will not vanish in this case, as they do in Eq. (2.13).

term in parentheses by

$$(1 - \tilde{\omega}^2 T^2) \rightarrow \frac{1}{1 + \tilde{\omega}^2 l^2/v^2} \rightarrow \frac{1}{1 + \tilde{\omega}^2 l_c^2/(6v^2)}, \quad (2.14)$$

which has the same shape for small T, but decreases smoothly to zero for large $\tilde{\omega} T$. Note that T^2 has been replaced by $(l/v)^2$ (where l depends on the impact parameter b) in the first expression and then by $l_0^2/6v^2$ to account for the average over impact parameter. l_c is a characteristic length assumed to be independent of j_i and j_f, and the factor 6 is appropriate only for an R^{-6} potential. This adiabatic correction factor is then squared and applied to the rate constants, resulting in the energy-corrected sudden (ECS) approximation (DePristo et al., 1979).

Because it is in close accord with considerable experimental data we now give mathematical expressions for the ECS scaling law. The correction factor in Eq. (2.14) is applied to both $k_{l \to 0}$ and $k_{j_i \to j_f}$ in the IOS scaling law to get the ECS scaling law. This results in the appearance of an adiabatic factor,

$$A_l^j = \frac{1 + \tau_l^2/6}{1 + \tau_j^2/6}, \quad (2.15)$$

in the IOS scaling law (Eq. (2.8)). The ECS scaling law then becomes (in a form consistent with detailed balance)

$$k_{j_i \to j_f}^{ECS} = (2j_f + 1) \exp\left(\frac{E_{j_i}^r - E_{j_>}^r}{kT}\right) \sum_l \begin{bmatrix} j_i & j_f & l \\ 0 & 0 & 0 \end{bmatrix}^2 (2l + 1)[A_{j_>}]^2 k_{l \to 0}. \quad (2.16)$$

B. Statistically Based Scaling Laws

In this section we discuss scaling laws in which 'trivial' phase space factors are divided out of the rate constants. Our discussion of such laws is based on the velocity averaged T-matrix expression (Messiah, 1962) for the rate constant:

$$k_{j_i \to j_f} = \frac{2\pi}{h} T_{i \to f}^2 \langle \rho_t(E^t - \Delta E) \rangle N(j_i \to j_f) \quad (2.17)$$

in which ρ_t is the density of final translational states at the final relative kinetic energy $E^t - \Delta E$, $\langle \square \rangle$ denotes an average over the initial translational energy (E^t) distribution, N is an angular momentum degeneracy factor, and $T_{i \to f}^2$ is the energy-averaged integral over solid angle of the magnitude squared of the T-matrix element, averaged over all *allowed* pairs of magnetic sublevels m_i, m_f.

By defining the dimensionless translational phase space factor

$$R(\Delta E) \equiv \frac{\langle \rho_t(E^t - \Delta E) \rangle}{\langle \rho_t(E^t) \rangle} = \frac{\langle (E^t - \Delta E)^{1/2} \rangle}{\langle (E^t)^{1/2} \rangle}, \quad (2.18)$$

we can rewrite Eq. (2.17) in the more convenient form

$$k_{j_i \to j_f} = C T_{i \to f}^2 R(\Delta E) N(j_i \to j_f) \tag{2.19}$$

where C is an unimportant constant. In the important case that the initial kinetic energy distribution is thermal, at temperature T, $R(\Delta E; T)$ has been analytically calculated (Procaccia and Levine, 1976a).

The spin degeneracy factor $N(j_i \to j_f)$ is usually taken to be

$$N_\Delta = 2j_f + 1, \tag{2.20}$$

the number of m sublevels in level j_f. This factor is appropriate if there is no restriction on how much m changes in a collision. An alternate spin degeneracy factor,

$$N_0 = \frac{2j_< + 1}{2j_i + 1}, \tag{2.21}$$

where $j_<$ is the lesser of j_i and j_f, assumes the selection rule $|\Delta m| = 0$. N_0 was found to be useful in correlating rate constants for Na_2^* colliding with Xe (Brunner et al., 1978, 1979).

A straightforward generalization of the m-randomizing N_Δ and the m-conserving N_0 assumes the selection rule $|\Delta m| \le \lambda$. The resulting spin degeneracy factor (Brunner, 1980; Brunner et al., 1981) is

$$N_\lambda(j_i \to j_f) = \frac{(2n_2 + 1)(2n_3 + 1) - \alpha(\alpha + 1)}{2j_i + 1}, \tag{2.22}$$

where of the three numbers $\{j_i, j_f, \lambda\}$, n_1 is the largest, n_3 the smallest, n_2 the middle one, and α is the greater of zero and $n_2 + n_3 - n_1$. Statistically based fits to experimental data with intermediate λ have been shown to be substantially better than those with either $\lambda = 0 (N_0)$ or $\lambda = \infty (N_\Delta)$ (Brunner, 1980; Brunner et al., 1981).

The basic assumption of statistically based scaling laws is that $k/(RN) \propto T^2$ is 'simpler' than k. An important class of statistically based scaling laws makes the key assumption that $T_{i \to f}^2$ depends only on $|\Delta E \equiv E_{j_f} - E_{j_i}|$ (and possibly on the vibrational quantum number or the initial kinetic energy E^t as well— these variables are assumed fixed in the present discussion). $T_{i \to f}^2$ must not depend explicitly on j_i or j_f; any dependence on these variables resides in the angular momentum statistical factor, N. Eq. (2.17) would then relate any two rate constants with the same $|\Delta E|$, since $k/(RN_\lambda) \propto T^2$ depends only on $|\Delta E|$. If we determine the curve $k/(RN_\lambda)$ versus $|\Delta E|$ by using data with any one j_i then we predict all the other data by the requirement that they fall on this curve. By changing λ one can vary the relationship between rate constants with the same $|\Delta E|$; this is essentially a change in the scaling law. In summary, the assumption that T^2 depends only on $|\Delta E|$ and the use of the N_λ statistical factor thus enables the prediction of the entire rate matrix from a row or column (since this contains an array of k's which samples all $|\Delta E|$); this results in a particular scaling law being built in, regardless of the details of the dependence of T^2 on $|\Delta E|$.

Although surprisal analysis is generally justified from information theory (Levine and Bernstein, 1975; Levine and Kinsey, 1979), it may also be justified from the T-matrix expression in Eq. (2.19). From this perspective RN_Δ is the 'prior' (Levine and Bernstein, 1975; Levine et al., 1976; Goldflam et al., 1977), since it is proportional to the final density of states; the surprisal is then defined as

$$S \equiv -\ln\left[\frac{k}{R(\Delta E)N_\Delta}\right], \qquad (2.23)$$

a logarithmic comparison of the observed rate constant with the prior. The more general spin degeneracy factor N_λ may be incorporated into standard surprisal analysis by considering the 'persuasion' (Levine and Kinsey, 1979),

$$P \equiv -\ln\left[\frac{k}{R(\Delta E)N_\lambda}\right], \qquad (2.24)$$

of k with respect to the 'theoretical distribution' RN_λ. Of course, it is unimportant whether one considers $k/(RN)$ or $-\ln[k/(RN)]$ to be the interesting quantity.

C. Fitting Laws

Scaling laws can seldom be used directly to fit data because there are so many basis rate constants that the number of free parameters in the fit becomes too large. Fitting laws represent the basis rate constants or T^2 by simple functions which have a small number of variable parameters. We present several important fitting laws which have been applied to RI data, in some cases with remarkable success.

1. Statistically Based Fitting Laws

We now discuss fitting laws which arise from statistically based scaling laws where we assume that T^2 depends only on $|\Delta E|$.

a. Statistical power gap (SPG) law. The statistical power gap (SPG) law assumed that $k/(RN) \propto |\Delta E|^{-\alpha}$, which is equivalent to the average T-matrix element depending on $|\Delta E|$ as a simple power law. This expression is written

$$k^{SPG}_{j_i \to j_f} = a|\Delta E/B|^{-\alpha}N_\lambda(j_i \to j_f)R(\Delta E), \qquad (2.25)$$

where B is the rotational constant, and a, α, and λ are variable parameters.

(Both ΔE and B are measured in cm^{-1} so that their ratio is dimensionless and the parameter a has the dimensions of a rate constant.) We call this expression the statistical power gap (SPG) law to emphasize its two most important features—the spin statistical factor N_λ, and the power-law dependence on the transferred energy.

The SPG formula was first successfully applied to Na_2^*–Xe data (Brunner et al., 1978, 1979), yielding fits to within the experimental error of $\simeq 15\%$. A variety of calculated rate constants were shown to fit the SPG formula much better than the exponential gap law (Pritchard et al., 1979). (See section b below.) Additional experimental work in Na_2^* colliding with eight targets (Brunner, 1980; Brunner et al., 1981), $Li_2^*(B\Pi)$–Li (Vidal, 1978, $Li_2^*(A)$–rare gas (Ottinger and Schroder, 1979), and $I_2^*(B)$–Xe (Dexheimer et al., 1981) have established the wide applicability of the SPG formula. Along with this empirical justification of SPG, Sec. II D.1 presents a crude calculation which predicts such a power-law behavior.

b. *Exponential gap law (EGL)*. A widely used fitting law for RI collisions (because it is the oldest fitting law) is the exponential gap law (Procaccia and Levine, 1976a, b)

$$k_{j_i \to j_f}^{EGL} = a \exp(-\theta|\Delta E|)R(\Delta E)N_\Delta, \tag{2.26}$$

where a and θ are free parameters. EGL was first proposed by Polanyi and Woodall (1972) who showed that the assumption that the RI cross section scaled like $\exp(-|\Delta E|)$ explained their observation on the relaxation of a non-thermal rotational distribution better than several considerably less reasonable assumptions. Subsequently EGL has been given theoretical respectability by surprisal theory (Levine et al., 1976; Levine and Bernstein, 1975; Levine and Kinsey, 1979) and quantum-mechanical considerations (Heller, 1977), although none of this work should be thought of as proving EGL. While the EGL fitting law has been found to work well for a few very light (hydrogenic) molecules (Green, 1979), it does not work for heavier, more anisotropic systems.

c. *Statistical exponential–power gap (SEPG) law*. Under certain circumstances the SPG law must be generalized by including an exponential factor with an additional free parameter. We term this the statistical exponential–power gap (SEPG-λ) law and write it as

$$k_{j_i \to j_f}^{SEPG} = a|\Delta E/B|^{-\alpha} \exp(-\theta|\Delta E|)R(\Delta E)N_\lambda(j_i \to j_f), \tag{2.27}$$

where the parameters a, α, θ, and λ are all varied in the fit.

A mathematically similar formula has been derived and successfully applied to N_2–Ar data by Sanctuary (1979) (see also Eu, 1974), and we shall discuss the physical justification for it in Sec. II D.

2. *Dynamically Based Fitting Laws (ECS-P and ECS-EP)*

We now turn to the discussion of fitting laws which are based on the dynamical scaling laws discussed in Sec. II A: the infinite-order sudden (IOS) and energy-corrected sudden (ECS) scaling laws.

The idea of a power law has been carried over for the basis rate constants $k_{l \to 0}$ needed when applying the IOS or the ECS scaling laws to our data. We use the expression

$$k^P_{l \to 0} = a[l(l + 1)]^{-\gamma} \tag{2.28}$$

which has an energy gap dependence similar to SPG-λ (i.e. $k^P \propto |\Delta E|^{-\gamma} \propto |l(l + 1)|^{-\gamma}$). We call the fitting law which results from the combination of Eq. (2.28) and the infinite-order sudden (IOS) scaling law IOS-P (Eq. (2.8)) to emphasize the power-law basis rates $k^P_{l \to 0}$. The IOS-P fitting law was first applied to Na_2^* colliding with He, H_2, CH_4, and N_2 (Wainger *et al.*, 1979; Brunner *et al.*, 1981). The formula worked quite well for the light targets He and H_2, as expected, but failed as the target mass increased.

If the energy-corrected sudden scaling law (Eq. (2.16)) is used instead of IOS, we call the resulting fitting law ECS-P. Like SPG, ECS-P has three parameters, a, γ, and l_c. ECS-P has been spectacularly successful in fitting experimentally determined sets of rate constants in a variety of systems as we shall document in Sec. III. The excellent ECS-P fits to high-quality data sets indicates that both the ECS scaling law and the power-law assumption for the basis rates work well.

In a computer program to fit data to the ECS-P formula (Smith and Pritchard, 1981), a large fraction of the time is spent calculating the many 3-j coefficients appearing in Eq. (2.16). One way to avoid this computer-time bottleneck is to use a recently discovered approximation to the ECS-P formula which will be described in the next section.

We also propose here a new fitting law which contains an exponential fall-off at large Δj. We assume that the basis rates obey the formula

$$k^{EP}_{l \to 0} = a[l(l + 1)]^{-\gamma} \exp\left(\frac{-l(l + 1)}{l^*(l^* + 1)}\right), \tag{2.29}$$

where a, γ, and l^* are free parameters. Combining this with the ECS scaling law defines the ECS-EP fitting law. The quantity l^* corresponds to the angular momentum transfer at which the decrease of the rates with increasing l changes from a power law to an exponential. Justification for this form will be given in Sec. II D.

3. *Connection Between Statistically and Dynamically Based Fitting Laws*

Although dynamically based fitting laws and statistically based fitting laws arise from totally different physical assumptions, it turns out that there is a great mathematical similarity between them. This similarity may be demonstrated more formally using a new sum rule for 3-j coefficients (Smith and Pritchard, 1981),

$$\sum_l (2l+1) \begin{bmatrix} j_i & j_f & l \\ 0 & 0 & 0 \end{bmatrix}^2 \frac{1}{l(l+1)} = |j_f(j_f+1) - j_i(j_i+1)|^{-1}. \qquad (2.30)$$

The right-hand side is proportional to $|E_{j_f} - E_{j_i}|^{-1} = |\Delta E|^{-1}$ while the left-hand side is the sum appearing in the IOS scaling law (Eq. (2.8)) with $k_{0 \to l} = [l(l+1)]^{-1} \propto |E_l - E_0|^{-1}$. In the limit $\Delta E/kT \ll 1$ (so that both R in Eq. (2.5) and the exponential factor in Eq. (2.8) go to 1), we see that the IOS-P fitting law with $\gamma = 1$ is equivalent to the SPG fitting law with $\alpha = 1$ and $N = N_\Delta$.

Of course, in many collisional systems the IOS scaling law does not work well and the more general ECS scaling law should be used. Furthermore, fits to data show that γ is seldom 1, though in most cases it lies in the range 0·9–1·2. For these more realistic cases, where Eq. (2.30) is not applicable, an analytical approximation has been developed (Smith and Pritchard, 1981) which allows one accurately to fit data to the ECS-P fitting law without tediously evaluating 3-j coefficients. This approximate formula is quite similar to Eq. (2.25), and thus the SPG formula can be thought of as a simple, useful approximation to the more complicated ECS-P fitting law.

4. Non-Thermal Velocity Distributions

The scaling and fitting laws which we have discussed to this point have been applicable to thermally averaged data, but have often been derived by averaging a result valid for some sort of monochromatic velocity distribution. When attempting to apply these laws to data with monochromatic distributions, it is therefore important that these distributions match those of the monochromatic form of that law; otherwise some additional assumptions must be made. (The main problem is whether it is the initial velocity, final velocity, or total energy which is assumed to be constant for rate constants with different j_i and j_f.)

The SPG/SEPG fitting laws are based on the T-matrix. For consistency with microreversibility, the total energy $E = E^t + E^r$ must be constant, and in general this is taken to be the same for all rates to be scaled. In order to apply these laws to data which have constant initial E^t (as is often true experimentally), we assume that the monochromatic T-matrix element $T_{i \to j}$ depends on $|\Delta E|$ and E^t. This assumption was made implicitly in Eq. 2.19 for thermally averaged distributions, and it permits the application of these laws to data with arbitrary velocity distributions for *each* $k_{j_i \to j_f}$.

We note that the SPG/SEPG laws may work quite well even if $T_{i \to j}$ does depend on E, at least in cases in which $T_{i \to j}$ is not known from some *a priori* information (e.g. theory). This is because the T^2 of Eq. (2.19) must then be found from the data (i.e. from k/RN) and is then the weighted average

$$\frac{\langle T^2_{i \to f} \rho_t(E^t_i - \Delta E)\rangle}{\langle \rho_t(E^t - \Delta E)\rangle}$$

in which the velocity dependence of $T^2_{i \to j}$ may partially cancel or may be unobservable on account of affecting all $T^2_{i \to j}$ in a similar manner. Similar considerations apply to the exponential gap law (Procaccia and Levine, 1976b)

The ECS scaling law (Eq. (2.16)), like the statistical scaling laws, relates thermally averaged rate constants. Its unaveraged form is somewhat peculiar in that it relates rate constants with constant energy above threshold, i.e. for $\Delta j < 0$ the translational energy is some constant T while for $\Delta j > 0$ the translational energy is $T + \Delta E$ (DePristo et al., 1979; Smith et al., 1981). Moreover, a somewhat unsymmetrical situation comes about because the v in Eq. (2.22) is taken to be the initial relative velocity of the system. The theory is also restricted to de-excitation in DePristo et al. (1979), but this limitation can be overcome by using microscopic reversibility. Therefore we shall not be able to apply fitting laws which contain the ECS scaling law to data with constant E or E^t.

D. Theoretical Prediction of Fitting Law Parameters

For the many different systems which are represented by the fitting laws discussed in the previous section, the number of parameters necessary to describe the rate constant matrix has been reduced from $\simeq 10^4$ to several. This dramatic simplification not only makes manifest the similarities and differences of these systems, but it serves as a stimulus to simple, general theories which are capable of predicting the fitting parameters directly. In this section we shall consider general theoretical approaches for determining the physically important parameters in fitting laws. We believe that the relatively little work which has been done in this area (relative, say, to approximation schemes for a full close-coupling treatment to inelastic collisions) already demonstrates that this approach is likely to be the most fruitful way to make predictions of RET rate constants for heavy systems.

1. Impulsive Classical Calculation for $k^{IC}_{l \to 0}$

The ECS scaling law successfully reduces the entire rate constant matrix to the basis rate constants $k_{l \to 0}$. In many cases the collision is quite sudden for the most important basis rate constants (say those with $l < 30$), facilitating simplified calculation. We shall now outline an approximate classical calculation of the basis rate constants which predicts that the basis rate constants will obey a power law and which expresses the parameters a and γ in

Eq. (2.8) in terms of parameters in an assumed potential (Brunner *et al.*, no date). (Brechignac *et al.*, 1981).

The major approximations used in this calculation are as follows:

   (i) Classical mechanics—ought to be valid for non-hydrogenic collisional systems where the rotational quantum numbers are large and the de Broglie wavelength is small.

  (ii) Straight-line trajectory—RI cross sections are large, and the most probable collisions are a 'soft' one with small l and correspondingly small scattering angle.

 (iii) Sudden approximation—assumes that the molecule does not rotate during the collision. In these calculations we assume that $j_i = 0$, so the sudden approximation ought to be reasonable.

 (iv) Simple long-range potential—we assume that

$$V = C_n r^{-n} P_2(\cos\theta), \tag{2.31}$$

where θ is the angle between the molecular axis and the atom and P_2 is the second Legendre polynomial in $\cos\theta$. Typically, van der Waals' forces would dominate at long distances and so $n = 6$.

  (v) Impulsive approximation—assumes that most of the torque applied to the molecule by the colliding atom occurs near the point of closest approach.

With these assumptions it is straightforward, though tedious, to calculate the cross section $Q_{0 \to L}$ for adding angular momentum L to an initially stationary rotor, and from that the desired rate constants (Brunner *et al.*, no date). Because of the classical nature of this calculation we use the notation that L is the continuously variable magnitude of the angular momentum measured in units of h, and $Q_{0 \to L}\,dL$ is the cross section for the final angular momentum being between L and $L + dL$. The result of this impulsive classical (IC) calculation is

$$Q_{0 \to L}^{IC} = a_n \left(\frac{C_n}{hv}\right)^{2/(n-1)} L^{-(n+1)/(n-1)} \tag{2.32}$$

where

$$a_n = \frac{2\pi}{(n-1)} \left(\frac{3\Gamma(1/2)\Gamma((n-1)/2)}{2\Gamma(n/2)}\right)^{2/(n-1)} \int_0^1 dw$$

$$\cdot \frac{(1-t)^{1/2} + (1+t)^{1/2}}{2^{3/2}t} w^{2/(n-1)} \tag{2.33}$$

and $t \equiv (1 - w^2)^{1/2}$. (The integral in the definition of the coefficient a_n can easily be done numerically.)

To compare this expression more easily with rate constant measurements we write

$$k_{l \to 0}^{IC} = \frac{k_{0 \to l}^{IC}}{2l + 1}$$

$$k_{0 \to l}^{IC} = [2(l + \tfrac{1}{2})]^{-1} \int_{l-1}^{l+1} dL \langle v Q_{0 \to L} \rangle_T$$

$$\simeq a_n (C_n/h)^{2/(n-1)} \langle v^{(n-3)/(n-1)} \rangle_T [l(l + 1)]^{-[n/(n-1)]} \qquad (2.34)$$

where we have averaged the rate over a thermal velocity distribution at temperature T, approximated

$$l^{-(n+1)/(n-1)} (l + \tfrac{1}{2})^{-1} \simeq [l(l + 1)]^{[-n/(n-1)]}$$

and have gone to a discrete angular momentum l by summing over all classical momenta L within h of l. (This procedure is appropriate for systems with homonuclear symmetry.) This equation gives values of a and γ for Eq. (2.28).

We now make some general comments regarding comparison of Eq. (2.34) with experiment, reserving more detailed discussion for the next chapter. The value of γ, the exponent of $l(l + 1)$, is $-l$ less a little more which depends on the potential. The $-l$ comes from the general nature of RI collisions, independent of n, and is generally in accord with the theoretical and experimental data sets investigated here in which γ varies from 0·9 to 1·4. Thermally averaged rate constants are predicted to be proportional to $(T/\mu_s)^{(n-3)/(2n-2)}(C_n)^{2/(n-1)}$. Note that the factors containing μ_s and C_n tend to oppose one another because heavier molecules have larger polarizabilities.

2. Deviations from Power Law

The conditions for the power-law RI model just described are quite restrictive, and it is important to consider, qualitatively at least, what sorts of deviations might be expected from it. We shall consider only deviations resulting from the presence of a spherically symmetric term $V_0(R)$ in the potential—deviations from the impulsive (and therefore rotationally sudden) collisions assumed are more difficult to treat.

The first possibility which might occur is that the presence of $V_0(R)$, especially one with a minimum, could cause the distance of closest approach, hence the point at which the impulsive torque is calculated, to have a non-uniform distribution with respect to impact parameter. This could cause systematic deviations from the power law, especially at the large l values corresponding to smaller R (where V_0 is stronger). If a $V_0(R)$ potential with an attractive well concentrated turning points near the minimum, it is conceivable that local structure, such as a bump, would arise in the rate constant as a function of l.

A certain breakdown of the power-law model arises because real potentials have a repulsive core so that $V_0(R)$ will limit the distance of closest approach to some $R_m > 0$. Since the anisotropy of the potential is finite at this point there must be a corresponding maximum l^* to the angular momentum which can be transferred. Classically $k_{l \to 0}$ will be zero for $l > l^*$, but quantum mechanically one might expect an exponentially decreasing $k_{l \to 0}$ reflecting the presence of tunnelling—indeed an exponential gap factor may be shown to arise in RI collisions in which the transition occurs inside the classical turning point (Sanctuary, 1979). It thus seems likely that $k_{l \to 0}$ will in general behave as a power law for small l, but as a decaying exponential beyond l^*.

We can represent such a decay by adding an exponential factor to our usual power law, obtaining

$$k_{l \to 0} = a[l(l + 1)]^{-\gamma} \exp\left(\frac{-l(l + 1)}{l^*(l^* + 1)}\right). \tag{2.35}$$

When we use it in fits we shall designate it EP (i.e. ECS-EP).

A more concrete argument about l^* may be based on a classical view of the collision. An impulsive collision with the repulsive core of the potential is basically a collision with a hard ellipsoidal shell, a view of RI collisions which has recently received a great deal of discussion in connection with measurements of differential RI cross sections (Serri et al., 1980; Bosanac, 1980; Beck et al., 1979; Bergmann et al., 1980). A hard ellipsoid has a fixed eccentricity, and consequently the transferred angular and linear momenta, hl and Δp, are proportional and satisfy

$$\frac{hl}{\Delta p} \le R_A$$

where R_A is the classical moment arm. This behaviour is observed experimentally (Serri, 1980). (For the model discussed above, $R_A \sim R_m V_2(R_m)/V_0(R_m)$.) It is clear that the maximum of Δp must be of order p_i, the initial momentum (if most of the translational energy is transferred to rotation, $p_f \ll p_i$). Therefore, one obtains a maximum for l,

$$l^* \le \pi R_A/h.$$

Transfer of $l > l^*$ is classically forbidden.

Whether or not one actually observes this predicted transition from power-law to exponential behaviour depends on whether l^* is energetically accessible. The initial translational energy is $E^t = p_i^2/2\mu_s$, and the final rotational energy is (for transfer of l^* angular momentum)

$$E_f^{r*} = Bl^*(l^* + 1)$$
$$= h^2 l^*(l^* + 1)/(2\mu_m r^2)$$
$$= (R_A/r)^2 p_i^2/\mu_m.$$

The observation of the transition from power-law to exponential decrease of $k_{l\to 0}$ is not possible unless $E_f^{t*} < E^t$, or equivalently if

$$(R_A/r)^2 \mu_s/\mu_m < 1.$$

In systems where R_A is defined, this criterion would seem to be a natural way to define weakly *vs.* strongly anisotropic atom–molecule potentials. This application of strong/weak to the anistropy is distinct from an earlier application of the term 'weak' to RI collision processes (Oka, 1973) which are governed by first-order perturbation theory.

3. Surprisal Synthesis

a. Surprisal synthesis via dynamical calculations. The prior rate constants of surprisal theory are determined by maximizing the entropy assuming only energy conservation. In surprisal theory, the deviation of the observed rate constants from the prior is attributed to additional constraints in the entropy maximization. Suppose the surprisal (see Eq. (2.23)) can be expanded in the series (Levine and Kinsey, 1979)

$$S = -\lambda_0 - \sum_{i=1}^{n} \lambda_i A_i \tag{2.36}$$

where the A_i's are the 'natural variables'. Then the experimental distribution can be reproduced by maximizing the entropy subject to the n constraints that the average value of A_i be $\langle A_i \rangle$, the average of A_i over the experimental distribution.

This expansion leads to a useful method (Procaccia and Levine; 1975, 1976b) for using the results of a dynamical calculation of the $\langle A_i \rangle$ to predict the complete experimental distribution. Suppose that one knows what the n natural variables are, but one does not know their average value. Any dynamical calculation which provides values for the $\langle A_i \rangle$'s can be used to generate complete predictions of the experimental distribution. For a more concrete example, suppose that the one natural variable is $\ln |\Delta E|$, a choice unsupported by physical intuition, but which leads to the SPG fitting law. Any dynamical calculation of $\langle \ln |\Delta E| \rangle$ can be used as the constraint in a maximal entropy procedure to predict the final state distribution.

Perhaps the most practicable way to implement this procedure is to use a classical trajectories computer program. There is strong evidence that classically computed averages are an excellent approximation to the exact quantal average (Levine, 1973; Miller and Raczkowski, 1973). By using only averages of observables one avoids the problem of going from a continuous classical distribution to a discrete quantized one. A final advantage of this approach is that many fewer trajectories are required to calculate an average, e.g. $\langle \ln|\Delta E| \rangle$, accurately than to calculate, with comparable accuracy, the entire distribution.

The surprisal synthesis described above obviously requires knowledge of the potential surface of the colliding bodies in order to carry out the dynamical calculation. It turns out that even in the absence of such knowledge—in fact, knowing only the total energy E and rotational constant B—attempts have been made to predict the parameter θ in the exponential gap law. This idea has motivated several papers by Procaccia and Levine (1975, 1976b, c).

They assume that the total energy transfer rate (or sum)

$$R(j_i) = \sum_{j_f}(E_{j_f}^r - E_{j_i}^r)k_{j_i \to j_f} \tag{2.37}$$

is linear in $[j_i(j_i + 1)]^{1/2}$. Under this assumption, it is possible to generate the parameter θ of the exponential gap law. The utility of this procedure is suspect because so few collisional systems are known which satisfy the exponential gap law (for RI collisions). Attempts by us to apply a similar technique to generate the parameter γ in the SPG fitting law were not successful.

Procaccia and Levine (1975, 1976b, c) have applied surprisal synthesis using the sum rule to several systems. For example in HF colliding with He the predictions of the surprisal synthesis (Procaccia and Levine, 1976b) are in good agreement with the close-coupled calculations of Collins and Lane (1975). Since this method requires knowing only B and E, the predictions of the sum rule synthesis are.

III. POWER-LAW FITS

In this section we present a collection of data sets in which the RI rates have been found to decrease as a power law of either the transferred energy ΔE (i.e. SPG-λ) or the transferred angular momentum Δj (i.e. ECS-P). In part III A we discuss sets of RI rate constants with only one initial translational energy distribution; these include thermally averaged rate constants from ordinary gas cell experiments, rate constants measured with a narrower velocity distribution, and theoretical calculations done with a monochromatic distribution. In part III B we will discuss systems for which rate constants are available over a range of relative velocities. We shall consider only the three-parameter fitting laws, ECS-P and SPG-λ, and shall not consider the several simpler two-parameter laws (SPG-0, SPG-Δ, and IOS-P) which are special cases of the previous two.

The fitting procedure that we used to obtain the results of this section (as well as Sec. IV) has been previously described (Brunner, 1980; Brunner et al., 1981). We shall now briefly discuss several quantities which will be useful in describing how well a fitting law works. The 'reduced chi-squared' χ^2/v, the sum of the squares of the relative deviations (difference between fit and datum) each divided by its associated error, all divided by v, the number of data points less the number of free parameters, is the standard measure of whether the fit agrees with the data to within error. $\chi^2/v \simeq 1$ indicates a fit 'to within error' while $\chi^2/v \gg 1$ indicates an inadequate fit. The problem with using χ^2/v exclusively is that two experiments, one with 5% error bars and the other with 50% errors bars, may each claim to verify some fitting law because $\chi^2/v \simeq 1$. Clearly the experiment with smaller error bars is a more precise verification, yet this information is not reflected in χ^2/v.

For this reason we have defined (Brunner, 1980; Brunner $et\ al.$, 1981) a weighted RMS fractional deviation (between fit and data), $D \equiv \langle (y - y_l)/y_l \rangle$, where y_l is the lth datum, y the fit formula prediction, and $\langle \square \rangle$ denotes a weighted RMS average. We can similarly define, for any set of data, an average error $E \equiv \langle \sigma_l/y_l \rangle$, where σ_l is the (one standard deviation) error of the lth datum. We usually express both E and D as a percentage so that we might describe some set of data as having an average relative error E of 9%, and say that a certain fit to these data has an average deviation D of 11%. By properly choosing the weighting function in the average (Brunner, 1980; Brunner $et\ al.$, 1981), one can show that the following relationship (good to the order of number of fit parameters divided by number of data) links χ^2/v, E, and D:

$$\chi^2/v \simeq D^2/E^2, \tag{3.1}$$

where N is the number of data points in the fit. D enables one to compare the degree of agreement of a fit with a database essentially independently of the errors; E enables one to describe the average error in a data set.

For a variety of reasons we are presenting both ECS-P fits and SPG-λ fits in spite of our predilection towards ECS-P. For the thermally averaged data in sec. III A, SPG-λ is far simpler to apply whereas ECS-P offers, in some cases, better fits to data sets with more than one j_i Brunner $et\ al.$, 1981. The question of whether ECS-P is superior to SPG-λ has been muted by the recent discovery of a mathematical connection between the two fitting laws (Smith and Pritchard, 1981). This connection explains why in many cases the predictions of ECS-P and SPG-λ are in close accord for thermally averaged rate constants despite their very different theoretical bases. (For the data sets with non-thermal velocity distributions SPG-λ and ECS-P relate sets of rate constants at different initial velocities, as we have discussed in Sec. II C.4, and it is not always possible to apply ECS-P to such data sets.)

Both SPG-λ and ECS-P are able to fit the data in this section much better than any previously proposed fitting law, e.g. the exponential gap law. We shall demonstrate not only the high quality of such fits, but also the excellent interpolating ability of these formulae.

The data in this section are restricted to those which satisfy a power law; we shall discuss other fitting laws (e.g. EGL or ECS-EP) in Sec. IV. We emphasize as one of the major points of this review that power-law behaviour appears to be fairly common; of the roughly 20 sets of high-quality RI rate constants which we have considered, about 70% follow power-law behaviour.

A. Rate Constants with a single Initial Kinetic Energy

In this section we discuss several types of data sets, all of which have a single distribution of initial kinetic energy; that is in contradistinction to the data of part III B where we consider systems which have rate constants measured over a range of velocities. In parts III A.1 to A.3 we discuss thermally averaged rate constants, so that both ECS-P and SPG-λ are directly applicable. In parts III A.4 to A.6 we consider results from colliding beam or beam–gas experiments for which the velocity resolution is somewhat narrower ($\Delta v/v \simeq$ 5–20%). For these data only SPG-λ fits were done, partly because these data sets were not large and had only one j_i, but also for the reasons discussed in Sec. II C.4.

1. I_2^*–Xe

Measurements (Dexheimer et al., 1981) of $I_2^*(B^3\Pi)$ colliding with Xe in a room-temperature cell (293 K) are an excellent example of power-law

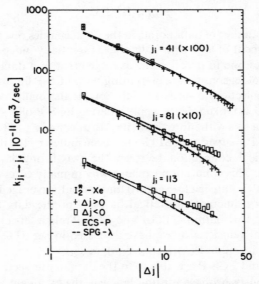

Fig. 1. ECS-P (solid line) and SPG-λ (dashed line) fits to experimental data of I_2^*–Xe (Dexheimer et al., 1981). The values of χ^2/v are 0·99 and 1·09 respectively, showing no preference for one law over the other, a conclusion confirmed by examination of the figure. While the average deviation of fits and data is 10%, the two fits differ by at most 5%.

TABLE II

System	Ref.	j_i range	$E(\%)$	ECS-P				SPG-λ		
				γ	$l_c(\text{Å})$	τ_{max}	$D(\%)$	$D(\%)$	α	λ
I_2^*–Xe	Dexheimer et al. (1981)	41–113		0·911	3·2	1·5	9	9	0·87	0·83
Li_2^*–Ar		11–27		1·11	6·8	2·5	14	16	1·00	17
Na_2^*–Xe	Brunner et al. (1981)	4–100	7·4	1·08	5·5	4·0	8	19	1·10	23
Na_2^*–Kr	Brunner et al. (1981)	4–100	6·5	1·14	5·3	3·5	11	12	1·24	36
Na_2^*–Ar	Brunner et al. (1981)	4–100	6·6	1·21	5·7		8	11	1·33	52
Na_2^*–Ne	Brunner et al. (1981)	16–66	5·7	0·99	4·4		6	7	1·02	57
Na_2^*–He	Brunner et al. (1981)	4–100	5·5	1·05	4·0	0·83	9	8	1·14	153
Li_2^*–Li	Vidal (1978)	24		1·33	4·8	1·3	12	12	1·64	300
Na_2–Ar	Serri et al. (1981)	7		n.a.	n.a.	n.a.	n.a.	28	1·19	300
LiH–He										
LiH–Ar										
LiH–HCN	Dagdigian and Alexander (1980)	1		1·38	0*	N	13	13	1·38	300*
Na_2^*–H₂	Brunner et al. (1981)	4–66	6·7	1·15	8·4	N	12	16	1·21	140
Na_2^*–N₂	Brunner et al. (1981)	4–66	7·2	1·17	5·5		13	15	1·26	61
Na_2^*–CH₄	Brunner et al. (1981)	4–66	6·6	1·29	6·0		11	15	1·47	122

*Constrained to this value.
N Could not be determined from data.
n.a. Not applicable.

behaviour. Accurate rate constants $(E = 9\%)$ have been measured with j_i between 41 and 113, and $|\Delta j|$ as great as 40. Rate constants are shown in Fig. 1 versus $|\Delta j|$ in a log–log plot. Results for both ECS-P and SPG-λ fits are tabulated in Table II.

For the ECS-P fit, shown as a solid line in Fig. 1, we vary the three parameters a, γ, and l_c in Eqs. (2.7), (2.28), and (2.15). (The apparently continuous nature of the fit is a result of connecting adjacent discrete points by line segments.) The value of γ is 0·91, somewhat lower than previous observations in other systems where γ has generally been in the range 1·0–1·3. $l_c = 3·2\text{Å}$ is physically reasonable for the length over which most of the interaction takes place. The quality of fit is excellent with $\chi^2/\nu = 0·99$ and an average fractional deviation $D = 10\%$.

The SPG-λ fit, shown as a dotted line in Fig. 1, is also excellent with $\chi^2/\nu = 1·09$ and $D = 10\%$; it can barely be distinguished from the ECS-P fit. Both fits correctly accounted for the fact that the $\Delta j < 0$ points lie on a different curve to the $\Delta j > 0$ points, that is both correctly account for the $+/-$ asymmetry. Both fitting laws also account for the decrease of the rate constant (at fixed Δj) with j_i. By way of comparison, the exponential gap law has $\chi^2/\nu = 10$ and average deviation D of 30%.

2. Li_2^*-Ar

Ottinger and Schroder (1979) have recently measured rate constants for RI collisions of $Li_2^*(A^1\Sigma)$ colliding with Ar at nine values of j_i. In Fig. 2 we plot $k/(RN_{\lambda = 17})$ versus $|\Delta E|$ for four different values of j_i. (We used all of the data (Ottinger and Schroder, 1979) in the fits, but omitted intermediate values of j_i for clarity in the figure.) Note that the data fall on a straight line, which indicates power-law behaviour in this log–log plot. Also, by choosing $\lambda = 17$ we find that data for all j_i fall on the same line, i.e. the statistical scaling portion of the fitting law works well. The straight line in Fig. 2 is the SPG-λ fit which has a D of 16%. The results of this fit, in Table III, show that α is 1·00, a typical value. The ECS-P fit to these data is marginally better with D of 14%.

3. Na_2^*-X

This system has been extensively studied, beginning with Na_2^*-Xe (Brunner et al., 1978, 1979) and subsequently Na_2^* colliding with the eight target gases Xe, Kr, Ar, Ne, He, H_2, CH_4, and N_2 (Wainger et al., 1979; Brunner, 1980; Brunner et al., 1981). For each target gas, approximately 60 rate constants were measured with $j_i = 4$, 16, 26, 38, 54, 66, and 100, and average errors $E \simeq 7\%$. $|\Delta j|$ varied from 2 to 28 and the dynamic range of the rate constants was 100:1.

a. Power-law fits and ECS-P interpolation. We now summarize the con-

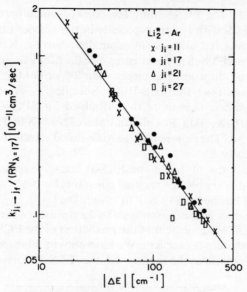

Fig. 2. SPG-λ fit to Li$_2^*$ ($A\Sigma$)–Ar data with $v_i = 15$ and j_i as shown. On this plot the SPG-λ fit is a straight line.

TABLE III

System	$\dfrac{k_{0\to2}^{IC}}{k_{0\to2}^{ECS}}$	$\dfrac{k_{0\to26}^{IC}}{k_{0\to26}^{ECS}}$	$D(\%)$
Na$_2$–Xe	1·32	0·73	31
Na$_2$–Kr	1·04	0·76	33
Na$_2$–Ar	0·96	0·96	14
Na$_2$–Ne	1·46	0·51	69
Na$_2$–He	0·88	0·40	120
Na$_2$–H$_2$	1·58	1·16	17
Na$_2$–N$_2$	1·42	1·17	14
Na$_2$–CH$_4$	0·99	1·44	17

clusions of Brunner *et al.* (1981), where several fitting laws were applied to these hundreds of high-quality data. The ECS-P fitting law was found to fit each one of the eight target gases with typical deviation $D \simeq 10\% (\chi^2/\nu \simeq 2)$, an unprecedented achievement considering the wide range of j_i and Δj. The power-law assumption for the basis rates (Eq. (2.28)) worked quite well and was not substantially improved upon by any of several generalizations including ECS-EP. The ECS scaling law worked much better than the IOS scaling law (i.e. ECS with $l_c = 0$), especially in the heavier target gases where χ^2/v for ECS-P was typically 40 times smaller than for IOS-P. The SPG-λ fitting law also worked well in many cases, but for heavy target gases at large j_i (where the reduced duration τ is larger than 2·5 or so) the SPG-λ fits were substantially worse than the ECS-P fits. Specifically, as one can see from Table II, the deviation D is more than doubled for SPG-λ compared with ECS-P for the Na_2^*–Xe data. For all examples of Na_2^*–X in Table III, γ is near 1·1 and l_c is near 5 Å. The exponential gap law failed utterly for these data with χ^2/v values typically 30.

We now discuss the utility of the ECS-P fitting law as an interpolating expression. In Fig. 3 are plotted the rate constants for Na_2^*–Kr at three values of j_i versus $|\Delta_j|$. The line is an ECS-P fit to *only* the $k_{4 \to 6}$, $k_{4 \to 22}$, and $k_{100 \to 102}$ data; these three data points are enough to determine the three parameters in the fit, a, γ, and l_c. The agreement of the prediction of the ECS-P formula with the other (unfitted) data is excellent. We have chosen this most extreme fitting procedure to dramatize the ability of the ECS-P formula to generate

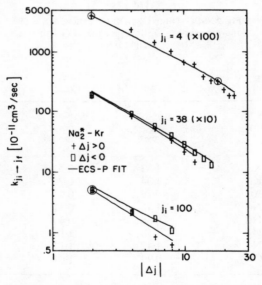

Fig. 3. ECS-P fit to Na_2^* ($A\Sigma$)–Kr data (Brunner *et al.*, 1981) with $v_i = 18$. The $j_i = 100$ rates lie almost an order of magnitude below the corresponding $j_i = 4$ (note powers of 10 displacement of curves for clarity); the ECS scaling attributes this to adiabaticity arising from collisions with a potential of effective range $l_c = 5·3$ Å. Only those data contained in the three circles were fitted: the excellent agreement with the rest of the data shows the interpolating ability of ECS-P.

'complete' rate constant matrices from a few (well-chosen) measured data—there can be no doubt that it also fits the several thousand unmeasured rate constants with $j_i < 100$ and $|\Delta j| = 20$ equally well.

b. Comparison with impulsive classical approximation. The impulsive classical calculation in Sec. II D.1 makes a prediction for the basis rate constants $k_{l \to 0}$. We obtain C_6 by estimating the polarizability of Na_2^* to be 300 Å^3 along and 40 Å^3 perpendicular to the internuclear axis (for a similar approach see Ottinger and Schroder, 1980), and use the known polarizabilities of the target gases (Karplus and Porter, 1970). Note that even if the Na_2^* polarizability is incorrect, the scaling of C_6 with different target gases should still be correct.

The results of this simple calculation are contained in Table III where we take the ratio of the result of the impulsive approximation $k_{0 \to l}^{IC}$ (Eq. (2.34)) to the $k_{0 \to l}^{ECS}$ determined by the ECS-P fit to our experimental data, for $l = 2$ and 26. We also tabulate the average deviation D between the calculated and the 'actual' (found via ECS-P fit) value for $k_{l \to 0}$ for all l between 2 and 26. The agreement is surprisingly good, especially if the lighter targets are excluded (long-range forces are well known to be less predominant in these systems). As a typical example of the agreement we have plotted in Fig. 4 the computed

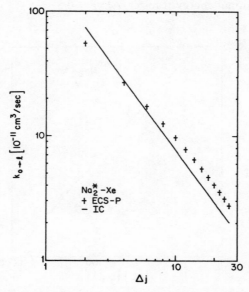

Fig. 4. Impulsive classical prediction of the basis rates $k_{l \to 0}$ for Na_2^*–Xe (solid line, no adjustable parameters) compared with the predictions of ECS-P(+) using the parameters found experimentally (+).

$k_{0 \to l}^{IC}$ and 'actual' $k_{0 \to l}^{ECS-P}$ for Na_2^*–Xe; the average deviation D is 31%. The theory is too large at small l and vice versa. For four out of the eight targets the agreement is better than 20%. The only evidence that the impulsive classical approximation misses significant physics is that for He and H_2 (both light, relatively unpolarizable targets where the assumption of a long-range potential and straight-line trajectory might be expected to fail) the theory misses in opposite directions: too low for He while it is too high for H_2, a discrepancy difficult to explain by invoking the internal degree of freedom of the H_2.

4. Na_2–Ar

A differential RI cross section experiment with two colliding supersonic beams has recently produced data with Δj as great as 80 (Serri *et al.*, 1980). We have fited the integrated cross sections of Serri *et al.* (1981) with $j_i = 7$ and $E^t = 2420 \, \text{cm}^{-1}$. Because of the narrow velocity range in the experiment we take the translational phase space factor to be $(1 - \Delta E/2420)^{1/2}$ (with ΔE in cm^{-1}). In Fig. 5 we have plotted $k/RN_\Delta)$ versus ΔE, where the data are observed to fall on a straight line. The power-law fit has a D of 28%, which is comparable to the average error E of these data. Note that these data extend to a very high fractional energy transfer, $\Delta E/E = 0.5$.

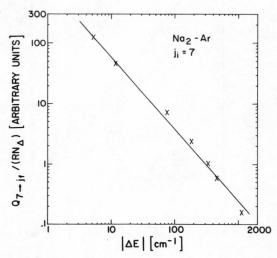

Fig. 5. SPG-Δ fit to molecular beam data for $Na_2(X\Sigma)$–Ar with $v_i = 0$ $j_i = 7$, and $E^t = 2420 \, \text{cm}^{-1}$. Note large range of Q/RN_Δ and ΔE. The Δj are 2, 4, 16, 28, 40, 48, and 80. The apparent negative curvature of the data is not significant since the fit has $\chi^2/\nu = $.

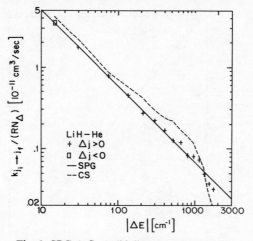

Fig. 6. SPG-Δ fit (solid line) to LiH–He data (Dagdigian and Wilcomb, 1980). The broken line results from a coupled states calculation (Jendrek and Alexander, 1980) using an *ab initio* potential (Silver, 1980). The calculation is discussed more fully in Sec. IV C.

5. *LiH–He, Ar*

Dagdigian and coworkers have used a beam–gas technique to measure cross sections for LiH ($v_i = 0$ and $j_i = 1$) with a number of collisional partners, including the rare gases Ar (Wilcomb and Dagdigian, 1977) and He (Dagdigian and Wilcomb, 1980). In this polar molecule, rates for both targets obey a power-law fit to within the estimated error. Fig. 6 shows an SPG-Δ fit to the LiH–He data (with only one j_i SPG-λ is unnecessary); a power-law fit to the LiH–Ar data is displayed in Pritchard *et al.* (1979). The χ^2/ν in Table II is near one for both targets, indicating a good fit.

6. *LiH–HCN*

Dagdigian and Alexander (1980) have measured integral RI cross sections Q for the collision of a LiH beam with HCN gas, an interesting combination since both molecules are quite polar. The dipole–dipole forces yield large total cross sections of $\simeq 1000\,\text{Å}^2$. In Fig. 7 we plot $Q/(RN_\Delta)$ versus ΔE, where we observe a straight line which decreases significantly faster than the previous data. Table II shows that D is 12%, quite comparable to the 9% (one standard deviation) error bars on these data. Also, notice that α is 1·39, much larger than the values for the previously described systems. An IOS-P fit to these data yields the same slope, $\gamma = 1·39$.

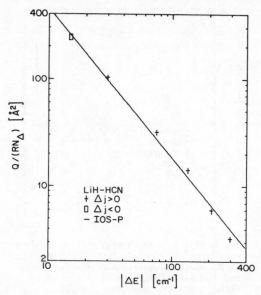

Fig. 7. SPG-Δ fit to LiH–HCN experimental data (Dagdigian and Alexander, 1980) taken with thermally distributed rotational distribution of HCN target. The final level of HCN was not determined, and ΔE is taken to be the energy transferred into the LiH (not necessarily from translation to rotation). These data yielded $\alpha = 1.39$, the largest exponent observed in any fit we have done.

7. Summary and Comparison with Other Data

We have found good power-law fits for all of the systems in parts III A.1 to A.6. Table II contains these results, along with results from a few other systems not discussed explicitly in the text. The existence of such a large class of RI systems which display power-law behaviour is in itself strong evidence for the fundamental correctness of the impulsive classical approximation. Further evidence comes from the fact that in such a diverse collection of collisional systems the ECS-P fitting parameter γ is so constant. This result can be explained by the impulsive sudden approximation which predicts that γ is $1 + 1/(n - 1)$; thus γ is only weakly dependent on the potential steepness and is independent of the potential's strength (i.e. the value of n).

In accord with this prediction γ should be largest for systems with longest range forces (i.e. $n < 6$). The LiH–HCN dipole–dipole interaction goes like R^{-3}, implying a γ of 1.5, while the fit to the data yields γ of 1.39 (the largest γ observed). In addition the system $\text{Li}_2^*(B^3\Pi)$–Li (Vidal, 1978) which (see complete discussion of these measurements in Sec. IV D) might be expected to have relatively long-range forces owing to the possibility of resonant excitation transfer also has a value of γ which is substantially larger than average. Finally, this theory is found approximately to predict the magnitude of the cross section for the Na_2^*-X data, as we demonstrated in Sec. III A.3b.

B. Velocity Dependence of RI Rates

While calculations generally are capable of finding the velocity dependence of RI cross sections (at a cost of more computer time), it is only recently that it has been possible to determine experimentally the velocity dependence of RI cross sections. For systems which obey a power law for fixed translational energy distributions, the velocity dimension presents two obvious challenges:

(i) Do the data in a particular system continue to display power-law behaviour at each initial translational energy?

(ii) If so, how do the parameters of the fitting law depend on velocity?

The only two experimental studies of the velocity dependence of RI collisions are on HF–Ar and Na_2^*–Xe (Smith et al., 1981). Both show that power-law fitting laws work well over the entire range of velocities studies. (The range was 2·1 in HF–Ar and 3·1 in Na_2^*–Xe.)

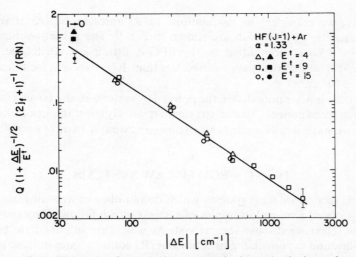

Fig. 8. SPG-Δ plot of cross sections measured using molecular beams for HF–Ar with $v_i = 1$, $j_i = 1$, at three different energies (in kcal mol^{-1}). $\Delta j > 0$ ($\Delta j < 0$) is indicated by open (solid) symbols. The small numbers indicate j_i and j_f. Except for $j_i = 1 \rightarrow 0$ there is little systematic change of cross section with E^t.

1. $HF-Ar$

Fig. 8 is plotted to display SPG fits as straight lines, and it is evident that the data not only fall on a straight line at each velocity but that it is the same straight line! The only possible deviation from this is the points for $j_i = 1 \to j_f = 0$ which show an increase at the lowest velocity. This deviation would cause γ (the exponent) to increase slightly at the lowest velocity.

2. Na_2^*-Xe

For Na_2-Xe (Smith *et al.*, 1981), data were taken with a variety of initial j_i, so that it is possible to determine the velocity dependence of all three ECS-P parameters. It was found that the exponent j_i increased $\sim 25\%$ over the velocity range, and that l_c increased by nearly a factor of 2 over this range. In addition it was found that the observed RI rates generally did not even increase linearly with velocity, indicating that the corresponding cross sections decreased. The designation of l_c as an effective range of the potential (Sec. II B) thus comes under some question, since it is difficult to imagine that the range should increase when the cross section decreases.

3. N_2-Ar

Lastly, we consider the calculations of RI rates in N_2-Ar (Pattengill and Bernstein, 1976), which are shown in Fig. 9. The subset of these with $|\Delta E|/E^t = 0.3$ has been fitted to the SPG-Δ fitting law (Pritchard *et al.*, 1979) and it was shown that γ varied less than 10% over a velocity range of $(0.9-4) \times 10^5$ cm s^{-1}.

In conclusion, we note that in the power-law systems studied over a range of velocities, the exponent γ tended to remain relatively constant. In addition, the cross sections generally tended to decrease somewhat with increasing velocity.

IV. NON-POWER-LAW SYSTEMS

In this section we treat systems which do not obey power-law fitting laws. We first observe several examples of systems which follows the exponential gap law. Next we discuss several systems which are intermediate between power-law and exponential gap behaviour: RI collision rates in these systems decrease as an exponential at large angular momentum transfer and as a power law at small transfer. We then discuss rates which display structure as a function of Δj. Finally we discuss systems in which there are alternations of the rate with respect to Δj.

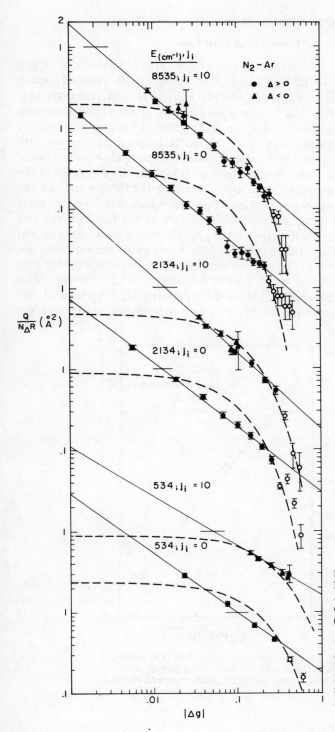

Fig. 9. SPG-Δ fit (solid line) with ordinate $\Delta g = \Delta E/E_{\text{total}}$ for classical calculations of N_2–Ar (Pattengill and Bernstein, 1976). Only closed points were fitted. The slopes are equal within combined errors. Horizontal lines show 1 A^2 point on fits. Dashed lines are EGL fits to each complete data set.

A. Exponential Gap Law

As might be expected from the arguments of Sec. II D.2, the exponential gap law (EGL) appears to be best satisfied in systems with relatively small anisotropy and low system mass, H_2–H, H_2–He, and HCl–Ar, being the best examples (Green, 1979). Fig. 10 shows this behaviour for H_2–H scattering with $E^t = 10^4 \, cm^{-1}$ (1·25 eV). This is a remarkable fit because the dynamic range is 10^6 (experimenters take envy!), by far the largest range for any data considered in this review. The fit also extends to a relatively high fractional energy transfer of 0·6. The average deviation is high by the standards of the preceding chapter, being 44% even after discarding the three worst points.

We know of no measurement of an RI process which definitively displays EGL behaviour, an ironic state of affairs in view of the fact that the first experimental evidence for the exponential gap law came from a measurement of RI processes (Polanyi and Woodall, 1972). These measurements were not truly level-to-level, and in our opinion were primarily important in that they showed that the rate decreased rapidly for larger energy transfer. (This experiment was not interpreted to show the superiority of an exponential over power law to represent this decrease.) Early experiments on I_2–rare gas

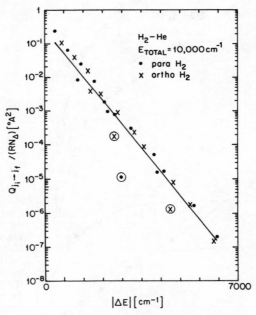

Fig. 10. H_2–H close-coupling calculations (Green, 1981). The straight line is a fit using the exponential gap law but with the circled points excepted from the fit.

(Steinfeld and Klemperer, 1965) were also claimed to verify the exponential gap law; however, we have obtained contradictory experimental evidence (Dexheimer *et al.*, 1982) and believe that the presence of multiple collisions (up to six per radiative lifetime) and both hyperfine and rotation-induced predissociation (Broyer *et al.*, 1975, 1976) invalidates the conclusion of this work. We are not aware of any experimental studies of level-to-level RI rates under single-collision conditions which are best fitted by EGL.

B. Exponential–Power Hybrids

In this subsection we discuss data sets which are fitted by the two hybrid exponential–power laws ECS-EP (the combination of ECS scaling (Eq. (2.16)) with the EP $k_{l \to 0}$ basis rates (Eq. (2.29)) and SEPG-λ (the statistical law with limited m-conservation (Eqs. (2.17) and (2.22)) with an exponential times a power law for the square of the T-matrix element (Eq. (2.27)). The scaling behaviour of EP systems is particularly revealing because EP fits have a definite kink or knee in the rate constant, and one can discriminate between ECS-EP and SEPG-λ by noting whether this feature holds its position more constant with transferred angular momentum or energy.

1. N_2-Ar

It is clear from Fig. 9 that the points with large energy transfer which were omitted from the power-law fit to the N_2-Ar theoretical data (Pattengill and Bernstein, 1976; Pritchard *et al.*, 1979) fall sharply below the power law, and are more nearly in accord with the dashed EGL fit. This type of behaviour is the hallmark of EP systems, and foreshadows the success of the EP fit in this system. While the pure power-law fits were applied only to a restricted range of ΔE, SEPG-λ fits all ΔE with the same (or better) quality of fit which we found using SPG-Δ on the earlier subset of these data. The results for the parameters are displayed in Table IV, and an example of such a fit is shown in Fig. 11.

We attempted to fit these data using ECS-EP by assuming that they were thermally averaged data appropriate for the temperature $T = 2k(E^t + E^r)/3$. Although these fits gave D somewhat larger than the SEPG-λ fits, we cannot draw the definite conclusion that the exponential cut-off depends on the energy transfer rather than the angular momentum transfer because the assumption of thermally averaged data is an extreme one for this data set which has data with a large ($> 70\%$) fraction of the energy in rotation.

2. I_2-He

A second example of EP behaviour is the system I_2-He (Dexheimer *et al.*, 1982). This system has the smallest ratio of atomic mass to molecular mass of

TABLE IV

System	Ref.	Energy (cm⁻¹)	i_i	τ_{max}	ECS-EP E(%)	a	-γ	l*	l_c(Å)	D(%)	SEPG-λ D(%)	a	-α	θ⁻¹(cm⁻¹)	λ	Comments
H₂H-H	Green[a]	10000	1–11	N	N	Fit did not converge.					45	0.19(1)	0*	455(4)	300*	EGL to para only
						Presumably D ≥ 100					75	0.37(4)	0*	362(5)	300*	EGL to ortho only
											70	0.30(2)	0*	385(3)	300*	EGL to all data
											68	876(53)	1.2(1)	492(11)	300*	SEPG-Δ to all data
N₂–Ar	Pattengill and Bernstein (1976)	2134	0–21	N	8.2	7.7	0.72(2)	21(1)	2(21)	22	16	10(1)	0.70(2)	862(50)	14(1)	ECS-EP and SEPG-λ
											23	20(1)	0.88(2)	0*	13(1)	SPG-λ
											40	0.73(2)	0*	121(2)	300*	EGL
I₂–He	Dexheimer et al. (1981)	290	41, 91	N	7.2	8.7(6)	0.73(2)	24.3(5)	0*	8.2	27	0.22(2)	0.53(3)	38(2)	300*	IOS-EP and SEPG-Δ
						8.7(6)	0.73(2)	24.5(5)	0.190	8.3	31	0.061(1)		25.1	ECS	ECS-EP
						39(1)	1.20(1)	0*	1(60)	40					300*	ECS-P and EGL
⁶Li⁷Li–Xe	Scott and Gottscho[b]		9	N	10						9.4	19(1)	0.91(1)	0*	12(2)	SPG-λ to even Δj
					10						7.2	175(22)	0.70(4)	700(240)	0*	SPG-O to even Δj
					17						27	0.28(2)	0.38(4)	0*	13(3)	SPG-λ to odd Δj
											26	2.5(3)	0.16(4)	700(300)	0*	SEPG-O to odd Δj
⁷Li₂–He			8	N	10						27	17(2)	0.94(3)	0*	14(2)	SPG-λ

N Not given or not determined by fit.
* Constrained to this value in fit.
[a] We thank S. Green for sending us unpublished data on H₂.
[b] We thank T. Scott and R. Gottscho for giving us unpublished data.

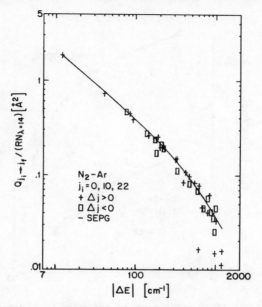

Fig. 11. SEPG-λ fit to N_2–Ar classical trajectories calculation (Pattengill and Bernstein, 1976) at $E = 2163$ cm^{-1}. Data with $j_i = 0$, 10, and 22 are not systematically different when $\lambda = 14$ and are all plotted using $+ (\Delta j > 0)$ and $\square$ ($\Delta j < 0$). Note downward curvature at large ΔE.

any system mentioned in this review and is therefore quite sudden; it is the only system for which IOS scaling had lower χ^2/ν than ECS (because ECS does not significantly reduce the χ^2 but has one extra free parameter). The data are plotted in Fig. 12 where a definite knee is evident in both the $j_i = 41$ and $j_i = 91$ curves. These knees appear at roughly the same value of Δj, and are thus better fitted by ECS-EP, the solid line, than by SEPG-λ, the dashed line (which would fit only if the knees occurred at the same energy transfer). Statistical measures of the quality of fit strongly support this assertion: whereas χ^2/ν is 1·3 for ECS-EP (with a corresponding $D = 13\%$), it is above 14 for SEPG-Δ (appropriate for this sudden system) and higher still for all other laws tried (SEPG-0, ECS-P, ECS, etc.).

The value of l^* obtained from the fit was 24, a little past the value of Δj which is subjectively located at the knee. Under the experimental conditions (temperature 270 K) this corresponds to an impact parameter $b \equiv 3$ Å using the semiclassical relationship $b = (\ell + \frac{1}{2})\lambda_{\text{deBroglie}}$. This value is larger than $(Q/\pi)^{1/2} = 1·5$ Å where Q is the sum of all quenching, vibrationally inelastic, and RI cross sections with $l > l^*$. Thus it appears that most of the collisions with $b < 3$ Å are RI collisions with $l < l^*$, as dictated by the lack of sufficient orbital

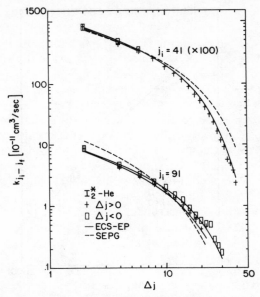

Fig. 12. Hybrid exponential–power fits to I_2–He (Dexheimer *et al.*, 1982) using both ECS-EP (solid line) and SEPG-Δ (dashed line). For clarity only the $\Delta j > 0$ predictions are shown for SEPG (the $\Delta j < 0$ prediction is very close to the SEPG curve shown, crossing it at $f_i = 2$ and 9 for $j_i = 91$).

angular momentum to cause transfer of more than l^* units of angular momentum to the I_2.

C. Bumps: LiH–He

Bumps or other structure observed in the Δj dependence of RI cross sections provide an opportunity to study the dynamics and the intermolecular potential in specific detail. We use the imprecise term 'bumps' intentionally as it can over local maxima and minima, and even the mild features associated with a region of positive second derivative in the $\ln k$ vs. $\ln \Delta j$ curve (which has negative or zero second derivatives with E, P, or EP behaviour). From the standpoint of scaling, the exact nature of the bump is unimportant; the key issue is whether the bump maintains its position with Δj or ΔE, for example, and how it moves with changes in velocity. From the standpoint of the potential, a bump is presumably revealing of significant departures from the smooth behaviour assumed in the classical impulsive approximation (which led to a corresponding structureless $k_{l \to 0}$). Relating an observed bump or decrease in $k_{l \to 0}$ to a particular feature of the potential is a difficult business, fraught with the perils of non-uniqueness, and beyond the scope of the present discussion.

Quite substantial local maxima have been calculated for $k_{j_i \to j_f}$ in LiH–He at fixed energy (Jendrek and Alexander, 1980). These maxima shift position dramatically with velocity; j_{bump} varies more rapidly than linearly with v, a real challenge for hard ellipse models which generally predict constant ratios of j to v. The features are so strong that they persist in spite of the average over relative velocities of the comparison experiment (Dagdigian and Wilcomb, 1980), as may be seen from Fig. 6. In view of the effort spent on both the potential (Silver, 1980) and the dynamics (Jendrek and Alexander, 1980) in this system, it is somewhat unsettling that the corresponding experiment (Dagdigian and Wilcomb, 1980) shows little hint of the predicted maximum (might it be beyond the measured j_f range?).

As an aside we note that this system is the classic example of scientists being so taken with the authority of detailed quantum-mechanical calculations that they cannot assess their experiments objectively. Dagdigian and Wilcomb (1980) claim that the experiment displays good agreement with the calculation and that 'an exponential or power-law dependence on the inelastic energy gap does not provide a suitable representation for the $[j_f]$ dependence of the LiH–He cross sections ...' In fact, as we indicated in Sec. III, χ^2/v is 0·76 for a power-law fit compared with 17 for the calculation. If the calculation is renormalized for best fit to the data, χ^2/v is 2·9, still outside the experimental error bars.

D. Alternations

The fitting laws that we have discussed in this review are characterized by a smooth change of the rate or cross section with Δj. Thus phenomena which are manifest as oscillations or alternations of the data present a special challenge for these fitting laws. In this subsection we discuss two different types of alternations which illustrate two different approaches to fitting data containing alternations. The two types of alternations are homonuclear propensity—the tendency for even Δj transitions to have larger RI rates than the adjacent odd Δj transitions—and $\Delta j +/-$ propensity—the tendency for RI rates which connect molecular levels with opposite $e–f$ symmetry to depend on whether e or f has higher energy, manifest as a strong preference for $\Delta j > 0$ vs. $\Delta j < 0$ (or vice versa). Two approaches to fitting data with these features are to fit the large and small rates independently or to average out the alternations; these approaches will be applied to the homonuclear and $+/-$ propensities respectively.

1. Homonuclear Propensity in $^6Li^7Li–X$

So far we have not commented on the well known selection rule that RI collisions involving collisions with homonuclear molecules always obey the selection rule $\Delta j =$ even. This may be viewed as resulting from the inability of RI collisions to change the nuclear spin symmetry of the diatomic (Herzberg, 1950; Gottscho, 1981) as a result of the invariance of the altom–diatom potential (exclusive of nuclear interaction) with respect to parity inversion of the diatom, or as the result of destructive interference of semiclassical amplitudes for scattering with a shift in the initial orientation angle of a homonuclear diatom (McCurdy and Miller, 1977) by π radians.

For systems in which the intermolecular potential has approximate parity invariance, or in which the centre of symmetry of the potential is not coincident with the centre of mass, the RI rates display an even/odd alteration reminiscent of the homonuclear selection rule, which has been termed 'homonuclear propensity'. These systems form an interesting challenge for fitting laws and may ultimately be very revealing about the interplay of molecular potentials and scaling behaviour because the even/odd differential enables one to isolate $P_{odd}(\cos\theta)$ and $P_{even}(\cos\theta)$ terms in the intermolecular potential, and to study their relative effects on the rates as the independent physical variables (e.g. Δj, j_i, and v_{rel}) are varied. This isolation is most easily

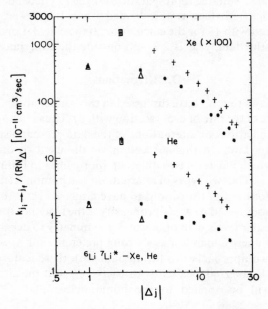

Fig. 13. Alternations in $^6Li^7Li$ in $v_i = 1$, $j_i = 9$ colliding with Xe (top) and He (bottom) (T. Scott and R. Gottscho, private communication). Even Δj is indicated by $+ (\Delta j > 0)$ and $\square$ $(\Delta j < 0)$ and odd Δj is indicated by $* (\Delta j > 0)$ and $\triangle (\Delta j < 0)$. Note that the alternations have a larger amplitude in He than in Xe.

achieved when one of the terms (e.g. P_1) is much smaller than the other (McCurdy and Miller, 1977).

We present in Fig. 13 results for $^6\text{Li}^7\text{Li}$ scattering from Xe and He. The alternation of rates with Δj is strongly evident in both systems, and is strongest at small $|\Delta j|$. It has been suggested (McCurdy and Miller, 1977) that there might be a change in the sign of the alternation at large Δj, but no hint of such behaviour is observed here.

When the rates for the even Δj processes are augmented by the average of the adjacent odd Δj RI rates, the resulting rates are fitted by the same (within error) power law which fits rates taken at a nearby j_i in $^7\text{Li}_2$ collisions.

2. $\Delta j +/-$ Propensity

A number of investigations (Ottinger *et al.*, 1970; Green and Zare, 1975; Lebed *et al.*, 1977) have dealt with the propensity of homonuclear molecules in π states to have a large asymmetry between rates for $+$ and $-$ Δj when Δj is odd. Although this at first may seem counter-intuitive, it violates no canons of

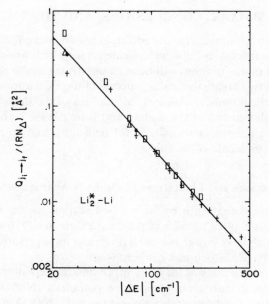

Fig. 14. $+/-\Delta$ propensity alternations in $\text{Li}_2^*(\text{B}^1\Pi_u)$–Li (Vidal, 1978) with $v_i = 71$, $j_i = 45$ symmetry type e or f. This SPG-type plot shows the data as $+(\Delta j > 0)$ and $\square$ $(\Delta j < 0)$. The averages of $e \rightarrow f$ and $f \rightarrow e$ (measured separately) rates are shown as $*(\Delta j > 0)$ and $\triangle(\Delta j < 0)$. These alternations do not persist beyond $\Delta j = 4$.

physical law (e.g. detailed balance) and simply reflects the fact that RI collisions with odd Δj in a homonuclear molecule can only occur between electronic levels of opposite intrinsic parity (so that overall parity is conserved); these levels can couple asymmetrically to the adjacent j levels. These electronic states are known as e and f levels: for a π potential one has its electron cloud predominantly parallel to J while in the other it is perpendicular—it is easy to see that these levels may be perturbed differentially by an approaching atom so that the e levels with odd j are all moved closer to $j f$ levels with even j above them, for example (Lebed *et al.*, 1977).

In fitting data which show this type of propensity, we have not explored the possibility of fitting the $\Delta j +$ and $\Delta j -$ data separately when Δj is odd, but rather have averaged them together. These average rates then seem to fall on the same curve with the even Δj data and we have simply fitted the $\Delta j = $ even and the average $\Delta j = $ odd RI rates with one curve. An example of this procedure (which also shows the unaveraged odd Δj data) is the Li_2^* $(B^1\Pi_u)$ data (Vidal, 1978) displayed in Fig. 14. Once the alternation has been removed, $Li_2^*(B^1\Pi)$–Li is a fine example of a power-law system.

V. SUMMARY, CONCLUSIONS, AND THE FUTURE

In this section we summarize the principal results and conclusions that we have reached in this review, and we indicate topics which we feel need further work or which in our opinion will become important in the near future. We concentrate on two areas: the scaling success of the ECS approximation, and the success of the impulsive classical calculations of the basis rates. We also consider the relationship of the scaling and fitting laws to the potential, to orientation and alignment transfer in RI collisions, and to other forms of inelastic atom–molecule collisions.

A. Successes and Limitations of the ECS Approximation

The ECS approximation is by far the best scaling law for RI data. It has proven to be better, often significantly, than the statistically based scaling laws for systems which have power-law behaviour, and its superiority is even wider in the system I_2^*–He which has EP behaviour.

The range and accuracy of the ECS approximation are quite impressive. It applies to systems with mainly repulsive potentials (Na_2^*–He, I_2^*–He), to systems with well depths probably several times E^t (Na_2^*–Xe), to quite polar systems (LiH–HCN, HF–Ar), and over a 3 to 1 range of relative velocities in the one system where such data exist (Na_2^*–Xe). For Na_2^*–rare gas systems it has been shown that the use of a multiparameter form for $k_{l \to 0}$ reduced the χ^2/ν to essentially unity (Brunner *et al.*, 1981), implying that the ECS scaling law introduced errors of less that 5%!

The ECS scaling law was significantly superior to the IOS scaling law in all but one of the systems studied I_2^*–He, where χ^2 was not significantly improved), and we strongly recommend the use of ECS instead of the widely used IOS scaling law whenever possible. Even if one must resort to IOS, use ECS to make an estimate of the error in IOS (this requires a guess of l_c). Finally we note that the correction term of ECS relative to IOS is of order $\tau^2/6$—thus it is apparent that the failure of IOS is due to the failure of the sudden approximation rather than to the breakdown of the energy sudden approximation ($\Delta E/E^t \ll 1$ used in most derivations of IOS.) This was confirmed in the Na_2^*–rare gas systems where $\Delta E/E^t$ was constant while IOS became worse with increasing τ in the heavier targets (Brunner et al., 1981).

We must also report the failures of ECS. The most glaring one is that it is not applicable to data sets where E^t is the same for all j_i or where $E^t + E^r$ is constant—this demands a generalization of the theory (we suggest, but have not had the conviction to implement, replacing v_{rel} in the upper level by its average value and assuming that the $k_{l \to 0}$ increase linearly with velocity). In addition it is untested for values of $\tau > 4$ and it fails completely to account for the increase in $k_{j_i \to j_f + \Delta}$ with j_i for constant Δ found in close-coupled calculations of RI in CO–He (Green and Thaddeus, 1976) at low temperatures—in this system E^r is several times E^t. Finally, it is doubtful that the H_2–H and H_2–He systems discussed in Sec. IV A could be fitted by any reasonable generalization of ECS to systems with constant total energy. It is not clear what happens to the basic angular momentum addition ideas which underlie IOS/ECS in these highly quantal systems.

Future work on the ECS should make it applicable to data with various velocity distributions, and should extend it to larger τ. This might also reveal its limits of applicability. It would be interesting to know if the IOS approximation to differential cross sections (which has recently been confirmed experimentally (Hefter et al., 1981) and theoretically Fitz and Kouri, 1979) has a similar ECS-type generalization (does l_c depend on scattering angle?). Similarly, a recent IOS expression which relates rates with vibrational inelasticity (Secrest and Liu, 1979) invites a similar generalization. It might be possible to push the ECS into regions where the potential has two discernible components (e.g. P_1 and P_2): might not the velocity dependence of even/odd Δj alternations reveal that these two parts of the potential have different l_c?

Finally, and most important, ECS cannot be regarded as a relatively complete theory until a prescription exists for finding l_c directly from the potential (rather than by fitting a good dynamical calculation).

B. Power Laws and Exponentials

The pure power-law systems discussed in Sec. II, in addition to those discussed in Brunner et al. (1981) and Pritchard et al. (1979), comprise a majority of all systems for which high-quality (average error $E < 10\%$) RI data exist. In these systems the power law (actually ECS-P or SPG-λ, each a power law plus a suitable scaling law) exhausts virtually all of the systematic information in the data as evidenced by the fact that the deviation between fit and data, D, generally lies in the 10–15% range, not much larger than the error in the data. An examination of the sources of the deviation between fit and data which caused χ^2/ν to average close to 2 (instead of 1) in Na_2^*–X systems revealed only two marginally significant systematic trends (Brunner et al., 1981): $k_{2 \to 0}$ was systematically overestimated by the power law (5%) and the power law often overestimated $k_{l \to 0}$ for the largest values of l, a trend which is similar to the behaviour of EP hybrids discussed in Sec. IV.

We now attempt to explain why power-law behaviour is observed in systems with such varied circumstances: for potentials whose well depths are both larger and smaller than E^t, in both strongly dipolar and homonuclear systems, in molecule–molecule systems, and over a range of more than 10 in E^t. Certainly the impulsive classical model summarized in Sec. II D must be a step in the right direction. While its assumptions are rather severe, they certainly do characterize RI collisions for small Δj: long range so that the dominant anisotropy is $R^{-n}P_2(\cos\theta)$ or $R^{-n}P_1(\cos\theta)$, collisions which are nearly impulsive, and which have large impact parameters so that neglect of the isotropic part of the potential is reasonable. There are difficulties with the model, though, since it is difficult to predict the C_n for the anisotropic term, and the model predicts that the exponent γ should not vary with relative velocity as it does (25% over a 3 to 1 velocity range) in Na_2^*–Xe (Smith et al., 1981). Obviously this simple approach should be extended if possible.

The existence of exponential–power systems discussed in Secs. IV B and C must also be accounted for dynamically. The knee in I_2^*–He observed in the EP systems affords an opportunity to probe the dynamics of RI in a general way by studying its movement as j_i, v_{rel}, and collision system are varied. It also presents a challenge to explain (or better yet predict) dynamically the origin and behaviour of this feature. The ideas advanced in Sec. II D.2 concerning tunnelling in those RI collisions with l exceeding the maximum l^* permitted by the limited anisotropy of the potential are suggested by the demonstrated ECS (as opposed to $|\Delta E|$) scaling of the knee in I_2^*–He and also by the observation of EP behaviour in systems with light reduced mass μ_s and with limited anisotropy. But these ideas cannot be regarded as substantiated either experimentally or theoretically without more work and without more examples. It might be possible to base an explanation of the transition from power law (at large range) to exponential (at short range) or the potential anisotropy on atom–molecule separation and different causes of EP behaviour may be important in different systems.

The existence of EP systems forms a logical bridge between EGL and power-law behaviour in RI collisions, but it raises new problems because EGL and power-law behaviour are explained by theories with quite different philosophical underpinnings. We have argued that there is a general dynamical explanation for the power law, and have given plausibility arguments for how EP behaviour might arise from realistic exceptions to the power law for harder collisions. These explanations are based in the mainstream of traditional molecular dynamics — long-range forces plus dynamical calculations — and therefore they pose a challenge to the relevance of surprisal theory with its information theoretic epistemology to RI collisions: if we see the progression from P to EP to EGL behaviour as a continuum associated with the progression to lighter systems with less long-range anisotropy (the picture which emerges strongly from this review), then is it not more philosophically pleasing to seek explanations of EGL in dynamical arguments? (Occum's razor is difficult to apply here because, while the EGL is predicted from fewer assumptions in surprisal theory, these assumptions must be made from a basis orthogonal to the dynamical one which works so well for clearly related phenomena.) It would seem that resolution of these difficulties could come either from a dynamical explanation of EGL or from an information theoretic treatment of P and EP behaviour. Obviously ability to predict the coefficients in the laws is a paramount discriminant.

The fact that the $P \rightarrow EP \rightarrow EGL$ progression accompanies the passage from heavy to light systems accounts for the fact that all experimentally studied systems have P or EP behaviour, while the theoretical studies all have EGL or EP behaviour. Clearly experimentalists should work towards lighter systems with more isotropic potentials and theorists should do the reverse.

C. Alignment and Orientation

Numerous recent studies of RI collisions using polarized light and polarization-sensitive detectors reveal the level-to-level rates for the transfer of orientation (first moment of m_j) and alignment (second moment of m_j) in RI collisions (Baylis, 1978; Rowe and McCaffery, 1979; Silvers et al., 1981; Brechignac et al., 1981; Kurzel and Steinfeld, 1972; Gottscho, 1981). So far this work has not been interpreted using the angular momentum addition ideas underlying IOS scaling, but it is clear that this is a fruitful approach. If these ideas are correct, then they will make correct predictions of these new types of rates without additional assumptions because under sudden conditions the direction of j_i is immaterial to the direction of angular momentum transfer I, permitting a prediction of the final alignment and orientation of j_f from the $k_{l \rightarrow 0}$ only. One would expect to see deviations from these predictions for less sudden collisions, where preferential alignment of I with respect to j_i might occur, for example.

Another topic which appears ripe for investigation is the relative orientation of I and the relative velocity. Experiments showing alignment of alkali dimers in supersonic expansions (Sinha *et al.*, 1974) suggest that such effects exist and similar effects have been observed in atom–atom scattering (Elbet *et al.*, 1975). Moreover, impulse (e.g. hard ellipse) models of RI collisions predict that I should be perpendicular to the direction of momentum transfer (Khare *et al.*, 1981a, b). We suggest that crossed molecular beam experiments should be conducted to study such effects in detail.

D. The End: Progress and Perspective

In the last half-decade a number of theoretical and experimental programs have produced reliable arrays of rate constants and cross sections for purely RI collisions in a wide variety of diatomic molecule–atom systems. We have shown that fitting laws exist which are capable of representing these rates in a majority of these systems with 10 to 25% accuracy, even when these rates extend over two orders of magnitude in size and over nearly comparable ranges of initial and transferred angular momentum. While we have indicated where there is room for improvement in the laws themselves and in their range of applicability, the most important thing is that these laws are presently capable of fitting the large majority of the atom–molecule systems for which extensive, high-quality data sets exist. For these systems, these laws represent a language for discussing the RI rates which is simultaneously short, easy to comprehend and communicate, and able to interpolate and extrapolate; it also forms the logical basis for comparing RI rates in different systems and for recognizing unusual behaviour of some of the rate constants. These laws are obviously well suited to representing the RI process in models of energy transfer in gaseous systems near equilibrium whether these systems are being studied to learn more about various types of collision processes (e.g. other types of inelastic collisions) or for more practical reasons (e.g. to make better lasers).

The success of the various fitting laws in describing the RI rates in a wide variety of systems enables us to draw some conclusions about the overall status of understanding RI collisions. To begin, it is apparent that we do not have a *general* understanding of rotationally inelastic collisions. Although there is a clear qualitative progression in the systems which are fitted by power laws, exponential–power laws, and exponential gap laws, we cannot predict which one to use in some unstudied system. (Even if we were given the potential, there is no way to determine which law to use except by examining the results of a full calculation.) We cannot explain all of the successful laws (i.e. ECS-P and EGL) from a unified perspective (i.e. dynamical or statistical). We can at best speculate on the extent to which the parameters in the fitting laws reflect general properties of RI collisions rather than system dependent details. There are some systems (e.g. CO–He) which cannot be fit by the laws described here. The need for theoretical effort of a general nature on RI collisions is obvious; the success of the general simple fitting laws described here may serve as a guide in such efforts, but it *must* serve as their stimulus.

We speculate that the ECS-EP fitting may prove to be extremely universal as thus far only H_2-He and CO-He can not be fit well by it. The value of the exponent, γ, is predicted to be close to $-1\cdot2$, and most observations support this. Thus γ appears to be universal – i.e. independent of the system. It appears from [Procaccia and Levine 1976d] and [Brunner *et al.*, 1981, Dexheimer *et al.*, 1981] respectively that values of θ in EGL and l in ECS-P do not depend on the target atom, but do depend on the molecule. Both theory [PRL 76d, Brunner *et al.*, 1982] and experiment [Brunner *et al.*, 1981] indicate that the overall magnitude of the rate constants depends on the particular system being studied, as will the critical 1* in the ECS-EP law.

Acknowledgements

We are deeply grateful to Tom Scott for the tremendous assistance that he has given us in the preparation of this manuscript and figures to go with it. We also express our gratitude to the Air Force Office of Scientific Research (AFOSR) for its long-standing support of our energy transfer research program.

References

M. H. Alexander (1979a), *J. Chem. Phys.*, **71**, 1683.

M. H. Alexander (1979b), *J. Chem. Phys.*, **71**, 5212.

M. H. Alexander and A. E. Depristo (1979), *J. Phys. Chem.*, **83**, 1499.

G. G. Balint-Kurti (1975), *The Theory of Rotationally Inelastic Molecular Collisions*, *Int. Rev. Sci.*, *MTP*, *Phys. Chem. Ser.* II, p. 283–326.

W. E. Baylis (1978), in *Progress in Atomic Spectroscopy*, part A, ed. W. Hanle and H. Kleinpoppen, Plenum, New York, p. 245.

D. Beck, V. Ross and W. Schepper (1979), *Z. Phys.*, **A293**, 107.

K. Bergmann, V. Hefter and J. Witt (1980), *J. Chem. Phys.*, **72**, 4777.

R. B. Bernstein (ed.) (1979), *Atom–Molecule Collision Theory*, Plenum Press, New York.

S. S. Bhattacharyya and A. S. Dickinson (1979), *J. Phys. B*, **12**, L521.

S. Bosanac (1980), *Phys. Rev.*, **A32**, 2617.

P. Brechignac, A. Picard-Bersellini and R. Charneau (1981), *J. Chem. Phys.*, to be published.

T. A. Brunner (1980), *PhD Thesis*, MIT, Cambridge, MA.

T. A. Brunner, R. D. Driver, N. Smith and D. E. Pritchard (1978), *Phys. Rev. Lett.*, **41**, 856.

T. A. Brunner, R. D. Driver, N. Smith and D. E. Pritchard (1979), *J. Chem. Phys.*, **70**, 4155.

T. A. Brunner, T. Scott and D. E. Pritchard, to be written.

T. A. Brunner, N. Smith, A. W. Karp and D. E. Pritchard (1981), *J. Chem. Phys.*, to be published.

M. Broyer, J. Vigue and J. C. Lehmann (1975), *J. Chem. Phys.*, **63**, 5428.

M. Broyer, J. Vigue and J. C. Lehmann (1976), *J. Chem. Phys.*, **64**, 4793.

S. I. Chu and A. Dalgarno (1975), *Proc. R. Soc. Lond. A*, **342**, 191.

L. A. Collins and N. F. Lane (1975), *Phys. Rev.*, **A12**, 811.

M. E. Coltrin and P. A. Marcus (1980), *J. Chem. Phys.*, **73**, 2179.

P. J. Dagdigian and M. H. Alexander (1980), *J. Chem. Phys.*, **72**, 6513.

P. J. Dagdigian and B. E. Wilcomb (1980), *J. Chem. Phys.*, **72**, 6462.

A. E. DePristo, S. D. Augustin, R. Ramaswamy and H. Rabitz (1979), *J. Chem. Phys.*, **71**, 850.

A. E. DePristo and H. Rabitz (1980), *J. Chem. Phys.*, **72**, 4685.

S. Dexheimer, M. Durand, T. Brunner and D. Pritchard (1982), in preparation.

A. S. Dickinson and D. Richards (1978), *J. Phys. B*, **20**, 3513.

J. L. Dunham (1932), *Phys. Rev.*, **41**, 721.

A. R. Edmonds (1974), *Angular Momentum in Quantum Mechanics*, Princeton University Press, Princeton, NJ, p. 31.

M. Elbel, H. Huhnermann, T. Meier and W. B. Schneider (1975), *Z. Phys.*, **275**, 339.

B. C. Eu (1974), *Chem. Phys.*, **5**, 95.

M. Faubel and J. P. Toennies (1977), *Adv. Atom. Mol. Phys.*, **13**, 229.

D. E. Fitz and D. J. Kouri (1979), *Chem Phys. Lett.*, **67**, 558.

R. B. Gaber, V. Buck and U. Buch (1980), *Phys. Rev. Lett.*, **44**, 1397.

R. Goldflam, S. Green and D. J. Kouri (1977a), *J. Chem. Phys.*, **67**, 4149.

R. Goldflam, D. J. Kouri and S. Green (1977b), *J. Chem. Phys.*, **67**, 5661.

R. Goldflam, D. J. Kouri, R. K. Preston and R. T. Pack (1977), *J. Chem. Phys.*, **66**, 2574.

R. A. Gottscho (1981), *Chem. Phys. Lett.*, submitted for publication.

S. Green (1979), *Chem. Phys.*, **40**, 1.

S. Green and P. Thaddeus (1976), *Astrophys. J.*, **205**, 766.

S. Green and R. N. Zare (1975), *Chem. Phys.*, **7**, 62.

E. F. Jendrek and M. H. Alexander (1980), *J. Chem. Phys.*, **72**, 6452.

U. Hefter, P. L. Jones, A. Matthews, J. Witt, K. Bergmann and R. Schinke (1981), *Phys. Rev. Letts.*, **46**, 915.

D. F. Heller (1977), *Chem. Phys. Lett.*, **45**, 64.

M. A. Henesian, R. L. Herbst and R. L. Byer (1976), *J. Appl. Phys.*, **47**, 1515.

G. Herzberg (1950), *Molecular Spectra and Molecular Structure*, vol. I, *Spectra of Diatomic Molecules*, Van Nostrand Reinhold, New York.

M. Karplus and R. M. Porter (1970), *Atoms and Molecules*, W. A. Benjamin, Reading, MA.

V. Khare (1978), *J. Chem. Phys.*, **68**, 4631.

V. Khare, D. J. Kouri and D. K. Hoffman (1981a), *J. Chem. Phys.*, **74**, 2413.

V. Khare, D. J. Kouri and D. K. Hoffman (1981b), *J. Chem. Phys.*, **74**, 2656.

R. B. Kurzel and J. I. Steinfeld (1972), *J. Chem. Phys.*, **56**, 5188.

I. V. Lebed, E. E. Nikitin and S. Ya. Umanskii (1977), *Opt. Spektrosk.*, **43**, 636.

R. J. LeRoy (1973), *Molecular Spectroscopy*, vol. 1, The Chemical Society, London.

R. D. Levine (1973), *J. Chem. Phys.*, **56**, 1633.

R. D. Levine and R. B. Bernstein (1975), in *Dynamics of Molecular Collisions*, ed. W. H. Miller, Plenum, New York.

R. D. Levine, R. B. Bernstein, P. Kahana, I. Procaccia and E. T. Upchurch (1976), *J. Chem. Phys.*, **64**, 796.

R. E. Levine and J. L. Kinsey (1979), in *Atom–Molecule Collision Theory*, ed. R. B. Bernstein, Plenum Press, New York, p. 301.

Kin-Wah Li, B. C. Eu and B. C. Sanctuary (1977), *Chem. Phys. Lett.*, **50**, 162.

C. W. McCurdy and W. H. Miller (1977), *J. Chem. Phys.*, **67**, 463.

A. Messiah (1962), *Quantum Mechanics*, Wiley, New York, p. 736.

W. H. Miller and A. W. Raczkowski (1973), *Faraday Discuss. Chem. Soc.*, **55**, 45.

M. H. Mittleman, J. L. Peacher and B. F. Rozsnyai (1968), *Phys. Rev.*, **176**, 180.
T. Mulloney and G. C. Schatz (1980), *Chem. Phys.*, **45**, 213.
T. Oka (1973), *Advances in Atomic and Molecular Physics*, vol. 9, ed. D. R. Bates, Academic Press, New York, p. 127.
Ch. Ottinger and M. Schroder (1980), *J. Phys. B.* **13**, 4163.
C. Ottinger, R. Velasco and R. N. Zare (1970), *J. Chem. Phys.*, **52**, 1636.
R. T. Pack (1975), *J. Chem. Phys.*, **62**, 3143.
M. D. Pattengill (1979), in *Atom–Molecule Collision Theory*, ed. R. B. Bernstein, Plenum Press, New York, chap. 9.
M. D. Pattengill and R. B. Bernstein (1976), *J. Chem. Phys.*, **65**, 4007. (We have fit only data calculated using the potential denoted *KDV.*)
J. C. Polanyi and K. B. Woodall (1972), *J. Chem. Phys.*, **56**, 1563.
D. E. Pritchard, N. Smith, R. D. Driver and T. A. Brunner (1979), *J. Chem. Phys.*, **70**, 2115.
I. Procaccia and R. D. Levine (1975), *J. Chem. Phys.*, **63**, 4261.
I. Procaccia and R. D. Levine (1976a), *Physica (Utrecht)*, A, **82**, 623.
I. Procaccia and R. D. Levine (1976b), *J. Chem. Phys.*, **64**, 808.
I. Procaccia and R. D. Levine (1976c), *Phys. Rev.*, A14, 1569.
I. Procaccia and R. D. Levine (1976d), *J. Chem. Phys.*, **65**, 495.
M. D. Rowe and A. J. McCaffery (1979), *Chem. Phys.*, **43**, 35, and references therein.
S. J. Silvers, R. A. Gottscho and R. W. Field (1981), *J. Chem. Phys.*, **74**, 6000.
B. C. Sanctuary (1979), *Chem. Phys. Lett.*, **62**, 378.
D. Secrest (1979), in *Atom–Molecule Collision Theory*, ed. R. B. Bernstein, Plenum Press, New York, chap. 8.
D. Secrest and C. S. Lin (1979), *J. Chem. Phys.*, **70**, 3420.
M. P. Sinha, C. D. Caldwell and R. M. Zare (1974), *J. Chem. Phys.*, **61**, 491.
N. Smith, T. A. Brunner and D. E. Pritchard (1981), *J. Chem. Phys.*, **74**, 467.
N. Smith and D. E. Pritchard (1981), *J. Chem. Phys.*, to be published.
D. M. Silver (1980), *J. Chem. Phys.*, **72**, 6445.
S. J. Silvers, R. A. Gottscho and R. W. Field (1981), *J. Chem. Phys.*, to be published.
J. I. Steinfeld and W. Klemperer (1965), *J. Chem. Phys.*, **42**, 3475.
W. C. Stwalley (1975), *J. Chem. Phys.*, **63**, 3062.
J. P. Toennies (1976), *Annu. Rev. Phys. Chem.*, **27**, 225.
Z. H. Top and D. J. Kouri (1978), *Chem. Phys.*, **37**, 265.
C. R. Vidal (1978), *Chem. Phys.*, **35**, 215.
J. Vigue and J. C. Lehmann (1980), unpublished results.
M. Wainger, I. Al-Agil, T. A. Brunner, A. W. Karp, N. Smith and D. E. Pritchard (1979), *J. Chem. Phys.*, **71**, 1977.
R. Wellegehausen, S. Shahdin, D. Friede and H. Willing (1977), *Appl. Phys.*, **13**, 97.
B. E. Wilcomb and P. J. Dagdigian (1977), *J. Chem. Phys.*, **67**, 3829.

AUTHOR INDEX

SUBJECT INDEX